LASERS

LASERS

PETER W. MILONNI
Los Alamos National Laboratory
Los Alamos, New Mexico

JOSEPH H. EBERLY
University of Rochester
Rochester, New York

WILEY

A WILEY-INTERSCIENCE PUBLICATION

JOHN WILEY & SONS

New York / Chichester / Brisbane / Toronto / Singapore

Library of Congress Cataloging in Publication Data:

Milonni, Peter W.
  Lasers / Peter W. Milonni and Joseph H. Eberly.
    p.   cm.

  ''A Wiley-Interscience publication.''
  Bibliography: p.
  ISBN 0-471-62731-3
  1. Lasers.   I. Eberly, Joseph H. II. Title.

QC688.M55 1988
621.36'6—dc19   87-26347
                                                            CIP

Printed in the United States of America

10 9 8 7 6 5 4 3 2 1

For our parents

PREFACE

In the ideal case a technical book covers its subject from all angles, providing both a connected chain of ideas and enough derivations to give the subject a firm basis, as well as historical digressions sufficient to exhibit the authors' scientific heroes in a favorable light. At the same time it should include homework problems not too difficult for the instructor and plenty of tables and charts summarizing all of the useful data of the field, but not too much of any of these to push the book's weight beyond the strength of the average student. But if such an ideal book exists, it does not deal with lasers.

The subject of lasers is now almost too big for any one book. The description of the laser as a solution looking for a problem was never really accurate. In the 1980s it is clear that the laser is the solution to a wide variety of problems. Some of them, such as holography and detached retinas, existed well before the laser was invented, and others, such as nonlinear optics, were more or less invented *by* the laser.

In writing this book we have tried to keep in mind a variety of possible readers. We think it is complete enough to be useful as an occasional reference even for many optical scientists and engineers who are already familiar with lasers. The book begins with elementary material and the mathematical level is not advanced, so it should also be suitable as a textbook in an introductory course.

There is too much material to be covered in a single term, so an instructor will be forced to make choices, and the book allows a number of options. It should be possible to make a survey of the entire subject by picking only the most important topics of each chapter, or to concentrate on four or five subject areas for detailed study. Among the areas suitable for detailed treatment are the rate equations of laser action, the classical theory of dispersion, absorption and emission, the theory of optical resonators, diffraction and Gaussian beams, the principles of optical coherence, introductory quantum physics, the wave mechanical basis of laser action, holography, wave mixing and nonlinear optics, and the operating principles of the most important laser types. There are problems and exercises included in every chapter after the first.

In a sense, the book comes as two independent half books. It is possible to begin reading at Chapter 10 and learn the most important details of laser operation and the nature of the most common laser systems, resonators, and applications, without looking at earlier chapters. To do so overlooks the fundamental basis for the parameters such as the gain coefficient and absorption cross section that govern laser operation. It also sidesteps questions underlying the origin of the population equations presented in Chapter 10. However, there may be compelling reasons for such an approach, including lack of time or an adequate earlier exposure to elementary dispersion theory, atomic physics, and quantum mechanics. In this case the first half of the book serves as a compact reference with the big advantage that it uses the same notation as the second half.

On the other hand, it is possible to stop reading the book after Chapter 9 and still have acquired a rather full introduction to laser physics, including the derivation of completely satisfactory semiclassical laser equations, an introduction to cavity modes, and a discussion of the evidence for photons. Such a reading could be used in a course which deals with atomic and quantum physics with the laser as a specific application in mind. There are other rearrangements of material that are also logical. For example, the chapters on laser resonators and on nonlinear optics are somewhat arbitrarily placed in the book. Both of them could be taken up very early, immediately after Chapter 2 if desired. Nonlinear optics does not really require quantum mechanics for a good practical understanding, and laser resonators require only Section 2.1 as an introductory review. Without difficulty optical resonators can be made the first serious topic considered in studying lasers. It is entirely possible that we have not even thought of the arrangement of topics that most people would consider optimal. Although both authors have taught courses based on the book, and parts of the book have been tested by others, neither author has yet had the pleasure of teaching a course requiring all 18 chapters.

Naturally the book contains the views of the authors. For these views we must take responsibility, even though they are due in part to colleagues who have repeatedly discussed and argued aspects of laser physics with us over a number of years. Although we have included a number of historical side notes, we have not mentioned scientific credits or priorities after about 1960. The history of laser science is a fascinating subject that deserves its own careful examination.

To an extent the contents of the book reflect our own research interests, and we express thanks to the organizations that have supported this research. We also owe a debt of gratitude to the people who worked with us to produce a final readable manuscript. These include Melissa Beatty, Bertha Marsh Griffin, Jean Eaton, Anne Schmidt, Wendy Bierman, Dotty Paine, Ellen Eckert, Leslie Smith and Michael Goggin, who contributed directly, and Connie Jones, who helped indirectly much more than she needed to. Most of the figures were drawn by Wayne Sheeler, and we appreciate his willingness to work to an unusual schedule without sacrificing care or good humor. Finally we thank Beatrice Shube. We hope the book comes up to her expectations.

P. W. MILONNI
J. H. EBERLY

Los Alamos, New Mexico
Rochester, New York

CONTENTS

Chapter 12   Multimode and Transient Oscillation   **365**

Chapter 13   Specific Lasers and Pumping Mechanisms   **411**

Chapter 16   Some Laser Applications   585

Chapter 17   Introduction to Nonlinear Optics   625

LASERS

1 INTRODUCTION TO LASER OPERATION

1.1 INTRODUCTION

The word *laser* is an acronym for the most significant feature of laser action: *light amplification by stimulated emission of radiation*. There are many different kinds of laser, but they all share a crucial element: each contains material capable of amplifying radiation. This material is called the gain medium, because radiation gains energy passing through it. The physical principle responsible for this amplification is called stimulated emission, and was discovered by Albert Einstein in 1916. It was widely recognized that the laser would represent a scientific and technological step of the greatest magnitude, even before the first one was constructed in 1960 by T. H. Maiman. The award of the 1964 Nobel Prize in physics to C. H. Townes, N. G. Basov, and A. M. Prokhorov carried the citation "for fundamental work in the field of quantum electronics, which has led to the construction of oscillators and amplifiers based on the maser–laser principle." These oscillators and amplifiers have since motivated and aided the work of thousands of scientists and engineers.

In this chapter we will undertake a superficial introduction to lasers, cutting corners at every opportunity. We will present an overview of the properties of laser light, with the goal of understanding what a laser is, in the simplest terms. We will introduce the theory of light in cavities and of cavity modes, and we will describe an elementary theory of laser action.

We can begin our introduction with Figure 1.1, which illustrates the four key elements of a laser. First, a collection of atoms or other material amplifies a light signal directed through it. This is shown in Figure 1.1*a*. The amplifying material is usually enclosed by a highly reflecting cavity that will hold the amplified light, in effect redirecting it through the medium for repeated amplifications. This refinement is indicated in Figure 1.1*b*. Some provision, as sketched in Figure 1.1*c*, must be made for replenishing the energy of the amplifier which is being converted to light energy. And some means must be arranged for extracting in the form of a beam at least part of the light stored in the cavity, perhaps as shown in Figure 1.1*d*. A schematic diagram of an operating laser embodying all these elements is shown in Figure 1.2.

It is clear that a well-designed laser must carefully balance gains and losses. It can be anticipated with confidence that every potential laser system will present its designer with more sources of loss than gain. Lasers are subject to the basic laws of physics, and every stage of laser operation from the injection of energy

1

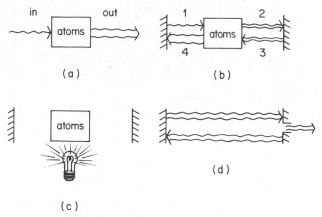

Figure 1.1 Basic elements of a laser.

into the amplifying medium to the extraction of light from the cavity is an opportunity for energy loss and entropy gain. One can say that the success of masers and lasers came only after physicists learned how atoms could be operated efficiently as thermodynamic engines.

One of the challenges in understanding the behavior of atoms in cavities arises from the strong feedback deliberately imposed by the cavity designer. This feedback means that a small input can be amplified in a straightforward way by the atoms, but not indefinitely. Simple amplification occurs only until the light field in the cavity is strong enough to affect the behavior of the atoms. Then the strength of the light as it acts on the amplifying atoms must be taken into account in determining the strength of the light itself. This sounds like circular reasoning, and in a sense it is. The responses of the light and the atoms to each other can become so strongly interconnected that they cannot be determined independently but only self-consistently. Strong feedback also means that small perturbations can be rapidly magnified. Thus it is accurate to anticipate that lasers are potentially highly

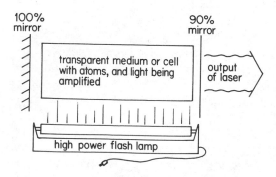

Figure 1.2 A complete laser system, showing elements responsible for energy input, amplification, and output.

erratic and unstable devices. In fact, lasers can provide dramatic exhibitions of truly chaotic behavior, and are the objects of fundamental study for this reason.

For our purposes lasers are principally interesting, however, when they operate stably, with well-determined output intensity and frequency as well as spatial mode structure. The self-consistent interaction of light and atoms is important for these properties, and we will have to be concerned with concepts such as gain, loss, threshold, steady state, saturation, mode structure, frequency pulling, and line-width.

In the next several sections we sketch properties of laser light, discuss modes in cavities, and give a theory of laser action. This theory is not really correct, but it is realistic within its own domain, and has so many familiar features that it may be said to be "obvious." It is also significant to observe what is not explained by this theory, and to observe the ways in which it is not fundamental but only em-pirical. These gaps and missing elements are an indication that the remaining 17 chapters of the book may also be necessary.

1.2 LASERS AND LASER LIGHT

Many of the properties of laser light are special or extreme in one way or another. In this section we provide a brief overview of these properties, contrasting them with the properties of light from more ordinary sources when possible.

Wavelength

Laser light is available in all colors from red to violet, and also far outside these conventional limits of the optical spectrum, as shown in Figure 1.3. Over a wide

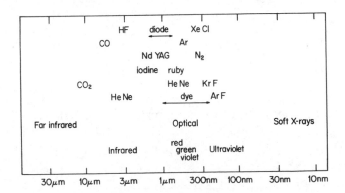

Figure 1.3 Near-optical regions of the electromagnetic spectrum, with types of lasers located approximately at their operating wavelength or wavelengths.

portion of the available range laser light is "tunable." This means that some lasers have the property of emitting light at any wavelength chosen within a range of wavelengths. Tunability is primarily a property of dye lasers. The longest laser wavelength can be taken to be in the far infrared, in the neighborhood of 100–500 μm. Devices producing coherent light at much longer wavelengths by the "maser–laser principle" are usually thought of as masers. The search for lasers with ever shorter wavelengths is probably endless. Coherent stimulated emission in the XUV or soft X-ray region (10–15 nm) has been reported. Appreciably shorter wavelengths, those characteristic of gamma rays, for example, may be quite difficult to reach.

Photon Energy

The energy of a laser photon is not different from the energy of an "ordinary" light photon of the same wavelength. A green–yellow photon, roughly in the middle of the optical spectrum, has an energy of about 2.5 eV (electron volts). This is the same as about 4×10^{-19} J (joules) $= 4 \times 10^{-12}$ ergs. It should be clear that electron volts are a much more convenient unit for laser photon energy than Joules or ergs. From the infrared to the X-ray region photon energies vary from about 0.01 eV to about 100 eV. For contrast, at room temperature the thermal unit of energy is $kT \approx \frac{1}{40}$ eV $= 0.025$ eV. This is two orders of magnitude smaller than the typical optical photon energy just mentioned, and as a consequence thermal excitation plays only a very small role in the physics of nearly all lasers.

Directionality

The output of a laser can consist of nearly ideal plane wave fronts. Only diffraction imposes a lower limit on the angular spread of a laser beam. The wavelength λ and the area A of the laser output aperture determine the order of magnitude of the beams's solid angle ($\Delta\Omega$) and vertex angle ($\Delta\theta$) of divergence (Figure 1.4) through the relation

$$\Delta\Omega \approx \frac{\lambda^2}{A} \approx (\Delta\theta)^2 \qquad (1.2.1)$$

This represents a very small angular spread indeed if λ is in the optical range, say

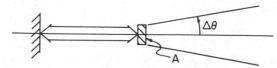

Figure 1.4 Sketch of a laser cavity showing angular beam divergence $\Delta\theta$ at the output mirror (area A).

500 nm, and A is macroscopic, say $(5\ \text{mm})^2$. In this example we compute $\Delta\Omega \approx$ $(500)^2 \times 10^{-18}\ \text{m}^2/(5^2 \times 10^{-6}\ \text{m}^2) = 10^{-8}$ steradians.

Monochromaticity

It is well known that lasers produce very pure colors. If they could produce exactly one wavelength, laser light would be fully monochromatic. This is not possible, in principle as well as for practical reasons. We will designate by $\Delta\lambda$ the range of wavelengths included in a laser beam of main wavelength λ. Similarly, the associated range of frequencies will be designated by $\Delta\nu$, the bandwidth. In the optical region of the spectrum we can take $\nu \approx 5 \times 10^{14}$ Hz (hertz, i.e., cycles per second). The bandwidth of sunlight is very broad, more than 10^{14} Hz. Of course filtered sunlight is a different matter, and with sufficiently good filters $\Delta\nu$ could be reduced a great deal. However, the cost in lost intensity would usually be prohibitive. (See subsection on spectral brightness below.) For lasers, a very low value of $\Delta\nu$ is 1 Hz, while a bandwidth around 100 Hz is spectroscopically practical in some cases (Figure 1.5). For $\Delta\nu \approx 100$ Hz the relative spectral purity of a laser beam is quite impressive: $\Delta\nu/\nu \approx 100/(5 \times 10^{14}) = 2 \times 10^{-13}$. This exceeds the spectral purity (Q factor) achievable in conventional mechanical and electrical resonators by many orders of magnitude.

Coherence Time

The existence of a finite bandwidth $\Delta\nu$ means that the different frequencies present in a laser beam can eventually get out of phase with each other. The time required for two oscillations differing in frequency by $\Delta\nu$ to get out of phase by a full cycle is obviously $1/\Delta\nu$. After this amount of time the different frequency components in the beam can begin to interfere destructively, and the beam loses "coherence." Thus $\Delta\tau = 1/\Delta\nu$ is called the beam's coherence time. This is a general definition, not restricted to laser light, but the extremely small values possible for $\Delta\nu$ in laser light make the coherence times of laser light extraordinarily long.

For example, even a "broadband" laser with $\Delta\nu \approx 1$ MHz has the coherence time $\Delta\tau \approx 1$ μsec. This is enormously longer than most "typical" atomic fluorescence lifetimes, which are measured in nanoseconds (10^{-9} sec). Thus even lasers that are not close to the limit of spectral purity are nevertheless effectively 100% pure on the relevant spectroscopic time scale. By way of contrast, sunlight

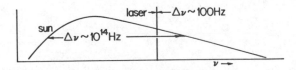

Figure 1.5 Spectral emission bands of the sun and of a representative laser, to indicate the much closer approach to purely monochromatic light achieved by the laser.

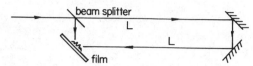

Figure 1.6 Sketch of a two-beam interferometer showing interference fringes obtained at the recording plane if the coherence length of the light is great enough.

has a bandwidth $\Delta \nu$ almost as great as its central frequency (yellow light, $\nu \approx 5 \times 10^{14}$ Hz). Thus for sunlight the coherence time is $\Delta \tau \approx 2 \times 10^{-15}$ sec, so short that unfiltered sunlight cannot be considered temporally coherent at all.

Coherence Length

The speed of light is so great that a light beam can travel a very great distance within even a short coherence time. For example, within $\Delta \tau \approx 1$ μsec light travels $\Delta z \approx (3 \times 10^8 \text{ m/sec}) \times (1 \text{ }\mu\text{sec}) \approx 300$ meters. The distance $\Delta z = c \Delta \tau$ is called the beam's coherence length. Only portions of the same beam that are separated by less than Δz are capable of interfering constructively with each other. No fringes will be recorded by the film in Figure 1.6, for example, unless $2L < c \Delta \tau = \Delta z$.

Spectral Brightness

A light beam from a finite source can be characterized by its beam divergence $\Delta \Omega$, source size (usually surface area A), bandwidth $\Delta \nu$, and spectral power density $P(\nu)$ (watts per hertz of bandwidth). From these parameters it is useful to determine the *spectral brightness* β_{ν} of the source, which is defined (Figure 1.7) to be the power flow per unit area, unit bandwidth, and steradian, namely $\beta_{\nu} = P_{\nu}/A \, \Delta \Omega \, \Delta \nu$. Notice that $P_{\nu}/A \, \Delta \nu$ is the spectral intensity, so β_{ν} can also be thought of as the spectral intensity per steradian.

For an ordinary *nonlaser optical source*, brightness can be estimated directly from the blackbody formula for $\rho(\nu)$, the spectral energy density (J/cm^3-Hz):

$$\rho(\nu) = \frac{8 \pi \nu^2}{c^3} \frac{h\nu}{e^{h\nu/kT} - 1} \tag{1.2.2}$$

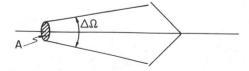

Figure 1.7 Geometrical construction showing the source area and emission solid angle appropriate to a discussion of spectral brightness.

The spectral intensity (watts/cm^2-Hz) is thus $c\rho$, and $c\rho/\Delta\Omega$ is the desired spectral intensity per steradian. Taking $\Delta\Omega = 4\pi$ for a blackbody, we have

$$\beta_\nu = \frac{2\nu^2}{c^2}\frac{h\nu}{e^{h\nu/kT}-1} \tag{1.2.3}$$

The temperature of the sun is about $T = 5800$ K $\approx 20 \times (300$ K$)$. Since the main solar emission is in the yellow portion of the spectrum, we can take $h\nu \approx 2.5$ eV. We recall that $kT \approx \frac{1}{40}$ eV for $T = 300$ K, so $h\nu/kT \approx 5$, giving $e^{h\nu/kT} \approx 150$ and finally

$$\beta_\nu \approx 1.5 \times 10^{-12} \text{ W/cm}^2\text{-sr-Hz} \quad \text{(sun)} \tag{1.2.4}$$

Several different estimates can be made for laser radiation, depending on the type of laser considered. Consider first a *low power He–Ne laser*. A power level of 1 mW is normal, with a bandwidth of around 10^4 Hz. From (1.2.1) we see that the product of beam cross-sectional area and solid angle is just λ^2, which for He–Ne light is $\lambda^2 \approx (6238 \times 10^{-8}$ cm$)^2 \approx 3.89 \times 10^{-9}$ cm^2. Combining these, we find

$$\beta_\nu \approx 25 \text{ W/cm}^2\text{-sr-Hz} \quad \text{(He–Ne laser)} \tag{1.2.5}$$

Another common laser is the *mode-locked neodymium-glass laser*, which can easily reach power levels around 10^4 MW. The bandwidth of such a laser is limited by the pulse duration $\tau_p \approx 30$ psec (30×10^{-12} sec), as follows. Since the laser's coherence time $\Delta\tau$ is equal to τ_p at most, its bandwidth is certainly greater than $1/\tau_p \approx 3.3 \times 10^{10}$ sec^{-1}. We convert from radians per second to cycles per second by dividing by 2π and get $\Delta\nu \approx 5 \times 10^9$ Hz. The wavelength of a Nd: glass laser is 1.06 μm, so $\lambda^2 \approx 10^{-8}$ cm^2. The result of combining these, again using $A \Delta\Omega = \lambda^2$, is

$$\beta_\nu \approx 2 \times 10^8 \text{ W/cm}^2\text{-sr-Hz} \quad \text{(Nd: glass laser)} \tag{1.2.6}$$

Recent developments have led to lasers with powers at the terawatt level (10^{12} W), so β_ν can be even several orders of magnitude larger.

It is clear that in terms of brightness there is practically no comparison possible between lasers and thermal light. Twenty orders of magnitude in brightness separate our sun from a mode-locked laser. This raises an interesting question of principle. Let us imagine a thermal light source filtered and collimated to the bandwidth and directionality of a He–Ne laser, and the He–Ne laser attenuated to the brightness level of the thermal light. The question is: could the two light beams with equal brightness, beam divergence, and bandwidth be distinguished in any way? The answer is that they could be distinguished, but not by any ordinary measurement of optics. Differences would show up only in the statistical fluctua-

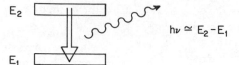

$h\nu \simeq E_2 - E_1$

Figure 1.8 Photon emission accompanying a quantum jump from level 2 to level 1.

tions in the light beam. These fluctuations can reflect the quantum nature of the light source, and are detected by photon counting.

Active Medium

The materials that can be used as the active medium of a laser are so varied that a listing is hardly possible. Gases, liquids, and solids of every sort have been made to *lase* (a verb contributed to science by the laser). The origin of laser photons, as shown in Figure 1.8, is most often in a transition between discrete upper and lower energy states in the medium, regardless of its state of matter. He–Ne, ruby, CO_2, and dye lasers are familiar examples, but exceptions are easily found: the excimer laser has an unbound lower state, the semiconductor diode laser depends on transitions between electron bands rather than discrete states, and understanding the free-electron laser does not require quantum states at all.

Type of Laser Cavity

All laser cavities share two characteristics that complement each other: (1) they are basically linear devices with one relatively long optical axis, and (2) the sides perpendicular to this axis are open, rather than closed by reflecting material as in a microwave cavity. There is no single best shape implied by these criteria, and in the case of ring lasers the long axis actually bends and closes on itself (Figure 1.9). Despite what may seem obvious, it is not always best to design a cavity with the lowest loss. In the case of Q switching an extra loss is temporarily introduced into the cavity for the laser to overcome, and very high-power lasers sometimes use mirrors that are deliberately designed to deflect light out of the cavity rather than contain it.

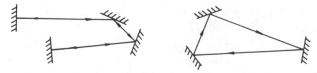

Figure 1.9 Two collections of mirrors making laser cavities, showing standing-wave and traveling-wave (ring) configurations on the left and right, respectively.

Applications of Lasers

There is apparently no end of possible applications of lasers. Many of the uses of lasers are well known by now to most people, such as for several kinds of surgical procedures, for holography, in ultrasensitive gyroscopes, to provide straight lines for surveying, in supermarket checkout scanners and compact disc players, for welding, drilling, and scribing, in compact death-ray pistols, and so on. (The sophisticated student knows, even before reading this book, that some "well-known" applications have never been realized outside the movie theater.)

1.3 LIGHT IN CAVITIES

In laser technology the terms *cavity* and *resonator* are used interchangeably. The theory and design of the cavity are important enough for us to devote all of Chapter 14 to them later in the book. In this section we will consider only a simplified theory of resonators, a theory that is certain to be at least partly familiar to most readers. This simplification allows us to introduce the concept of cavity modes, and to infer certain features of cavity modes that remain valid in more general circumstances. We also describe the great advantage of open, rather than closed, cavities for optical radiation.

We will consider only the case of a rectangular "empty cavity" containing radiation but no matter, as sketched in Figure 1.10. The assumption that there is radiation but no matter inside the cavity is obviously an approximation if the cavity is part of a working laser. This approximation is used frequently in laser theory, and it is accurate enough for many purposes because laser media are usually only sparsely filled with "active" atoms or molecules.

In Chapter 2 full solutions for the electric field in the cavity are given. For example, the z dependence of the x component of the field takes the form

$$E_x(z) = A \sin k_z z \tag{1.3.1}$$

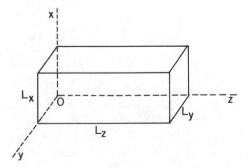

Figure 1.10 A rectangular cavity with side lengths L_x, L_y, L_z.

However, here we are interested only in the simplest features of the cavity field, and these can be obtained easily by physical reasoning.

The electric field should vanish at the walls of the cavity. It will do so if we fit exactly an integer number of half wavelengths into the cavity along each of its axes. This means, for example, that $L = n(\lambda/2)$ along the z axis, where $n = 1$, $2, \ldots$, is a positive integer, and L is the cavity length. If we use the relation between wave vector and wavelength, $k = 2\pi/\lambda$, this is the same as

$$k_z = \frac{\pi}{L} n \qquad (1.3.2)$$

for the z component of the wave vector. By substitution into the solution (1.3.1) we see that (1.3.2) is sufficient to guarantee that the required boundary condition is met, i.e., that $E_x(z) = 0$ for both $z = 0$ and $z = L$. Similar conditions apply to the x and y dependences, and to the other components of the wave vector, when we fit integer numbers of half waves along the x and y axes:

$$k_x = \frac{\pi}{L} l, \qquad l = 1, 2, \ldots \qquad (1.3.3)$$

$$k_y = \frac{\pi}{L} m, \qquad m = 1, 2, \ldots \qquad (1.3.4)$$

Note that we have taken the same length L for the x and y dimensions as well.

We can combine these three components of the wave vector to get an expression for the mode frequencies that are possible in the cavity. We use the familiar relations

$$\nu = \frac{c}{\lambda} = \frac{kc}{2\pi} \qquad (1.3.5)$$

and

$$k = |\mathbf{k}| = (k_x^2 + k_y^2 + k_z^2)^{1/2} \qquad (1.3.6)$$

to obtain the following set of allowed frequencies for a cubical cavity:

$$\nu = \nu_{lmn} = \frac{c}{2L} (l^2 + m^2 + n^2)^{1/2} \qquad (1.3.7)$$

An important distinction between masers and lasers can be explained on the basis of this formula for cavity frequencies. In the case of a maser (which typically operates at macroscopic wavelengths, $\lambda \approx 1$ cm) the cavity itself can be constructed on the scale of the wavelengths of interest. Consider, for example, a cubical cavity with

$$L = 1 \text{ cm}$$

From (1.3.7), then, the wavelengths of the cavity modes are

$$\lambda_{lmn} = \frac{c}{\nu_{lmn}} = \frac{2}{\sqrt{l^2 + m^2 + n^2}} \text{ cm} \qquad (1.3.8)$$

Thus the wavelengths of the principal (lowest) modes of the cavity are

$$\lambda_{110} = \lambda_{101} = \lambda_{011} = \frac{2}{\sqrt{2}} \text{ cm} \approx 1.41 \text{ cm}$$

$$\lambda_{111} = \frac{2}{\sqrt{3}} \text{ cm} \approx 1.15 \text{ cm}$$

$$\lambda_{210} = \lambda_{201} = \lambda_{012} = \lambda_{021} = \lambda_{102} = \lambda_{120} = \frac{2}{\sqrt{5}} \text{ cm} \approx 0.89 \text{ cm}$$

$$\lambda_{211} = \lambda_{121} = \lambda_{112} = \frac{2}{\sqrt{6}} \text{ cm} \approx 0.82 \text{ cm}$$

and so on. (We do not list wavelengths of "modes" for which any two of the integers l, m, n are zero, because the field then vanishes. See Chapter 2.)

The point of this explicit catalog of modes is to show that the mode wavelengths are all on the order of the cavity size, $L = 1$ cm, and that they are also separated from each other by amounts not too much smaller than the cavity size. (See Figure 1.11.) If we want to have coherent—ideally single-mode—maser emission in the wavelength range around 1 cm, such a cavity is just right. Even if the maser gain medium is fairly broadband, able to amplify radiation over an unusually wide band of frequencies, say $\Delta\lambda/\lambda = 0.01$ around $\lambda = 1.41$ cm, the cavity will nevertheless give single-mode output at $\lambda = 1.41$ cm, because there are no other modes within 1% of the 1.41-cm mode. The distance between the modes is too large for any other mode to experience significant gain. In the case of microwave wavelengths, therefore, single-mode oscillation is easily accomplished by a propitious choice of the cavity dimensions.

In the case of optical or near-infrared *laser* wavelengths, however, this obviously cannot be done very easily. Optical cavity dimensions are enormously larger than an optical wavelength ($\lambda \approx 5 \times 10^{-5}$ cm). In this case, Eq. (1.3.7)

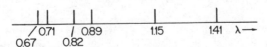

Figure 1.11 Wavelengths corresponding to the lowest modes of a cavity with $L_x = L_y = L_z = 1$ cm.

implies there can be a very large number of modes within the frequency range for which the active medium amplifies the field. The output laser radiation will then contain contributions from many different modes, differing in both their frequencies and their spatial distributions. Furthermore the relative proportions of the various modes might vary in time in a practically unpredictable way as a result of various perturbations such as mirror vibrations. Such unpredictable small variations would make the laser much "noisier" than we would like for many important applications.

It is not difficult to calculate the number of modes within a given frequency range when the cavity dimensions are much larger than the wavelength. The calculation is given in the next chapter. It is found that the number ΔN_ν of possible field modes (of a cavity of volume V) in the frequency interval from ν to $\nu + \Delta\nu$ is

$$\Delta N_\nu = \frac{8\pi\nu^2}{c^3} V \, \Delta\nu \tag{1.3.9}$$

Conversely, the interval $\Delta\nu$ between adjacent modes (i.e., for $\Delta N_\nu = 1$) is given by

$$\Delta\nu = \frac{c^3}{8\pi\nu^2 V} = \left(\frac{\lambda^3}{V}\right)\frac{\nu}{8\pi} \tag{1.3.10}$$

The modes are obviously extremely closely spaced if λ is optical and V is macroscopic. In a typical case ($\lambda \approx 600$ nm and $V \approx 1$ cm^3) we have

$$\Delta\nu \approx \frac{100}{8\pi} \text{ Hz} \approx 4 \text{ Hz} \tag{1.3.11}$$

Consider an example. In a 6328-Å He–Ne laser ($\nu = 4.7 \times 10^{14}$ Hz) the width of the gain curve (the frequency region over which stimulated emission is feasible) is about 1500 MHz. In this frequency interval a 1-cm^3 cavity has about 400 million modes available. A laser using this cavity would not produce anything resembling single-mode radiation.

It would appear from (1.3.10) and (1.3.11) that an extremely large number of modes must be present at optical frequencies. This is not the case. One way to reduce the mode density is to reduce the *dimensionality* of the cavity. A way to do this was suggested independently by Schawlow and Townes, R.H. Dicke, and Prokhorov in 1958: use an *open* resonator consisting of two parallel mirrors, as in Figure 1.1*d*.

The idea behind this suggestion is intuitively clear. A wave traveling at an angle to the axis joining the two mirrors will escape from the resonator after a few reflections. It does not represent a cavity mode. Also, it will not complete many

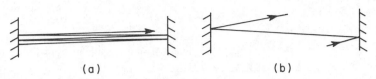

Figure 1.12 Sketch illustrating the advantage of a one-dimensional cavity. Stable modes are associated only with beams that are retroreflected many times.

round-trip passes through the gain medium, and therefore will not have a chance to be amplified very much. A wave traveling nearly exactly along the axis, however, will continue to bounce back and forth between the mirrors. It can set up a standing-wave pattern and does represent a cavity mode. Also, such a wave mode continues to extract energy from the gain medium as a result of stimulated emission. If the gain is large enough, exceeding a certain threshold value, laser radiation will build up in the form of such waves traveling back and forth along the axis. This is illustrated in Figure 1.12.

The shift to a one-dimensional cavity has a dramatic effect on mode spacing. Roughly speaking, in (1.3.10) the factor λ^3/V can be interpreted as $(\lambda/L)^3$, where L is a typical cavity side length. In a one-dimensional cavity we can anticipate that λ/L appears in the appropriate mode-spacing formula instead of $(\lambda/L)^3$. Thus we expect a mode spacing on the order of $\Delta\nu \approx (\lambda/L)(\nu/8\pi)$. A careful analysis (see Chapter 2) confirms this estimate with a slight change in the numerical factor 8π. It is found that in a one-dimensional cavity the successive modes are separated in frequency by

$$\Delta\nu = \frac{c}{2L} = \frac{\lambda}{L}\frac{\nu}{2} \tag{1.3.12}$$

which gives, in our He–Ne example, if we take $L = 50$ cm as is typical,

$$\Delta\nu = \frac{3 \times 10^{10}\ \text{cm/sec}}{(2)\,(50\ \text{cm})} = 300\ \text{MHz} \tag{1.3.13}$$

for the separation in frequency of adjacent resonator modes. As indicated in Figure 1.13, the number of possible frequencies that can lase is therefore at most

$$\frac{1500\ \text{MHz}}{300\ \text{MHz}} = 5$$

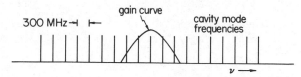

Figure 1.13 Mode frequencies separated by 300 MHz corresponding to a 50-cm one-dimensional cavity. A 1500-MHz gain curve overlaps only five modes.

The maximum number of possible modes, including polarization, is therefore 10, considerably smaller than the estimate of 4×10^8 modes obtained for three-dimensional cavities.

These results do not include the effects of diffraction of radiation at the mirror edges. Diffraction determines the x, y dependence of the field, which we have ignored completely. Accurate calculations of resonator modes, including diffraction, are often done with computers. Such calculations were first made in 1961 for the plane-parallel resonator of Figure 1.12 with either rectangular or circular mirrors. Actually lasers are seldom designed with flat mirrors. Laser resonator mirrors are usually spherical surfaces, for reasons to be discussed in Chapter 14. A great deal about laser cavities can nevertheless be understood without worrying about diffraction or mirror shape. In particular, *for most practical purposes, the mode-frequency spacing is given accurately enough by* (1.3.12).

1.4   LIGHT EMISSION AND ABSORPTION IN QUANTUM THEORY

The modern interpretation of light emission and absorption was first proposed by Einstein in 1905 in his theory of the photoelectric effect. Einstein assumed the difference in energy of the electron before and after its photoejection to be equal to the energy $h\nu$ of the photon absorbed in the process.

This picture of light absorption was extended in two ways by Bohr: to apply to atomic electrons that are not ejected during photon absorption but instead take on a higher energy within their atom, and to apply to the reverse process of photon emission, in which case the energy of the electron should decrease. These extensions of Einstein's idea fitted perfectly into Bohr's quantum-mechanical model of an atom in 1913. This model was the first to suggest that electrons are restricted to a certain fixed set of orbits around the atomic nucleus. This set of orbits was shown to correspond to a fixed set of allowed electron energies. The idea of a "quantum jump" was introduced to describe an electron's transition between two allowed orbits.

The amount of energy involved in a quantum jump depends on the quantum system. Atoms have quantum jumps whose energies are typically in the range 1–6 eV, as long as an outer-shell electron is doing the jumping. This is the ordinary case, so atoms usually absorb and emit photons in or near the optical region of the spectrum. Jumps by inner-shell atomic electrons require much more energy and are associated with X-ray photons. On the other hand, quantum jumps among the so-called Rydberg energy levels, those outer-electron levels lying far from the ground level and near to the ionization limit, involve only a small amount of energy, corresponding to far-infrared or even microwave photons.

Molecules have vibrational and rotational degrees of freedom whose quantum jumps are smaller (perhaps much smaller) than the quantum jumps in free atoms, and the same is often true of jumps between conduction and valence bands in semiconductors. Many crystals are transparent in the optical region, which is a sign that they do not absorb or emit optical photons, because they do not have quantum energy levels that permit jumps in the optical range. However, colored crystals such as ruby have impurities that do absorb and emit optical photons. These impurities are frequently atomic ions, and they have both discrete energy levels and broad bands of levels that allow optical quantum jumps (ruby is a good absorber of green photons and so appears red).

1.5 EINSTEIN THEORY OF LIGHT–MATTER INTERACTIONS

The atoms of a laser undergo repeated quantum jumps and so act as microscopic transducers. That is, each atom accepts energy and jumps to a higher orbit as a result of some input or "pumping" process, and converts it into other forms of energy—e.g., into light energy (photons)—when it jumps to a lower orbit. At the same time, each atom must deal with the photons that have been emitted earlier and reflected back by the mirrors. These prior photons, already channeled along the cavity axis, are the origin of the stimulated component to the atom's emission of subsequent photons.

In Figure 1.14 we indicate some ways in which energy conversion can occur. For simplicity we focus our attention on quantum jumps between two energy levels, 1 and 2, of an atom. The five distinct energy-conversion diagrams of Figure 1.14 are interpreted as follows:

- *a.* Absorption of an increment $\Delta E = E_2 - E_1$ of energy from the pump; the atom is raised from level 1 to level 2. In other words, an electron in the atom jumps from an inner orbit to an outer orbit.
- *b.* Spontaneous emission of a photon of energy $h\nu = E_2 - E_1$; the atom jumps down from level 2 to the lower level 1. The process occurs "spontaneously" without any external influence.
- *c.* Stimulated emission; the atom jumps down from energy level 2 to the lower level 1, and the emitted photon of energy $h\nu = E_2 - E_1$ is an exact replica of a photon already present. The process is induced, or stimulated, by the incident photon.
- *d.* Absorption of a photon of energy $h\nu = E_2 - E_1$; the atom jumps up from

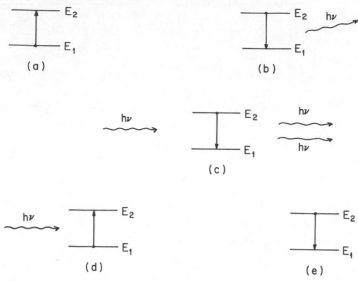

Figure 1.14 Energy conversion processes in a lasing atom or molecule: (*a*) absorption of energy $\Delta E = E_2 - E_1$ from the pump; (*b*) spontaneous emission of a photon of energy ΔE; (*c*) stimulated emission of a photon of energy ΔE; (*d*) absorption of a photon of energy ΔE; (*e*) nonradiative deexcitation.

    level 1 to the higher level 2. As in *c*, the process is induced by the incident photon.

e. Nonradiative deexcitation; the atom jumps down from level 2 to the lower level 1, but no photon is emitted so the energy $E_2 - E_1$ must appear in some other form (e.g., increased vibrational or rotational energy in the case of a molecule, or increased translational energy in the case of either an atom or a molecule).

    All these processes occur in the gain medium of a laser. Lasers are often classified according to the nature of the pumping process (*a*) which is the source of energy for the output laser beam. In electric-discharge lasers, for instance, the pumping occurs as a result of collisions of electrons in a gaseous discharge with the atoms (or molecules) of the gain medium. In an optically pumped laser the pumping process is the same as the absorption process (*d*), except that the pumping photons are supplied by a lamp or perhaps another laser.

    Our quantum picture is consistent with a highly simplified description of laser action. Suppose that lasing occurs on the transition defined by levels 1 and 2 of Figure 1.14. In the most favorable situation the lower level of this lasing transition is rapidly deexcited either by spontaneous emission or by nonradiative deexcitation. In this case, before it can absorb a laser photon of energy $h\nu = E_2 - E_1$ and jump to the upper level 2 of the lasing transition, the atom will drop from level 1 to some lower level. We will assume for simplicity that the rate of deexcitation of the lower level 1 is so large that the number of atoms remaining in that level is negligible compared with the number in level 2; this is a reasonably good approximation for many lasers. Under this approximation laser action can be described in terms of two "populations": the number n of atoms in the upper level 2, and the number q of photons in the laser cavity.

The number of laser photons in the cavity changes for two main reasons:

   i. Laser photons are continually being added because of stimulated emission.
  ii. Laser photons are continually being lost because of mirror transmission, scattering or absorption at the mirrors, etc.

Thus we can write a (provisional) equation for the rate of change of the number of photons, incorporating the gain and loss described in (i) and (ii) as follows:

$$\frac{dq}{dt} = anq - bq \qquad (1.5.1)$$

That is, the rate at which the number of laser photons changes is the sum of two separate rates: the rate of increase (amplification or gain) due to stimulated emission, and the rate of decrease (loss) due to imperfect mirror reflectivity.

As Eq. (1.5.1) indicates, the gain of laser photons due to stimulated emission is not only proportional to the number n of atoms in level 2, but also to the number q of photons already in the cavity. The efficiency of the stimulated emission process depends on the type of atom used and other factors. These factors are reflected in the size of the amplification or *gain coefficient a*. The rate of loss of laser photons is simply proportional to the number of laser photons present.

We can also write a provisional equation for n. Both stimulated and spontaneous emission cause n to decrease (in the former case in proportion to q, in the latter case not), and the pump causes n to increase at some rate we denote by p. Thus we write

$$\frac{dn}{dt} = -anq - fn + p \qquad (1.5.2)$$

Note that the first term appears in both equations, but with opposite signs. This reflects the central role of stimulated emission, and shows that the decrease of n (excited atoms) due to stimulated emission corresponds precisely to the increase of q (photons).

Equations (1.5.1) and (1.5.2) are equations describing laser action. They show how the numbers of lasing atoms and laser photons in the cavity are related to each other. They do not indicate what happens to the photons that leave the cavity, or what happens to the atoms when their electrons jump to some other level. Above all, they do not tell how to evaluate the coefficients a, b, f, p. They must be taken only as provisional equations, not well justified although intuitively reasonable.

It is important to note that neither equation (1.5.1) nor (1.5.2) can be solved independently of the other. That is, (1.5.1) and (1.5.2) are *coupled* equations. The coupling is due physically to stimulated emission: the lasing atoms of the gain medium can increase the number of photons via stimulated emission, but by the same process the presence of photons will also decrease the number of atoms in the upper laser level. This coupling between the atoms and the cavity photons is indicated schematically in Figure 1.15.

We also note that Eqs. (1.5.1) and (1.5.2) are nonlinear. The nonlinearity (the product of the two variables nq) occurs in both equations and is another manifestation of stimulated emission. No established systematic methods exist for solving nonlinear differential equations, and there is no known general solution to these laser equations. However, they have a number of well-defined limiting cases of some practical importance, and some of these do have known solutions. The most important case is steady state.

In steady state we can put both dq/dt and dn/dt equal to zero. Then (1.5.1) reduces to

$$n = b/a \equiv n_t \qquad (1.5.3)$$

which can be recognized as a *threshold* requirement on the number of upper-level atoms. That is, if $n < b/a$ then $dq/dt < 0$ and the number of photons in the cavity decreases, terminating laser action. The steady state of (1.5.2) also has a direct interpretation. From $dn/dt = 0$ and $n = n_t = b/a$ we find

$$q = \frac{p}{b} - \frac{f}{a} \qquad (1.5.4)$$

This equation establishes a threshold for the pumping rate, since the number of photons q can't be negative. Thus the minimum or threshold value of p compatible with steady-state operation is found by putting $q = 0$:

$$p_t = \frac{fb}{a} = fn_t \qquad (1.5.5)$$

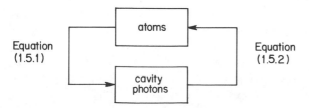

Figure 1.15 Self-consistent pair of laser equations.

In words, the threshold pumping rate just equals the loss rate per atom times the number of atoms present at threshold.

In Section 8.5 and in the four chapters beginning with Chapter 10 we will return to a discussion of laser equations. We will deal there with steady state as well as many other aspects of laser oscillation in two-level, three-level, and four-level quantum systems.

1.6 SUMMARY

The theory of laser action and description of cavity modes presented in this chapter can be regarded only as caricatures. In common with all caricatures, they display outstanding features of their subject boldly and simply. All theories of laser action must address the questions of *gain*, *loss*, *steady state*, and *threshold*. The virtues of our caricatures in addressing these questions are limited. They do not even suggest matters such as *linewidth*, *saturation*, *output power*, *mode locking*, *tunability*, and *stability*.

Obviously, one must not accept a caricature as the truth. Concerning the many aspects of the truth that are distorted or omitted by these first discussions, it will take much of this book to get the facts straight. This is not only a matter of dealing with details within the caricatures, but also with concepts that are larger than the caricatures altogether.

One should ask whether lasers are better described by photons or electric fields. Also, is Einstein's theory always satisfactory, or does Schrödinger's wave equation play a role? Are Maxwell's equations for electromagnetic waves significant? The answer to these questions is no, yes, yes. Laser theory is usually based on Schrödinger's and Maxwell's equations, neither of which was needed in this chapter.

From a different point of view another kind of question is equally important in trying to understand what a laser is. For example, why were lasers not built before 1960? Are there any rules of thumb that can predict, approximately and without detailed calculation, how much one can increase the output power or change the operating frequency? What are the most sensitive design features of a gas laser? a chemical laser? a semiconductor laser? Is a laser essentially quantum-mechanical, or can classical physics explain all the important features of laser operation?

It will not be possible to give detailed answers to all of these questions. However, these questions guide the organization of the book, and many of them are addressed individually. In the following chapters the reader should encounter the concepts of physics and engineering that are most important for understanding laser action in general, and that provide the background for pursuing further questions of particular theoretical or practical interest.

2 CLASSICAL DISPERSION THEORY

2.1 INTRODUCTION

An understanding of lasers requires some knowledge of the way in which light and matter interact. Many features of this interaction may be explained using a simple classical model. In this chapter and the next we describe this model and use it to introduce some concepts that are especially important for lasers.

We assume that the reader has previously encountered Maxwell's equations, at least briefly, and understands that they provide the most fundamental description of electric and magnetic fields. For a neutral dielectric medium (one with no free charges) Maxwell's equations are

$$\nabla \cdot \mathbf{D} = 0 \qquad (2.1.1)$$

$$\nabla \cdot \mathbf{B} = 0 \qquad (2.1.2)$$

$$\nabla \times \mathbf{E} = -\frac{\partial \mathbf{B}}{\partial t} \qquad (2.1.3)$$

$$\nabla \times \mathbf{H} = \frac{\partial \mathbf{D}}{\partial t} \qquad (2.1.4)$$

We will be interested only in nonmagnetic media, for which

$$\mathbf{B} = \mu_0 \mathbf{H} \qquad (2.1.5)$$

where $\mu_0 = 4\pi \times 10^{-7} \ \text{N/A}^2$. The electric displacement $\mathbf{D}$ is defined as

$$\mathbf{D} = \epsilon_0 \mathbf{E} + \mathbf{P} \qquad (2.1.6)$$

where $1/4\pi\epsilon_0 = 8.9874 \times 10^9 \ \text{N-m}^2/\text{C}^2$ and the polarization $\mathbf{P}$ is the electric dipole moment per unit volume of the medium. $\mathbf{P}$ is the only term in the Maxwell equations relating directly to the medium.

Applying the curl operation to both sides of Eq. (2.1.3), we obtain

$$\nabla \times (\nabla \times \mathbf{E}) = -\nabla \times \frac{\partial \mathbf{B}}{\partial t} = -\frac{\partial}{\partial t}(\nabla \times \mathbf{B}) \qquad (2.1.7)$$

Now we use the general identity (see Problem 2.1)

$$\nabla \times (\nabla \times \mathbf{E}) = \nabla(\nabla \cdot \mathbf{E}) - \nabla^2\mathbf{E} \qquad (2.1.8)$$

of vector calculus, together with (2.1.5) and the Maxwell equation (2.1.4), to write

$$\nabla(\nabla \cdot \mathbf{E}) - \nabla^2\mathbf{E} = -\mu_0 \frac{\partial^2\mathbf{D}}{\partial t^2} \qquad (2.1.9)$$

Finally we use the definition (2.1.6) of $\mathbf{D}$ and rearrange terms:

$$\nabla^2\mathbf{E} - \nabla(\nabla \cdot \mathbf{E}) - \frac{1}{c^2}\frac{\partial^2\mathbf{E}}{\partial t^2} = \frac{1}{\epsilon_0 c^2}\frac{\partial^2\mathbf{P}}{\partial t^2} \qquad (2.1.10)$$

Here we have used the fact that

$$\epsilon_0\mu_0 = \frac{1}{c^2} \qquad (2.1.11)$$

where $c = 2.998 \times 10^8$ m/sec is the velocity of light in vacuum.[1]

Equation (2.1.10) is a partial differential equation with independent variables x, y, z, and t. It tells us how the electric field depends on the electric dipole moment density $\mathbf{P}$ of the medium. We will be particularly interested in *transverse* fields (sometimes called *solenoidal* or *radiation* fields). Such fields satisfy

$$\nabla \cdot \mathbf{E} = 0 \qquad (2.1.12)$$

Transverse fields therefore satisfy the inhomogeneous wave equation

$$\nabla^2\mathbf{E} - \frac{1}{c^2}\frac{\partial^2\mathbf{E}}{\partial t^2} = \frac{1}{\epsilon_0 c^2}\frac{\partial^2\mathbf{P}}{\partial t^2} \qquad (2.1.13)$$

This is the fundamental electromagnetic field equation for our purposes. In order to make any use of it we must somehow specify the polarization $\mathbf{P}$. This cannot be done solely within the framework of the Maxwell equations, for $\mathbf{P}$ is a property of the material medium that the field $\mathbf{E}$ propagates in. We need to know how a dipole moment density $\mathbf{P}$ is produced in the medium. For this purpose we will introduce in the next section a theory of dielectric media.

First, however, we will finish this section with a discussion of solutions to the homogeneous (free-space) wave equation, which applies when there is no polar-

---

1. For convenience we include inside the cover of the book a table of physical constants that appear frequently in our study of lasers.

ization present. Laser resonator theory is based on the free-space wave equation and free-space solutions. Such solutions are useful, even though lasers do not operate in vacuum, because most laser media are optically homogeneous.

We will consider only the case of a rectangular cavity, as sketched in Figure 1.10. We also imagine we have perfectly reflecting walls; then the components of the electric field parallel to the walls must vanish on the walls. The electric field inside the cavity satisfies the wave equation

$$\nabla^2 \mathbf{E} - \frac{1}{c^2} \frac{\partial^2}{\partial t^2} \mathbf{E} = 0 \tag{2.1.14}$$

For a monochromatic field of angular frequency $\omega = 2\pi\nu$, we may use the complex-field representation (where the physical electric field is understood to be the real part of the right-hand side):

$$\mathbf{E}(\mathbf{r}, t) = \mathbf{E}_0(\mathbf{r})\, e^{-i\omega t} \tag{2.1.15}$$

and (2.1.14) becomes

$$\nabla^2 \mathbf{E}_0(\mathbf{r}) + k^2 \mathbf{E}_0(\mathbf{r}) = 0, \qquad k \equiv \omega/c \tag{2.1.16}$$

That is,

$$(\nabla^2 + k^2)\, E_{0x}(\mathbf{r}) = 0 \tag{2.1.17}$$

and likewise for the y and z components.

To solve (2.1.17), it is convenient to use the method of separation of variables. Separation means that the solution is written in a factored form:

$$E_{0x}(x, y, z) = F(x)\, G(y)\, H(z) \tag{2.1.18}$$

and then inserted in (2.1.17). After carrying out the differentiations required by $\nabla^2 = \partial^2/\partial x^2 + \partial^2/\partial y^2 + \partial^2/\partial z^2$, we divide through by the product FGH and obtain

$$\frac{1}{F}\frac{d^2 F}{dx^2} + \frac{1}{G}\frac{d^2 G}{dy^2} + \frac{1}{H}\frac{d^2 H}{dz^2} + k^2 = 0 \tag{2.1.19}$$

Since each of the first three terms on the left side is a function of a different independent variable, Eq. (2.1.19) [and hence (2.1.17)] can only be true for all x, y, and z if each term is separately constant, i.e.,

$$\frac{1}{F}\frac{d^2F}{dx^2} = -k_x^2 \qquad (2.1.20\text{a})$$

$$\frac{1}{G}\frac{d^2G}{dy^2} = -k_y^2 \qquad (2.1.20\text{b})$$

$$\frac{1}{H}\frac{d^2H}{dz^2} = -k_z^2 \qquad (2.1.20\text{c})$$

with

$$k_x^2 + k_y^2 + k_z^2 = k^2 \qquad (2.1.21)$$

The boundary condition that the tangential component of the electric field vanishes on the cavity walls means that

$$E_{0x}(x, y = 0, z) = E_{0x}(x, y = L_y, z) = 0 \qquad (2.1.22\text{a})$$

and

$$E_{0x}(x, y, z = 0) = E_{0x}(x, y, z = L_z) = 0 \qquad (2.1.22\text{b})$$

or

$$G(0) = G(L_y) = 0 \qquad (2.1.23\text{a})$$

$$H(0) = H(L_z) = 0 \qquad (2.1.23\text{b})$$

A solution of (2.1.20b) satisfying the boundary condition $G(0) = 0$ is

$$G(y) = \sin k_y y \qquad (2.1.24)$$

In order to satisfy $G(L_y) = 0$ as well, we must have $\sin k_y L_y = 0$, or in other words

$$k_y L_y = m\pi, \qquad m = 0, 1, 2, \ldots \qquad (2.1.25\text{a})$$

In exactly the same way we find that solutions of Eq. (2.1.20c) satisfying (2.1.23b) are only possible if

$$k_z L_z = n\pi, \qquad n = 0, 1, 2, \ldots \qquad (2.1.25\text{b})$$

Finally, consideration of the equations for the y and z components of $\mathbf{E}_0(\mathbf{r})$, together with the appropriate boundary conditions, shows that allowed solutions for $\mathbf{E}_0(\mathbf{r})$ must satisfy (2.1.25a), (2.1.25b), and

$$k_x L_x = l\pi, \qquad l = 0, 1, 2, \ldots \tag{2.1.25c}$$

The full solutions for the components of $\mathbf{E}(\mathbf{r}, t)$ satisfying Maxwell's equations and the boundary conditions inside the cavity are (Problem 2.2)

$$E_x(x, y, z, t) = A_x e^{-i\omega t} \cos \frac{l\pi x}{L_x} \sin \frac{m\pi y}{L_y} \sin \frac{n\pi z}{L_z} \tag{2.1.26a}$$

$$E_y(x, y, z, t) = A_y e^{-i\omega t} \sin \frac{l\pi x}{L_x} \cos \frac{m\pi y}{L_y} \sin \frac{n\pi z}{L_z} \tag{2.1.26b}$$

$$E_z(x, y, z, t) = A_z e^{-i\omega t} \sin \frac{l\pi x}{L_x} \sin \frac{m\pi y}{L_y} \cos \frac{n\pi z}{L_z} \tag{2.1.26c}$$

where the coefficients A_x, A_y, and A_z must satisfy the condition (Problem 2.3)

$$\frac{l}{L_x} A_x + \frac{m}{L_y} A_y + \frac{n}{L_z} A_z = 0 \tag{2.1.27}$$

implied by the Maxwell equation $\nabla \cdot \mathbf{E} = 0$, valid in the empty cavity.

From Eqs. (2.1.21) and (2.1.25) we have

$$k^2 = \pi^2 \left(\frac{l^2}{L_x^2} + \frac{m^2}{L_y^2} + \frac{n^2}{L_z^2} \right) \tag{2.1.28}$$

The possible modes of the rectangular closed cavity have allowed frequencies determined by (2.1.28) and $k = \omega/c = 2\pi\nu/c$ [recall (2.1.16)]:

$$\nu = \nu_{lmn} = \frac{c}{2} \left(\frac{l^2}{L_x^2} + \frac{m^2}{L_y^2} + \frac{n^2}{L_z^2} \right)^{1/2} \tag{2.1.29}$$

This formula is a generalization of (1.3.7), in which case the cavity was assumed cubical ($L_x = L_y = L_z = L$). The same mode-density formulas given in Chapter 1 are valid here, independent of cavity shape, as long as L_x, L_y, L_z are all much larger than $\lambda = c/\nu$. That is, the number of modes in the frequency interval $[\nu, \nu + d\nu]$ is

$$dN_\nu = \frac{8\pi\nu^2}{c^3} V \, d\nu \tag{2.1.30a}$$

and in the wavelength range $d\lambda$ the number is (Problem 2.4)

$$dN_\lambda = 8\pi \left(\frac{V}{\lambda^3} \right) \frac{d\lambda}{\lambda} \tag{2.1.30b}$$

• It is not difficult to derive the important mode-density formulas for dN_ν and dN_λ given above. First of all, the number of modes available in a cavity is infinite. This is clear because in (2.1.29), for example, an infinite number of values are permitted for any of the three mode indices l, m, n. However, the number of modes *whose frequency lies in the neighborhood $d\nu$ of a given value ν* is finite. This number is related to the number of modes *whose frequency is less than ν*, and it is this number we will determine first.

The number of modes we want is the number of terms in the triple sum:

$$N = \sum_l \sum_m \sum_n \tag{2.1.31}$$

where the upper limits on the sums are determined by the maximum frequency to be included. The simplest approach to this problem is to stipulate that the cavity is much larger than a typical wavelength (obviously true for realistic cavities and optical wavelengths). Then the discrete nature of the sum is not important and we can rewrite the sum as a triple integral:

$$N = \int dl \int dm \int dn \tag{2.1.32}$$

In addition, for a large cavity the shape is not very important in determining the *number* of modes (although critical for the spatial characteristics of the modes, of course), so for our present purpose we can just as well assume the simplest shape—a cube with sides equal to L. For a cubical cavity (2.1.29) becomes

$$\left(\frac{2L}{c}\right)^2 \nu^2 = l^2 + m^2 + n^2 \tag{2.1.33}$$

It is a useful trick to regard the triplet (l, m, n) as the components of a fictitious vector **q**:

$$\mathbf{q} = \mathbf{i}l + \mathbf{j}m + \mathbf{k}n \tag{2.1.34a}$$

with magnitude

$$\mathbf{q}^2 = q^2 = l^2 + m^2 + n^2 \tag{2.1.34b}$$

Then the triple integral can be denoted

$$N = \int\int\int d^3q \tag{2.1.35}$$

Equation (2.1.33) indicates that ν depends only on the length, but not the orientation, of the vector **q**. Thus we rewrite the mode integral in spherical coordinates:

$$N = \int\int\int q^2 \, dq \, \sin\theta_q \, d\theta_q \, d\phi_q \tag{2.1.36}$$

and carry out the integrations to get:

$$N = \frac{4\pi}{8} \int q^2 \, dq$$

$$= \frac{4\pi}{8} \frac{q^3}{3} \tag{2.1.37}$$

Here the factor 4π is the result of the angular integration (4π total solid angle) and the $\frac{1}{8}$ is due to the restriction on the original integers l, m, n to be positive, so that only the vectors $\mathbf{q}$ in the positive octant of the integration (2.1.35) should be counted as corresponding to physical modes.

In (2.1.37) q is the length of the vector $\mathbf{q}$ compatible with the given frequency ν. From (2.1.33) it is clear that $q = (2L/c)\,\nu$, so we finally get

$$N_\nu = \frac{\pi}{6} \left(\frac{2L}{c}\right)^3 \nu^3$$

$$= 4\pi \frac{\nu^3}{3c^3} V \tag{2.1.38}$$

where $V = L^3$ is the cavity volume.

Since our derivation of (2.1.38) did not take account of the polarization of the cavity modes, we are still free to choose independently any *two* polarizations (see Problem 2.3). Thus we have

$$N_\nu = \frac{8\pi\nu^3}{3c^3} V \quad \text{(all polarizations)} \tag{2.1.39}$$

for the number of possible cavity modes of frequency less than ν, counting all polarizations.

The number of possible field modes in the frequency interval from ν to $\nu + d\nu$ is therefore

$$dN_\nu = \frac{8\pi\nu^2}{c^3} V \, d\nu \tag{2.1.40}$$

which is exactly the formula used in Section 1.3 and reproduced in (2.1.30a). The formula for the density dN_λ given in (2.1.30b) is obtained from dN_ν by application of the relation $\nu = c/\lambda$ (see Problem 2.4). •

2.2 THE ELECTRON OSCILLATOR MODEL

In classical physics the motion of a particle is described by Newton's second law. For a charged particle in an electromagnetic field the force referred to in Newton's second law is the Lorentz force

$$\mathbf{F} = e(\mathbf{E} + \mathbf{v} \times \mathbf{B})$$ (2.2.1)

where e and $\mathbf{v}$ are the charge and velocity, respectively, of the particle.

We are interested in the atoms that play roles in laser action. These are atoms in dielectrics such as crystals, vapors, glasses, and liquids. Atoms are neutral objects, but (2.2.1) applies to the individual protons and electrons in these atoms. The interaction of atomic protons and electrons with light can be treated very accurately in most cases with classical laws and concepts. The quantum-mechanical basis for our classical treatment, and quantum corrections to several classical formulas, are discussed in Chapter 6.

The electron has mass m_e and charge e (a negative number), and the oppositely charged core of the atom ("nucleus") has mass m_n and charge $-e$. The nucleus exerts a binding force $\mathbf{F}_{en}$ on the electron, depending on the relative separation $\mathbf{r}_{en}$ $= \mathbf{r}_e - \mathbf{r}_n$, as shown in Figure 2.1. The electron also exerts a force $\mathbf{F}_{ne}$ on the nucleus, and according to Newton's third law,

$$\mathbf{F}_{ne}(\mathbf{r}_{en}) = -\mathbf{F}_{en}(\mathbf{r}_{en})$$ (2.2.2)

The Newton equations of motion for the electron and nucleus are therefore

$$m_e \frac{d^2\mathbf{r}_e}{dt^2} = e\mathbf{E}(\mathbf{r}_e, t) + \mathbf{F}_{en}(\mathbf{r}_{en})$$ (2.2.3a)

and

$$m_n \frac{d^2\mathbf{r}_n}{dt^2} = -e\mathbf{E}(\mathbf{r}_n, t) + \mathbf{F}_{ne}(\mathbf{r}_{en})$$ (2.2.3b)

In writing Eqs. (2.2.3) we have dropped the magnetic contribution to the Lorentz force. Because optical phenomena do not normally involve relativistic particle velocities, we can safely disregard the magnetic force hereafter.

The interaction of electromagnetic fields with charges is mainly determined by

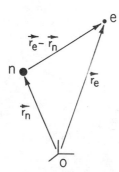

Figure 2.1 The position vectors $\mathbf{r}_e$ and $\mathbf{r}_n$ of the electron and nucleus, measured from some origin O. By Newton's third law the force $\mathbf{F}_{en}(\mathbf{r}_e - \mathbf{r}_n)$ exerted by the nucleus on the electron is equal in magnitude but opposite in direction to the force $\mathbf{F}_{ne}(\mathbf{r}_e - \mathbf{r}_n)$ exerted by the electron on the nucleus.

the acceleration of the charges. The nucleus is so massive compared to an electron that its acceleration is generally negligible. In this case only the electron equation is needed. The binding force $\mathbf{F}_{en}$ is strong enough to restrict the atomic electrons to small excursions about the (approximately stationary) nucleus. Thus we can write $\mathbf{r}_e = \mathbf{r}_n + \mathbf{x}$, where $\mathbf{x}$ is an atomic dimension ($|\mathbf{x}| \lesssim 10$ Å $= 1$ nm $= 10^{-9}$ m, 1 nanometer). The electric radiation field varies on the scale of an optical wavelength ($\lambda \approx 6000$ Å $= 600$ nm for yellow light) and is not sensitive to variations as small as $|\mathbf{x}|$, so we have $\mathbf{E}(\mathbf{r}_e, t) \approx \mathbf{E}(\mathbf{r}_n, t)$.

Within these approximations it is easy to see that the effective interaction equation replacing Eqs. (2.2.3) is

$$m \frac{d^2\mathbf{x}}{dt^2} = e\mathbf{E}(\mathbf{R}, t) + \mathbf{F}_{en}(\mathbf{x}) \qquad (2.2.4)$$

Here we have dropped the subscript e from the electron mass and have written $\mathbf{R}$ for the position of the stationary nucleus. Actually, $\mathbf{x}$ and $\mathbf{R}$ are the relative coordinate and center-of-mass coordinate of the electron–nucleus pair, and m is the associated reduced mass. These terms are defined in the black-dot section below. For our purposes it is accurate to continue to think of $\mathbf{R}$ as the position of the nucleus and m as the electron mass. Only in exceptional cases in which the two charges have nearly equal mass, such as the positronium atom (an atom in which the nucleus is a positron, i.e., an anti-electron, rather than a proton), would significant corrections be required.

Note that the electric force in (2.2.4) can be written in terms of a potential $V(\mathbf{x}, \mathbf{R}, t)$. If we define

$$V(\mathbf{x}, \mathbf{R}, t) = -e\mathbf{x} \cdot \mathbf{E}(\mathbf{R}, t) \qquad (2.2.5)$$

then we have

$$eE(\mathbf{R}, t) = -\nabla_\mathbf{x} \left[-e\mathbf{x} \cdot \mathbf{E}(\mathbf{R}, t) \right]$$
$$= -\nabla_\mathbf{x} V(\mathbf{x}, \mathbf{R}, t) \qquad (2.2.6)$$

where $\nabla_\mathbf{x}$ indicates the gradient with respect to the coordinate $\mathbf{x}$. The potential V is proportional to $e\mathbf{x} = \mathbf{p} = e(\mathbf{r}_e - \mathbf{r}_n)$, which is the *electric dipole moment* of the atom. Thus the equation of motion (2.2.4) is frequently identified as belonging to the *dipole approximation*. Recall that the main approximation is to take $\mathbf{E}$ to depend only on the nuclear position $\mathbf{R}$. This was justified by the tightness of the atomic binding, which restricts $\mathbf{x}$ to a range much smaller than the wavelength of the radiation described by $\mathbf{E}$.

- The center of mass of the electron–nucleus system is defined to be

$$\mathbf{R} = \frac{m_e \mathbf{r}_e + m_n \mathbf{r}_n}{M} \tag{2.2.7a}$$

where $M = m_e + m_n$ and $\mathbf{x}$ is the electron coordinate relative to the nucleus,

$$\mathbf{x} = \mathbf{r}_{en} \tag{2.2.7b}$$

in terms of which

$$\mathbf{r}_e = \mathbf{R} + \frac{m_n}{M} \mathbf{x} \tag{2.2.8a}$$

$$\mathbf{r}_n = \mathbf{R} - \frac{m_e}{M} \mathbf{x} \tag{2.2.8b}$$

Then Eqs. (2.2.3) may be written as

$$m_e \frac{d^2\mathbf{R}}{dt^2} + m \frac{d^2\mathbf{x}}{dt^2} = e\mathbf{E}\left(\mathbf{R} + \frac{m_n}{M} \mathbf{x}, t\right) + \mathbf{F}_{en}(\mathbf{x}) \tag{2.2.9a}$$

$$m_n \frac{d^2\mathbf{R}}{dt^2} - m \frac{d^2\mathbf{x}}{dt^2} = -e\mathbf{E}\left(\mathbf{R} - \frac{m_e}{M} \mathbf{x}, t\right) + \mathbf{F}_{ne}(\mathbf{x}) \tag{2.2.9b}$$

where

$$m = \frac{m_e m_n}{M} = \frac{m_e m_n}{m_e + m_n} \tag{2.2.10}$$

is called the *reduced mass* of the electron–nucleus system.

By adding and subtracting Eqs. (2.2.9), and using (2.2.2), we obtain the equations of motion

$$M \frac{d^2\mathbf{R}}{dt^2} = e\left[\mathbf{E}\left(\mathbf{R} + \frac{m_n}{M} \mathbf{x}, t\right) - \mathbf{E}\left(\mathbf{R} - \frac{m_e}{M} \mathbf{x}, t\right)\right] \tag{2.2.11a}$$

and

$$m \frac{d^2\mathbf{x}}{dt^2} = \frac{e}{2}\left[\mathbf{E}\left(\mathbf{R} + \frac{m_n}{M} \mathbf{x}, t\right) + \mathbf{E}\left(\mathbf{R} - \frac{m_e}{M} \mathbf{x}, t\right)\right]$$

$$+ \mathbf{F}_{en}(\mathbf{x}) + \tfrac{1}{2}(m_n - m_e)\frac{d^2\mathbf{R}}{dt^2} \tag{2.2.11b}$$

Equation (2.2.11a) describes the motion of the center of mass of the atom. In the absence of an external field, the center of mass moves with constant velocity. Equation (2.2.11b) describes the motion of the relative coordinate $\mathbf{x}$ of the electron–nucleus system.

We have already mentioned that optical radiation is characterized by wavelengths that are a few thousand Ångstrom units (1 Å $= 10^{-10}$ m) or larger, and the electron–nucleus separations in atoms are typically only 1–10 Å in size. The extreme disparity of these sizes is the basis of a fundamental approximation called the dipole approximation. The dipole approximation arises from the leading terms of a Taylor series expansion of the type

$$F(X + \delta X) = F(X) + \delta X F'(X) + \tfrac{1}{2}(\delta X)^2 F''(X) + \cdots$$

applied to the electric field vectors in (2.2.11). The vector analog of the Taylor series gives

$$\mathbf{E}\left(\mathbf{R} - \frac{m_e}{M}\mathbf{x}, t\right) = \mathbf{E}(\mathbf{R}, t) - \frac{m_e}{M}\mathbf{x} \cdot \nabla \mathbf{E}(\mathbf{R}, t) + \cdots \qquad (2.2.12a)$$

and

$$\mathbf{E}\left(\mathbf{R} + \frac{m_n}{M}\mathbf{x}, t\right) = \mathbf{E}(\mathbf{R}, t) + \frac{m_n}{M}\mathbf{x} \cdot \nabla \mathbf{E}(\mathbf{R}, t) + \cdots \qquad (2.2.12b)$$

and if we retain only the first two terms in the Taylor series (2.2.12), then Eqs. (2.2.11) become

$$M\frac{d^2\mathbf{R}}{dt^2} \approx e\mathbf{x} \cdot \nabla \mathbf{E}(\mathbf{R}, t) \qquad (2.2.13a)$$

$$m\frac{d^2\mathbf{x}}{dt^2} \approx e\mathbf{E}(\mathbf{R}, t) + \left(\frac{m_n - m_e}{M}\right) e\mathbf{x} \cdot \nabla \mathbf{E}(\mathbf{R}, t) + \mathbf{F}_{en}(\mathbf{x}) \qquad (2.2.13b)$$

We have already mentioned that the vector

$$\mathbf{p} = e\mathbf{x} \qquad (2.2.14)$$

is the electric dipole moment of the electron–nucleus pair. In terms of $\mathbf{p}$ Eq. (2.2.13a) is

$$M\frac{d^2\mathbf{R}}{dt^2} = \mathbf{p} \cdot \nabla \mathbf{E}(\mathbf{R}, t) = \mathbf{F} = -\nabla_{\mathbf{R}}V(\mathbf{x}, \mathbf{R}, t) \qquad (2.2.15)$$

where $\nabla_{\mathbf{R}}$ denotes the gradient with respect to the coordinate $\mathbf{R}$ and

$$V(\mathbf{x}, \mathbf{R}, t) = -\mathbf{p} \cdot \mathbf{E}(\mathbf{R}, t) \qquad (2.2.16)$$

is the potential energy of an electric dipole $\mathbf{p}$ at the point $\mathbf{R}$ in an electric field. It was already identified in (2.2.5) above. Finally, we retain only the leading $\mathbf{E}$ term on the right-hand side of (2.2.13b) and obtain

$$m\frac{d^2\mathbf{x}}{dt^2} \approx e\mathbf{E}(\mathbf{R}, t) + \mathbf{F}_{en}(\mathbf{x}) = -\nabla_{\mathbf{x}}V + \mathbf{F}_{en}(\mathbf{x})$$

$$(\text{electric-dipole approximation}) \qquad (2.2.17)$$

which is Eq. (2.2.4) again, this time with m, $\mathbf{x}$ and $\mathbf{R}$ carefully defined. ●

For most of our purposes we can assume that the center-of-mass motion of the atom is unaffected by the field, so that we can ignore (2.2.15). However, this is possible only because we are interested mainly in explaining laser action, which depends mostly on internal transitions within atoms or molecules, transitions based on the relative coordinate $\mathbf{x}$. For other purposes Eq. (2.2.15) is essential. For example, the important topics of laser trapping and laser cooling depend directly on the effects produced by laser light on the atomic center of mass.

In order to proceed with (2.2.4) or (2.2.17) it is necessary to know $\mathbf{F}_{en}(\mathbf{x})$. For reasons that only quantum theory can explain (see Chapter 6), the classical theory satisfactorily treats many important features of the interaction of light with matter by adopting an *ad hoc* hypothesis about $\mathbf{F}_{en}$ due to H. A. Lorentz (around 1900). This hypothesis states that an electron in an atom responds to light as if it were bound to its atom or molecule by a simple spring. As a consequence the electron can be imagined to oscillate about the nucleus.

This *electron oscillator model* is often called the *Lorentz model* of an atom. It is not really a model of an atom as such, but rather a model of the way an atom responds to a perturbation. The Lorentz model was developed before atoms were understood to have massive nuclei. Lorentz simply asserted that each electron in an atom has a certain equilibrium position when there are no external forces. Under the influence of an electromagnetic field, the electron experiences the Lorentz force and is displaced from its equilibrium position, and according to Lorentz "the displacement will immediately give rise to a new force by which the particle is pulled back towards its original position, and which we may therefore appropriately distinguish by the name of *elastic force*."[2] Lorentz's assertion is equivalent to the replacement $\mathbf{F}_{en}(\mathbf{x}) \rightarrow -k_s\mathbf{x}$, where k_s is the "spring constant" associated with the hypothetical elastic force. This leads to the equation

$$m\frac{d^2\mathbf{x}}{dt^2} = e\mathbf{E}(\mathbf{R}, t) - k_s\mathbf{x}$$

(Lorentz-model fundamental equation)   (2.2.18a)

or

$$\left(\frac{d^2}{dt^2} + \omega_0^2\right)\mathbf{x} = \frac{e}{m}\mathbf{E}(\mathbf{R}, t),$$ (2.2.18b)

where we have defined the electron's natural oscillation frequency $\omega_0 = \sqrt{k_s/m}$.

It is easy to see how a nonzero dipole moment density $\mathbf{P}$ such as is required in the wave equation (2.1.13) arises in the electron oscillator model. Let us continue to suppose that each atom of the medium has only one electron. When a field is applied, each atom's electron is displaced by some $\mathbf{x}$ from its original position (Figure 2.2). Thus, according to (2.2.14), each atom has a dipole moment $\mathbf{p} =$

---

2. H. A. Lorentz, *The Theory of Electrons* (Dover, New York, 1952), p. 9.

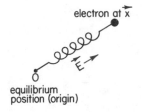

electron at $\vec{x}$

equilibrium
position (origin)

Figure 2.2 The electron oscillator (Lorentz) model of an atomic electron. An applied field displaces the electron from its equilibrium position. The atom reacts as though the electron were a charged mass on a spring.

*e*x. (The electron charge *e* is negative, so that **p** points from the negative charge to the positive.) If the density of atoms is denoted *N*, then the density of dipole moments is *N* times the individual dipole moment of each atom. That is, we have the polarization density

$$\mathbf{P} = N\mathbf{p} = Ne\mathbf{x} \qquad (2.2.19)$$

induced in the medium by the field.

   This expression for **P** closes a circle. We now have a model for the interaction between electromagnetic radiation and a material medium consisting of one-electron atoms. The Maxwell equation (2.1.13) tells us how the electric field **E** depends upon the dipole moment density **P** of the medium. Newton's equation (2.2.18) tells us how the electron displacement **x** depends upon the electric field **E**. And Eq. (2.2.19) connects these basic equations by relating **P** to **x**. The electron oscillator model thus ties together the Maxwell equations with Newton's law of motion. Solutions of these coupled equations will provide the model's predictions about the mutual interaction of light and matter.

2.3   REFRACTIVE INDEX AND POLARIZABILITY

In many applications it is necessary to know the effect on a bound electron of a monochromatic field, i.e., a field varying purely sinusoidally in time with the single frequency ω. It will often be convenient for our purposes to consider a monochromatic *plane* wave, i.e., a monochromatic wave having the same value across a plane perpendicular to the direction (z, say) of propagation. We will furthermore deal initially with a linearly polarized wave. In this case the electric field at the position of the atom has the form

$$\mathbf{E}(z, t) = \hat{\varepsilon} E_0 \cos(\omega t - kz) \qquad (2.3.1)$$

where k is an undetermined constant. We assume that the wave amplitude E_0 does not depend on t or z. The unit vector $\hat{\varepsilon}$ specifies the polarization of the wave and lies in the xy plane, normal to the z axis (i.e., the wave is "transverse"). The period of the wave described by (2.3.1) is $2\pi/\omega$, and so the frequency ν is $\omega/2\pi$. Finally we recall that the wavelength, or spatial period, of the wave is $\lambda = 2\pi/k$.

The electric field (2.3.1) is transverse, satisfying condition (2.1.12):

$$\nabla \cdot \mathbf{E} = \frac{\partial}{\partial x} E_x + \frac{\partial}{\partial y} E_y + \frac{\partial}{\partial z} E_z = 0 \tag{2.3.2}$$

because it does not depend upon x or y and because $E_z = 0$ (because $\varepsilon_z = 0$). If $\mathbf{E}$ is also to be a solution of Maxwell's equations in *vacuum*, it must satisfy the wave equation (2.1.13) with $\mathbf{P} = 0$:

$$\nabla^2 \mathbf{E} - \frac{1}{c^2} \frac{\partial^2 \mathbf{E}}{\partial t^2} = 0 \tag{2.3.3}$$

Suppose we try to satisfy this wave equation with the plane wave given in Eq. (2.3.1). Then since $\nabla^2 \mathbf{E} = \partial^2 \mathbf{E}/\partial z^2 = -k^2 \mathbf{E}$ and $\partial^2 \mathbf{E}/\partial t^2 = -\omega^2 \mathbf{E}$ in this case, we must have

$$\left(-k^2 + \frac{\omega^2}{c^2} \right) \hat{\boldsymbol{\varepsilon}} E_0 \cos (\omega t - kz) = 0 \tag{2.3.4}$$

Except in the trivial and physically uninteresting case $E_0 = 0$, we can satisfy this equation only if the constant k satisfies

$$k^2 = \frac{\omega^2}{c^2} \tag{2.3.5}$$

This relation between k and ω is an example of a dispersion relation. We have the two possibilities $k = \pm \omega/c$, or

$$\mathbf{E}(z, t) = \hat{\boldsymbol{\varepsilon}} E_0 \cos \omega(t \pm z/c).$$

The choices $k = \omega/c$ or $k = -\omega/c$ correspond respectively to waves propagating in the positive or negative z direction with the phase velocity

$$v_p = \frac{\omega}{|k|} = c \tag{2.3.6}$$

This merely tells us something we already know: the propagation velocity of an electromagnetic wave in vacuum is c, the (vacuum) speed of light, and (2.3.5) is sometimes called the vacuum dispersion relation.

Now let us consider the case of a material medium of propagation, where the right-hand side of Eq. (2.1.13) is not zero. According to the electron oscillator model of a medium of one-electron atoms, $\mathbf{P}$ is given by Eq. (2.2.19), with the electron displacement $\mathbf{x}$ satisfying the equation of motion (2.2.18). If the electric field (2.3.1) is to be a solution of the coupled Maxwell–Newton equations it must be the driving field in the Newton equation (2.2.18):

$$\frac{d^2x}{dt^2} + \omega_0^2 x = \hat{\epsilon} \frac{e}{m} E_0 \cos(\omega t - kz) \qquad (2.3.7)$$

This equation has the solution[3]

$$x = \hat{\epsilon} \left(\frac{eE_0/m}{\omega_0^2 - \omega^2} \right) \cos(\omega t - kz) \qquad (2.3.8)$$

It is sometimes convenient to write this solution in the form

$$p = ex = \alpha E \qquad (2.3.9)$$

where we define the electronic *polarizability*

$$\alpha(\omega) = \frac{e^2/m}{\omega_0^2 - \omega^2} \qquad (2.3.10)$$

as the ratio of the induced dipole moment **p** of the atom to the electric field **E** that produces this dipole moment.

Thus we have the dipole moment density

$$\mathbf{P} = N\mathbf{p} = N\alpha(\omega)\mathbf{E} = \hat{\epsilon} \left(\frac{Ne^2/m}{\omega_0^2 - \omega^2} \right) E_0 \cos(\omega t - kz) \qquad (2.3.11)$$

This is the polarization predicted by the electron oscillator model when there is a field (2.3.1) in the medium. This solution for the polarization provides the source term on the right-hand side of the Maxwell equation (2.1.13). In order that we have a consistent theory, the electric field (2.3.1) appearing in (2.3.11) must satisfy (2.1.13). Using the assumed plane-wave, monochromatic solution (2.3.1) for **E** in (2.1.13), together with the above results for **P**, we obtain the consistency condition

$$\left(-k^2 + \frac{\omega^2}{c^2} \right) \hat{\epsilon} E_0 \cos(\omega t - kz) = -\frac{\omega^2}{c^2} \frac{N\alpha(\omega)}{\epsilon_0} \hat{\epsilon} E_0 \cos(\omega t - kz) \qquad (2.3.12)$$

3. This is the particular solution. The homogeneous solution has been omitted for reasons that are explained in Section 3.3.

To satisfy this equation k must satisfy a more general dispersion relation:

$$k^2 = \frac{\omega^2}{c^2}\left(1 + \frac{N\alpha(\omega)}{\epsilon_0}\right)$$

$$= \frac{\omega^2}{c^2} n^2(\omega) \tag{2.3.13}$$

where we have defined

$$n^2(\omega) = 1 + \frac{N\alpha(\omega)}{\epsilon_0} \tag{2.3.14a}$$

or

$$n(\omega) = \left(1 + \frac{N\alpha(\omega)}{\epsilon_0}\right)^{1/2}$$

$$= \left(1 + \frac{Ne^2/m\epsilon_0}{\omega_0^2 - \omega^2}\right)^{1/2} \tag{2.3.14b}$$

The plane monochromatic wave (2.3.1) is therefore a solution of the Maxwell equation (2.3.3) *in vacuum* if

$$k^2 = \frac{\omega^2}{c^2} \quad \text{(vacuum)} \tag{2.3.15}$$

but it can be a solution *in a material medium* only if

$$k^2 = n^2(\omega)\frac{\omega^2}{c^2} \quad \text{(material medium)} \tag{2.3.16}$$

In the second case the phase velocity of the wave is

$$v_p = \frac{\omega}{|k|} = \frac{c}{n(\omega)} \tag{2.3.17}$$

which identifies $n(\omega)$ as the refractive index of the medium for light of frequency ω.

The result (2.3.14) for the refractive index applies to a medium of one-electron atoms. This restriction may be removed. If we assume that the Z electrons in an atom respond independently of one another to an imposed field, then the displacement $\mathbf{x}_i$ of each electron satisfies (2.2.18) with perhaps a different "spring constant" k_i for each electron. The solution is given by (2.3.8):

$$\mathbf{x}_i = \hat{\varepsilon}\left(\frac{eE_0/m}{\omega_i^2 - \omega^2}\right)\cos(\omega t - kz) \qquad (2.3.18)$$

where $\omega_i = (k_i/m)^{1/2}$ is the natural oscillation frequency of the ith electron. The dipole moment (2.3.9) of the atom is replaced by

$$\mathbf{p} = \sum_{i=1}^{Z} e\mathbf{x}_i \qquad (2.3.19)$$

where the summation is over the displacements of the Z electrons in the atom. From (2.3.9), therefore, we have the multielectron polarizability

$$\alpha(\omega) = \sum_{i=1}^{Z} \frac{e^2/m}{\omega_i^2 - \omega^2} \qquad (2.3.20)$$

and the dipole density

$$\mathbf{P} = \hat{\varepsilon}N\left(\sum_{i=1}^{Z} \frac{e^2/m}{\omega_i^2 - \omega^2}\right)E_0\cos(\omega t - kz) \qquad (2.3.21)$$

It follows that the generalization of (2.3.14) for a medium of Z-electron atoms is

$$n(\omega) = \left(1 + \frac{N}{\epsilon_0}\alpha(\omega)\right)^{1/2}$$

$$= \left(1 + \frac{N}{\epsilon_0}\sum_{i=1}^{Z}\frac{e^2/m}{\omega_i^2 - \omega^2}\right)^{1/2} \qquad (2.3.22a)$$

A further generalization occurs if more than one species of atom is present. If species a has Z_a electrons with resonant frequencies labeled ω_{ai}, then we have

$$n(\omega) = \left(1 + \sum_a \frac{N_a}{\epsilon_0}\sum_{i=1}^{Z_a}\frac{e^2/m}{\omega_{ai}^2 - \omega^2}\right)^{1/2} \qquad (2.3.22b)$$

where N_a is the density of species a. In the case of gases, $n(\omega) \approx 1$ for optical frequencies and we can expand the square root in (2.3.22b) using the binomial formula $(1 + x)^m = 1 + mx + m(m - 1)x^2/2 + \cdots$. If we retain only the first-order correction to the zero-order result $n(\omega) = 1$, we find

$$n(\omega) = 1 + \sum_a \frac{N_a}{2\epsilon_0}\sum_{i=1}^{Z_a}\frac{e^2/m}{\omega_{ai}^2 - \omega^2} \qquad (2.3.23)$$

This approximation implies that the index of a mixture of gases is the proportionate sum of their individual indices (see Problem 2.5).

TABLE 2.1   The Maxwell-Newton Equations of the Electron Oscillator Model

$$\nabla \cdot \mathbf{D} = 0, \qquad \mathbf{D} = \epsilon_0 \mathbf{E} + \mathbf{P}$$

$$\nabla \cdot \mathbf{B} = 0$$

$$\nabla \times \mathbf{E} = -\frac{\partial \mathbf{B}}{\partial t}$$

$$\nabla \times \mathbf{B} = \mu_0 \frac{\partial \mathbf{D}}{\partial t}$$

$$\mathbf{P} = Ne \sum_{i=1}^{Z} \mathbf{x}_i$$

$$\frac{d^2 \mathbf{x}_i}{dt^2} + \omega_i^2 \mathbf{x}_i = \frac{e}{m} \mathbf{E}$$

Using Eqs. (2.3.1), (2.3.13), and (2.3.21), we easily obtain from the Maxwell equations the corresponding solutions for $\mathbf{B}$ and $\mathbf{D}$. In Table 2.1 we collect the coupled Maxwell–Newton equations of the electron oscillator model, and in Table 2.2 we display the particular solutions we have found for these coupled equations. For definiteness we have taken the polarization unit vector $\hat{\varepsilon}$ to be $\hat{\mathbf{x}}$, the unit vector in the x direction. The reader may easily verify that Table 2.2 provides a solution to the equations of Table 2.1.

It is sometimes useful to use the electric susceptibility $\chi(\omega)$ instead of the polarizability $\alpha(\omega)$. Since $\chi(\omega)$ is defined by the relation

$$\mathbf{P} = \epsilon_0 \chi \mathbf{E} \tag{2.3.24}$$

the connection with $\alpha(\omega)$ is easily determined, by comparison with (2.3.11), to be

$$\chi(\omega) = N\alpha(\omega)/\epsilon_0 \tag{2.3.25}$$

Apart from the dimensional factor $1/\epsilon_0$ (which is not even present in cgs units) the only physical difference between χ and α is that χ is directly associated with a medium (proportional to N) and α is a single-oscillator characteristic. Expressions for quantities associated with an optical medium are usually somewhat simpler to express in terms of χ; for example, Eqs. (2.3.14) for the index of refraction can be rewritten as

$$n^2(\omega) = 1 + \chi(\omega) \tag{2.3.26a}$$

TABLE 2.2   A Solution of the Maxwell–Newton Equations

$$\mathbf{E} = \hat{\mathbf{x}} \, E_0 \cos \omega\left(t \pm n(\omega)\frac{z}{c}\right), \qquad E_0 = \text{constant}$$

$$\mathbf{D} = n^2(\omega) \, \mathbf{E}$$

$$\mathbf{B} = \mp \frac{n(\omega)}{c} \, \hat{\mathbf{y}} \, E_0 \cos \omega\left(t \pm n(\omega)\frac{z}{c}\right)$$

$$\mathbf{P} = \epsilon_0 \, [n^2(\omega) - 1] \, \mathbf{E}$$

$$\mathbf{x}_i = \frac{e/m}{\omega_i^2 - \omega^2} \, \mathbf{E}$$

$$n(\omega) = \left(1 + \frac{N}{\epsilon_0} \sum_{i=1}^{Z} \frac{e^2/m}{\omega_i^2 - \omega^2}\right)^{1/2}$$

or

$$n(\omega) = \left[1 + \chi(\omega)\right]^{1/2} \tag{2.3.26b}$$

In the Lorentz model, $\chi(\omega)$ is given by

$$\chi(\omega) = \frac{Ne^2}{m\epsilon_0} \sum_{i=1}^{Z} \frac{1}{\omega_i^2 - \omega^2} \tag{2.3.27}$$

• The atoms or molecules of a medium do not form a continuum, but have empty spaces between them. As a result, there is a difference between the "mean" field satisfying the wave equation (2.1.13) and the actual field acting on a given atom. In most cases the only practical consequence of this difference is that we must modify the expression (2.3.22) for the refractive index to read

$$\frac{n^2(\omega) - 1}{n^2(\omega) + 2} = \frac{N\alpha(\omega)}{3\epsilon_0} \tag{2.3.28}$$

The origin of this "Lorentz–Lorenz relation" is discussed in many textbooks on electromagnetic theory. Note that when the refractive index is close to unity, so that $n^2(\omega) + 2 \approx 3$, the Lorentz–Lorenz relation reduces to (2.3.22) or (2.3.26). •

2.4 THE CAUCHY FORMULA

We will now compare the formula (2.3.22) for the refractive index with experimental data. Such data often show an increase of the refractive index with increasing frequency (decreasing wavelength). This is familiar from the way a glass prism disperses (separates) white light into its spectral components, with violet being deviated more than red. An example of such "normal" dispersion appears in Figure 2.3, which gives the refractive index of He gas over a range of wavelengths extending from the infrared to the ultraviolet.

To compare the dispersion formula (2.3.22) predicted by the electron oscillator model with experimental results, we first write it in terms of wavelength, $\lambda = 2\pi c / \omega$ (This is convenient because one normally measures the wavelength of optical radiation rather than the frequency.):

$$n^2(\lambda) = 1 + \frac{Ne^2}{4\pi^2\epsilon_0 mc^2} \sum_{i=1}^{z} \frac{\lambda_i^2 \lambda^2}{\lambda^2 - \lambda_i^2} \tag{2.4.1}$$

where λ_i is a wavelength associated with the natural oscillation frequency ω_i of the electron oscillator model by the definition

$$\lambda_i = \frac{2\pi c}{\omega_i} \tag{2.4.2}$$

It follows immediately from (2.4.1) that if the refractive index at wavelength λ exceeds unity, the sum must be positive; if $n(\lambda) < 1$, then the sum is negative. Most transparent materials we encounter daily (e.g., air, water, glass) have refractive indices greater than unity at visible wavelengths. We therefore conclude that in most transparent media the electrons have oscillation wavelengths λ_i that are less than optical wavelengths, which lie roughly between 4000 and 7000 Å, 400–700 nm. From Eq. (2.4.1), we not only have $n(\lambda) > 1$ but also $dn/d\lambda < 0$ for $\lambda_i < \lambda$, consistent with the frequent observation mentioned above. Of course,

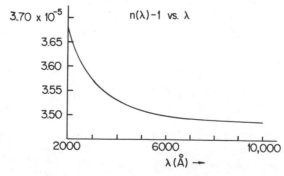

Figure 2.3 Refractive index of helium at standard temperature and pressure.

in order that $n(\lambda) > 1$ it is really only necessary that at least one of the λ_i must be less than λ; we are postulating here that all of the λ_i are less than visible wavelengths λ.

If $\lambda_i < \lambda$, we may again use the binomial series expansion

$$(1 + x)^m = 1 + mx + \tfrac{1}{2} m(m - 1) x^2 + \cdots \qquad (2.4.3)$$

which is valid for $|x| < 1$, to write

$$\frac{\lambda^2}{\lambda^2 - \lambda_i^2} = \left(1 - \frac{\lambda_i^2}{\lambda^2} \right)^{-1} = 1 + \frac{\lambda_i^2}{\lambda^2} + \left(\frac{\lambda_i^2}{\lambda^2} \right)^2 + \cdots \qquad (2.4.4)$$

in Eq. (2.4.1). If λ_i^2 / λ^2 is very small, i.e.,

$$\lambda_i^2 \ll \lambda^2 \qquad (2.4.5)$$

we can approximate (2.4.4) by the first two terms of the expansion, and write

$$n^2(\lambda) \approx 1 + \frac{Ne^2}{4\pi^2 \epsilon_0 mc^2} \sum_{i=1}^{z} \lambda_i^2 \left(1 + \frac{\lambda_i^2}{\lambda^2} \right) \qquad (2.4.6)$$

as an approximation to (2.4.1). If we suppose further that

$$\left| n^2(\lambda) - 1 \right| \ll 1 \qquad (2.4.7)$$

we may again use (2.4.3) to obtain the approximation $(1 + x)^{1/2} \approx 1 + x/2$ for $x \ll 1$, and write the refractive index given by (2.4.6) as

$$n(\lambda) \approx 1 + \frac{Ne^2}{8\pi^2 \epsilon_0 mc^2} \sum_{i=1}^{z} \lambda_i^2 \left(1 + \frac{\lambda_i^2}{\lambda^2} \right) \qquad (2.4.8)$$

This is a good approximation to $n(\lambda)$ when the two conditions (2.4.5) and (2.4.7) are met.

Equation (2.4.8) explains why the refractive indices of many gases have long been known to follow the Cauchy formula

$$n(\lambda) - 1 = A\left(1 + \frac{B}{\lambda^2} \right) \qquad (2.4.9)$$

where A and B are constants. Such an empirical relation was proposed by Cauchy in 1830, before the electromagnetic theory of light. For He gas at standard temperature and pressure (STP), for example, $A = 3.48 \times 10^{-5}$ and $B = 2.3 \times 10^{-11}$ cm^2. Now we notice that (2.4.8) is of precisely the form (2.4.9), with

$$A = \frac{Ne^2}{8\pi^2\epsilon_0 mc^2} \sum_{i=1}^{Z} \lambda_i^2 \qquad (2.4.10)$$

and

$$B = \frac{\sum_{i=1}^{Z} \lambda_i^4}{\sum_{i=1}^{Z} \lambda_i^2} \qquad (2.4.11)$$

This suggests that gases obeying the relation (2.4.9) satisfy the condition (2.4.5) assumed in the derivation of (2.4.8).

A more quantitative check of the electron oscillator model requires a knowledge of the natural wavelengths λ_i. The values of the λ_i are not given by the Lorentz oscillator model, since they are only a different manifestation of the hypothetical elastic force constants k_i. However, the λ_i can be determined empirically from dispersion data. Consider He gas at standard temperature and pressure. Helium is known to absorb strongly radiation of wavelength $\lambda \approx 584$ Å. Suppose we take this to be the natural wavelength λ_i of each electron in the electron oscillator model of He (atomic number $Z = 2$). Then

$$\sum_{i=1}^{Z} \lambda_i^2 = 2(584 \times 10^{-8} \text{ cm})^2 = 6.82 \times 10^{-11} \text{ cm}^2$$

$$\sum_{i=1}^{Z} \lambda_i^4 = 2(584 \times 10^{-8} \text{ cm})^4 = 2.33 \times 10^{-21} \text{ cm}^4$$

An ideal gas at STP has about 2.69×10^{19} atoms (or molecules) per cm^3 (Problem 2.7). Therefore we compute from (2.4.10) the value $A \approx 8.23 \times 10^{-5}$. This is more than twice the tabulated value, given above, that has been measured for He. For the coefficient B we get $\approx 3.42 \ 10^{-11} \text{ cm}^2$, about 50% greater than the tabulated value.

This example brings out a general feature of our electron oscillator model: it is often in close *qualitative* accord with our observations but not in detailed quantitative agreement. Not infrequently it provides, as in this example, reasonable orders of magnitude. Before quantum mechanics, the electron oscillator model was modified *ad hoc* in order to bring its numerical predictions in line with experimental results. The nature of this modification will be discussed in Section 3.7, and in Chapter 6 we will interpret this modification from the standpoint of the quantum theory of the interaction of light with matter.

It is also interesting to consider the case in which the frequency of the radiation is much greater than the natural oscillation frequencies of the medium, i.e., the natural wavelengths are relatively very large: $\lambda_i \gg \lambda$. The simplest example occurs for free electrons, for which there is no elastic binding force. The natural oscillation frequencies ω_i are then zero, and the dispersion formula (2.3.22) reduces to

$$n(\omega) = \left(1 - \frac{Ne^2}{\epsilon_0 m\omega^2}\right)^{1/2} \tag{2.4.12}$$

where N is now the density of free electrons. In some cases this result is known to be fairly accurate; it is applicable, for instance, to the upper atmosphere, where ultraviolet solar radiation produces free electrons by photoionization (Problem 2.8). Another example is the refraction of X-rays by glass. In this case the natural oscillation frequencies of the medium are not zero, but are much less than X-ray frequencies. In both of these examples the refractive index is less than one. Thus X-rays propagating in glass have a phase velocity greater than c.

• Equation (2.4.12) for free electrons shows that the refractive index can even be a pure *imaginary* number. This occurs whenever $\omega < \omega_p$, where

$$\omega_p = \left(\frac{Ne^2}{m\epsilon_0}\right)^{1/2} \tag{2.4.13}$$

is called the plasma frequency. In terms of the plasma frequency the refractive index (2.4.12) of a free-electron gas may be written as

$$n(\omega) = \left(1 - \frac{\omega_p^2}{\omega^2}\right)^{1/2} = \frac{1}{\omega}(\omega^2 - \omega_p^2)^{1/2} \tag{2.4.14}$$

To understand the physical significance of an imaginary refractive index, let us consider again the solutions of the Maxwell–Newton equations we have displayed in Table 2.2. To be specific, the solution corresponding to plane-wave propagation in the positive z direction is

$$\mathbf{E}(z, t) = \hat{x}E_0 \cos \omega\left(t - n(\omega)\frac{z}{c}\right) \tag{2.4.15}$$

This is not feasible if $n(\omega)$ is imaginary. Since the electric field must obviously be a real quantity we can instead write

$$\mathbf{E}(z, t) = \text{Re}\,[\hat{x}E_0\, e^{-i\omega[t - n(\omega)z/c]}] \tag{2.4.16}$$

where Re means "real part of". This reduces to (2.4.15) when $n(\omega)$ is real. For the free-electron case $n(\omega)$ is given by (2.4.14) and we have therefore

$$\mathbf{E}(z, t) = \text{Re}\,[\hat{x}E_0\, e^{-i\omega t}\, e^{i(\omega^2 - \omega_p^2)^{1/2}z/c}] \tag{2.4.17}$$

When $\omega < \omega_p$ we have

$$(\omega^2 - \omega_p^2)^{1/2} = i(\omega_p^2 - \omega^2)^{1/2} = icb(\omega) \tag{2.4.18}$$

with $b(\omega) = (1/c)(\omega_p^2 - \omega^2)^{1/2} > 0$. Equation (2.4.17) then gives

$$\mathbf{E}(z, t) = \hat{\mathbf{x}} E_0 \, e^{-b(\omega)z} \cos \omega t \tag{2.4.19}$$

indicating that the electric field decreases exponentially with penetration depth in the medium. It follows from Table 2.2 that the magnetic field is attenuated in the same manner.

The field (2.4.19) is not a propagating wave. Thus when the field frequency ω is less than the plasma frequency ω_p, the free-electron gas will not support a propagating mode of the field. The field is therefore reflected. This applies, for instance, to the propagation of radio waves in the earth's atmosphere. High-frequency FM radio waves are not reflected by the ionosphere ($\omega > \omega_p$), whereas the lower-frequency AM waves are. AM radio broadcasts therefore reach more distant points on the earth's surface (see Problem 2.9). •

2.5 ELECTRIC DIPOLE RADIATION

According to Eq. (2.3.8), a monochromatic electric field forces an electron into oscillation at the frequency of the field. The field thus induces an oscillating dipole moment in an atom, and according to (2.1.13) this dipole can serve as the source of a new electric field. The electromagnetic field radiated by an oscillating electric dipole is discussed in textbooks on electromagnetic theory.[4] We will only discuss the most important results needed for our purposes. In the next section we will use these results to treat the scattering of light by atoms.

Suppose we have a time-dependent electric dipole moment

$$\mathbf{p} = \hat{\mathbf{x}} p(t) \tag{2.5.1}$$

where the unit vector $\hat{\mathbf{x}}$ specifies the fixed direction in which the dipole is assumed to point. Then at the position $\mathbf{r}$ measured from the dipole the electric and magnetic field vectors are[4]

$$\mathbf{E}(\mathbf{r}, t) = \frac{1}{4\pi\epsilon_0} \left[3(\hat{\mathbf{x}} \cdot \hat{\mathbf{r}})\hat{\mathbf{r}} - \hat{\mathbf{x}} \right] \left[\frac{1}{r^3} p\left(t - \frac{r}{c}\right) + \frac{1}{cr^2} \frac{d}{dt} p\left(t - \frac{r}{c}\right) \right]$$

$$+ \frac{1}{4\pi\epsilon_0} \left[(\hat{\mathbf{x}} \cdot \hat{\mathbf{r}})\hat{\mathbf{r}} - \hat{\mathbf{x}} \right] \frac{1}{c^2 r} \frac{d^2}{dt^2} p\left(t - \frac{r}{c}\right) \tag{2.5.2a}$$

$$\mathbf{B}(\mathbf{r}, t) = \frac{1}{4\pi\epsilon_0 c} (\hat{\mathbf{x}} \times \hat{\mathbf{r}}) \left[\frac{1}{cr^2} \frac{d}{dt} p\left(t - \frac{r}{c}\right) + \frac{1}{c^2 r} \frac{d^2}{dt^2} p\left(t - \frac{r}{c}\right) \right] \tag{2.5.2b}$$

4. See, for example, R. P. Feynman, R. B. Leighton, and M. Sands, *The Feynman Lectures on Physics* (Addison-Wesley, Reading, Mass., 1964), vol. 2, Section 21-4, or J. D. Jackson, *Classical Electrodynamics* (Wiley, New York, 1975), or J. B. Marion, *Classicial Electromagnetic Radiation* (Academic Press, New York, 1965).

where the unit vector $\hat{\mathbf{r}}$ points from the dipole to $\mathbf{r}$, and r is the magnitude of $\mathbf{r}$ (i.e., $\mathbf{r} = \hat{\mathbf{r}}r$). Note that, owing to the finite velocity c of light, the electric and magnetic fields at $\mathbf{r}$ at time t depend upon the dipole moment at the earlier (retarded) time $t - r/c$.

The simplest example of (2.5.2) is the electrostatic case in which $p(t)$ is time-independent. In this case the fields (2.5.2) reduce to

$$\mathbf{E}(\mathbf{r}, t) = [3(\hat{\mathbf{x}} \cdot \hat{\mathbf{r}})\hat{\mathbf{r}} - \hat{\mathbf{x}}] \frac{p}{4\pi\epsilon_0 r^3} \tag{2.5.3a}$$

$$\mathbf{B}(\mathbf{r}, t) = 0 \tag{2.5.3b}$$

The reader may recall that these electrostatic dipole fields are derivable independently of the general formulas (2.5.2), simply by considering the static (Coulomb) fields of two point charges q and $-q$ separated by a distance d small compared with r; the electric dipole moment is $\mathbf{p} = q\mathbf{d}$ and the magnetic dipole moment is zero. However, we are interested now in the field (2.5.2) radiated by a *time-dependent* dipole moment, specifically the dipole moment induced in an atom by a time-dependent external field.

In particular, we are interested in the rate at which an oscillating electric dipole radiates electromagnetic energy. We will calculate this rate from the fields (2.5.2). To do this we use the *Poynting vector* of an electromagnetic field (recall Problem 2.6):

$$\mathbf{S} = \mathbf{E} \times \mathbf{H} = \epsilon_0 c^2 \mathbf{E} \times \mathbf{B} \tag{2.5.4}$$

The Poynting vector of the field (2.5.2) gives the rate at which electromagnetic energy crosses a unit area. Its SI units are W/m^2 but W/cm^2 is more commonly used in laser physics. The total rate at which electromagnetic energy is radiated may be obtained by integrating the normal (radially outward) component of $\mathbf{S}$ over a large sphere of radius R centered at the dipole. Since the differential element of area on a sphere is given in terms of the polar angle θ and the azimuthal angle ϕ by

$$dA = R^2 \sin\theta \, d\theta \, d\phi \tag{2.5.5}$$

the radiated power (written Pwr) at the spherical surface is

$$\text{Pwr} = R^2 \int_0^{2\pi} d\phi \int_0^{\pi} d\theta \sin\theta \, \mathbf{S}(R, \theta, \phi) \cdot \hat{\mathbf{R}} \tag{2.5.6}$$

From (2.5.2) and the definition (2.5.4) of the Poynting vector, we see that $\mathbf{S}(R, \theta, \phi)$ has a term varying as R^{-2}, and other terms falling off as R^{-3}, R^{-4}, and R^{-5}. The terms decreasing faster than R^{-2} (so-called near-zone and induction-zone fields) do not contribute to outward power flow at any value of R, since they have zero time average. [In many texts it is pointed out that these terms make no contribution to the net energy flow at infinity, obtained by integrating the Poynting flux over a large sphere of radius R_0 surrounding the dipole:

$$\int_{R_0} \mathbf{S} \cdot \mathbf{n} \, d(\text{area}) = 0,$$

in the limit $R_0 \to \infty$. This is also true.] Only the term in $\mathbf{S}$ falling off as R^{-2} contributes to the net rate of electromagnetic energy flow. This term arises from the so-called "radiation zone" fields:

$$\mathbf{E}(\mathbf{R}, t) = \frac{1}{4\pi\epsilon_0 c^2 R} [(\hat{\mathbf{x}} \cdot \hat{\mathbf{R}})\hat{\mathbf{R}} - \hat{\mathbf{x}}] \frac{d^2}{dt^2} p\left(t - \frac{R}{c}\right) \qquad (2.5.7a)$$

$$\mathbf{B}(\mathbf{R}, t) = \frac{1}{4\pi\epsilon_0 c^3 R} [\hat{\mathbf{x}} \times \hat{\mathbf{R}}] \frac{d^2}{dt^2} p\left(t - \frac{R}{c}\right) \qquad (2.5.7b)$$

which follow from (2.5.2) when R is so large that

$$|p| \ll \frac{R}{c}\left|\frac{dp}{dt}\right| \quad \text{and} \quad \left|\frac{dp}{dt}\right| \ll \frac{R}{c}\left|\frac{d^2 p}{dt^2}\right| \qquad (2.5.8)$$

In the radiation zone the electric and magnetic fields (2.5.7) are orthogonal to each other and to $\hat{\mathbf{R}}$.

The Poynting vector in the radiation zone defined by (2.5.8) follows from (2.5.4) and (2.5.7):

$$\mathbf{S} = \frac{\epsilon_0 c}{(4\pi\epsilon_0)^2}\left(\frac{1}{c^2 R}\right)^2 [(\hat{\mathbf{x}} \cdot \hat{\mathbf{R}})\hat{\mathbf{R}} - \hat{\mathbf{x}}] \times [\hat{\mathbf{x}} \times \hat{\mathbf{R}}]$$

$$\times \left[\frac{d^2}{dt^2} p\left(t - \frac{R}{c}\right)\right]^2 \qquad (2.5.9)$$

The vector cross product in this equation may be evaluated using the identity

$$\mathbf{A} \times (\mathbf{B} \times \mathbf{C}) = \mathbf{B}(\mathbf{A} \cdot \mathbf{C}) - \mathbf{C}(\mathbf{A} \cdot \mathbf{B}) \qquad (2.5.10)$$

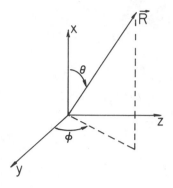

Figure 2.4 The polar angle θ in Eqs. (2.5.11) and (2.5.12) is measured from the dipole axis.

to obtain

$$
\begin{aligned}
[(\mathbf{x} \cdot \hat{\mathbf{R}})\hat{\mathbf{R}} - \hat{\mathbf{x}}] \times [\hat{\mathbf{x}} \times \hat{\mathbf{R}}] &= [(\hat{\mathbf{x}} \cdot \hat{\mathbf{R}})\hat{\mathbf{R}} \times (\hat{\mathbf{x}} \times \hat{\mathbf{R}}) - \hat{\mathbf{x}} \times (\hat{\mathbf{x}} \times \hat{\mathbf{R}})] \\
&= \hat{\mathbf{R}}[1 - (\hat{\mathbf{R}} \cdot \hat{\mathbf{x}})^2] \\
&= \hat{\mathbf{R}} \sin^2 \theta \tag{2.5.11}
\end{aligned}
$$

where θ is the angle between the dipole axis and the radius vector $\mathbf{R}$. (Figure 2.4). Thus

$$
\mathbf{S} = \hat{\mathbf{R}} \left(\frac{\sin^2 \theta}{16\pi^2 \epsilon_0 c^3 R^2} \right) \left[\frac{d^2}{dt^2} p \left(t - \frac{R}{c} \right) \right]^2 \tag{2.5.12}
$$

which has the angular dependence (or "radiation pattern") depicted in Figure 2.5. The radiation is greatest in directions normal to the dipole axis ($\theta = \pi/2$), and falls to zero in the two directions parallel to the axis ($\theta = 0, \pi$).

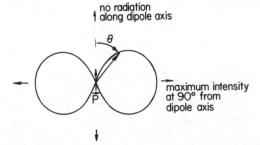

Figure 2.5 The radiation pattern ($\sin^2 \theta$) of a linear electric dipole oscillator. The three-dimensional pattern is obtained by rotating the two-lobe pattern in the figure about the dipole (vertical) axis.

The total rate at which electromagnetic energy flows out from the dipole at radius R may now be calculated using (2.5.6) and (2.5.12):

$$
\begin{aligned}
\text{Pwr} &= R^2 \int_0^{2\pi} d\phi \int_0^{\pi} d\theta \, \sin\theta \, \hat{\mathbf{R}} \cdot \mathbf{S} \\
&= R^2 \int_0^{2\pi} d\phi \int_0^{\pi} d\theta \, \sin\theta \left(\frac{\sin^2\theta}{16\pi^2\epsilon_0 c^3 R^2} \right) \left[\frac{d^2}{dt^2} p\left(t - \frac{R}{c} \right) \right]^2 \\
&= \frac{1}{16\pi^2\epsilon_0 c^3} \left[\frac{d^2}{dt^2} p\left(t - \frac{R}{c} \right) \right]^2 \int_0^{2\pi} d\phi \int_0^{\pi} d\theta \, \sin^3\theta \\
&= \left(\frac{1}{4\pi\epsilon_0} \right) \frac{2}{3c^3} \left[\frac{d^2}{dt^2} p\left(t - \frac{R}{c} \right) \right]^2
\end{aligned}
\tag{2.5.13}
$$

The rate at which the dipole at the center of our fictitious sphere radiates electromagnetic energy is therefore

$$
\text{Pwr} = \left(\frac{1}{4\pi\epsilon_0} \right) \frac{2}{3c^3} \left(\frac{d^2 p}{dt^2} \right)^2
\tag{2.5.14}
$$

• This result is similar to a general (nonrelativistic) formula for the power radiated by a particle of charge q and acceleration $\mathbf{a}$,

$$
\text{Pwr} = \left(\frac{1}{4\pi\epsilon_0} \right) \frac{2q^2 a^2}{3c^3}
\tag{2.5.15}
$$

From this formula an estimate can be made of the length of time an oscillator *not* driven by an external field, but simply freely oscillating, can be expected to take to convert its oscillation energy to radiation.

If the radiative "lifetime" of an oscillator obtained in this way is designated τ, then $1/\tau$ is the radiation rate, and one finds (see Problem 2.10)

$$
\frac{1}{\tau} = \frac{1}{4\pi\epsilon_0} \frac{2e^2\omega_0^2}{3mc^3}
\tag{2.5.16}
$$

It is interesting that the ad hoc replacement of half the oscillator's energy (the potential energy) by half the lowest energy of a quantum oscillator (see Section 4.5) turns (2.5.16) into a quantum-mechanical formula for the spontaneous emission rate. That is, if we use $\frac{1}{2}m\omega_0^2 x^2 = \frac{1}{2}(\hbar\omega_0/2)$ to eliminate m from (2.5.16), we find

$$
\frac{1}{\tau} = \frac{1}{4\pi\epsilon_0} \frac{4e^2 x^2 \omega_0^3}{3\hbar c^3}
\tag{2.5.17}
$$

This is the correct quantum-mechanical rate for spontaneous radiative decay, if the oscillator amplitude x is interpreted appropriately as a coordinate "matrix element" (see Eq. (7.6.5)) •

2.6 RAYLEIGH SCATTERING

Using the solutions of Table 2.2, we may write the dipole moment induced in an atom by a monochromatic plane wave traveling in the z direction as

$$\mathbf{p} = e \sum_{i=1}^{z} \mathbf{x}_i = \frac{\mathbf{P}}{N} = \hat{\mathbf{x}}\alpha(\omega)E_0 \cos\omega\left(t - \frac{z}{c}\right) \tag{2.6.1}$$

where $\alpha(\omega)$ is the atomic dipole polarizability, given in (2.3.20). We assume the atom to be in a dilute host medium such as a gas for which $n(\omega) = 1$ is a satisfactory approximation. We will now apply the results we have obtained for electric dipole radiation to the case where $\mathbf{p}$ is given by (2.6.1). That is, we will calculate the power radiated by any atom as a result of its having an oscillating dipole moment induced by a monochromatic field.

For the dipole moment (2.6.1) we have

$$\frac{d^2\mathbf{p}}{dt^2} = -\omega^2\alpha(\omega)\hat{\mathbf{x}}E_0 \cos\omega\left(t - \frac{z}{c}\right) \tag{2.6.2}$$

According to (2.5.14), therefore, electromagnetic energy is radiated at the rate

$$\text{Pwr} = \frac{\omega^4}{6\pi\epsilon_0 c^3} \alpha^2(\omega)E_0^2 \cos^2\omega\left(t - \frac{z}{c}\right) \tag{2.6.3}$$

This is the instantaneous power radiated by a single dipole. It has been obtained by applying the general result (2.5.14) of electromagnetic theory to the solutions of the coupled Maxwell–Newton equations given in Table 2.2.

Using the fields $\mathbf{E}$ and $\mathbf{B}$ of Table 2.2 in (2.5.4), we obtain the Poynting vector [recall in this case we have assumed $n(\omega) = 1$]

$$\mathbf{S} = \epsilon_0 c \, \hat{\mathbf{z}}E_0^2 \cos^2\omega\left(t - \frac{z}{c}\right) \tag{2.6.4a}$$

Therefore

$$E_0^2 \cos^2\omega\left(t - \frac{z}{c}\right) = \frac{1}{\epsilon_0 c}|\mathbf{S}| \tag{2.6.4b}$$

and so the power (2.6.3) radiated by the dipole may be written as

$$\text{Pwr} = \frac{8\pi\omega^4/c^4}{3(4\pi\epsilon_0)^2} \alpha^2(\omega) \, |\mathbf{S}| \tag{2.6.5}$$

The radiation from an atom having an oscillating electric dipole moment goes off in all directions in accordance with the angular distribution (2.5.12). This means there is a spatial redistribution or *scattering* of electromagnetic radiation. The incident plane wave still travels in the z direction, but with diminished amplitude because of the scattering. The scattering of radiation out of the incident wave is indicated pictorially in Figure 2.6.

Equation (2.6.5) shows that the power radiated by an atom is linearly proportional to $|\mathbf{S}|$, the rate per unit area at which the field driving the dipole transports electromagnetic energy. Since Pwr has the dimensions of energy per unit time, the constant of proportionality in (2.6.5) has the dimensions of an area. This area is called the *scattering cross section*, and denoted $\sigma(\omega)$. It is given by

$$\sigma(\omega) = \frac{8\pi}{3}\omega^4 \left(\frac{\alpha(\omega)}{4\pi\epsilon_0 c^2}\right)^2 \tag{2.6.6}$$

Note that cross section, like polarizability, is a single-atom parameter. If there is a collection of scattering dipoles, making up a continuous medium, then the cross section can be expressed in terms of the medium's index of refraction by the use of (2.3.22): $\epsilon_0[n^2(\omega) - 1] = N\alpha(\omega)$. Then (2.6.6) becomes

$$\sigma(\omega) = \frac{8\pi\omega^4}{3c^4} \left(\frac{n^2(\omega) - 1}{4\pi N}\right)^2 \tag{2.6.7}$$

The relation (radiated power) = (scattering cross section) × (field power per unit area incident on the dipole) means that the radiated power is the same as if the scattering atom (dipole) had a cross-sectional area σ, and *all* of the incident field power intercepted by this area were scattered. There is no actual geometrical object the same size as this cross section; it is merely a convenient measure of the effectiveness of the scatterer.

By scattering the incident light, the dipoles cause the light beam to be attenuated

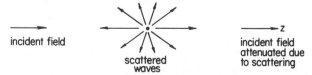

incident field

scattered
waves

incident field
attenuated due
to scattering

Figure 2.6 An incident field induces an oscillating dipole moment in an atom, which then radiates. The dipole radiation field is the field scattered by the atom.

as it propagates. The degree of attenuation can be calculated as follows. Suppose there are N atoms (scatterers) per unit volume. According to (2.6.5) and (2.6.6), each one radiates electromagnetic energy at the rate σI, where the intensity I (typically measured in W/cm^2) of the incident wave is defined as the cycle-averaged magnitude of the Poynting vector:

$$I(t, z) = \frac{1}{T} \int_t^{t+T} |\mathbf{S}(t', z)| \, dt' \qquad (2.6.8a)$$

where $T = 2\pi/\omega$. Therefore, in the case of a monochromatic plane wave $I(t, z) \rightarrow I(z)$. From (2.6.4) we find this intensity to be given by

$$I = \tfrac{1}{2} \epsilon_0 c |E_0|^2 \qquad (2.6.8b)$$

Consider, as in Figure 2.7, an imaginary slab at position z in the medium, drawn so that it is perpendicular to the direction z of propagation of the wave. Let the energy density of the wave at the entrance face of the slab be denoted $u(z)$. As a result of scattering of electromagnetic radiation into directions of propagation other than z, the energy density $u(z + \Delta z)$ of the wave at the exit face of our imaginary slab must be less than $u(z)$. We may write

$$u(z + \Delta z) - u(z) \approx -N[\sigma(\omega)I(z)] (\Delta z/c) \qquad (2.6.9)$$

The right-hand side of this expression is simply the number of atoms per unit volume, times the power scattered by each atom, times the time $\Delta t = \Delta z/c$ taken by the wave to cross the slab drawn in Figure 2.7. The approximation (2.6.9) is valid provided Δz is small enough that I remains relatively constant over the width of the slab, so that the same radiated power $\sigma I(z)$ may be associated with every atom in the slab. We divide by Δz in (2.6.9) and then take the limit $\Delta z \rightarrow 0$ to ensure the validity of this approximation. We obtain the differential equation

$$\frac{du}{dz} = -\frac{N\sigma(\omega) \, I(z)}{c} \qquad (2.6.10)$$

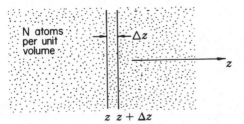

Figure 2.7 An imaginary thin slab of width Δz in the scattering medium.

where as usual

$$\frac{du}{dz} = \lim_{\Delta z \to 0} \frac{u(z + \Delta z) - u(z)}{\Delta z} \tag{2.6.11}$$

In many contexts in physics it is helpful to recognize that intensity I is simply energy density u times wave propagation velocity. This can be checked in the electromagnetic case by using the solutions in Table 2.2 to compute the energy density u [see Problem 2.6] for comparison with the intensity (magnitude of Poynting vector) obtained in (2.6.4). The result for the case $n = 1$ is

$$I(z) = cu(z) \tag{2.6.12}$$

A simple analogy to this result is a stream of pellets moving with speed v. If ρ is the number of pellets per unit volume, then the "intensity," or flux, of the stream is

$$\Phi = \rho v \tag{2.6.13}$$

This is the number of pellets crossing a unit area per unit time. In Eq. (2.6.12) the intensity I and energy density u of the electromagnetic wave play the roles of Φ and ρ, respectively, and the light velocity c plays the role of v.

It follows from (2.6.10) and (2.6.12) that

$$\frac{dI}{dz} = -N\sigma(\omega) I(z) \tag{2.6.14}$$

The solution to this simple differential equation is

$$I(z) = I_0 e^{-a_s(\omega)z} \tag{2.6.15}$$

This gives the intensity at a distance z into the medium, given the "initial condition" that the intensity at $z = 0$ is I_0. The initial condition may refer to the actual entrance to the scattering medium, or merely some convenient reference plane inside the medium. We have defined

$$a_s(\omega) = N\sigma(\omega) \tag{2.6.16}$$

which is called the *extinction coefficient due to scattering*. The length a_s^{-1} is the distance of propagation for which the intensity decreases by a factor e^{-1} as a result of scattering. The relation (2.6.16) provides a general connection between the density N of scatterers, the cross section σ associated with each scatterer, and the extinction coefficient due to scattering. An expression of the same form as (2.6.15) also describes the attenuation of the flux of a beam of particles in a medium of scatterers.

Using (2.6.7) and (2.6.16), we may conveniently express the extinction coefficient in terms of the refractive index:

$$a_s(\omega) = \left(\frac{\omega}{c}\right)^4 \frac{\left[n^2(\omega) - 1\right]^2}{6\pi N} \qquad (2.6.17)$$

Equation (2.6.17) was derived by Lord Rayleigh (J.W. Strutt) in 1899. Rayleigh scattering is characterized by the extinction coefficient (2.6.17).

• There is a conceputal inconsistency in our (and most textbooks') treatment of attenuation here and our solution of the Maxwell–Newton equations earlier. Recall that in Table 2.2 we have given the fully exact solutions for the electromagnetic fields **E** and **B** propagating through a continuous medium of atomic dipoles characterized by a real index of refraction $n(\omega)$. The solutions given there show no evidence of scattered light or of attenuation of the incident beam. This is correct. In fact there is no attenuation when a light beam passes through such an ideal collection of dipoles.

However, an important assumption of our idealized derivation is usually violated for atmospheric light propagation. This assumption is that the density of scatterers N is constant. If the density were actually constant, the scattered fields from individual atoms or molecules would actually cancel out in every direction except the forward z direction, with the solutions of Table 2.2 as the result.

When N is not constant in either space or time the side-scattered fields do not cancel out, and appreciable side scattering can occur, as Rayleigh's formula predicts. However, since the assumption of constant N was used in Rayleigh's derivation [and in ours in the substitution between (2.6.6) and (2.6.7)], that leaves open the matter of explaining why (2.6.17) is correct. This is a significant problem of statistical physics, first solved by Smoluchowski and Einstein in the first decade of the century, which we need not discuss further here. •

The frequencies of the scattered and incident fields are equal in Rayleigh scattering. This is an inevitable consequence of the electron oscillator model. The induced dipole moment oscillates at the frequency of the driving (incident) field, and therefore radiates at this frequency. Indeed, this feature of Rayleigh scattering is predicted by *any* model in which the induced dipole moment is linearly proportional to the imposed field.

The ω^4 (i.e., $1/\lambda^4$) dependence of the extinction coefficient (2.6.17) for Rayleigh scattering means that the amount of scattering increases sharply with increasing frequency. (The refractive index n generally varies much more slowly with ω than ω^4.) Rayleigh used (2.6.17) to explain why the sky is blue and the sunset is red. When we look at the sky away from the sun on a sunny day, we see light that has been scattered by air molecules exposed to sunlight. This scattered light is predominantly blue because the high-frequency components of the visible solar radiation are scattered more strongly than the low-frequency components. The sunset, however, is reddish because the sunlight has traveled a sufficient distance through the earth's atmosphere that much of the high-frequency component has been scattered away.

Consider the Rayleigh scattering of visible radiation by molecules in the earth's atmosphere. Taking $\lambda = 6000$ Å, and $n \approx 1.0003$ for the refractive index of air at optical frequencies, we find from (2.6.17) that

$$a_s^{-1} \approx 4.4 \times 10^{-13} \, N \, \text{cm} \tag{2.6.18}$$

where N is the number of air molecules in a cubic centimeter. Assuming an ideal gas at STP, we have $N \approx 2.69 \times 10^{19}$ (Problem 2.7) and therefore

$$a_s^{-1} \approx 118 \text{ km} \tag{2.6.19}$$

for the distance in which 6000-Å radiation is attenuated by a factor $e^{-1} \approx 37\%$ at sea level. Rayleigh compared such calculations with astronomers' estimates for the transmission of stellar radiation through the earth's atmosphere. He drew the important conclusion that scattering of light by molecules alone, without suspended particles (dust), is strong enough to cause the blue sky, which he poetically called the "heavenly azure" (see also Problem 2.12).

This explanation of the blue sky suggests, in fact, that the sky should be violet, since violet light should be scattered more strongly than blue. One reason the sky appears blue rather than violet is that the eye is more sensitive to blue. Furthermore the solar spectrum is not uniform, but has somewhat less radiation at the shorter visible wavelengths.

2.7 POLARIZATION BY RAYLEIGH SCATTERING

A less obvious characteristic of skylight (in contrast to direct sunlight) is that it is polarized. This effect, which was observed in 1811 by D.F. Arago, is easily observed with Polaroid sunglasses. The extent of polarization appears to be strongest from directions near 90° to the direction of the sun from the observer. It has been observed that bees and certain other insects are sensitive to the polarization of light and use it for navigation. Human eyes, of course, are not directly sensitive to polarization.

To understand the polarization of light by Rayleigh scattering, let us return again to the electric field (2.5.7a) in the radiation zone of an oscillating dipole. The unit vector $\hat{x}$, defining the direction of the dipole moment, also defines the polarization of the incident field (propagating in the z direction). $\hat{R}$ is the unit vector pointing from the dipole to the point of observation. If we observe the scattered field at right angles to the "plane of incidence" defined by the directions of polarization and propagation, we see from (2.5.7a) that it will be polarized in the x direction, since

$$(\hat{\mathbf{x}} \cdot \hat{\mathbf{R}})\hat{\mathbf{R}} - \hat{\mathbf{x}} = (\hat{\mathbf{x}} \cdot \hat{\mathbf{y}})\hat{\mathbf{y}} - \hat{\mathbf{x}} = -\hat{\mathbf{x}} \tag{2.7.1}$$

when $\hat{\mathbf{R}} = \hat{\mathbf{y}}$ (Figure 2.8). If instead the incident field driving the dipole is polarized in the y direction, there is no scattered field in the y direction, since

$$(\hat{\mathbf{y}} \cdot \hat{\mathbf{R}})\hat{\mathbf{R}} - \hat{\mathbf{y}} = (\hat{\mathbf{y}} \cdot \hat{\mathbf{y}})\hat{\mathbf{y}} - \hat{\mathbf{y}} = 0 \tag{2.7.2}$$

The light from the sun is unpolarized. In fact virtually all nonlaser sources, including fluorescent and incandescent lamps, emit unpolarized light. In such sources, thermal energy is converted via collisions into internal atomic energy. According to the electron oscillator model, this internal energy appears as electron oscillations. We may imagine a thermal source of radiation to be a large collection of dipole oscillators (atoms) which each radiate, independently of one another, the power (2.5.14). At any instant the dipole moment of each atom oscillates along some axis, thus radiating a field with a polarization determined by this axis [Eq. (2.5.7a)]. The total electric field at any frequency is the sum of the fields from the individual atoms. This (total) radiated field at a given point has a direction determined by the vector addition of the individual atoms' fields. However, this direction varies rapidly in time. Each individual atom radiates for only a short time, typically about ten nanoseconds, before losing the internal energy picked up in a collision (Problem 2.10). After such an interval a different subset of atoms contributes to the total field, which will then point in a different direction. The field from such a source is "unpolarized" because we cannot detect the rapid variations of the electric field direction.

The direction of the dipole moment driven by an unpolarized wave propagating in the z direction (Figure 2.8) will therefore be varying rapidly in the xy plane. Equations (2.7.1) and (2.7.2) show that the dipole radiates in the y direction only when its oscillation has a nonzero component in the x direction (Figure 2.8), and

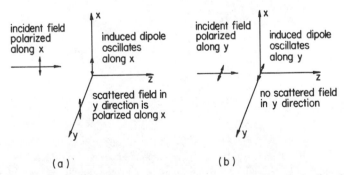

(a)   (b)

Figure 2.8 An incident field propagating in the z direction with polarization (*a*) along *x*, in which case the field scattered in the y direction is also polarized along *x*, and (*b*) along *y*, in which case there is no scattering in the y direction.

in that case the radiation is polarized in the x direction. Rayleigh scattering thus produces polarized light. This explains the polarization of skylight.

The theory of scattering becomes much more complicated when the particle dimensions are not negligible compared with the wavelength. In this case the light scattered at 90° is not completely polarized. And as the particle size increases, the scattering cross section becomes less sensitive to the wavelength; the radiation scattered from white light becomes "whiter" as the particle size increases. This explains why clouds, consisting of water droplets suspended in air, are white.

2.8   THE TYNDALL–RAYLEIGH EXPERIMENT

All of these features of light scattering may be observed beautifully in a simple experiment that the reader can perform with readily available materials (Figure 2.9). A few spoonfuls of photographic fixing powder (sodium thiosulfate) are dissolved in a beaker or small tank of water. The addition of about 100 ml of dilute sulfuric acid causes small grains of sulfur to precipitate out of solution after a few minutes. In the initial stages of precipitation these grains are very small. The scattered light has a faintly bluish hue, gradually becoming a deeper blue (the "blue sky"). The scattered light viewed at 90° (Figure 2.9) is observed to be strongly polarized. The light transmitted through the tank has a yellowish and eventually a strongly reddish hue (the "sunset"). After several more minutes the scattered light is not blue but white ("clouds"), and it is no longer strongly polarized. At this stage we are observing light scattering from sulfur grains that have grown to a size comparable to or larger than optical wavelengths.

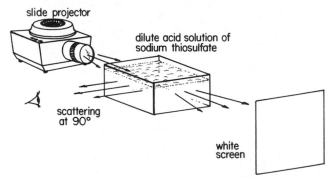

Figure 2.9 Experiment demonstrating the effect of particle size in light scattering. In the initial stage of precipitation the sulfur particles suspended in sodium thiosulfate solution are small compared with an optical wavelength. The tank takes on a blue color, the light scattered at 90° is strongly polarized, and on the white screen one observes the "red sunset." After a few minutes the particles have grown larger than a wavelength. Then the tank has a cloudy appearance and the light scattered at 90° is no longer strongly polarized.

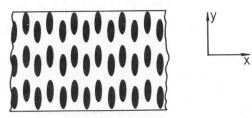

Figure 2.10 A birefringent material in which the molecules are aligned along the y axis.

Such an experiment was performed by John Tyndall in 1869. Rayleigh (1881) found that the ''hypo'' (sodium thiosulfate) solution demonstrates the scattering effects especially well.

It should be borne in mind, of course, that the notion of ''color'' is a subjective one. The theory of color vision has attracted the attention of many physicists, including Isaac Newton, Thomas Young, H. Helmholtz, J.C. Maxwell, Rayleigh, and E.H. Land.[5] Even now, however, color vision is not completely understood. Although it would be inappropriate for us to explore the subject here, it may be worthwhile to mention that in dim light we see nearly in ''black and white.'' This explains, for example, why we do not see the brilliant colors of the Crab nebula that are apparent in photographs taken with long exposure times. Another example is the night sky itself. It looks black instead of blue because of its low intensity; a photograph taken with a long exposure time shows it to be ''really'' blue.

2.9 BIREFRINGENCE

We have tacitly assumed that the elastic restoring force acting on a bound electron is independent of the electron displacement x. In general, however, the ''spring constant'' k_s may depend on the direction of x, in which case the medium is said to be optically anisotropic. In such a medium the refractive index will be different for different polarizations of the field propagating in the medium. This is called *birefringence*. All transparent crystals with noncubic lattice structure, such as calcite, quartz, ice, and sugar, are birefringent.

It is easy to understand how such anisotropy might arise. In particular, imagine a long, rod-shaped molecule in which the elastic restoring force is different for electron displacements parallel and perpendicular to the axis. There will be different refractive indices for different directions of field polarization in a medium in which there is some degree of molecular alignment.

Consider a birefringent material in which the molecules are aligned along the y axis, as illustrated in Figure 2.10. This preferred direction may be called the *optic axis*. Waves that are linearly polarized in the x or y directions will propagate with different refractive indices n_x or n_y, respectively. That is, a field

5. E.H. Land, Experiments in Color Vision, *Scientific American*, **200**, May 1959, p. 84.

$$\mathbf{E}(z,\, t) = \hat{\mathcal{E}} E_0 \cos\,(\omega t - kz)$$

$$= \hat{\mathcal{E}} E_0 \cos \omega \left(t - \frac{nz}{c} \right) \qquad (2.9.1)$$

in the medium will propagate with refractive index $n = n_x$ if $\hat{\mathcal{E}} = \hat{\mathbf{x}}$, or $n = n_y$ if $\hat{\mathcal{E}} = \hat{\mathbf{y}}$. For a general linearly polarized wave in which

$$\hat{\mathcal{E}} = \epsilon_x \hat{\mathbf{x}} + \epsilon_y \hat{\mathbf{y}} \qquad (2.9.2)$$

and $\epsilon_x^2 + \epsilon_y^2 = 1$, we have

$$\mathbf{E}(z,\, t) = \epsilon_x \hat{\mathbf{x}} E_0 \cos \omega \left(t - \frac{n_x z}{c} \right) + \epsilon_y \hat{\mathbf{y}} E_0 \cos \omega \left(t - \frac{n_y z}{c} \right) \quad (2.9.3)$$

If the field

$$\mathbf{E}(0,\, t) = \epsilon_x \hat{\mathbf{x}} E_0 \cos \omega t + \epsilon_y \hat{\mathbf{y}} E_0 \cos \omega t \qquad (2.9.4)$$

is incident at the face $z = 0$ of the material, the field at $z = l$ in the material will be

$$\mathbf{E}(l,\, t) = \epsilon_x \hat{\mathbf{x}} E_0 \cos \omega \left(t - \frac{n_x l}{c} \right) + \epsilon_y \hat{\mathbf{y}} E_0 \cos \omega \left(t - \frac{n_y l}{c} \right) \quad (2.9.5)$$

Thus the x and y polarization components of the field, since they propagate with different phase velocities, develop a phase difference

$$\phi(l) = \frac{\omega l}{c} (n_y - n_x) \qquad (2.9.6)$$

Therefore we may write (2.9.5) as

$$\mathbf{E}(l,\, t) = \epsilon_x \hat{\mathbf{x}} E_0 \cos\,(\omega t - k_x l) + \epsilon_y \hat{\mathbf{y}} E_0 \cos\,[\omega t - k_x l - \phi(l)] \quad (2.9.7)$$

where $k_x = n_x \omega / c$ and $\phi(l)$ [Eq. (2.9.6)] is the phase difference of the x and y components after a propagation distance l.

This phase difference is used, for instance, in the construction of a *quarter-wave plate*. A quarter-wave plate is a sheet of birefringent material of just the right thickness l that $|\phi(l)| = \pi/2$, i.e., the phase difference between waves linearly polarized parallel and perpendicular to the optic axis is a quarter of a cycle. In other words, the optical path difference $l|n_y - n_x|$ is a quarter of the wavelength under consideration. If light incident on a quarter-wave plate is linearly polarized

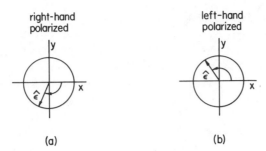

right-hand
polarized

left-hand
polarized

(a)

(b)

Figure 2.11 The x and y components of a circularly polarized field propagating in the z direction trace out a circle: (*a*) right-hand circular polarization; (*b*) left-hand circular polarization.

at 45° to the optic axis, so that $|\epsilon_x| = |\epsilon_y|$, then the transmitted field has the form [Eq. (2.9.7)]

$$\mathbf{E}(l, t) = \frac{1}{\sqrt{2}} E_0 \left[\hat{\mathbf{x}} \cos (\omega t - k_x l) \pm \hat{\mathbf{y}} \sin (\omega t - k_x l) \right] \qquad (2.9.8)$$

since $|\epsilon_x| = |\epsilon_y| = 1/\sqrt{2}$. The x and y components of this electric field trace out a circle, as indicated in Figures 2.11 *a* and *b*. In other words, the transmitted field is circularly polarized. It is called right-hand circularly polarized if $\mathbf{E}$ rotates clockwise to an observer viewing the oncoming light.[6] Thus the $-$ and $+$ signs in (2.9.8) correspond to right- and left-hand circular polarization, respectively.

In general a linearly polarized wave incident on a quarter-wave plate will be converted to an elliptically polarized wave. Linear and circular polarization are special cases of elliptical polarization.

Many isotropic media can be made birefringent by the application of an electric field. Consider, for instance, a liquid consisting of long molecules having permanent dipole moments. The existence of a permanent dipole moment implies a certain asymmetry, namely a preponderance of positive charge at one end of the molecule, and negative charge at the other. Because of collisions, the molecules are randomly oriented, and so the liquid will be macroscopically isotropic, and will not exhibit birefringence. An applied electric field, however, will tend to align the molecules, creating an anisotropy and making the liquid birefringent. This creation of birefringence by an applied electric field is called the *electro-optic effect*.

In the *Kerr electro-optic effect* the induced optic axis is parallel to the applied field, and the difference in refractive indices for light polarized parallel and perpendicular to the optic axis is proportional to the square of the applied field. Kerr cells of liquid or solid transparent media can be used together with polarizers to

---

6. This is the convention traditional in optics. It is unnatural in the sense that the right-hand rule applied to the **k** vector would suggest the opposite.

transmit or block light, depending on whether an electric field is applied. Such "light switches" have many uses in laser technology. (See Section 12.5.)

In certain crystals the induced optic axis is perpendicular to the applied electric field, and the difference in refractive indices for light polarized parallel and perpendicular to the optic axis is linearly proportional to the applied field. This electro-optic effect is called the *Pockels effect*. Pockels cells have uses similar to Kerr cells.

In certain crystals the molecules are aligned in such a way that light polarized in one direction is transmitted, whereas light polarized in the perpendicular direction is strongly absorbed. Such materials are said to be *dichroic*. (Some such materials appear colored in white light because the effect is wavelength-dependent, whence the term dichroic.) They are used, of course, as polarizers. If a wave linearly polarized in some direction $\hat{\varepsilon}$ is normally incident on a dichroic sheet in which the transmitting direction is $\hat{x}$, the transmitted wave will be linearly polarized in the x direction, and will be diminished in amplitude by the factor $\hat{\varepsilon} \cdot \hat{x}$. The intensity is thus diminished by $(\hat{\varepsilon} \cdot \hat{x})^2 = \cos^2 \theta$; this is called Malus's law. Similarly, unpolarized light incident on the polarizer will be reduced in intensity by 50%.

The most common types of polarizer are Polaroid filters, invented around 1926 by Edwin Land (the inventor of the Polaroid Land camera). One type of Polaroid filter consists of a plastic sheet in which are embedded needle-like crystals of dichroic herapathite (quinine sulfate periodide). The most common Polaroid used today is made by dipping a plastic (whose long molecules have been aligned by stretching) in iodine, which makes the plastic dichroic.

• The dichroic property of herapathite was known long before Land, but the crystals were very fragile and difficult to grow in sizes large enough to be useful. The essence of Land's idea was to embed the tiny crystals in a plastic which was stretched while soft to align them. The discovery of herapathite is rather interesting. Land writes:[7] "In the literature are a few pertinent high spots in the development of polarizers, particularly the work of William Bird Herapath, a physician in Bristol, England, whose pupil, a Mr. Phelps, had found that when he dropped iodine into the urine of a dog that had been fed quinine, little scintillating green crystals formed in the reaction liquid. Phelps went to his teacher, and Herapath then did something which I think was curious under the circumstances; he looked at the crystals under a microscope and noticed that in some places they were light where they were overlapped and in some places they were dark. He was shrewd enough to recognize that here was a remarkable phenomenon, a new polarizing material." •

We have assumed in our discussion of birefringence that the optic axis is parallel to the surface on which the field is incident. In general there will be a direction of propagation in which the refractive index is independent of the polarization; this direction defines the optic axis. (Note that the optic axis is not really a single axis, but rather refers to two opposite directions within the birefringent material.) The

7. E.H. Land, *Journal of the Optical Society of America* **41**, 957 (1951).

material may have a single optic axis, in which case it is called *uniaxial*, or it may have two optic axes, in which case it is called *biaxial*. Here we will consider only the simpler, uniaxial case.

If the direction of field propagation is not parallel to the optic axis of a uniaxial crystal, there will be different refractive indices for different field polarizations. In fact only two types of wave can propagate in a uniaxial crystal, namely waves linearly polarized perpendicular to the plane formed by the optic axis and the direction of incidence, and waves linearly polarized parallel to this plane. Waves of the first type are called *ordinary*, whereas those of the second type are called *extraordinary*. Figure 2.12 illustrates what is "ordinary" and "extraordinary" about these waves. In Figure 2.12a, a normally incident field is linearly polarized in a direction perpendicular to the plane formed by the optic axis and the direction of incidence. In this case the field simply passes through the crystal in the expected or "ordinary" way, i.e., according to Snell's law. In Figure 2.12b, however, the field is linearly polarized parallel to the plane defined by the optic axis and the direction of incidence. In this case something unexpected or "extraordinary" happens: the wave is deflected at the boundaries and the rays emergent at the exit face are displaced with respect to the incident rays.

This means that if an anisotropic crystal such as Iceland spar ($CaCO_3$) is laid over a small dot on a piece of paper, we will see a double image of the dot. This phenomenon of *double refraction* (sometimes called *anomalous refraction*) was noted 400 years ago by European sailors visiting Iceland.

The refractive index for ordinary waves is denoted n_0, and is independent of the direction of propagation. The refractive index $n_e(\theta)$ for extraordinary waves, however, depends on the direction of propagation (θ) relative to the optic axis. If n_e denotes the extraordinary index for propagation normal to the optic axis (i.e., $n_e = n_e(\theta = \pi/2)$), then the following relation holds for n_0, n_e, and $n_e(\theta)$:

$$\frac{1}{n_e(\theta)^2} = \frac{\cos^2 \theta}{n_0^2} + \frac{\sin^2 \theta}{n_e^2} \tag{2.9.9}$$

Values of n_0 and n_e for different materials are tabulated in various handbooks.

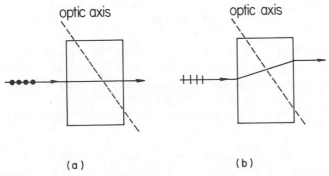

(a)   (b)

Figure 2.12 Ordinary (a) and extraordinary (b) waves in a uniaxial birefringent crystal.

If unpolarized light is incident on a doubly refracting crystal, it is separated into ordinary and extraordinary waves which are linearly polarized orthogonally to each other. This splitting of a light beam into two orthogonally polarized beams in Iceland spar (calcite) was observed by Arago and A.J. Fresnel early in the nineteenth century. Since their polarizations are orthogonal, the two beams do not interfere. This led Young (1817), and later Fresnel, to propose that light waves are transverse, for the absence of interference could not be explained if light waves were longitudinal.

2.10 SUMMARY

Equation (2.3.22) relates a property of a macroscopic medium, the refractive index, to a microscopic characteristic of the medium, the natural oscillation frequencies ω_i of bound electrons. The latter cannot be calculated with any classical model. However, this result for the refractive index is a major success of the electron oscillator model; as we have shown in Sections 2.4–2.9 it explains some important general features of optical dispersion. These features cannot be understood solely on the basis of the Maxwell equations without "atomistic" model assumptions about the medium of propagation.

The classical oscillator model of individual bound electrons was used extensively before the development of quantum mechanics. It was applied by Lorentz to the theory of nearly all the optical phenomena known at the turn of the century. Before Lorentz, physicists generally believed that electromagnetic fields were inextricably associated with continuous distributions of matter. The great contribution of Lorentz was to attribute to the electromagnetic field a truly independent existence of its own, in which, according to Einstein,

> . . .The only connection between the electromagnetic field and ponderable matter arises from the fact that elementary electric charges are rigidly attached to atomistic matter. For the latter, Newton's law of motion holds.
>
> Upon this simplified foundation Lorentz based a complete theory of all electromagnetic phenomena known at the time, including those of the electrodynamics of moving bodies. It is a work of such lucidity, consistency, and beauty as has only rarely been obtained in an empirical science.

In later chapters we will deal with concepts like "oscillator strengths" and "negative oscillators" which, though basically of a quantum-mechanical nature, reflect even by their names the conceptual framework established by Lorentz.

PROBLEMS

2.1   Write out the x component of the left and right sides of the vector identity given in (2.1.8) and show that they are equal. Remember that $(\nabla \times \mathbf{V})_x = \partial V_z / \partial y - \partial V_y / \partial z$ for any vector $\mathbf{V}$.

2.2   Verify by substitution in (2.1.14) that the z component of $\mathbf{E}$ given in (2.1.26c) is a solution of the free-space wave equation, assuming that l, m, n obey (2.1.28).

2.3   Show that the condition given in (2.1.27) is necessary if the electric field solutions given in (2.1.26) are to be "transverse," i.e., if they are to satisfy $\nabla \cdot \mathbf{E} = 0$ as assumed in (2.1.12). Note that (2.1.27) implies that only two of the amplitudes A_x, A_y, A_z can be chosen independently. This is the reason a transverse field has only two independent polarizations.

2.4   Show that the formulas for mode densities dN_ν and dN_λ, given in (2.1.30), are related through the identity $\nu = c/\lambda$.

2.5   (a) Show that the refractive index of a mixture of gases is

$$n(\omega) = \sum_i f_i n_i(\omega)$$

where $n_i(\omega)$ is the index of the ith species and f_i is its fractional concentration (number of atoms of species i divided by the total number).

   (b) Using the refractive indices $n_{O_2} = 1.000272$ and $n_{N_2} = 1.000297$ of STP oxygen and nitrogen at 5890 Å, estimate the refractive index of STP air at this wavelength.

2.6   Consider a medium with $n = 1$. The Poynting vector $\mathbf{S} = \mathbf{E} \times \mathbf{H}$ is associated with electromagnetic energy flow. Compute $\mathbf{S}$ from the solutions for $\mathbf{E}$ and $\mathbf{B} = \mu_0 \mathbf{H}$ given in Table 2.2. Also compute the electromagnetic energy density $u = \frac{1}{2}(\epsilon_0 E^2 + B^2/\mu_0)$. By comparing $\mathbf{S}$ and u, determine the velocity of energy flow in this case.

2.7   Verify that the number of molecules per cubic centimeter in an ideal gas at standard temperature and pressure is about 2.69×10^{19}.

2.8   Assume that in the earth's ionosphere the refractive index for 100 MHz radio waves is 0.90, and that free electrons make the greatest contribution to the index.

   (a) Estimate the number of electrons per cubic centimeter.

   (b) Why is the contribution of positively charged ions to the refractive index much smaller?

2.9   (a) Choose an AM and an FM radio station in your area and compare their frequencies with the plasma frequency of the ionosphere.

   (b) Why are automobile antennas oriented vertically rather than horizontally?

2.10* Consider the electron oscillator model for the case in which there is *no field* acting on the atom. Suppose that at $t = 0$ an electron is given the displacement x_0 from equilibrium, and the velocity v_0.

(a) Show that the electron coordinate $x(t)$ is given by

$$x(t) = x_0 \cos \omega_0 t + \frac{v_0}{\omega_0} \sin \omega_0 t$$

(b) What is the total (kinetic plus potential) energy of the electron?

(c) Using the formula (2.5.14), derive an expression for the rate at which the oscillating electron radiates electromagnetic energy. Give the rate averaged over times long compared with the period of oscillation.

(d) Show that the electron can be expected to radiate away most of its energy in a time

$$\tau = 4\pi\epsilon_0 \left(\frac{2e^2 \omega_0^2}{3mc^3} \right)^{-1}$$

This is the classical picture of "spontaneous emission," which we consider in Chapter 7.

(e) Estimate numerically the "radiative lifetime" τ found in part (d) for the case of an electron oscillating at an optical frequency $\nu_0 (= \omega_0 / 2\pi)$.

2.11 Show that the scattering cross section for radiation of frequency ω much greater than the natural oscillation frequency ω_0 is given by the Thomson formula

$$\sigma(\omega \gg \omega_0) \approx \frac{8\pi}{3} r_0^2 = \frac{8\pi}{3} \left(\frac{e^2}{4\pi\epsilon_0 mc^2} \right)^2$$

where $r_0 = e^2 / 4\pi\epsilon_0 mc^2$ is called the "classical electron radius," What is the magnitude of r_0?

2.12 A typical He–Ne laser operating at 6328 Å contains about five times as much He as Ne, with a total pressure of about one Torr. The length of the gain cell is about 50 cm. Estimate the fraction of laser radiation intensity lost due to Rayleigh scattering in passing a billion times through the gain cell. (Note: For STP Ne the constants in (2.4.9) are $A = 6.66 \times 10^{-5}$ and $B = 2.4 \times 10^{-11}$ cm^2.) This illustrates the fact that Rayleigh scattering is usually very weak in gas laser media.

*Starred problems are somewhat more difficult.

3 CLASSICAL THEORY OF ABSORPTION

3.1 INTRODUCTION

Most objects around us are not self-luminous but are nevertheless visible because they scatter the light that falls upon them. Most objects are *colored*, however, because they absorb light, not simply because they scatter it. The colors of an object typically arise because materials selectively absorb light of certain frequencies, while freely scattering or transmitting light of other frequencies. Thus if an object absorbs light of all visible frequencies, it is black. An object is red if it absorbs all (visible) frequencies except those our eyes perceive to be "red" (wavelengths roughly between about 6300 and 6800 Å), and so on.[1]

The physics of the absorption process is simplest in well-isolated atoms. These are found most commonly in gases. White light propagating through a gas is absorbed at the resonance frequencies of the atoms or molecules, so that one observes gaps in the wavelength distribution of the emerging light. On a spectrogram these gaps appear as bright lines on the dark, exposed background. The gaps, shown as lines in Figure 3.1, correspond to the absorption of sunlight by the atmosphere *of the sun* before the light reaches the earth. The absorbed energy is partially converted into heat (translational kinetic energy of the atoms) when excited atoms (or molecules) which have absorbed radiation collide with other particles. The absorbed radiation is also partially reradiated in all directions at the frequency of the absorbed radiation. This is called resonance radiation, or resonance fluorescence. When the pressure of the gas is increased, collisions may rapidly convert the absorbed radiation into heat before it can be reradiated. In this case the resonance radiation is said to be quenched.

Most atoms have electronic resonance frequencies in the ultraviolet, although resonances in the visible and infrared are not uncommon. Sodium, for instance, has strong absorption lines in the yellow region at 5890 and 5896 Å, the Fraunhofer "D lines," and their position is indicated in Figure 3.1.

Electronic resonances in molecules also tend to lie in the ultraviolet. We have "white" daylight because the atmosphere, consisting mostly of N_2 and O_2, does not absorb strongly at visible frequencies.

In molecules the separate atoms act approximately as if they were connected to each other by springs, so that entire atoms vibrate back and forth. Atoms are of

---

1. The principal features of the electromagnetic spectrum for our purposes are summarized in Table 3 inside the cover of the book.

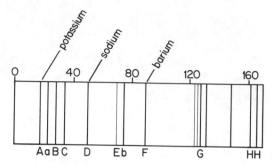

Figure 3.1 Absorption lines of the sun's atmosphere. The Fraunhofer D lines of sodium at 5890 and 5896 Å are not resolved in this sketch.

course much more massive (by 10^3–10^5 times) than electrons, and the natural vibrations of molecules are consequently slower. We can estimate, on the basis of this mass difference (Problem 3.1), that molecular vibration frequencies should lie in the infrared portion of the electromagnetic spectrum.

A molecule as a whole can also rotate; the resonance frequencies associated with molecular rotations lie in the microwave portion of the spectrum. Molecules therefore typically have resonances in the ultraviolet, infrared, and microwave regions of the spectrum.

Absorption in liquids and solids is much more complicated than in gases. In liquids and amorphous solids such as glass, the absorption lines have such large widths that they overlap. Water, for example, is obviously transparent in the visible, but absorbs in the near infrared, i.e., at infrared wavelengths not far removed from the visible. Its absorption curve is wide enough, in fact, that it extends into the red edge of the visible. (Figure 3.2) The weak absorption in the red portion of

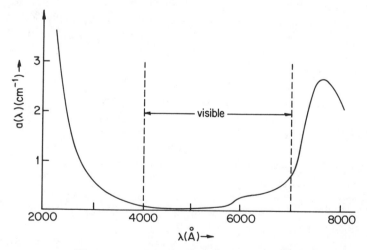

Figure 3.2 Absorption coefficient of water.

the visible spectrum explains why things appear green when one is sufficiently submerged under water.

A broad absorption curve covering all visible wavelengths except those in a particular narrow band is characteristic of the molecules of a dye. The absorbed radiation is converted into heat before it can be reradiated. Such broad absorption curves and fast quenching rates require the high molecular number densities of liquids and solids.

In metals some of the atomic electrons are able to move freely about under the influence of an electromagnetic field. The fact that metals contain these "free" electrons explains, of course, why they are good conductors of electricity. In the *free-electron approximation* we may apply the dispersion formula (2.4.12). The plasma frequency ω_p for metals is usually in the ultraviolet (Problem 3.2). Thus visible frequencies ($\omega < \omega_p$) cannot penetrate into the metal. They are completely reflected, just as AM radio waves are reflected by the ionosphere. This strong reflection gives metals their shine. In a metal like gold there is also absorption, associated with the electrons that remain bound to atoms, and it is this that gives the metal a characteristic color.

In a solid that is a good electrical insulator, the electrons are tightly bound, and consequently the natural oscillation frequencies are high, typically corresponding to wavelengths less than 4000 Å. An insulator, therefore, is usually transparent in the visible but opaque in the ultraviolet. In semiconductors the natural oscillation frequencies are smaller. Silicon, for example, absorbs visible wavelengths (it is black), but transmits radiation of wavelength greater than one micron (1 micron $= 1 \ \mu m$).

Lattice defects (deviations from periodicity) can substantially modify the absorption spectra of crystalline solids. Ruby, for instance, is corundum (Al_2O_3) with an occasional (roughly 0.05% by weight) random substitution of Cr^{+3} ions in place of Al^{+3}. The chromium ions absorb green light and thus ruby is pink, in contrast to the transparency of pure corundum.

The variety of natural phenomena resulting from the selective absorption of certain wavelengths and the transmission of others is too broad to treat here. We mention only one important example, the "greenhouse effect."[2] Visible sunlight is transmitted by the earth's atmosphere and heats (by absorption) both land and water. The warmed earth's surface is a source of thermal radiation, the dominant emission for typical ambient temperatures being in the infrared. This infrared radiation, however, is strongly absorbed by CO_2 and H_2O vapor in the earth's atmosphere, preventing its rapid escape into space. Without this effect, the earth would be a much colder place. An increased burning of fossil fuels could conceivably enhance the greenhouse effect by increasing the level of CO_2 in the atmosphere.

---

2. The term "greenhouse effect" is actually a misnomer, originating in the observation that the glass in a greenhouse, which is transparent in the visible but opaque to the infrared, plays an absorptive role similar to that of CO_2 and H_2O in the earth's atmosphere. This effect, however, does not contribute significantly to the warming of the air inside a real greenhouse. A real greenhouse mainly prevents cooling by wind currents. This point was demonstrated experimentally by R. W. Wood (1909), although the contrary misconception persists even among scientists.

3.2 ABSORPTION AND THE LORENTZ MODEL

The strength of an electromagnetic field will be reduced in transit through a material medium if the atoms (or molecules) of the medium can absorb radiant energy. More commonly than not, in a wide variety of materials, absorption can be explained by the assumption that the Lorentz electron oscillators introduced in Chapter 2 are subject to a frictional force. The origin of a "frictional" force is itself a subject for discussion, which will be found in Section 3.9. For the moment, however, we will take a frictional force for granted, and explore its consequences.

We simply amend the Newton force law (2.2.18) to read

$$m \frac{d^2\mathbf{x}}{dt^2} = e\mathbf{E}(\mathbf{R}, t) - k_s\mathbf{x} + \mathbf{F}_{\text{fric}} \tag{3.2.1}$$

and we make the simplest assumption compatible with the idea of frictional drag:

$$\mathbf{F}_{\text{fric}} = -b\mathbf{v} = -b \frac{d\mathbf{x}}{dt} \tag{3.2.2}$$

Then the Newton equation of motion (2.3.7) for an electron oscillator in a linearly polarized monochromatic plane wave takes the form

$$\frac{d^2\mathbf{x}}{dt^2} + 2\beta \frac{d\mathbf{x}}{dt} + \omega_0^2\mathbf{x} = \hat{\varepsilon} \frac{e}{m} E_0 \cos(\omega t - kz) \tag{3.2.3}$$

where for later convenience we have defined

$$\beta = \frac{b}{2m}$$

As in Chapter 2 we have introduced the natural oscillation frequency

$$\omega_0 = \left(\frac{k_s}{m}\right)^{1/2} \tag{3.2.4}$$

associated with Lorentz's elastic force.

If there is no applied field, Eq. (3.2.3) becomes

$$\frac{d^2\mathbf{x}}{dt^2} + 2\beta \frac{d\mathbf{x}}{dt} + \omega_0^2\mathbf{x} = 0 \tag{3.2.5}$$

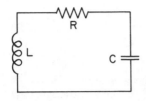

Figure 3.3 An *LRC* circuit. The charge on the capacitor obeys the equation of motion (3.2.6) for a damped oscillator.

This is the equation describing a damped oscillator. A well-known example is an *LRC* circuit (Figure 3.3), where the charge q on the capacitor satisfies the equation

$$\frac{d^2q}{dt^2} + \frac{R}{L}\frac{dq}{dt} + \frac{1}{LC}q = 0 \qquad (3.2.6)$$

In this case the natural oscillation frequency and the damping rate are determined by the fundamental parameters of the circuit:

$$\omega_0 = \left(\frac{1}{LC}\right)^{1/2} \qquad (3.2.7a)$$

and

$$\beta = \frac{R}{2L} \qquad (3.2.7b)$$

The solution of the differential equation (3.2.6) is

$$q(t) = (A\cos\omega_0't + B\sin\omega_0't)\,e^{-\beta t} \qquad (3.2.8a)$$

where

$$\omega_0' = (\omega_0^2 - \beta^2)^{1/2} \qquad (3.2.8b)$$

Under most conditions of interest the oscillator will be significantly underdamped [see Eq. (3.3.10)] and we can replace ω_0' by ω_0. Since (3.2.6) is a second-order linear differential equation, its solution has two constants of integration which are determined by the initial conditions for $q(t)$ and $dq(t)/dt$. We have denoted these two constants by A and B.

If the LRC circuit is driven by a sinusoidal emf (Figure 3.4),

$$V(t) = V_0\cos(\omega t - \theta) \qquad (3.2.9)$$

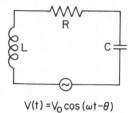

V(t) = $V_0 \cos(\omega t - \theta)$

Figure 3.4 An *LRC* circuit with a sinusoidal emf. The charge on the capacitor obeys the equation of motion (3.2.10) for a sinusoidally driven, damped oscillator.

then q satisfies the *forced-oscillator* equation

$$\frac{d^2 q}{dt^2} + 2\beta \frac{dq}{dt} + \omega_0^2 q = \frac{V_0}{L} \cos(\omega t - \theta) \qquad (3.2.10)$$

where ω_0 and β are given by (3.2.7). This is just a scalar version of the electron oscillator vector equation (3.2.3), with L corresponding to the electron's mass/charge ratio:

$$L = \frac{m}{e} \qquad (3.2.11)$$

and θ corresponding to the field phase at the position of the atom:

$$\theta = kz = \frac{2\pi z}{\lambda} \qquad (3.2.12)$$

In contrast to the homogeneous solution (3.2.8a), which decays to zero, the solution to the forced-oscillator equation (3.2.10) is a steady sinusoidal oscillation with an amplitude depending on ω and ω_0. The amplitude has a maximum when $\omega \approx \omega_0$, and one says that the circuit of Figure 3.4 exhibits a resonance. From (3.2.7a) we see that this resonance condition is met when the capacitance is

$$C = \frac{1}{\omega^2 L} \qquad (3.2.13)$$

When the resonance condition is approached by tuning the capacitance to the resonance value (3.2.13), the amplitude of the oscillating current in the circuit increases dramatically, as shown in Figure 3.5. This resonant enhancement is used in simple radio receivers, where a variable capacitor permits tuning to various broadcast frequencies.

The interaction of an atom with a monochromatic field is similarly enhanced when

$$\omega = \omega_0 \qquad (3.2.14)$$

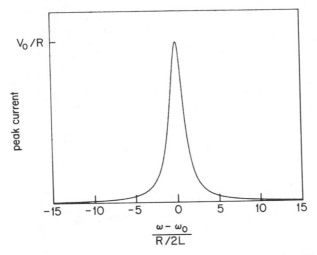

Figure 3.5 The amplitude of the oscillating current in an *LRC* circuit with emf (3.2.9). The current oscillation is at the driving frequency ω, and has maximum amplitude when the resonance condition $\omega = \omega_0 = (1/LC)^{1/2}$ is satisfied.

i.e., when the frequency of the field coincides with a natural oscillation frequency of a bound electron. This enhancement of the interaction is already implied by our result (2.3.14b) for the refractive index. However, that result is obviously undefined if $\omega = \omega_0$. A frictional force in the electron oscillator model allows us to understand formulas like (2.3.14b) even for $\omega = \omega_0$, while also providing the physical mechanism for the absorption of electromagnetic energy.

3.3 COMPLEX POLARIZABILITY AND INDEX OF REFRACTION

The equation (3.2.3) for the electron oscillator with damping is most easily solved by first writing it in complex form:

$$\frac{d^2x}{dt^2} + 2\beta \frac{dx}{dt} + \omega_0^2 x = \hat{\varepsilon} \frac{e}{m} E_0 e^{-i(\omega t - kz)} \tag{3.3.1}$$

where we follow the convention of writing $E_0 \cos(\omega t - kz)$ as $E_0 e^{-i(\omega t - kz)}$. This means that $x(t)$ in Eq. (3.3.1) is also regarded mathematically as a complex quantity in our calculations, but only its real part is physically meaningful. In other words, we may defer the process of taking the real part of (3.3.1) until *after* our calculations, at which point the real part of our solution for $x(t)$ is the (real) electron displacement. This approach is standard in solving linear equations, but there are pitfalls that can arise in nonlinear problems. [See Chapter 17, where modifications of Eq. (3.3.1) are used in an introduction to nonlinear optics.]

We solve (3.3.1) by temporarily writing

$$\mathbf{x}(t) = \mathbf{a} \, e^{-i(\omega t - kz)} \qquad (3.3.2)$$

and after inserting this in (3.3.1) we obtain

$$(-\omega^2 - 2i\beta\omega + \omega_0^2) \, \mathbf{a} = \hat{\varepsilon} \, \frac{e}{m} \, E_0 \qquad (3.3.3)$$

Therefore the assumed solution (3.3.2) satisfies Eq. (3.3.1) if

$$\mathbf{a} = \frac{-\hat{\varepsilon}(e/m) \, E_0}{\omega^2 - \omega_0^2 + 2i\beta\omega} \qquad (3.3.4)$$

and the physically relevant solution is therefore

$$\mathbf{x}(t) = \mathrm{Re} \left(\frac{\hat{\varepsilon}(e/m) \, E_0 \, e^{-i(\omega t - kz)}}{\omega_0^2 - \omega^2 - 2i\beta\omega} \right) \qquad (3.3.5)$$

Note that (3.3.5) actually gives only the steady-state solution of (3.3.1). Any solution of the homogeneous version of (3.3.1) can be added to (3.3.5), and the total will still be a solution of (3.3.1). The homogeneous version is

$$\frac{d^2 \mathbf{x}_{\mathrm{hom}}}{dt^2} + 2\beta \frac{d\mathbf{x}_{\mathrm{hom}}}{dt} + \omega_0^2 \mathbf{x}_{\mathrm{hom}} = 0 \qquad (3.3.6)$$

and its general solution is an obvious vectorial extension of (3.2.8a):

$$\mathbf{x}_{\mathrm{hom}} = [\mathbf{A} \cos \omega_0' t + \mathbf{B} \sin \omega_0' t] \, e^{-\beta t} \qquad (3.3.7)$$

where again

$$\omega_0' = (\omega_0^2 - \beta^2)^{1/2} \approx \omega_0 \qquad (3.3.8)$$

We will usually neglect the homogeneous part of the full solution to (3.3.1). This is obviously an approximation. The approximation is however an excellent one whenever

$$t \gg 1/\beta \qquad (3.3.9)$$

Under this condition, $e^{-\beta t} \ll 1$ and we can safely neglect the homogeneous component (3.3.7) because it makes only a short-lived transient contribution to the solution.

Even though the homogeneous damping time, or lifetime $\tau_0 = 1/\beta$, is very short, it is not the shortest time in the problem. Typically the oscillation periods $T = 2\pi/\omega$ and $T_0 = 2\pi/\omega_0$ associated with the natural oscillation frequency ω_0 or the forcing frequency ω are very much shorter. In the case of ordinary optically transparent materials such as atomic vapors, glasses, and many crystals and liquids, both ω_0 and ω are typically in the neighborhood of 10^{15} sec^{-1}, and β falls in a wide range of much smaller frequencies:

$$\beta \approx 10^6\text{--}10^{12} \text{ sec}^{-1} \ll \omega_0, \omega \tag{3.3.10}$$

Relations (3.3.9) and (3.3.10), taken together, imply that times of physical interest must be much longer than an optical period:

$$t \gg \beta^{-1} \gg \omega_0^{-1}, \omega^{-1} \tag{3.3.11}$$

That is, steady-state solutions of (3.3.1) are valid for times that are many periods of oscillator vibration ($T_0 = 2\pi/\omega_0$) and forced vibration ($T = 2\pi/\omega$) removed from $t = 0$, but they cannot be used to predict the oscillator's response within the first few cycles after $t = 0$. This is, however, a restriction of no real significance in optical physics, as it is equivalent to

$$t \gg 10^{-15} \text{ sec } (=10^{-3} \text{ ps}) \tag{3.3.12}$$

This is a time span one or two orders of magnitude smaller than can presently be resolved optically.

The steady-state solution (3.3.5) is very close to the solution (2.3.8) for the undamped oscillator. It implies that the electric field induces in an atom a dipole moment $\mathbf{p} = e\mathbf{x}$, or $\mathbf{p} = \Sigma_j e\mathbf{x}_j$ in the case of many electrons:

$$\mathbf{p} = \text{Re} \left(\hat{\varepsilon} \frac{e^2}{m} \frac{E_0 e^{-i(\omega t - kz)}}{\omega_0^2 - \omega^2 - 2i\beta\omega} \right) \tag{3.3.13}$$

or

$$\mathbf{p} = \text{Re} \left(\hat{\varepsilon} \frac{e^2}{m} E_0 e^{-i(\omega t - kz)} \sum_{j=1}^{z} \frac{1}{\omega_j^2 - \omega^2 - 2i\beta_j\omega} \right)$$

The real part can be found explicitly to be

$$\mathbf{p} = \hat{\varepsilon} \frac{e^2}{m} \left(\frac{(\omega_0^2 - \omega^2) E_0 \cos(\omega t - kz) + 2\beta\omega E_0 \sin(\omega t - kz)}{(\omega_0^2 - \omega^2)^2 + 4\beta^2\omega^2} \right) \tag{3.3.14}$$

with a corresponding expression for a multielectron system.

Because of the frictional damping (i.e., because $\beta \neq 0$) the dipole moment no longer oscillates completely in phase with the electric field as it did in (2.3.8). The new term proportional to $\sin(\omega t - kz)$ signifies the existence of a phase lag in the dipole response. Thus there is no single real polarizability coefficient that can be identified as the ratio of the dipole moment and the electric field strength.

It is possible nevertheless, and generally very convenient, to introduce a complex polarizability. This is done by recognizing that (3.3.2) can be used to define a complex dipole moment $\mathbf{p}$:

$$\mathbf{p} = e\mathbf{x} = e\mathbf{a}\, e^{-i(\omega t - kz)} \tag{3.3.15}$$

The complex polarizability α is defined by the relation between complex moment and complex field:

$$\mathbf{p} = \alpha(\omega)\,\hat{\varepsilon} E_0\, e^{-i(\omega t - kz)} \tag{3.3.16}$$

In the present case, by comparing (3.3.13) and (3.3.16) we easily identify the complex polarizability of a Lorentzian atom to be

$$
\begin{aligned}
\alpha(\omega) &= \frac{e^2/m}{\omega_0^2 - \omega^2 - 2i\beta\omega} \\[2mm]
&= \frac{e^2}{m}\frac{\omega_0^2 - \omega^2 + 2i\beta\omega}{\left(\omega_0^2 - \omega^2\right)^2 + 4\beta^2\omega^2}
\end{aligned}
\tag{3.3.17}
$$

or in the case of many electrons,

$$
\begin{aligned}
\alpha(\omega) &= \sum_{j=1}^{Z} \frac{e^2/m}{\omega_j^2 - \omega^2 - 2i\beta_j\omega} \\[2mm]
&= \sum_{j=1}^{Z} \frac{e^2}{m}\frac{\omega_j^2 - \omega^2 + 2i\beta_j\omega}{\left(\omega_j^2 - \omega^2\right)^2 + 4\beta_j^2\omega^2}
\end{aligned}
\tag{3.3.18}
$$

Given the complex polarizability (3.3.17) or (3.3.18), the complex polarization density is

$$\mathbf{P} = N\mathbf{p} = N\alpha(\omega)\,\hat{\varepsilon} E_0\, e^{-i(\omega t - kz)} \tag{3.3.19}$$

Using this polarization density in the wave equation (2.1.13), together with the complex form of the assumed solution (2.3.1), we obtain

$$
\begin{aligned}
\left(-k^2 + \frac{\omega^2}{c^2}\right) &\hat{\varepsilon}\, E_0\, e^{-i(\omega t - kz)} \\[2mm]
&= -\frac{\omega^2}{c^2}\frac{N\alpha(\omega)}{\epsilon_0}\,\hat{\varepsilon}\, E_0\, e^{-i(\omega t - kz)}
\end{aligned}
\tag{3.3.20}
$$

Therefore k must satisfy the dispersion relation

$$k^2 = \frac{\omega^2}{c^2}\left(1 + \frac{N\alpha(\omega)}{\epsilon_0}\right)$$

$$= \frac{\omega^2}{c^2}\,n^2(\omega), \tag{3.3.21}$$

just as in Eq. (2.3.13).

In this case, because $\alpha(\omega)$ is complex the refractive index is also a complex number:

$$n^2(\omega) = 1 + \frac{Ne^2/m\epsilon_0}{\omega_0^2 - \omega^2 - 2i\beta\omega}$$

$$= 1 + \frac{Ne^2}{m\epsilon_0}\,\frac{\omega_0^2 - \omega^2 + 2i\beta\omega}{\left(\omega_0^2 - \omega^2\right)^2 + 4\beta^2\omega^2}$$

$$= \left[n_R(\omega) + in_I(\omega)\right]^2 \tag{3.3.22}$$

The most important consequence of these results is that the electric field in the medium behaves differently from the field discussed in Chapter 2 because $n(\omega)$ is now complex:

$$\mathbf{E}(z,\,t) = \hat{\mathbf{\varepsilon}}E_0\,e^{-i(\omega t\,-\,kz)}$$

$$= \hat{\mathbf{\varepsilon}}E_0\,e^{-i\omega[t-n(\omega)z/c]}$$

$$= \hat{\mathbf{\varepsilon}}E_0\,e^{-[n_I(\omega)]\omega z/c}\,e^{-i\omega\{t-[n_R(\omega)]z/c\}} \tag{3.3.23}$$

Note that $\mathbf{E}(z,\,t)$ is no longer purely oscillatory. Due to $n_I(\omega)$, the field decays with increasing distance of propagation. Since the intensity is proportional to the square of the (real) electric field [recall Eq. (2.6.4) and (2.6.8)], the intensity shows exponential decay with z:

$$I_\omega(z) = I_\omega(0)\left(e^{-[n_I(\omega)]\omega z/c}\right)^2 = I_0\,e^{-a(\omega)z} \tag{3.3.24}$$

where we call $a(\omega)$ the *absorption coefficient* or *extinction coefficient*:

$$a(\omega) = 2[n_I(\omega)]\omega/c$$

$$= \frac{2Ne^2}{\epsilon_0 mc}\sum_j \frac{\beta_j\omega^2}{\left(\omega_j^2 - \omega^2\right)^2 + 4\beta_j^2\omega^2} \tag{3.3.25}$$

As in (2.3.23) we have used $n \approx 1$. This is a very important result, and we will return to it shortly.

The phase velocity of the wave (3.3.23) is $c/n_R(\omega)$. The real part of the com-

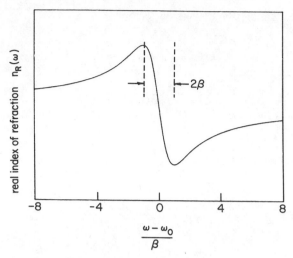

Figure 3.6 Anomalous dispersion curve for a collision-broadened absorption line.

plex refractive index is therefore what would ordinarily be called the "refractive index." This refractive index is plotted versus frequency in Figure 3.6. On the low-frequency side of each resonance frequency, $n_R(\omega)$ increases with increasing frequency, i.e., we have "normal dispersion" (Section 2.4). However, when ω gets within β_j of ω_j, $n_R(\omega)$ begins *decreasing* with increasing frequency. This decrease continues until ω is more than β_j from ω_j on the high-frequency side, whereupon it again increases with increasing frequency. Because most media show normal dispersion at optical frequencies, the negative slope of the dispersion curve near an absorption line was historically termed *anomalous dispersion*.

• Anomalous dispersion was observed by R. W. Wood in 1904. Wood studied the dispersion of light at frequencies near the sodium D lines (5890 and 5896 Å). The basic idea of Wood's experiment is sketched in Figure 3.7. Light enters a tube in which sodium vapor is produced by heating sodium. The vapor pressure decreases upwards in the tube, so that for normal dispersion the light would be bent downward, in the direction of greater density

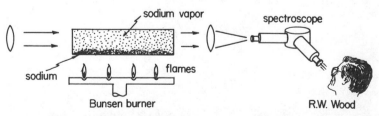

Figure 3.7 One of R. W. Wood's experiments on anomalous dispersion in sodium vapor.

and refractive index. The vapor thus acts as a kind of prism. The light emerging at the other end of the tube is focused onto the entrance slit of a spectroscope. Wood writes:

> On heating the tube, the sodium prism deviates the rays of different wave-length up or down by different amounts, curving the spectrum into two oppositely directed branches. The spectrum on the green side of the D lines will be found to bend down in the spectroscope, which means that the rays are deviated upwards in passing through the sodium tube, since the spectroscope inverts the image of its slit. This means that this phase velocity is greater in the sodium vapor than in vacuo, or the prism acts for these rays like an air prism immersed in water. The red and orange region is deviated in the opposite direction; these rays are therefore retarded by the vapor.

In other words, the refractive index on the low-frequency side of resonance was observed to be greater than unity, whereas on the high-frequency side it was less than unity. This is the behavior shown in Figure 3.6. In fact Wood's measured curve of refractive index versus frequency showed exactly the "anomalous dispersion" form predicted by the electron oscillator model. •

3.4 POLARIZABILITY AND INDEX OF REFRACTION NEAR A RESONANCE

Most of the time we will be primarily interested in the response of the dipoles that are very nearly resonant with an applied field. These dipoles will usually be a small minority of the dipoles present. The sharpness of their resonant response (recall Figure 3.5) makes them particularly important. However, the other dipoles in the far off-resonant "background" can be so numerous that they also make a significant contribution to the polarizability and index of refraction, and we cannot overlook them.

Equation (3.3.18) shows that the polarizability is additive over all dipole response frequencies. Thus we can write

$$\alpha(\omega) = \alpha_b(\omega) + \alpha_r(\omega) \qquad (3.4.1)$$

where α_b and α_r are the contributions from "background" and "resonant" dipoles, respectively. The background dipoles may reside in an actual host material, in which the atoms with the resonant dipoles are embedded, or they may be dipoles associated with nonresonant oscillations in the same atoms as the resonant dipoles. In either event, the relations (3.4.1) and (3.3.21) imply

$$n^2(\omega) = 1 + \sum_i \frac{N_{bi}}{\epsilon_0} \alpha_{bi}(\omega) + \frac{N_r}{\epsilon_0} \alpha_r(\omega) \qquad (3.4.2)$$

where we have indicated a sum over all background species.

The first two terms in (3.4.2) determine $n_b(\omega)$, the index of refraction of the background or host material. Thus we will write

$$n^2(\omega) = n_b^2(\omega) + \frac{N_r \alpha_r(\omega)}{\epsilon_0}$$

$$= n_b^2(\omega) \left(1 + \frac{N_r \alpha_r(\omega)}{n_b^2(\omega) \epsilon_0} \right)$$

$$= n_b^2(\omega) \left(1 + \frac{N_r \alpha_r(\omega)}{\epsilon_b(\omega)} \right) \tag{3.4.3}$$

where $\epsilon_b = n_b^2 \epsilon_0$ is the dielectric permittivity of the background. If the resonant atoms are present in a monatomic beam, then the background material is vacuum or nearly so and the background contributions can largely be ignored. Even in an atomic vapor n_b can be taken to be unity to three or four significant figures. However, in laser physics, the background material is frequently a solid or liquid. For example, the ruby laser operates because of dipoles associated with chromium ions thinly dispersed throughout a solid lattice (the crystal called corundum), and the dye molecules of a dye laser are dissolved in a liquid solvent (for example ethanol). Then n_b is significantly different from unity, typically in the range 1.3–2.0. We will write n_b in place of $n_b(\omega)$ hereafter because the resonances of the background are typically in the infrared or ultraviolet and n_b is effectively constant at optical frequencies.

The resonant dipoles do not make a correspondingly large contribution, since they are usually present in such small concentrations. The concentration of the chromium ions in ruby, for example, may be only 10^{19} per cm^3 or even less, much smaller than typical solid densities. As a consequence the last term in (3.4.3) is typically much smaller than unity. Then the total index of refraction can be expressed compactly as follows:

$$n(\omega) = n_b \left(1 + \frac{N_r \alpha_r(\omega)}{\epsilon_b} \right)^{1/2}$$

$$\approx n_b + \frac{N_r \alpha_r(\omega)}{2 n_b \epsilon_0} \tag{3.4.4}$$

where we have again used $\epsilon_b = n_b^2 \epsilon_0$ after expanding the square root and keeping only the first term in the binomial series $(1 + x)^{1/2} = 1 + x/2 + x^2/8 + \cdots$.

Now we must consider what we mean by "near to resonance." Note in (3.3.17) that when $\beta = 0$ the imaginary part of $\alpha(\omega)$ vanishes and the real part reduces to (2.3.10). In any event, if ω is far enough from the resonance frequencies ω_j, we can put $\beta = 0$ without affecting the result appreciably. It should be clear then that "far from resonance" is only a relative term, relative to the damping coefficient β. For any resonance frequency ω_j, then, "far from resonance" means

$$|\omega_j - \omega| \gg \beta_j \tag{3.4.5a}$$

and "near to resonance" means

$$|\omega_j - \omega| \le \beta_j \qquad (3.4.5b)$$

A significant contribution to $\alpha(\omega)$ can come from a resonance if the associated β is small enough. Suppose there is one frequency $\omega_j = \omega_0$ close enough to ω to satisfy (3.4.5b) and all others satisfy the off-resonance condition (3.4.5a). For clarity we will label the resonant damping coefficient β without a subscript. Then we can write

$$\alpha_r(\omega) = \frac{e^2}{m} \frac{1}{\omega_0^2 - \omega^2 - 2i\beta\omega}$$

The resonant part of $\alpha(\omega)$ can be written in a still simpler form if ω is close enough to ω_0 to justify the approximation

$$|\omega_0 - \omega| \ll \omega, \omega_0 \qquad (3.4.6)$$

which is always guaranteed in practice whenever the earlier approximation $|\omega_0 - \omega| \le \beta$ is valid. In this case we can write

$$\omega_0^2 - \omega^2 = (\omega_0 + \omega)(\omega_0 - \omega) \approx 2\omega(\omega_0 - \omega), \qquad (3.4.7)$$

and under this condition we have

$$\alpha_r(\omega) = \frac{e^2/2m\omega}{\omega_0 - \omega - i\beta}. \qquad (3.4.8)$$

When the field frequency ω is far removed from *all* the resonance frequencies ω_j of the medium, the complex polarizability (3.3.17) reduces to the real polarizability (2.3.10). In this case the refractive index predicted by the electron oscillator model has been discussed in Chapter 2. For frequencies ω near to any of the ω_j, however, the friction coefficient β becomes important. For example, it is just because β is not zero that the refractive index does not become infinite whenever $\omega = \omega_j$, as is (erroneously) predicted by (2.3.14).

The real and imaginary parts of the index of refraction can now be identified easily, using (3.4.8) for $\alpha_r(\omega)$, and we find

$$n_R(\omega) = n_{bR} + \frac{Ne^2}{4n_{bR}\epsilon_0 m\omega} \frac{\omega_0 - \omega}{(\omega_0 - \omega)^2 + \beta^2} \qquad (3.4.9)$$

$$n_I(\omega) = n_{bI} + \frac{Ne^2}{4n_{bR}\epsilon_0 m\omega} \frac{\beta}{(\omega_0 - \omega)^2 + \beta^2} \qquad (3.4.10)$$

Here we have written n_{bR} and n_{bI} for the real and imaginary parts of n_b. Also we have assumed $n_{bR} \gg n_{bI}$. Finally, by comparison with (3.3.25) and (3.4.10), we obtain the absorption coefficient due to the resonance frequency ω_0:

$$a(\omega) = 2n_I(\omega)\, \omega/c$$

$$= a_b(\omega) + \frac{Ne^2}{2n_{bR}\epsilon_0 mc} \frac{\beta}{(\omega_0 - \omega)^2 + \beta^2} \qquad (3.4.11)$$

where $a_b(\omega) = 2n_{bI}\, \omega/c$ is the background absorption coefficient.

3.5 LORENTZIAN ATOMS AND RADIATION IN CAVITIES

The Newton–Lorentz equation for the response of an atomic dipole to an applied radiation field was given in (3.2.3) under the assumption that the radiation took the form of a traveling wave. That is, in complex notation, the electric field was assumed to have the form

$$\mathbf{E}(z, t) = \pmb{\varepsilon} E_0\, e^{-i(\omega t - kz)} \qquad (3.5.1)$$

This is not appropriate for dipoles in cavities, where the electric field takes the form of a standing wave:

$$\mathbf{E}(z, t) = \pmb{\varepsilon} E_n \sin k_n z\, e^{-i\omega t} \qquad (3.5.2)$$

where

$$k = k_n = n\pi/L, \qquad n = 1, 2, 3, \ldots \qquad (3.5.3)$$

as we indicated in Eq. (1.3.2) and derived in Section 2.1.

In this section we will examine the polarizability of atoms exposed to a standing-wave field, and the radiation emitted by these atoms into the cavity. The principal consequences are a new expression for the relation between k and ω and the discovery that a classical laser of Lorentz dipoles can't work.

In free space [recall (2.3.13)] we specified ω and used the coupled Maxwell–Newton equations to find $k = k(\omega)$, and this dispersion relation defined the index of refraction: $n(\omega) = k(\omega)c/\omega$, as in (3.3.21). In a cavity, we specify the cavity length L which first determines the wave vector $k = k_n = n\pi/L$ [recall (1.3.2)], but not the frequency ω. We will use the coupled Maxwell–Newton equations to find $\omega = \omega(k_n) \neq \omega_n$. That is, we will find that the presence of dipoles in the cavity will bias ω, the actual oscillation frequency of the field, away from the natural frequency of the cavity mode, $\omega_n = n\pi c/L$.

First we rewrite (3.2.3) using (3.5.2) and obtain

$$\frac{d^2\mathbf{x}}{dt^2} + 2\beta \frac{d\mathbf{x}}{dt} + \omega_0^2\mathbf{x} = \hat{\varepsilon}\frac{e}{m} E_n \sin k_n z\, e^{-i\omega t} \qquad (3.5.4)$$

The solution of this equation is the obvious analog of (3.3.2):

$$\mathbf{x} = \mathbf{a} \sin k_n z\, e^{-i\omega t} \qquad (3.5.5)$$

where the amplitude **a** can be found easily by substitution into (3.5.4). It satisfies (3.3.3) exactly. In other words the atomic polarizability $\alpha(\omega)$ remains as derived in (3.3.17), even though the atoms are in a standing wave.

Next we determine the field amplitude. The appropriate Maxwell wave equation for a cavity is the same as (2.1.13), except that cavity losses can be included by adding an ohmic current $\mathbf{J} = \sigma \mathbf{E}$ to the right side of (2.1.4). Then we obtain

$$\left(\frac{\partial^2}{\partial z^2} - \frac{\sigma}{\epsilon_0 c^2}\frac{\partial}{\partial t} - \frac{1}{c^2}\frac{\partial^2}{\partial t^2} \right) \mathbf{E}(z, t) = \frac{1}{\epsilon_0 c^2}\frac{\partial^2}{\partial t^2} \mathbf{P}(z, t) \qquad (3.5.6)$$

Here the second term represents the effect of ohmic losses, such as would be due to a finite conductivity σ (Problem 3.3). This is a common method for modeling cavity losses in laser theory.

The polarization is defined to be $\mathbf{P} = Ne\mathbf{x}$, as before, so we can use (3.5.5) to evaluate the derivatives on the right side of (3.5.6) and use (3.5.2) for computing the derivatives on the left side. After differentiating we can cancel the common factor $\hat{\varepsilon} E_n \sin k_n z\, e^{-i\omega t}$ on both sides to get:

$$-k_n^2 + i\left(\frac{\omega\sigma}{\epsilon_0 c^2} \right) + \left(\frac{\omega}{c} \right)^2 = -\left(\frac{\omega}{c} \right)^2 \frac{Ne^2/\epsilon_0 m}{\omega_0^2 - \omega^2 - 2i\beta\omega}$$

Now we use $k_n = \omega_n/c$, and the near-resonance approximation (3.4.7) twice:

$$\omega_0^2 - \omega^2 \approx 2\omega(\omega_0 - \omega) \qquad (3.5.7a)$$

$$\omega^2 - \omega_n^2 \approx 2\omega(\omega - \omega_n) \qquad (3.5.7b)$$

to get

$$\omega - \omega_n + i\frac{\sigma}{2\epsilon_0} = -\frac{Ne^2}{4\epsilon_0 m}\frac{\omega_0 - \omega + i\beta}{(\omega_0 - \omega)^2 + \beta^2}$$

$$= \tfrac{1}{2}(\delta + ig)c \qquad (3.5.8)$$

where we have defined g and δ as abbreviations for:

$$g = -\frac{Ne^2}{2\epsilon_0 mc} \frac{\beta}{(\omega_0 - \omega)^2 + \beta^2} \tag{3.5.9}$$

and

$$\delta = -\frac{Ne^2}{2\epsilon_0 mc} \frac{\omega_0 - \omega}{(\omega_0 - \omega)^2 + \beta^2}$$

$$= \frac{g(\omega_0 - \omega)}{\beta} \tag{3.5.10}$$

We note immediately that all reference to the field amplitude has dropped out in the step from (3.5.6) to the solution (3.5.8), as it did in the similar step between Eqs. (3.3.20) and (3.3.21). What remains is the consistency condition (3.5.8) on the parameters of the interaction. That is, (3.5.8) is the dispersion relation for the cavity.

Let us now solve for g and δ. By matching imaginary parts of (3.5.8) we quickly determine

$$g = \sigma/\epsilon_0 c \tag{3.5.11}$$

Next we look at the real parts of (3.5.8). With the aid of (3.5.10) we find the simple relation

$$\omega_n - \omega = \frac{gc}{2\beta}(\omega - \omega_0) \tag{3.5.12}$$

We can interpret this second relation as a condition on the oscillation frequency ω. Note that if ω is below the cavity frequency ω_n, the left side of (3.5.12) is positive and the right side shows that ω must then lie above the atomic dipole frequency ω_0. Conversely, if ω is above ω_n, then it must be below ω_0. In other words, no matter whether $\omega_0 > \omega_n$ or $\omega_n > \omega_0$, the operating frequency lies between the cavity frequency and the dipole frequency. This is called *frequency pulling*; the interaction with the atomic dipoles pulls the electric field frequency away from the free-space cavity frequency and toward the dipole frequency. An explicit solution of (3.5.12) is

$$\omega = \frac{\beta\omega_n + (gc/2)\omega_0}{\beta + gc/2}$$

$$\approx \omega_n + \frac{gc}{2\beta}(\omega_0 - \omega_n) \quad (\beta \gg gc/2) \tag{3.5.13}$$

It is possible to give a physical interpretation to the equation for g as well. If we were to allow E_n in Eq. (3.5.2) to be time-dependent, then upon substitution into Maxwell's wave equation (3.5.6) we would obtain a differential equation for $E_n(t)$ instead of the consistency equation (3.5.11). We would find that $E_n(t)$ grows exponentially in time if $g > \sigma/\epsilon_0 c$. Thus gc is the classical *gain coefficient* for the interaction of radiation with atomic dipoles in a cavity (Problem 3.4).

We could go on and formulate immediately a classical theory of laser action. For example, the equality in (3.5.11) gives the value of $g = \sigma/\epsilon_0 c$ at which amplification is first possible. This is the *threshold gain*, usually denoted g_t. Unfortunately, none of this is realistic because (3.5.11) cannot be satisfied. That is, from (3.5.9) we see immediately that g is intrinsically negative. Radiation in the cavity will only be damped and never amplified by classical dipoles. A classical laser theory based on the linear electron oscillator model is not possible.

The negative value of g is inherent in the classical theory. It requires a quantum-mechanical treatment of the light–matter interaction to understand how g can be made positive. Apart from this detail, it is remarkable how much of the present classical formulation survives the transition to quantum theory. For example, except for its sign, the form of the gain coefficient will turn out to be exactly correct. The frequency-pulling equation (3.5.12) is exactly correct as it stands. The threshold condition (3.5.11) is correct. We will find how to make g positive in Chapter 7, and in so doing will find other missing elements of laser theory, such as saturation and power broadening.

3.6 THE ABSORPTION COEFFICIENT

We can associate the energy absorbed from an electromagnetic wave by an atom with the work done by the wave on the Lorentzian oscillators. In classical mechanics the rate at which work is done on an atom when a force $\mathbf{F}$ is exerted on it is $dW\,W_A/dt = \mathbf{F} \cdot \mathbf{v}$. In the electron oscillator model the force exerted on an electron by the monochromatic field (2.3.1) is simply the Lorentz force appearing on the right side of (2.2.18):

$$\mathbf{F}_{em} = e\hat{\varepsilon}E_0 \cos(\omega t - kz) \tag{3.6.1}$$

in which case we can write

$$\frac{dW\,W_A}{dt} = \mathbf{E} \cdot \frac{d\mathbf{p}}{dt} \tag{3.6.2}$$

This expression does not lead to energy absorption by the oscillator if $\mathbf{p}$ is in phase with $\mathbf{E}$. In this section we focus attention on an oscillator near to resonance, for which $\alpha(\omega)$ has a significant imaginary (quadrature) part and for which energy absorption does occur. We can use (3.4.8) to obtain

$$\mathbf{p} = \hat{\boldsymbol{\varepsilon}} \, \frac{e^2}{2m\omega} \, \frac{1}{\omega_0 - \omega - i\beta} \, E_0 \, e^{-i(\omega t - kz)} \tag{3.6.3a}$$

which corresponds to the (real) physical moment:

$$\mathbf{p}(t) = \hat{\boldsymbol{\varepsilon}} e \left[U \cos(\omega t - kz) - V \sin(\omega t - kz) \right] \tag{3.6.3b}$$

The coefficients U and V are easily found by computing the real part of (3.6.3a) and comparing with (3.6.3b):

$$U = + \frac{eE_0}{2m\omega} \, \frac{\omega_0 - \omega}{\left(\omega_0 - \omega\right)^2 + \beta^2} \tag{3.6.4}$$

and

$$V = - \frac{eE_0}{2m\omega} \, \frac{\beta}{\left(\omega_0 - \omega\right)^2 + \beta^2} \tag{3.6.5}$$

The corresponding solution obtained without damping would have no quadrature component corresponding to V. The existence of the quadrature component is critical to our discussion of absorption, as we now demonstrate. From Eq. (3.6.3b) we obtain

$$\frac{d\mathbf{p}}{dt} = -\omega \hat{\boldsymbol{\varepsilon}} e \left[U \sin(\omega t - kz) + V \cos(\omega t - kz) \right] \tag{3.6.6}$$

Therefore the rate at which the dipole energy changes is given by

$$\frac{dW_A}{dt} = -eE_0 \left[\omega U \sin(\omega t - kz) \cos(\omega t - kz) + \omega V \cos^2(\omega t - kz) \right]$$

$$= e\omega E_0 \left[-\tfrac{1}{2} U \sin(2\omega t - 2kz) - V \cos^2(\omega t - kz) \right] \tag{3.6.7}$$

Notice that the dipole's energy gain has two distinct contributions. The first term oscillates extremely rapidly and is zero on average, and thus does not give rise to any permanent change in energy. The second term, however, is always positive-definite and corresponds to a steady decrease in field energy with time. Then the rate of change of electromagnetic field energy, equal and opposite to dW_A/dt, is effectively governed by the second term alone:

$$\frac{dW_{em}}{dt} = e\omega E_0 V \cos^2 (\omega t - kz)$$

$$= -\frac{e^2}{2m} \frac{\beta}{(\omega_0 - \omega)^2 + \beta^2} E_0^2 \cos^2 (\omega t - kz) \qquad (3.6.8)$$

where we have used the expression (3.6.5) for V.

Thus we may express the rate (3.6.8) at which electromagnetic energy is absorbed by an atom in terms of the magnitude of the Poynting vector at the atom [recall (2.6.4)]:

$$\frac{dW_{em}}{dt} = -\frac{e^2}{2\epsilon_0 mc} \frac{\beta}{(\omega_0 - \omega)^2 + \beta^2} |\mathbf{S}| \qquad (3.6.9)$$

This result is similar to (2.6.5). Both equations show that dW_{em}/dt is proportional to $|\mathbf{S}|$. Of course (2.6.5) gives the rate of change of electromagnetic energy in a light beam due to scattering, whereas (3.6.9) gives the rate due to absorption.

The similarity of (3.6.9) to (2.6.5) means that we may define an *absorption cross section*:

$$\sigma(\omega) = \frac{e^2}{2\epsilon_0 mc} \frac{\beta}{(\omega_0 - \omega)^2 + \beta^2} \qquad (3.6.10)$$

We may follow the same steps, leading from (2.6.6) to the extinction coefficient (2.6.16) due to scattering, to obtain the extinction coefficient due to absorption in a medium of N atoms per unit volume:

$$a(\omega) = N\sigma(\omega) = \frac{Ne^2}{2\epsilon_0 mc} \frac{\beta}{(\omega_0 - \omega)^2 + \beta^2} \qquad (3.6.11)$$

This extinction coefficient is usually called simply the *absorption coefficient*. The intensity of the incident wave after propagating a distance z into the absorbing medium is

$$I_\omega(z) = I_\omega(0) e^{-a(\omega)z} \qquad (3.6.12)$$

just as in the case (2.6.15) when the incident wave is attenuated because of scattering.

Equation (3.6.12) is identical to (3.3.24). We have simply obtained the same physical result for the field attenuation due to absorption using two approaches. In the first approach, leading to (3.3.24), absorption was associated with the imaginery part of the complex refractive index. In this section we have obtained the same result via the rate at which a single atom absorbs energy from the field. The two approaches are equivalent. Keep in mind, however, that the absorption coefficient derived here is physically distinct from the extinction coefficient due to scattering derived in Section 2.6. Both lead to exponential attenuation of intensity, and the total extinction coefficient includes both.

The absorption coefficient is often written in terms of the circular frequency ν,

$$\nu = \omega/2\pi \tag{3.6.13}$$

rather than the angular frequency ω. From (3.6.11), therefore,

$$a(\nu) = \frac{Ne^2}{4\pi\epsilon_0 mc} \frac{\delta\nu_0}{(\nu - \nu_0)^2 + \delta\nu_0^2} \tag{3.6.14}$$

where

$$\nu_0 = \omega_0/2\pi \tag{3.6.15}$$

and

$$\delta\nu_0 = \beta/2\pi \tag{3.6.16}$$

The absorption coefficient (3.6.14) is frequently written in the form

$$a(\nu) = \frac{Ne^2}{4\epsilon_0 mc} L(\nu) \tag{3.6.17}$$

where the lineshape function $L(\nu)$ is defined by

$$L(\nu) = \frac{\delta\nu_0/\pi}{(\nu - \nu_0)^2 + \delta\nu_0^2} \tag{3.6.18}$$

This is called the *Lorentzian function*, and is plotted in Figure 3.8.

The Lorentzian function is a mathematically idealized lineshape in several respects. We have already shown that it is the near-resonance approximation to the more complicated function (3.3.25). The Lorentzian function is defined mathematically for negative frequencies, even though they have no physical significance. It is exactly normalized to unity when integrated over all frequencies, as is easily checked:

$$\int_{-\infty}^{\infty} d\nu \, L(\nu) = \frac{\delta\nu_0}{\pi} \int_{-\infty}^{\infty} \frac{d\nu}{(\nu - \nu_0)^2 + \delta\nu_0^2} = 1 \tag{3.6.19}$$

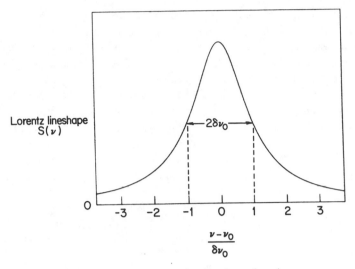

Figure 3.8 The Lorentzian lineshape function.

and the normalization is approximately the same when only the physical, positive frequencies are used. The approximation is excellent for $\delta\nu_0 \ll \nu_0$ [recall (3.3.10)]. In other words, the contribution of the unphysical negative frequencies is negligible because the linewidth is negligible compared with the resonance frequency, and in this sense $L(\nu)$ is physically as well as mathematically normalized to unity.

The maximum value of $L(\nu)$ occurs at the resonance $\nu = \nu_0$:

$$L(\nu)_{max} = L(\nu_0) = \frac{1}{\pi\delta\nu_0} \tag{3.6.20}$$

At $\nu = \nu_0 \pm \delta\nu_0$ we have

$$L(\nu_0 \pm \delta\nu_0) = \frac{1}{2\pi\delta\nu_0} = \frac{1}{2}L(\nu)_{max} \tag{3.6.21}$$

Because of this property, $2\delta\nu_0$ is called the width of the Lorentzian function or the *full width at half maximum* (FWHM), and $\delta\nu_0$ is called the *half width at half maximum* (HWHM). The Lorentzian function is fully specified by its width (FWHM or HWHM) and the frequency ν_0 where it peaks. The absorption coefficient is greatest at resonance, where

$$a(\nu = \nu_0) = \frac{Ne^2}{4\pi\epsilon_0 mc\delta\nu_0} \tag{3.6.22}$$

and decreases to half this resonance value when the field is "detuned" from resonance by the half width $\delta\nu_0$ of the Lorentzian function.

Our classical theory thus predicts that the absorption is strongest when the frequency of the light equals one of the natural oscillation frequencies of the bound electrons. Far out in the wings of the Lorentzian, where $|\nu - \nu_0| \gg \delta\nu_0$, there is very little absorption. A knowledge of the width $\delta\nu_0$ is therefore essential to a quantitative interpretation of absorption data. In order to determine the numerical magnitude of $\delta\nu_0$ in a given situation, we must consider in some detail the physical origin of this absorption width. This we do in Section 3.9.

We shall see later that $a(\nu)$ does not always have the Lorentzian form (3.6.17). However, it will always be possible to write the absorption coefficient as

$$a(\nu) = a_t S(\nu) \tag{3.6.23}$$

where the lineshape function $S(\nu)$, whatever its form, is normalized to unity:

$$\int_0^\infty d\nu S(\nu) = 1 \tag{3.6.24}$$

With this normalization it follows that

$$\int_0^\infty d\nu a(\nu) = a_t \int_0^\infty d\nu S(\nu) = a_t \tag{3.6.25}$$

The integrated absorption coefficient a_t is convenient because it is independent of the lineshape function $S(\nu)$, which may vary with parameters like pressure, temperature, etc. It thus provides a measure of the inherent absorbing strength of the atoms.

3.7 OSCILLATOR STRENGTH

Even more than the integrated absorption coefficient, the integrated absorption cross section, namely

$$\sigma_t = a_t/N \tag{3.7.1}$$

is a convenient measure of absorption, because it characterizes the inherent absorbing strength of a single atom. From (3.6.23) and (3.6.17) we see that

$$a_t = \frac{Ne^2}{4\epsilon_0 mc} \tag{3.7.2}$$

and therefore

$$\sigma_t = \frac{e^2}{4\epsilon_0 mc} \tag{3.7.3}$$

The numerical value of σ_t is easily computed to be approximately 2.65×10^{-2} cm^2-sec^{-1}. This is a universal value, and according to our theory is applicable to absorption by any atomic material.

Extensive experimental absorption data exist for atomic hydrogen. For example, it is known to absorb strongly at a wavelength of about 1216 Å, with an integrated absorption coefficient of about 1.1×10^{-2} cm^2-sec^{-1}. Thus our classical electron oscillator theory gives a reasonable order of magnitude, although it is far from being quantitatively accurate. However, atoms do not absorb at only one wavelength. Table 3.1 lists some wavelengths at which atomic hydrogen absorbs radiation. Our classical theory gives an integrated absorption cross section (3.7.3) which is independent of ν_0, so that the same numerical value (2.65×10^{-2} cm^2-sec^{-1}) should apply to every wavelength listed in Table 3.1. The second column of Table 3.1 lists the observed integrated cross sections of these absorption lines, while the third column gives the ratio of the observed value for each line to the result (3.7.3) of the classical theory. We see that the classical result comes close to the integrated absorption cross section only for the 1216-Å line.

Before the advent of the quantum theory, this quantitative failure of the classical theory was sidestepped by writing the integrated absorption cross section of a one-electron atom as

$$\sigma_t = \frac{e^2}{4\epsilon_0 mc} f \tag{3.7.4}$$

where the parameter f is called the "oscillator strength," and its values are given by the third column of Table 3.1. In other words, the classical theory was patched up by assigning a different "oscillator strength" to each natural oscillation fre-

TABLE 3.1 Some Integrated Cross Sections of Atomic Hydrogen

Wavelength (Å)	σ_t (actual) (cm^2-sec^{-1})	$f = \dfrac{\sigma_t \text{ (actual)}}{\sigma_t \text{ (classical theory)}}$
1216	1.10×10^{-2}	0.416
1026	2.10×10^{-3}	0.079
973	7.69×10^{-4}	0.029
950	3.71×10^{-4}	0.014
938	2.07×10^{-4}	0.0078
931	1.27×10^{-4}	0.0048

quency. In fact the integrated absorption cross section for any atom could be written in the form (3.7.4) by making the *ad hoc* replacement

$$\frac{e^2}{m} \rightarrow \frac{e^2}{m} f \tag{3.7.5}$$

wherever the quantity on the left appeared. In this way the Lorentz theory (of both absorption and the refractive index) was brought into detailed numerical agreement with experimental results. We will include f in most classical formulas hereafter. Like the natural oscillation frequencies, however, the oscillator strengths had to be taken as empirical parameters, without a theoretical basis. Quantum theory removes both of these defects of Lorentz's model.

3.8 ABSORPTION OF BROADBAND LIGHT

The rate at which the energy W_A of an atom increases due to absorption of electromagnetic energy may be obtained from (3.6.2) or (3.6.9):

$$\frac{dW_A}{dt} = \frac{-dW_{em}}{dt} = \frac{\pi e^2 f}{2\epsilon_0 mc} \frac{\beta/\pi}{\left(\omega - \omega_0\right)^2 + \beta^2} I$$

$$= \frac{\pi e^2 f}{2\epsilon_0 mc} \left(\frac{1}{2\pi} S(\nu)\right) I_\nu \tag{3.8.1}$$

where we have added the subscript ν to remind us that I_ν refers to the intensity of monochromatic radiation at the frequency ν.

Equation (3.8.1) gives the rate of increase of the energy of an atom due to absorption from a monochromatic field of frequency ν. In reality, of course, the applied field will not be perfectly monochromatic. Hereafter we will indicate explicitly the dependence of field quantities on the frequency: $W_{em} \rightarrow W_{em}^\nu$ and $I \rightarrow I_\nu$. The change in atomic energy is due to the action of all the frequency components:

$$\left(\frac{dW_A}{dt}\right)_{total} = \sum_\nu \left(\frac{-dW_{em}^\nu}{dt}\right)$$

$$= \frac{e^2 f}{4\epsilon_0 mc} \sum_\nu S(\nu) I_\nu \tag{3.8.2}$$

In many cases of interest the field is composed of a continuous range of frequencies, and the summation in (3.8.2) must be replaced by an integral:

$$\left(\frac{dW_A}{dt}\right)_{total} \rightarrow \frac{e^2 f}{4\epsilon_0 mc} \int_0^\infty S(\nu)\, I(\nu)\, d\nu \qquad (3.8.3)$$

where $I(\nu)\, d\nu$ is the intensity of radiation in the frequency band from ν to $\nu + d\nu$.

It is convenient to define a spectral energy density $\rho(\nu)$, such that $\rho(\nu)\, d\nu$ is the electromagnetic energy per unit volume in the same frequency band (Figure 3.9). The total electromagnetic energy per unit volume is then

$$\int_0^\infty \rho(\nu)\, d\nu = \frac{1}{c} \int_0^\infty I(\nu)\, d\nu \qquad (3.8.4)$$

Clearly (3.8.3) may be rewritten

$$\left(\frac{dW_A}{dt}\right)_{total} = \frac{e^2 f}{4\epsilon_0 m} \int_0^\infty S(\nu)\, \rho(\nu)\, d\nu \qquad (3.8.5)$$

We can now define "broadband" light as follows. Whenever the spectral energy density $\rho(\nu)$ is a flat, almost constant function of ν compared with the atomic lineshape function $S(\nu)$, we can write

$$\int_0^\infty d\nu\, S(\nu)\, \rho(\nu) \approx \rho(\nu_0) \int_0^\infty d\nu\, S(\nu)$$

$$= \rho(\nu_0) \qquad (3.8.6)$$

If $\rho(\nu)$ is perfectly constant, then of course (3.8.6) is an equality. Whether $\rho(\nu)$ is flat enough in its variation to justify the approximation (3.8.6) depends on the lineshape function $S(\nu)$. The narrower the width of $S(\nu)$, the easier it is to satisfy (3.8.6). When this approximation is valid we may say that we have broadband light and broadband absorption, as opposed to the opposite extreme of narrow-band (i.e., monochromatic) absorption. Both extremes are limiting cases of (3.8.5).

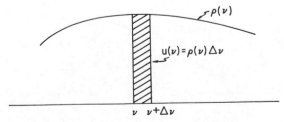

Figure 3.9 The spectral energy density $\rho(\nu)$ is defined so that $u(\nu) = \rho(\nu)\, \Delta\nu$ is the electromagnetic energy per unit volume in the narrow frequency interval from ν to $\nu + \Delta\nu$.

Therefore, the energy absorption rate for an atom exposed to broadband radiation is

$$\frac{dW_A}{dt} = \frac{e^2 f}{4\epsilon_0 m} \rho(\nu_0) \tag{3.8.7}$$

We see that for broadband absorption the rate at which the energy of the atom increases is completely independent of the form of the lineshape function $S(\nu)$, and is simply proportional to the spectral energy density of the field at the dipole's natural oscillation frequency ν_0.

In Table 3.2 we collect the most important results of this section. For simplicity we omit the background refractive index from the equations. This is always an excellent approximation for gaseous media, where n_b is close to unity. However, for solid media the index must be included. We return to this point in our discus-

TABLE 3.2   Results of the Classical Theory of Absorption by a Medium with N Atoms per Unit Volume.

Energy Absorption Rate of an Atom

$$\frac{dW_A}{dt} = \frac{e^2 f}{4\epsilon_0 m} \int_0^\infty d\nu\, S(\nu)\, \rho(\nu) \qquad (f = \text{oscillator strength})$$

$$\approx \frac{e^2 f}{4\epsilon_0 mc} S(\nu)\, I_\nu \qquad\qquad \text{(narrowband radiation)}$$

$$\approx \frac{e^2 f}{4\epsilon_0 m} \rho(\nu_0) \qquad\qquad \text{(broadband radiation)}$$

Lineshape Function

$S(\nu)$ peaks at the resonance frequency $\nu = \nu_0$ and

$$\int_0^\infty d\nu\, S(\nu) = 1$$

Attenuation of Intensity for Radiation of Frequency ν

$$I_\nu(z) = I_\nu(0)\, e^{-a(\nu)z}$$

$$a(\nu) = \frac{Ne^2 f}{4\epsilon_0 mc} S(\nu) \qquad \text{(absorption coefficient)}$$

$$a_t = \frac{Ne^2 f}{4\epsilon_0 mc} \qquad\qquad \text{(integrated absorption coefficient)}$$

The oscillator strength f has been included by making the replacement (3.7.5), $e^2/m \rightarrow e^2 f/m$.

sion of laser gain in Chapter 10. We have furthermore refrained from specifying the form of the lineshape function $S(\nu)$; the question of different lineshapes is taken up in the following sections. The equations of Table 3.2 are valid for any lineshape function.

3.9 COLLISIONS AND "FRICTION" IN THE LORENTZ MODEL

In the preceding sections of this chapter we have shown that light is strongly absorbed when it is nearly resonant with one of the natural oscillation frequencies of the molecules of a medium, and that absorption is due to "frictional" processes that damp out dipole oscillations. We have also shown that any frictional force in the Newton equation of an electron oscillator leads to a broadened absorption line, the lineshape being Lorentzian. We did not, however, give any fundamental explanation for the existence of frictional processes. We will now approach the question of absorption and lineshape from a more fundamental viewpoint, focusing our attention on "line broadening" mechanisms in gases, in order to answer the question of the origin of the frictional coefficient β.

It is a well-known result of experiment that, for sufficiently large pressures, the width of an absorption line in a gas increases as the pressure increases. This broadening is due to collisions of the molecules and is therefore called collision broadening, or sometimes pressure broadening. Collision broadening is the most important line-broadening mechanism in gases at atmospheric pressures, and is often dominant at much lower pressures as well. We will begin our study by considering the details of collision broadening.

Our treatment of collision broadening will follow the original approach of Lorentz. We will find, for instance, that a kind of frictional force arises naturally as a result of collisions, and that the damping rate β can be interpreted as simply the collision rate.

Let us consider the effect of collisions on an atom in the electric field of a laser beam. We imagine collisions to occur in billiard-ball fashion, each collision lasting for a time that is very short compared with the time between collisions. We suppose that, immediately prior to a collision, the active electrons in an atom are oscillating along the axis defined by the field polarization, as indicated by (3.3.13). During a collision, the interaction between the two atoms causes a reorientation of the axes of oscillation. Since each atom in a gas may be bombarded by other atoms from any direction, we can assume that *on the average* all orientations of the displacements and velocities of the atomic electrons are equally probable following a collision. This is the assumption made by Lorentz. It is an assumption about the statistics of a large number of collisions, rather than about the details of a single collision.

Consider a gas of atoms at a given time t. Most atoms are not at this moment involved in a collision. Consider in particular those atoms that underwent their most recent collision at the earlier time t_1. According to our (Lorentz's) assumption, the average of the electron displacements and velocities for these atoms vanished at the time t_1, since all orientations of displacement and velocity vectors were equally probable immediately after their collision. We assume that the electrons in those atoms that had their last collision at time t_1 obey the Newton–Lorentz equation (2.3.7), which we write here in complex notation:

$$\frac{d^2\mathbf{x}}{dt^2} + \omega_0^2 \mathbf{x} = \hat{\boldsymbol{\varepsilon}} \frac{e}{m} E_0 e^{-i(\omega t - kz)} \tag{3.9.1}$$

The electron displacement for a dipole satisfying this equation is obtained by combining the homogeneous and particular solutions in such a way that $\mathbf{x}(t)$ obeys the initial conditions

$$\mathbf{x}(t_1) = \left(\frac{d\mathbf{x}}{dt}\right)_{t=t_1} = 0 \tag{3.9.2}$$

Note that these are initial conditions applying to the "average" atom, since we have assumed that all displacements and velocities are equally likely after a collision. The corresponding solution to (3.9.1) will be written

$$\mathbf{x}(t; t_1) = \hat{\boldsymbol{\varepsilon}} \frac{eE_0/m}{\omega_0^2 - \omega^2} \left[e^{-i\omega t} - \frac{1}{2}\left(1 + \frac{\omega}{\omega_0}\right) e^{-i\omega_0(t-t_1)} e^{-i\omega t_1} \right.$$
$$\left. - \frac{1}{2}\left(1 - \frac{\omega}{\omega_0}\right) e^{i\omega_0(t-t_1)} e^{-i\omega t_1} \right] e^{ikz} \tag{3.9.3}$$

It is easy to verify that (3.9.3) is the desired solution, by checking that it satisfies both (3.9.1) and the initial conditions (3.9.2). This solution will now be taken to represent the average atom. It has this average significance even if it is not applicable to any one of the atoms individually.

We wish to calculate the average electron displacement at time t for atoms in the gas, no matter when their last collision. We can obtain this by summing (3.9.3) over all possible t_1. We only need to know (at time t) the fraction $df(t, t_1)$ of atoms for which the last collision occurred between t_1 and $t_1 + dt_1$. We show below that this is given by

$$df(t, t_1) = e^{-(t-t_1)/\tau} \frac{dt_1}{\tau} \tag{3.9.4}$$

where τ is the mean time between collisions. The average electron displacement $\langle \mathbf{x}(t) \rangle$ for any atom at time t is therefore obtained by multiplying (3.9.3) by the fraction (3.9.4) of atoms to which it applies, and then summing (integrating) over all possible values of earlier times t_1:

$$\langle \mathbf{x}(t) \rangle = \int_{-\infty}^{t} \mathbf{x}(t; t_1) \, df(t, t_1)$$

$$= \hat{\varepsilon} \frac{eE_0/m}{\omega_0^2 - \omega^2} e^{-i(\omega t - kz)} \left(\frac{1}{\tau}\right) \int_{-\infty}^{t} dt_1 \, e^{-(t-t_1)/\tau}$$

$$\times \left[1 - \frac{1}{2}\left(1 + \frac{\omega}{\omega_0}\right) e^{-i(\omega_0 - \omega)(t - t_1)} - \frac{1}{2}\left(1 - \frac{\omega}{\omega_0}\right) e^{i(\omega_0 + \omega)(t - t_1)} \right]$$

$$(3.9.5)$$

The required integrals are

$$\int_{-\infty}^{t} dt_1 \, e^{-(t-t_1)/\tau} = \tau \qquad (3.9.6a)$$

$$\int_{-\infty}^{t} dt_1 \, e^{-i(\omega_0 - \omega)(t - t_1)} e^{-(t-t_1)/\tau} = \frac{-i}{\omega_0 - \omega - i/\tau} \qquad (3.9.6b)$$

$$\int_{-\infty}^{t} dt_1 \, e^{i(\omega_0 + \omega)(t - t_1)} e^{-(t-t_1)/\tau} = \frac{i}{\omega_0 + \omega + i/\tau} \qquad (3.9.6c)$$

The average electron displacement is therefore given by

$$\langle \mathbf{x}(t) \rangle = \hat{\varepsilon} \frac{eE_0/m}{\omega_0^2 - \omega^2} e^{-i(\omega t - kz)}$$

$$\times \left[1 + \frac{i}{2\tau} \frac{1 + \omega/\omega_0}{\omega_0 - \omega - i/\tau} - \frac{i}{2\tau} \frac{1 - \omega/\omega_0}{\omega_0 + \omega + i/\tau} \right]$$

$$= \frac{\hat{\varepsilon}(eE_0/m) \, e^{-i(\omega t - kz)}}{\omega_0^2 - \omega^2 - 2i\omega/\tau + 1/\tau^2} \qquad (3.9.7)$$

and the corresponding polarizability is

$$\alpha(\omega) = \frac{e^2/m}{\omega_0^2 - \omega^2 - 2i\omega/\tau + 1/\tau^2}. \qquad (3.9.8)$$

Note that, except for the term $1/\tau^2$, this is the same as (3.3.17) if we identify the frictional coefficient β with the collision rate $1/\tau$.

The main conclusion to be drawn from our collision analysis is obvious. Given the strong inequality

$$\omega\tau \gg 1 \qquad (3.9.9)$$

which implies that the mean time between collisions is much longer than an optical period ($\sim 10^{-15}$ sec), and which is an excellent approximation in practice, the last term in the denominator of (3.9.8) can be dropped. Then the effect of collisions is exactly the same as the effect of a frictional damping force if we let

$$\beta = \frac{1}{\tau} = \text{collision rate} \qquad (3.9.10)$$

However, we must not lose sight of the statistical nature of our treatment of collisions. We should really say that a frictional term in the Newton equation is justified by the effects of collisions *on the average*. Thus we can give up the artificial notion of friction at the atomic level, but still use all of the results derived from it, if we reinterpret $x(t)$, U, V, and W_A in Sections 3.5 and 3.6 as *average* values in the sense of (3.9.5). We are thus led to regard the results of Table 3.2 with the Lorentzian lineshape function (3.6.18) as the consequences of collision broadening. The width (HWHM) of this collision-broadened lineshape function is

$$\delta\nu_0 = \frac{\beta}{2\pi} = \frac{1}{2\pi\tau} \qquad (3.9.11)$$

The damping term we introduced empirically earlier in (3.2.3) may now be interpreted as the damping of the average electron displacement, i.e.,

$$\frac{d^2}{dt^2}\langle \mathbf{x}\rangle + 2\beta\frac{d}{dt}\langle \mathbf{x}\rangle + \omega_0^2\langle \mathbf{x}\rangle = \hat{\boldsymbol{\epsilon}}\frac{e}{m}E_0\,e^{-i(\omega t - kz)} \qquad (3.9.12)$$

Collision broadening is often described equivalently in terms of a "dephasing" of the electron oscillators, as follows. Immediately after a collision the phase of the electron's oscillation has no correlation with the precollision phase. Collisions have the effect of "interrupting" the phase of oscillation, leading to an overall decay of the average electron displacement from equilibrium (Figure 3.10). The damping rate in (3.9.12) is sometimes called a "dephasing" rate, in order to distinguish it from an "energy decay" rate. The latter would appear as a frictional term in the equation of motion of each electron oscillator as well as in the average equation. In the absence of any inelastic collisions to decrease the energy of the electron oscillators, each oscillator would satisfy the Newton equation (2.3.7) with no damping term. Due to elastic collisions, i.e., collisions which only interrupt the phase of oscillation but do not produce any change in energy, the *average* electron displacement follows equation (3.9.12), which includes damping.

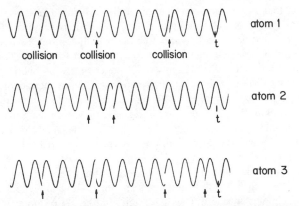

Figure 3.10 Electron oscillations in three different atoms in a gas. Collisions completely interrupt the phase of the oscillation. The average electron displacement associated with all the atoms in the gas therefore decays to zero at a rate given by the inverse of the mean collision time.

• To complete our derivation of (3.9.7), we must prove our assertion (3.9.4). The mean time between collisions, τ, is obviously an average; a given atom certainly does not have collisions in evenly spaced intervals of time τ. We can only say that the *probability* of any given atom having a collision in a small time interval Δt is given by Δt times the mean number of collisions per unit time, $1/\tau$. If at time T there are $\eta(T)$ atoms which have not yet had a collision since the time $T = 0$, then the number of "collisionless" atoms at time $T + \Delta T$ is

$$\eta(T + \Delta T) = \eta(T) - \eta(T)\frac{\Delta T}{\tau} \tag{3.9.13}$$

In words, $\eta(T)$ decreases by the amount $\eta(T)\,\Delta T/\tau$, which is the number of atoms collisionless since the time $T = 0$, times the probability that any one such atom will have a collision in the time interval ΔT. Thus

$$\frac{\eta(T + \Delta T) - \eta(T)}{\Delta T} = -\frac{\eta(T)}{\tau} \tag{3.9.14}$$

The limit $\Delta T \to 0$ gives the simple differential equation

$$\frac{d\eta(T)}{dT} = -\frac{1}{\tau}\eta(T) \tag{3.9.15}$$

with the solution

$$\eta(T) = \eta_0 e^{-T/\tau} \tag{3.9.16}$$

Since η_0 is the total number of atoms in the gas, the quantity

$$P(T) = e^{-T/\tau} \tag{3.9.17}$$

is the probability that a given atom has had no collision for a time T. At $T = 0$, when we begin "looking," this probability is unity. For $T \gg \tau$ it is very small, because in all likelihood the atom will have a collision before many collision times τ have elapsed.

The probability that a given atom will have no collision for a time T, and then have a collision within a time interval dT, is just the product of the probabilities for these two "events", i.e., $P(T) \, dT/\tau$. This is just the fraction $df(T)$, of the total number of atoms, that can be expected to have their *next* collision within the time interval from T to $T + dT$, after we begin "looking" at $T = 0$. If we imagine a movie showing the movements and collisions of the atoms, we can run our film backwards in time and the collisions will exhibit the same statistical behavior. And we will observe the same statistical behavior regardless of where we begin looking.

The atoms at time t that had their last collisions in the interval from t_1 to $t_1 + dt_1$ will be just those having their *next* collision in the same interval when we look at the gas backwards in time beginning at time t. Thus the fraction $df(t, t_1)$ of atoms at time t that had their last collision in an interval dt_1 of $t_1 < t$ will be the same as the fraction of atoms at time t which will have their next collision in an interval dt_1 of t_1 when the film is run backwards. This is just the probability $P(T) \, dt_1/\tau$ found above, with $T = t - t_1$. Thus

$$df(t, t_1) = P(t - t_1) \frac{dt_1}{\tau} = e^{-(t-t_1)/\tau} \frac{dt_1}{\tau} \qquad (3.9.18)$$

which is the same as (3.9.4). •

3.10   COLLISION CROSS SECTIONS

We have shown in the preceding section that collisions, on average, can produce the same effect as frictional damping on an electron oscillator. The damping rate β can be identified with the collision rate $1/\tau$. Therefore the magnitude of $1/\tau$ is of direct significance for realistic estimates of line broadening.

The collision rate $1/\tau$ may be expressed in terms of the number density N of atoms, the collision cross section σ between atoms, and the average relative velocity $\bar{v}$ of the atoms. Imagine some particular atom to be at rest and bombarded by a stream of identical atoms of velocity $\bar{v}$. If the number of atoms per unit volume in the stream is N, then the number of collisions per unit time undergone by the atom at rest is $N\sigma\bar{v}$, where the area σ is the *collision cross section* between the atom at rest and an atom in the stream. The number of collisions per second is the same as if all the stream atoms within a cross-sectional area σ collide with the stationary atom. The idea here is exactly the same one used to define scattering and absorption cross sections for incident light.

According to the kinetic theory of gases, an atom of mass m_x has an rms velocity

$$v_{\text{rms}} = \left(\frac{8kT}{\pi m_x} \right)^{1/2} \qquad (3.10.1)$$

in a gas in thermal equilibrium at temperature T, where k is Boltzmann's constant.

To obtain the average relative velocity $\bar{v}_{rel}$ of colliding atoms of masses m_x and m_y in the gas, we replace m_x in (3.10.1) by the reduced mass

$$\mu_{x,y} = \frac{m_x m_y}{m_x + m_y} = \left(\frac{1}{m_x} + \frac{1}{m_y}\right)^{-1} \tag{3.10.2}$$

Thus

$$\bar{v}_{rel} = \left[\frac{8kT}{\pi}\left(\frac{1}{m_x} + \frac{1}{m_y}\right)\right]^{1/2} \tag{3.10.3}$$

It is convenient to express this in terms of the atomic (or molecular) weights M_x and M_y:

$$\bar{v}_{rel} = \left[\frac{8RT}{\pi}\left(\frac{1}{M_x} + \frac{1}{M_y}\right)\right]^{1/2} \tag{3.10.4}$$

where R, the universal gas constant, is Boltzmann's constant times Avogadro's number. The collision rate for molecules of type x is therefore

$$\frac{1}{\tau} = \sum_Y N(Y)\sigma(X, Y)\,\bar{v}_{rel}(X, Y)$$

$$= \sum_Y N(Y)\sigma(X, Y)\left[\frac{8RT}{\pi}\left(\frac{1}{M_x} + \frac{1}{M_y}\right)\right]^{1/2} \tag{3.10.5}$$

where the sum is over all species y, including x.

The important "unknowns" in the expression (3.10.5) are the collision cross sections $\sigma(X, Y)$. It often happens that these are not known very accurately. They are difficult to derive theoretically, and experimental determinations are not always unambiguous. The simplest approximation to the cross section is the "hard-sphere" approximation. We write

$$\bar{\sigma}(X, Y) = \frac{\pi}{4}(d_x + d_y)^2 \tag{3.10.6}$$

where d_x and d_y are the "hard-sphere" molecular diameters, estimates of which are sometimes tabulated. $\bar{\sigma}(X, Y)$ is just the area of a circle of diameter $d_x + d_y$, just what we would expect if the molecules acted like spheres of diameters d_x and d_y. For CO_2, for example, the hard-sphere diameter is about 4.00 Å. From (3.10.6), therefore, the hard-sphere cross section for two CO_2 molecules is $\bar{\sigma}(CO_2, CO_2) = 5.03 \times 10^{-15}$ cm^2. For a gas of pure CO_2 at $T = 300$ K we find the average relative velocity of two colliding CO_2 molecules to be $\bar{v}_{rel} = 5.37 \times 10^4$ cm/sec. The collision rate (3.10.5) in the "hard-sphere" approximation is therefore

$$\frac{1}{\tau} = N(5.03 \times 10^{-15} \text{ cm}^2)(5.37 \times 10^4 \text{ cm/sec})$$

$$= 2.70 \times 10^{-10} N/\text{sec} \tag{3.10.7}$$

where N is the number of CO_2 molecules in a cubic centimeter. For an ideal gas we calculate (Problem 3.5)

$$N = 9.65 \times 10^{18} \frac{P(\text{Torr})}{T} \tag{3.10.8}$$

where $P(\text{Torr})$ is the pressure in Torr (1 atmosphere = 760 Torr) and T is the temperature (K). From (3.10.7), finally, the collision rate for a gas of CO_2 at 300 K is

$$\frac{1}{\tau} = 8.69 \times 10^6 P(\text{Torr}) \text{ sec}^{-1} \tag{3.10.9}$$

Thus at a pressure of 1 atmosphere (760 Torr) we calculate the collision rate

$$\frac{1}{\tau} = 6.60 \times 10^9 \text{ sec}^{-1} \tag{3.10.10}$$

and from (3.9.11) the collision-broadened linewidth

$$\delta\nu_0 = 1.05 \times 10^9 \text{ Hz} \tag{3.10.11}$$

The actual collision-broadened linewidths can be larger, by as much as an order of magnitude or more, than those calculated in the hard-sphere approximation. The value calculated above, however, is reasonable, and it allows us to point out some general features of the collision-broadened linewidths. First we note that the collision rate (3.10.10) is very much smaller than an optical frequency, as assumed in (3.9.9). The linewidth $\delta\nu_0$ is thus also orders of magnitude less than an optical frequency. This explains why we can speak of absorption "lines" in a gas, even though the absorption occurs over a band of frequencies: the band has a width ($\sim 2\delta\nu_0$) that is very small compared with the resonance frequency ν_0.

From (3.10.9) we note that the linewidth is linearly proportional to the pressure. For this reason, experimental results for collision-broadened linewidths are often reported in units such as MHz-Torr^{-1}. The linewidth calculated above, for instance, may be expressed as 1.38 MHz-Torr^{-1} at 300 K.

Our treatment of collision broadening only highlights some general features of a complex subject. In actual calculations we prefer always to use measured values of the collision-broadened linewidths. We note parenthetically that, for the 10.6-μm CO_2 laser line, the linewidth (1.38 MHz-Torr^{-1}) computed above is about three times smaller than the experimentally determined value. It is possible to calculate these widths more accurately, but this will not concern us.

3.11 DOPPLER BROADENING

The Doppler effect was demonstrated for sound waves in 1845 by C. H. D. Buys Ballot, who employed trumpeters performing in a moving train to demonstrate it. The mathematician C. J. Doppler had predicted the effect in 1842. His prediction applied also to light, although Maxwell's electromagnetic theory of light waves was still nearly a quarter of a century away.

Let us consider again a gaseous medium, this time only very weakly influenced by collisions (i.e., β is very small). Every electron oscillator will thus undergo practically undamped oscillation at the field frequency. Nevertheless we will show that, because of the Doppler effect, an absorption line is broadened and its width can be much larger than β. We will find that the lineshape associated with the Doppler effect is not the Lorentzian function (3.6.18), but rather the Gaussian function given in (3.11.6) below.

To an atom moving with velocity $v \ll c$ away from a source of radiation of frequency v, the frequency of the radiation appears to be shifted:

$$v' = v \left(1 - \frac{v}{c}\right) \tag{3.11.1}$$

This is the Doppler effect. It implies that a source of radiation (e.g., a laser) exactly resonant in frequency with an absorption line of a stationary atom will not be in resonance with the same absorption line in a moving atom, and the frequency offset is $\delta v = (v/c)v$. Similarly, a nonresonant absorption line of an atom may be brought *into* resonance with the field as a result of atomic motion. Since the atoms in a gas exhibit a wide variety of velocities, a broad range of different effective resonance frequencies will be associated with a given absorption line. In other words, the absorption line is broadened because of the Doppler effect. The absorption line is thus said to be Doppler-broadened.

For a gas in thermal equilibrium at the temperature T, the fraction $df(v)$ of atoms having velocities between v and $v + dv$ along any one axis is given by the (one-dimensional) Maxwell–Boltzmann distribution,

$$df(v) = \left(\frac{m_x}{2\pi kT}\right)^{1/2} e^{-m_x v^2/2kT} dv \tag{3.11.2}$$

Here again k is the Boltzmann constant and m_x is the mass of an atom or molecule. Because we have assumed that collisions are almost negligible, an atom with resonance frequency ν_0 and velocity v moving away from the source of radiation will only absorb radiation very near to (within $\Delta\nu = \beta/2\pi$ of) the frequency

$$\nu = \nu_0\left(1 + \frac{v}{c}\right) \tag{3.11.3}$$

The fraction of atoms absorbing within the frequency interval from ν to $\nu + d\nu$ is thus equal to the fraction of atoms with velocity in the interval from v to $v + dv$. From (3.11.3) we have

$$v = \frac{c}{\nu_0}(\nu - \nu_0) \tag{3.11.4}$$

and $dv = (c/\nu_0)\, d\nu$. Using (3.11.2) we can determine that this fraction is

$$df_\nu(v) = \left(\frac{m_x}{2\pi kT}\right)^{1/2} e^{-m_x c^2(\nu - \nu_0)^2/2kT\nu_0^2}\left(\frac{c}{\nu_0} d\nu\right) \tag{3.11.5}$$

Since the absorption rate at frequency ν must be proportional to $df_\nu(v)$, we may write the Doppler lineshape function as

$$S(\nu) = \left(\frac{m_x c^2}{2\pi kT\nu_0^2}\right)^{1/2} e^{-m_x c^2(\nu - \nu_0)^2/2kT\nu_0^2} \tag{3.11.6}$$

Since (3.11.2) was normalized to unity when integrated over velocity, (3.11.6) is normalized to unity with respect to the frequency offset (or "detuning") $\nu - \nu_0$, as required by the definition of lineshape function.

By direct computation using (3.11.6) we find

$$\int_0^\infty d\nu\, S(\nu) = S(\nu_0)\int_0^\infty d\nu\, e^{-m_x c^2(\nu - \nu_0)^2/2kT\nu_0^2}$$

$$= S(\nu_0)\int_{-\nu_0}^\infty d\mu\, e^{-m_x c^2\mu^2/2kT\nu_0^2}$$

$$\approx S(\nu_0)\int_{-\infty}^\infty d\mu\, e^{-m_x c^2\mu^2/2kT\nu_0^2}$$

$$= S(\nu_0) \left[\frac{\nu_0}{c} \left(\frac{2\pi kT}{m_x} \right)^{1/2} \right]$$

$$= 1 \qquad (3.11.7)$$

We have used the excellent approximation $kT \ll m_x c^2$ to replace the lower limit of the integral by $-\infty$. Thus we may write

$$S(\nu) = \frac{c}{\nu_0} \left(\frac{m_x}{2\pi kT} \right)^{1/2} e^{-m_x c^2 (\nu - \nu_0)^2 / 2kT\nu_0^2} \qquad (3.11.8)$$

It is convenient to define

$$\delta \nu_D = 2 \frac{\nu_0}{c} \left(\frac{2kT}{m_x} \ln 2 \right)^{1/2} \qquad (3.11.9)$$

in terms of which

$$S(\nu) = \frac{1}{\delta \nu_D} \left(\frac{4 \ln 2}{\pi} \right)^{1/2} e^{-4(\nu - \nu_0)^2 \ln 2 / \delta \nu_D^2}, \qquad (3.11.10)$$

and we recognize that $\delta \nu_D$ is the width (FWHM) of the Doppler absorption curve, since

$$S(\nu_0 \pm \tfrac{1}{2} \delta \nu_D) = S(\nu_0) e^{-\ln 2} = \tfrac{1}{2} S(\nu_0) \qquad (3.11.11)$$

$\delta \nu_D$ is commonly called the *Doppler width* (Figure 3.11). The Doppler width is also often defined in terms of the $1/e$ point of the curve, rather than the half-maximum point. Sometimes it is defined as the half width at half maximum (HWHM) rather than the FWHM. Thus one finds formulas in the literature differing by factors of 2, ln 2, etc. It is important to keep these possible differences in mind when comparing calculations.

The peak of the Doppler curve at $\nu = \nu_0$ has the value

$$S(\nu_0) = \frac{1}{\delta \nu_D} \left(\frac{4 \ln 2}{\pi} \right)^{1/2} \qquad (3.11.12)$$

where $S(\nu_0)$ is evidently the peak value of $S(\nu)$, for which $\nu = \nu_0$. $S(\nu_0)$ is determined from the normalization condition (3.11.7).

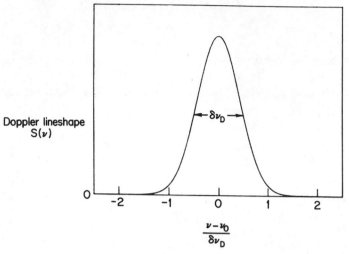

Figure 3.11 The Doppler lineshape function.

In terms of the molecular weight M_x and the wavelength $\lambda_0 = c/\nu_0$ of the absorption line, the Doppler width is

$$\delta\nu_D = \frac{2}{\lambda_0}\left(\frac{2RT}{M_x}\ln 2\right)^{1/2}$$

$$= 2.15 \times 10^{12}\left[\frac{1}{\lambda_0}\left(\frac{T}{M_x}\right)^{1/2}\right] \tag{3.11.13}$$

where λ_0 is expressed in angstroms, M_x in grams, and T in kelvins. In these same units the formula

$$\frac{\delta\nu_D}{\nu_0} \approx 7.16 \times 10^{-7}\left(\frac{T}{M_x}\right)^{1/2} \tag{3.11.14}$$

for the ratio of the Doppler width to the resonance frequency is also useful.

The Doppler width depends only on the transition frequency, the gas temperature, and the molecular weight of the absorbing species. It is therefore much simpler to calculate than the collision-broadened width, which involves the collision cross section. As an example, consider the 6328-Å line of Ne in the He–Ne laser. Since $M_{Ne} \approx 20.18$ g for Ne, we obtain from (3.11.13) the Doppler width

$$\delta \nu_D \approx 1500 \text{ MHz} \qquad (3.11.15)$$

for $T = 400$ K. For the 10.6-μm line of CO_2 and the same temperature, however, we find a much smaller Doppler width:

$$\delta \nu_D \approx 61 \text{ MHz} \qquad (3.11.16)$$

3.12 THE VOIGT PROFILE

Doppler broadening is an example of what is called *inhomogeneous* broadening. The term inhomogeneous means that individual atoms within a collection of otherwise identical atoms do not have the same resonant response frequencies. Thus atoms in the collection can show resonant response over the available range of frequencies. This is true even though the atoms are nominally identical. In the Doppler case this is because individual (nominally identical) atoms can have different velocities. These different velocities serve as tags or labels for the individual atoms, and any discussion of the behavior of a sample of such atoms must take account of all the velocity labels.

There are other possible inhomogeneities that have the same effect as the Doppler distribution of velocities. For example, impurity atoms embedded randomly in a crystal are subjected to different local crystal fields due to strains and defects. These have the effect of shifting the resonance frequency of each atom slightly differently. The distribution of such shifts acts very much like the Doppler distribution, and gives rise to an inhomogeneous broadening of the absorption line associated with the nominally identical impurity atoms subjected to different local fields in the crystal. This type of random strain broadening is present in the Cr^{3+} line associated with ruby laser light, for example.

The line broadening associated with collisions is different, and is called *homogeneous*. This is because each atom can itself absorb light over a range of frequencies, due to the interruptions of its dipole oscillations by collisions. Since the collisional history of every atom is assumed to be the same, no greater collisional broadening is associated with the collection of atoms than is associated with an individual atom.

In general we cannot characterize an absorption lineshape of a gas as a pure collision-broadened Lorentzian or a pure Doppler-broadened Gaussian. Both phase-interrupting collisions and the Doppler effect may play a role in determining the lineshape. We will now derive the Voigt profile, describing the absorption lineshape when both collision broadening and Doppler broadening must be taken into account.

Equation (3.6.18) gives the collision-broadened lineshape for each atom in the gas. If an atom has a velocity component v moving away from the source of light of frequency $\nu \approx \nu_0$, its absorption curve is Doppler-shifted to

$$S(v, \nu) = \frac{(1/\pi)\delta\nu_0}{\left(\nu_0 - \nu + v\nu/c\right)^2 + \delta\nu_0^2} \tag{3.12.1}$$

In other words, the peak absorption for this atom will occur at the field frequency ν such that (3.11.3) is satisfied:

$$\nu \approx \nu_0 + \frac{\nu_0 v}{c} \tag{3.12.2}$$

The lineshape function for the gas is thus obtained by integrating over the velocity distribution (3.11.2):

$$S(\nu) = \int_{-\infty}^{\infty} dv \, S(v, \nu) \left(\frac{M_x}{2\pi RT}\right)^{1/2} e^{-M_x v^2/2RT}$$

$$= \left(\frac{M_x}{2\pi RT}\right)^{1/2} \frac{\delta\nu_0}{\pi} \int_{-\infty}^{\infty} \frac{dv \, e^{-M_x v^2/2RT}}{\left(\nu_0 - \nu + \nu_0 v/c\right)^2 + \delta\nu_0^2}$$

$$= \frac{1}{\pi^{3/2}} \frac{b^2}{\delta\nu_0} \int_{-\infty}^{\infty} \frac{dy \, e^{-y^2}}{\left(y + x\right)^2 + b^2} \tag{3.12.3}$$

where we have made the change of variables (Problem 3.6)

$$x = (4 \ln 2)^{1/2} \frac{\nu_0 - \nu}{\delta\nu_D} \tag{3.12.4}$$

and we have defined

$$b = (4 \ln 2)^{1/2} \frac{\delta\nu_0}{\delta\nu_D} \tag{3.12.5}$$

The lineshape function (3.12.3) is called the *Voigt profile*.

In the case when the applied field is tuned exactly to the resonance frequency ν_0, we have $x = 0$ and therefore

$$S(\nu_0) = \frac{b^2}{\pi^{3/2} \delta\nu_0} \int_{-\infty}^{\infty} \frac{dy \, e^{-y^2}}{y^2 + b^2} \tag{3.12.6}$$

The integral may be looked up in a table of integrals. It is found that

$$\int_{-\infty}^{\infty} \frac{dy \, e^{-y^2}}{y^2 + b^2} = \frac{\pi}{b} e^{b^2} \, \text{erfc} \, (b) \qquad (3.12.7)$$

where

$$\text{erfc} \, (b) = \frac{2}{\pi^{1/2}} \int_{b}^{\infty} du \, e^{-u^2} \qquad (3.12.8)$$

is the *complementary error function*. From (3.12.6) and (3.12.7), therefore, the lineshape function for the resonance frequency $\nu = \nu_0$ has the value

$$S(\nu_0) = \frac{b^2}{\pi^{3/2} \delta\nu_0} \frac{\pi}{b} e^{b^2} \, \text{erfc} \, (b)$$

$$= \frac{b}{\pi^{1/2} \delta\nu_0} e^{b^2} \, \text{erfc} \, (b)$$

$$= \left(\frac{4 \ln 2}{\pi}\right)^{1/2} \frac{1}{\delta\nu_D} e^{b^2} \, \text{erfc} \, (b) \qquad (3.12.9)$$

This function is plotted for several values of the parameter b in Figure 3.12.

$S(\nu_0)$ depends strongly on the ratio of the linewidths for collision and Doppler broadening. When the collision width $\delta\nu_0$ is much greater than the Doppler width $\delta\nu_D$, we have $b \gg 1$. For large values of b it is known that

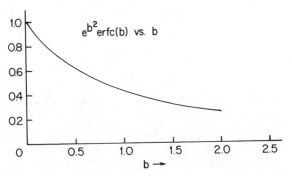

Figure 3.12 The function $e^{b^2}\text{erfc} \, (b)$.

$$e^{b^2} \, \text{erfc} \, (b) \approx \frac{1}{\pi^{1/2} b} \quad (b \gg 1) \tag{3.12.10}$$

In this "collision-broadened limit," therefore, we have from (3.12.9) the result

$$S(\nu_0) \approx \frac{1}{\pi \delta \nu_0} \quad (b \gg 1) \tag{3.12.11}$$

which is exactly the result (3.6.20) for the case of pure collision broadening. In the limit in which the Doppler width is much greater than the collision broadened width, on the other hand, we have $b \ll 1$, in which case the function

$$e^{b^2} \, \text{erfc} \, (b) \approx 1 \quad (b \ll 1) \tag{3.12.12}$$

Then from (3.12.9) we have

$$S(\nu_0) \approx \frac{1}{\delta \nu_D} \left(\frac{4 \ln 2}{\pi} \right)^{1/2} \quad (b \ll 1) \tag{3.12.13}$$

which is the result (3.11.12) for pure Doppler broadening. The limits $\delta \nu_0 \gg \delta \nu_D$ and $\delta \nu_0 \ll \delta \nu_D$ thus reproduce the results for pure collision broadening and pure Doppler broadening, respectively. In general, for arbitrary values of b, $S(\nu_0)$, given by (3.12.9), must be evaluated by using tables of erfc (b).

For the general case of arbitrary values of both the parameter b and the detuning parameter x, the lineshape function $S(\nu)$ given by Eq. (3.12.3) must be evaluated from tabulated values of the more complicated function

$$\int_{-\infty}^{\infty} \frac{dy \, e^{-y^2}}{(y + x)^2 + b^2} = \frac{\pi}{b} \, \text{Re} \left(\frac{i}{\pi} \int_{-\infty}^{\infty} \frac{dy \, e^{-y^2}}{x + y + ib} \right)$$

$$= \frac{\pi}{b} \, \text{Re} \, w(x + ib) \tag{3.12.14}$$

where w is the "error function of complex argument." Numerical values are tabulated in various mathematical handbooks.[3]

In Table 3.3 we summarize our results for collision broadening and Doppler broadening, as well as the more general case of the Voigt profile. Tables 3.2 and 3.3 together summarize the results of our classical theory of absorption. With slight

3. See, for example, M. Abramowitz and I. A. Stegun, *Handbook of Mathematical Functions* (Dover, New York, 1971), pp. 325–328.

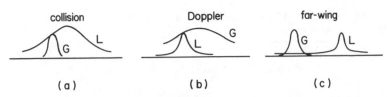

Figure 3.13 Sketch of factors in the integrand of (3.12.3) in three limiting cases: (*a*) collision-broadened limit, (*b*) Doppler-broadened limit, (*c*) far-wing limit.

modifications, these formulas are basically the same as those given by the quantum theory of absorption. The case of *gain*, or "negative absorption," is also described by very similar formulas. Thus the results of our classical theory will prove to be far more relevant to the operation of lasers than one might at first suspect. Indeed, we will refer back to Tables 3.2 and 3.3 in our study of lasers.

• Without going to numerical tables, and even without a study of the asymptotic properties of $w(x + ib)$, it is possible to evaluate the Voigt integral (3.12.3) in several limits because both factors in the integrand are normalized lineshapes themselves. There are three limits of interest, as shown in Figure 3.13.

Collisional Limit ($\delta\nu_0 \gg \delta\nu_D$): In this case $S(\nu, v)$ is very broad and slowly varying compared to the narrow Gaussian velocity distribution (Fig. 3.13*a*). Since the Gaussian is normalized to unity it acts like the delta function $\delta(v)$, and the Voigt integral reduces to the result $S(\nu) = S(\nu, v = 0)$, which is just the original collisional Lorentzian lineshape given in (3.6.18).

Doppler Limit ($\delta\nu_D \gg \delta\nu_0$): In this case the reverse is true (Fig. 3.13*b*), and the collisional function $S(\nu, v)$ acts like the delta function $\delta(\nu_0 - \nu + \nu v/c)$. Thus the Voigt integral gives back the Gaussian function (3.11.10). Except at high pressures or in cases where the Doppler distribution is altered by beam collimation it is usually valid to assume that the inequality $\delta\nu_D \gg \delta\nu_0$ is accurate and the Doppler limit applies.

Far-Wing Limit ($|\nu - \nu_0| \gg \delta\nu_D, \delta\nu_0$): This case refers to the spectral region far from line center, far outside the halfwidths of either the collisional or Doppler factors in the Voigt integrand. Thus the integrand is the product of two peaked functions. Each peak falls in the remote wing of the other function (see Fig. 3.13*c*). Here the qualitative difference between Gaussian and Lorentzian functions is significant. The Gaussian is much more compact. It falls to zero much more rapidly than the Lorentzian. Because the Lorentzian's wings are falling relatively slowly, as $1/v^2$ for large v, it still has nonzero value at the position of the Gaussian peak. However, the value of the Gaussian function is effectively zero by comparison near the Lorentzian's peak. Thus the contribution of the Gaussian function in the Lorentzian wing is much greater than that of the Lorentzian function in the Gaussian wing, and the Voigt integral can be replaced by (3.6.18) in its far wing:

$$S(\nu) \rightarrow \frac{\delta\nu_0/\pi}{(\nu - \nu_0)^2} \qquad (13.12.15)$$

This result is anomalous in the sense that the lineshape behaves like a Lorentzian in the far wing even if the broadening is principally Doppler, not collisional ($\delta\nu_D \gg \delta\nu_0$). •

TABLE 3.3   Collision, Doppler, and Voigt Lineshapes

Collision-Broadening Lineshape

$$S(\nu) = \frac{(1/\pi)\,\delta\nu_0}{(\nu - \nu_0)^2 + \delta\nu_0^2}$$

$$\delta\nu_0 = \frac{\text{collision rate}}{2\pi}$$

Doppler-Broadening Lineshape

$$S(\nu) = \frac{0.939}{\delta\nu_D}\, e^{-2.77(\nu - \nu_0)^2/\delta\nu_D^2}$$

$$\delta\nu_D = 2.15 \times 10^6 \left[\frac{1}{\lambda_0}\left(\frac{T}{M}\right)^{1/2}\right] \text{MHz}$$

T = gas temperature (K)

M = molecular weight of absorber (g)

λ_0 = wavelength (Å) of absorption line

Voigt Lineshape

$$S(\nu) = \frac{0.939}{\delta\nu_D}\, \text{Re}\, w(x + ib)$$

$$x = 1.67\,\frac{\nu_0 - \nu}{\delta\nu_D}$$

$$b = 1.67\,\frac{\delta\nu_0}{\delta\nu_D}$$

w = error function of complex argument

3.13   EXAMPLE: ABSORPTION BY SODIUM VAPOR

Let us consider an example of the use of Tables 3.2 and 3.3. Consider the 5890-Å absorption line of sodium vapor at 300 K. The Doppler width is

$$\delta\nu_D \approx 2.15 \times 10^{12}\left[\frac{1}{5890}\left(\frac{300}{23}\right)^{1/2}\right] \text{Hz}$$

$$\approx 1300 \text{ MHz} \tag{3.13.1}$$

since the atomic weight of sodium is $M_{\text{Na}} = 23$ g. From tabulated data we can estimate the collision broadening linewidth of the 5890-Å line in pure sodium vapor at 300 K to be

$$\delta\nu_0 \approx 1700P(\text{Torr}) \text{ MHz} \tag{3.13.2}$$

The ratio b of Table 3.3 is therefore

$$b \cong 2.2P(\text{Torr}) \tag{3.13.3}$$

If $P(\text{Torr})$ is less than, say, about 0.1 Torr, we are in the "Doppler regime." In this case the absorption coefficient for narrowband light exactly resonant with the 5890-Å absorption line is found from Table 3.2 at the end of Section 3.8 to be

$$a(\nu_0) = \frac{e^2 f}{4\epsilon_0 mc} N S(\nu_0)$$

$$= \frac{e^2 f}{4\epsilon_0 mc} N \frac{1}{\delta\nu_D} \left(\frac{4 \ln 2}{\pi}\right)^{1/2} \tag{3.13.4}$$

For the sodium D lines the oscillator strength—the factor f—is of order unity. In fact the 5890- and 5896-Å lines have oscillator strengths of 0.355 and 0.627, respectively. From (3.13.4), (3.13.1), and (3.10.8), therefore, we obtain

$$a(\nu_0) \approx 2.2 \times 10^5 P(\text{Torr}) \text{ cm}^{-1} \tag{3.13.5}$$

which is valid provided the pressure is small enough that Doppler broadening prevails. For narrowband light of frequency ν not necessarily equal to ν_0 we have

$$a(\nu) \approx 2.2 \times 10^5 P(\text{Torr}) e^{-4(\nu - \nu_0)^2 \ln 2/\delta\nu_D^2} \text{ cm}^{-1} \tag{3.13.6}$$

In Figure 3.14 we have plotted the transmission coefficient

$$\frac{I_\nu(z)}{I_\nu(0)} = e^{-a(\nu)z} \tag{3.13.7}$$

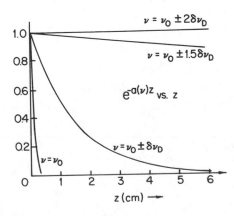

Figure 3.14 Transmission coefficient for 5890-Å radiation in sodium vapor at $T = 300$ K, $P = 5 \times 10^{-5}$ Torr. In this case the absorption line is Doppler-broadened, with $\delta\nu_0 \approx 1300$ MHz. The four curves illustrate the high selectivity of the absorption process.

for nearly monochromatic light of slightly different frequencies. It is evident that the transmission coefficient has an extremely strong dependence on the detuning $\nu - \nu_0$. The detuning of the field frequency ν by just a very small fraction of ν from the resonance frequency ν_0 results in a very sharp increase in the transmission coefficient. Similar results apply at higher pressures, where collision broadening becomes important.

PROBLEMS

3.1   Assume the "spring constants" k for the binding of electrons in atoms are approximately the same as those for the binding of atoms in molecules. If $\nu \approx 5 \times 10^{14}$ Hz is a typical electronic oscillation frequency, estimate the range of frequencies typical of atomic vibrations in molecules, given typical electron–atom mass differences. Does your estimate indicate that molecular vibrations lie in the infrared region of the spectrum?

3.2   The atomic weight of lithium is 6.939 g, and the density of lithium is 0.534 g/cm^3. Assuming each lithium atom contributes one electron to the "free-electron gas," calculate the plasma frequency ν_p. For what wavelengths would you expect lithium to be transparent? (Note: The transparency of the alkali metals in the ultraviolet was discovered by R. W. Wood in 1933.)

3.3   The addition of an ohmic current density to Maxwell's equations leads to the wave equation (3.5.6). Show this by adding $\mathbf{J} = \sigma\mathbf{E}$ to the right side of (2.1.4) and then retracing the derivation of the wave equation (2.1.13).

3.4*   Derive the equation for classical "laser amplification" by substituting (3.5.2) into (3.5.6), allowing E_n to be time-dependent: $E_n = E_n(t)$. It is sufficient to assume that $E_n(t)$ is slowly varying so that terms proportional to d^2E_n/dt^2 can be discarded.

   (a) Obtain the equation for dE_n/dt.

   (b) Use the approximations and abbreviations given in Eqs. (3.5.7)–(3.5.10) to show that $d|E_n|^2/dt = 0$ if $g = \sigma/\epsilon_0 c$.

   (c) Sketch on one graph the behavior of $E_n(t)$ vs. t obtained from the solution of the equation for $d|E_n|^2/dt$ under the (unrealistic) assumption that $g = 2\sigma/\epsilon_0 c$, and the (more realistic) assumption that $g = -2\sigma/\epsilon_0 c$.

3.5   Show that the number of atoms (or molecules) per cm^3 of an ideal gas at pressure P and temperature T is given by (3.10.8).

3.6   **a.**   Verify Eq. (3.12.3).

   b.   Using Eqs. (3.4.9) and (3.4.11), show that in the absence of background atoms

$$n_R(\nu) - 1 = \frac{\lambda_0}{4\pi} \frac{\nu_0 - \nu}{\delta \nu_0} a(\nu), \qquad \lambda_0 = \frac{c}{\nu_0}$$

This equation relates the refractive index near a collision-broadened absorption line to the absorption coefficient.

3.7* Although the relation derived in Problem 3.6 applies to the case of collision broadening, a similar relation holds more generally. Show in the case of a Voigt profile that

$$n_R(\nu) - 1 = \frac{\lambda_0}{4\pi} \frac{\mathrm{Im}\, w(x + ib)}{\mathrm{Re}\, w(x + ib)} a(\nu)$$

where w, x, and b are defined in Section 3.12.

[Note: The relation between the refractive index and the absorption coefficient (or, equivalently, between the real and imaginary parts of the complex refractive index) is a special case of the so-called *Kramers–Kronig relations*. Such relations may be derived on very general grounds based on causality.]

3.8 Estimate the absorption coefficient for 5890-Å radiation in sodium vapor containing 2.7×10^{12} atoms$/$cm^3 at 200°C. [See J. E. Bjorkholm and A. Ashkin, *Phys. Rev. Lett.* **32**, 129 (1974)].

3.9 The CO_2 molecule has strong absorption lines in the neighborhood of $\lambda = 10\ \mu$m. Assuming that the cross sections of CO_2 molecules with N_2 and O_2 molecules are $\sigma(CO_2, N_2) = 120\ \text{Å}^2$ and $\sigma(CO_2, O_2) = 95\ \text{Å}^2$, estimate the collision-broadened linewidth for CO_2 in the earth's atmosphere. (Note: since the concentration of CO_2 is very small compared with N_2 and O_2 in air, you may assume that only N_2–CO_2 and O_2–CO_2 collisions contribute to the linewidth.) Compare this with the Doppler width.

3.10 Consider the absorption coefficient $a(\nu_0)$ of a pure gas precisely at resonance. Show that $a(\nu_0)$ is proportional to the number density of atoms when the absorption line is Doppler-broadened, but is independent of the number density when the pressure is sufficiently large that collision broadening is dominant.

4 ATOMS, MOLECULES, AND SOLIDS

4.1 INTRODUCTION

It is frequently said that quantum physics began with Max Planck's discovery of the correct blackbody radiation formula in 1900. But quantum mechanics did not come easily; it was more than a quarter of a century before Planck's formula could be derived from a satisfactory theory of quantum mechanics. Nevertheless, once formulated, quantum mechanics answered so many questions that it was adopted and refined with remarkable speed between 1925 and 1930. By 1930 there were new and successful quantum theories of atomic and molecular structure, electromagnetic radiation, electron scattering, and thermal, optical, and magnetic properties of solids.

Lasers can be understood without a detailed knowledge of the quantum theory of matter. However, several consequences of the quantum theory are essential. This chapter provides a review of the results of quantum theory applied to simple models of atoms, molecules, and semiconductors.

4.2 THE BOHR ATOM

In 1913 Niels Bohr discovered a way to use Planck's radiation constant h in a radically new, but still mostly classical, theory of the hydrogen atom. Bohr's theory was the first quantum theory of atoms. Its importance was recognized immediately, even though it raised as many questions as it answered.

One of the most important questions it answered had to do with the Balmer formula:

$$\lambda = \frac{bn^2}{n^2 - 4} \tag{4.2.1}$$

This relation had been published in 1885 by Johann Jacob Balmer, a German school teacher. Balmer pointed out that if b were given the value 3645.6, then λ equaled the wavelength (measured in Ångstrom units) of a line in the hydrogen spectrum.[1] This was true for $n = 3$, 4, 5, and 6 (and possibly for higher integers as well, but no measurements existed to confirm or deny the possibility).

---

1. Historically, the term spectral "line" arose because lines appeared as images of slits in spectrometers.

For almost thirty years the Balmer formula was a small oasis of regularity in the chaotic field of spectroscopy—the science of measuring and cataloging the wavelengths of radiation emitted and absorbed by different elements and compounds. Unfortunately the Balmer formula could not be explained, or applied to any other element, or even applied to other known wavelengths emitted by hydrogen atoms. It might well have been a mere coincidence, without any significance.

Bohr's model of the hydrogen atom not only explained the Balmer formula, but also gave scientists their first glimpse of atomic structure. Three quarters of a century later, it still serves as the basis for most scientists' working picture of an atom.

Bohr adopted the nuclear model that had been successful in explaining Rutherford's scattering experiments with alpha particles between 1910 and 1912. In other words, Bohr assumed that almost all the mass of a hydrogen atom is concentrated in a positively charged nucleus, allowing most of the atomic volume free for the motion of the much lighter electron. The electron was assumed attracted to the nucleus by the Coulomb force law governing opposite charges (Figure 4.1). In magnitude this force is

$$F = \frac{1}{4\pi\epsilon_0} \frac{e^2}{r^2} \tag{4.2.2}$$

Bohr also assumed that the electron travels in a circular orbit about the massive nucleus. Moreover he assumed the validity of Newton's laws of motion for the orbit. Thus, in common with every planetary body in a circular orbit, the electron was assumed to experience an inward (centripetal) acceleration of magnitude

$$a = v^2/r \tag{4.2.3}$$

Newton's second law of motion, $\mathbf{F} = m\mathbf{a}$, then gives

$$\frac{mv^2}{r} = \frac{1}{4\pi\epsilon_0} \frac{e^2}{r^2} \tag{4.2.4}$$

which is the same as saying that the electron's kinetic energy, $T = \frac{1}{2} mv^2$, is half

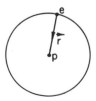

Figure 4.1 The electron in the Bohr model is attracted to the nucleus with a force of magnitude $F = e^2/4\pi\epsilon_0 r^2$.

as great as the magnitude of its potential energy, $V = -e^2/4\pi\epsilon_0 r$. In the Coulomb field of the nucleus the electron's total energy is therefore

$$E = T + V = -\frac{1}{4\pi\epsilon_0}\frac{e^2}{2r} \tag{4.2.5}$$

These results are familiar consequences of Newton's laws. Bohr then introduced a single, radical, unexplained restriction on the electron's motion. He asserted that only certain circles are actually used by electrons as orbits. These orbits are the ones that permit the electron's angular momentum L to have one of the values

$$L = n\frac{h}{2\pi} \tag{4.2.6}$$

where n is an integer ($n = 1, 2, 3, \ldots$) and h is the constant of Planck's radiation formula:

$$h \approx 6.625 \times 10^{-34} \text{ J-sec} \tag{4.2.7}$$

With the definition of angular momentum for a circular orbit,

$$L = mvr \tag{4.2.8}$$

it is easy to eliminate r between (4.2.4) and (4.2.8) and find

$$v = \frac{1}{4\pi\epsilon_0}\frac{e^2}{L} = \frac{1}{4\pi\epsilon_0}\frac{2\pi e^2}{nh} \tag{4.2.9}$$

Then the combination of (4.2.4) and (4.2.5), namely

$$E = -mv^2/2 \tag{4.2.10}$$

together with (4.2.9), gives the famous Bohr formula for the allowed energies of a hydrogen electron:

$$E_n = -\left(\frac{1}{4\pi\epsilon_0}\right)^2\frac{me^4}{2n^2h^2} \tag{4.2.11}$$

This can be seen, by comparison with (4.2.5), to be the same as a formula for the allowed values of electron orbital radius:

$$r_n = 4\pi\epsilon_0\frac{n^2h^2}{me^2} \tag{4.2.12}$$

In both (4.2.11) and (4.2.12) we have adopted the modern notation for Planck's constant:

$$\hbar = \frac{h}{2\pi} \approx 1.054 \times 10^{-34} \text{ J-sec} \tag{4.2.13}$$

The first thing to be said about Bohr's model and his unsupported assertion (4.2.6) is that they were not contradicted by known facts about atoms, and, for small values of n, the allowed radii defined by (4.2.12) are numerically about right.[2] For example, the smallest of these radii (conventionally called "the Bohr radius" and denoted a_0) is:

$$a_0 = r_1 = 4\pi\epsilon_0 \frac{\hbar^2}{me^2} \approx 0.53 \text{ Å} \tag{4.2.14}$$

This might have been an accident without further consequences. Since no way existed to measure such small distances with any precision, Bohr needed a connection between (4.2.12) and a possible laboratory experiment. A second unsupported assertion supplied the connection.

Bohr's second assertion was that the atom was stable when the electron was in one of the permitted orbits, but that jumps from one orbit to another were possible if accompanied by light emission or absorption. To be specific, Bohr combined earlier ideas of Planck and Einstein and stated that a jump from a higher to a lower orbit would find the decrease of the electron's energy transformed into a quantum of radiation that would be emitted in the process. In other words, Bohr postulated that

$$(\Delta E)_{n,n'} = h\nu = \text{energy of emitted photon} \tag{4.2.15}$$

Here $(\Delta E)_{n,n'}$ denotes the energy lost by the electron in switching orbits from r_n to $r_{n'}$, and ν is the frequency of the photon emitted in the process (Figure 4.2).

The relation (4.2.15) led immediately to a connection between Bohr's theory and all the spectroscopic data known for atomic hydrogen. By using (4.2.11) for two different orbits, i.e., for two different values of the integer n, we easily find for the energy decrement $(\Delta E)_{n,n'}$ the expression

$$E_n - E_{n'} = \left(\frac{1}{4\pi\epsilon_0}\right)^2 \frac{me^4}{2\hbar^2} \left(\frac{1}{n'^2} - \frac{1}{n^2}\right) \tag{4.2.16}$$

---

2. Lord Rayleigh (1890) was able to estimate molecular dimensions by dropping olive oil onto a water surface. Assuming that an oil drop spreads until it forms a layer one molecule thick (a *molecular monolayer*), he could give a reasonable estimate of a molecular diameter from the area of the layer and the volume of the original drop.

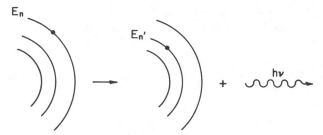

Figure 4.2 A radiative transition of an atomic electron in the Bohr model.

Furthermore, the connection between the frequency and wavelength of a light wave is

$$\lambda = c/\nu \tag{4.2.17}$$

Thus Bohr's statement (4.2.15) and his energy formula (4.2.11) are actually equivalent to the postulate that *all* spectroscopic wavelengths of light associated with atomic hydrogen fit the formula

$$\lambda = \frac{hc}{\Delta E_{n,n'}} = (4\pi\epsilon_0)^2 \frac{4\pi\hbar^3 c}{me^4} \frac{n^2 n'^2}{n^2 - n'^2} \tag{4.2.18}$$

where n and n' are integers to be chosen, but where all the other parameters are fixed.

It is obvious that if $n' = 2$ and $n > 2$, then (4.2.18) becomes

$$\lambda = (4\pi\epsilon_0)^2 \frac{16\pi\hbar^3 c}{me^4} \frac{n^2}{n^2 - 4} \tag{4.2.19}$$

which is exactly the Balmer formula (4.2.1). The numerical value of the product of the coefficients in (4.2.19) is just what Balmer had said the constant b was, 28 years earlier. Bohr's expression (4.2.18) was quickly found, for values of n' not equal to 2, to agree with other wavelengths associated with hydrogen but which had not fitted the Balmer formula (Problem 4.1).

Bohr's theory opened a new viewpoint on atomic spectroscopy. All observed spectroscopic wavelengths could be interpreted as evidence for the existence of certain allowed electron orbits in all atoms, even if formulas corresponding to Bohr's (4.2.18) were not known for any atom but hydrogen.

In particular, it seemed probable that the resonance frequencies postulated in the classical electron oscillator model of an atom were to be associated somehow with the "transition frequencies" ν of Bohr's formula (4.2.15). In this way the most useful features of the classical Lorentz theory survived the quantum revolution unchanged. How this is possible, in view of the obvious fact that the assumed Coulomb force (4.2.2) between electron and nucleus is not the harmonic-oscillator force appearing in (2.2.18), will become clear in Chapter 6.

It is easy to have second thoughts about Bohr's model, no matter how successful it is. For example, one can ask why (4.2.6) does not include the possibility $n = 0$. There is no apparent reason why zero angular momentum must be excluded, except that the energy formula (4.2.11) is not defined for $n = 0$. This point, whether physical significance can be assigned to the orbit with zero angular momentum, cannot be clarified within the Bohr theory, and it proved puzzling to physicists for more than a decade until quantum mechanics was developed. In a similar fashion, one can ask how Bohr's results are modified by relativity. The kinetic energy formula used above, $T = \frac{1}{2} mv^2$, is certainly nonrelativistic. Again, this point was not fully answered until the development of quantum mechanics.

• Relativistic corrections to Newtonian physics become important when particle velocities approach the velocity of light. If v is the velocity of a particle, then typically the first correction terms are found to be proportional to $(v/c)^2$, where c is the velocity of light. The value of $(v/c)^2$ can easily be estimated within the Bohr theory. It follows from (4.2.4) that

$$\left(\frac{v}{c}\right)^2 = \frac{1}{4\pi\epsilon_0} \frac{e^2}{rmc^2}$$

(4.2.20)

By inserting r from (4.2.12) and taking the square root, the ratio v/c can be found for any of the allowed orbits:

$$\frac{v}{c} = \frac{1}{4\pi\epsilon_0} \frac{e^2}{n\hbar c}$$

(4.2.21)

Equation (4.2.21) shows that the largest velocity to be expected in the Bohr atom is associated with the lowest orbit, $n = 1$. The ratio of this maximum velocity to the velocity of light is given by the remarkable (dimensionless) combination of electromagnetic and quantum-mechanical constants, $e^2/4\pi\epsilon_0\hbar c$. The numerical value of this parameter is easily found:

$$\frac{1}{4\pi\epsilon_0} \frac{e^2}{\hbar c} = \frac{(1.602 \times 10^{-19} \text{ C})^2}{(4\pi)(8.854 \times 10^{-12} \text{ C/V-m})(1.054 \times 10^{-34} \text{ J-sec})(2.998 \times 10^8 \text{ m/sec})}$$
$$= 0.007297 \ (= 1/137.04)$$

(4.2.22)

The value found in (4.2.22) is small enough that corrections to the Bohr model from relativistic effects are of the relative order of magnitude 10^{-4} or smaller, and thus completely negligible in most circumstances (Problem 4.2). Spectroscopic measurements, however, are commonly accurate to five significant figures. Arnold Sommerfeld, in the period 1915–1920, studied the relativistic corrections to Bohr's formulas, and showed that they accounted accurately for some of the fine details or *fine structure* in observed spectra. For this reason the parameter $e^2/4\pi\epsilon_0\hbar c$ is called *Sommerfeld's fine structure constant*.

The fine-structure constant appears so frequently in expressions of atomic radiation physics that it is very useful to remember its numerical value. Because the value given in (4.2.22) is very nearly equal to $1/137$, it is in this form that its value is memorized by physicists. •

4.3 QUANTUM STATES AND DEGENERACY

In the Bohr model a state of the electron is characterized by the "quantum number" n. Everything the model can say about the allowed states of the electron is given in terms of n.

The full quantum theory of the hydrogen atom also yields the allowed energies (4.2.11). However, in the quantum theory a state of the electron is characterized by other quantum numbers in addition to the *principal quantum number n* appearing in (4.2.11). The results of the quantum theory for the hydrogen atom, in addition to (4.2.11), are mainly the following:

i. For each principal quantum number n (= 1, 2, 3, . . .) there are n possible values of the *orbital angular momentum quantum number l*. The allowed values of l are 0, 1, 2, . . . $n - 1$. Thus for $n = 1$ we can have only $l = 0$, whereas for $n = 2$ we can have $l = 0$ or 1, etc.

ii. For each l there are $2l + 1$ possible values of the *magnetic quantum number* m. The possible values of m are $-l$, $-l + 1$, . . . , $-1, 0, 1$, . . . , $l - 1$, l.

iii. In addition to orbital angular momentum, an electron also carries an intrinsic angular momentum, which is called simply *spin*. The spin of an electron always has magnitude $\frac{1}{2}$ (in units of $\hbar$). But in any given direction the electron spin can be either "up" or "down"; that is, quantum theory says that when the component of electron spin along any direction is measured, we will always find the spin angular momentum to be along that direction or opposite to it. Because of this, an electron state must also be labeled by an additional quantum number m_s, called the *spin magnetic quantum number*, whose only possible values are $\pm\frac{1}{2}$.

Thus, for a given n, there are n possible values of l, and for each l there are $2l + 1$ possible values of m, for a total of

$$\sum_{l=0}^{n-1} (2l + 1) = n^2 \tag{4.3.1}$$

states. And each of these states is characterized further by m_s, which may be $+\frac{1}{2}$ or $-\frac{1}{2}$. Therefore there are $2n^2$ states associated with each principal quantum number n. In contrast to the Bohr model, in which an allowed state of the electron in the hydrogen atom is characterized by n, quantum theory characterizes each allowed state by the four quantum numbers n, l, m, and m_s; and since the electron energy depends only on n [recall (4.2.11)], there are $2n^2$ states with the same energy for every value of n. These $2n^2$ states are called *degenerate states*, or are said to be degenerate in energy.

Historical designations for the orbital angular momentum quantum numbers are still in use:

$l = 0$ designates the so-called s orbital

$l = 1$ p orbital

$l = 2$ d orbital

$l = 3$ f orbital

$l = 4$ g orbital

.   .

.   .

.   .

The first three letters came from the words "sharp," "principal," and "diffuse," which described the character of atomic emission spectra in a qualitative way long before quantum theory showed that they could be associated systematically with different orbital angular momentum values for an electron in the atom.

4.4   THE PERIODIC TABLE

Although hydrogen is the only atom for which explicit expressions like (4.2.11) can be written down, we can nevertheless understand the gross features of the periodic table of the elements. That is, we can understand the chemical regularity, or periodicity, that occurs as the atomic number Z increases. The key to this understanding is the *exclusion principle* of Wolfgang Pauli (1925), which forbids two electrons from occupying the same quantum-mechanical state. The Pauli exclusion principle may be proved only at an advanced level which is well beyond the scope of this book. We will simply accept it as a fundamental truth.

But the Pauli principle alone is not sufficient for an understanding of the periodic table. We must also deal with the electron–electron interactions in a multi-electron atom. These interactions present us with an extremely complicated many-body problem that has never been solved. A useful approximation, however, is to assume that each electron moves independently of all the others; each electron is thought of as being in a spherically symmetric potential $V(r)$ due to the Coulomb field of the nucleus plus the $Z - 1$ other electrons. In this independent-particle approximation an electron state is still characterized by the four quantum numbers (n, l, m, m_s), as in the case of hydrogen. However, in this case the simple energy formula (4.2.11) does not apply, and in particular the energy depends on both n and l (but not m or m_s).

The simplest multielectron atom, of course, is helium, in which there are $Z = 2$ electrons. The lowest-energy state for each electron is characterized in the independent-particle approximation by the quantum numbers $n = 1$, $l = 0$, $m = 0$, and $m_s = \pm\frac{1}{2}$. Since the energy depends now on both n and l, we can label this particular *electron configuration* as $1s$, a shorthand notation meaning $n = 1$ and $l = 0$. Both electrons are in the *shell* $n = 1$, one having spin up $(m_s = \frac{1}{2})$, the other spin down $(m_s = -\frac{1}{2})$. Since the number 2 is the maximum number of electrons allowed by the Pauli exclusion principle for the $1s$ configuration, we say that the $1s$ shell is completely filled in the helium atom.

In the case $Z = 3$ (lithium), there is one electron left over after the $1s$ shell is filled. The next allowed electron configuration is $2s$ $(n = 2, l = 0)$, and one of the electrons in lithium is assigned to this configuration. Since the $2s$ configuration can accommodate two electrons, the $2s$ *subshell* in lithium is only partially filled. The next element is beryllium, with $Z = 4$ electrons, and in this case the $2s$ subshell is completely filled, there being two electrons in this "slot."

For $Z = 5$ (boron), the added electron goes into the $2p$ configuration $(n = 2, l = 1)$. This configuration can accommodate $2(2l + 1) = 6$ electrons. Thus there are five other elements (C, N, O, F, and Ne) in which the outer subshell of electrons corresponds to the configuration $2p$. The eight elements lithium through neon, for which the outermost electrons belong to the $n = 2$ shell, constitute the first full row of the periodic table.

In Table 4.1 we list the first 36 elements and their electron configurations. The configurations are assigned in a similar manner as done above for $Z = 1$ to 10. Also listed in Table 4.1 is the ionization energy, defined as the energy required to remove one electron from the atom. For hydrogen the ionization energy W_I may be calculated from Eq. (4.2.11) with $n = 1$, i.e., W_I is just the binding energy of the electron in ground-state hydrogen:

$$W_I = |E_1| = \left(\frac{1}{4\pi\epsilon_0}\right)^2 \frac{me^4}{2\hbar^2} = 2.17 \times 10^{-18} \text{ J} \qquad (4.4.1)$$

Such small energies are usually expressed in units of electron volts (eV), an electron volt being the energy acquired by an electron accelerated through a potential difference of 1 volt:

$$1 \text{ eV} = (1.602 \times 10^{-19} \text{ C})(1 \text{ V})$$

$$= 1.602 \times 10^{-19} \text{ J} \qquad (4.4.2)$$

TABLE 4.1 Electron Configurations and Ionization Energies of the Elements of Atomic Number Z = 1 to 36

Z	Element	W_I (eV)	1s	2s	2p	3s	3p	3d	4s	4p
1	H	13.6	1							
2	He	24.6	2							
3	Li	5.4	2	1						
4	Be	9.3	2	2						
5	B	8.3	2	2	1					
6	C	11.3	2	2	2					
7	N	14.5	2	2	3					
8	O	13.6	2	2	4					
9	F	17.4	2	2	5					
10	Ne	21.6	2	2	6					
11	Na	5.1	2	2	6	1				
12	Mg	7.6	2	2	6	2				
13	Al	6.0	2	2	6	2	1			
14	Si	8.1	2	2	6	2	2			
15	P	10.5	2	2	6	2	3			
16	S	10.4	2	2	6	2	4			
17	Cl	13.0	2	2	6	2	5			
18	Ar	15.8	2	2	6	2	6			
19	K	4.3	2	2	6	2	6		1	
20	Ca	6.1	2	2	6	2	6		2	
21	Sc	6.5	2	2	6	2	6	1	2	
22	Ti	6.8	2	2	6	2	6	2	2	
23	V	6.7	2	2	6	2	6	3	2	
24	Cr	6.8	2	2	6	2	6	5	1	
25	Mn	7.4	2	2	6	2	6	5	2	
26	Fe	7.9	2	2	6	2	6	6	2	
27	Co	7.9	2	2	6	2	6	7	2	
28	Ni	7.6	2	2	6	2	6	8	2	
29	Cu	7.7	2	2	6	2	6	10	1	
30	Zn	9.4	2	2	6	2	6	10	2	
31	Ga	6.0	2	2	6	2	6	10	2	1
32	Ge	7.9	2	2	6	2	6	10	2	2
33	As	9.8	2	2	6	2	6	10	2	3
34	Se	9.7	2	2	6	2	6	10	2	4
35	Br	11.8	2	2	6	2	6	10	2	5
36	Kr	14.0	2	2	6	2	6	10	2	6

The ionization energy of hydrogen is therefore

$$W_I = \frac{2.17 \times 10^{-18} \text{ J}}{1.602 \times 10^{-19} \text{ J/eV}}$$

$$= 13.6 \text{ eV}$$

The ionization energy of a hydrogen atom in any state (n, l, m, m_s) is likewise $(13.6 \text{ eV})/n^2$.

The elements He, Ne, Ar, and Kr of Table 4.1 are chemically inactive. We note that each of these atoms has a completely filled outer shell. Evidently an atom with a filled outer shell of electrons tends to be "satisfied" with itself, having very little proclivity to share its electrons with other atoms (i.e., to join in chemical bonds). However, a filled outer subshell does not necessarily mean chemical inertness. Beryllium, for instance, has a filled $2s$ subshell, but it is not inert. Furthermore, even the noble gas Kr is not entirely inert.

The alkali metals Li, Na, and K of Table 4.1 have only one electron in an outer subshell, and, because of the low ionization energies of these elements, their outer electrons are weakly bound. These elements are highly reactive; they will readily give up their "extra" electron. On the other hand, the halogens F, Cl, and Br are one electron short of a filled outer subshell. These atoms will readily take another electron, and so they too are quite reactive chemically.

The characterization of atomic electron states in terms of the four quantum numbers n, l, m, and m_s, together with the Pauli exclusion principle, thus allows us to understand why Na is chemically similar to K, Mg is chemically similar to Ca, and so forth. These chemical periodicities, according to which the periodic table is arranged, are consequences of the way electrons fill in the allowed "slots" when they combine with nuclei to form atoms.

Of course there is a great deal more that can be said about the periodic table. For a rigorous treatment of atomic structure, we must again refer the reader to textbooks on atomic physics. As mentioned earlier, however, we can understand lasers without a detailed understanding of atomic and molecular physics.

4.5 MOLECULAR VIBRATIONS

As in the case of atoms, there are only certain allowed energy levels for the electrons of a molecule. Quantum jumps of electrons in molecules are accompanied by the emission of photons that typically belong to the ultraviolet region of the electromagnetic spectrum. For our purposes the electronic energy levels of molecules are quite similar to those of atoms.

However, in contrast with atoms, molecules have vibrational and rotational as well as electronic energy. This is because the relative positions and orientations of the individual atomic nuclei in molecules are not absolutely fixed. The energies associated with molecular vibrations and rotations are also quantized, i.e., restricted to certain allowed values. In this section and the next we will discuss the main features of molecular vibrational and rotational energy levels. Transitions between vibrational levels lie in the infrared portion of the electromagnetic spectrum, whereas rotational spectra are in the microwave region. Some of the most powerful lasers operate on molecular vibrational–rotational transitions.

Consider the simplest kind of molecule, namely a diatomic molecule such as O_2, N_2, or CO. There is a molecular binding force that is responsible for holding the two atoms together. To a first (and often very good) approximation the binding force is linear, so that the potential energy function is

$$V(x) = \tfrac{1}{2} k (x_2 - x_1 - x_0)^2 = \tfrac{1}{2} k (x - x_0)^2 \qquad (4.5.1)$$

where k is the "spring constant," $x = x_2 - x_1$ is the distance between the two nuclei, and x_0 is the internuclear separation for which the spring force

$$F = -k (x - x_0) \qquad (4.5.2)$$

vanishes (Figure 4.3). In other words, if the separation x is greater than x_0, the binding force is attractive and brings the nuclei closer; if x is less than x_0, the force is repulsive. The separation $x = x_0$ is therefore a point of stable equilibrium. The origin of the binding force is quantum-mechanical; we will not attempt to explain it, but will simply accept the result (4.5.1) and consider its consequences.

For simplicity let us assume that the nuclei can move only in one dimension. The total energy of a diatomic system (i.e., the sum of kinetic and potential energies) is then

$$E = \tfrac{1}{2} m_1 \dot{x}_1^2 + \tfrac{1}{2} m_2 \dot{x}_2^2 + \tfrac{1}{2} k (x_2 - x_1 - x_0)^2 \qquad (4.5.3)$$

where the dots denote differentiation with respect to time, i.e., $\dot{x} = dx/dt$. As in Section 2.2 it is convenient to define the total mass

$$M = m_1 + m_2 \qquad (4.5.4)$$

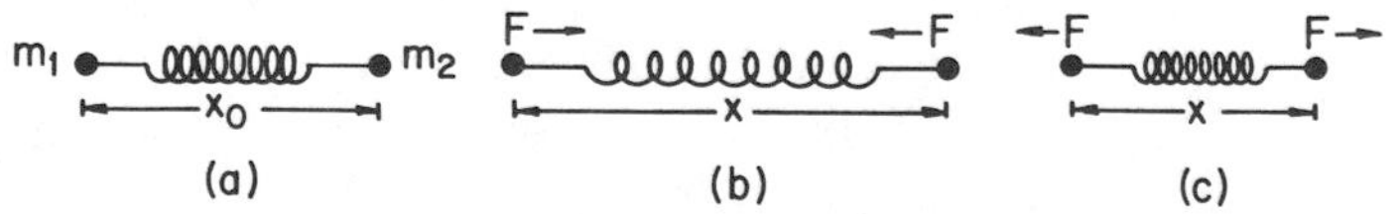

(a)   (b)   (c)

Figure 4.3 (a) When the two nuclei of a diatomic molecule are separated by the equilibrium distance x_0 there is no force between them. If their separation x is larger than x_0, there is an attractive force (b), whereas when x is less than x_0 the force is repulsive (c). The internuclear force is approximately harmonic, i.e., springlike.

and the reduced mass

$$m = m_1 m_2 / M \qquad (4.5.5)$$

of the system, as well as the center-of-mass coordinate

$$X = \frac{m_1 x_1 + m_2 x_2}{M} \qquad (4.5.6)$$

In terms of these quantities we may write (4.5.3) as (Problem 4.4)

$$E = \tfrac{1}{2} M \dot{X}^2 + \tfrac{1}{2} m \dot{x}^2 + \tfrac{1}{2} k (x - x_0)^2 \qquad (4.5.7)$$

The first term is just the kinetic energy associated with the center-of-mass motion. We ignore it and focus our attention on the internal vibrational energy

$$E = \tfrac{1}{2} m \dot{x}^2 + \tfrac{1}{2} k(x - x_0)^2 \qquad (4.5.8)$$

The vibrational motion of a diatomic molecule must clearly be one-dimensional, and so we lose nothing in the way of generality by restricting ourselves to one-dimensional vibrations from the start [Eq. (4.5.3)].

The quantum mechanics of the motion associated with the energy formula (4.5.8) has much in common with that for the hydrogen-atom electron. The most important result is that the allowed energies E of the oscillator are also quantized. The quantized energies are given by (Appendix 5.B)

$$E_n = \hbar\omega(n + 1/2), \qquad n = 0, 1, 2, 3, \ldots \qquad (4.5.9)$$

with

$$\omega = \sqrt{k/m} \qquad (4.5.10)$$

This formula is clearly quite different from Bohr's formula for hydrogen. The quantum-mechanical energy spectrum for a harmonic oscillator is simply a ladder of evenly spaced levels separated by $\hbar\omega$ (Figure 4.4). The ground level of the oscillator corresponds to $n = 0$. However, an oscillator in its ground level is not at rest at its stable equilibrium point $x = x_0$. Even the lowest possible energy of a quantum-mechanical oscillator has finite kinetic and potential energy contributions. At zero absolute temperature, where classically all motion ceases, the quantum-mechanical oscillator still has a finite energy $\tfrac{1}{2} \hbar\omega$. For this reason the energy $\tfrac{1}{2} \hbar\omega$ is called the zero-point energy of the harmonic oscillator.

Of course real diatomic molecules are not perfect harmonic oscillators, and their

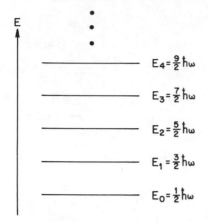

$E_4 = \frac{9}{2}\hbar\omega$

$E_3 = \frac{7}{2}\hbar\omega$

$E_2 = \frac{5}{2}\hbar\omega$

$E_1 = \frac{3}{2}\hbar\omega$

$E_0 = \frac{1}{2}\hbar\omega$ **Figure 4.4** The energy levels of a harmonic oscillator form a ladder with rung spacing $\hbar\omega$.

vibrational energies do not satisfy (4.5.9) precisely. Figure 4.5 shows the sort of potential energy function $V(x)$ that describes the bonding of a real diatomic molecule. The Taylor series expansion of the function $V(x)$ about the equilibrium point x_0 is

$$V(x) = V(x_0) + (x - x_0)\left(\frac{dV}{dx}\right)_{x=x_0} + \frac{1}{2}(x - x_0)^2\left(\frac{d^2V}{dx^2}\right)_{x=x_0}$$

$$+ \frac{1}{6}(x - x_0)^3\left(\frac{d^3V}{dx^3}\right)_{x=x_0} + \ldots \tag{4.5.11}$$

Here $V(x_0)$ is a constant, which we put equal to zero by shifting the origin of the

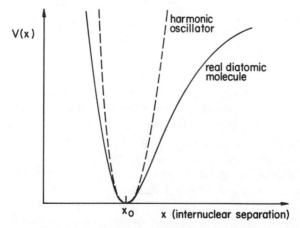

Figure 4.5 The potential energy function of a real diatomic molecule is approximately like that of a harmonic oscillator for values of x near to x_0.

energy scale. Also $(dV/dx)_{x=x_0} = 0$, because, by definition, $x = x_0$ is the equilibrium separation, at which the potential energy is a minimum. Furthermore (d^2V/dx^2) at $x = x_0$ is positive if x_0 is a point of stable equilibrium (Figure 4.5). Thus we can replace (4.5.11) by

$$V(x) = \tfrac{1}{2} k (x - x_0)^2 + A(x - x_0)^3 + B(x - x_0)^4 + \cdots \quad (4.5.12)$$

where $A, B, \ldots$ are constants and $k = (d^2V/dx^2)_{x=x_0}$.

From (4.5.11) we can conclude that *any* potential energy function describing a stable equilibrium [i.e., $(dV/dx)_{x=x_0} = 0$, $(d^2V/dx^2)_{x=x_0} > 0$] can be approximated by the harmonic-oscillator potential (4.5.1) for small enough displacements from equilibrium. Of course, what is "small" is determined by the constants A, $B, \ldots$ in (4.5.12), i.e., by the shape of the potential function $V(x)$. If the terms involving third and/or higher powers of $x - x_0$ in (4.5.12) are not negligible, however, we have what is called an *anharmonic* potential. The energy levels of an anharmonic oscillator do not satisfy the simple formula (4.5.9).

Real diatomic molecules have vibrational spectra that are usually only slightly anharmonic. In conventional notation the vibrational energy levels of diatomic molecules are written in the form

$$E_v = hc\omega_e\left[(v + \tfrac{1}{2}) - x_e(v + \tfrac{1}{2})^2 + y_e(v + \tfrac{1}{2})^3 + \cdots\right]$$

where

$$v = 0, 1, 2, 3, \ldots \quad (4.5.13)$$

and ω_e is in units of "wave numbers," i.e., cm^{-1}; $c\omega_e$ is the same as $\omega/2\pi = \nu$ in this notation. If the anharmonicity coefficients $x_e, y_e, \ldots$ are all zero, we recover the harmonic-oscillator spectrum (4.5.9). The numerical values of $\omega_e, x_e, y_e, \ldots$ are tabulated in the literature.[3] Values of ω_e, x_e, and y_e are given for several diatomic molecules in Table 4.2. The deviations from perfect harmonicity are small until v becomes large, i.e., until we climb fairly high on the vibrational ladder. The level spacing decreases as v increases, in contrast to the even spacing of the ideal harmonic oscillator (Figure 4.6).

• As a simple check on our theory, consider the two molecules hydrogen fluoride (HF) and deuterium fluoride (DF). These molecules differ only to the extent that D has a neutron and proton in its nucleus and H has only a proton. Since neutrons have no effect on molecular bonding, we expect HF and DF to have the same potential function V(x) and therefore the same "spring constant" k. According to (4.5.10), therefore, we should have

$$\frac{\omega_e^{HF}}{\omega_e^{DF}} = \left(\frac{m^{DF}}{m^{HF}}\right)^{1/2} \quad (4.5.14)$$

3. A standard source is G. Herzberg, *Spectra of Diatomic Molecules* (Van Nostrand, Princeton, N.J., 1964).

TABLE 4.2   Vibrational Constants of the Ground Electronic State for a Few Diatomic Molecules

Molecule	ω_e (cm^{-1})	x_e	y_e
H_2	4395.24	0.0268	6.67×10^{-5}
O_2	1580.36	0.00764	3.46×10^{-5}
CO	2170.21	0.00620	1.42×10^{-5}
HF	4138.52	0.0218	2.37×10^{-4}
HCl	2989.74	0.0174	1.87×10^{-5}

where m^{DF} and m^{HF} are the *reduced* masses of DF and HF, respectively, so that

$$\frac{m^{DF}}{m^{HF}} = \frac{m^D m^F}{m^D + m^F} \Big/ \frac{m^H m^F}{m^H + m^F}$$

$$\approx \frac{(2)(19)}{2 + 19} \times \frac{1 + 19}{(1)(19)}$$

$$\approx 1.90 \tag{4.5.15}$$

From the value of ω_e for HF given in Table 4.2, therefore, we calculate

$$\omega_e^{DF} \approx (4138.52)(1.90)^{-1/2} = 2998.64 \text{ cm}^{-1} \tag{4.5.16}$$

and indeed this result is very close to the tabulated value $\omega_e = 2998.25$ cm^{-1} for DF. •

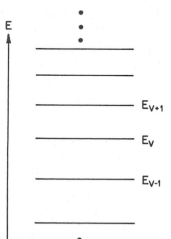

Figure 4.6 The vibrational energy level spacing of a real (anharmonic) diatomic molecule decreases with increasing vibrational energy.

4.6 MOLECULAR ROTATIONS

The rotations of a diatomic molecule can be understood in two stages. First we imagine the molecule to be a dumbbell consisting of two masses, m_1 and m_2, held together by a (massless) rigid rod of length x_0 (Figure 4.7). The dumbbell can rotate about its center of mass. The moment of inertia I is mx_0^2, where m is the reduced mass (4.5.5) and x_0 is the distance separating the masses m_1 and m_2. If the angular velocity of rotation (radians per second) is ω_R, the angular momentum and kinetic energy are respectively

$$L = I\omega_R \quad \text{(magnitude of angular momentum vector)} \quad (4.6.1)$$

and

$$E = \tfrac{1}{2} I\omega_R^2 = \frac{L^2}{2I} \quad (4.6.2)$$

These classical formulas are the starting point of a quantum-mechanical treatment of the rigid dumbbell, just as similar classical formulas underlie treatments of the hydrogen atom and the vibrations of molecules. It is found that the rotational energy (4.6.2) of the molecule has the allowed values

$$E_J = \frac{\hbar^2}{2I} J(J + 1), \quad J = 0, 1, 2, \ldots \quad (4.6.3)$$

Actual diatomic molecules are, of course, not rigid dumbbells. In particular, the masses m_1 and m_2 do not stay a fixed distance x_0 apart. As the molecule rotates, the centrifugal force tends to increase the separation of the two masses, and therefore also the moment of inertia I. This decreases the rotational energy, the more so as the rate of rotation (i.e., J) increases. In the notation of molecular spectroscopy this is accounted for by writing

$$E_J = hcBJ(J + 1) - hcDJ^2(J + 1)^2 \quad (4.6.4)$$

where the J-independent quantities B and D have units of wave numbers.

The fact that the molecule can vibrate also tends to increase the effective mo-

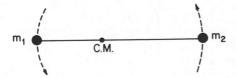

Figure 4.7 A dumbbell rotating about an axis through its center of mass serves as a classical model for the rotations of a diatomic molecule.

ment of inertia, the more so as the vibrational quantum number v increases. This is accounted for by writing

$$B = B_e - \alpha_e(v + \tfrac{1}{2}) \tag{4.6.5}$$

where B_e and α_e (in cm^{-1}) are independent of v and J. The rotational energy levels associated with the vibrational level v of a diatomic molecule are therefore written as

$$E_J(v) = hc\big[B_e - \alpha_e(v + \tfrac{1}{2})\big] J(J + 1) - hcDJ^2(J + 1)^2 \tag{4.6.6}$$

Higher-order corrections are necessary in general to explain the fine details of the rotational energy spectrum of a diatomic molecule. However, (4.6.6) is often accurate enough for practical purposes, and in fact the term involving D is often negligible.

The constants B_e, α_e, ... for different molecules are tabulated in the spectroscopic literature. The constants B_e and α_e for several molecules are given in Table 4.3. For our purposes it will suffice to make the rigid-dumbbell approximation and write

$$E_J(v) \approx E_J \approx hcB_eJ(J + 1) \tag{4.6.7}$$

• Once again it is possible to check our theory with an example. A comparison of Eqs. (4.6.3) and (4.6.7) shows that the rotational constant B_e of a diatomic molecule should be inversely proportional to its moment of inertia $I = mx_0^2$. Since the equilibrium separation x_0 is determined primarily by chemical (i.e., electromagnetic) forces, we expect that it should be practically the same for the two molecules HF and DF. Thus we expect

$$\frac{B_e^{HF}}{B_e^{DF}} = \frac{m^{DF}}{m^{HF}} = 1.90 \tag{4.6.8}$$

TABLE 4.3   Rotational Constants of the
Ground Electronic State for the Molecules
Listed in Table 4.2

Molecule	$B_e(\text{cm}^{-1})$	$\alpha_e(\text{cm}^{-1})$
H_2	60.81	2.993
O_2	1.44567	0.01579
CO	1.9314	0.01749
HF	20.939	0.770
HCl	10.5909	0.3019

It follows from the data in Table 4.3, therefore, that the rotational constant for DF should be $B_e^{DF} \approx 11.02$ cm^{-1}, in excellent agreement with the value 11.007 cm^{-1} tabulated by Herzberg. •

In summary, with every electronic state are associated vibrational constants ω_e, x_e, y_e, ... and rotational constants B_e, α_e, ... In Tables 4.2 and 4.3 the vibrational–rotational constants are given for the ground (lowest-energy) electronic state.

4.7 EXAMPLE: CARBON DIOXIDE

In our treatment of molecular vibrations and rotations we have only considered the relatively simple case of diatomic molecules. Rather than now discussing general polyatomic molecules, which are more complicated but fundamentally much the same as diatomics, we will consider only the specific case of the carbon dioxide molecule. We choose this example because the CO_2 laser is perhaps the most important molecule laser.

Carbon dioxide is a linear triatomic molecule (Figure 4.8). Such a molecule has three so-called *normal modes* of vibration, shown in Figure 4.8*b*. For obvious reasons these are called the asymmetric stretch, bending, and symmetric stretch modes. With each of these normal modes is associated a characteristic frequency of vibration. Each mode of vibration has a ladder of allowed energy levels asso-

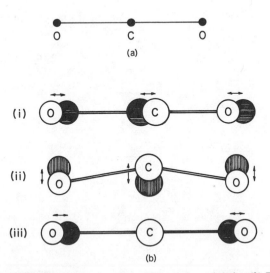

Figure 4.8 (*a*) Carbon dioxide (CO_2) is a linear triatomic molecule. (*b*) The normal modes of vibration of the CO_2 molecule: (i) The asymmetric stretch mode, (ii) the bending mode, and (iii) the symmetric stretch mode.

ciated with it, as in the case of a diatomic molecule (which has only one normal mode of vibration). The vibrational energy levels of the molecule may therefore be labeled by three integers v_1, v_2, and v_3 ($= 0, 1, 2, 3, \ldots$), and we have approximately

$$E(v_1, v_2, v_3) = hc\omega_e^{(1)}(v_1 + \tfrac{1}{2})$$

$$+ hc\omega_e^{(2)}(v_2 + \tfrac{1}{2})$$

$$+ hc\omega_e^{(3)}(v_3 + \tfrac{1}{2}) \qquad (4.7.1)$$

where $\omega_e^{(1)}$, $\omega_e^{(2)}$, and $\omega_e^{(3)}$ are the normal-mode frequencies in units of wave numbers. In reality each normal mode is slightly anharmonic, but the harmonic-oscillator approximation will suffice for our purposes. For CO_2 the normal-mode frequencies are

$$\omega_e^{(1)} = \omega(\text{symmetric stretch}) \approx 1388 \text{ cm}^{-1} \qquad (4.7.2a)$$

$$\omega_e^{(2)} = \omega(\text{bending}) \approx 667 \text{ cm}^{-1}. \qquad (4.7.2b)$$

$$\omega_e^{(3)} = \omega(\text{asymmetric stretch}) \approx 2349 \text{ cm}^{-1} \qquad (4.7.2c)$$

The first few vibrational energy levels ($v_1 v_2 v_3$) of the CO_2 molecule are indicated in Figure 4.9.

Since CO_2 is a linear molecule, its rotational energy spectrum has the same

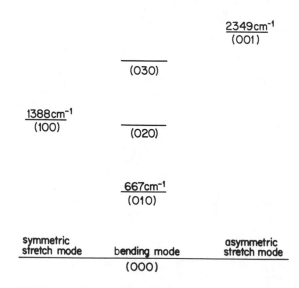

Figure 4.9 The first few vibrational energy levels of the CO_2 molecule.

character as that for diatomic molecules. The CO_2 rotational energy levels are thus given to a good approximation by Eq. (4.6.7):

$$E_J = hcB_eJ(J + 1), \qquad J = 0, 1, 2, \ldots \qquad (4.7.3)$$

where the rotational constant B_e for the CO_2 molecule is

$$B_e = 0.39 \text{ cm}^{-1}. \qquad (4.7.4)$$

4.8 ENERGY BANDS IN SOLIDS

In a gas the average distance between molecules (or atoms) is large compared with molecular dimensions. In liquids and solids, however, the intermolecular distance is comparable to a molecular diameter (Problem 4.5). Consequently the intermolecular forces are roughly comparable in strength to the interatomic bonding forces in the molecules. The molecules in liquids and solids are thus influenced very strongly by their neighbors.

What is generally called "solid-state physics" is mostly the study of crystalline solids, i.e., solids in which the molecules are arranged in a regular pattern called a crystal lattice. It is in this restricted sense that solid-state physics is a well-developed field with so many applications in modern technology.

The central result of the quantum theory of crystalline solids is that the discrete energy levels of the individual molecules are split into *energy bands*, each containing many closely spaced levels (Figure 4.10). Between these allowed energy bands are forbidden energy gaps.

We can reach a crude understanding of the formation of energy bands by be-

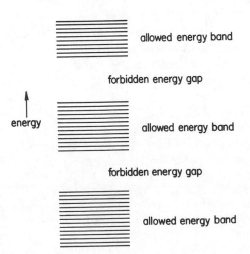

Figure 4.10 In a crystalline solid the allowed electron energy levels occur in *bands* of closely spaced levels. Between these allowed energy bands are forbidden energy gaps.

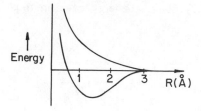

Figure 4.11 Energy levels of a pair of identical atoms as a function of their separation R. At large separations the energy levels are those of the individual atoms. As the atoms are brought close together, each level splits into two levels. The sketch above is typical of the splitting of each energy level of the isolated atoms.

ginning with the case of two identical atoms. When the atoms are far apart and effectively noninteracting, their electron configurations are identical. As they are brought closer together, however, the electrons of each begin to feel the presence of the other atom, and the allowed energy levels become those of the two atoms as a whole. Because of the Pauli exclusion principle, no two electrons of this two-atom system can occupy the same state. Nature's way of obeying the Pauli principle is to split each energy level of the individual atoms into two distinct energy levels.[4] The difference between these two new energy levels depends upon the interatomic spacing (Figure 4.11).

Similarly, if we begin with N atoms, each level of each atom is split into N levels as the atoms are brought close enough together. The difference between the highest and the lowest of these N levels depends on the interatomic distances, amounting typically to several electron volts for atomic spacings of a few angstroms, typical of solids. Now if we increase N, keeping the interatomic spacing fixed as in a crystalline solid, the total energy spread of the N levels stays about the same, but the levels become more densely spaced. For the large values of N typical of a solid (say, something like 10^{23} atoms/cm^3), each set of N levels thus becomes in effect a continuous energy band, as in Figure 4.12.

The chemical and optical properties of atoms are determined primarily by their outer electrons. In solids, similarly, many important properties are determined by the electrons in the highest energy bands, the bands evolving out of the higher states of the individual atoms. Consider, for instance, a solid in which the highest occupied energy band is only partially filled, as illustrated in Figure 4.13a. In an applied electric field the electrons in this band can readily take up energy and move up within the band. A solid whose highest occupied band is only partially filled is a good *conductor* of electricity.

Now consider a solid whose highest occupied band is completely filled with electrons, as illustrated in Figure 4.13b. In this case it is quite difficult for an electron to move because all the energetically allowed higher states in the band already have their full measure of electrons permitted by the Pauli principle.

---

4. These two energy levels correspond to symmetric and antisymmetric spatial wave functions for the electrons, with correspondingly antisymmetric and symmetric spin eigenfunctions. The twofold exchange degeneracy in the case of widely separated atoms is broken when their wave functions begin to overlap.

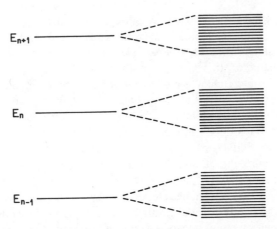

Figure 4.12 In a crystalline solid energy bands are formed from the energy levels of the isolated atoms.

Therefore a solid whose highest occupied energy band is filled will be an electrical *insulator*; in other words, its electrons will not flow freely when an electric field is applied. Implicit in this definition of an insulator is the assumption that the forbidden energy gap between the highest filled band and the next allowed energy band, denoted E_g in Fig. 4.13, is large compared with the amount of energy an electron can pick up in the applied field.

Solids in which this band gap is not so large are called *semiconductors*. Their band structure is indicated in Figure 4.13 c. At the absolute zero of temperature the *valence* band of a semiconductor is completely filled, whereas the *conduction* band, the next allowed energy band, is empty. At room temperature, however,

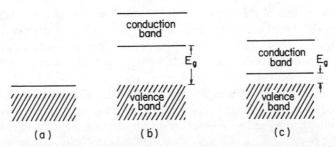

Figure 4.13 In a good conductor of electricity (a), the highest occupied band is only partially filled with electrons, whereas in a good insulator (b) it is filled. In (b) the energy gap E_g between the valence band and the conduction band is large. In the case (c) of a semiconductor, however, this gap is small, and electrons in the valence band can easily be promoted to the conduction band.

electrons in the valence band may have enough thermal energy to cross the narrow energy gap and go into the conduction band. Thus diamond, which has a band gap of about 7 eV, is an insulator, whereas silicon, with a band gap of only about 1 eV, is a semiconductor. In a metallic conductor, by contrast, there is no band gap at all; the valence and conduction bands are effectively overlapped.

This characterization of solids as insulators, conductors, and semiconductors is obviously more descriptive than explanatory. To understand *why* a given solid is an insulator, conductor, or semiconductor, we must consider the nature of the forces binding the atoms (or molecules) together in the solid.

In *covalent solids* the atoms are bound by the sharing of outer electrons in partially filled configurations. In a true covalent solid, there are no free electrons, and so such solids do not conduct electricity very well. The covalent bonds tend to hold the atoms tightly together, thus causing covalent solids to be rather hard and have high melting points. A good example is diamond, in which carbon atoms are arranged in a lattice such that each atom is at the center of a tetrahedron formed by its four nearest neighbors.

In *ionic solids* such as NaCl, the binding is produced by electrostatic forces between oppositely charged ions. The reason for the binding is the same as in ionic molecules. In NaCl, for instance, the energy required to remove the $3s$ electron from Na and transfer it to Cl, to form Na^+ and Cl^-, is less than the electrostatic energy of attraction between the ions. Here again there are no free electrons available to conduct heat or electricity, and so ionic solids are not good conductors.

The so-called *molecular solids*, which include many organic compounds, are also poor conductors. In such solids the binding is due to the very weak van der Waals forces, which were originally postulated by J. D. van der Waals (1837) in order to explain deviations from the ideal gas law. The van der Waals energy of attraction between two molecules ordinarily varies as the inverse sixth power of the distance between the molecules. Because of the weakness of the van der Waals interaction, molecular solids are much easier to deform or compress than covalent or ionic solids.

Of course electrical technology as we know it would be impossible without *metallic solids*, which are good conductors of electricity. In a metallic solid the electrons are not all tightly bound at crystal lattice sites. Some of the electrons are free to move over large distances in the metal, much as atoms move freely in a gas. This occurs because metals are formed from atoms in which there are one, two, or occasionally three outer electrons in unfilled configurations. The binding is associated with these weakly held electrons leaving their parent ions and being shared by all the ions, and so we can regard metallic binding as a kind of covalent binding. We can also think of the positive ions as being held in place because their attraction to the "electron gas" exceeds their mutual repulsion.

It is sometimes a useful approximation to regard the conduction electrons of a metal as completely free to move about. We have already mentioned this free-electron approximation in Section 3.1, in connection with metallic shine. Of course conduction electrons are not really completely free, as evidenced by the fact that even the very best conductors, copper, silver, and gold, have a finite resistance to the flow of electricity.

It should be emphasized that many solids do not fit so neatly into the covalent, ionic, molecular, or metallic categories. Furthermore many important properties of various solids are determined by imperfections such as impurities and dislocations in the crystal lattice. Steel, for instance is much harder than pure iron because of the small amount of carbon that was mixed into the iron melt. Impurities can also determine the color of a crystal, as in the case of ruby (Section 3.1). We will shortly discuss how the addition of certain impurities in semiconductors is responsible for modern electronic technology.

4.9 SEMICONDUCTORS

A semiconductor is distinguished from an insulator by the fact that the band gap between the valence and conduction bands is small, about 2 eV or less. The important semiconductors silicon and germanium, for instance, are covalent solids with band gaps of about 1.12 and 0.67 eV, respectively, at 300 K. At very low temperatures they are insulators, but the conductivity increases rapidly with increasing temperature, because valence-band electrons can be thermally excited into the conduction band. This increase of conductivity with increasing temperature is an important distinction between semiconductors and metals.[5] It was known to Faraday and other physicists early in the nineteenth century, but was explained only when quantum mechanics and the band theory of solids were developed.

Another way to promote electrons into the conduction band of a semiconductor is by absorption of radiation, if the energy of an incident photon exceeds the gap energy E_g (i.e., $h\nu > E_g$, where ν is the frequency of the radiation). This photoconductive effect is quite similar to the photoelectric effect, except that electrons are not actually released from the surface of the material. Photoconductive cells, in which the current in an electric circuit is controlled by the intensity of incident light, have applications similar to photoelectric cells (exposure meters in photography, automatic door openers, etc.), except that they require an auxiliary voltage supply to move electrons that have been put into the conduction band.

By far the most important means of producing conduction electrons in a semiconductor is by doping it with a certain type of impurity. Tiny junctions of differently doped semiconductors are the basis not only for transistors, but also for light-emitting diodes and the increasingly important diode lasers. In order to understand the operation of such devices, however, we must first discuss the concept of a *hole* in the valence band of a semiconductor.

5. Perhaps it is worth emphasizing what a tremendous range of electrical conductivities is found in different materials. A good insulator might have a resistivity (the inverse of the conductivity) of 10^{22} Ω-cm, whereas a metal may have a value 10^{-6} Ω-cm. Room-temperature resistivities of semiconductors, by contrast, are typically somewhere between 10^{-3} and 10^9 Ω-cm. This range of about 28 orders of magnitude is often cited as one of the broadest variations of any physical parameter.

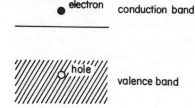

conduction band

valence band

Figure 4.14 In going from the valence band to the conduction band, an electron leaves a hole in the valence band.

The basic idea is very simple. If an electron somehow goes from the valence band to the conduction band, it leaves a hole—the absence of an electron—in the valence band (Figure 4.14). That is, a hole corresponds to the absence of an electron from an otherwise filled valence band. It turns out to be very useful to think of a hole as a particle like an electron. The removal of an electron increases the charge of the valence band, so clearly a hole must be a positively charged "particle," with charge opposite to that of an electron.

Consider a piece of a semiconductor in which electron–hole pairs have been created in some way (i.e., electrons have been put into the conduction band, leaving holes in the valence band). If we connect it with wires to the terminals of a battery, there will be a flow of current, since there are "free" electrons in the conduction band ready to respond to an externally applied field. These electrons drift in the direction shown in Figure 4.15, from the negative electrode to the positive. The net effect, as seen from the outside, is that electrons enter the semiconductor from the right and exit to the left. However, this is not the whole story, for electrons in the valence band are also affected by the potential difference. Specifically, an electron to the right of the hole indicated in Figure 4.15 can fall into the hole; that is, it will go into the state previously occupied by another electron. In doing so, it leaves a hole at the site it left, which can now be filled by another electron. This electron drift in the valence band constitutes a current in the same direction (left to right, by convention, in Figure 4.15) as the current of the electrons in the conduction band. Equivalently, we can view the situation as one in

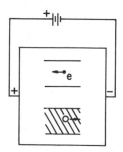

Figure 4.15 When a potential difference is applied to two ends of a semiconductor, electrons in the valence and conduction bands drift from the negative side to the positive. In the valence band, the effect is equivalent to the drift of positively charged holes from the positive side to the negative. The total current can therefore be attributed to electrons in the conduction band and holes in the valence band.

which electrons in the conduction band are moving from right to left, while holes in the valence band are moving from left to right. In other words, we can describe the charge motion in the conduction band in terms of electrons, and that in the valence band in terms of holes, and *both electrons and holes contribute to the total current*.

By doping a semiconductor with a certain kind of impurity, we can arrange for a current in the semiconductor to be due predominantly to either electrons or holes. In the former case the semiconductor is called *n-type* (because electrons are negatively charged) and in the latter it is called *p-type* (because holes have positive charge). To see how this works, we will consider the example of silicon doped with phosphorus.

In pure silicon each atom shares its four valence electrons in the unfilled $3s3p$ subshell (Table 4.1) to form covalent bonds with its four nearest neighbors. Each silicon atom needs four more electrons to complete the sp configuration, and by sharing electrons in this way it comes closer to having a filled outer subshell. The crystal structure is that of diamond, with each silicon atom at the center of a regular tetrahedron (pyramid) and its four nearest neighbors at the vertices. This structure is a consequence of the fact that the bonds associated with shared electrons are spaced as far from each other as possible at equal angles from each atom. It is useful to represent the situation in the schematic, two-dimensional form of Figure 4.16.

In its pure form silicon has a very low conductivity at room temperature, because so few electrons can be thermally excited across the 1.12-eV energy gap. Under ordinary circumstances the current passed is so small as to be practically useless. To pass useful current we must find a way to get more electrons into the conduction band, or holes into the valence band.

Suppose that one of the silicon atoms in the crystal is replaced by an atom of phosphorus, which has *five* electrons in the unfilled $3s3p$ subshell. Four of these can contribute to the covalent bonding of the crystal, as indicated in Figure 4.17, but there is one left-over electron which cannot take part in the bonding. This fifth electron is very loosely bound, and so is free to move through the crystal when an electric field is applied. In other words, if we add a small amount of phosphorus

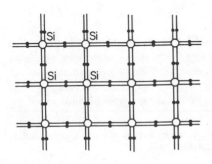

Figure 4.16 Schematic illustration of the covalent bonding in silicon, in which each atom shares its four valence electrons with its four nearest neighbors.

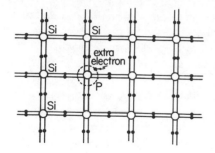

Figure 4.17 If a silicon atom is replaced by a phosphorus atom in Figure 4.16, there is an extra electron left over which cannot take part in the covalent bonding. This electron is very loosely bound and therefore available for the conduction of electric current.

to a silicon melt, the crystal that forms will be an *n*-type semiconductor.[6] We can also make an *n*-type semiconductor by doping silicon with other pentavalent elements like arsenic and antimony (Table 4.1).

Imagine instead that we replace a silicon atom by an atom of boron, which has *three* electrons in an unfilled outer shell. In this case there is one electron short of the four needed to join in complete covalent bonding in the host silicon lattice. Thus if boron shares its three valence electrons with neighboring silicon atoms, there will be a missing bond in the crystal, as indicated in Figure 4.18. This missing electron is a hole that can be filled by an electron that happends to be nearby. But when that electron fills the hole, it leaves another hole, which can be filled by another electron, and so in an electric field we get a migration of *holes* (or equivalently, of course, a migration of electrons in the opposite direction). In other words, by doping silicon with boron we can create a *p*-type semiconductor. Other trivalent elements like aluminum or gallium are also suitable dopants for this purpose.

As a matter of terminology, a dopant that produces an n-type semiconductor is called a *donor*, because it donates electrons to the conduction band. A dopant that produces a *p*-type semiconductor is called an *acceptor*, because it puts holes in the valence band, i.e., it accepts electrons to fill the missing slots. In either case, of

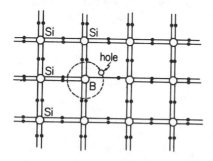

Figure 4.18 If a silicon atom in Figure 4.16 is replaced by a boron atom, there will be a missing bond because there is one electron short of the number necessary for complete bonding. The missing electron is represented by a hole.

6.   The proportion of dopant must be small in order to preserve the integrity of the host crystal lattice, since the dopant by itself forms its own crystal lattice structure.

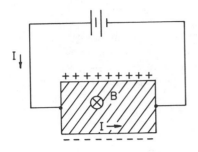

Figure 4.19 Experimental arrangement to determine the sign of the charge carriers (Hall effect). A magnetic field **B** is applied in the direction into the page. The top and bottom of the sample become charged + and −, respectively, if the carriers are positive, and − and +, respectively, if the carriers are negative.

course, the crystal remains charge-neutral. Note also that the added impurities produce either electrons or holes, but no electron–hole pairs as in, for instance, photoconductivity. Thus in an *n*-type semiconductor any current is due predominantly to electrons, whereas in *p*-type material it is due to holes.

Figure 4.19 shows an experiment that can distinguish between *n*-type and *p*-type semiconductors. Two ends of the material are connected to battery terminals to produce a current, and we also apply a magnetic field **B** at right angles to the current. The magnetic field exerts a force $q\mathbf{v} \times \mathbf{B}$ on particles of charge q moving with velocity **v**. Regardless of the sign of q, this magnetic force is upward for the arrangement shown (Problem 4.6). Therefore the top will become positively charged or negatively charged, depending on the sign of the charge carriers. This displacement of charge creates an electric force on the charge carriers, and this electric force opposes the magnetic force. This is called the *Hall effect*. In equilibrium these vertical electric and magnetic forces exactly cancel each other and the current flows horizontally. By measuring with a voltmeter the potential difference between the top and bottom of the sample, we can determine whether the charge carriers are positive or negative, and therefore whether the semiconductor is *p*-type or *n*-type.

Actually some metals, such as beryllium, also exhibit an "anomalous Hall effect" in which the dominant charge carriers are positive. This is because beryllium has a filled $2s$ subshell in which the holes happen to be much more mobile than the $2p$ electrons. The important point, again, is that it is very convenient to think in terms of electrons and holes, even though the real charge carriers are of course the electrons.

• The extra electron indicated in Figure 4.17 is not actually free but is attracted by the positively charged impurity ion. The electron–ion system is thus analogous to the hydrogen atom. It might be expected, therefore, that there are electron orbits about the ion with allowed energies given by the Bohr formula (4.2.11). In particular, the energy of the lowest-energy state should be

$$E_1 = -\left(\frac{1}{4\pi\epsilon_0}\right)^2 \frac{me^4}{2\hbar^2} \qquad (4.9.1)$$

and an energy $|E_1| = 13.6$ eV should be required to free the extra electron and make it available for conduction.

This argument overestimates the binding energy of the extra electron for two reasons. First, the free-space permittivity ϵ_0 in (4.9.1) should be replaced by the material dielectric constant ϵ. Second, it turns out that the band theory ascribes to electrons (and holes) a certain *effective mass* m^*; because the electron is in the periodic potential of the crystal and not in free space, it acts *as if* its mass were smaller, say m^*. The value of m^* depends on the energy of the electron within an energy band, and can also vary with direction within the crystal (Section 5.5). Thus we should replace ϵ_0 by ϵ and m by m^* in (4.9.1):

$$E_1 = - \left(\frac{\epsilon_0}{\epsilon}\right)^2 \left(\frac{m^*}{m}\right) \left(\frac{1}{4\pi\epsilon_0}\right)^2 \frac{me^4}{2\hbar^2}$$

$$= - \left(\frac{\epsilon_0}{\epsilon}\right)^2 \left(\frac{m^*}{m}\right) (13.6 \text{ eV}) \tag{4.9.2}$$

Using the values $\epsilon = 11.8\,\epsilon_0$ and $m^* = 0.26\,m$ for silicon, we obtain

$$E_1 = -0.025 \text{ eV} \tag{4.9.3}$$

Therefore the extra electron is in fact bound very weakly, requiring only an energy of about 0.025 eV to put it into the conduction band. This energy is so small that thermal excitation is enough to free some extra electrons in *n*-type doped silicon.

The Bohr levels of the extra electrons represent new energy levels not found in the pure semiconductor. These levels are called *donor levels*. Because they are small and negative, the donor levels lie just below the bottom of the conduction band.

The hole produced by an acceptor impurity as in Figure 4.18 is likewise bound, and it also requires only a small amount of energy to be freed. Its Bohr energy levels, called *acceptor levels*, lie just above the top of the valence band.

In summary, the doping of a semiconductor with a donor or acceptor does not by itself produce conduction-band electrons or valence-band holes, as assumed in our discussion based on Figures 4.17 and 4.18. However, the energy required to "ionize" these donors or acceptors is so small that thermal excitation at moderate temperatures will do the job. •

Another important semiconductor, germanium, is similar to silicon in that it is tetravalent. It may therefore be doped in the same ways to produce *n*-type and *p*-type materials. In addition to such elemental semiconductors are "III–V" binary semiconductors like gallium arsenide or indium antimonide, in which trivalent and pentavalent atoms share in covalent bonding as a result of unfilled *sp* subshells (Problem 4.7).

4.10   SEMICONDUCTOR JUNCTIONS

Semiconductors of either *n* type or *p* type are not by themselves very useful. They are, after all, just second-class conductors. Their great utility is realized only when they are brought together to form junctions. All of semiconductor technology is

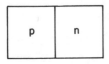

Figure 4.20 A *pn* junction is formed by joining *p*-type and *n*-type semiconductor materials.

based ultimately on the properties of the *pn* junction, the joining of an *n*-type material to a *p*-type material (Figure 4.20).

We will imagine the junction to be abrupt, although of course it cannot be so on the atomic scale. Junctions can be made in practice from a single crystal, one region of which has been made *p*-type whereas an adjoining region has been made *n*-type. The boundary between the two regions can be made very narrow, typically less than a micron (1 micrometer = 10^{-4} cm), and the sharp-boundary idealization is therefore a reasonable approximation.

In a *p*-type material the negative acceptor ions are fixed in position and the holes are mobile, whereas in an *n*-type material the positive donor ions are fixed and the electrons are mobile. Figure 4.21 illustrates what happens at a *pn* junction. Electrons from the *n* side are attracted by the positive holes at the boundary and drift over into the *p* side, while holes on the *p* side are attracted by electrons and drift across the boundary to the *n* side. When an electron meets a hole it falls into it, becoming part of a covalent bond. This diffusion and annihilation of mobile charge carriers thus produces a *depletion region* at the boundary between the *p*- and *n*-type materials. The depletion region is short of both electrons and holes, consisting mainly of negative acceptor ions on the *p* side and positive donor ions on the *n* side (Figure 4.21). Because of this charge separation, there is a static electric field pointing from the *n* side to the *p* side. This field opposes further diffusion of electrons and holes, and thus keeps the depletion region confined to a narrow layer at the boundary. In other words, there is some voltage drop, which we call V_0, in going from the *n* side to the *p* side (Figure 4.22).

In thermal equilibrium there is a diffusion of individual electrons and holes across the junction, but no net flow of current. Holes on the *n* side, for instance,

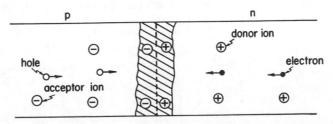

Figure 4.21 At a *pn* junction, electrons from the *n* side are attracted to the *p* side, and holes from the *p* side are attracted to the *n* side. When electron–hole pairs meet they are "annihilated" as the electron becomes part of a covalent bond. This results in a depletion region at the boundary, which is a region in which there are very few electrons or holes.

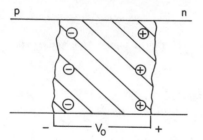

Figure 4.22 The charge separation due to donor and acceptor ions in the depletion layer of a *pn* junction results in a potential difference V_0.

have no difficulty dropping down the potential-energy hill and going over to the *p* side. Holes on the *p* side, on the other hand, have to cross the potential-energy barrier eV_0 to get to the *n* side (here e is understood to be positive). According to statistical mechanics the fraction of holes able to cross the barrier is given by the Boltzmann factor $\exp(-eV_0/kT)$, where k is Boltzmann's constant ($k \approx 1.38 \times 10^{-23}$ J/K $\approx 1/40$ eV/300K) and T is the absolute temperature. If the current due to holes diffusing from the *n* side to the *p* side is to be exactly balanced by the hole current in the opposite direction, therefore, we must have

$$N_p(n \text{ side}) = N_p(p \text{ side}) \, e^{-eV_0/kT} \qquad (4.10.1)$$

where N_p denotes the number of holes per unit volume. Equation (4.10.1) shows, as expected, that there is a greater density of holes on the *p* side than on the *n* side. The same reasoning leads to the relation

$$N_n(p \text{ side}) = N_n(n \text{ side}) \, e^{-eV_0/kT} \qquad (4.10.2)$$

for the number density N_n of electrons. According to these relations, the product $N_n N_p$ is the same for the two sides of the junction.

Now suppose that the *p* and *n* sides are connected to the terminals of a battery. If the *p* side is connected to the positive terminal and the *n* side to the negative terminal, the junction is said to be *forward-biased* (Figure 4.23a). Since the de-

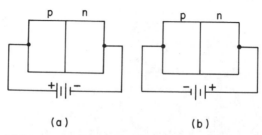

(a) (b)

Figure 4.23 Forward-biased (*a*) and reverse-biased (*b*) *pn* junctions.

pletion layer is much more resistive to current than the bulk regions, most of the applied voltage V is dropped across the depletion layer. In other words, the applied voltage has the effect of lowering the barrier voltage V_0 to $V_0 - V$. We will assume for the present that V is small compared with V_0.

This lowering of the barrier potential by a forward-biased voltage does not affect very much the diffusion current of holes from the n side to the p side, because with or without it they have no potential-energy barrier to cross. Their diffusion current therefore is unaffected by the applied voltage. The hole current from the p side to the n side, however, will increase with the applied voltage, because now a fraction $e^{-(V_0-V)/kT}$ of them are able to cross over. For the net current of holes diffusing from the p side to the n side we therefore have

$$I_p \propto N_p(p \text{ side}) \, e^{-e(V_0-V)/kT} - N_p(n \text{ side})$$

$$= N_p(n \text{ side}) \, e^{eV/kT} - N_p(n \text{ side})$$

$$= N_p(n \text{ side}) \, (e^{eV/kT} - 1) \qquad (4.10.3)$$

where in the second line we have used (4.10.1). In other words, we have the relation

$$I_p = I_{p0} \, (e^{eV/kT} - 1) \qquad (4.10.4)$$

for the net hole current flowing from the p side to the n side under forward bias, where I_{p0} is the hole diffusion current from the n side to the p side. A similar expression is obtained for the net electron current under forward biasing:

$$I_n = I_{n0}(e^{eV/kT} - 1) \qquad (4.10.5)$$

where I_{n0} is the electron diffusion current flowing from the p side to the n side. The total current flowing from the p side to the n side of a forward-biased pn junction is therefore

$$I = I_p + I_n = I_0(e^{eV/kT} - 1) \qquad \text{(forward bias)} \qquad (4.10.6)$$

where $I_0 = I_{p0} + I_{n0}$ is called the *saturation current* of the junction.

Suppose instead that the leads are reversed, as in Figure 4.23b. In this case the junction is said to be *reverse-biased*. The barrier potential is now increased from V_0 to $V_0 + V$, and the net current is obtained simply by changing V to $-V$ in (4.10.6):

$$I = I_0(e^{-eV/kT} - 1) \qquad \text{(reverse bias)} \qquad (4.10.7)$$

Equations (4.10.6) and (4.10.7) give the so-called current–voltage (IV) characteristics of an ideal pn junction (Figure 4.24). Under forward biasing the current

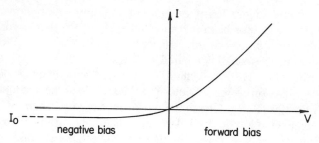

Figure 4.24 Current–voltage characteristics of an ideal *pn* junction.

increases rapidly (exponentially) with increasing voltage. Under reverse bias the current *saturates* with increasing voltage to the value I_0. Typical barrier voltages V_0 in silicon and germanium diodes are roughly on the order of half a volt. Saturation current densities are extremely small, perhaps less than 10^{-10} A/cm^2; for a typical junction area of 10^{-2} cm^2, this amounts to a saturation current I_0 of less than 10^{-12} A. Formulas can be derived to estimate V_0 and I_0 as a function of carrier concentrations and other parameters, but we will not take the time to do so.

The key property of a *pn* junction, therefore, is that it will conduct current in one direction but not the other. That is, it can act as a diode, a sort of automatic switch that closes a circuit when voltage is applied in a forward sense, but blocks the flow of current otherwise. This diode can be used as a rectifier, converting ac current to dc current.

• Real semiconductor diodes do not display exactly the same *IV* characteristics as the idealized diode we have considered. For one thing, we have ignored electron–hole recombination within the depletion layer. This and other effects may be taken into account by replacing (4.10.6) by

$$I = I_0(e^{eV/\beta kT} - 1) \qquad (4.10.8)$$

where the "ideality factor" β is a dimensionless parameter between 1 and 2, depending on T.

Furthermore, at large reverse-bias voltages a real diode no longer blocks the flow of current. At a certain "breakdown" voltage there is a sudden jump in the reverse current. One reason for this is that high-energy charge carriers colliding with atoms in the crystal lattice can ionize them, producing more charge carriers, which lead to further ionization and therefore increasing the current. This is called *avalanche breakdown*. Another mechanism for reverse-current generation is the *Zener effect*, in which electrons undergo a quantum-mechanical "tunneling" from the *p* side to the *n* side. •

An important example of a *pn* junction is the *light-emitting diode*, or *LED*. Because of its similarity in principle to the diode laser, we will use the remainder of this section to introduce the LED.

In our discussion of electrons and holes we have mentioned several times that an electron can "fall into" a hole, meaning that the electron can replace the missing electron represented by the hole. In doing so the electron becomes part of a covalent bond. This simultaneous annihilation of an electron and a hole is called *recombination*.

Now if an electron from the conduction band recombines with a hole in the valence band, it loses energy, having been free ($E > 0$) and then becoming bound ($E < 0$). There are two ways in which this energy can be discarded by the electron. One way is for it to appear as heat in the form of vibrations of the crystal atoms about their equilibrium positions. Another way is *radiative recombination*, in which the electron transition is accompanied by the emission of a photon (Figure 4.25). This is analogous to (spontaneous) emission by an atom. The emitted photon has a frequency ν satisfying $h\nu = E_f - E_i$, where E_i and E_f are the energies of the initial and final states, respectively. In a transition from the conduction band to the valence band, the minimum value of $E_f - E_i$ is clearly the gap energy E_g. The maximum photon wavelength in interband radiative recombination is in turn given by

$$\lambda_{max} = \frac{c}{\nu_{min}} = \frac{hc}{E_g} \tag{4.10.9}$$

For silicon, therefore, with a gap energy of 1.12 eV, $\lambda_{max} \cong 11,000$ Å. For germanium, with $E_g = 0.67$ eV, $\lambda_{max} \cong 19,000$ Å.

The light from an LED is produced by radiative recombination of electrons and holes injected across the junction of a forward-biased *pn* diode. The electrons drifting from the *n* side to the *p* side recombine radiatively with holes, and the holes drifting from the *p* side to the *n* side recombine radiatively with electrons. On either side, of course, the emitted photons are produced when an electron makes a downward transition from a state of energy E_f to one of energy E_i. Not every recombination of an electron with a hole will be radiative, because there are competing nonradiative recombinations in which the energy lost by the electron appears in the form of crystal lattice vibrations. Although there are LEDs in which nearly every recombination process is radiative, other factors come into play to

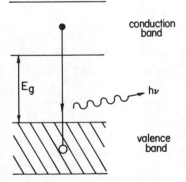

Figure 4.25 Radiative recombination process in which the electron–hole annihilation is accompanied by the emission of a photon.

substantially limit the actual device efficiency. One of the more important of these factors is discussed below.

Interband recombination radiation has a distribution of wavelengths associated with the thermal distribution of electron energy within the conduction band. The maximum wavelength λ_{max} given by (4.10.9), however, provides a good estimate of the peak wavelength. Thus we can deduce from it that LEDs made from Si or Ge junctions will not generate much visible radiation.

Actually there are other types of radiative recombination that produce longer wavelengths than the interband maximum λ_{max}. As noted in the preceding section, there are donor levels and acceptor levels associated with the impurities of a doped semiconductor, and these levels lie just below the bottom of the conduction band and just above the top of the valence band, respectively. Radiative recombination processes involving these impurity levels produce radiation of wavelengths $\lambda > \lambda_{max}$, as is clear from Figure 4.26. In part (a) of the figure we indicate an interband radiative recombination transition, i.e., a transition of an electron from the conduction band to the valence band. Part (b) shows a transition from a donor level to the valence band, while (c) shows a transition from the conduction band to an acceptor level. Finally we show in (d) a transition from a donor level to an acceptor level. Processes (b)–(d) obviously lead to wavelengths greater than the interband process (a), and so LED wavelengths are often *greater* than the interband maximum (4.10.9). Because the differences $E_c - E_d$ and $E_a - E_v$ are small compared with E_g, however, (4.10.9) provides a good estimate of the sort of wavelength that can be expected of a given semiconductor.

The question of wavelength is obviously an important one if an LED is to be used for visual display purposes. Silicon and germanium, for instance, are eminently useful electronically because of the relative ease with which they can be doped and fabricated as diodes, but their band gaps are too small to make them useful as LEDs for visible radiation. (Moreover, Si and Ge are radiatively too inefficient to be used in LEDs. This is because they are *indirect-band-gap* semiconductors, for which the interband radiative recombination rates are very low. GaAs, by contrast, is a *direct-band-gap* semiconductor, and is consequently a much more efficient radiator. *Indirect-band-gap* materials can be used in LEDs if there are efficient radiative pathways in addition to interband recombination.)

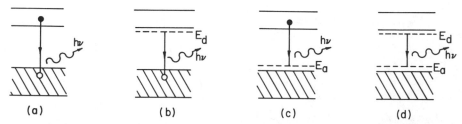

Figure 4.26 Radiative recombination involving a transition from (a) the conduction band to the valence band, (b) a donor level to the valence band, (c) the conduction band to an acceptor level, and (d) a donor level to an acceptor level.

In addition to having a band gap large enough to produce visible radiation, a semiconductor to be used in an LED must of course have both p-type and n-type forms that can be made by suitable doping. As a rule of thumb, large-gap materials tend to have high melting points, making doping of a melt more difficult, and furthermore they tend to have low conductivities even when doped. Among the more commonly used LED materials is gallium arsenide (GaAs), with a band gap of 1.44 eV (and therefore $\lambda_{max} \approx 8610$ Å). Depending on the dopant, the dominant radiative-recombination transition may be interband (Figure 4.26a) or from the conduction band to an acceptor level (Figure 4.26c).

Even if every charge carrier injected across the junction gave rise to an emitted photon, the efficiency of an LED would still be much less than 100%. An important reason for this is a phenomenon well known in classical optics: total internal reflection. For a quick review of this effect, recall that the refraction of light at an interface of two media is governed by Snell's law:

$$n_1 \sin \theta_1 = n_2 \sin \theta_2 \qquad (4.10.10)$$

Here n_1 and n_2 are the refractive indices on the two sides of the interface, and θ_1 and θ_2 are the corresponding angles of incidence, as in Figure 4.27a. Now if $n_1 > n_2$, it is possible for light propagating from medium 1 to medium 2 to be reflected back into medium 1 instead of penetrating the interface and going into medium 2. This total internal reflection occurs at angles of incidence θ_1 greater than the critical value θ_c for which $\theta_2 = 90°$:

$$n_1 \sin \theta_c = n_2 \sin 90° = n_2$$

or

$$\theta_c = \sin^{-1}\left(\frac{n_2}{n_1}\right) \qquad (4.10.11)$$

This is illustrated in Figure 4.27b.

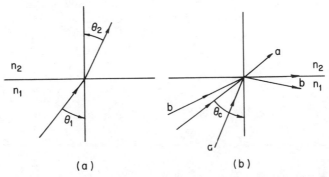

Figure 4.27 (a) Geometry for Snell's law if $n_1 < n_2$. (b) Total internal reflection occurs at the interface if $n_1 > n_2$ and the angle of incidence exceeds the critical angle θ_c given by (4.10.11).

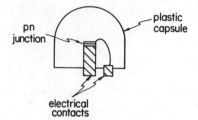

Figure 4.28 Total internal reflection in an LED is minimized by surrounding the emitting junction with a plastic material of higher refractive index than the emitting material, and shaping the plastic enclosure into the form of a hemispherical dome to reduce total internal reflection at the plastic–air interface.

In an LED, any light emerging from either the *p* side or the *n* side, propagating into air, is passing from a medium of higher index to a medium of lower index. This means that light approaching the LED–air interface at an angle greater than the critical angle θ_c will be reflected back into the LED instead of emerging as useful output radiation. In fact the refractive indices of LED materials are often quite large, making the critical angle for total internal reflection rather small. In GaAs, for instance, $n \approx 3.6$, so that $\theta_c \approx 16°$ for the GaAs–air interface.

The deleterious effect of total internal reflection is minimized in the common LED design shown in Fig. 4.28. The junction is enclosed in a plastic case of refractive index $n \approx 1.5$. This reduces the effect of total internal reflection at the emitting surface, because the critical angle (4.10.11) is increased over the value approximate to a diode-air interface. Of course there is still total internal reflection at the plastic–air interface, but this is minimized by shaping the plastic into the form of a hemispherical dome. With this geometry, most of the light rays at the plastic–air interface have angles of incidence less than the critical angle for total internal reflection. Of course an alternative design is to simply make the emitting LED surface dome-shaped, but this is more expensive than the plastic enclosure.

Light-emitting diodes have replaced incandescent lamps in applications demanding compactness, low power consumption, and a high degree of reliability. For such purposes they are used either singly or in arrays. In the latter case, a pattern or message is conveyed when certain of the LEDs are switched on. A familiar example is the digital display used in things like clock radios and calculators. These commonly employ the seven-segment display shown in Fig. 4.29, in which each segment is an individual LED. The numerals 0 through 9 are displayed by turning on only certain of these LEDs at a time.

Figure 4.29 Seven-segment display format used with LEDs and LCDs. The ten digits 0 through 9 may be displayed by lighting selected segments.

• In applications in which only a very small amount of power from a small battery is available, such as in digital wrist watches and many pocket calculators, the *liquid-crystal display*, or LCD, is used instead of the LED. LCDs consume less power because they do not generate any light of their own, but use ambient light. Their operation is based on the properties of certain organic liquids of rod-shaped molecules. The molecules can take on certain organized relative alignments (whence the term liquid "crystal"), in such a way that the polarization of an incident light wave is rotated by 90° in passing through the LCD cell. The cell is a liquid-crystal layer sandwiched between two clear plates whose inner surfaces are coated with a transparent conducting material arranged in a certain pattern. When a voltage is applied between the plates, the molecular alignment is altered and the polarization of incident light is no longer rotated by 90°. By using orthogonally oriented polarizing sheets in front of and behind the cell, and a mirror at the back, we can arrange for incident light to be reflected when there is no applied voltage, but for no light to be reflected from those areas where there is an applied voltage. Then we see the black-and-white alphanumeric display patterns that have become so common in recent years. •

The transistor is the most important application of semiconductor junctions, consisting basically of two adjacent *pn* junctions (*pnp* or *npn*) such that the middle layer is very thin. The operation of a transistor may be understood within the electron–hole framework we have used to discuss LEDs. Because they are not as directly relevant to laser technology as LEDs, however, we will not take the time here for such a discussion.[7]

4.11 SUMMARY

In this chapter we have introduced some aspects of atoms, molecules, and solids that will be important for the remaining chapters. The most important aspect is the restriction of internal energies to a fixed set of allowed values.

We discussed the Bohr model of the hydrogen atom for three reasons. First, it was the first view of an atom that incorporated any quantum-mechanical features (the postulated discrete values of orbital angular momentum). Second, it is still the model that most scientists and engineers use to *think* about atoms, although the mathematical machinery of quantum mechanics is needed to *calculate* about atoms. And third, its main results were correct. That is, it gives the right expression (4.2.11) for the energy levels of hydrogen, and Bohr's interpretation of atomic spectral lines on the basis of electron jumps between these levels was the key insight showing that atoms are not classical objects.

Obviously hydrogen is atypical in many ways. After more than half a century of quantum mechanics, hydrogen is still the only element whose exact energy levels can be written down explicitly. Nevertheless, the results of the quantum theory of the hydrogen atom are useful in understanding the structure of other atoms. We have seen, for instance, how these results, together with the Pauli

7. See, for instance, R. P. Feynman, R. B. Leighton, and M. Sands, *The Feynman Lectures on Physics* (Addison-Wesley, Reading, Mass., 1965), Volume III, Section 14–6.

principle and the independent-particle approximation, explain the chemical regularities in the periodic table of the elements.

In much of laser physics it is sufficient to regard any atom as simply a "black box" for electrons, with the special property that the electrons inside can only be in certain energy slots. We can adopt a similar view for the electronic structure of molecules.

Of course we cannot ignore the fact that molecules can also vibrate and rotate. However, the most important vibrational and rotational characteristics of molecules are, for us, very similar to their electronic characteristics. First and foremost among these similarities, of course, is the restriction of the vibrational and rotational energies to a fixed set of allowed values. Just as an electron can jump to a higher (or lower) energy level with the absorption (or emission) of a photon, so too can the molecule as a whole "jump" to a different vibrational or rotational state with the simultaneous absorption (or emission) of a photon. In fact the electronic, vibrational, and rotational states of a molecule can *all* change as the molecule absorbs or emits a photon. Molecular spectra are more complicated than atomic spectra, but this simply means that the "black box" we call a molecule is more complicated on the inside than an atom.

The properties of solids are determined to a large extent by the outer electron orbitals of its constituent atoms or molecules. In crystalline solids the allowed electron energies are spread into energy bands as a consequence of the orderly and periodic arrangement of atoms in a crystal lattice. The concept of energy bands provides a satisfactory interpretation of insulators, conductors, and semiconductors.

Semiconductor junctions are an especially interesting and important application of the quantum mechanics (band theory) of solids. In particular, the concept of a missing electron or hole, as a sort of particle in its own right, greatly facilitates our understanding of semiconductor junctions. The basic *pn* junction acts as a diode, passing a current when it is forward-biased but not when it is reverse-biased. Light-emitting diodes are important not only in alphanumeric displays, but also as the gain cells of diode lasers.

The existence of atomic and subatomic particles, as the basic building blocks of matter in all its forms, is arguably the most basic and significant discovery of post-Newtonian science. A strong argument can also be made that the most far-reaching technological developments of the past half century have involved the controlled manipulation of quantum states of these particles. Among these developments are nuclear power sources and transistor-based computer technology. The laser is another example. In this case populations of excited atomic and molecular states are created and controlled to generate light.

PROBLEMS

4.1   **(a)** Equation (4.2.18) with $n' = 2$ and $n = 3, 4, 5, \cdots$ gives the *Balmer series* of the hydrogen spectrum. In what region of the electromagnetic spectrum (e.g., infrared, visible, ultraviolet) are the wavelengths of the Balmer series?

(b) Equation (4.2.18) with $n' = 1$ and $n = 2, 3, 4, \cdots$ gives the *Lyman series* of hydrogen. In what region of the spectrum are the wavelengths of the Lyman series?

(c) Equation (4.2.18) with $n' = 3$ and $n = 4, 5, 6, \cdots$ gives the *Paschen series* of hydrogen. In what region of the spectrum are these wavelengths?

4.2 Assuming the result (4.2.22) is of the correct order of magnitude for an electron in any atom, justify the approximation of neglecting the magnetic force in the electron oscillator model of Chapters 2 and 3.

4.3 **(a)** Consider a head-on collision of two ground-state hydrogen atoms, one at rest and the other having a kinetic energy of 13 eV. On the basis of the conservation of energy and linear momentum, show that the atoms must still be in their ground states after collision.

(b) Show that the moving atom in part (a) must actually have a kinetic energy 50% greater than the ionization energy of hydrogen in order that one of the atoms might be excited above the ground state after collison.

4.4 Verify Eq. (4.5.7).

4.5 Given the fact that the molecular weight of water is 18, estimate the average distance between two water molecules in ice.

4.6 Show that the magnetic force acting on the charge carriers in the Hall-effect experiment of Figure 4.19 is upward, regardless of whether the charges are positive or negative.

4.7 Assuming for GaAs a dielectric constant $\epsilon = 13.0 \, \epsilon_0$ and an effective mass $m^* = 0.07 \, m$, estimate the energy required to ionize donor impurities.

5 THE SCHRÖDINGER EQUATION

5.1 INTRODUCTION

We emphasized in the preceding chapter that the energies of bound states of electrons in atoms are restricted to a discrete set of allowed values. This restriction is an important feature of any quantum-mechanical discussion of atoms. However, it does not by itself capture the only difference between classical and quantum physics. Quantum theory deals with physical reality from a statistical viewpoint. This means that quantum theory predicts only probabilities, average values, correlations, and related statistical concepts. These must be interpreted appropriately, as we will show.

5.2 THE WAVE FUNCTION

In classical physics the concepts of wave and particle are very different. In 1923 Louis de Broglie opened the way for quantum mechanics when he suggested that a particle can be "associated with" a wave. On the basis of some complicated and speculative reasoning, he argued that for a particle moving with linear momentum $\mathbf{p}$, the associated wave should have wavelength λ given by

$$\lambda = h/|\mathbf{p}| \tag{5.2.1}$$

where h is Planck's constant. For a while de Broglie's hypothesis did not attract much attention because there was no experimental evidence of wavelike behavior of particles.

However, in 1925 Erwin Schrödinger put forth an equation for a wave $\Psi(\mathbf{r}, t)$ associated with a particle. He applied this equation to an electron in a Coulomb potential, and derived the correct energy levels of the hydrogen atom. The Schrödinger equation for the wave function is

$$i\hbar \frac{\partial \Psi}{\partial t} = H\Psi \tag{5.2.2}$$

where H is the *Hamiltonian* function or, more loosely, the energy function of the particle. The Hamiltonian is expressed in terms of the particle coordinates (x, y,

z) and momenta (p_x, p_y, p_z). For instance, for a particle of mass m moving in the x direction and in the absence of any forces, the kinetic energy is $\frac{1}{2}mv_x^2$. However, since $mv_x = p_x$, the Hamiltonian expression for the kinetic energy is

$$H = p_x^2/2m$$

and (5.2.2) becomes

$$ih\frac{\partial \Psi(x,t)}{\partial t} = \frac{p_x^2}{2m}\Psi(x,t) \tag{5.2.3}$$

The difference between the classical Hamiltonian or energy, and the Hamiltonian $p_x^2/2m$ appearing in the Schrödinger equation (5.2.3), is that in the latter H is interpreted as an *operator* acting on Ψ. Specifically, for one-dimensional motion the momentum p_x appearing in (5.2.3) is interpreted in quantum mechanics as the combination of differentiation with respect to the coordinate x and multiplication by $\hbar/i$:

$$p_x \rightarrow \frac{\hbar}{i}\frac{\partial}{\partial x} \equiv \mathsf{p}_x \tag{5.2.4}$$

Since the momentum now has a nonclassical attribute (differentiation), we signal this with a sans serif p. In (5.2.3) and (5.2.2) we require the square of p_x, which in quantum mechanics means applying the p operation twice:

$$p_x^2 \rightarrow \mathsf{p}_x^2 = \left(\frac{\hbar}{i}\frac{\partial}{\partial x}\right)^2 = -\hbar^2\frac{\partial^2}{\partial x^2} \tag{5.2.5}$$

Thus the Schrödinger equation (5.2.2) for a free particle of mass m moving along the x axis is

$$i\hbar\frac{\partial \Psi}{\partial t} = -\frac{\hbar^2}{2m}\frac{\partial^2\Psi}{\partial x^2} \tag{5.2.6}$$

The generalization of (5.2.4) to three dimensions is

$$\mathsf{p} \rightarrow \frac{\hbar}{i}\left(\hat{\mathbf{x}}\frac{\partial}{\partial x} + \hat{\mathbf{y}}\frac{\partial}{\partial y} + \hat{\mathbf{z}}\frac{\partial}{\partial z}\right) = \frac{\hbar}{i}\nabla \tag{5.2.7}$$

Similarly, (5.2.5) generalizes to

$$\mathsf{p}^2 = \frac{\hbar}{i}\nabla \cdot \frac{\hbar}{i}\nabla = -\hbar^2\nabla^2 \tag{5.2.8}$$

and the Schrödinger equation for a free particle in three dimensions is

$$ i\hbar \, \frac{\partial \Psi}{\partial t} = -\frac{\hbar^2}{2m} \nabla^2 \Psi \tag{5.2.9}$$

It is important to realize that quantum mechanics and the Schrödinger equation did not follow deductively from other considerations. Like Newtonian mechanics, quantum mechanics is simply a compact embodiment of laws of nature discovered over many years in a wide variety of observations and experiments. Quantum mechanics can be said to represent a deeper understanding of these laws than Newtonian mechanics because it contains all Newtonian mechanics as a well-defined limiting case (roughly speaking, the limit of macroscopic masses, energies, and angular momenta, where $\hbar$ is negligible).

Another important point is that $\Psi(\mathbf{r}, t)$ is a *scalar* wave. That is, at each point $\mathbf{r}$ in space and at any time t, $\Psi(\mathbf{r}, t)$ is simply a number. However, Ψ is a *complex* function of $\mathbf{r}$ and t, which means that it does not correspond directly to any physical quantity, i.e., to anything we can measure in terms of real numbers.

In order to establish a connection between the Schrödinger equation and the de Broglie wave hypothesis, let us try to satisfy (5.2.9) by assuming a solution that is clearly wavelike:

$$ \Psi(\mathbf{r}, t) = A \, e^{i(\mathbf{k} \cdot \mathbf{r} - \omega t)} \tag{5.2.10}$$

where A is some fixed complex number and where the wave vector $\mathbf{k}$ and frequency ω must be interpreted in a way appropriate to the particle of mass m described by (5.2.9). Using this guess in (5.2.9), we obtain the equation

$$ i\hbar(-i\omega) A \, e^{i(\mathbf{k} \cdot \mathbf{r} - \omega t)} = -\frac{\hbar^2}{2m} (-k^2) A \, e^{i(\mathbf{k} \cdot \mathbf{r} - \omega t)} \tag{5.2.11}$$

or the following dispersion relation for the particle wave:

$$ \hbar\omega = \frac{\hbar^2 k^2}{2m} \quad \text{or} \quad k = \frac{\sqrt{2m\omega}}{\hbar} \tag{5.2.12}$$

Recalling that a monochromatic electromagnetic wave of angular frequency ω is associated with photons of energy $\hbar\omega$, we suppose by analogy that the wave function (5.2.10) is associated with the motion of a particle of energy

$$ E = \hbar\omega \tag{5.2.13}$$

But since $E = p^2/2m$ for a classical free particle of mass m and linear momentum $\mathbf{p}$, we evidently have from (5.2.12) the relation

$$\frac{p^2}{2m} = \frac{\hbar^2 k^2}{2m} \tag{5.2.14}$$

or

$$|\mathbf{p}| = \hbar |\mathbf{k}| \tag{5.2.15}$$

and since the wavelength of the plane wave (5.2.10) is given by

$$\lambda = 2\pi / |\mathbf{k}| \tag{5.2.16}$$

it follows from (5.2.15) that

$$\lambda = \frac{2\pi\hbar}{|\mathbf{p}|} = \frac{h}{|\mathbf{p}|} \tag{5.2.17}$$

which is just the de Broglie relation (5.2.1).

This "derivation" of the de Broglie relation shows that (5.2.1) and (5.2.10) are consistent. Furthermore, in 1926 the experiments of C. Davisson and L. H. Germer revealed that electrons can be diffracted by crystal lattices in much the same way that X-rays are. And a year later G. P. Thomson demonstrated the diffraction of electrons[1] by thin films and provided added support for the de Broglie relation (5.2.1). Still, this evidence does not at all indicate the nature of the "association" of the wave function Ψ with the particle. In fact quantum mechanics was successfully applied for several years before it was accepted that Ψ has a purely statistical significance, according to the 1926 proposal of Max Born.

According to the Born interpretation, the quantity $|\Psi(\mathbf{r}, t)|^2 \, d^3r$ is the *probability* that the particle will be found within the infinitesimal volume element $d^3r = dx \, dy \, dz$ centered at point $\mathbf{r}$. Loosely speaking, Ψ provides us with a way to say where the particle is *likely* to be found. Since the particle associated with Ψ is somewhere, its total probability must be unity, meaning that the wave function must satisfy the "normalization" condition

$$\int_{\text{all space}} |\Psi(\mathbf{r}, t)|^2 \, d^3r = 1 \tag{5.2.18}$$

Furthermore we can easily write expressions for any number of related statistical properties. For example, the probability that the particle will be found within a

---

1. Recent beautiful experiments of H. Rauch demonstrate many wave aspects of neutrons. It has been accepted for more than 50 years now that all particles have wave properties.

given volume V is

$$\int_V \left| \Psi(\mathbf{r}, t) \right|^2 d^3r \le 1 \qquad (5.2.19)$$

Obviously, given (5.2.18), and since V is only a finite region of space, the integral (5.2.19) lies between 0 and 1.

We write the average or expected value, or "expectation value" as it is called in quantum theory, with angular brackets. Then, according to the probability interpretation of $|\Psi|^2$, the average position is given by

$$\langle \mathbf{r} \rangle = \int_{\text{all space}} \mathbf{r} \left| \Psi(\mathbf{r}, t) \right|^2 d^3r$$

$$= \int_{\text{all space}} \Psi^*(\mathbf{r}, t) \, \mathbf{r} \, \Psi(\mathbf{r}, t) \, d^3r \qquad (5.2.20)$$

Thus if $|\Psi(\mathbf{r}, t)|^2$ is sharply peaked at the particular point in space $\mathbf{r}_0$, then the expectation value of $\mathbf{r}$ is close to $\mathbf{r}_0$. It is not always the case that the probability distribution is in fact sharply peaked. One therefore must consider the dispersion of values of $\mathbf{r}$, that is, the amount by which $\mathbf{r}$ can be expected to deviate from its expectation value $\langle \mathbf{r} \rangle$. There are various ways to measure this deviation from the mean. The conventional way is to use the average (i.e., expectation value) of the *square* of $\mathbf{r} - \langle \mathbf{r} \rangle$. (The average of $\mathbf{r} - \langle \mathbf{r} \rangle$ itself is useless because it is identically zero.) This is called the *variance* of $\mathbf{r}$, and is denoted $\langle \Delta r^2 \rangle$:

$$\begin{aligned} \langle \Delta r^2 \rangle &= \langle (\mathbf{r} - \langle \mathbf{r} \rangle)^2 \rangle \\ &= \langle r^2 - 2\mathbf{r} \cdot \langle \mathbf{r} \rangle + \langle \mathbf{r} \rangle^2 \rangle \\ &= \langle r^2 \rangle - 2\langle \mathbf{r} \rangle^2 + \langle \mathbf{r} \rangle^2 \\ &= \langle r^2 \rangle - \langle \mathbf{r} \rangle^2 \end{aligned} \qquad (5.2.21)$$

This is equivalent to

$$\langle \Delta r^2 \rangle = \langle \Delta x^2 \rangle + \langle \Delta y^2 \rangle + \langle \Delta z^2 \rangle \qquad (5.2.22)$$

where

$$\begin{aligned} \langle \Delta x^2 \rangle &= \langle (x - \langle x \rangle)^2 \rangle \\ &= \langle x^2 \rangle - \langle x \rangle^2 \end{aligned} \qquad (5.2.23)$$

defines the variance of the Cartesian components of $\mathbf{r}$. The variance can also be called the *mean square deviation from the mean*. Thus the rms (root mean square) deviation of x, to be denoted Δx, is just the square root of the variance:

$$(x - \langle x \rangle)_{rms} = \Delta x = \langle \Delta x^2 \rangle^{1/2} \qquad (5.2.24)$$

One should realize that there are many circumstances in which a given quantum-mechanical probability is unity. This is not the same as knowing all about a particle, however. For instance, it is possible for the particle's wave function to be such that $\Delta x = 0$, in which case the particle is certainly located at $x = \langle x \rangle$. It is also possible that $\Delta p_x = 0$. But there is no wave function for which both Δx and Δp_x are simultaneously zero. In fact, it may be shown quite generally that

$$\Delta x \, \Delta p_x \geqq \frac{\hbar}{2} \qquad (5.2.25a)$$

$$\Delta y \, \Delta p_y \geqq \frac{\hbar}{2} \qquad (5.2.25b)$$

$$\Delta z \, \Delta p_z \geqq \frac{\hbar}{2} \qquad (5.2.25c)$$

In other words, we cannot predict exactly *both* members of the "conjugate pairs" (x, p_x), (y, p_y) and (z, p_z). This aspect of quantum-mechanical indeterminism is called the *Heisenberg uncertainty principle* and is discussed in texts on quantum mechanics.

• Einstein never accepted quantum mechanics as a complete theory of physical reality, at least not as he perceived that reality. His remark that "God does not play dice" conveyed his displeasure with indeterminism. Nevertheless, his epic debates with Bohr, especially during the early years of quantum mechanics (1927–1936), contributed greatly to a physical understanding of the theory. It is ironic that some of the other key contributors to quantum mechanics, including Planck and de Broglie, later became dissatisfied with it. But no one has ever devised a better alternative, and quantum mechanics over the years has been so successful in explaining and predicting the results of so many experiments that today its validity is hardly questioned. •

In a book such as this we cannot discuss all the subtle implications of quantum mechanics. For the most part we will only need some results of the Schrödinger equation, typically for problems involving a particle with kinetic energy $T = p^2/2m$ and potential energy V. Then the total energy of the particle is $E = p^2/2m + V$, and so the Hamiltonian in the Schrödinger equation is the differential operator obtained by the replacement $\mathbf{p} \rightarrow \mathbf{p}$, namely

$$H = -\frac{\hbar^2}{2m} \nabla^2 + V \qquad (5.2.26)$$

That is, the Schrödinger equation for the wave function of a particle of mass m in a potential V is

$$ i\hbar \frac{\partial \Psi}{\partial t} = \left[-\frac{\hbar^2}{2m} \nabla^2 + V \right] \Psi \qquad (5.2.27) $$

The problems of interest are of two types, depending on whether V is time-dependent or time-independent. We consider first potentials that do not depend on time.

5.3 STATIONARY STATES

The simplest potential V that is independent of time is $V = 0$. This is the potential for a free particle, and the corresponding Schrödinger equation (5.2.9) has the solution (5.2.10):

$$ \Psi(\mathbf{r}, t) = A\, e^{i(\mathbf{k}\cdot\mathbf{r} - \omega t)} $$
$$ = A\, e^{i\mathbf{k}\cdot\mathbf{r}}\, e^{-iEt/\hbar} \qquad (5.3.1) $$

where $E = \hbar\omega$, as we have already shown. The second form of this solution is appropriate, with some modification, to a much wider class of potentials.

Consider any time-independent potential $V(\mathbf{r})$ that can restrict a particle to bounded orbits. Two examples familiar from classical physics are:

$$ V = \frac{-1}{4\pi\epsilon_0} \frac{e^2}{|\mathbf{r}|} \qquad \text{(Coulomb potential)} \qquad (5.3.2) $$

$$ V = \tfrac{1}{2} kx^2 \qquad \text{(one-dimensional harmonic oscillator)} \qquad (5.3.3) $$

The corresponding Schrödinger equation (5.2.27) has the factored solution

$$ \Psi(\mathbf{r}, t) = \Phi(\mathbf{r})\, e^{-iEt/\hbar} \qquad (5.3.4) $$

as one can easily verify by direct substitution. The only condition is that the space-dependent function $\Phi(\mathbf{r})$ must satisfy the equation

$$ \left(-\frac{\hbar^2}{2m} \nabla^2 + V(\mathbf{r}) \right) \Phi(r) = E\, \Phi(r) \qquad (5.3.5a) $$

which is the same as

$$ H\Phi(\mathbf{r}) = E\, \Phi(\mathbf{r}) \qquad (5.3.5b) $$

Both forms of (5.3.5) are referred to as the *time-independent Schrödinger equation*.

From (5.3.4) it follows that

$$\left|\Psi(\mathbf{r}, t)\right|^2 = \left|\Phi(\mathbf{r})\right|^2 \left|e^{-iEt/\hbar}\right|^2 \tag{5.3.6}$$

and since E, t, and $\hbar$ are represented by real, not complex, numbers, we have $|\exp(-iEt/\hbar)|^2 = 1$, and therefore

$$\int \left|\Phi(\mathbf{r})\right|^2 d^3r = 1 \tag{5.3.7}$$

Thus the probability distribution (5.3.6) is independent of time. For this reason a wave function satisfying (5.3.4) is called a *stationary state* of the system.

A stationary state of the system is a state of definite energy. That is, a measurement of the energy of the system will yield with certainty the value E appearing in (5.3.5). This follows from the fact that the dispersion ΔH is zero. To see this, note first that the expectation value of the energy, when (5.3.4) is satisfied, is just E:

$$
\begin{aligned}
\langle H \rangle &= \int \Psi^*(\mathbf{r}, t)\, H\Psi(\mathbf{r}, t)\, d^3r \\
&= \int \Psi^*(\mathbf{r}, t)\, E\Psi(\mathbf{r}, t)\, d^3r \\
&= E \int \Psi^*(\mathbf{r}, t)\, \Psi(\mathbf{r}, t)\, d^3r = E
\end{aligned}
\tag{5.3.8}
$$

where we have used the normalization condition (5.2.18) and the fact that E is just a number, not an operator, and can be pulled outside the integral. Furthermore

$$\langle H^2 \rangle = \int \Psi^*(\mathbf{r}, t)\, H^2\Psi(\mathbf{r}, t)\, d^3r \tag{5.3.9}$$

and, using (5.3.5b), we have

$$
\begin{aligned}
H^2\Psi &= H(H\Psi) \\
&= H(E\Psi) \\
&= E(H\Psi) \\
&= E^2\Psi
\end{aligned}
\tag{5.3.10}
$$

so that

$$\langle H^2 \rangle = \int \Psi^*(\mathbf{r}, t) \, E^2 \Psi(\mathbf{r}, t) \, d^3r$$

$$= E^2 \int \Psi^*(\mathbf{r}, t) \, \Psi(\mathbf{r}, t) \, d^3r = E^2 \qquad (5.3.11)$$

Therefore

$$\Delta H = (\Delta H)_{\text{rms}} = \left(\langle H^2 \rangle - \langle H \rangle^2 \right)^{1/2}$$

$$= (E^2 - E^2)^{1/2} = 0 \qquad (5.3.12)$$

and our assertion that $\Delta H = 0$ for a stationary state is thus proven. In fact the converse may also be shown to be true: if $\Delta H = 0$ then the system is in a stationary state.

5.4 ATOMIC WAVE FUNCTIONS AND THEIR PROPERTIES

In the case of a particle in a potential $V(\mathbf{r})$, the Hamiltonian is given by Eq. (5.2.26), and the time-independent Schrödinger equation is given in (5.3.5):

$$\left(-\frac{\hbar^2}{2m} \nabla^2 + V(\mathbf{r}) \right) \Phi(\mathbf{r}) = E \, \Phi(\mathbf{r})$$

or

$$\nabla^2 \Phi(\mathbf{r}) + \frac{2m}{\hbar^2} \left[E - V(\mathbf{r}) \right] \Phi(\mathbf{r}) = 0 \qquad (5.4.1)$$

This is a very complicated partial differential equation. However, for any potential energy function $V(\mathbf{r})$, this equation has physically allowable solutions only for a restricted set of values of the real number E. These allowed values of E are just the allowed energies of the particle in the potential $V(\mathbf{r})$. This is how the allowed energies of a system are obtained in quantum mechanics. For instance, if $V(\mathbf{r})$ is the Coulomb potential $(-e^2/4\pi\epsilon_0 r)$, the allowed values of E are given by the Bohr formula (4.2.11), as shown by Schrödinger. Another example is the harmonic-oscillator potential, $V(x) = \frac{1}{2} kx^2$, in which case the allowed energies are given by (4.5.9).

The $\mathbf{r}$ dependences of the functions $\Phi(\mathbf{r})$ show how the quantum system arranges itself in the potential $V(\mathbf{r})$. When we have solved (5.4.1) for all Φ's corresponding to all E's, we can say we have obtained full knowledge of the spatial structure of the quantum system. Multielectron generalizations of (5.4.1) are the subject of continuing study in atomic structure research. The determination of the structure of most atoms remains an extremely difficult and unsolved problem, but the nature of the structure equation itself, Eq. (5.4.1) in our study, allows certain important properties of the functions $\Phi(\mathbf{r})$ to be known even if the explicit form of the Φ's is not known.

Let us label the energies E for which (5.4.1) has a solution by the index n, and arrange the energies from lowest to highest, E_1, E_2, E_3, $\cdots$, E_n, $\cdots$. The solutions for $\Phi(\mathbf{r})$ that correspond to these energies can be listed similarly: $\Phi_1(\mathbf{r})$, $\Phi_2(\mathbf{r})$, $\cdots$, $\Phi_n(\mathbf{r})$ $\cdots$. The set of energies E_n and the set of functions $\Phi_n(\mathbf{r})$ are different for every potential function $V(\mathbf{r})$. Sometimes it is necessary to use a labeling method more complex than a single integer n. The first few members of the sets $\{E_n\}$ and $\{\Phi_n(\mathbf{r})\}$ associated with hydrogen and with a one-dimensional harmonic oscillator, i.e., the sets associated with $V(r) = -e^2/4\pi\epsilon_0 r$ and $V(x) = \frac{1}{2}kx^2$, are listed at the end of the chapter in two appendices.

A remarkable and extremely useful property of these sets $\{\Phi_n(\mathbf{r})\}$ is that they are *complete*. By this one means that any function of $\mathbf{r}$, say $F(\mathbf{r})$, can be written as a sum of the members of one of these sets with appropriate numerical coefficients:

$$F(\mathbf{r}) = \sum_n a_n \Phi_n(\mathbf{r}) \tag{5.4.2}$$

What is equally remarkable is that the various coefficients are unique to $F(\mathbf{r})$. In other words, a set of coefficients $\{a_n\}$ always exists to permit (5.4.2) to be written, and there is only one such set for every function $F(\mathbf{r})$. Clearly, the set of coefficients $\{a_n\}$ is in this way *equivalent* to $F(\mathbf{r})$, since $F(\mathbf{r})$ can be reconstructed from the a_n via (5.4.2) and the functions $\Phi_n(\mathbf{r})$.

The completeness of a set $\{\Phi_n(\mathbf{r})\}$ is similar to the completeness of the set of unit vectors $\mathbf{i}$, $\mathbf{j}$, $\mathbf{k}$. Every vector $\mathbf{A}$ can be resolved into its x, y, and z components and then reconstructed by using the three unit vectors:

$$\mathbf{A} = \mathbf{i}A_x + \mathbf{j}A_y + \mathbf{k}A_z$$
$$= A_x\mathbf{i} + A_y\mathbf{j} + A_z\mathbf{k} \tag{5.4.3}$$

In the second line we have written the reconstruction in reverse to emphasize the similarity to (5.4.2). That is, $\mathbf{i}$, $\mathbf{j}$, and $\mathbf{k}$ contain the vector properties of $\mathbf{A}$ just as the $\Phi_n(\mathbf{r})$ contain the functional properties of $F(\mathbf{r})$. The components A_x, A_y, and A_z merely indicate how much of each vector property is needed to make up $\mathbf{A}$, and the a_n indicate how much of each function $\Phi_n(\mathbf{r})$ is needed to reconstruct the full $F(\mathbf{r})$. Of course every vector $\mathbf{A}$ has a unique x component, so there is only one possible coefficient A_x that multiplies $\mathbf{i}$ in (5.4.3). This makes it easier to see that the set of a_n in (5.4.2) should be unique; the a_n are just the "components" of the "vector" $F(\mathbf{r})$ along the "axes" defined by the set of $\Phi_n(\mathbf{r})$.

The analogy between (5.4.2) and (5.4.3) can be taken even further. Recall that the unit vectors $\mathbf{i}$, $\mathbf{j}$ and $\mathbf{k}$ are perpendicular (orthogonal) to each other:

$$\mathbf{i} \cdot \mathbf{j} = \mathbf{i} \cdot \mathbf{k} = \mathbf{j} \cdot \mathbf{k} = 0$$

The center dot indicates the scalar product, the projection of one vector on another. Since $\mathbf{i}$, $\mathbf{j}$, and $\mathbf{k}$ are orthogonal vectors, they obviously have zero projections on each other. The same is true of the $\Phi_n(\mathbf{r})$. They are also orthogonal to each other. Of course they are not vectors in the same way as $\mathbf{i}$, $\mathbf{j}$ and $\mathbf{k}$, so the nature of a mutual projection (scalar product) of two Φ's is different. We write $\langle \Phi_m | \Phi_n \rangle$ instead of $\Phi_m \cdot \Phi_n$, and the projection means an integral in this case (Problem 5.2):

$$\langle \Phi_m | \Phi_n \rangle = \int_{\text{all space}} d^3r\, \Phi_m^*(\mathbf{r})\, \Phi_n(\mathbf{r}) = 0 \quad (m \neq n) \quad (5.4.4)$$

Note that the first member of the projection is complex-conjugated before the integration is carried out. This is a reminder that quantum-mechanical functions [recall (5.2.10)] are generally complex.

One last analogy remains between the unit vectors $\mathbf{i}$, $\mathbf{j}$ and $\mathbf{k}$ and the functions $\Phi_n(\mathbf{r})$. The unit vectors are normalized:

$$\mathbf{i} \cdot \mathbf{i} = \mathbf{j} \cdot \mathbf{j} = \mathbf{k} \cdot \mathbf{k} = 1 \quad\quad\quad (5.4.5)$$

Their self-projections are all the same and equal to 1. The $\Phi_n(\mathbf{r})$ also have unit self-projections:

$$\langle \Phi_n | \Phi_n \rangle = \int d^3r\, \Phi_n^*(\mathbf{r})\, \Phi_n(\mathbf{r}) = 1 \quad\quad\quad (5.4.6)$$

It is common to use the Kronecker delta symbol to compress (5.4.4) and (5.4.6) into a single statement:

$$\langle \Phi_m | \Phi_n \rangle = \int d^3r\, \Phi_m^*(\mathbf{r})\, \Phi_n(\mathbf{r}) = \delta_{mn} \quad\quad\quad (5.4.7)$$

Here, as usual, we have defined

$$\delta_{mn} = \begin{cases} 1 \text{ if } & m = n \\ 0 \text{ if } & m \neq n \end{cases} \quad\quad\quad (5.4.8)$$

In summary, the members of the set of functions $\{\Phi_n(\mathbf{r})\}$ that represent all possible solutions of the time-independent Schrödinger equation (5.4.1) are *complete, orthogonal,* and *normalized.* One frequently uses the term *orthonormal* to express the combination of orthogonality and unit normalization.

In obtaining the solution of the time-independent Schrödinger equation (5.4.1), one solves the structure problem for the potential $V(r)$. This means that one has determined the spatial form of the electron wave function $\Phi_n(\mathbf{r})$ for each allowed energy E_n. The classical analog is the determination of all electron orbits, one for each energy E. Each $\Phi_n(\mathbf{r})$ is different, so each solution corresponds to a different arrangement in space of electronic probability. In the case of hydrogen [i.e., $V(r) = -e^2/4\pi\epsilon_0 r$], for the lowest values of E_n (the most tightly bound Bohr orbits) the corresponding Φ_n's are nearly, but not exactly, zero for $r \gg a_0$. That is, there is almost no probability that an electron in a low-lying orbit will be found farther than a few Bohr radii from the nucleus. In general, the probability distributions obtained from $|\Phi_n(\mathbf{r})|^2$ depend on x, y, and z, or r, θ, ϕ in polar coordinates. Appendix 5.*A* at the end of the chapter shows that the hydrogen wave functions for *s* states (zero orbital angular momentum) depend only on r, and not on θ or ϕ. In the notation of App. 5.*A*, the integral (5.2.18) becomes

$$\int_{\substack{\text{all}\\\text{space}}} \left|\Psi(\mathbf{r}, t)\right|^2 d^3r = \int_{\substack{\text{all}\\\text{space}}} \left|\Phi(\mathbf{r})\right|^2 d^3r$$

$$= \int_0^\infty R_{n0}^2(r)\, r^2\, dr = 1 \qquad \text{(for } s \text{ states)} \quad (5.4.9)$$

Figure 5.1 plots the radial probability density $r^2 R_{n0}^2(r)$ vs r for the three lowest *s* states and one higher state. One can see only vestiges of classical orbits in the graphs. Each state has a "preferred" radius where the electron is more likely to be found, and these radii increase with orbital quantum number (with energy), as Bohr's original (mostly classical) atomic model predicted.

In addition to the three-dimensional hydrogen potential $V(r) = -e^2/4\pi\epsilon_0 r$, there are many other potentials that are important in atomic, molecular, and laser physics. We cannot treat all of them, of course. We will mention briefly only the one-dimensional harmonic-oscillator potential $V(x) = \frac{1}{2}kx^2$. It plays an important role in molecular structure and also in laser resonators.

Appendix 5.*B* contains a description of harmonic-oscillator functions $\Phi_n(x)$, and a short list of the ones corresponding to the lowest allowed energies. Figure 5.2 shows the probability density $|\Phi_n(x)|^2$ for the lowest five states ($n = 0$, $\cdots$, 4). The classical behavior of an oscillator becomes more evident for higher n values (in accordance with Bohr's *correspondence principle*). The increasing tendency for $|\Phi_n(x)|^2$ to avoid the region around $x = 0$ for higher n values corresponds to the relatively long time a classical oscillator spends near the turning points of its motion.

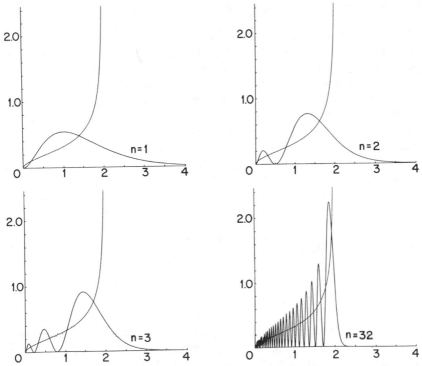

Figure 5.1 The radial probability density $r^2 R_{n0}^2(r)$ vs. r (in units of a_0, the Bohr radius) for four s states of hydrogen. The steeply rising curve common to all four graphs is the probability distribution for classical motion with zero angular momentum in a Coulomb potential. Clearly, the quantum probability density oscillates more closely about this curve as the principal quantum number gets larger. This can be considered an illustration of Bohr's "correspondence principle." From E. G. P. Rowe, *European Journal of Physics* **8**, 81 (1987).

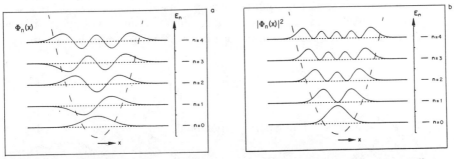

Figure 5.2 (*a*) The wave function $\Phi_n(x)$ vs. x for the lowest five states of a one-dimensional harmonic oscillator. (*b*) The probability density $|\Phi_n(x)|^2$ vs. x for a linear harmonic oscillator. In all cases the dashed curves indicate the turning points of classical motion with the same energy. From S. Brandt and H. D. Dahmen, *The Picture Book of Quantum Mechanics* (Wiley, New York, 1985).

The most quantum-mechanical electron state is in this sense the most tightly bound one, the ground state. Its probability curve shows no preference for the edges. Note that for all of the probability curves, but more so for the ground state, there is a small probability to be outside the potential well altogether. This is, of course, completely impossible in classical physics. A classical harmonic oscillator is never found beyond the turning points of its motion, whereas the probability is 0.16 that a quantum oscillator in its ground state is beyond the classical limit (Problem 5.3).

In another sense the ground state is the *least* quantum-mechanical stationary state of the harmonic oscillator. It may be shown that the *n*th stationary state has the uncertainty product

$$\Delta x \, \Delta p_x = (n + \tfrac{1}{2}) \, \hbar \tag{5.4.10}$$

According to the Heisenberg uncertainty relation (5.2.25), therefore, the ground state has the smallest possible *x-p* uncertainty product allowed by quantum mechanics, and in this sense is the most nearly "classical" stationary state. Such states of minimal uncertainty product are called *coherent states*, and for the harmonic oscillator it is clear from (5.4.10) that the ground state is the only stationary state that is also a coherent state. In general, of course, the coherent states may be constructed from linear superpositions of stationary states.

5.5   ENERGY BANDS AND WAVE FUNCTIONS FOR SOLIDS

In Section 4.8 we used the quantum-mechanical result that the allowed electron energies in crystalline solids occur in bands, with forbidden energy gaps between these bands. In this section we will use a simple one-dimensional model of a solid to show how this band structure arises in quantum mechanics. This will also serve as an example of a full solution to the one-dimensional Schrödinger equation

$$\frac{d^2\Phi}{dx^2} + \frac{2m}{\hbar^2}\left[E - V(x)\right]\Phi = 0 \tag{5.5.1}$$

For the case of an electron in a periodic potential,

$$V(x + d) = V(x) \tag{5.5.2}$$

where the distance d is the lattice spacing in our one-dimensional model. Note that (5.5.2) implies that $V(x + nd) = V(x)$, where n is any integer, so in our model the solid is infinitely long.

If $V(x)$ were identically zero, we could satisfy (5.5.1) with the free-particle plane-wave solution

$$\Phi(x) = u e^{ikx} \qquad (5.5.3)$$

with u some constant (complex) amplitude and k such that [Eq. (5.2.12)]

$$E = \frac{\hbar^2 k^2}{2m} \qquad (5.5.4)$$

For a potential which is not identically zero, and which satisfies (5.5.2), it is natural to try to satisfy the Schrödinger equation (5.5.1) with a wave function of the form

$$\Phi(x) = u(x) e^{ikx} \qquad (5.5.5a)$$

with k a real number and $u(x)$ now not a constant, but a function with the periodicity of the potential:

$$u(x + d) = u(x) \qquad (5.5.5b)$$

Indeed, it may be shown that a solution of the Schrödinger equation, with a potential satisfying (5.5.2), *must* be of the form (5.5.5). This statement is Floquet's theorem, and in solid-state physics it is called *Bloch's theorem*. We will use it in our treatment of the one-dimensional solid.

Different models of a one-dimensional solid are characterized by different choices of the potential $V(x)$ satisfying (5.5.2), but the most important results are insensitive to the specific $V(x)$ chosen. A particularly simple choice is the series of "square wells" shown in Figure 5.3. This will serve as a crude idealization of the sort of potential encountered by an electron in a crystal lattice, each square well representing the effect of an atom at a lattice site. The lattice spacing in Figure 5.3 is $a + b$, with a the width of each potential well.

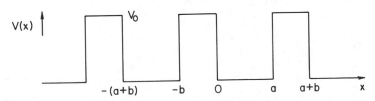

Figure 5.3 A model of the potential $V(x)$ encountered by an electron in a one-dimensional crystal lattice.

We consider first the solution in the "unit cell" $0 < x < a + b$. For $0 < x < a$, $V(x) = 0$, and

$$\Phi(x) = Ae^{i\alpha x} + Be^{-i\alpha x} \qquad (5.5.6)$$

where A and B are constants and

$$\alpha = \left(\frac{2mE}{\hbar^2}\right)^{1/2} \qquad (5.5.7)$$

This solution is just a sum of free-particle, plane-wave solutions, one wave propagating to the right and the other to the left. This is the most general possible solution of (5.5.1) with $V = 0$. Similarly, for $a < x < a + b$, where $V(x) = V_0$, the general solution of (5.5.1) is[2]

$$\Phi(x) = Ce^{\beta x} + De^{-\beta x} \qquad (5.5.8)$$

where C and D are constants and

$$\beta = \left(\frac{2m}{\hbar^2}(V_0 - E)\right)^{1/2} \qquad (5.5.9)$$

According to Bloch's theorem the wave function must have the form (5.5.5). We therefore use the solutions (5.5.6) and (5.5.8) to identify the function $u = \Phi e^{-ikx}$:

$$u(x) = A\,e^{i(\alpha-k)x} + B\,e^{-i(\alpha+k)x}, \qquad 0 < x < a \qquad (5.5.10a)$$
$$u(x) = C\,e^{(\beta-ik)x} + D\,e^{-(\beta+ik)x}, \qquad a < x < a + b \quad (5.5.10b)$$

The wave equation (5.5.1) is a second-order differential equation. As such it demands that both Φ and $d\Phi/dx$ must be continuous functions of x, because a function that is differentiable must be continuous. This means that u and du/dx must also be continuous functions of x. In particular, continuity of u and du/dx at $x = 0$ require the following relations among A, B, C, and D in (5.5.10):

$$A + B = C + D \qquad (5.5.11a)$$
$$i(\alpha - k)A - i(\alpha + k)B = (\beta - ik)C - (\beta + ik)D \quad (5.5.11b)$$

2. Since we will be interested in the case $V_0 > E$, in which β is a real number, we have written (5.5.8) in terms of real exponentials. If $V_0 < E$, then β is purely imaginary and (5.5.8) is a sum of two plane waves, as in (5.5.6) for the case $V_0 = 0$.

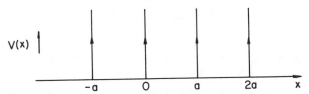

$V(x)$

$-a$ 0 a $2a$ x

Figure 5.4 The Kronig–Penney model for the potential energy $V(x)$ for an electron in a one-dimensional crystal lattice. This is the limit of the potential of Figure 5.3 for the case in which V_0 is very large and b is very small, such that $V_0 b$ is a finite number. In this limit the lattice spacing is a.

Now $u(x)$ must, according to Bloch's theorem, have the periodicity of the potential. Thus u and du/dx must have the same values at $x = a$ as at $x = -b$. This condition of periodicity requires that

$$A\, e^{i(\alpha-k)a} + B\, e^{-i(\alpha+k)a}$$
$$= C\, e^{-(\beta-ik)b} + D\, e^{(\beta+ik)b} \qquad (5.5.11c)$$

$$i(\alpha - k)\, A\, e^{i(\alpha-k)a} - i(\alpha + k)\, B\, e^{-i(\alpha+k)a}$$
$$= (\beta - ik)\, C\, e^{-(\beta-ik)b} - (\beta + ik)\, D\, e^{(\beta+ik)b} \qquad (5.5.11d)$$

The conditions (5.5.11) are four linear, homogeneous, algebraic equations for the four "unknowns" A, B, C, D. A trivial, uninteresting solution is $A = B = C = D = 0$. In order for a nontrivial solution to exist, the 4×4 determinant of the coefficients must vanish. After some simple but tedious algebra, we find that this condition for a nontrivial solution takes the form[3]

$$\frac{\beta^2 - \alpha^2}{2\alpha\beta}\, \sinh \beta b\, \sin \alpha a + \cosh \beta b\, \cos \alpha a = \cos k\,(a + b) \qquad (5.5.12)$$

Since a and b are fixed in our model of a one-dimensional solid, this equation imposes a relation among α, β, and k. As we now show, this relation gives rise to allowed energy bands separated by forbidden energy gaps.

It is convenient to consider a special case of (5.5.12) in which V_0 is very large and b is very small (Figure 5.4). Specifically, we take $V_0 \to \infty$ and $b \to 0$ in such a way that $V_0 b$ remains a finite number. Since $\beta^2 \sim V_0$ for large V_0, this limit is such that $\beta^2 b$ has a finite limit as $\beta \to \infty$ and $b \to 0$. For convenience we denote this limiting value $2P/a$, which is the same as defining

---

3. See, for instance, C. Kittel, *Introduction to Solid State Physics* (Wiley, New York, 1957), Chapter 11.

$$P = \lim_{\beta \to \infty} \lim_{b \to 0} (\tfrac{1}{2} \beta^2 ab) \qquad (5.5.13)$$

Since $\beta b = (1/\beta)(\beta^2 b)$, it follows that $\beta b \to 0$ in this limit. Thus

$$\lim_{\beta \to \infty} \lim_{b \to 0} \cosh \beta b = \lim_{x \to 0} \cosh x = 1 \qquad (5.5.14a)$$

and similarly

$$\lim_{\beta \to \infty} \lim_{b \to 0} \frac{\beta^2 - \alpha^2}{2\alpha\beta} \sinh \beta b = \frac{1}{\alpha a} \lim_{\beta \to \infty} \lim_{b \to 0} \frac{\beta^2 ab}{2} \frac{\sinh \beta b}{\beta b}$$

$$= \frac{1}{\alpha a} \lim_{\beta \to \infty} \lim_{b \to 0} \frac{\beta^2 ab}{2} = \frac{P}{\alpha a} \qquad (5.5.14b)$$

since

$$\lim_{x \to 0} \frac{\sinh x}{x} = 1$$

This limit of $\beta^2 \to \infty$ and $b \to 0$, such that $\beta^2 b$ stays finite, is called the *Kronig–Penney model*. It is useful as a simplification of (5.5.12). With (5.5.14), the condition (5.5.12) in this limit reduces to

$$P \frac{\sin \alpha a}{\alpha a} + \cos \alpha a = \cos ka \qquad (5.5.15)$$

If P is very small, the first term on the left may be neglected, and (5.5.15) becomes $\cos \alpha a = \cos ka$, or $\alpha = k$ (except possibly for a trivial shift of $2\pi/a$). Using the definition (5.5.7) of α, we see that $\alpha = k$ gives the free-particle E–k relation (5.5.4).

If P is very large, on the other hand, then (5.5.15) can only make sense when $(\sin \alpha a)/\alpha a$ is very small. In the limit $P \to \infty$ then we must have

$$\alpha a = n\pi, \qquad n = \pm 1, \pm 2, \pm 3, \ldots \qquad (5.5.16)$$

From the expression (5.5.7) for α, this condition is seen to restrict the electron energy to one of the values

$$E_n = \frac{n^2 \pi^2 \hbar^2}{2ma^2}, \qquad n = 1, 2, 3, \ldots \qquad (5.5.17)$$

These allowed energies are those for an electron in a single, infinitely deep ($V_0 \rightarrow \infty$) square well of width a (Problem 5.5). They may be regarded heuristically as the allowed levels of an electron in an isolated "atom" in the present model.

From the discussion in Section 4.8 we expect these discrete energy levels to broaden into bands when the atoms are brought together to form a crystal lattice. In Figure 5.5 we plot the left-hand side of Eq. (5.5.15) for a case in which P has been arbitrarily chosen to be 1.0. Obviously those values of αa for which this function exceeds unity do not allow (5.5.15) to be satisfied, since $|\cos ka| \leq 1$ for all (real) values of ka. Those values of α for which $|(P \sin \alpha a)/\alpha a + \cos \alpha a| > 1$ define the *forbidden* values of E via the relation (5.5.7):

$$E = \frac{\hbar^2 \alpha^2}{2m} = \frac{\pi^2 \hbar^2}{2ma^2} \left(\frac{\alpha a}{\pi}\right)^2 \qquad (5.5.18)$$

On the other hand, those values of αa for which $|P(\sin \alpha a/\alpha a) + \cos \alpha a| \leq 1$ permit (5.5.15) to be satisfied for real values of k, as required by Bloch's theorem, and the corresponding energies (5.5.18) are the *allowed* electron energies.

In Figure 5.6 we plot the allowed energies, in units of $\pi^2 \hbar^2/2ma^2$, for the example $P = 1.0$ of Figure 5.5. These are the results for a particularly simple one-dimensional model crystal. The theory for a three-dimensional crystal lattice leads similarly, from the periodicity of the potential, to a band structure for the allowed electron energies. This result is brought out by the one-dimensional model we have considered, and so we will not pursue a more complicated, albeit more realistic, three-dimensional model.

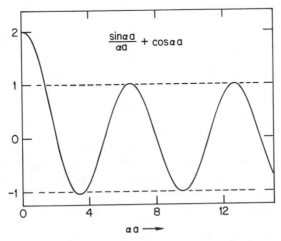

Figure 5.5 Plot of the left side of Eq. (5.5.15) for the Kronig–Penney model when $P = 1.0$.

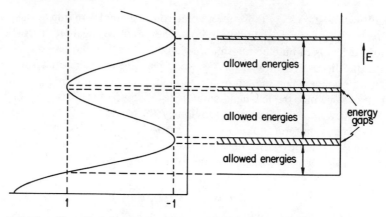

Figure 5.6 The allowed energies given by the Kronig–Penney model for $P = 1.0$. These energies are given by (5.5.18) for those values of $a\alpha$ for which Eq. (5.5.15) allows $|\cos ka| \leqslant 1$. The allowed energies appear in bands.

• We can obtain Bloch's theorem (5.5.5) as follows. First, the periodicity of the potential in (5.5.1) suggests immediately that the probability distribution $|\Phi(x)|^2$ for the electron should also be periodic:

$$\left|\Phi(x + d)\right|^2 = \left|\Phi(x)\right|^2 \tag{5.5.19}$$

which means that

$$\Phi(x + d) = C\Phi(x) \tag{5.5.20}$$

with

$$\left|C\right|^2 = 1 \tag{5.5.21}$$

Note that (5.5.20) implies that

$$\Phi(x + nd) = C^n\Phi(x) \tag{5.5.22}$$

Our one-dimensional crystal is assumed to be infinitely long; this is implicit in the statement (5.5.2) of the periodicity of the potential. The underlying assumption, of course, is that there are enough atoms in a real crystal to make the model of an infinite lattice a reasonable one. That is, "edge effects" in a real crystal are assumed to be very small. In this vein it is also reasonable to suppose there is some integer N, perhaps very large, such that

$$\Phi(x + Nd) = C^N\Phi(x) = \Phi(x) \tag{5.5.23}$$

It can be assumed that the distance Nd, after which the wave function repeats itself according to (5.5.23), is large enough that the assumption (5.5.23) will not affect any physical predictions of the model. In other words, "edge effects" associated with the artificial *periodic boundary condition* (5.5.23) do not have any real physical consequences.

Equation (5.5.23) implies that $C^N = 1$, which means that C must be one of the Nth roots of unity:

$$C = e^{2\pi i M/N}, \quad M = 0, 1, 2, \ldots, N - 1 \tag{5.5.24}$$

It then follows from (5.5.20) that $\Phi(x)$ must have the form

$$\Phi(x) = e^{2\pi i M x/Nd} u(x) = e^{ikx} u(x) \tag{5.5.25a}$$

with $k = 2\pi M/Nd$ and

$$u(x + d) = u(x) \tag{5.5.25b}$$

That is, Eqs. (5.5.20) and (5.5.24) are satisfied when $\Phi(x)$ has the form (5.5.25). Bloch's theorem is easily extended to the case of a three-dimensional lattice. •

It is also instructive to plot E vs. k, as shown in Figure 5.7 for the example $P = 1.0$. We also show for comparison the free-particle E–k relation (5.5.4). In the E–k curve the energy gaps occur at those values of k for which the right-hand side of (5.5.15) is ± 1, i.e., for

$$k = \frac{n\pi}{a}, \quad n = \pm 1, \pm 2, \pm 3, \ldots \tag{5.5.26}$$

This may be understood physically as follows. The wave function

$$\Psi(x, t) = \Phi(x) e^{-iEt/\hbar} = u(x) e^{i(kx - Et/\hbar)} \tag{5.5.27}$$

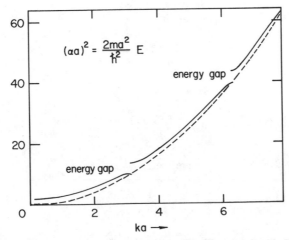

Figure 5.7 Plot of E vs. k for the Kronig–Penney model with $P = 1.0$.

associated with an electron propagating down the lattice has wavelength $\lambda = 2\pi/|k|$ associated with the plane-wave factor

$$e^{i(kx-\omega t)} = e^{i(px-Et)/\hbar} \tag{5.5.28}$$

If $2a = n\lambda$, where $n = 1, 2, 3, \ldots$, the spacing between the potential barriers in Figure 5.4 is an integral number of half wavelengths. This means that the waves reflected from these barriers are all in phase and interfere constructively. In other words, when

$$|k| = \frac{2\pi}{\lambda} = \frac{n\pi}{a}, \qquad n = 1, 2, 3, \ldots \tag{5.5.29}$$

the wave (5.5.27) is strongly reflected and "forbidden" from propagating unhindered down the lattice. This is why the energies associated with the k values (5.5.29), or equivalently (5.5.26), are forbidden.

• In Section 4.9 we invoked the concept of an *effective mass*, m^*, of an electron in a crystal lattice. To see how this concept arises, suppose that a force F acts on the electron. The rate of change of the electron energy as a result of this force is

$$\frac{dE}{dt} = Fv \tag{5.5.30}$$

where v is the electron velocity. Now the force equals the rate of change of the momentum $p = \hbar k$:

$$F = \frac{dp}{dt} = \hbar\frac{dk}{dt} \tag{5.5.31}$$

Thus

$$\frac{dE}{dt} = \hbar v\frac{dk}{dt} \tag{5.5.32}$$

or

$$v = \frac{1}{\hbar}\frac{(dE/dt)}{(dk/dt)} = \frac{1}{\hbar}\frac{dE}{dk} \tag{5.5.33}$$

The acceleration of the electron is therefore

$$a = \frac{dv}{dt} = \frac{1}{\hbar}\frac{d}{dt}\left(\frac{dE}{dk}\right)$$

$$= \frac{1}{\hbar}\frac{dk}{dt}\frac{d}{dk}\left(\frac{dE}{dk}\right) = \frac{F}{\hbar^2}\frac{d^2E}{dk^2} \tag{5.5.34}$$

This equation has the form $F = m*a$, which defines the effective mass as

$$m^* = \hbar^2 \left(\frac{d^2E}{dk^2}\right)^{-1} \tag{5.5.35}$$

For some purposes the electron in the lattice behaves *as if* its mass were m^*, as in the example of Section 4.9 for the calculation of donor ionization energy. The value of the effective mass is determined by the E–k curve according to (5.5.35).

The physical basis for effective mass can be understood as follows. The total force acting on the electron is the external force F_{ext} plus the force F_{crys} due to the atoms of the crystal lattice. This total force equals ma. In our derivation above, however, the force F is only F_{ext}; F_{crys} is accounted for only indirectly via the E–k relation for the electron in the crystal. Thus m^* arises from the proportionality of the electron acceleration to the *external* force. •

5.6 SUMMARY

In this chapter we have introduced some basic ideas of quantum mechanics that are needed for a proper understanding of the interaction of light with matter. The results will be used in the following chapters, where we discuss the emission and absorption of radiation.

One of the most significant characteristics of electrons in matter is that they have only certain allowed energies—discrete levels in atoms and molecules, and bands in crystalline solids. The vibrations and rotations of molecules are also quantized. These features are beautifully accounted for by quantum mechanics in the form of the time-independent Schrödinger equation.

In the study of lasers we are primarily interested in transitions between these allowed energy levels and bands. Because of the fundamentally indeterminate, or probabilistic, nature of quantum mechanics, we can only speak in general of probabilities for a system to be found in one of its stationary states. Similarly, we can only calculate probabilities for transitions, or "quantum jumps," between these stationary states when a system is perturbed by some outside influence like an electromagnetic field. For this purpose our primary tool is the *time-dependent* Schrödinger equation, and Chapter 6 begins with that topic.

APPENDIX 5.A WAVE FUNCTIONS FOR THE ELECTRON IN A HYDROGEN ATOM

n, l, m are the usual hydrogenic quantum numbers (recall Section 4.3), and the wave functions are usually written as a product:

$$\Phi_{nlm}(\mathbf{r}) = R_{nl}(r)\, Y_{lm}(\theta,\,\phi)$$

In particular we have

$$1s \text{ state: } \Phi_{100}(\mathbf{r}) = R_{1,0}(r)\, Y_{0,0}(\theta, \phi) \tag{5.A.1}$$

$$2s \text{ state: } \Phi_{200}(\mathbf{r}) = R_{2,0}(r)\, Y_{0,0}(\theta, \phi) \tag{5.A.2}$$

$$2p \text{ states: } \Phi_{210}(\mathbf{r}) = R_{2,1}(r)\, Y_{1,0}(\theta, \phi) \tag{5.A.3}$$

$$\Phi_{21\pm1}(\mathbf{r}) = R_{2,1}(r)\, Y_{1,\pm1}(\theta, \phi) \tag{5.A.4}$$

The radial and angular factors R_{nl} and Y_{lm} are given explicitly below. They are normalized separately:

$$\int_0^\infty \left| R_{nl}(r) \right|^2 r^2\, dr = 1 \tag{5.A.5}$$

$$\int_0^{2\pi} \int_0^\pi \left| Y_{lm}(\theta, \phi) \right|^2 \sin\theta\, d\theta\, d\phi = 1 \tag{5.A.6}$$

A few radial wave functions $R_{nl}(r)$ for hydrogen are

$$R_{1,0}(r) = 2a_0^{-3/2}\, e^{-r/a_0} \tag{5.A.7}$$

$$R_{2,0}(r) = 2(2a_0)^{-3/2} \left(1 - \frac{r}{2a_0} \right) e^{-r/2a_0} \tag{5.A.8}$$

$$R_{2,1}(r) = \frac{2}{\sqrt{3}} (2a_0)^{-3/2} \left(\frac{r}{2a_0} \right) e^{-r/2a_0} \tag{5.A.9}$$

$$R_{3,0}(r) = 2(3a_0)^{-3/2} \left[1 - \frac{2r}{3a_0} + \frac{2}{3}\left(\frac{r}{3a_0} \right)^2 \right] e^{-r/3a_0} \tag{5.A.10}$$

$$R_{3,1}(r) = \frac{4\sqrt{2}}{3} (3a_0)^{-3/2} \left[\frac{r}{3a_0} \left(1 - \frac{r}{6a_0} \right) \right] e^{-r/3a_0} \tag{5.A.11}$$

$$R_{3,2}(r) = \frac{2}{3}\sqrt{\frac{2}{5}} (3a_0)^{-3/2} \left(\frac{r}{3a_0} \right)^2 e^{-r/3a_0} \tag{5.A.12}$$

In all of these formulas a_0 is the Bohr radius:

$$a_0 = (4\pi\epsilon_0) \frac{\hbar^2}{me^2} \approx 0.53 \text{ Å} \quad \text{(Bohr radius)} \tag{5.A.13}$$

A few angular wave functions (spherical harmonics) $Y_{lm}(\theta, \phi)$ are

$$Y_{0,0} = \left(\frac{1}{4\pi}\right)^{1/2} \tag{5.A.14}$$

$$Y_{1,-1} = \left(\frac{3}{8\pi}\right)^{1/2} \sin\theta \, e^{-i\phi} \tag{5.A.15}$$

$$Y_{1,0} = \left(\frac{3}{4\pi}\right)^{1/2} \cos\theta \tag{5.A.16}$$

$$Y_{1,1} = -\left(\frac{3}{8\pi}\right)^{1/2} \sin\theta \, e^{i\phi} \tag{5.A.17}$$

$$Y_{2,-2} = \left(\frac{15}{32\pi}\right)^{1/2} \sin^2\theta \, e^{-2i\phi} \tag{5.A.18}$$

$$Y_{2,-1} = \left(\frac{15}{8\pi}\right)^{1/2} \sin\theta \cos\theta \, e^{-i\phi} \tag{5.A.19}$$

$$Y_{2,0} = \left(\frac{5}{16\pi}\right)^{1/2} (3\cos^2\theta - 1) \tag{5.A.20}$$

$$Y_{2,1} = -\left(\frac{15}{8\pi}\right)^{1/2} \sin\theta \cos\theta \, e^{i\phi} \tag{5.A.21}$$

$$Y_{2,2} = \left(\frac{15}{32\pi}\right)^{1/2} \sin^2\theta \, e^{2i\phi} \tag{5.A.22}$$

APPENDIX 5.B WAVE FUNCTIONS FOR A PARTICLE IN THE POTENTIAL $V(x) = \frac{1}{2} kx^2$ (ONE-DIMENSIONAL HARMONIC OSCILLATOR)

The harmonic-oscillator scale length is $x_{HO} = (\hbar/m\omega)^{1/2}$, so we introduce the dimensionless coordinate $\xi = x/x_{HO}$, in terms of which the Schrödinger equation (5.4.1) becomes

$$\left(\frac{d^2}{d\xi^2} - \xi^2 + 2\epsilon\right) \Phi(\xi) = 0 \tag{5.B.1}$$

where $\epsilon = E/\hbar\omega$, the energy in units of the quantum energy $\hbar\omega$. The allowed values of E are

$$E_n = (n + \tfrac{1}{2})\hbar\omega, \qquad n = 0, 1, 2, \ldots \tag{5.B.2}$$

and the corresponding wave functions are usually written

$$\Phi_n(\xi) = N_n e^{-\xi^2/2} H_n(\xi) \tag{5.B.3}$$

where N_n is a constant and $H_n(\xi)$ is one of a set of polynomials called *Hermite polynomials*. A list of the first few Hermite polynomials follows:

$$H_0(\xi) = 1 \tag{5.B.4}$$

$$H_1(\xi) = 2\xi \tag{5.B.5}$$

$$H_2(\xi) = 4\xi^2 - 2 \tag{5.B.6}$$

$$H_3(\xi) = 8\xi^3 - 12\xi \tag{5.B.7}$$

$$H_4(\xi) = 16\xi^4 - 48\xi^2 + 12 \tag{5.B.8}$$

If we take

$$N_n = (2^n n! \sqrt{\pi})^{-1/2} \tag{5.B.9}$$

then Φ_n satisfies

$$\langle \Phi_n | \Phi_n \rangle = \int_{-\infty}^{\infty} \Phi_n^*(\xi) \, \Phi_n(\xi) \, d\xi = 1 \tag{5.B.10}$$

Note that "all space" for a one-dimensional harmonic oscillator is the entire (positive and negative) x axis, so the limits on the normalization integral are $\pm\infty$.

PROBLEMS

5.1 Show that a linear superposition (with constant coefficients) of two solutions of the Schrödinger equation (5.3.5) is also a solution of (5.3.5).

5.2 Verify Eq. (5.4.4) explicitly for the projection of the $1s$ and $2s$ states of hydrogen (see Appendix 5.A).

5.3 (a) Show that the classical turning points of a harmonic oscillator of energy E are at $x = \pm \sqrt{2E/m\omega^2}$.

(b) Verify the value of the probability of finding x outside the classical turning points given at the end of Section 5.4. Also find the probability that x lies beyond two times the turning point distance.

5.4 Determine how much electron probability lies outside $r = a_0$ for the $1s$ and $2s$ wave functions of hydrogen.

5.5 Consider a particle of mass m in an infinitely deep, one-dimensional square well of width a. Between the walls the particle is free ($V = 0$), but because it cannot penetrate the walls the wave function must vanish at $x = 0$ and $x = a$ and for all x outside those limits. Show that the normalized stationary-state wave functions are given by

$$\Phi_n(x) = \left(\frac{2}{a}\right)^{1/2} \sin\left(\frac{n\pi x}{a}\right), \qquad n = 1, 2, 3, \ldots$$

with corresponding allowed energies

$$E_n = \frac{n^2\pi^2\hbar^2}{2ma^2}$$

5.6 For the lowest two wave functions given in Problem 5.5, compute Δx and Δp_x and show that the Heisenberg uncertainty principle is satisfied. For which of the two wave functions is the "quantum limit" $\Delta x \, \Delta p_x = \hbar/2$ more closely approached? Is the result as you expected? Why?

5.7 For two functions Φ_n and Φ_m (n and m arbitrary) of the class given in Problem 5.5, verify (5.4.7) explicitly.

5.8 Substitute Φ_n from (5.B.3) into (5.B.1) to find the equation satisfied by the Hermite polynomial $H_n(\xi)$. We will find in Chapter 14 that this equation also plays a role in laser resonator theory.

6 THE TIME-DEPENDENT SCHRÖDINGER EQUATION

6.1 INTRODUCTION

The stationary-state solutions $\Phi_n(\mathbf{r})$ obtained for a potential $V(\mathbf{r})$ determine the state probability distributions of particles in specific energy states. The orbit associated with any given energy E_n is not like a classical orbit; it is not a trajectory in space. Instead it is a spread-out region of more or less concentrated probability. The orbit is static; it is a probability distribution fixed in time.

Energy can be imparted to or taken from a quantum system only if the system can jump from one energy E_m to another energy E_n, i.e., only if it can change its orbit. A change from one orbit to another can occur if an external time-dependent force $\mathbf{F}_{\text{ext}}$ acts on the quantum system.

We can associate this force with a new potential energy in the usual way:

$$\mathbf{F}_{\text{ext}}(\mathbf{r}, t) = -\nabla V_{\text{ext}}(\mathbf{r}, t).$$

Then the system's total energy function (Hamiltonian) can be altered to account for the new force by adding V_{ext}:

$$H = H_a + V_{\text{ext}}(\mathbf{r}, t) \tag{6.1.1}$$

Here we are using the subscript a to mean atomic, and by H_a we mean the Hamiltonian function given in (5.3.5) which describes an atomic particle bound by the potential $V(\mathbf{r})$. The new total Hamiltonian $H(\mathbf{p}, \mathbf{r}, t)$ incorporates the particle's kinetic energy $\mathbf{p}^2/2m$ and both the static binding potential $V(\mathbf{r})$ and the time-dependent external potential $V_{\text{ext}}(\mathbf{r}, t)$. The Schrödinger equation becomes

$$\left(-\frac{\hbar^2}{2m} \nabla^2 + V(\mathbf{r}) + V_{\text{ext}}(\mathbf{r}, t) \right) \Psi(\mathbf{r}, t) = i\hbar \frac{\partial \Psi}{\partial t} \tag{6.1.2}$$

6.2 TIME-DEPENDENT SOLUTIONS

Since H is now time-dependent, the time-dependent part of the wave function of the system cannot be factored as in (5.3.4). This is discussed in Problem 6.1.

Another method of solution was proposed by P. A. M. Dirac around 1926. It makes use of the "completeness" of the orbital functions $\Phi_n(\mathbf{r})$, discussed in Section 5.4. Because the set $\Phi_n(\mathbf{r})$ is complete, any function can be written in terms of the Φ_n's, for example $\Psi(\mathbf{r}, t)$. Following Dirac, we write the exact (still unknown) time-dependent wave function as a sum of Φ_n's:

$$\Psi(\mathbf{r}, t) = \sum_n a_n \Phi_n(\mathbf{r}) \qquad (6.2.1)$$

However, since Ψ changes in time, the specific set of coefficients a_n that can be used to reconstitute Ψ out of Φ_n's must also change in time. That is, the a's are also time-dependent.

Since Schrödinger's equation (6.1.2) is responsible for the time dependence of Ψ, it is also indirectly responsible for the time dependence of the a's. We can determine the equations for the a's as follows. First, we apply (6.1.2) to Ψ written as in (6.2.1):

$$\sum_n a_n [H_a + V_{\text{ext}}] \Phi_n(\mathbf{r}) = \sum_n i\hbar \frac{\partial a_n}{\partial t} \Phi_n(\mathbf{r}) \qquad (6.2.2)$$

Next we use the fact that the set $\Phi_n(\mathbf{r})$ is the solution set for the time-independent Schrödinger equation (5.3.5). This means that $H_a \Phi_n = E_n \Phi_n$, so (6.2.2) can be rewritten as

$$\sum_n a_n [E_n + V_{\text{ext}}] \Phi_n(\mathbf{r}) = \sum_n i\hbar \frac{\partial a_n}{\partial t} \Phi_n(\mathbf{r}) \qquad (6.2.3)$$

Next we take the projection of both sides of (6.2.3) along the function $\Phi_m(\mathbf{r})$. Here we understand projection in the sense explained in Section 5.4 [recall (5.4.7)]:

$$\langle \Phi_m | \Phi_n \rangle = \int_{\substack{\text{all} \\ \text{space}}} \Phi_m^*(\mathbf{r}) \Phi_n(\mathbf{r}) \, d^3r = \delta_{mn} \qquad (6.2.4)$$

Thus we find

$$i\hbar \dot{a}_m = E_m a_m + \sum_n a_n \int \Phi_m^*(\mathbf{r}) V_{\text{ext}} \Phi_n(\mathbf{r}) \, d^3r \qquad (6.2.5)$$

where we have used a dot to indicate time derivative and the definition (5.4.8) to evaluate the sums

$$\sum_n a_n E_n \langle \Phi_m | \Phi_n \rangle = a_m E_m$$

$$\sum_n i\hbar \dot{a}_n \langle \Phi_m | \Phi_n \rangle = i\hbar \dot{a}_m \qquad (6.2.6)$$

The integration in (6.2.5) cannot yet be carried out, because the $\mathbf{r}$ dependence of V_{ext} has not been specified. However, the integral can be abbreviated conveniently as

$$V_{mn}(t) = \int \Phi_m^*(\mathbf{r}) \, V_{\text{ext}}(\mathbf{r}, t) \, \Phi_n(\mathbf{r}) \, d^3r \qquad (6.2.7)$$

For reasons that are explained in the next black-dot section, V_{mn} is called the *matrix element* of V_{ext} between the states m and n of the atomic system. In terms of this matrix element, Eq. (6.2.5) takes a more compact form:

$$i\hbar \dot{a}_m = E_m a_m + \sum_n V_{mn}(t) \, a_n \qquad (6.2.8)$$

This equation, as well as (6.1.2), is frequently called the *time-dependent Schrödinger equation*. This is reasonable, since knowledge of the a's, obtained by solving (6.2.8), can be used as in (6.2.1) to reconstitute the full $\Psi(\mathbf{r}, t)$.

The a's themselves are called *probability amplitudes*. This name follows from the normalization property of Ψ. According to the probability interpretation of quantum mechanics we must have

$$\int_{\substack{\text{all} \\ \text{space}}} \Psi^*(\mathbf{r}, t) \, \Psi(\mathbf{r}, t) \, d^3r = 1 \qquad (6.2.9)$$

Therefore (6.2.1) implies

$$\int \left(\sum_m a_m \Phi_m \right)^* \left(\sum_n a_n \Phi_n \right) d^3r = 1$$

$$= \sum_m a_m^* \sum_n a_n \langle \Phi_m | \Phi_n \rangle$$

$$= \sum_m \sum_n a_m^* a_n \delta_{mn}$$

$$= \sum_m |a_m|^2 = 1 \qquad (6.2.10)$$

It is natural to identify each term in (6.2.10) as an orbital probability. That is, the squared magnitude $|a_m|^2$ is the probability that the quantum system (for example, the atomic electron) is in its mth orbit. The term probability amplitude is then used for a_m itself.

There is a significant shift in viewpoint between (6.2.9) and (6.2.10) even though they express the same normalization. Recall that $|\Psi(\mathbf{r}, t)|^2 \, d^3r$ is the electron probability assigned to the differential volume element d^3r. There is no information about orbitals in this assignment, and indeed, many or all of the orbitals may make a contribution to the probability within d^3r. On the other hand,

$|a_m|^2$ plays the opposite role. It is the probability that the electron is in the mth orbital, without providing any information about the spatial location of the electron in that orbit. In laser physics information about orbital occupation by the atomic electron is much more useful than information about its spatial location. For this reason we will concentrate completely on solutions of the second form of the time-dependent Schrödinger equation, given in (6.2.8).

• Let us write out the equations for the coefficients a_n in (6.2.8) in order, i.e.,

$$i\hbar\dot{a}_1 = E_1 a_1 + V_{11} a_1 + V_{12} a_2 + V_{13} a_3 + \cdots$$

$$i\hbar\dot{a}_2 = E_2 a_2 + V_{21} a_1 + V_{22} a_2 + V_{23} a_3 + \cdots$$

$$i\hbar\dot{a}_3 = E_3 a_3 + V_{31} a_1 + V_{32} a_2 + V_{33} a_3 + \cdots$$

$$\vdots$$

$$(6.2.11)$$

We see that they can be written as a single *matrix equation*:

$$i\hbar\underline{\dot{\Psi}} = \underline{H}\underline{\Psi} \tag{6.2.12}$$

where

$$\underline{\Psi} = \begin{bmatrix} a_1 \\ a_2 \\ a_3 \\ \vdots \end{bmatrix} \tag{6.2.13}$$

and

$$\underline{H} = \begin{bmatrix} E_1 + V_{11} & V_{12} & V_{13} & \cdots \\ V_{21} & E_{22} + V_{22} & V_{23} & \cdots \\ V_{31} & V_{32} & E_3 + V_{33} & \cdots \\ \vdots & \vdots & \vdots & \end{bmatrix} \tag{6.2.14}$$

This matrix form of the Schrödinger equation is the origin of the term "matrix element" for V_{nm}. In this form $\underline{H}$ is called the *Hamiltonian matrix* and $\underline{\Psi}$ the *state vector*. Heisenberg's original approach to quantum mechanics (1925) was through such matrices. Physical observables were represented by Hermitian matrices, whose matrix elements satisfy the relation $V_{nm}^* = V_{mn}$. It was not immediately appreciated that Heisenberg's "matrix mechanics" is equivalent to Schrödinger's "wave mechanics". •

6.3 TWO-STATE QUANTUM SYSTEMS AND SINUSOIDAL EXTERNAL FORCES

According to Bohr's description of quantum jumps, an atom can increase its energy by jumping from an orbit with energy E to one with higher energy E' if a photon of frequency ω is simultaneously absorbed, where $\hbar\omega = E' - E$. The reverse process is associated with the emission of a photon of frequency ω.

We associate photons of frequency ω with an electromagnetic wave of the same frequency. According to our analysis in Section 2.2, an external electromagnetic field interacts with an electron via the time-dependent potential

$$V_{\text{ext}}(\mathbf{r}, \mathbf{R}, t) = -e\mathbf{r} \cdot \mathbf{E}(\mathbf{R}, t)$$

We showed in Section 2.2 that $\mathbf{r}$ is the relative electron–nuclear distance and $\mathbf{R}$ is the location of the center of atomic mass. We will begin by considering a monochromatic plane wave for $\mathbf{E}$:

$$\mathbf{E}(\mathbf{R}, t) = \hat{\mathbf{\varepsilon}}E_0 \cos(\mathbf{k}\cdot\mathbf{R} - \omega t)$$

$$= \tfrac{1}{2}\hat{\mathbf{\varepsilon}}E_0\, e^{i(\mathbf{k}\cdot\mathbf{R} - \omega t)} + \text{c.c.}$$

$$\rightarrow \tfrac{1}{2}\hat{\mathbf{\varepsilon}}E_0 e^{-i\omega t} + \text{c.c.} \tag{6.3.1}$$

where c.c. means complex conjugate and for convenience $\mathbf{R}$ has been put at the origin. This form of V_{ext} is the result of the dipole approximation, which is highly accurate when applied to optical transitions in atoms.

The implication of Bohr's rule for quantum jumps is that only pairs of energy levels in the atom that are separated by $\Delta E = \hbar\omega$ are affected by radiation present at frequency ω. Therefore, we will begin our study by restricting our attention to just two of the electronic energy levels. These are shown in Figure 6.1 and designated 1 and 2, with energies E_1 and E_2, such that $\Delta E = E_2 - E_1 = \hbar\omega$.

For such a two-state system the expression (6.2.1) is simply

$$\Psi(\mathbf{r}, t) = a_1(t)\,\Phi_1(\mathbf{r}) + a_2(t)\,\Phi_2(\mathbf{r}) \tag{6.3.2}$$

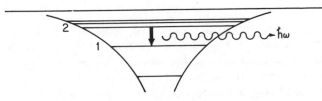

Figure 6.1 Energy levels of a hypothetical atom. Radiation of angular frequency ω is nearly resonant with the $E_1 \rightarrow E_2$ transition.

and the Schrödinger equation (6.2.8) reduces to

$$i\hbar \dot{a}_1(t) = E_1 a_1(t) + V_{11} a_1(t) + V_{12} a_2(t) \qquad (6.3.3a)$$

$$i\hbar \dot{a}_2(t) = E_2 a_2(t) + V_{21} a_1(t) + V_{22} a_2(t) \qquad (6.3.3b)$$

Level 1 may be the ground level but need not be. In most cases of interest the parity selection rule (see Problem 6.5) requires the diagonal matrix elements V_{11} and V_{22} of the interaction to be zero. Then

$$i\hbar \dot{a}_1(t) = E_1 a_1(t) + V_{12} a_2(t) \qquad (6.3.4a)$$

$$i\hbar \dot{a}_2(t) = E_2 a_2(t) + V_{21} a_1(t) \qquad (6.3.4b)$$

Equations (6.3.4) give the time variation of the probability amplitudes a_1 and a_2 for the two-state system. If the two-state model is a reasonable approximation we can assume that the system has negligible probability of being in any state other than Φ_1 or Φ_2. In other words, the probability that the system will be found in one or the other of these two states is unity at any time:

$$\left| a_1(t) \right|^2 + \left| a_2(t) \right|^2 = 1 \qquad (6.3.5)$$

This is the two-state version of (6.2.10).

Equations (6.3.4) show how the 1–2 and 2–1 matrix elements of V_{ext} are involved in changes in the amplitudes $a_1(t)$ and $a_2(t)$. From (6.2.7) and (6.3.1) we can express these matrix elements more explicitly as

$$V_{12}(t) = -e\mathbf{r}_{12} \cdot \tfrac{1}{2}(\hat{\mathbf{\varepsilon}} E_0 e^{-i\omega t} + \text{c.c.}) \qquad (6.3.6)$$

$$V_{21}(t) = -e\mathbf{r}_{21} \cdot \tfrac{1}{2}(\hat{\mathbf{\varepsilon}} E_0 e^{-i\omega t} + \text{c.c.}) \qquad (6.3.7)$$

where, for example, the 1–2 matrix element of $\mathbf{r}$ is defined by

$$\mathbf{r}_{12} \equiv \int \Phi_1^*(\mathbf{r})\, \mathbf{r}\, \Phi_2(\mathbf{r})\, d^3r \qquad (6.3.8)$$

Note that $\mathbf{r}_{12}$ is generally a complex-valued vector because the Φ's may be complex. The numerical value of $\mathbf{r}_{12}$ depends on the wave functions Φ_1 and Φ_2, so the size of the matrix elements V_{12} and V_{21} must be expected to vary from atom to atom. As a typical magnitude (associated with an optical transition to or from an atomic ground state), one can expect $|\mathbf{r}_{12}|$ to differ from the Bohr radius $a_0 \approx \tfrac{1}{2}$ Å by less than a factor of 10. Listings of the radial and angular parts of the coordinate matrix element r_{12} are given in Appendix 6.A for various transitions in hydrogen. Notice that the matrix element tends to have a significantly larger magnitude if the n values are different by only one unit.

With $V_{12}(t)$ and $V_{21}(t)$ given by Eqs. (6.3.6) and (6.3.7) we can insert them in Eq. (6.3.4). It is convenient to adopt several conventions at the same time. We will work with frequencies instead of energies, so we divide through by $\hbar$ and define

$$\omega_{21} = \frac{E_2 - E_1}{\hbar} \tag{6.3.9}$$

$$\chi_{21} = e(\mathbf{r}_{21} \cdot \hat{\boldsymbol{\varepsilon}}) \frac{E_0}{\hbar} \tag{6.3.10a}$$

$$\chi_{12} = e(\mathbf{r}_{12} \cdot \hat{\boldsymbol{\varepsilon}}) \frac{E_0}{\hbar} \tag{6.3.10b}$$

Note that even though $\mathbf{r}_{12} = \mathbf{r}_{21}^*$ (see Problem 6.2), we cannot write $\chi_{12} = \chi_{21}^*$, since $\hat{\boldsymbol{\varepsilon}} E_0$ may be complex (for circularly polarized radiation, for example). Evidently χ is the field–atom interaction energy in frequency units. It is also known as the "Rabi frequency," as we explain below. Also, we now set the arbitrary zero of energy at E_1, so $E_2 \to E_2 - E_1 = \hbar\omega_{21}$. Then Eqs. (6.3.4) become

$$i\dot{a}_1 = -\tfrac{1}{2}(\chi_{12}e^{-i\omega t} + \chi_{21}^* e^{i\omega t}) a_2 \tag{6.3.11a}$$

$$i\dot{a}_2 = \omega_{21}a_2 - \tfrac{1}{2}(\chi_{21}e^{-i\omega t} + \chi_{12}^* e^{i\omega t}) a_1 \tag{6.3.11b}$$

In the absence of any radiation field ($\chi = 0$) we find $a_1(t) = a_1(0)$ from (6.3.11a) and $a_2(t) = a_2(0)\exp[-i\omega_{21}t]$ from (6.3.11b). In the presence of a nearly resonant radiation field (6.3.1) oscillating at frequency $\omega \approx \omega_{21}$ we adopt similar trial solutions

$$a_1(t) = c_1(t) \tag{6.3.12a}$$

$$a_2(t) = c_2(t) e^{-i\omega t} \tag{6.3.12b}$$

and find these equations for $c_1(t)$ and $c_2(t)$:

$$i\dot{c}_1 = -\tfrac{1}{2}(\chi_{12} e^{-2i\omega t} + \chi_{21}^*) c_2 \tag{6.3.13a}$$

$$i\dot{c}_2 = (\omega_{21} - \omega) c_2 - \tfrac{1}{2}(\chi_{21} + \chi_{12}^* e^{2i\omega t}) c_1 \tag{6.3.13b}$$

Equations (6.3.13) are more useful because of their isolation of the $\exp[\pm 2i\omega t]$ terms. These terms oscillate so rapidly compared with every other time variation in the equations that they can be assumed to average to zero over any realistic time interval. In this way it is argued that they can simply be discarded. This is known as the *rotating-wave approximation* (abbreviated RWA in the optical resonance literature). It leads to these elementary working equations:

$$i\dot{c}_1 = -\tfrac{1}{2}\chi^* c_2 \tag{6.3.14a}$$

$$i\dot{c}_2 = \Delta c_2 - \tfrac{1}{2}\chi c_1 \tag{6.3.14b}$$

where we have dropped the subscript 21 from χ_{21} and have introduced Δ to stand for the atom–field frequency offset, or *detuning*:

$$\chi = \chi_{21} = (e\mathbf{r}_{21} \cdot \hat{\mathbf{\varepsilon}}) \frac{E_0}{\hbar} \tag{6.3.15a}$$

$$\Delta = \omega_{21} - \omega \tag{6.3.15b}$$

If $\hat{\mathbf{\varepsilon}}E_0$ is a constant vector χ can be taken to be a purely real number. This can be arranged by the right choice of phases of the wave functions Φ_1 and Φ_2 (see Problem 6.3). Unless the context indicates otherwise (see Sec. 8.2), we will assume this has been done.

The great advantage of Eqs. (6.3.14) is their relative simplicity. The smallness of the coefficients Δ and χ (compared with ω and ω_{21}) shows that the c's are "slow" variables (compared with the a's). They contain the essential physics once the rapid oscillations associated with the frequencies ω and ω_{21} are removed by the rotating-wave approximation. The solutions for the c's are easily found (see Problem 6.4):

$$c_1(t) = \left(\cos \frac{\Omega t}{2} + i\frac{\Delta}{\Omega} \sin \frac{\Omega t}{2} \right) e^{-i\Delta t/2} \tag{6.3.16a}$$

$$c_2(t) = \left(i\frac{\chi}{\Omega} \sin \frac{\Omega t}{2} \right) e^{-i\Delta t/2} \tag{6.3.16b}$$

Here we have adopted the phase choice to make χ real. Also, we have assumed the atom to be in state 1 initially: $c_1(0) = 1$, $c_2(0) = 0$, and we have introduced the generalized Rabi frequency

$$\Omega = (\chi^2 + \Delta^2)^{1/2} \tag{6.3.17}$$

which reduces to the ordinary Rabi frequency χ at exact resonance ($\Delta = 0$). For this reason χ is sometimes called the resonance Rabi frequency.

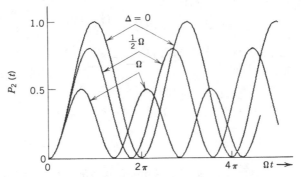

Figure 6.2 Plot of the upper-state probability $P_2(t)$ given by Eq. (6.3.18). Note that larger detuning Δ corresponds to higher frequency of Rabi oscillation but lower amplitude.

The corresponding probabilities $P_1(t) = |a_1(t)|^2$ and $P_2(t) = |a_2(t)|^2$ are

$$P_1(t) = \frac{1}{2}\left[1 + \left(\frac{\Delta}{\Omega}\right)^2\right] + \frac{1}{2}\left(\frac{\chi}{\Omega}\right)^2 \cos\Omega t \qquad (6.3.18a)$$

$$P_2(t) = \frac{1}{2}\left(\frac{\chi}{\Omega}\right)^2[1 - \cos\Omega t], \qquad (6.3.18b)$$

The justification for defining Ω and χ exactly as we have and calling them (instead of $\Omega/2$ or 2Ω) "the" Rabi frequencies is evident in Figure 6.2, where $P_2(t)$ is plotted. It is clear that Ω is precisely the frequency at which probability oscillates between levels 1 and 2. It is easy to check that $P_1(t) + P_2(t) = 1$ for all t, so P_1 simply oscillates at the same frequency with the opposite phase from $P_2(t)$.

6.4 QUANTUM MECHANICS AND THE LORENTZ MODEL

In Chapters 2 and 3 we discussed Lorentz's classical electron oscillator model for the interaction of light with matter. Although this model is completely classical, it offers, as we saw, good explanations for a wide variety of phenomena. In this section we will explain, from the viewpoint of quantum mechanics, why the Lorentz model is so successful and also why the oscillator strength f must be introduced.

The basic dynamical variable for an atomic electron in the Lorentz model is its displacement **x**. In order to establish the connection between the Lorentz model and the quantum-mechanical theory of an atomic electron, let us consider the corresponding quantum displacement, i.e., the expectation value $\langle \mathbf{r} \rangle$ in our two-state atom. By definition (5.2.20) this expectation value is

$$\langle \mathbf{r} \rangle = \int \Psi^*(\mathbf{r}, t) \, \mathbf{r} \Psi(\mathbf{r}, t) \, d^3r \tag{6.4.1}$$

For the two-state atom, $\Psi(\mathbf{r}, t)$ is given by the linear superposition (6.3.2) of the two stationary states. Thus

$$\langle \mathbf{r} \rangle = \int (a_1^* \Phi_1^* + a_2^* \Phi_2^*) \, \mathbf{r}(a_1 \Phi_1 + a_2 \Phi_2) \, d^3r$$

$$= |a_1|^2 \mathbf{r}_{11} + |a_2|^2 \mathbf{r}_{22} + a_1^* a_2 \mathbf{r}_{12} + a_2^* a_1 \mathbf{r}_{21} \tag{6.4.2}$$

This complicated expression can be simplified. We noted in the preceding section that for atoms we can take $V_{11} = V_{22} = 0$, which means $\mathbf{r}_{11} = \mathbf{r}_{22} = 0$. Thus we have

$$\langle \mathbf{r} \rangle = \mathbf{r}_{12} a_1^* a_2 + \mathbf{r}_{21} a_2^* a_1$$

$$= \mathbf{r}_{12} a_1^* a_2 + \text{c.c.} \tag{6.4.3}$$

Since we are looking for a quantum analog to the Lorentzian physics of Chapters 2 and 3, we look for the equation of motion obeyed by the quantum expectation value $\langle \mathbf{r} \rangle$. For simplicity we consider linear polarization, so that $\hat{\varepsilon} E_0$ is real. From equations (6.3.4) we easily compute (retaining the choice of phases that leads to real-valued $\mathbf{r}_{21} \cdot \hat{\varepsilon} E_0$ and V_{21})

$$\hbar \frac{d}{dt} (a_1^* a_2) = -i(E_2 - E_1) a_1^* a_2 - i V_{21} (|a_1|^2 - |a_2|^2) \tag{6.4.4}$$

and

$$\hbar^2 \frac{d^2}{dt^2} (a_1^* a_2) = -(E_2 - E_1)^2 a_1^* a_2 - (E_2 - E_1) V_{21} (|a_1|^2 - |a_2|^2)$$

$$- i\hbar \frac{d}{dt} [V_{21} (|a_1|^2 - |a_2|^2)]. \tag{6.4.5}$$

Therefore, since we have real $\mathbf{r}_{12}$ we can write

$$\langle \mathbf{r} \rangle = \mathbf{r}_{12} (a_1^* a_2 + a_1 a_2^*) \tag{6.4.6}$$

and we can combine (6.4.5) with its complex conjugate equation to get

$$\left(\frac{d^2}{dt^2} + \omega_0^2 \right) \langle \mathbf{r} \rangle = +\frac{2e\omega_0}{\hbar} \mathbf{r}_{12} (\mathbf{r}_{21} \cdot \mathbf{E}) (|a_1|^2 - |a_2|^2) \tag{6.4.7}$$

Here we have adopted the classical notation ω_0 for the transition frequency ω_{21}.

There is a close similarity of (6.4.7) to the Lorentz equation (2.2.18) for the classical electron displacement **x**:

$$\left(\frac{d^2}{dt^2} + \omega_0^2\right)\mathbf{x} = \frac{e}{m}\mathbf{E} \tag{6.4.8}$$

To interpret the differences between (6.4.7) and (6.4.8) we first recall the special circumstances for which the Lorentz model was invented (around 1900). The phenomena Lorentz sought to explain involved only natural light (from the sun), or light from man-made thermal sources (lamps). The spectral intensity (W/cm²-Hz) of any such radiation is weak (recall Section 1.2). This suggests that we focus our attention on the quantum equation (6.4.7) in the case in which the excited-state occupation probability is close to zero. [Sufficient conditions for this are worked out later, in Section 7.5.] In our two-state atom this means $|a_2|^2 \ll 1$ and therefore $|a_1|^2 \approx 1$, which gives

$$\left|a_1(t)\right|^2 - \left|a_2(t)\right|^2 \approx 1 \tag{6.4.9}$$

Under such circumstances of low excitation, we can approximate (6.4.7) by the equation

$$\left(\frac{d^2}{dt^2} + \omega_0^2\right)\langle\mathbf{r}\rangle = \frac{2e\omega_0}{\hbar}\mathbf{r}_{12}(\mathbf{r}_{21}\cdot\mathbf{E}) \tag{6.4.10}$$

This equation is still different from the Lorentz-model equation (6.4.8), but only in the constants on the right-hand side. Let us look at it more closely in a particular example.

Suppose the electric field points in the z direction ($\mathbf{E} = \hat{z}E$), so that

$$\left(\frac{d^2}{dt^2} + \omega_0^2\right)\langle\mathbf{r}\rangle = \frac{2e\omega_0}{\hbar}\mathbf{r}_{12}z_{21}E \tag{6.4.11}$$

and let the atomic states 1 and 2 be the 100 and 210 ($1s$ and $2p$) states of hydrogen. Then from Appendix 5.A we have

$$\Phi_1(\mathbf{r}) = R_{1,0}(r)\,Y_{0,0}(\theta, \phi) \tag{6.4.12}$$

$$\Phi_2(\mathbf{r}) = R_{2,1}(r)\,Y_{1,0}(\theta, \phi) \tag{6.4.13}$$

[This choice of states is discussed below. Spin does not play any role in this calculation. It may be assumed that $m_s = +\frac{1}{2}$ (or $-\frac{1}{2}$) for both states.] Now we can evaluate the three components of $\mathbf{r}_{12}$.

The components of **r** in Cartesian coordinates are shown in Figure 6.3 to be related to those in spherical coordinates by

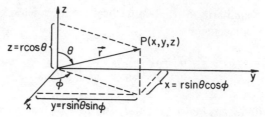

Figure 6.3 Cartesian components of the radius vector **r** in relation to spherical coordinates.

$$x = r \sin \theta \cos \phi \qquad (6.4.14a)$$

$$y = r \sin \theta \sin \phi \qquad (6.4.14b)$$

$$z = r \cos \theta \qquad (6.4.14c)$$

We begin our evaluations with the z component:

$$z_{21} = \int \int \int \Phi_2^*(\mathbf{r})\, z \Phi_1(\mathbf{r})\, d^3r$$

$$= \int_0^\infty r^3 R_{2,1}^*(r)\, R_{1,0}(r)\, dr \int_0^{2\pi} \int_0^\pi Y_{1,0}^*(\theta, \phi) \sin \theta \cos \theta\, Y_{0,0}(\theta, \phi)\, d\theta\, d\phi$$

$$(6.4.15)$$

The radial and angular integrations are traditionally considered separately. We consider the radial part first, and denote it r_{21}:

$$r_{21} = (2a_0^{-3/2}) \frac{2}{\sqrt{3}} (2a_0)^{-3/2} \int_0^\infty \left(\frac{r}{2a_0} \right) \exp\left(-r/2a_0 \right) r^3 e^{-r/a_0}\, dr$$

With a change of variable and integral tables one finds

$$\int_0^\infty r^4 \exp\left(-3r/2a_0 \right) dr = \left(\frac{2a_0}{3} \right)^5 \int_0^\infty x^4 e^{-x}\, dx = 4! \left(\frac{2a_0}{3} \right)^5$$

and therefore

$$r_{21} = 1.29 a_0 \qquad (6.4.16)$$

Note that we have verified the first line of Table 6.1 in Appendix 6.A. The angular part is also easily worked out. We will denote it by $\hat{z}_{21}$, where the circumflex indicates that the magnitude r has been taken out and we are considering the z

component of the *unit vector* $\hat{\mathbf{r}} = \hat{\mathbf{x}} \sin \theta \cos \phi + \hat{\mathbf{y}} \sin \theta \sin \phi + \hat{\mathbf{z}} \cos \theta$. We find the value

$$
\hat{z}_{21} = \sqrt{\frac{1}{4\pi}\frac{3}{4\pi}} \int_0^{2\pi} d\phi \int_0^{\pi} \cos^2 \theta \sin \theta \, d\theta
$$

$$
= \frac{\sqrt{3}}{4\pi} 2\pi \int_{-1}^{1} x^2 \, dx
$$

$$
= \sqrt{\tfrac{1}{3}} \tag{6.4.17}
$$

The combination of (6.4.16) and (6.4.17) then gives

$$
z_{21} = r_{21}\hat{z}_{21} = 0.745a_0 \tag{6.4.18}
$$

Proceeding in exactly the same fashion, we find that x_{21} and y_{21} vanish for the states (6.4.12) and (6.4.13). It is easy to see why: x and y have ϕ dependences of $\cos \phi$ and $\sin \phi$, respectively. Since the integral from 0 to 2π of either $\cos \phi$ or $\sin \phi$ is zero, $x_{21} = y_{21} = 0$. Note that we have verified the first line of Table 6.2 in Appendix 6.A.

Equation (6.4.11) may therefore be simplified:

$$
\left(\frac{d^2}{dt^2} + \omega_0^2\right) \langle \mathbf{r} \rangle = \frac{2e\omega_0}{\hbar} \hat{\mathbf{z}} z_{12} z_{21} E \tag{6.4.19}
$$

$$
= \frac{2e\omega_0}{\hbar} (z_{12})^2 \mathbf{E} \tag{6.4.20}
$$

Now we observe that if in the classical formula (6.4.8) we make the replacement

$$
\frac{e}{m} \rightarrow \frac{2e\omega_0}{\hbar} z_{12}^2 \tag{6.4.21}
$$

or equivalently

$$
\frac{e^2}{m} \rightarrow \frac{2e^2\omega_0}{\hbar} z_{12}^2
$$

$$
= \frac{e^2}{m} \left[\frac{2m\omega_0}{\hbar} z_{12}^2 \right] \tag{6.4.22}
$$

then *Lorentz's classical equation for* $\mathbf{x}$ *is exactly the same as the quantum-mechanical equation for* $\langle \mathbf{r} \rangle$ *under the ground-state approximation* (6.4.9).

Comparing (6.4.22) with (3.7.5), it is evident that they agree if we identify the factor in brackets in the former with the oscillator strength factor f:

$$f = \frac{2m\omega_0}{\hbar} z_{12}^2$$

Since f values have been known from absorption and dispersion experiments for many years, this identification can easily be tested. Using our calculated result (6.4.18) and the expression $\omega_0 = (E_2 - E_1)/\hbar$ for the Bohr transition frequency, we compute from (6.4.22) the (dimensionless) numerical value

$$f = 0.416 \tag{6.4.23}$$

Comparison with the first entry of Table 3.1 confirms that 0.416 is indeed the oscillator strength of the $n = 1 \rightarrow n = 2$ transition of hydrogen. This shows that we have not only demonstrated the validity in quantum theory of the Lorentz model under conditions of low excitation ($|a_2|^2 \ll 1$), but we have also derived an expression for the oscillator strength of the transition in terms of fundamental atomic parameters. We note that the presence of $\hbar$ indicates the quantum nature of f.

Our example gives the impression that the z direction plays a special role. This is not actually the case. We can correct the misimpression as follows. First we note that the classical atom, according to (6.4.8), is free to respond to the electric field, no matter in what direction $\mathbf{E}$ points. On the other hand, according to (6.4.10) the quantum mechanical atom responds only to the component of the field parallel to the matrix element vector $\mathbf{r}_{21}$. That is, the quantum mechanical atom appears to have an internal or intrinsic sense of direction.

In fact a quantum mechanical atom does have a sense of direction if it is exposed to orienting or aligning forces such as from static external electric or magnetic fields. External forces have the effect of destroying ("lifting") the degeneracy of states mentioned in Section 4.3. When the degeneracy is lifted, each set (n, l, m) of state quantum numbers refers to a distinct value of energy. Conversely, in the absence of external alignment or orientation (in an atom in free space) each transition is degenerate since all possible m's associated with the same l denote states with the same energy. Thus, to compare a quantum mechanical atom with a classical atom in free space requires that the degeneracy of the initial and final states be recognized. Figure 6.4 shows the states with differing m values contributing to the 1s–2p transition chosen in (6.4.12) and (6.4.13) above. Figure 6.5 shows the corresponding situation for a p–d transition.

We see now that the right-hand side of (6.4.10) correctly refers to an atom in free space only if the three possible m values ($+1$, 0, -1) of the 2p state are all included. As it happens, for an $s \rightarrow p$ transition, if the field is polarized in the z direction as assumed in the calculation, the $m = \pm 1$ contributions to the result

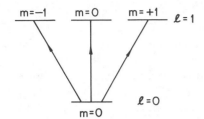

Figure 6.4 Absorptive transitions between an s level ($l = 0$) and a p level ($l = 1$). The transition $m = 0 \rightarrow 0$ occurs in a linearly polarized field, whereas the $m = 0 \rightarrow +1$ and $m = 0 \rightarrow -1$ transitions occur in left and right circularly polarized fields, respectively.

turn out to be zero [see Problem 6.6a]. This is why the calculation reached the correct result, given in (6.4.23).

On the other hand, if the field had not been chosen in the z direction, then the $m = \pm 1$ terms would have contributed enough to the result to produce the same final number. This follows from the important quantum mechanical result

$$\sum_{m's} |x_{12}|^2 = \sum_{m's} |y_{12}|^2 = \sum_{m's} |z_{12}|^2 \qquad (6.4.24)$$

where the summations are over all of the m values for levels 1 and 2. The usual expression for the oscillator strength f recognizes these and other symmetry principles [see Problem 6.8]. As a result f is conventionally written in terms of the isotropic combination $|x_{12}|^2 + |y_{12}|^2 + |z_{12}|^2 = \mathbf{r}_{12} \cdot \mathbf{r}_{21}$, as follows:

$$f = (2m\omega_{21}/3\hbar) \sum_m \mathbf{r}_{12} \cdot \mathbf{r}_{21} \qquad (6.4.25)$$

where the sum is over the m values of the *final* state of the transition. Note that the final state is unambiguously Φ_2 in our example because of our assumption about level probabilities in (6.4.9).

This quantum-mechanical validation of the classical Lorentz model is little short of wonderful. We have shown that, under conditions of low excitation probability, an atomic electron responds to an electric field exactly as if it were bound by a spring to the nucleus. The classical oscillation frequency ω_0 corresponds to a Bohr transition frequency, just as we surmised in Chapters 2 and 3. And if we want the

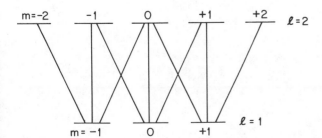

Figure 6.5 Transitions between p and d levels.

classical Lorentz model to agree in quantitative detail with quantum mechanics, we simply introduce the oscillator strength as in (3.7.5).

We have now justified the Lorentz model quantum-mechanically in a special case, but have included only two atomic levels in our calculations. It is not difficult to justify the Lorentz model using the full time-dependent Schrödinger equation (6.2.8) for the probability amplitudes of *all* atomic levels, assuming that the atom remains with high probability in the ground level. Actually, however, it is usually not really essential, under conditions of low excitation probability, to include more than the ground and first excited levels of an atom. The reason for this is made clear by inspection of Table 3.1 for the oscillator strengths of hydrogen: the transition involving the $n = 1$ and $n = 2$ energy levels has a much larger oscillator strength than other transitions involving the ground level. Thus our two-level approach, including only $n = 1$ and $n = 2$ levels, is a reasonable approximation under conditions of high ground-state probability.

6.5 DENSITY MATRIX AND COLLISIONAL RELAXATION

The identification of quantum and classical electron displacements $\langle \mathbf{r} \rangle$ and $\mathbf{x}$ allows the classical Lorentz theory to be put in its correct perspective, as we showed in the preceding section. The same identification assists us with our development of the quantum theory of absorption and emission, because it correctly suggests, via Eq. (6.4.3), that the combinations $a_1^* a_2$ and $a_1 a_2^*$ are more useful than either a_1 or a_2 alone.

We will pursue this approach by obtaining the equations of motion for these combination variables, except that we will focus on the related but simpler quantities c_1 and c_2 defined in (6.3.12). First we adopt a conventional notation and use the Greek letter ρ (rho) to define

$$\rho_{12} \equiv c_1 c_2^* \tag{6.5.1a}$$

$$\rho_{21} \equiv c_2 c_1^* \tag{6.5.1b}$$

$$\rho_{11} \equiv c_1 c_1^* = |c_1|^2 \tag{6.5.1c}$$

$$\rho_{22} \equiv c_2 c_2^* = |c_2|^2 \tag{6.5.1d}$$

The ρ's are elements of the so-called "density matrix" of the atom, as we explain briefly at the end of this section. However, independent of this terminology, it is clear that ρ_{11} and ρ_{22} are just new ways to write the levels' occupation probabilities. The physical meanings of ρ_{12} and ρ_{21} are related to the electron displacement vector through (6.4.3) and (6.3.12), so we can think of ρ_{21} as the complex amplitude of the electron's displacement $\langle \mathbf{r} \rangle$.

By using equations (6.3.14) repeatedly we can easily derive the following equations for the ρ's (see Problem 6.9):

$$\dot{\rho}_{12} = i\Delta\rho_{12} + i\frac{\chi^*}{2}(\rho_{22} - \rho_{11}) \tag{6.5.2a}$$

$$\dot{\rho}_{21} = -i\Delta\rho_{21} - i\frac{\chi}{2}(\rho_{22} - \rho_{11}) \tag{6.5.2b}$$

$$\dot{\rho}_{11} = -\frac{i}{2}(\chi\rho_{12} - \chi^*\rho_{21}) \tag{6.5.2c}$$

$$\dot{\rho}_{22} = \frac{i}{2}(\chi\rho_{12} - \chi^*\rho_{21}) \tag{6.5.2d}$$

The solutions of these equations can be constructed from the solutions for $c_1(t)$ and $c_2(t)$ given in (6.3.16).

However, the equations themselves are not yet in their most useful form. This is because they do not reflect the existence of relaxation processes such as collisions. The same statistical principles employed to treat collisions in Chapter 3 will be used again here. There will be one added complication compared with the classical case, originating with the population variables ρ_{22} and ρ_{11}, which have no classical counterparts.

First we will concentrate on the electron's complex displacement variable ρ_{21}, and on one type of collision, namely purely elastic collisions, which do not affect the populations ρ_{11} and ρ_{22}. If the radiation field present is steady, then $\chi = $ constant. We require the solution for $\rho_{21}(t)$ that vanishes at an earlier time t_1, which (as in Chapter 3) we associate with the time of the atom's most recent collision. We assume that collisions are frequent, so that $t - t_1$ is short enough to neglect changes in $\rho_{22} - \rho_{11}$. The required solution is:

$$\rho_{21}(t)\Big|_{t_1} = -\frac{\chi(\rho_{22} - \rho_{11})}{2\Delta}\left(1 - e^{-i\Delta(t-t_1)}\right). \tag{6.5.3}$$

This can be checked by substitution in (6.5.2b), remembering to hold χ and $\rho_{22} - \rho_{11}$ constant. Next we average this solution over all possible earlier times t_1 at which a collision might have occurred, using the familiar expression (3.9.4) for the probability df that the collision occurred in the time between t_1 and $t_1 + dt_1$:

$$df(t, t_1) = e^{-(t-t_1)/\tau}\frac{dt_1}{\tau} \tag{6.5.4}$$

The result is

$$\langle \rho_{21}(t) \rangle = -\frac{\chi(\rho_{22} - \rho_{11})}{2\Delta\tau}\int_{-\infty}^{t} dt_1\, e^{-(t-t_1)/\tau}\left(1 - e^{-i\Delta(t-t_1)}\right)$$

$$= -\frac{\chi(\rho_{22} - \rho_{11})}{2}\frac{1}{\Delta - i/\tau} \tag{6.5.5}$$

• It is instructive to compare (6.5.5) in detail with its classical counterpart, which is (3.9.7). Since $\langle \mathbf{x} \rangle$ was written there in complex form, the correspondence is $\langle \mathbf{x} \rangle_{cl} \rightarrow \langle \mathbf{r} \rangle_{qm} = 2\mathbf{r}_{12} \langle \rho_{21} \rangle \, e^{-i(\omega t - kz)}$. As we have seen above, we should also take $\rho_{22} \rightarrow 0$ and $\rho_{11} \rightarrow 1$ for a classical comparison. Furthermore the equations for c_1 and c_2, and thus for ρ_{21}, were written in the rotating-wave approximation, which assumes the near-resonance condition $\omega_0 \approx \omega$, or $\omega_0^2 - \omega^2 \approx 2\omega(\omega_0 - \omega) = 2\omega\Delta$. With these adjustments, to ensure similarity of assumptions, we find

$$\langle \mathbf{x} \rangle_{cl} \rightarrow \frac{e}{2m\omega} \frac{\hat{\boldsymbol{\varepsilon}} E_0 \, e^{-i(\omega t - kz)}}{\Delta - i/\tau + 1/2\omega\tau^2} \tag{6.5.6}$$

$$\langle \mathbf{r} \rangle_{qm} \rightarrow \frac{e\mathbf{r}_{12}(\mathbf{r}_{21} \cdot \hat{\boldsymbol{\varepsilon}})}{\hbar} \frac{E_0 \, e^{-i(\omega t - kz)}}{\Delta - i/\tau} \tag{6.5.7}$$

We can drop the final term in the denominator of (6.5.6) because it has already been discarded in Chapter 3. If we replace ω by ω_{21}, as the resonance approximation permits, then the only other difference between the classical and quantum expressions is the difference between $(e/2m\omega_{21})\hat{\boldsymbol{\varepsilon}}$ and $e\mathbf{r}_{12}(\mathbf{r}_{21} \cdot \hat{\boldsymbol{\varepsilon}})/\hbar$, which we determined in Section 6.4 to be just a factor of f, the oscillator strength. Thus the quantum and classical theories are again found to be in complete accord, given the universal use of the oscillator-strength factor in classical formulas—again, *if* we put all of the atomic population in the ground level, $\rho_{22} - \rho_{11} = -1$. •

The same result (6.5.5), obtained by a collision average, can also be reached by a simple modification of the original equation of motion. It can be checked (Problem 6.10) that collisions are already included if we rewrite the ρ_{21} and ρ_{12} equations as follows:

$$\dot{\rho}_{12} = -\left(\frac{1}{\tau} - i\Delta\right)\rho_{12} + i\frac{\chi^*}{2}(\rho_{22} - \rho_{11}) \tag{6.5.8a}$$

$$\dot{\rho}_{21} = -\left(\frac{1}{\tau} + i\Delta\right)\rho_{21} - i\frac{\chi}{2}(\rho_{22} - \rho_{11}) \tag{6.5.8b}$$

As in the classical case, we cannot apply these equations any longer to an individual atom. Instead they represent an "average" atom in the sense of the collision average in (6.5.5). We have omitted averaging brackets $\langle \cdots \rangle$ for notational convenience, as in the classical discussion.

Note that equations (6.5.8) can be read as if the average atom's ρ_{12} and ρ_{21} variables undergo change for two reasons. That is, we can interpret (6.5.8b) as the result of adding two independent rates of change:

$$\rho_{21} = (\dot{\rho}_{21})_{\text{elastic collisions}} + (\dot{\rho}_{21})_{\text{Schrödinger equation}} \tag{6.5.9}$$

where

$$(\dot{\rho}_{21})_{\text{elastic collisions}} = -\frac{1}{\tau}\rho_{21} \tag{6.5.10a}$$

and

$$\left(\dot{\rho}_{21} \right)_{\text{Schrödinger equation}} = -i\Delta\rho_{21} - i\frac{\chi}{2}\left(\rho_{22} - \rho_{11} \right) \qquad (6.5.10b)$$

Such an interpretation will be very helpful in dealing with the effect of collisions on the level populations ρ_{22} and ρ_{11}, for which there are no classical analogs.

The collision rate $1/\tau$ appearing in (6.5.8) is often referred to as the atomic dipole's "dephasing" rate. To understand this we can recall the discussion in Section 3.9, following Eq. (3.9.12). It was assumed there that the orientation of both $\mathbf{x}$ and $d\mathbf{x}/dt$ was random for each dipole after a collision, and thus zero on average. No assumption was made about $(\mathbf{x})^2$ or $(d\mathbf{x}/dt)^2$. That is, both classically and quantum-mechanically we have been discussing only the effects of energy-nonchanging (elastic) collisions. However, inelastic collisions can also occur, in which the electron can change its energy level.

To account for inelastic collisions we simply assert that their effect is to knock population out of levels 1 and 2 into other unspecified levels of the atom at the fixed rates Γ_1 and Γ_2. At the same time we can include the effect of spontaneous photon emission as a special type of "collision" that transfers population between the two specified levels, from 2 to 1. Following Einstein's notation we will denote the spontaneous emission rate by A_{21}. Then we write, in analogy to (6.5.9),

$$\dot{\rho}_{22} = \left(\dot{\rho}_{22} \right)_{\text{collisions}} + \left(\dot{\rho}_{22} \right)_{\text{spontaneous emission}}$$

$$+ \left(\dot{\rho}_{22} \right)_{\text{Schrödinger equation}} \qquad (6.5.11)$$

where

$$\left(\dot{\rho}_{22} \right)_{\text{collisions}} = -\Gamma_2\rho_{22} \qquad (6.5.12a)$$

$$\left(\dot{\rho}_{22} \right)_{\text{spontaneous emission}} = -A_{21}\rho_{22} \qquad (6.5.12b)$$

$$\left(\dot{\rho}_{22} \right)_{\text{Schrödinger equation}} = \frac{i}{2}\left(\chi\rho_{12} - \chi^*\rho_{21} \right) \qquad (6.5.12c)$$

In a similar vein we write the separate contributions to $\dot{\rho}_{11}$:

$$\left(\dot{\rho}_{11} \right)_{\text{collisions}} = -\Gamma_1\rho_{11} \qquad (6.5.13a)$$

$$\left(\dot{\rho}_{11} \right)_{\text{spontaneous emission}} = +A_{21}\rho_{22} \qquad (6.5.13b)$$

$$\left(\dot{\rho}_{11} \right)_{\text{Schrödinger equation}} = -\frac{i}{2}\left(\chi\rho_{12} - \chi^*\rho_{21} \right) \qquad (6.5.13c)$$

Note that the contribution from spontaneous emission to $\dot{\rho}_{11}$ is positive, and just equal to the negative contribution to $\dot{\rho}_{22}$, on the assumption that the atom makes a jump from level 2 to level 1 while emitting a photon spontaneously.

As a result of these collisional and spontaneous contributions we obtain the following equations for the level populations:

$$\dot{\rho}_{11} = -\Gamma_1\rho_{11} + A_{21}\rho_{22} - \frac{i}{2}(\chi\rho_{12} - \chi^*\rho_{21}) \qquad (6.5.14a)$$

$$\dot{\rho}_{22} = -(\Gamma_2 + A_{21})\rho_{22} + \frac{i}{2}(\chi\rho_{12} - \chi^*\rho_{21}) \qquad (6.5.14b)$$

Again, for notational convenience, we do not include brackets $\langle \cdots \rangle$. However, these equations must also be understood as applying only in an average sense to the atoms under consideration.

Finally we must return to the elastic-collision-averaged ρ_{12} and ρ_{21} equations. What is the effect of inelastic collisions on them? A simple answer is based on the obvious relation $|\rho_{12}| = (\rho_{11}\rho_{22})^{1/2}$, which holds before collision averaging. It is a direct consequence of the definitions (6.5.1). This relation says that the effect of collisions on the *magnitude* of ρ_{12}, as distinct from the effect on its *phase*, is directly related to the effect on the level populations in a specific way. That is, if inelastic collisions alone cause ρ_{11} and ρ_{22} to decay, i.e.,

$$\rho_{11}(t)\big|_{\text{collisions}} = \rho_{11}(0)\,e^{-\Gamma_1 t} \qquad (6.5.15a)$$

$$\rho_{22}(t)\big|_{\text{collisions}} = \rho_{22}(0)\,e^{-\Gamma_2 t} \qquad (6.5.15b)$$

which are the solutions to (6.5.12a) and (6.5.13a), then inelastic collisions alone cause $|\rho_{12}(t)|$ to decay as

$$\begin{aligned}
\big|\rho_{12}(t)\big| &= \big[\rho_{11}(t)\,\rho_{22}(t)\big]^{1/2} \\
&= \big[\rho_{11}(0)\,\rho_{22}(0)\,\exp\big[-(\Gamma_1 + \Gamma_2)t\big]\big]^{1/2} \\
&= \big|\rho_{12}(0)\big|\,\exp\left(-\frac{\Gamma_1 + \Gamma_2}{2}t\right) \qquad (6.5.16)
\end{aligned}$$

In words, the effect on ρ_{12} of inelastic collisions alone is to add an extra decay rate to the elastic collision decay rate $1/\tau$. This added rate is just $(\Gamma_1 + \Gamma_2)/2$, one-half the sum of the population decay rates for ρ_{11} and ρ_{22}.

Thus we write our final equations for ρ_{12} and ρ_{21} averaged over both elastic and inelastic collisions (and including spontaneous emission) in the form

$$\dot{\rho}_{12} = -(\beta - i\Delta)\rho_{12} + i\frac{\chi^*}{2}(\rho_{22} - \rho_{11}) \qquad (6.5.17a)$$

$$\dot{\rho}_{21} = -(\beta + i\Delta)\rho_{21} - i\frac{\chi}{2}(\rho_{22} - \rho_{11}) \qquad (6.5.17b)$$

where β is the total relaxation rate:

$$\beta = \frac{1}{\tau} + \tfrac{1}{2}(\Gamma_1 + \Gamma_2 + A_{21}) \qquad (6.5.18)$$

Only the first term in β refers to elastic ("soft" or "dephasing") collisions, but it is often dominant. It is usually likely that an atom suffers many distant soft dephasing collisions for every close collision that is hard enough to cause population changes. Thus, to a good approximation in many cases

$$\beta \approx \frac{1}{\tau} \gg \tfrac{1}{2}(\Gamma_1 + \Gamma_2 + A_{21}) \qquad (6.5.19)$$

To a surprising degree, laser action of the usual kind depends very strongly on this inequality. We will require (6.5.19) in the next chapter.

The effects of collisional dephasing relaxation can be illustrated in detail by integrating the coupled equations for the ρ's (see Problem 6.11). In Figure 6.6 we show the solutions for a wide range of parameters. We have chosen a special case that is free of complications. We take $\Gamma_1 = \Gamma_2 = 0$ (no transfer of probability to levels other than 1 and 2), and we take $\Delta = 0$ (exact resonance). Since $\Gamma_1 = \Gamma_2 = 0$, we have $d\rho_{11}/dt + d\rho_{22}/dt = 0$. Thus, $\rho_{11} + \rho_{22} = 1$ (conservation of probability), and it is enough to determine either ρ_{11} or ρ_{22}. Actually it is most convenient to deal with the inversion $\rho_{22} - \rho_{11}$, since it enters Eqs. (6.5.17) naturally. Furthermore, Eqs. (6.5.17) show that at exact resonance $\rho_{21} + \rho_{12}$ is coupled only to itself and plays no role in the dynamics, so we can pay attention solely to the difference, $\rho_{12} - \rho_{21}$, which in any event is the variable that couples directly to ρ_{11} and ρ_{22}, as Eqs. (6.5.14) make clear.

Thus we can focus on two real variables

$$v = i(\rho_{21} - \rho_{12}) \qquad (6.5.20a)$$

$$w = \rho_{22} - \rho_{11} \qquad (6.5.20b)$$

which obey the equations (at resonance, and in the absence of the Γ_1 and Γ_2 collisions and for real χ)

$$\dot{v} = -\beta v + \chi w \qquad (6.5.21a)$$

$$\dot{w} = -A_{21}(1 + w) - \chi v \qquad (6.5.21b)$$

These equations can be obtained directly from (6.5.14) and (6.5.17) and the definitions in (6.5.20). They are discussed further in Chapter 8. Of course, following (6.5.18) and the absence of Γ_1 and Γ_2, we have $\beta = 1/\tau + A_{21}/2$.

The solutions shown in Figure 6.6 are chosen to illustrate the influence of elastic

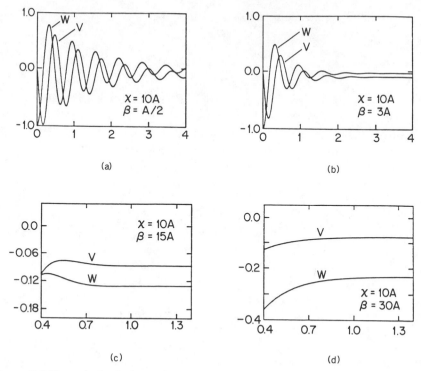

Figure 6.6 Numerical solutions of the v, w equations (6.5.21) for a range of collisional damping rates. Note scale changes.

collisions. As the elastic-collision rate $1/\tau$ increases from zero, the damping parameter β also increases and the oscillatory (so-called "coherent") response of the atom to the applied radiation changes to nonoscillatory ("incoherent") decay. Note the changes in scale needed in the figure to make evident the different types of response.

• The notation used for the ρ's suggests that they are the elements of a 2×2 matrix:

$$\underline{\rho} = \begin{bmatrix} \rho_{11} & \rho_{12} \\ \rho_{21} & \rho_{22} \end{bmatrix} \qquad (6.5.22)$$

This is indeed the case, and quantum statistical mechanics is devoted to the study of such matrices. They were introduced into quantum theory independently by L. D. Landau and J. von Neumann before 1930. For historical reasons $\underline{\rho}$ is called the *density matrix* of the system, and in this case $\underline{\rho}$ is the density matrix of a two-level atom.

The density matrix is a generalization of a related 2×2 matrix

$$
\begin{bmatrix} c_1 c_1^* & c_1 c_2^* \\ c_2 c_1^* & c_2 c_2^* \end{bmatrix}
\tag{6.5.23}
$$

and the two are occasionally confused. Note that they are *not* the same matrix, despite the original definitions in (6.5.1): $\rho_{11} = c_1 c_1^*$, $\rho_{12} = c_1 c_2^*$, etc. This is because the ρ's are now understood to refer to collision averages of $c_1 c_1^*$, etc. Thus the equations (6.5.14) and (6.5.17) for the elements of the density matrix cannot be obtained from simpler equations for c_1 and c_2 separately. [The reader is challenged to try to construct equations for c_1 and c_2 that can be used to obtain (6.5.14) and (6.5.17).] This is the most important sense in which the cc^* combinations are more physical than c's and c^*'s alone.

The existence of the matrix (6.5.22) establishes a definite meaning to the terms "diagonal" and "off-diagonal". Obviously ρ_{11} and ρ_{22} are the elements on the diagonal, and ρ_{12} and ρ_{21} are the off-diagonal elements. This terminology is frequently applied to the damping rates. Referring to Eqs. (6.5.14), we see that Γ_1 and $\Gamma_2 + A_{21}$ can be called diagonal damping rates, and from Eqs. (6.5.17) we see that β is the off-diagonal damping rate. A fundamental relation, obtained from (6.5.18), is illustrated by the inequality

$$
\beta \geq \tfrac{1}{2} (\Gamma_1 + \Gamma_2 + A_{21})
\tag{6.5.24}
$$

As we have seen, because the off-diagonal elements ρ_{12} and ρ_{21} have a complex phase as well as a magnitude, they are susceptible to purely phase-destructive, as well as population-changing, relaxation. •

APPENDIX 6.A   MATRIX ELEMENTS OF THE ELECTRON COORDINATE r FOR LOW-LYING STATES OF HYDROGEN

The coordinate vector $\mathbf{r}$ can be written $\mathbf{r} = r\hat{\mathbf{r}}$, where $r = |\mathbf{r}|$ is the magnitude of $\mathbf{r}$, and $\hat{\mathbf{r}} = \hat{\mathbf{x}} \sin \theta \cos \phi + \hat{\mathbf{y}} \sin \theta \sin \phi + \hat{\mathbf{z}} \cos \theta$ is the radial unit vector (recall Figure 6.3). The matrix element of $\mathbf{r}$ between two states Φ_1 and Φ_2 is defined by (6.3.8):

$$
\mathbf{r}_{12} = \iiint \Phi_1^*(\mathbf{r})\, \mathbf{r}\, \Phi_2(\mathbf{r})\, d^3 r
\tag{6.A.1}
$$

We separate the integral into its radial and angular parts, according to the corresponding separation of the wave functions of Appendix 5.A. That is,

$$
\mathbf{r}_{12} = r_{12} \hat{\mathbf{r}}_{12}
\tag{6.A.2}
$$

where r_{12} is the matrix element of r, and $\hat{\mathbf{r}}_{12}$ is the matrix element of the radial unit vector $\hat{\mathbf{r}}$. In terms of hydrogen wave functions we have

TABLE 6.1 Radial Coordinate Matrix Elements for Atomic Hydrogen

Transition	r_{12}/a_0	r_{12}^2/a_0^2
$1s$–$2p$	1.29	1.66
$1s$–$3p$	0.517	0.267
$1s$–$4p$	0.305	0.093
$2s$–$3p$	3.07	9.4
$2s$–$4p$	1.28	1.64
$2p$–$3s$	0.95	0.9
$2p$–$3d$	4.75	22.5
$2p$–$4d$	1.71	2.92

TABLE 6.2 Cartesian Components of Angular Matrix Elements

Angular Matrix Element	x Component	y Component	z Component
.... s–p Transitions			
$(00\|\hat{\mathbf{r}}\|10)$	0	0	$\sqrt{\frac{1}{3}}$
$(00\|\hat{\mathbf{r}}\|11)$	$-\sqrt{\frac{1}{6}}$	$-i\sqrt{\frac{1}{6}}$	0
$(00\|\hat{\mathbf{r}}\|1-1)$	$\sqrt{\frac{1}{6}}$	$-i\sqrt{\frac{1}{6}}$	0
.... p–d Transitions			
$(10\|\hat{\mathbf{r}}\|20)$	0	0	$\sqrt{\frac{4}{15}}$
$(10\|\hat{\mathbf{r}}\|21)$	$-\sqrt{\frac{1}{10}}$	$-i\sqrt{\frac{1}{10}}$	0
$(10\|\hat{\mathbf{r}}\|2-1)$	$\sqrt{\frac{1}{10}}$	$-i\sqrt{\frac{1}{10}}$	0
$(10\|\hat{\mathbf{r}}\|22)$	0	0	0
$(10\|\hat{\mathbf{r}}\|2-2)$	0	0	0
$(11\|\hat{\mathbf{r}}\|20)$	$\sqrt{\frac{1}{30}}$	$-i\sqrt{\frac{1}{30}}$	0
$(11\|\hat{\mathbf{r}}\|21)$	0	0	$\sqrt{\frac{1}{5}}$
$(11\|\hat{\mathbf{r}}\|2-1)$	0	0	0
$(11\|\hat{\mathbf{r}}\|22)$	$-\sqrt{\frac{1}{5}}$	$-i\sqrt{\frac{1}{5}}$	0
$(11\|\hat{\mathbf{r}}\|2-2)$	0	0	0
$(1-1)\|\hat{\mathbf{r}}\|20)$	$-\sqrt{\frac{1}{30}}$	$-i\sqrt{\frac{1}{30}}$	0
$(1-1)\|\hat{\mathbf{r}}\|21)$	0	0	0
$(1-1)\|\hat{\mathbf{r}}\|2-1)$	0	0	$\sqrt{\frac{1}{5}}$
$(1-1)\|\hat{\mathbf{r}}\|22)$	0	0	0
$(1-1)\|\hat{\mathbf{r}}\|2-2)$	$\sqrt{\frac{1}{5}}$	$-i\sqrt{\frac{1}{5}}$	0

$$r_{12} = \int R^*_{n_1 l_1}(r)\, r R_{n_2 l_2}(r)\, r^2\, dr \tag{6.A.3}$$

$$\hat{\mathbf{r}}_{12} = \int\int Y^*_{l_1 m_1}(\theta,\,\phi)\, \hat{\mathbf{r}}\, Y_{l_2 m_2}(\theta,\,\phi)\, \sin\theta\, d\theta\, d\phi \tag{6.A.4}$$

The radial part depends only on principal quantum numbers and orbital angular momenta (n, l), and we find the values in Table 6.1. For the vector part of $\mathbf{r}_{12}$ we use the notation

$$\hat{\mathbf{r}}_{12} = (l_1 m_1 | \hat{\mathbf{r}} | l_2 m_2) \tag{6.A.5}$$

and we give the Cartesian components of the matrix element separately in Table 6.2.

PROBLEMS

6.1 Show that the factorization (5.3.4) does not work when H is time-dependent. That is, show that the resulting equation for $\Phi(\mathbf{r})$ depends on t, and so $\Phi(\mathbf{r})$ itself must depend on t, contrary to the factorization assumption.

6.2 From the definition of $\mathbf{r}_{12}$ in (6.3.8), show that $(\mathbf{r}_{12})^* = \mathbf{r}_{21}$.

6.3 Every solution $\Phi(\mathbf{r})$ of the Schrödinger equation (5.4.1) remains a solution when multiplied by a constant K, and it remains normalized according to (5.4.6) if K is a pure phasor: $K = e^{i\mu}$. In this sense every $\Phi(\mathbf{r})$ has arbitrary complex phase that can be adjusted for convenience. Assume that an initial phase choice for the wave functions Φ_1 and Φ_2 (perhaps from a table such as given in Appendix 5.A) leads to the complex matrix element $V_{12} = \alpha - i\beta$ (where α and β are real).
(a) Replace Φ_1 by $K\,\Phi_1$. Find the value of K that makes V_{12} real.
(b) What is the new purely real value of V_{12}?

6.4 (a) Find the second-order differential equations satisfied by the probability amplitudes c_1 and c_2 by differentiation and substitution between Eqs. (6.3.14).
(b) Write the general solution for $c_2(t)$ in terms of $\sin(\Omega t/2)$ and $\cos(\Omega t/2)$, and fix the coefficients to fit the initial condition $c_1(0) = 0$, $c_2(0) = 1$.
(c) The initial condition specified in (b) is opposite to the one used to obtain the solutions (6.3.16) in the text. Comment on the differences (if any) between (6.3.16) and the solutions obtained in (b).

6.5 Use the definition of $\mathbf{r}_{12}$ in (6.3.8) to show that $\mathbf{r}_{11} = 0$. To obtain this result you must assume that $|\Phi_1(\mathbf{r})|^2$ is an even function of $\mathbf{r}$. More precisely, $\Phi_1(\mathbf{r})$ must have a definite *parity*, i.e., $\Phi_1(-\mathbf{r})$ is identically the same as either $\Phi_1(\mathbf{r})$ (even parity) or $-\Phi_1(\mathbf{r})$ (odd parity).

6.6   (a) Evaluate z_{12} for the transition $(1, 0, 0) \rightarrow (2, 1, \pm 1)$ in hydrogen.

(b) Evaluate $\hat{\varepsilon} \cdot \mathbf{r}_{12}$ for the $2s$–$4p$ transition in hydrogen, assuming that $\hat{\varepsilon}$ represents circular polarization: $\hat{\varepsilon} = (\hat{x} \pm i\hat{y})/\sqrt{2}$.

6.7   (a) Verify the spherical symmetry expressed in (6.4.24) when applied to p–d transitions (see Figure 6.5), by evaluating $|\hat{x}_{12}|^2$, $|\hat{y}_{12}|^2$, and $|\hat{z}_{12}|^2$, summed over all m's in each case.

(b) Determine the value of the oscillator strength for the $2p$–$3d$ transition in hydrogen.

6.8   Let $\hat{\mathbf{r}}_{12} \equiv (l_1 m_1 |\hat{\mathbf{r}}| l_2 m_2)$, as in Appendix 6.A, and denote by $(\hat{\mathbf{r}}_{12})_i$ the three Cartesian components of the radial unit vector, i.e., $(\hat{\mathbf{r}}_{12})_1 = \hat{x}_{12}$, $(\hat{\mathbf{r}}_{12})_2 = \hat{y}_{12}$, $(\hat{\mathbf{r}}_{12})_3 = \hat{z}_{12}$. Problem 6.7 illustrated that $\sum_{m_1} \sum_{m_2} |(\hat{\mathbf{r}}_{12})_i|^2 =$ constant (*independent of i*). This can be interpreted as spherical symmetry. There are two other symmetries of $\hat{\mathbf{r}}_{12}$. Consider p–d transitions, as in Figure 6.5, and use Table 6.2 to evaluate

(a) $\sum_i \sum_{m_2} |(\hat{\mathbf{r}}_{21})_i|^2$ and show that it has the same value for all m_1's.

(b) $\sum_i \sum_{m_1} |(\hat{\mathbf{r}}_{12})_i|^2$ and show that it has the same value for all m_2's.

The ratio of the values found in (a) and (b) should be $(2l_1 + 1)/(2l_2 + 1)$, the ratio g_1/g_2 of the level degeneracies.

6.9   Derive Eqs. (6.5.2a) and (6.5.2d) from Eqs. (6.3.14).

6.10   Solve Eq. (6.5.8b) by assuming that χ and $\rho_{22} - \rho_{11}$ are constant. Show that the result approaches (6.5.5) in the limit $t \gg \tau$.

6.11   Find the steady-state values of ρ_{21} and ρ_{22} predicted by Eqs. (6.5.14) and (6.5.17). Note that the true steady-state values must be zero, due to "leakage" from levels 1 and 2 into other atomic levels via the collision rates Γ_1 and Γ_2. Assume that this leakage occurs very slowly, and thus put $\Gamma_1 = \Gamma_2 = 0$. Assume $\rho_{11} + \rho_{22} = 1$ initially, and note that $\rho_{11} + \rho_{22}$ is a constant [add Eqs. (6.5.14) to check].

7 EMISSION AND ABSORPTION AND RATE EQUATIONS

7.1 INTRODUCTION

The results of this chapter are based on the fact that the collisional relaxation rate $1/\tau$, or the "off-diagonal" rate β, is usually very fast. For most considerations β is much faster than the rate at which external forces cause electrons to jump between atomic energy levels. For this reason the quasisinusoidal oscillations of atomic probability shown in Figure 6.6 are usually well damped, and the result of an external force such as $\mathbf{F}_{ext} = -e\mathbf{E}$ is only to produce a gradual increase or decrease in probability. The rate of this increase or decrease is what we call the absorption rate or stimulated emission rate. In this chapter we will determine the fundamental origin of these rates and their close association with Einstein's B coefficient and the optical absorption cross section.

The simplification that occurs because β is large is not the first example we have found of the benefits that can follow from recognizing a disparity of time scales. Laser physics is characterized by many different time scales, and when a variable is responding on two time scales at once, the faster process can usually be averaged. That is, fast phenomena can often be treated with sufficient accuracy in an average sense.

We can recall two previous instances of this. In our treatment of collisions in Chapters 3 and 6 we used the extremely short time of a collisional impact to justify overlooking the details of a collision trajectory in favor of simply setting the electron coordinate and velocity equal to zero (on average) following a collision. Also, in our derivation of the fundamental quantum amplitude equations in Chapter 6 we encountered time-dependent factors such as $\exp(\pm 2i\omega t)$, and we discarded them, arguing that they average to zero very rapidly.

7.2 STIMULATED ABSORPTION AND EMISSION RATES

When the inequality (6.5.19) is satisfied, the diagonal and off-diagonal components of the density matrix relax at such different rates that the off-diagonal elements rapidly come to a quasisteady state determined by the instantaneous values of the slowly changing diagonal elements ρ_{11} and ρ_{22}. When this is the case, the off-diagonal variables can be completely eliminated, and so-called rate equations are obtained for the diagonal elements alone, as follows.

The quasisteady-state values of ρ_{12} and ρ_{21} are found by putting $\dot{\rho}_{12} = \dot{\rho}_{21} = 0$ and solving (6.5.17) to obtain

$$\rho_{12} = \frac{i\chi^*/2}{\beta - i\Delta} (\rho_{22} - \rho_{11}) \tag{7.2.1a}$$

$$\rho_{21} = \frac{-i\chi/2}{\beta + i\Delta} (\rho_{22} - \rho_{11}) \tag{7.2.1b}$$

These are called the adiabatic solutions for ρ_{21} and ρ_{12}. Obviously these solutions are not truly constant, but the solutions will be very good approximations if $\rho_{22} - \rho_{11}$ changes very slowly compared with β. It is said that ρ_{12} and ρ_{21} *adiabatically follow* the inversion $\rho_{22} - \rho_{11}$ when solutions (7.2.1) are valid. Thus the inequality (6.5.19) is said to be the condition for *adiabatic following* to occur. Adiabatic following is shown in Figure 6.6d.

The adiabatic approximation is usually used for the purpose of *adiabatic elimination*, which is accomplished by using the adiabatic solutions for ρ_{12} and ρ_{21} in the population equations (6.5.14). We need only the difference $\chi\rho_{12} - \chi^*\rho_{21}$:

$$\chi\rho_{12} - \chi^*\rho_{21} = \frac{i|\chi|^2\beta}{\Delta^2 + \beta^2} (\rho_{22} - \rho_{11}) \tag{7.2.1c}$$

Equations (6.5.14) for ρ_{11} and ρ_{22} then become

$$\dot{\rho}_{11} = -\Gamma_1\rho_{11} + A_{21}\rho_{22} + \frac{|\chi|^2\beta/2}{\Delta^2 + \beta^2} (\rho_{22} - \rho_{11}) \tag{7.2.2a}$$

$$\dot{\rho}_{22} = -(\Gamma_2 + A_{21})\rho_{22} - \frac{|\chi|^2\beta/2}{\Delta^2 + \beta^2} (\rho_{22} - \rho_{11}) \tag{7.2.2b}$$

Now all references to the "off-diagonal" variables ρ_{12} and ρ_{21} have been eliminated, and the two populations ρ_{11} and ρ_{22} are coupled only to each other. Equations (7.2.2) are examples of so-called *population rate equations*.

Note that every term in Eqs. (7.2.2) has a natural interpretation. The terms involving Γ's or A_{21} have been explained already in Section 6.5, and are "spontaneous" in the sense that they do not depend on the existence of an external light field. The terms appearing as a result of adiabatic elimination of ρ_{21} and ρ_{12} are new and correspond to "induced" or "stimulated" processes in the sense that they are proportional to $\chi^2 \propto E_0^2$ and therefore proportional to the intensity of light falling on the atom. If there is no light falling on the atom, these terms vanish. The sign of each of the various induced terms makes it clear whether it corresponds to an absorption or an emission process. Therefore, we can now identify the *absorption rate* and the *stimulated emission rate* of an atom in a light field:

$$\text{absorption rate} = \frac{\left|e\mathbf{r}_{21}\cdot\hat{\epsilon}E_0/\hbar\right|^2 \beta/2}{\left(\omega_{21} - \omega\right)^2 + \beta^2} \qquad (7.2.3a)$$

$$\text{stimulated emission rate} = \frac{\left|e\mathbf{r}_{21}\cdot\hat{\epsilon}E_0/\hbar\right|^2 \beta/2}{\left(\omega_{21} - \omega\right)^2 + \beta^2} \qquad (7.2.3b)$$

The expressions given in (7.2.3) are *state-specific*, and are appropriate for non-degenerate transitions. In the presence of initial- and final-state degeneracy it is necessary to modify the underlying population rate equations (7.2.2). We will do this in Appendix 7.A. In the meantime it is worth noting that, according to (7.2.3) the absorption rate between states 1 and 2 is exactly the same as the stimulated emission rate between states 2 and 1. This is an example of microscopic reversibility.

Notice that in (7.2.2) and (7.2.3) we have the square magnitude $|\chi|^2$. The resulting expressions are correct for the induced transition rates for an arbitrary choice of wave-function phase, not just the choice that makes χ real. Thus, any arbitrary or conventional phase choice (such as is found in a table of wave-functions) may be used for $\mathbf{r}_{21}$ in these formulas. Furthermore, the expressions (7.2.3) can be simplified in many cases when for a variety of practical reasons—unpolarized radiation, rotational or collisional disorientation, etc.—it is the orientational average of $|\chi|^2$ that is relevant. It is not difficult to compute the required average and show that (see Problem 7.1):

$$\left\langle \left|e\mathbf{r}_{21}\cdot\hat{\epsilon}E_0\right|^2 \right\rangle_{\text{orientation}} \equiv \left\langle \left|\mu E_0\right|^2 \right\rangle$$

$$= \tfrac{1}{3} e^2\mathbf{r}_{21}\cdot\mathbf{r}_{12}\left|E_0\right|^2 = \tfrac{1}{3} D^2\left|E_0\right|^2 \qquad (7.2.4)$$

Here we have defined $\mathbf{D}$ and μ as the complex dipole moment and its projection on $\hat{\epsilon}$:

$$\mathbf{D} = e\mathbf{r}_{21} \quad \text{and} \quad \mu = \hat{\epsilon}\cdot\mathbf{D} \qquad (7.2.5)$$

In terms of cartesian components we also have:

$$D^2 = \left|\mathbf{D}\right|^2 = e^2\mathbf{r}_{21}\cdot\mathbf{r}_{12} = e^2\left(\left|x_{12}\right|^2 + \left|y_{12}\right|^2 + \left|z_{12}\right|^2\right) \qquad (7.2.6)$$

We will use the abbreviation R_{12} for the induced transition rates (7.2.3). Note that $R_{12} = R_{21}$. Thus, we can write

$$R_{12} = \frac{\left|\mu E_0/\hbar\right|^2 \beta/2}{(\omega_{21} - \omega)^2 + \beta^2}$$

$$= \frac{D^2 |E_0|^2}{6\hbar^2} \frac{\beta}{\Delta^2 + \beta^2} \qquad \text{(orientation-averaged)} \qquad (7.2.7)$$

These expressions will be useful later in deriving the quantum-mechanical formula for the atomic absorption cross section $\sigma(\Delta)$. We will show in (7.4.2) that R_{12} and σ are related by

$$R_{12} = \sigma(\Delta) \, \Phi \qquad (7.2.8)$$

where Φ is the photon flux (photons/cm²-sec) associated with the light field E_0. For the moment we will just use (7.2.8) to simplify the population rate equations.

7.3 POPULATION RATE EQUATIONS

In studying the population rate equations (7.2.2) we will first multiply them by N, the density of atoms. The products $N\rho_{11}$ and $N\rho_{22}$ are then interpreted as the densities of atoms in levels 1 and 2:

$$N_1 = N\rho_{11}$$

$$N_2 = N\rho_{22} \qquad (7.3.1)$$

and the rate equations (7.2.2) become

$$\dot{N}_1 = -\Gamma_1 N_1 + A_{21} N_2 + \sigma\Phi \, (N_2 - N_1) \qquad (7.3.2a)$$

$$\dot{N}_2 = -\Gamma_2 N_2 - A_{21} N_2 - \sigma\Phi \, (N_2 - N_1) \qquad (7.3.2b)$$

By adding these two equations we find that the rate of change of the total population density is always negative:

$$\dot{N}_1 + \dot{N}_2 = -\Gamma_1 N_1 - \Gamma_2 N_2 \qquad (7.3.3)$$

which indicates that inelastic collisions will eventually take all of the atoms out of levels 1 and 2 into other atomic levels. Equation (7.3.3) implies that, for long times, $N_1 = N_2 = 0$. However Γ_1 and Γ_2 are frequently small relative to A_{21} and $\sigma\Phi$. Then the long-time limit is reached very slowly, and the intermediate-time behavior is of the most interest. In this case we can simply ignore the slow influence of Γ_1 and Γ_2 and write simpler rate equations:

$$\dot{N}_1 = A_{21}N_2 + \sigma\Phi\,(N_2 - N_1) \tag{7.3.4a}$$

$$\dot{N}_2 = -A_{21}N_2 - \sigma\Phi\,(N_2 - N_1) \tag{7.3.4b}$$

Now it is obvious that $\dot{N}_1 + \dot{N}_2 = 0$, so we have

$$N_1 + N_2 = N = \text{constant} \qquad \text{(no inelastic collisions)} \tag{7.3.5}$$

which allows N_1 to be eliminated: $N_1 = N - N_2$. All of the physics is then in the N_2 equation:

$$\begin{aligned}
\dot{N}_2 &= -A_{21}N_2 - \sigma\Phi(2N_2 - N) \\
&= -(A_{21} + 2\sigma\Phi)N_2 + \sigma\Phi N
\end{aligned} \tag{7.3.6}$$

which has the solution (assuming Φ to be constant)

$$N_2(t) = \left[N_2(0) - \frac{N\sigma\Phi}{A_{21} + 2\sigma\Phi} \right] e^{-(A_{21} + 2\sigma\Phi)t} + \frac{N\sigma\Phi}{A_{21} + 2\sigma\Phi} \tag{7.3.7}$$

Several examples of this solution are shown in Figure 7.1. Several special cases of (7.3.7) are important in their own right (see also Problem 7.2). When there is no radiation field present ($\Phi = 0$) we have the result of purely spontaneous decay out of level 2:

$$N_2(t) = N_2(0)\, e^{-A_{21}t} \qquad \text{(spontaneous decay)} \tag{7.3.8}$$

Even if a radiation field is present, most light sources are weak, and a common special case corresponds to $\sigma\Phi \ll A_{21}$, with the upper level initially empty, $N_2(0) = 0$. Then we find

$$N_2(t) = \frac{N\sigma\Phi}{A_{21}} \left[1 - e^{-A_{21}t} \right] \qquad \text{(weak excitation)} \tag{7.3.9}$$

In this case there is a very small steady-state excited population:

$$N_2(t) \rightarrow \frac{N\sigma\Phi}{A_{21}} \ll N \tag{7.3.10}$$

and most of the atoms stay in their ground levels. This is the same case in which $\rho_{22} - \rho_{11} \approx -1$, and the Lorentz classical theory is valid [recall (6.4.9)].

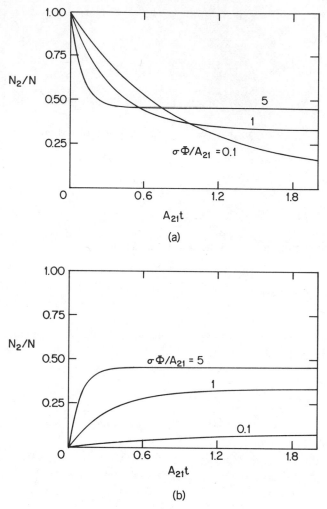

(a)

(b)

Figure 7.1 The upper-level population as a function of time, as predicted by the rate equations (7.3.4). Part (*a*) shows the decrease of population if $N_2(0) = N$, and part (*b*) shows the increase of population if $N_2(0) = 0$. Note that in every case the upper-level population satisfies $N_2 < N/2$ at long enough times.

On the other hand, (7.3.7) shows if the radiation present is sufficiently intense that $\sigma\Phi \gg A_{21}$, then we have a fully saturated atomic transition:

$$N_2(t) = \left[N_2(0) - \frac{N}{2} \right] e^{-2\sigma\Phi t} + \frac{N}{2}$$

$$\rightarrow \frac{N}{2} \quad \text{(saturated steady state)} \qquad (7.3.11)$$

Saturation of a two-level transition means that the population is approximately equally divided between the levels. Examples are shown in Figure 7.1a and b and in Sec. 7.5. This illustrates an important general result: rate equations permit half, at most, of the two-level population to remain in or to be excited into level 2. This result has an important consequence for laser operation: To achieve a steady-state positive inversion *some pumping mechanism other than radiative excitation of the 1–2 transition* must be introduced.

There is so-called *power broadening* associated with the steady-state solutions for the populations. Consider (7.3.7) in the limit $(A_{21} + 2\sigma\Phi)t \gg 1$. In this case we get

$$N_2(\infty) = \frac{N\sigma\Phi}{A_{21} + 2\sigma\Phi} \qquad (7.3.12)$$

Power broadening is made evident by rearranging (7.3.12) as follows:

$$\frac{N_2(\infty)}{N} = \frac{\sigma(\Delta)\,\Phi/A_{21}}{1 + 2\sigma(\Delta)\,\Phi/A_{21}}$$

$$= \frac{\chi^2\beta/2A_{21}}{\Delta^2 + \beta^2 + \chi^2\beta/A_{21}} \text{ (steady state)} \qquad (7.3.13)$$

where the last step uses (7.2.7) and (7.2.8). The form of (7.3.13) is obviously a Lorentzian function of Δ. It is shown in Fig. 7.2. Since χ^2 is proportional to E_0^2, it is also obvious that the half-width of the Lorentzian is dependent on the intensity (or power) of the incident light field. The "power-broadened" half-width is easily seen to be

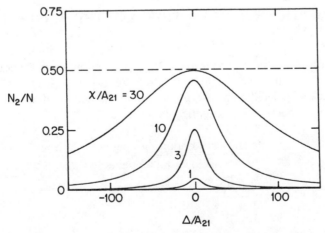

Figure 7.2 The upper level population in steady state, as a function of detuning, for four different Rabi frequencies χ. Power broadening is clearly evident in the higher curves.

$$\Delta_{1/2} = \left(\beta^2 + \frac{\chi^2 \beta}{A_{21}} \right)^{1/2}$$

$$= \beta \left(1 + \frac{2\sigma(\Delta=0)\Phi}{A_{21}} \right)^{1/2} \tag{7.3.14}$$

This power-dependent width is evident in Figure 7.2.

• Another example of saturation is already exhibited by the solution (6.3.18b) for the upper level population. This solution is valid in the absence of any damping (i.e., for times so short that the adiabatic approximation is not valid, but long enough to contain many cycles Ωt):

$$\rho_{22}(t) = \frac{1}{2} \left(\frac{\chi}{\Omega} \right)^2 (1 - \cos \Omega t) \tag{7.3.15}$$

$$\bar{\rho}_{22} = \frac{\chi^2/2}{\Delta^2 + \chi^2} \qquad \text{(time average)} \tag{7.3.16}$$

Equation (7.3.16) shows that the time average of ρ_{22} is proportional to a Lorentzian function of Δ, with its maximum value $\frac{1}{2}$ at $\Delta = 0$. The half-maximum of the inversion is at $\Delta = \chi$, and the power-broadening half-width $\Delta_{1/2}$ is obviously given here by

$$\Delta_{1/2} = \chi \tag{7.3.17}$$

Note that (7.3.17) and (7.3.14) do not agree, and they should not. They represent two physically distinct kinds of saturation, appropriate at very short times and very long times, respectively. For further discussion of saturation, see Section 7.5. •

7.4 ABSORPTION CROSS SECTION AND THE EINSTEIN B COEFFICIENT

The quantum mechanical expression for the absorption cross section associated with the 1–2 transition can be obtained from (7.2.8) almost by inspection. The transition rate associated with absorption is R_{12}. The definition of absorption cross section given in Section 3.6 is applicable:

$$\sigma_{\text{abs}} = \frac{\text{energy absorption rate per atom in level 1}}{\text{incident radiant energy flux}}$$

$$= \frac{\hbar\omega R_{12}}{\frac{1}{2} c\epsilon_0 E_0^2} \tag{7.4.1}$$

Note that since $\frac{1}{2} c\epsilon_0 E_0^2 = I = \hbar\omega\Phi$, where Φ is the photon flux (photons/cm²-sec), we can also convert (7.4.1) into a usefully compact expression for R_{12}:

$$R_{12} = \sigma_{\text{abs}} \Phi \tag{7.4.2}$$

We anticipated this result in Eq. (7.2.8). The expression (7.4.1) for σ can be simplified easily using (7.2.8), and we find

$$\sigma = \frac{D^2 \omega}{3\epsilon_0 \hbar c} \frac{\beta}{(\omega_{21} - \omega)^2 + \beta^2} \tag{7.4.3}$$

in analogy to the classical formula (3.6.10).

The absorption cross section σ takes a special form at resonance and in the absence of collisions, i.e., when $\omega_{21} = \omega$ and $\beta = A_{21}/2$. Consider (7.4.3) in this limit:

$$\sigma \rightarrow \frac{2D^2 \omega}{3\epsilon_0 \hbar c A_{21}} \tag{7.4.4}$$

Now insert the value for A_{21} given in (7.6.5) in the section on spontaneous emission. One finds the extremely simple formula

$$\sigma = \lambda^2/2\pi \quad \text{(on resonance, natural broadening)} \tag{7.4.5}$$

This cross section depends only on the light wavelength, and is completely independent of the properties of the absorbing atom. Its value is the largest possible for σ and is obviously easy to estimate, given λ. For an optical wavelength around 600 nm we obtain

$$\sigma_{\text{max}} \approx 6 \times 10^{-10} \text{ cm}^2 \tag{7.4.6}$$

A typical optical cross section is six orders of magnitude smaller than this.

In the quantum theory the extinction coefficient a is related to the cross section by the density of absorbing atoms through the defining relation

$$a = N\sigma \tag{7.4.7}$$

just as in classical theory. Thus the quantum expression for a is

$$a = \frac{N|\mu|^2 \omega}{\epsilon_0 \hbar c} \frac{\beta}{\Delta^2 + \beta^2}$$

$$= \frac{ND^2 \omega}{3\epsilon_0 \hbar c} \frac{\beta}{(\omega_{21} - \omega)^2 + \beta^2} \quad \text{(orientation-averaged)} \tag{7.4.8}$$

Values for a vary enormously, given the variability of both σ and N. We can regard

$a \sim 0.01\text{--}100 \text{ cm}^{-1}$ as the range of main interest in laser physics. The so-called absorption length or Beer length a^{-1} then lies between $100 \ \mu\text{m}$ and 1 m.

We have shown that the absorption rate coefficient associated with the 1–2 transition can be written several ways, for example,

$$R_{12} = \sigma \Phi = \sigma I / \hbar \omega \qquad (7.4.9)$$

However, these all refer to absorption of monochromatic incident light that is nearly resonant with the atomic transition frequency ($\omega \approx \omega_{21}$). As in the classical case, we can generalize the absorption formula so that it applies to broadband incident light. In the quantum case one result is an expression for the Einstein B coefficient.

Following the same line adopted in our classical discussion (Section 3.8), we take account of the dependence of σ and Φ on the frequency $\nu = \omega / 2\pi$, and then sum over ν to include all frequencies; e.g., $R_{12} = \sigma \Phi \rightarrow \Sigma \sigma(\nu) \ \Phi_\nu$. We rewrite (7.4.3) as

$$\sigma = \frac{D^2 \omega}{3\epsilon_0 \hbar c} \frac{2\pi \ \delta\nu_{21}}{(2\pi)^2 (\nu - \nu_{21})^2 + (2\pi)^2 (\delta\nu_{21})^2}$$

$$= \frac{D^2 \omega}{6\epsilon_0 \hbar c} L(\nu), \qquad (7.4.10)$$

where $L(\nu)$ is the normalized Lorentzian atomic lineshape function discussed in Chapter 3 with $2\pi \delta\nu_{21} = \beta$:

$$L(\nu) = \frac{\delta\nu_{21}/\pi}{(\nu - \nu_{21})^2 + (\delta\nu_{21})^2} \qquad (7.4.11)$$

For an arbitrary (possibly not Lorentzian) atomic lineshape function $S(\nu)$ the total transition rate R_{12} can be written, using $I_\nu = h\nu \Phi_\nu$,

$$R_{12} = \frac{D^2 \omega}{6\epsilon_0 \hbar c} \sum_\nu \frac{S(\nu) \ I_\nu}{h\nu} \qquad \text{(many discrete frequencies)} \qquad (7.4.12)$$

When a continuous band of frequencies contributes to (7.4.12), then the intensity I_ν at frequency ν must be identified with $I(\nu) \ d\nu$, where $I(\nu)$ is the density-in-frequency of intensity (W/cm^2-Hz) over the interval $[\nu, \nu + d\nu]$, and the sum must be replaced by an integral. It is conventional to introduce the spectral energy density ρ by $I(\nu) = c\rho(\nu)$, and in terms of $\rho(\nu)$ the continuous limit is

$$R_{12} = \frac{D^2 \omega}{6\epsilon_0 \hbar} \int_0^\infty S(\nu) \frac{\rho(\nu)}{h\nu} d\nu \qquad \text{(continuous band of frequencies)} \qquad (7.4.13)$$

The broadband limit is reached whenever $\rho(\nu)$ is a broad function of ν by comparison with $S(\nu)$. Since $S(\nu)$ is normalized to unit area, when it is sharply peaked it is equivalent to a Dirac delta function. For example, in the Lorentzian case we have the well-known representation (when $\delta\nu_{21} \to 0$)

$$\frac{\delta\nu_{21}/\pi}{(\nu - \nu_{21})^2 + (\delta\nu_{21})^2} \to \delta(\nu - \nu_{21}) \tag{7.4.14}$$

In the broadband limit the integral in (7.4.13) can be evaluated with the delta-function approximation:

$$R_{12} = \frac{D^2}{6\epsilon_0\hbar^2}\rho(\nu_{21}) \quad \text{(broadband limit)}$$

$$= B\rho(\nu_{21}) \tag{7.4.15}$$

The coefficient of $\rho(\nu_{21})$ in the transition-rate formula is the B coefficient introduced by Einstein on empirical grounds in 1916. An explicit expression for it was not available until a decade later, when it could be derived from wave mechanics. From (7.4.15) we have

$$B = \frac{D^2}{6\epsilon_0\hbar^2} \tag{7.4.16}$$

Thermal radiation is usually an excellent example of broadband light. Unfiltered blackbody radiation, for example, has a bandwidth comparable to its peak frequency [recall Section 1.2]. Einstein's interest in Planck's formula and the extensive experimental work on blackbody radiation around 1900 made it natural for him to develop a theory of transition rates from the broadband viewpoint. One is now much more likely to encounter very narrowband laser light in the laboratory. The formula (7.4.15) must not be confused with its narrowband counterpart (7.4.9), i.e., the one-frequency limit of (7.4.12):

$$R_{12} = \frac{D^2}{6\epsilon_0\hbar^2 c}S(\nu)\,I_\nu \quad \text{(narrowband limit)} \tag{7.4.17}$$

Of course, if the B coefficient is viewed just as a collection of atomic constants, it can also be used in the narrowband rate formula if $\nu \approx \nu_{21}$:

$$R_{12} = \frac{1}{c}BS(\nu)\,I_\nu \tag{7.4.18}$$

Finally, by comparing (7.4.18) with the original expression $R_{12} = \sigma\Phi$, and using $h\nu\Phi_\nu = I_\nu$, we can obtain an expression for σ in terms of the B coefficient:

$$\sigma(\nu) = \frac{h\nu}{c} BS(\nu) \qquad (7.4.19)$$

and, by using (7.4.16) and (7.6.5) below, an expression in terms of the A coefficient:

$$\sigma(\nu) = \frac{\lambda^2 A_{21}}{8\pi} S(\nu) \qquad (7.4.20)$$

It is helpful in remembering the differences between the narrowband and broadband limits to keep in mind that each formula, (7.4.15) and (7.4.17), explicitly contains the lineshape function that is broader, ρ or S as the case may be, but it is evaluated at the peak frequency of the narrower. That is, in the broadband formula (7.4.15) the broad ρ appears, but it is evaluated at ν_{21}. In the narrowband formula (7.4.17) the relatively broad S appears, but it is evaluated at ν.

A more thorough discussion of thermal radiation and the B coefficient and its relation to the A coefficient, including degeneracy factors, is given in Appendix 7.A.

7.5 STRONG FIELDS AND SATURATION

The two steady-state formulas (7.3.13) for N_2 provide more than the expression (7.3.14) for the power-broadened half-width. They provide a criterion by which one can determine whether or not a radiation field is "strong." We will call a field *strong* if it saturates N_2, that is, if $N_2 \sim N/2$. This criterion can be expressed in different ways. For example, the field is strong if

$$\Phi \gg A_{21}/2\sigma \qquad (7.5.1)$$

or if

$$\chi^2 \beta/A_{21} \gg \Delta^2 + \beta^2 \qquad (7.5.2)$$

which come from the two equivalent formulas in (7.3.13).

The first of these criteria is of greater practical use, since A_{21} and σ are both tabulated and accessible to direct measurement. The ratio $A_{21}/2\sigma$ is minimized for pure radiative broadening at exact resonance. In this (very) special case values in the neighborhood of

$$\sigma \sim 10^{-10} \text{ cm}^2, \qquad A_{21} \sim 10^8 \text{ sec}^{-1}$$

can be used, with the result that (for allowed atomic ground-state transitions) the saturation flux

$$\Phi^{\text{sat}} = A_{21}/2\sigma \qquad (7.5.3)$$

has the value

$$\Phi^{\text{sat}} \sim 10^{18} \text{ photons/cm}^2\text{-sec} \qquad \text{(on resonance)}$$

The corresponding saturation intensity

$$I_{\text{sat}} = \hbar\omega\Phi^{\text{sat}}$$

$$= \hbar\omega A_{21}/2\sigma \qquad (7.5.4)$$

has a value in the neighborhood of

$$I^{\text{sat}} \sim 0.3 \text{ W/cm}^2 \qquad \text{(on resonance)}$$

since $\hbar\omega \sim$ 2–4 eV for an optical-frequency photon and 1 eV $= 1.6 \times 10^{-19}$ J. However, collisional and Doppler broadening and other factors can easily reduce σ by six or eight orders of magnitude, even on resonance, leading to the estimate

$$I^{\text{sat}} \sim 1 \text{ MW/cm}^2 \qquad (7.5.5)$$

In some specialized cases (e.g., multiphoton ionization) the strong-field threshold can be much higher:

$$I^{\text{sat}} \sim 10^6 \text{ MW/cm}^2$$

or more. Thus, while there is a definite formula, namely (7.5.1), which determines whether a radiation field is "strong" or not, the value of intensity that is "strong" can vary by very many orders of magnitude, depending on the situation.

Finally we observe that these strong-field saturation criteria also provide another way to state the limit of validity of the classical Lorentz theory. To see this we compute the steady-state dipole moment of the atomic electron. According to (6.4.3), and using (6.3.12) and (6.5.1), the electron's complex displacement amplitude is $2\mathbf{r}_{12}\rho_{21}e^{-i\omega t}$. From (7.2.1), (7.3.1), (7.3.12) and $N_1 + N_2 = N$, we compute the steady-state complex dipole amplitude

$$2e\mathbf{r}_{12}\rho_{21} = \frac{1}{1 + 2\sigma\Phi/A_{21}} \frac{e\mathbf{r}_{12}(e\mathbf{r}_{21} \cdot \hat{\boldsymbol{\varepsilon}})}{\hbar} \frac{1}{\Delta - i\beta} E_0 \qquad (7.5.6)$$

Assuming an atom in free space [recall Section 6.4], the oscillator strength f converts this expression to

$$2e\mathbf{r}_{12}\rho_{21} = \frac{1}{1 + 2\sigma\Phi/A_{21}} \frac{e^2 f}{2m\omega} \frac{1}{\Delta - i\beta} \hat{\boldsymbol{\varepsilon}} E_0 \qquad (7.5.7)$$

and it is easy to check that the complex dipole moment $2er_{12}\rho_{21}$ agrees with the classical expression (3.6.3) under two conditions:

i . the oscillator strength f must be inserted into the classical formula, and

ii. the radiation field must be far below saturation level,

$$\Phi \ll A_{21}/2\sigma \equiv \Phi^{\text{sat}} \qquad (7.5.8a)$$

Thus under these two conditions the classical Lorentzian theory is adequate and quantum-mechanical corrections are unnecessary. Of course we recognize the first condition from our comparison of classical and quantum electron displacement formulas in Section 6.4. What is new here is condition ii, which replaces the relatively vague statement (6.4.9), namely $\rho_{11} - \rho_{22} \approx 1$, with a quantitative statement about radiation intensity, i.e., $\Phi \ll \Phi^{\text{sat}}$, or

$$I \ll \frac{\hbar\omega A_{21}}{2\sigma} \equiv I^{\text{sat}} \qquad (7.5.8b)$$

7.6 SPONTANEOUS EMISSION AND THE EINSTEIN A COEFFICIENT

An atom in an excited state does not remain in that state indefinitely. It will eventually drop to a state of lower energy, even in the absence of any field or other atoms. This is known from experiments in which collisions are very infrequent, and each atom is effectively free of any external influence. Associated with this decay of an excited state is the *spontaneous emission* of a photon. Spontaneous emission would occur even for a single excited atom in a perfect vacuum. We have used the symbol A_{21} to indicate this process in the rate equations of Section 7.3.

Most of the light around us is ultimately the result of spontaneous emission. The phenomenon goes by various names, depending on the context. The term *luminescence*, for instance, is often used to describe spontaneous emission from atoms or molecules excited by some means other than by heating. If excitation occurs in an electric discharge such as a spark, the term *electroluminescence* is used. If the excited states are produced as a by-product of a chemical reaction, the emission is called *chemiluminescence*, or if this occurs in a living organism, *bioluminescence*. The luminescence of fireflies, for example, results from a complicated sequence of chemical reactions. *Fluorescence* refers to spontaneous emission from an excited state produced by the absorption of light. *Phosphorescence* describes the situation in which the emission persists long after the exciting light is shut off, and is associated with a metastable (long-lived) level, as illustrated in Figure 7.3. Phosphorescent materials are used, for instance, in toy figurines that glow in the dark.

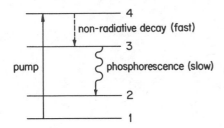

Figure 7.3 A model of a phosphorescent molecular system. The molecule is pumped to level 4 by absorption of radiation, and then decays to level 3. Level 3 is metastable, i.e., it decays very slowly. As a result the molecule continues to fluoresce long after the source of radiation has been shut off.

In most situations a given excited level has several or many spontaneous decay channels. Thus, the general case is somewhat more complex than our notation A_{21} implies. For example, the exponential decay law,

$$N_2(t) = N_2(0)\, e^{-A_{21}t} \qquad (7.6.1)$$

indicates that the population of the upper level, labeled 2, decays to zero with the characteristic time constant $\tau_2 = 1/A_{21}$. However, if level 2 has other decay channels, they will obviously shorten the effective lifetime of level 2 and this expression for τ_2 will be incomplete.

According to quantum mechanics, the effective spontaneous radiative lifetime of level 2 is determined by the sum of all possible radiative channel rates:

$$A_2 = \sum_m A_{2m} \qquad (7.6.2)$$

and the correct expression for upper-state lifetime is given by

$$\tau_2 = \frac{1}{A_2} = \frac{1}{\sum_m A_{2m}} \qquad (7.6.3)$$

where the summation is over all states m with energy E_m lower than the energy of level 2 (see Figure 7.4). Numerical values of the "*A* coefficients" A_{nm} for atomic transitions are usually included in tables of oscillator strengths. Radiative lifetimes of excited atomic states are typically on the order of 10–100 nsec.

Spontaneous emission is a purely quantum-mechanical phenomenon in respect to its subtle statistical features. No theory based entirely on classical concepts has ever been made fully consistent with all of the facets of spontaneous emission. Nevertheless there are some common principles at work in the classical and quantum pictures of the emission of radiation. These stem ultimately from the fact that Maxwell's differential equations of electromagnetism are fully valid quantum-mechanically as well as classically. The classical principle that radiation from a non-relativistic source depends on the square of dipole acceleration [recall Eq. (2.5.14)] is reflected in the quantum theory. The rate A_{nm} for spontaneous emission on the n–m transition involves the square of the dipole matrix element $\mathbf{r}_{nm}$. Moreover $\mathbf{r}_{nm}$

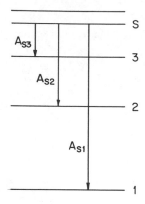

Figure 7.4 An atomic state s may make spontaneous transitions to lower states 1, 2, 3, . . . with rates A_{s1}, A_{s2}, ... The total spontaneous decay rate of state s is $A_s = A_{s1} + A_{s2} + \cdots$.

is involved in spontaneous emission in the same way that it is involved in stimulated emission, via the projection of $e\mathbf{r}_{nm}$ on the polarization vector of the radiation: $|e\mathbf{r}_{nm}\cdot\hat{\varepsilon}|^2$. An important difference is that neither the direction of emission $\hat{\mathbf{k}}$ of a spontaneous photon, nor the direction $\hat{\varepsilon}_k$ of its polarization, is predictable in advance. Thus the square of the matrix element must be integrated over all the directions $\hat{\mathbf{k}}$ and summed over the polarization directions.

The complete quantum-mechanical expression for the spontaneous emission rate can be shown to be

$$A_{nm} = \frac{1}{4\pi\epsilon_0}\frac{e^2\omega_{nm}^3}{2\pi\hbar c^3}\sum_{\text{pol}}\int |\mathbf{r}_{nm}\cdot\hat{\varepsilon}_k|^2 \sin\theta_k\, d\theta_k\, d\phi_k$$

$$= \frac{1}{4\pi\epsilon_0}\frac{e^2\omega_{nm}^3}{2\pi\hbar c^3} 2 \times 4\pi\, \tfrac{1}{3}\big[|x_{nm}|^2 + |y_{nm}|^2 + |z_{nm}|^2\big] \quad (7.6.4)$$

The evaluation of the sum and integral can be worked out easily (see Problem 7.3), but the various factors can be understood directly. The factor 4π represents the integration over the angles of $\hat{\mathbf{k}}$. The $\tfrac{1}{3}$ and the sum of squares of the components of the matrix element express the spherical averaging of the integration, and the factor 2 comes from the polarization sum—the transverse character of the emitted light restricts the polarization vector $\hat{\varepsilon}_k$ to two components perpendicular to $\hat{\mathbf{k}}$. The final, and customary, expression for the spontaneous emission rate is

$$A_{nm} = \frac{1}{4\pi\epsilon_0}\frac{4}{3}\frac{D_{nm}^2\omega_{nm}^3}{\hbar c^3} = \frac{D_{nm}^2\omega_{nm}^3}{3\pi\epsilon_0\hbar c^3} \quad (7.6.5)$$

In Gaussian cgs units the expression is the same except that the factor $1/4\pi\epsilon_0$ is removed. Here D_{nm}^2 is defined as in (7.2.6):

$$D_{nm}^2 = e^2\mathbf{r}_{nm}\cdot\mathbf{r}_{mn} = e^2|\mathbf{r}_{nm}|^2 \quad (7.6.6)$$

We should keep in mind the quantum-mechanical origin of spontaneous emission. Suppose we apply (7.6.1) to a single atom. According to quantum mechanics, this formula gives only the *probability* at time t that the atom will be found in state 2. Without a measurement we cannot say with certainty whether the atom at time t is in state 2 or in some lower-energy state, nor can we predict exactly when a quantum jump to a lower state will occur.

Under most circumstances in gaseous media the broadening of spectral lines is due mainly to collisions (pressure broadening) and atomic motion (Doppler broadening). However, even if these effects could be neglected, a spectral line would still have a nonvanishing "natural" width. This natural line broadening is associated with spontaneous emission.

If A_2 and A_1 are the spontaneous emission rates of the upper and lower states, respectively, of a transition, then the natural radiative lineshape function $S(\nu)$ given by quantum mechanics is the Lorentzian function

$$S(\nu) = \frac{\delta\nu_{\mathrm{rad}}/\pi}{(\nu - \nu_0)^2 + \delta\nu_{\mathrm{rad}}^2} \qquad (7.6.7)$$

where the *natural linewidth* $\delta\nu_{\mathrm{rad}}$ is

$$\delta\nu_{\mathrm{rad}} = \frac{1}{4\pi}(A_2 + A_1) \qquad (7.6.8)$$

and A_2 and A_1 are the total spontaneous emission rates, as given by (7.6.2). In particular, A_2 includes A_{21}, the rate of spontaneous emission on the transition 2 → 1. Thus if state 1 is the ground state, and state 2 the first excited state, then $A_2 = A_{21}$ and $A_1 = 0$. Then $\delta\nu_{\mathrm{rad}} = A_{21}/4\pi$ or, in terms of angular frequency, $\delta\omega_{\mathrm{rad}} = 2\pi\,\delta\nu_{\mathrm{rad}} = A_{21}/2$.

In a gas the total homogeneous linewidth of a transition is $\delta\nu_0 = \delta\nu_c + \delta\nu_{\mathrm{rad}}$, where $\delta\nu_c$ and $\delta\nu_{\mathrm{rad}}$ are respectively the collision-broadened and natural linewidths. As the pressure goes to zero, $\delta\nu_c$ also goes to zero and the only contribution to the homogeneous linewidth is the natural radiative width. If furthermore the inhomogeneous (Doppler) broadening is negligible, then the natural line broadening associated with spontaneous emission is the only line-broadening mechanism; the spectral line is then homogeneously broadened, with lineshape function $S(\nu)$ given by (7.6.7). Unlike other line-broadening mechanisms, spontaneous emission cannot ordinarily be reduced by changing certain variables like pressure or temperature in an experiment. This is why radiative line broadening associated with spontaneous emission is called natural line broadening. The term "natural" is meant to imply that the linewidth $\delta\nu_{\mathrm{rad}}$ is immutable, i.e., that $\delta\nu_{\mathrm{rad}}$ is fundamentally the smallest possible linewidth that can be realized in any experiment. This is not strictly true, however. For instance, we will see that laser radiation can be narrower than the "natural" linewidth of the laser transition. The term *radiative broadening* is therefore better for the line broadening due to spontaneous emission.

7.7 THERMAL EQUILIBRIUM RADIATION

In thermal equilibrium the populations N_1 and N_2 of atoms in states 1 and 2, respectively, are related by (abbreviating $E_2 - E_1 = h\nu_0$)

$$\frac{(N_2)_T}{(N_1)_T} = e^{-(E_2 - E_1)/kT} = e^{-h\nu_0/kT} \tag{7.7.1}$$

where $k = 1.38 \times 10^{-23}$ J/K [$k \approx (1/40)$ eV/300K] and T is the absolute temperature. Equation (7.7.1) follows from the principles of statistical mechanics. Specifically, it is a consequence of the fact that, in thermal equilibrium, the occupation probability of a state of energy E is proportional to the Boltzmann factor $e^{-E/kT}$.

Because there is some nonvanishing probability for each atom to be in the upper state of the $2 \rightarrow 1$ transition, there will be spontaneous emission in thermal equilibrium, and furthermore the spontaneously emitted radiation can stimulate the emission of more radiation at frequency ν_0. Of course, this thermal radiation can also be absorbed, and *in thermal equilibrium the absorption and emission processes must balance each other in just such a way that the Boltzmann condition* (7.7.1) *is satisfied*. Applying this condition to the steady-state populations obtainable from (7.3.4), we have (again writing R_{12} for $\sigma\Phi$)

$$\frac{(N_2)_T}{(N_1)_T} = \frac{R_{12}}{A + R_{12}} = \frac{1}{A/R_{12} + 1} = e^{-h\nu_0/kT}$$

or

$$R_{12} = \frac{A}{e^{h\nu_0/kT} - 1} \tag{7.7.2}$$

This relation between the stimulated and spontaneous emission rates R_{12} and A_{21} provides an expression for the spectral energy density $\rho(\nu_0)$ of thermal radiation at the transition frequency ν_0: (7.4.15) and (7.7.2) together imply that

$$\rho(\nu_0) = \frac{A/B}{e^{h\nu_0/kT} - 1} \tag{7.7.3}$$

The ratio A/B of the Einstein A and B coefficients follows easily from (7.4.16) and (7.6.5):

$$\frac{A}{B} = \frac{2\hbar\omega_0^3}{\pi c^3} = \frac{8\pi h\nu_0^3}{c^3} \tag{7.7.4}$$

Thus (7.7.3) implies the *Planck spectrum*

$$\rho(\nu_0) = \frac{8\pi h\nu_0^3/c^3}{e^{h\nu_0/kT} - 1} \qquad (7.7.5)$$

This formula is the correct expression for the temperature dependence of thermal radiation, and it was the first formula in which the quantum constant h appeared. Planck fixed the value of h empirically, in order for the formula (7.7.5) to agree with experimental data on thermal radiation. Note that the requirement that the Boltzmann condition (7.7.1) be fulfilled—that is, the requirement of thermal equilibrium—leads to a *universal* spectral density function (7.7.5). In other words, the Planck spectrum is independent of the specific atomic or molecular properties of the system, such as transition dipole moment, linewidth, oscillator strength, etc., and depends only on its temperature.

The Boltzmann condition (7.7.1) implies that $(N_2)_T < (N_1)_T$ for any finite temperature T: in thermal equilibrium any pair of levels will have more atoms in the lower than the upper. This means that an externally applied radiation field will lose energy to the atoms. It is convenient to define an ideal *blackbody* as an object that absorbs *all* the radiation, of *any* frequency, incident upon it. In such a black-body the absorption and emission of radiation are exactly balanced in a steady state of thermal equilibrium, and any radiation incident upon its surface would be completely absorbed.

Although no perfect blackbody is known to exist, it is possible to construct an excellent approximation to an ideal blackbody surface. Consider a cavity inside a metal block, with a small hole drilled through to provide an opening to the outside, as illustrated in Figure 7.5. Any radiation incident on the hole from the outside is repeatedly reflected within the cavity, and eventually absorbed, so that the amount of incident radiation escaping back through the hole to the outside is negligibly small. The hole itself thus acts as the "surface" of a blackbody. Furthermore a small amount of equilibrium thermal radiation inside the cavity, produced by spontaneous and stimulated emission from the cavity walls, can escape through the hole to the outside. This radiation escaping through the hole is a sampling of the

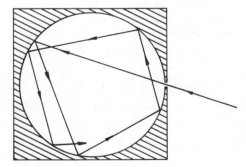

Figure 7.5 A cavity inside a metal block kept at constant temperature. A small hole allows radiation to enter the cavity, and the radiation is diffused by repeated internal scattering. The hole itself, therefore, acts as the surface of a blackbody.

thermal radiation inside the cavity, and therefore the spectral density of this "cavity radiation" should satisfy the Planck formula. If the block is placed inside an oven it can be kept in thermal equilibrium at some fixed temperature T. The earliest accurate measurements of the spectrum of such cavity radiation were carried out by O. Lummer and E. Pringsheim in 1899. These and other measurements, particularly those of H. Rubens and F. Kurlbaum, motivated Planck to reconsider the existing theory of thermal radiation, and led to his announcement of formula (7.7.5) at the 19 October 1900 session of the Prussian Academy of Science.

• From a historical viewpoint the blackbody problem is valuable not only because of intrinsic interest in thermal radiation, but also because it illustrates a failure of classical physics. It is therefore worthwhile to consider briefly a classical approach to the blackbody problem, using the electron oscillator model of an atom.

In thermal equilibrium the rate at which an atom absorbs radiation from the field in the cavity must exactly balance the rate at which it gives energy back to the field. The classical energy absorption rate for an electron oscillator of natural frequency ν_0 is given in Table 3.2. For broadband radiation, such as thermal radiation, we find, without f,

$$\frac{dW_A}{dt} = \frac{e^2}{4\epsilon_0 m} \rho(\nu_0) \tag{7.7.6}$$

The rate at which the oscillator radiates electromagnetic energy is given by (2.5.15):

$$(Pwr) = \frac{e^2 \mathbf{a}^2}{6\pi\epsilon_0 c^3} \tag{7.7.7}$$

where $\mathbf{a}$ is the acceleration of the electron. For an oscillation at frequency ν_0 we have

$$\mathbf{a} = -(2\pi\nu_0)^2 \mathbf{x} \tag{7.7.8}$$

and so

$$(Pwr) = \left(\frac{8\pi^3 e^2 \nu_0^4}{3\epsilon_0 c^3}\right) \mathbf{x}^2 \tag{7.7.9}$$

According to a general principle of classical statistical mechanics, the average energy of a one-dimensional harmonic oscillator in thermal equilibrium is kT, and this energy is distributed equally between the kinetic and potential energies of the oscillator. Thus our classical three-dimensional electron oscillators must have an average potential energy $\frac{3}{2} kT$:

$$\left(\tfrac{1}{2} m\omega_0^2 \mathbf{x}^2\right)_{avg} = 2\pi^2 m\nu_0^2 \left(\mathbf{x}^2\right)_{avg} = \tfrac{3}{2} kT$$

or

$$\left(\mathbf{x}^2\right)_{avg} = \frac{3kT}{4\pi^2 m\nu_0^2} \tag{7.7.10}$$

Therefore in thermal equilibrium the average power radiated by an electron oscillator is found from (7.7.9) and (7.7.10) to be

$$(Pwr)_{\text{avg}} = \frac{2\pi e^2 \nu_0^2 kT}{m\epsilon_0 c^3} \tag{7.7.11}$$

The spectral density of thermal equilibrium radiation follows by equating (7.7.6) and (7.7.11):

$$\rho(\nu_0) = \frac{8\pi\nu_0^2}{c^3} kT \tag{7.7.12}$$

This prediction of classical physics for the spectrum of thermal radiation is called the *Rayleigh–Jeans law*. It obviously differs from the (correct) Planck formula (7.7.5), but it does agree with the Planck law in the high-temperature limit $kT \gg h\nu_0$.

Aside from its disagreement with the Planck formula, there is an obvious flaw in the classical law (7.7.12): the total electromagnetic energy density implied by the Rayleigh–Jeans law is

$$\int_0^\infty \rho(\nu)\, d\nu = \frac{8\pi kT}{c^3} \int_0^\infty \nu^2\, d\nu \quad \text{(classical)} \tag{7.7.13}$$

which is obviously infinite. This infinity was historically referred to as the "ultraviolet catastrophe" of classical physics. •

The quantum theory of thermal radiation predicts a total electromagnetic energy density given by integrating (7.7.5), which leads to

$$\int_0^\infty \rho(\nu)\, d\nu = \frac{8\pi^5 k^4}{15c^3 h^3} T^4 \quad \text{(quantum-mechanical)} \tag{7.7.14}$$

The total intensity radiated by a blackbody is then [cf. Eq. (3.8.4)]

$$I_{\text{total}} = \int_0^\infty I(\nu)\, d\nu = \frac{c}{4} \int_0^\infty \rho(\nu)\, d\nu = \frac{2\pi^5 k^4}{15c^2 h^3} T^4 = \sigma T^4 \tag{7.7.15}$$

where $\sigma = 5.67 \times 10^{-8}$ J-m^{-2}-sec^{-1}-K^{-4} = 5.67×10^{-12} W/cm^2-K^{-4} is the *Stefan–Boltzmann constant*. The new factor $\frac{1}{4}$ multiplying the integral in (7.7.15) arises for two reasons. First, a factor $\frac{1}{2}$ arises because, along any axis through the blackbody, there are equal intensities of radiation propagating in opposite directions, but in (7.7.15) we are only interested in radiation propagating outward through the surface. A second factor of $\frac{1}{2}$ arises because the average component of light velocity normal to the surface is

$$(\cos \theta)_{avg} = \frac{1}{2\pi} \int_0^{2\pi} \int_0^{\pi/2} \cos \theta \sin \theta \, d\theta \, d\phi = \frac{1}{2} \qquad (7.7.16)$$

Many sources of radiation have spectral characteristics approximating those of an ideal blackbody. Stars, for instance, are certainly not perfect blackbodies, but they come sufficiently close to the ideal that we can estimate their surface temperatures by fitting their spectra to Planck's law (Figure 7.6). In particular, the peak emission wavelength λ_{max} of a blackbody at temperature T (K) is given by (Problem 7.4)

$$\lambda_{max} = \frac{2.898 \times 10^7}{T} \, \text{Å} \qquad (7.7.17)$$

Thus the sun, which has a spectral output approaching that of a blackbody at 5800 K, has a peak emission wavelength $\lambda_{max} \approx 5000$ Å. Its total intensity at the earth's surface is about $0.14 \, \text{W/cm}^2 = 1.4 \, \text{kW/m}^2$ (Problem 7.5). Equation (7.7.15) is consistent with the observation that the color of hot bodies shifts to shorter wavelengths with increasing temperature (e.g., "white hot" is hotter than "red hot"). This shift is evident in Figure 7.7.

For most thermal sources of light, the dominant radiative process is spontaneous emission. This can be seen by considering the ratio of spontaneous and stimulated emission rates for atoms in the upper state of some transition:

$$\frac{AN_2}{R_{12}N_2} = \frac{A}{R_{12}} = \frac{A}{B\rho(\nu_0)} = e^{h\nu_0/kT} - 1 \qquad (7.7.18)$$

where we have used (7.7.3). For wavelengths in the visible, and for temperatures less than several tens of thousands of kelvins, the ratio (7.7.18) is much greater than unity. For the solar temperature $T = 5800$ K, for instance, and $\lambda = 5000$ Å, the ratio is about 142. Thus we can infer that more than 99% of the light from the sun is due to spontaneous rather than stimulated emission of radiation. This dom-

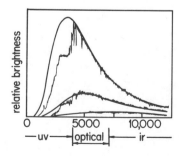

Figure 7.6 Comparison of blackbody (smooth curves) and stellar emission spectra for two temperatures, 8000K and 5800K [after L.U.M. Protheroe, E.R. Capriotti and T.H. Newsome, *Exploring the Universe*, third edition (Merrill, Columbus, 1984)].

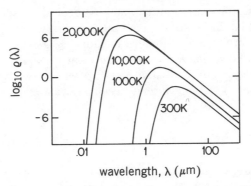

Figure 7.7 $\text{Log}_{10}\,\rho(\lambda)$ vs. λ for an ideal blackbody radiator at four different temperatures.

inance of spontaneous emission is responsible for the considerable degree of temporal incoherence of light from thermal sources in comparison with laser light, which is generated by stimulated emission.

7.8 PHOTONS PER MODE

In Section 2.1 we showed, for a cavity of dimensions much larger than a wavelength, that $N(\nu)\,d\nu = (8\pi\nu^2/c^3)\,d\nu$ is the number of modes per unit volume in the frequency band $[\nu, \nu + d\nu]$. Thus the Planck formula (7.7.5) has the form

$$\rho(\nu_0)\,d\nu = \big(N(\nu_0)\,d\nu\big)\,q(\nu_0)\,h\nu_0$$

$$= \big(\text{number of modes per unit volume in}$$

$$\text{the frequency band } [\nu_0, \nu_0 + d\nu]\big)$$

$$\times\,(\text{number of photons in mode of frequency } \nu_0)$$

$$\times\,(\text{energy } h\nu_0 \text{ of a photon of frequency } \nu_0) \qquad (7.8.1)$$

This identifies

$$q(\nu_0) = \frac{1}{e^{h\nu_0/kT} - 1} \qquad (7.8.2)$$

as the (average) number of photons in a mode of frequency ν_0 in the case of thermal radiation. Thus, from (7.7.3), the stimulated emission rate may be written

$$R_{12} = B\rho(\nu_0) = A\left(e^{h\nu_0/kT} - 1\right)^{-1} = Aq(\nu_0) \qquad (7.8.3)$$

In other words, the stimulated emission rate is equal to the spontaneous emission rate times the number of photons at the transition frequency. Although this result has been inferred for the case of thermal radiation, it is more generally valid, and may be stated as follows: *the rate of stimulated emission into any mode of the field is equal to the spontaneous emission rate into the mode, times the expectation value of the number of photons already occupying that mode.*

APPENDIX 7.A   RELATIONS AMONG A AND B COEFFICIENTS

In Table 7.1 below we list some important relations among A and B coefficients and the oscillator strength of a transition. We include in these relations the possible degeneracies of the upper and lower levels of the transition, i.e., the possibility that there may be g_2 states of the same energy E_2 and g_1 states of the same energy E_1.

Until now we have almost ignored the possibility of level degeneracy in our discussion of emission and absorption processes. To see how expressions such as those for absorption and stimulated emission first given in (7.2.3) are altered by degeneracy, we return to the population rate equations (7.2.2), where we now interpret ρ_{11} as $\rho_{m_1 m_1}$ and ρ_{22} as $\rho_{m_2 m_2}$ (since they were derived for a specific state-to-state transition). Furthermore, we will rewrite (7.2.2a) and (7.2.2b) as

$$\dot{\rho}_{11} = -\Gamma_1 \rho_{11} + A(m_1, m_2)\rho_{22} + R(m_1, m_2)(\rho_{22} - \rho_{11}) \quad (7.A.1a)$$

$$\dot{\rho}_{22} = -\Gamma_2 \rho_{22} - A(m_1, m_2)\rho_{22} - R(m_1, m_2)(\rho_{22} - \rho_{11}) \quad (7.A.1b)$$

to emphasize the specific state labels appropriate to R and A:

$$R_{12} \rightarrow R(m_1, m_2) = \frac{|\chi|^2 \beta/2}{(\omega_{21} - \omega)^2 + \beta^2} \quad (7.A.2a)$$

$$A_{21} \rightarrow A(m_1, m_2) = \frac{D^2 \omega_{21}^3}{3\pi\epsilon_0 \hbar c^3} \quad (7.A.2b)$$

where D and χ refer specifically to the $m_1 \rightarrow m_2$ transition.

In the presence of degeneracy several modifications of the derivation of the equations are required. For example, the spontaneous decay of ρ_{22} is not via a single channel to a single state m_1, but to all lower states m_1, so we must sum A_{21} over all m_1. Similarly, the increase in ρ_{11} due to stimulated emission from level 2 is due to all states m_2, not just to one of them, etc. If we concentrate first on the probabilities associated with the states of level 2, and consider separately the emission and absorption processes, we can write

$$[\dot{\rho}_{22}]_{abs} = \sum_{m_1} R(m_1, m_2)\, \rho_{11} \quad (7.A.3a)$$

$$[\dot{\rho}_{22}]_{\text{em}} = -\sum_{m_1} [A(m_1, m_2) + R(m_1, m_2)] \rho_{22} \qquad (7.A.3b)$$

Now we also sum over m_2 to obtain the rates of change of the full level probability for level 2, $P_2 = \sum_{m_2} \rho_{22}$. For example, the counterparts of (7.A.3a) and (7.A.3b) are

$$[\dot{P}_2]_{\text{abs}} = \sum_{m_1, m_2} R(m_1, m_2) \rho_{11}$$

$$[\dot{P}_2]_{\text{em}} = -\sum_{m_1, m_2} [A(m_1, m_2) + R(m_1, m_2)] \rho_{22}$$

These expressions can be simplified in two situations. The first occurs if R has been orientationally averaged, as in (7.2.7). Then when summed over one of the m's it becomes independent of the other [recall Problem 6.8]. The second situation occurs in the case of so-called "natural excitation" of the atom.[1] In this case all of the states of a given level are equally populated, so ρ_{11} and ρ_{22} are independent of m_1 and m_2 and thus equal to P_1/g_1 and P_2/g_2, respectively, where g_j is the degeneracy of level j. In both situations we obtain

$$[\dot{P}_2]_{\text{abs}} = \sum_{m_1 m_2} R(m_1, m_2) \frac{P_1}{g_1} \qquad (7.A.4a)$$

and

$$[\dot{P}_2]_{\text{em}} = -\sum_{m_1 m_2} [A(m_1, m_2) + R(m_1, m_2)] \frac{P_2}{g_2} \qquad (7.A.4b)$$

Einstein identified his famous A and B coefficients in the course of his derivation of the Planck radiation formula for blackbody radiation. He wrote the equivalent of Eqs. (7.A.4) as

$$[\dot{P}_2]_{\text{abs}} = B(1 \rightarrow 2) \rho(\nu) P_1 \qquad (7.A.5a)$$

$$[\dot{P}_2]_{\text{em}} = -[A(2 \rightarrow 1) + B(2 \rightarrow 1) \rho(\nu)] P_2 \qquad (7.A.5b)$$

1. By natural excitation one means that the conditions of preparation are sufficiently isotropic that the numbers of atoms in each of the different degenerate states of the same level are equal. In this respect a remark by Condon and Shortley is revealing: "If the excitation occurs in some definitely non-isotropic way, as by absorption from a unidirectional beam of light . . . , large departures from natural excitation may be produced. The study of such effects raises a whole complex of problems somewhat detached from the main body of spectroscopy." [From *The Theory of Atomic Spectra*, by E. U. Condon and G. H. Shortley (Cambridge University Press, London, 1935), p. 97.] Whether laser spectroscopy is considered to be "somewhat detached" or not from the main body of spectroscopy is only a matter of viewpoint, of course. However, the student must be aware that formulas for absorption and dispersion found in pre-laser reference material may not be directly applicable to atoms prepared by laser radiation, which is definitely not isotropic.

where $\rho(\nu)$ is the spectral energy density of the radiation. Blackbody radiation is isotropic, so Einstein was dealing with a situation of "natural excitation," and therefore the formulas (7.A.4) can be compared with their counterparts in (7.A.5), with the resulting identifications

$$A(2 \rightarrow 1) = \frac{1}{g_2} \sum_{m_1, m_2} A(m_1, m_2) \qquad (7.A.6a)$$

$$B(1 \rightarrow 2) \rho(\nu) = \frac{1}{g_1} \sum_{m_1, m_2} R(m_1, m_2) \qquad (7.A.6b)$$

$$B(2 \rightarrow 1) \rho(\nu) = \frac{1}{g_2} \sum_{m_1, m_2} R(m_1, m_2) \qquad (7.A.6c)$$

There are several consequences of this identification. First, the B coefficients for emission and absorption are directly related through the respective level degeneracies:

$$g_1 B(1 \rightarrow 2) = g_2 B(2 \rightarrow 1) \qquad (7.A.7)$$

Second, at steady state, when emission and absorption are balanced, we can solve for the ratio of the level probabilities:

$$\frac{P_2}{P_1} = \frac{B(1 \rightarrow 2) \rho(\nu)}{A(2 \rightarrow 1) + B(2 \rightarrow 1) \rho(\nu)} \qquad (7.A.8)$$

If the steady state is one of thermal equilibrium, then the probability ratio is determined by the Boltzmann factor to be

$$\left(\frac{P_2}{P_1} \right)_T = \frac{g_2}{g_1} e^{-h\nu/kT} \qquad (7.A.9)$$

where $h\nu = E_2 - E_1$. The combination of these last three expressions reproduces Einstein's derivation of the Planck formula:

$$\rho(\nu) = \frac{A(2 \rightarrow 1)/B(2 \rightarrow 1)}{e^{h\nu/kT} - 1} \qquad (7.A.10)$$

If the high-temperature limit is to agree with the classical result of Rayleigh and Jeans given in (7.7.12), we must have

$$\frac{A(2 \rightarrow 1)}{B(2 \rightarrow 1)} = \frac{8\pi h\nu^3}{c^3} \qquad (7.A.11)$$

At this point we will convert our state-summed A and B coefficients to a more compact notation:

$$A(2 \rightarrow 1) \rightarrow A(2,1) \qquad\qquad (7.A.12a)$$

$$B(2 \rightarrow 1) \rightarrow B(2,1) \qquad\qquad (7.A.12b)$$

$$B(1 \rightarrow 2) \rightarrow B(1,2) \qquad\qquad (7.A.12c)$$

and the order of the subscripts will denote the direction of the transition. The relations of these coefficients to each other, as well as to oscillator strengths, etc., are displayed systematically in Table 7.1.

Another useful measure of the strength of a transition is the *line strength*. To introduce the notion of line strength, consider first a transition between two *nondegenerate* atomic states. The rate of spontaneous emission at the $2 \rightarrow 1$ transition is given by Eq. (7.6.5):

$$A_{21} = \frac{\omega_{21}^3}{3\pi\epsilon_0\hbar c^3}\,|\mathbf{D}|^2$$

$$= \frac{16\pi^3\nu_{21}^3}{3\epsilon_0 hc^3}\,|\mathbf{D}|^2$$

where $\mathbf{D} = e\mathbf{r}_{21}$ is the electric dipole matrix element of the transition [recall (7.2.5)]. In this nondegenerate case the quantity defined to be the line strength is identical to $|\mathbf{D}|^2$:

$$S_{12} = |\mathbf{D}|^2 \quad \text{(nondegenerate case)} \qquad\qquad (7.A.13)$$

However, if the levels are degenerate, we define the line strength as the sum of $|\mathbf{D}|^2$ over all degenerate states:

$$S(1,2) \equiv \sum_{m_2}^{g_2} \sum_{m_1}^{g_1} |\mathbf{D}_{m_2 m_1}|^2 \qquad\qquad (7.A.14)$$

The line strength so defined has the useful property of being symmetric with respect to the levels 1 and 2, just as D^2 is symmetric in the states m_1 and m_2, so we can write it either as $S(1,2)$ or $S(2,1)$. The line strength is often given in terms of the "atomic unit" for electric dipole moment:

$$(ea_0)^2 = 7.188 \times 10^{-59}\ \text{C}^2\text{-m}^2$$

$$= 6.459 \times 10^{-36}\ \text{esu}^2\text{-cm}^2 \quad \text{(cgs units)} \qquad\qquad (7.A.15)$$

The oscillator strength can be written in terms of the line strength for the (absorptive) $1 \to 2$ transition:

$$f(1,2) = \frac{4\pi m\nu_{21}}{3\hbar e^2}\frac{1}{g_1}S(1,2) \qquad (7.A.16)$$

Recall that f was introduced in Chapter 3 as a "weighting factor" necessary to bring the classical theory of absorption into quantitative agreement with experiment [Eq. (3.7.5)].

Not surprisingly, the classical theory of dispersion discussed in Chapter 2 is similarly modified by quantum theory. In particular, the polarizability $\alpha(\omega)$ of a one-electron atom, given in the classical electron oscillator model by Eq. (2.3.20), is given in quantum theory by the formula

$$\alpha_i(\omega) = \frac{e^2}{m}\sum_j \frac{f(i,j)}{\omega_{ji}^2 - \omega^2} \qquad (7.A.17)$$

TABLE 7.1   Useful Relations among Emission and Absorption Parameters for a Transition of Frequency $\nu_{21} = (E_2 - E_1)/h$

	$A(2,1)$	$B(1,2)$	$f(1,2)$	$S(1,2)$
$A(2,1)$	1	$\dfrac{g_1}{g_2}\dfrac{8\pi h\nu_{21}^3}{c^3}$	$\dfrac{g_1}{g_2}\dfrac{2\pi e^2\nu_{21}^2}{\epsilon_0 mc^3}$	$\dfrac{1}{g_2}\dfrac{16\pi^3\nu_{21}^3}{3\epsilon_0 hc^3}$
$B(1,2)$	$\dfrac{g_2}{g_1}\dfrac{c^3}{8\pi h\nu_{21}^3}$	1	$\dfrac{e^2}{4\epsilon_0 mh\nu_{21}}$	$\dfrac{1}{g_1}\dfrac{2\pi^2}{3\epsilon_0 h^2}$
$f(1,2)$	$\dfrac{g_2}{g_1}\dfrac{\epsilon_0 mc^3}{2\pi e^2\nu_{21}^2}$	$\dfrac{4\epsilon_0 mh\nu_{21}}{e^2}$	1	$\dfrac{1}{g_1}\dfrac{8\pi^2 m\nu_{21}}{3e^2 h}$
$S(1,2)$	$g_2\dfrac{3\epsilon_0 hc^3}{16\pi^3\nu_{21}^3}$	$g_1\dfrac{3\epsilon_0 h^2}{2\pi^2}$	$g_1\dfrac{3e^2 h}{8\pi^2 m\nu_{21}}$	1

The entries in this table relate the row coefficients to the column coefficients. Thus

$$A(2,1) = \frac{g_1}{g_2}\frac{8\pi h\nu_{21}^3}{c^3}B(1,2), \text{ etc.}$$

In c.g.s. units the same relations hold with ϵ_0 replaced by $1/4\pi$. In addition, the following relations should be noted:

$$g_2 B(2,1) = g_1 B(1,2)$$

$$g_2 f(2,1) = -g_1 f(1,2)$$

$$S(2,1) = S(1,2)$$

The B coefficients are defined such that $B(2,1)\rho(\nu_{21})$ and $B(1,2)\rho(\nu_{21})$ are the stimulated emission and absorption rates on the transition for broadband radiation of spectral energy density $\rho(\nu)$, and $\rho(\nu)$ is defined so that $\rho(\nu)\,d\nu$ is the electromagnetic energy per unit volume in the frequency interval $[\nu, \nu + d\nu]$.

for an atom in level i. In this formula the summation is over all levels j connected to level i by the oscillator strength $f(i,j)$ and the Bohr frequency $\nu_{ji} = \omega_{ji}/2\pi$.

In the limit $\omega^2 \gg \omega_{ji}^2$, where the field frequency is much larger than the atomic transition frequencies, we have

$$\alpha_i(\omega) \approx -\frac{e^2}{m\omega^2} \sum_j f(i,j) \tag{7.A.18}$$

In this high-frequency limit we might expect the polarizability to be accurately given by the classical, free-electron theory in which there are no restrictions on electron energy, and no discrete Bohr frequencies. That is, we might expect (7.A.18) to reduce to the classical free-electron polarizability[2] $\alpha(\omega) = -e^2/m\omega^2$, which requires that

$$\sum_j f(i,j) = 1 \tag{7.A.19}$$

Equation (7.A.19) is in fact correct, and may be derived quantum-mechanically. It is called the Thomas–Reiche–Kuhn sum rule, or simply the *f sum rule*.

In Table 3.1 we listed some oscillator strengths of atomic hydrogen for transitions involving the ground level $1s$. These oscillator strengths are for the transitions $1s \rightarrow np$, beginning with $1s \rightarrow 2p$, which has an oscillator strength of 0.416. The sum of the oscillator strengths for the six transitions listed is about 0.551. In fact the sum of *all* the oscillator strengths for transitions from $1s$ to higher-lying, bound p states is only about 0.565, which would appear to contradict the *f* sum rule. The reason for this discrepancy is that in the *f* sum rule we must also include all unbound states j connected to the state i. Such bound–free, or *continuum*, transitions contribute $1.0 - 0.565 = 0.435$ to the sum (7.A.19) for the ground state of hydrogen.

Thus far we have defined oscillator strengths only for absorptive transitions. For an emissive transition $2 \rightarrow 1$ $(E_2 > E_1)$, we define the oscillator strength $f(2,1)$ by switching the labels 1 and 2 in (7.A.16):

$$f(2,1) = \frac{4\pi m \nu_{12}}{3\hbar e^2} \frac{1}{g_2} S(2,1)$$

$$= -\frac{4\pi m \nu_{21}}{3\hbar e^2} \frac{1}{g_2} S(1,2) \tag{7.A.20}$$

since $\nu_{12} = -\nu_{21}$ and $S(1,2) = S(2,1)$. Thus we have

$$g_2 f(2,1) = -g_1 f(1,2) \tag{7.A.21}$$

2.   The classical free-electron polarizability follows from (2.3.22) and (2.4.12).

relating the oscillator strengths for absorption ($f(1,2)$, a positive number) and emission ($f(2,1)$, a negative number) between levels of energy E_1 and E_2. If level i is an excited state, the summation in the f sum rule (7.A.19) must include all absorptive and emissive transitions involving level i, i.e.,

$$\sum_{j>i} f(i,j) + \sum_{j<i} f(i,j) = 1$$

or

$$\sum_{j>i} f(i,j) - \sum_{j<i} |f(i,j)| = 1 \qquad (7.A.22)$$

In Section 6.4 we calculated the oscillator strength for the $n = 1 \rightarrow 2$ transition of hydrogen. We can in fact calculate the oscillator strengths for all transitions of hydrogen. However, it is not possible at present to calculate exactly the oscillator strengths of any multielectron atom because, as discussed in Section 5.4, the stationary-state wave functions are not known exactly. This prevents an exact calculation of the matrix elements of the electric dipole moment

$$\mathbf{D} = \sum_{i} e\, \mathbf{r}_i \qquad (7.A.23)$$

where the sum is over all the electrons of the atom. Nevertheless considerable progress has been made, both experimentally and by approximate theoretical methods, in the determination or estimation of oscillator strengths of multielectron atoms. Useful tabulations of many oscillator strengths and line strengths have been available for decades, and additions are still being made.[3]

• The f sum rule (7.A.22) for a one-electron atom played an important role in the development of quantum theory. Historically, downward transitions associated with stimulated emission and negative oscillator strengths were referred to in terms of "negative oscillators." Consider the refractive index $n(\omega)$ of a monatomic gas, which is given by the formula

$$n(\omega) = 1 + \frac{1}{2\epsilon_0} \sum_{i} N_i \alpha_i(\omega) \qquad (7.A.24)$$

for $n(\omega) \approx 1$, where N_i is the number density of atoms in level i and $\alpha_i(\omega)$ is the polarizability of an atom in level i. From (7.A.17) we have

$$n(\omega) = 1 + \frac{e^2}{2m\epsilon_0} \sum_{i} \sum_{j} \frac{N_i f(i,j)}{\omega_{ji}^2 - \omega^2} \qquad (7.A.25)$$

---

3. See, for instance, W. L. Wiese, M. W. Smith, and B. M. Glennon, *Atomic Transition Probabilities* (U.S. Government Printing Office, Washington, D.C., 1966).

In particular, the contribution to the index by the 1–2 transition is

$$n(\omega)_{12} = 1 + \frac{e^2}{2m\epsilon_0} \left(\frac{N_1 f(1,2)}{\omega_{21}^2 - \omega^2} + \frac{N_2 f(2,1)}{\omega_{12}^2 - \omega^2} \right)$$

$$= 1 + \frac{e^2 f(1,2)/2m\epsilon_0}{\omega_{21}^2 - \omega^2} \left(N_1 - \frac{g_1}{g_2} N_2 \right) \qquad (7.A.26)$$

where we have related $f(2,1)$ to $f(1,2)$ by the equation (7.A.21). We see from this result that the excited-level population N_2 of a transition contributes to $n(\omega)$ a term of opposite sign to the contribution from the lower-level population. This is the origin of the term "negative oscillator" for the excited-state contributions to the refractive index.

This contribution of negative oscillators to the refractive index was studied experimentally by R. Ladenburg and H. Kopfermann in 1928. They studied the variation of refractive index with electric current of a discharge tube filled with neon. According to our theory, for $\omega < \omega_{21}$ in (7.A.26), with neither 1 nor 2 the ground level, $n(\omega)_{12}$ should initially increase with increasing current, because electron–atom collisions produce atoms in excited level 1. With further increase of the current, however, the rate of growth of $n(\omega)_{12}$ with current should decrease, because excited level 2 then has appreciable population N_2, and acts as a negative oscillator. This is just the sort of behavior observed by Ladenburg and Kopfermann, thus confirming the role of "negative" oscillators. •

The relations we have obtained among the A and B coefficients, the oscillator strengths, and the line strengths of transitions may be generalized to molecular as well as atomic transitions. However, because molecules have vibrational and rotational energy levels in addition to electronic energy levels, the situation is much more complicated than in the case of atoms. Since the details of molecular transition strengths are not crucial to our study of lasers, we will not take up the subject in this book except for some brief remarks in the following chapters. In Chapter 13 we calculate the absorption and "gain" coefficients for vibrational–rotational transitions as an illustration of the kind of considerations involved in the treatment of molecular transitions.

PROBLEMS

7.1 Let $\mathbf{r}_{21}$ and $\hat{\epsilon}$ be complex vectors, and let $\hat{\epsilon}$ be a unit vector in the complex sense: $\hat{\epsilon} \cdot \hat{\epsilon}^* = 1$. Designate $\mathbf{r}_{21} = \mathbf{p} + i\mathbf{q}$ and $\hat{\epsilon} = \boldsymbol{\alpha} + i\boldsymbol{\beta}$, where $\mathbf{p}$, $\mathbf{q}$, $\boldsymbol{\alpha}$, and $\boldsymbol{\beta}$ are purely real vectors.

(a) Show that $\alpha^2 + \beta^2 = 1$.

(b) Write $|\mathbf{r}_{21} \cdot \hat{\epsilon}|^2$ in terms of $\mathbf{p}$, $\mathbf{q}$, $\boldsymbol{\alpha}$, and $\boldsymbol{\beta}$.

(c) Compute the average of $(\mathbf{p} \cdot \boldsymbol{\alpha})^2$ over all relative $\mathbf{p}$–$\boldsymbol{\alpha}$ angles. Note that the spherical average of $\cos^2 \theta$ is *not* $\frac{1}{2}$.

(d) Extrapolate from (c), using (b) as well, to evaluate the average of $|\mathbf{r}_{12} \cdot \boldsymbol{\epsilon}|^2$.

7.2 (a) Find the steady-state solutions for N_1 and N_2 from Eqs. (7.3.4).

(b) From N_1 and N_2 obtain the steady-state inversion density $N_2 - N_1$, and give its limiting values for very high and very low incident light flux.

7.3* Evaluate the integral and sum given in (7.6.4). This can be done by first expressing $(\mathbf{r} \cdot \hat{\mathbf{\varepsilon}}_k)^2$ as a function of the angles θ_k and ϕ_k of the wave vector $\mathbf{k}$. One way to do this is to do the sum first. Note that, if $\hat{\mathbf{\varepsilon}}(1)$ and $\hat{\mathbf{\varepsilon}}(2)$ are the two independent polarization vectors (perpendicular to $\hat{\mathbf{k}}$ and to each other), then $\mathbf{r}$ can be written

$$\mathbf{r} = \hat{\mathbf{\varepsilon}}(1)\left(\mathbf{r} \cdot \hat{\mathbf{\varepsilon}}(1)\right) + \hat{\mathbf{\varepsilon}}(2)\left(\mathbf{r} \cdot \hat{\mathbf{\varepsilon}}(2)\right) + \hat{\mathbf{k}}(\mathbf{r} \cdot \hat{\mathbf{k}}).$$

That is, $\hat{\mathbf{\varepsilon}}(1)$, $\hat{\mathbf{\varepsilon}}(2)$, and $\hat{\mathbf{k}}$ are a *complete set*. Thus $\Sigma_{\text{pol}}\,(\mathbf{r} \cdot \hat{\mathbf{\varepsilon}}_k)^2 = \Sigma_{j=1,2}\,[\mathbf{r} \cdot \hat{\mathbf{\varepsilon}}(j)]^2 = [\mathbf{r} \cdot \hat{\mathbf{\varepsilon}}(1)]^2 + [\mathbf{r} \cdot \hat{\mathbf{\varepsilon}}(2)]^2 = r^2 - (\mathbf{r} \cdot \hat{\mathbf{k}})^2 = r^2\,(1 - \cos^2\theta_k)$.

7.4 Derive the Wien displacement law (7.7.17) for the peak emission wavelength of a blackbody at temperature T.

7.5 Calculate the total intensity, over all wavelengths, of light from the sun (a) at the earth's surface and (b) at the surface of Pluto. Assume that the sun is a blackbody of temperature 5800 K and that the orbital distances are $R_E = 150 \times 10^6$ km and $R_P = 5900 \times 10^6$ km.

7.6 Estimate the temperature of a blacktop road on a sunny day. Assume the asphalt is a perfect blackbody.

8 SEMICLASSICAL RADIATION THEORY

8.1 INTRODUCTION

Chapters 6 and 7 were based on the so-called *semiclassical* theory of light-matter interactions, without using that name. The semiclassical theory ignores the quantum-mechanical nature of the electromagnetic field while using the full regalia of the Schrödinger equation to determine the behavior of the matter. The result is a theory that is ultimately inconsistent, because quantum fluctuations associated with the electromagnetic field do have an effect on atomic behavior, but the theory refuses to recognize these effects except in isolated instances. For example, the semiclassical theory recognizes spontaneous emission only to the extent of assigning an empirical coefficient A_{21} to the spontaneous rate. The full quantum theory of radiation not only shows how to derive the value of A_{21}, but also shows how the existence of spontaneous emission is connected to the existence of atom-field fluctuations.

Nevertheless, the justification for the semiclassical theory is extremely strong in a wide domain. This domain embraces all of atomic radiation theory where the numbers of photons are much larger than unity. In most laser modes the number of photons is practically astronomical, and effects due to field quantization are insignificant so long as attention remains on stimulated processes. On the other hand, it is quite possible to use laser fields to probe the subtle correlations and fluctuations inherent in quantum theory, but in order to do so an observation must be undertaken of a quantity that is sensitive to single-photon differences. Some such observations are discussed in Chapter 9. They are rarely, if ever, directly important for an understanding of laser operation.

In this chapter we will complete our development of the semiclassical theory by allowing the atomic dipoles to serve as sources for radiation. This will require the use of Maxwell's wave equation, in parallel with the classical discussions in Chapters 2 and 3. We will use a variant of the density-matrix equations called the optical Bloch equations, and derive the coupled Maxwell-Bloch equations. One result will be an elementary semiclassical theory of laser action.

8.2 OPTICAL BLOCH EQUATIONS

Our description of a two-level atom's response to near-resonant radiation has been based on the density-matrix equations (6.5.14) and (6.5.17). An equivalent set of vector equations frequently used to treat two-level atoms (and also multilevel atoms in some special cases) is named for F. Bloch, who recognized their utility in describing magnetic resonance phenomena in the 1940s. The first proposal to use a similar formalism in optical resonance was advanced by R. P. Feynman, F. L. Vernon, and R. W. Hellwarth in 1957.

In their simplest form the optical Bloch equations are equivalent to the density-matrix equations (6.5.2). The Bloch formalism uses the conservation of probability, $\rho_{11} + \rho_{22} = 1$, to replace the four density-matrix elements with three (real) variables:

$$u = \rho_{21} + \rho_{12} \tag{8.2.1a}$$

$$v = i(\rho_{21} - \rho_{12}) \tag{8.2.1b}$$

$$w = \rho_{22} - \rho_{11} \tag{8.2.1c}$$

The equations obeyed by u, v, and w can be obtained by substituting equations (8.2.1) into (6.5.2). The results are (taking the Rabi frequency χ to be real)

$$\frac{du}{dt} = -\Delta v \tag{8.2.2a}$$

$$\frac{dv}{dt} = \Delta u + \chi w \tag{8.2.2b}$$

$$\frac{dw}{dt} = -\chi v \tag{8.2.2c}$$

The usefulness of the u, v, w equations comes from the fact that they satisfy a single vector equation whose solutions are easy to interpret. In a fictitious space with unit vectors $\hat{1}$, $\hat{2}$, $\hat{3}$ (analogous to $\hat{x}$, $\hat{y}$, $\hat{z}$) we define a coherence vector **S** (sometimes called the pseudospin vector or simply the Bloch vector):

$$\mathbf{S} = \hat{1}u + \hat{2}v + \hat{3}w \tag{8.2.3}$$

and a ''torque'' vector or ''axis'' vector **Q**:

$$\mathbf{Q} = -\hat{1}\chi + \hat{3}\Delta \tag{8.2.4}$$

Then it can be shown by considering each component separately that the vector equation

$$\frac{d\mathbf{S}}{dt} = \mathbf{Q} \times \mathbf{S} \tag{8.2.5}$$

is exactly equivalent to the three equations (8.2.2). For example, the $\hat{2}$ (analogous to $\hat{y}$) component of (8.2.5) is

$$\frac{dv}{dt} = (\mathbf{Q} \times \mathbf{S})_2$$

$$= Q_3 S_1 - Q_1 S_3$$

$$= \Delta u + \chi w$$

which is the same as (8.2.2b).

The role of $\mathbf{Q}$ as a "torque" or "axis" vector follows from the vector form of (8.2.5). Since $\mathbf{Q} \times \mathbf{S}$ is perpendicular to $\mathbf{S}$ (by definition of the vector cross product), the effect of $\mathbf{Q}$ is only to rotate $\mathbf{S}$ about the direction of $\mathbf{Q}$. It cannot lengthen or shorten $\mathbf{S}$. It is easy to confirm that the magnitude $S^2 = \mathbf{S} \cdot \mathbf{S}$ is constant:

$$\frac{dS^2}{dt} = 2\mathbf{S} \cdot \frac{d\mathbf{S}}{dt}$$

$$= 2\mathbf{S} \cdot (\mathbf{Q} \times \mathbf{S})$$

$$= 0 \qquad (8.2.6)$$

• The constant magnitude of $\mathbf{S}$ has a physical meaning. Recall the definitions (8.2.3) and (8.2.1):

$$S^2 = u^2 + v^2 + w^2$$

$$= (\rho_{21} + \rho_{12})^2 - (\rho_{21} - \rho_{12})^2 + (\rho_{22} - \rho_{11})^2$$

$$= 4\rho_{21}\rho_{12} + \rho_{22}^2 - 2\rho_{22}\rho_{11} + \rho_{11}^2 \qquad (8.2.7)$$

This expression can be reduced further by using the original definitions of the ρ's in (6.5.1): $\rho_{12} = c_1 c_2^*$, etc. We find

$$S^2 = 4 c_2 c_1^* c_1 c_2^* + (c_2 c_2^*)^2 - 2 c_2 c_2^* c_1 c_1^* + (c_1 c_1^*)^2$$

$$= (c_2 c_2^* + c_1 c_1^*)^2$$

$$= 1^2 = 1 \qquad (8.2.8)$$

The unit length of the Bloch vector is thus seen to be equivalent (in the absence of collisions) to the conservation of probability in the two-level atom. •

Since $S^2 = 1$, the tip of the Bloch vector lies on the surface of a unit sphere. Only the angles of the vector change with time. In effect, we have now shown that the time evolution of the two-level density matrix (with elements ρ_{21}, etc.) is equivalent to changes in the orientation of **S**, as sketched in Figure 8.1.

Let us consider the orientations of the Bloch vector in the fictitious $\hat{\mathbf{1}}$-$\hat{\mathbf{2}}$-$\hat{\mathbf{3}}$ space. First we note from (8.2.1c) that increasing or decreasing the degree of inversion $\rho_{22} - \rho_{11}$ corresponds to moving up or down the "vertical" $\hat{\mathbf{3}}$ axis. When the Bloch vector points straight up and $w = 1$, then obviously $\rho_{22} = 1$ and $\rho_{11} = 0$, and the atomic population is entirely in the upper level. Thus the north pole of the unit sphere ("Bloch sphere") corresponds to a fully inverted atom. By the same reasoning, at the south pole $w = -1$ and the atom is in its lower level.

The rotation angles of the Bloch vector also have direct physical interpretations. At resonance ($\Delta = 0$) the dynamical evolution consists entirely of rotation about the $\hat{\mathbf{1}}$ axis, since we have

$$\mathbf{Q} \rightarrow -\hat{\mathbf{1}}\chi \quad \text{(on resonance)} \tag{8.2.9}$$

Then we can characterize the Bloch vector by a single parameter, namely Θ, the rotation angle about the $\hat{\mathbf{1}}$ axis:

$$v = -\sin \Theta \tag{8.2.10a}$$

$$w = -\cos \Theta \tag{8.2.10b}$$

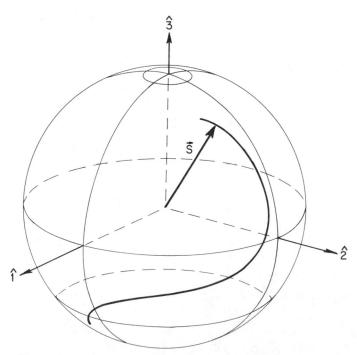

Figure 8.1 Bloch vector and its trajectory on the Bloch sphere.

where Θ is measured from the *south* pole, since that is the normal initial state, corresponding to the atom in its lower level (Figure 8.2).

Substitution of (8.2.10) into (8.2.2c) gives an equation for Θ:

$$\frac{d\Theta}{dt} = \chi \qquad (8.2.11)$$

which leads to the obvious solutions

$$\Theta = \chi t \qquad (8.2.12)$$

and

$$v = -\sin \chi t \qquad (8.2.13a)$$

$$w = -\cos \chi t \qquad (8.2.13b)$$

Agreement with the solutions found earlier in Eqs. (6.3.18) is worth checking in the same limit $\Delta = 0$. For example, since $\rho_{11} + \rho_{22} = 1$, we can write

$$\rho_{11} = \frac{1 - w}{2} \qquad (8.2.14a)$$

$$\rho_{22} = \frac{1 + w}{2} \qquad (8.2.14b)$$

and combine this with (8.2.13b) to obtain

$$\rho_{11} = \tfrac{1}{2}(1 + \cos \chi t) \qquad (8.2.15a)$$

$$\rho_{22} = \tfrac{1}{2}(1 - \cos \chi t) \qquad (8.2.15b)$$

which are the same as (6.3.18a) and (6.3.18b) on resonance.

Even if the Rabi frequency χ is not constant, the solutions (8.2.10) are the same, but $\Theta(t)$ is then given by

$$\Theta(t) = \int_0^t \chi(t') \, dt' \qquad (8.2.16)$$

where

$$\chi(t) = \frac{e(\mathbf{r}_{21} \cdot \hat{\mathbf{\varepsilon}}) E_0(t)}{\hbar} = \frac{\mu E_0(t)}{\hbar} \qquad (8.2.17a)$$

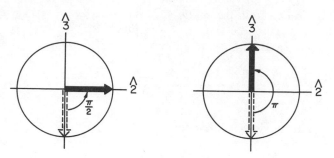

Figure 8.2 Effect of $\pi/2$ and π pulses on the Bloch vector for an on-resonance atom.

In this way the Bloch-vector formalism connects a property of the atoms, namely the Bloch-vector rotation angle on resonance, directly with a property of the incident radiation field, namely the time integral of the field amplitude.

The integral on the right side of (8.2.16) is called the *area* of the pulse. This name comes because any integral can be viewed as an area, as in Figure 8.3. The solutions (8.2.10) make it clear that pulses with areas equal to multiples of π are special. For example, in Figure 8.2 a π pulse turns Θ through 180° about the $\hat{1}$ axis, and thus inverts the atomic population from the lower to the upper level (Fig. 8.2). Much more surprising is the action of a (stronger) 2π pulse. It rotates the resonant Bloch vector through 360° and thus returns the atom exactly to its initial state. The same is true of 4π, 6π, ... pulses. There is therefore a set of pulses that *have no effect* on the resonant atoms when propagating in an absorbing medium at resonance. *Area* is a pulse parameter completely overlooked by conventional optical spectroscopy, but one with fundamental significance.

If E_0 is allowed to be time-dependent we gain the ability to treat the interaction of atoms with light *pulses* as well as with steady light beams. At the same time two working assumptions must be modified. First, the argument employed to justify the rotating wave approximation in going from (6.3.13) to (6.3.14) fails if χ itself contributes rapid temporal variation. Thus we must assume that $E_0(t)$ is a slowly varying time function in a sense made precise in (8.3.2) below. Second, if not just the amplitude but the phase of E_0 changes in time we can no longer assume that an adjustment of the wave function phase will make $\chi(t)$ real. In order to

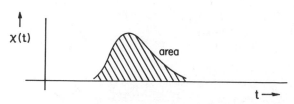

Figure 8.3 The area of a laser pulse according to (8.2.16).

recognize this important shift in assumptions we now change our symbol for the electric field's complex amplitude: $E_0(t) \rightarrow \mathcal{E}(t)$, and the Rabi frequency becomes

$$\chi(t) = \mu \mathcal{E}(t)/\hbar \tag{8.2.17b}$$

For complex $\chi(t)$ the Bloch equations take a slightly more general form, which is given in (8.2.18). In the remainder of the book we will allow for complex $\mathcal{E}$ and complex χ.

Equations (8.2.2) and (8.2.5) are no more than Schrödinger's equation written in another convenient form. Bloch's name should actually be associated with these equations only after relaxation processes are taken into account in a particular way. The Bloch equations, including relaxation, were originally written in the context of spin resonance. In that case the Bloch vector **S** is actually the magnetic spin vector. Relaxation processes for spins commonly include phase-interrupting collisions of the kind we have considered in Section 6.5, but not inelastic decay to other levels such as we have associated with the rates Γ_1 and Γ_2.

Elastic collisions and spontaneous emission relaxation are easily taken into account by reference to (6.5.14) and (6.5.17). The ρ equations can be converted easily into new u, v, w equations, again using the definitions (8.2.1). These are properly called Bloch equations, which we write in complex form to allow for complex χ:

$$\dot{u} - i\dot{v} = -(\beta + i\Delta)(u - iv) - i\chi w \tag{8.2.18a}$$

$$\dot{w} = -\frac{1}{T_1}(1 + w) + \frac{i}{2}\left[\chi(u + iv) - \chi^*(u - iv)\right] \tag{8.2.18b}$$

Here we have used $1/T_1$ to indicate a total rate due to both spontaneous emission and any other inelastic decay processes connecting levels 1 and 2, represented by the rate Γ_{12}. Therefore β and $1/T_1$ are now understood to be given by

$$\beta = \frac{1}{\tau} + \frac{1}{2T_1} \tag{8.2.19a}$$

$$\frac{1}{T_1} = A_{21} + \Gamma_{21} \tag{8.2.19b}$$

In Bloch's original notation β was designated $1/T_2$, and T_1 and T_2 were called the longitudinal and transverse lifetimes of the spin. The terms longitudinal and transverse refer to polar and equatorial directions on the Bloch sphere, directions in magnetic resonance experiments either along, or transverse to, the static magnetic field (conventionally the z axis). In view of (8.2.19) one has the frequently quoted inequality

$$T_2 \leq 2T_1 \qquad (8.2.20)$$

between the transverse and longitudinal lifetimes.

8.3 MAXWELL–BLOCH EQUATIONS

The interaction of light and atoms has two sides. As we emphasized in Chapter 1, this interaction should ideally be dealt with self-consistently. We have not yet done this in the quantum framework, having taken the field to be a fixed monochromatic wave for the most part.

Maxwell's wave equation (2.1.13) remains valid in quantum mechanics. Given a source polarization, even one described quantum-mechanically, we can solve the wave equation to find $\mathbf{E}(\mathbf{r}, t)$. As usual, we will assume that only the z coordinate will be significant, and that the field will be almost monochromatic. That is, we will assume that $\mathbf{E}$ can be conveniently written

$$\mathbf{E}(\mathbf{r}, t) = \mathbf{\hat{\varepsilon}}\mathcal{E}(z, t) \, e^{-i(\omega t - kz)} \qquad (8.3.1)$$

where the real part is understood to be the physical electric field, and where $\mathcal{E}(z, t)$ is the unknown complex amplitude to be determined. An additional assumption implied by the form assumed in (8.3.1) is that the amplitude $\mathcal{E}(z, t)$ varies slowly compared with the carrier wave $e^{-i(\omega t - kz)}$. This justifies inequalities such as

$$\left| \frac{\partial \mathcal{E}}{\partial z} \right| \ll k |\mathcal{E}|$$

$$\left| \frac{\partial^2 \mathcal{E}}{\partial z^2} \right| \ll k \left| \frac{\partial \mathcal{E}}{\partial z} \right|$$

$$\left| \frac{\partial \mathcal{E}}{\partial t} \right| \ll \omega |\mathcal{E}| \qquad (8.3.2)$$

In physical terms these inequalities state that $\mathcal{E}(z, t)$ represents a smooth enough pulse in both space and time. This restriction is not severe, since it would be violated only if $\mathcal{E}(z, t)$ represented a pulse shorter than a few optical periods ($\sim 10^{-15}$ sec) in time or a few wavelengths ($\sim 1 \, \mu$m) in space.

In parallel with (8.3.1) and (8.3.2), we make similar assumptions about the polarization density arising from whatever atoms are present. The semiclassical theory uses the quantum expectation value for the polarization density, namely $Ne\langle \mathbf{r} \rangle$. In complex form analogous to (8.3.1) we have

$$\mathbf{P}(z, t) = 2Ne\mathbf{r}_{12} \, a_1^* a_2$$

$$= 2Ne\mathbf{r}_{12}\rho_{21}(z, t)e^{-i(\omega t - kz)} \qquad (8.3.3)$$

where the real part is the physical polarization. We have used (6.4.3), (6.3.12), and (6.5.1). The slowly varying character of $\mathcal{E}(z, t)$ is also imputed to ρ_{21}:

$$\left|\frac{\partial \rho_{21}}{\partial t}\right| \ll \omega |\rho_{21}|$$

$$\left|\frac{\partial^2 \rho_{21}}{\partial t^2}\right| \ll \omega \left|\frac{\partial \rho_{21}}{\partial t}\right| \tag{8.3.4}$$

etc.

The wave equation for one spatial propagation direction (the z direction) is

$$\left(\frac{\partial^2}{\partial z^2} - \frac{1}{c^2}\frac{\partial^2}{\partial t^2}\right) \mathbf{E}(z, t) = \frac{1}{\epsilon_0 c^2}\frac{\partial^2}{\partial t^2} \mathbf{P}(z, t) \tag{8.3.5}$$

After substituting (8.3.1) and (8.3.3) into (8.3.5), and making use of (8.3.2) and (8.3.4), we keep the largest terms (lowest-order derivatives) on each side. After projecting both sides on $\hat{\boldsymbol{\varepsilon}}^*$ and using $\hat{\boldsymbol{\varepsilon}}^* \cdot \hat{\boldsymbol{\varepsilon}} = 1$ we obtain the wave equation

$$\left(\frac{\partial}{\partial z} + \frac{\partial}{\partial ct}\right) \mathcal{E}(z, t) = \frac{ik}{\epsilon_0} N\mu^* \rho_{21}(z, t) \tag{8.3.6}$$

where we have used the convenient abbreviation [recall (7.2.5)]

$$\mu^* = e(\mathbf{r}_{12} \cdot \hat{\boldsymbol{\varepsilon}}^*) \tag{8.3.7}$$

for the projection of the transition dipole moment on the direction of polarization. When the relations (8.2.1) for the Bloch variables u and v are introduced, we find

$$\left(\frac{\partial}{\partial z} + \frac{\partial}{\partial ct}\right) \mathcal{E}(z, t) = \frac{ik}{2\epsilon_0} N\mu^* (u - iv) \tag{8.3.8}$$

Both (8.3.6) and (8.3.8) are known as the reduced wave equation, or the wave equation in the slowly-varying-envelope approximation. Equations (8.2.18) and (8.3.8) together are said to be the coupled *Maxwell–Bloch equations* (Problem 8.4).

With these equations we have a quantum theory that can be treated self-consistently. That is, the coupled Maxwell–Bloch equations allow the atoms and the field to influence each other mutually, and the theory treats this mutual interaction at a fundamental level (Figure 8.4). We already saw in Chapter 1 an earlier example of such a mutual interaction [recall Figure 1.15], but there the theory was completely empirical. We now reexamine some earlier results, including those of Chapter 1, from our present, more satisfactory foundation.

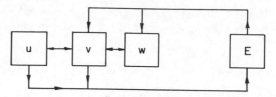

Figure 8.4 The mutual interactions embodied in the Maxwell–Bloch equations. The coupling is much more intricate than in the conventional rate-equation theory illustrated in Figure 1.15.

8.4   LINEAR ABSORPTION AND AMPLIFICATION

First we determine what effect other atoms, e.g., the collision partners of our two-level atoms, have on the Maxwell equation. These atoms also have dipole moments and give rise to an added polarization density. Thus we add the term

$$\frac{ik}{\epsilon_0} \overline{N}\, \overline{\mu}{}^* \overline{\rho}_{21}(z, t) \tag{8.4.1}$$

to the right side of (8.3.6), where the overbars denote background-atom parameters. These atoms are far from resonance and come to steady state extremely quickly, so we can use the adiabatic result (7.2.1) for $\overline{\rho}_{21}$. We can safely assume that the background atoms are at most only very slightly excited, so that $\overline{\rho}_{22} = 0$ and $\overline{\rho}_{11} = 1$. Thus we have

$$\begin{aligned}
\overline{\rho}_{21} &= \frac{i\overline{\chi}/2}{\overline{\beta} + i\,\overline{\Delta}} \\[2mm]
&= \frac{i(\overline{\chi}/2)\,(\overline{\beta} - i\overline{\Delta})}{\overline{\Delta}{}^2 + \overline{\beta}{}^2}
\end{aligned} \tag{8.4.2}$$

where $\overline{\chi} = \overline{\mu}\mathcal{E}/\hbar$. This expression for $\overline{\rho}_{21}$ gives

$$\begin{aligned}
\frac{ik}{\epsilon_0} \overline{N}\, \overline{\mu}{}^* \overline{\rho}_{21} &= \frac{\overline{N}\, \overline{\mu}{}^* \omega}{2\epsilon_0 c} \frac{\overline{\mu}\, \mathcal{E}}{\hbar} \frac{i\overline{\Delta} - \overline{\beta}}{\overline{\Delta}{}^2 + \overline{\beta}{}^2} \\[2mm]
&= -\frac{1}{2}\Big(\overline{a} - i\overline{\delta}\Big)\mathcal{E}
\end{aligned} \tag{8.4.3}$$

Here we have defined

$$\bar{a} = \frac{\bar{N}\,|\bar{\mu}|^2\omega}{\epsilon_0\hbar c}\,\frac{\bar{\beta}}{\bar{\Delta}^2 + \bar{\beta}^2} \tag{8.4.4}$$

$$\bar{\delta} = (\bar{\Delta}/\bar{\beta})\,\bar{a} = \frac{\bar{N}\,|\bar{\mu}|^2\omega}{\epsilon_0\hbar c}\,\frac{\bar{\Delta}}{\bar{\Delta}^2 + \bar{\beta}^2} \tag{8.4.5}$$

where $\bar{a}$ is seen by comparison with (7.4.8) to be the extinction coefficient of the background atoms. It is also the imaginary part of the index of refraction for light transmitted through the background atoms alone. [Recall (3.4.11) with the oscillator strength f included.] Similarly $\bar{\delta}$ can be recognized by comparison with (3.4.9) to correspond to the background correction to the real part of the index of refraction.

The effect of the background atoms (collision partners, etc.) on the slowly varying Maxwell equation (8.3.6) is therefore simply to add two terms to the left side:

$$\left(\frac{\partial}{\partial z} + \frac{\bar{a}}{2} - \frac{i\bar{\delta}}{2} + \frac{\partial}{\partial ct}\right)\mathcal{E}(z, t) = \frac{ik}{\epsilon_0}N\mu^*\,\rho_{21}(z, t) \tag{8.4.6}$$

Now consider the Maxwell–Bloch equations in steady state. We discard the time derivative in (8.4.6) and use the quasisteady solution (7.2.1b) for ρ_{21}. All of the steps from (8.3.8) to (8.4.6) apply as well to ρ_{21} as to $\bar{\rho}_{21}$, except that we cannot assume $\rho_{22} = 0$ and $\rho_{11} = 1$ for on-resonant atoms. This minor distinction is easily accounted for, and we can write (8.4.6) as

$$\left(\frac{\partial}{\partial z} + \frac{\xi - i\eta}{2}\right)\mathcal{E}(z) = 0 \tag{8.4.7}$$

where we have introduced the temporary abbreviation

$$\frac{\xi - i\eta}{2} = \frac{1}{2}\left(\bar{a} - i\bar{\delta} + (a - i\delta)(\rho_{11} - \rho_{22})\right) \tag{8.4.8}$$

Let us multiply (8.4.7) by $\mathcal{E}^*$ and add the complex conjugate equation to get

$$\mathcal{E}^*\left(\frac{\partial}{\partial z} + \frac{\xi - i\eta}{2}\right)\mathcal{E} + \mathcal{E}\left(\frac{\partial}{\partial z} + \frac{\xi + i\eta}{2}\right)\mathcal{E}^* = 0 \tag{8.4.9}$$

Since $\partial|\mathcal{E}|^2/\partial z = \mathcal{E}^* (\partial\mathcal{E}/\partial z) +$ c.c., this is the same as

$$\left(\frac{\partial}{\partial z} + \xi\right) |\mathcal{E}|^2 = 0 \qquad (8.4.10)$$

and since $|\mathcal{E}|^2$ is proportional to I, we have

$$\frac{\partial I}{\partial z} = -\xi I \qquad (8.4.11)$$

Equation (8.4.11) is the same as Eq. (2.6.14) and has the same exponential decay solution

$$I(z) = I(0) \, e^{-a_s z} \qquad (8.4.12)$$

as was given in (2.6.15), if we identify ξ with the extinction coefficient a_s:

$$a_s \leftrightarrow \xi = \bar{a} + a(\rho_{11} - \rho_{22}) \qquad (8.4.13)$$

In fact, we find much more than a simple identification of coefficients. From the Maxwell–Bloch result (8.4.11) we can draw two conclusions about light propagation in a medium of atoms both near to resonance and far off resonance (background atoms):

i. If the resonant atoms are all in their ground states ($\rho_{11} \approx 1$, $\rho_{22} \approx 0$), then the classical law of exponential extinction is valid, and $a_s = \bar{a} + a$. That is, both on-resonance and off-resonance atoms contribute alike to the attenuation of the field.

ii. If the nearly resonant atoms are in their excited states ($\rho_{11} \approx 0$, $\rho_{22} \approx 1$), then the solution is still exponential in form but with $a_s = \bar{a} - a$. It is quite possible that $a \gg \bar{a}$, since the detuning $\bar{\Delta}$ appearing in the denominator of the expression for $\bar{a}$ in (8.4.4) is by assumption very large, and the detuning Δ in the corresponding expression for a is small or even zero. Under these conditions a_s is negative, and describes not attenuation but amplification (Figure 8.5).

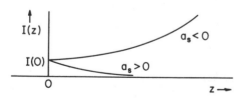

Figure 8.5 Amplification or attenuation of an incident pulse.

It is clear that in case (ii) the the possibility of laser action is foretold. Equation (8.4.13) also gives clear instructions on how to go about obtaining the right conditions for it. One need only (!) ensure that the near-resonant atoms are sufficiently excited by some means, so that

$$\rho_{22} - \rho_{11} > \bar{a}/a \qquad (8.4.14)$$

All formulas for critical or threshold inversion governing laser action derive from the principles leading to (8.4.14). Mirror losses can, in practice, be more significant than attenuation due to absorption or scattering in the laser medium, but they lead to similar results, as we will find in Chapter 10. The methods for achieving threshold inversion are many. They vary greatly with type of laser, and are discussed in Chapter 13.

8.5 SEMICLASSICAL LASER THEORY

Laser action based on inversion is a quantum effect, as (8.4.14) suggests. The need for a positive inversion cannot be satisfied in a classical system, for which the concept of inversion does not exist. We present the elements of a quantum-mechanical laser theory in this section. It will not be a full theory for many reasons, but it will be complete enough to correct the flaws in the classical theory of the laser, given in Section 3.5.

We must first recall the expression given in (1.3.1) and (3.5.2) for the electric field in a laser cavity. The allowed wave vectors and mode functions are determined by the cavity length:

$$k_m = m\pi/L \qquad (8.5.1)$$

so we have $\mathbf{E}(z, t) = \Sigma_m \mathbf{E}_m(z, t)$, where

$$\mathbf{E}_m = \hat{\boldsymbol{\varepsilon}}_m \, \mathcal{E}_m(t) \sin k_m z \, e^{-i\omega t} \qquad (8.5.2)$$

is the electric field of the mth mode of the cavity. Note that the complex mode amplitude $\mathcal{E}_m(t)$ does not depend on z, since the cavity mode function $\sin k_m z$ is assumed to express the z dependence fully. The frequency ω of laser oscillation is not known initially, although it will be close to one of the cavity mode frequencies: $\omega \approx \omega_m$.

It is convenient to express the polarization's z dependence in terms of cavity mode functions as well. That is, we will write $\mathbf{P}(z, t) = \Sigma_m \mathbf{P}_m(z, t)$, where

$$\mathbf{P}_m = 2N e \mathbf{r}_{12} \rho_{21}^{(m)}(z, t) \sin k_m z \, e^{-i\omega t} \qquad (8.5.3)$$

with the real part representing the physical polarization [recall (3.5.5)]. Now we follow the procedure of Section 3.5 and substitute (8.5.2) and (8.5.3) into the second-order wave equation (3.5.6) to find a "reduced" equation for the mode amplitude $\mathcal{E}_m$. This equation is the cavity analog of (8.4.6); losses are effected by the conductivity in the cavity instead of by background absorption due to other atoms. [In practice the cavity "conductivity" is due to mirror losses, and background absorption is also present but usually smaller. This is discussed in Chapter 10.] The reduced (i.e., first-order) equation for $\mathcal{E}_m$ is

$$
\left(-i(\omega - \omega_m) + \frac{\sigma}{2\epsilon_0} + \frac{\partial}{\partial t} \right) \mathcal{E}_m(t) \sin k_m z
$$

$$
= \frac{i\omega}{\epsilon_0} N\mu^* \rho_{21}^{(m)}(z, t) \sin k_m z \qquad (8.5.4)
$$

We have used the near-resonance approximation

$$
(k_m c)^2 - \omega^2 = (k_m c + \omega)(k_m c - \omega) \approx 2\omega(\omega_m - \omega)
$$

where

$$
\omega_m = k_m c = m\pi c/L \qquad (8.5.5)
$$

We will adopt an approximation already implied by (8.5.4), namely that different cavity modes are not coupled. They are actually coupled through the z dependence of $\rho_{21}^{(m)}(z, t)$, but in many lasers this coupling is not important. We will replace $\rho_{21}^{(m)}(z, t)$ by its cavity-average value and then simply cancel the two $\sin k_m z$ factors in (8.5.4). We will also drop the superscripts (m) for ease of writing, to obtain the single-mode reduced Maxwell equation

$$
\left(-i(\omega - \omega_m) + \frac{\sigma}{2\epsilon_0} + \frac{\partial}{\partial t} \right) \mathcal{E}_m(t) = \frac{i\omega}{\epsilon_0} N\mu^* \rho_{21}(t) \qquad (8.5.6)
$$

We will treat only quasisteady-state laser operation here, so the adiabatic approximation can be used for ρ_{21}:

$$
\rho_{21} = \frac{-i\chi_m/2}{\beta + i\Delta} (\rho_{22} - \rho_{11}) \qquad (8.5.7)
$$

where

$$
\chi_m = \mu \mathcal{E}_m/\hbar \qquad (8.5.8)
$$

The reduced wave equation becomes

$$\left(-i(\omega - \omega_m) + \frac{\sigma}{2\epsilon_0} + \frac{\partial}{\partial t}\right)\mathcal{E}_m = \frac{i\omega}{\epsilon_0} N\mu^* \frac{-i\chi_m/2}{\beta + i\Delta}(\rho_{22} - \rho_{11})$$

$$= \frac{N|\mu|^2\omega}{2\epsilon_0\hbar}\frac{\beta - i\Delta}{\Delta^2 + \beta^2}\mathcal{E}_m(t)\left[\rho_{22}(t) - \rho_{11}(t)\right]$$

(8.5.9)

This is the quantum-mechanical analog of the equation required in Problem 3.4.
Let us now adopt the common notation [recall (7.3.1)]

$$N_1(t) = N\rho_{11}(t), \qquad N_2(t) = N\rho_{22}(t) \qquad (8.5.10)$$

where N_1 and N_2 are the densities (atoms per cm^3) of active atoms in levels 1 and
2, and satisfy equations (7.3.2):

$$\dot{N}_1 = -\Gamma_1 N_1 + A_{21}N_2 + \sigma(\Delta)\Phi_m(N_2 - N_1) \qquad (8.5.11)$$

$$\dot{N}_2 = -(\Gamma_2 + A_{21})N_2 - \sigma(\Delta)\Phi_m(N_2 - N_1) \qquad (8.5.12)$$

The reduced wave equation (8.5.9) can then be written

$$\frac{\partial}{\partial t}\mathcal{E}_m = \frac{1}{2}\left(\frac{-\sigma}{\epsilon_0} + 2i(\omega - \omega_m) + c(g - i\delta)\right)\mathcal{E}_m \qquad (8.5.13)$$

where

$$g = \frac{|\mu|^2\omega}{\epsilon_0\hbar c}\frac{\beta}{\Delta^2 + \beta^2}(N_2 - N_1)$$

$$= \sigma(\Delta)(N_2 - N_1) \qquad (8.5.14)$$

$$\delta = \frac{\omega_{21} - \omega}{\beta}g \qquad (8.5.15)$$

(The dependence of the cross section $\sigma(\Delta)$ on detuning should be enough to prevent confusion with the conductivity σ.)

Equation (8.5.13) is a fundamental equation of semiclassical laser theory for
the laser field. It is coupled to equations for the atomic medium through the dependence of g on $N_2 - N_1$. The most important aspects of laser behavior are
associated with steady-state or cw (continuous wave) operation. In steady state the
terms on the right side of (8.5.13) must cancel:

$$-i(\omega - \omega_m) + \frac{\sigma}{2\epsilon_0} = \frac{c}{2}(g - i\delta) \tag{8.5.16}$$

Notice that this is exactly Eq. (3.5.8).

We have here obtained a full correction of the classical steady-state "laser theory" of Chapter 3. To help explain the correction, we have copied the notation of (3.5.9) and (3.5.10). The values of the coefficients are however taken from (8.5.9). The physical interpretation of an equation such as (8.5.13) has already been given in Section 3.5 in connection with the laser model there. That is, g is the gain coefficient, and the threshold gain condition is

$$g = \sigma/\epsilon_0 c \quad \text{(threshold)} \tag{8.5.17}$$

The important difference with the classical model is that a positive sign for g is now possible. In fact, just as we expect, the gain becomes positive whenever the inversion $N_2 - N_1$ is positive. As mentioned in Section 3.5, it is not necessary to reverse the sign of β to obtain positive gain. Of course (8.5.17) is also the cavity threshold analog of the amplification threshold $\xi = 0$ implied by (8.4.13): $a(\rho_{22} - \rho_{11}) = \bar{a}$.

The frequency pulling relation remains exactly as found in (3.5.12), and the lasing frequency ω obeys

$$\omega_m - \omega = \frac{gc}{2\beta}(\omega - \omega_{21}), \tag{8.5.18a}$$

or

$$\omega = \left(\omega_m + \frac{gc}{2\beta}\omega_{21}\right)\bigg/\left(1 + \frac{gc}{2\beta}\right)$$

$$\simeq \omega_m + \frac{gc}{2\beta}(\omega_{21} - \omega_m) \tag{8.5.18b}$$

when $gc/2\beta \ll 1$. Note that in this case the assumption made above (8.5.5) is well justified.

We have thus validated the *form* of the classical model of Chapter 3, with a correction factor $N_2 - N_1$ to the gain formula. Of course this correction to the gain has major physical consequences. Now what can be said about the "Einstein" laser model of Section 1.5?

The Einstein model may be based directly on Eqs. (8.5.12) and (8.5.13). We can show this in three steps. First we multiply (8.5.13) by $\mathcal{E}_m^*$ and add the complex conjugate equation to get

$$\frac{1}{c}(\mathcal{E}_m^* \dot{\mathcal{E}}_m + \dot{\mathcal{E}}_m^* \mathcal{E}_m) = \left(g - \frac{\sigma}{\epsilon_0 c}\right)\mathcal{E}_m^* \mathcal{E}_m$$

This is the same as

$$\frac{d|\mathcal{E}_m|^2}{dt} = c\sigma(\Delta)\,(N_2 - N_1)|\mathcal{E}_m|^2 - \frac{\sigma}{\epsilon_0}|\mathcal{E}_m|^2 \qquad (8.5.19)$$

where we have used (8.5.14)

The second step is to recognize that (8.5.12) has no term representing the rate of increase of population of level 2 due to a pumping mechanism, and to add such a term. At the same time let us assume that the pump is able to maintain $N_2 \gg N_1$ constantly. Then we can ignore N_1 compared with N_2, and rewrite (8.5.12) as

$$\frac{dN_2}{dt} = -(\Gamma_2 + A_{21})N_2 - \sigma(\Delta)\,\Phi N_2 + K \qquad (8.5.20)$$

where K is the pumping rate.

The final step is to multiply (8.5.20) by the cavity volume V in order to convert the density of upper-level atoms into the number of such atoms: $N_2 V = n_2$. And we multiply (8.5.19) by $\frac{1}{2}\epsilon_0 V / \hbar\omega$ to convert $|\mathcal{E}_m|^2$ into the number of photons in the cavity, q. Also $\Phi V/c$ is another expression for q. With these factor adjustments we can rewrite (8.5.19) as

$$\frac{dq}{dt} = \frac{c\sigma(\Delta)}{V}\,n_2 q - \frac{\sigma}{\epsilon_0}\,q \qquad (8.5.21)$$

where we have ignored n_1 compared with n_2; and we can rewrite (8.5.20) as

$$\frac{dn_2}{dt} = -\frac{c\sigma(\Delta)}{V}\,q\,n_2 - (\Gamma_2 + A_{21})n_2 + KV \qquad (8.5.22)$$

It should be clear that these last two equations exactly reproduce Eqs. (1.5.1) and (1.5.2) if we identify the parameters from Chapter 1 as follows:

$$a = \frac{c\sigma(\Delta)}{V} \qquad (8.5.23)$$

$$b = \frac{\sigma}{\epsilon_0} \qquad (8.5.24)$$

$$f = \Gamma_2 + A_{21} \qquad (8.5.25)$$

$$p = KV \qquad (8.5.26)$$

In this way we see that the rate-equation model is rederived. Of course, despite the formal parallels in the equations, only here in Chapter 8 have we really made a derivation. This is especially so regarding the identification formulas (8.5.23) and (8.5.25), which supply for the first time microscopic interpretations of the empirical parameters a and f of Chapter 1.

The principal message of this section is not difficult to find. It is simply that the framework of the coupled atom–light equations that we have called the Maxwell–Bloch equations is sufficiently simple to show the essential elements of laser theory without much work. It is, however, remarkable that these simple equations are also complete enough to explain the empirical model of Chapter 1 and the Lorentz model "laser" of Chapter 3, and thus to unify them, despite their obvious differences. The same message applies, in a more general sense of course, to this entire chapter. The wave-mechanical theory of atomic transitions can be made both tractable and relevant to laser theory, with only a little attention to appropriate models and approximations.

Despite these successes, the task of explaining what lasers are like under a variety of operating conditions, with attention to important details omitted here, remains for Chapters 10–14.

PROBLEMS

8.1   Find the solutions for u, v, and w from the Bloch equations (8.2.2) in the following separate special cases:

(a) Exact resonance ($\Delta = 0$).

(b) Zero field strength ($\chi = 0$).

In both cases, take the initial values to be u_0, v_0, and w_0, and assume that Δ and χ are constants.

8.2*   Find the solutions for u, v, and w from the damped Bloch equations (8.2.18) in the special case of exact resonance ($\Delta = 0$). Assume that χ is a real constant, and that the atom is initially in the upper level.

8.3   Find the steady-state solutions for u, v, and w from the damped Bloch equations (8.2.18). Assume that χ is a real constant. Make two plots on the same graph of w vs. Δ/β for $\chi = \beta/2$ and $\chi = 2\beta$, assuming $\beta = 1/T_1$. Compare the halfwidths of the two curves.

8.4*   A famous solution of the Maxwell–Bloch equations represents an optical "soliton," a pulse traveling without change of shape. This soliton pulse must be short enough, in practice, to permit losses to be ignored in both the Maxwell and the Bloch equations. The solution for the field envelope is

$$\chi(t, z) = \frac{2}{\tau} \operatorname{sech}\left(\zeta/\tau\right)$$

where $\zeta = t - z/V$, and V is the soliton velocity.

(a) Compute the soliton's total pulse area.

(b) The solutions for u and v for an on-resonant atom, corresponding to this χ solution, are

$$u(t) = 0$$
$$v(t) = 2 \, \text{sech} \, (\zeta/\tau) \, \tanh \, (\zeta/\tau)$$

Use these solutions, and Eq. (8.2.2b), to find the solution for $w(t)$. Compute $u^2 + v^2 + w^2$.

(c) Use the u, v, and χ solutions in (8.3.8) to find an expression for the soliton velocity V.

9 WAVE–PARTICLE DUALITY OF LIGHT

9.1 WHAT IS A PHOTON?

One of the long-standing questions about light is whether it consists of particles or waves. Newton, around 1700, thought it consisted of particles. The term *photon* is the modern name[1] for a particle of light, and we have used it confidently in several contexts in Chapters 1–8. However, in the period 1700–1850, as a result of careful experimentation on the interference and diffraction of light, Newton's particles of light were abandoned in favor of a wave picture championed by Thomas Young and others. The wave picture was brilliantly formulated mathematically by J. C. Maxwell at roughly the time of the American Civil War. The experiments of Heinrich Hertz (1887) gave convincing evidence for the validity of Maxwell's electromagnetic theory of light. Hertz confirmed that accelerated charges radiate electromagnetic waves, and that these waves have essentially the same characteristics as visible light.

By the 1890s the wave theory based on Maxwell's equations was used by Lorentz and others to explain, at least semiquantitatively, nearly all the known optical phenomena (recall Chapters 2 and 3). Many scientists held that Newton's theories of mechanics and gravitation, together with Maxwell's theory of electromagnetism, contained the fundamental laws of the universe, and that in principle everything about Nature might one day be understood in terms of them.

But there were some experimental results that did not seem to fit into the picture. One was the spectrum of blackbody radiation, which led to the light-quantum hypotheses of Planck (1900) and Einstein (1905). These ideas were used by Bohr in his theory of the hydrogen atom (1913), and eventually led to quantum mechanics and the view that nature is fundamentally statistical rather than deterministic.

According to quantum mechanics the wave and particle views of light are both oversimplifications. Radiation and matter have both wave and particle attributes, or a "wave–particle duality."

What, then, is a photon? By considering a few "thought experiments" we will try to explain the essence of the answer given by quantum mechanics. The answer is subtle and also very beautiful. The thought experiments described below deal with photon polarization, photon-induced recoil, and photon interference.

---

1. The term *photon* was coined in 1926 by the chemist G. N. Lewis. Before 1926 physicists referred to "quanta" of light.

9.2 PHOTON POLARIZATION: ALL OR NOTHING

Consider the situation shown in Figure 9.1. A plane electromagnetic wave of frequency ν is incident upon a Polaroid sheet oriented so that it passes radiation of polarization along the x direction, but absorbs radiation of the orthogonal y polarization. The incident light is assumed to be linearly polarized at an angle θ with respect to the x direction (i.e., the electric field vector at any point on the wave oscillates along a line at an angle θ with respect to the x axis). According to the classical law of Malus, a fraction $\cos^2 \theta$ of the incident intensity will pass through the sheet. This is the prediction of classical physics.

Now suppose that the incident light is reduced in intensity so that only a single photon (of energy $h\nu$) is incident upon the polarizer. In this case the law of Malus fails: the incident photon is not split by the polarizer. Instead we find that the *entire* photon either passes through the polaroid or is absorbed. It is "all or nothing" when a single photon is incident on the polaroid. The energy $h\nu$ is evidently an indivisible unit of energy for radiation of frequency ν.

If we repeat this one-photon experiment many times, always with the same source and arrangement, we find that sometimes the photon passes through the sheet and sometimes it does not. If θ is zero, however, we find that the incident photon always passes through the sheet, whereas if $\theta = 90°$ no photon ever passes through. Repetition of the experiment many times reveals that $\cos^2 \theta$ is the *probability* that a photon polarized at an angle θ to the polaroid axis will pass through.

The situation here is akin to coin flipping. When we say that the probability of

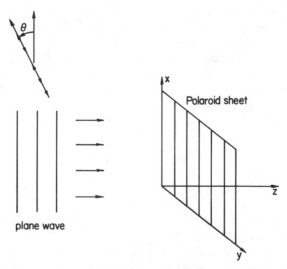

Figure 9.1 A plane monochromatic wave is incident upon a Polaroid sheet. The field is assumed to be linearly polarized at an angle θ with respect to the polarizing axis of the sheet.

getting either heads or tails is $\frac{1}{2}$, we mean that in a large number of tosses heads and tails will turn up approximately the same number of times. In any one toss of the coin, however, we get either heads or tails, just as our one-photon experiment either records an entire photon or nothing. Quantum mechanics asserts that the statistical aspect of our one-photon experiment is a fundamental characteristic of Nature.

Suppose the incident field in our experiment has an energy corresponding to n photons. Since each photon has the probability $\cos^2\theta$ of passing through the sheet, we expect on the average that $n \cos^2\theta$ photons will be transmitted. As n increases, the deviation of the number of transmitted photons from the average $n \cos^2\theta$ becomes smaller relative to the average number. Thus it becomes increasingly more accurate to say that a fraction $\cos^2\theta$ of the incident field intensity is passed by the filter. In other words, *we approach the classical law of Malus when the number of incident photons is large.* Most optical experiments involve enormous numbers of photons (Problem 9.1). In such experiments with polarized light, deviations from the law of Malus go unnoticed. In other words, the "all or nothing" nature of photon polarization may be ignored for all practical purposes in such experiments.

Our discussion has assumed an ideal photon source, one that produces a single photon on demand. We have also assumed an ideal photon counter. Such an ideal device would count every incident photon, and it would give no spurious counts when there are no incident photons. Furthermore it would respond immediately to the incident signal. Needless to say, there is no such perfect detector. Nevertheless, available detectors, including the human eye, come fairly close to the ideal. We give a brief discussion of photon detectors in the appendix to this chapter.

9.3 FAILURE OF CLASSICAL WAVE THEORY: RECOIL AND COINCIDENCE

A photon of radiation of frequency ν carries not only an energy $h\nu$ but also a linear momentum of magnitude

$$p = h\nu/c \qquad (9.3.1)$$

In other words, the linear momentum of radiation of frequency ν is quantized in indivisible units of magnitude $h\nu/c$. Conservation of linear momentum demands that an atom that undergoes spontaneous emission must recoil with a linear momentum of magnitude (9.3.1). According to quantum mechanics we cannot predict exactly in which direction the photon will be emitted, and therefore we cannot predict the direction of atomic recoil (Figure 9.2).

Experiments with beams of excited atoms confirm the recoil associated with spontaneous emission (Figure 9.3). In fact the recoil of a spontaneously emitting

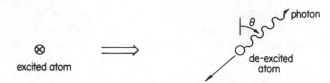

Figure 9.2 Conservation of linear momentum implies that an atom recoils when it undergoes spontaneous emission. The direction of photon emission (and atomic recoil) is not predictable.

atom was inferred by O. R. Frisch in 1933, and has in recent years been confirmed with greater accuracy.

It is not surprising that an atom recoils when it emits a photon. It is like the recoil a person feels upon firing a rifle, the person and the bullet corresponding to the atom and the photon, respectively. However, this recoil of a spontaneously emitting atom is not accounted for in the classical wave theory of radiation. Figure 9.4 shows why. Classical theory treats spontaneous emission as a smooth process (recall Problem 2.10), with radiation being continuously emitted more or less in all directions, according to the radiation pattern of the source. (We discussed the pattern of dipole radiation, for example, in Section 2.5.) Classical radiation from an isolated system like an atom is also characterized by inversion symmetry. That is, the intensity of spontaneous radiation emitted in the x direction is equal to the intensity emitted in the $-x$ direction, and so on for every direction. It is obvious that an emitter of this type suffers equal and opposite recoil forces along every axis, and does not recoil at all. So classical wave theory predicts no recoil in spontaneous emission, in contradiction to experimental results.

Stimulated emission and absorption also impart a recoil momentum to an atom (Figure 9.5). In these cases, of course, the direction of atomic recoil follows exactly from the direction of propagation of the incident radiation. It is interesting that the Doppler effect in the emission or absorption of radiation by a moving atom may be understood as a consequence of the fact that photons carry energy $h\nu$ and linear momentum $h\nu/c$ (Problem 9.3).

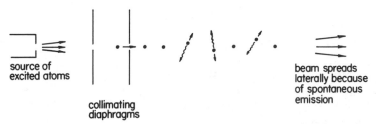

Figure 9.3 A well-collimated atomic beam of excited atoms will spread laterally because of the recoil associated with spontaneous emission.

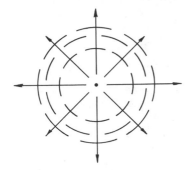

Figure 9.4 A source emitting a spherical wave cannot recoil, because the spherical symmetry of the wave prevents it from carrying any linear momentum from the source.

• The classic experimental manifestation of the expression for photon momentum given in (9.3.1) is the Compton effect, as discussed in textbooks on modern physics. A. H. Compton's experiments were reported in 1923. Einstein had already inferred in 1917 that photon momentum causes an atom to recoil when it undergoes spontaneous emission. He did this by a careful analysis of thermal equilibrium between radiation and matter, showing that the Planck distribution *requires* atomic recoil in spontaneous emission.

It is interesting to remember that in his special theory of relativity Einstein had even earlier (in 1905) proposed the equation

$$E = \sqrt{m^2c^4 + p^2c^2} \qquad (9.3.2)$$

connecting the energy E and linear momentum p of a body of rest mass m. For a photon, $m = 0$ and $E = h\nu$, so (9.3.1) then follows from Einstein's 1905 relation. Nevertheless, the equation $p = h\nu/c$ for a photon was not stated by Einstein (or anyone else) before his paper of 1917 on "The Quantum Theory of Radiation." •

Consider another situation involving an atom and a single photon. Take a single excited atom A and two identical photon detectors B and C, as illustrated in Figure 9.6a. We imagine that B and C are perfect detectors in that each registers a count if and only if a photon of radiation is incident upon it.

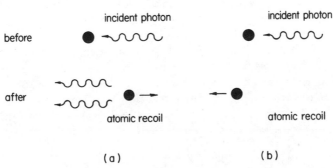

Figure 9.5 Atomic recoil associated with (*a*) stimulated emission and (*b*) absorption.

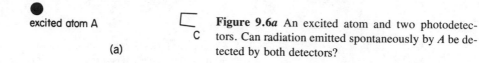

excited atom A

(a)

Figure 9.6a An excited atom and two photodetectors. Can radiation emitted spontaneously by A be detected by both detectors?

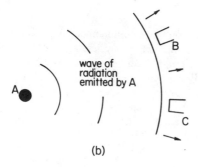

wave of
radiation
emitted by A

(b)

Figure 9.6b According to classical radiation theory the field emitted by A can be measured at both B and C.

When A undergoes spontaneous emission, it need not register a count in either B or C, because the emitted photon may simply go off in some other direction. But there is some probability that a count is registered in B or C. We pose the following question: is there any possibility that a coincidence occurs, i.e., that the emitted radiation from A can be detected at *both* B and C?

The answer given by the quantum theory of radiation is clear: No. For A emits a single photon, and a photon is an indivisible unit of energy that can trigger the emission of only one photoelectron. Therefore the radiation from A can register a count at either B or C, or neither, but never at both B and C.

Now think of the radiation from A as a classical electromagnetic wave. As such it has nonzero values over a certain region of space (and time). In particular, as illustrated in Figure 9.6b, the field propagates outwards from A and eventually reaches both B and C, so that there are measurable electric and magnetic fields at both B and C. And these fields can, according to classical electromagnetic theory, trigger photoelectric counts at both B and C. Thus there is a clear disagreement here between the classical and quantum theories of radiation.

Experiment supports the quantum theory of radiation: the radiation emitted in a single atomic transition cannot be split into more than one unit or excite more than one detector.

9.4   WAVE INTERFERENCE AND PHOTONS

Photons are energy quanta of the electromagnetic field, carrying energy $h\nu$ and linear momentum $h\nu/c$. Photons also have intrinsic angular momentum ("spin"), but we will not be concerned with photon spin in this book. Photons are often pictured as particles, like bullets or billiard balls, but having no mass and moving at the speed of light. This particle picture certainly helps us to understand atomic recoil in spontaneous emission, and phenomena like Compton scattering, but what have these effects to do with the vast body of evidence for the wave nature of light? To see how quantum mechanics reconciles the wave and particle aspects of light, we will consider the example of the Young two-slit experiment. We will assume that the slits are illuminated by a plane monochromatic wave. As illustrated in Figure 9.7, the slits in this case produce a well-known interference pattern on a screen behind the slits.

The location of the intensity maxima is easily found from the condition for constructive interference of the two waves emerging from the slits: their path difference $s_2 - s_1$ must be an integral number of wavelengths. This gives the condition

$$y_n^{(max)} = n\,\frac{\lambda D}{d}, \qquad n = 0, \pm 1, \pm 2, \ldots \qquad (9.4.1a)$$

for the intensity maxima on the observation screen. When the path difference

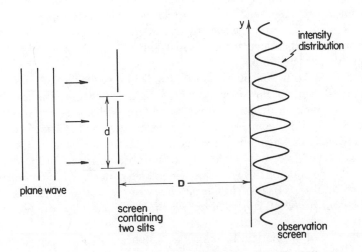

Figure 9.7 The Young two-slit interference experiment.

$s_2 - s_1$ is an odd integral number of half wavelengths, however, we get complete destructive interference:

$$y_n^{(min)} = \left(n + \tfrac{1}{2}\right) \frac{\lambda D}{d}, \qquad n = 0, \pm 1, \pm 2, \ldots \qquad (9.4.1b)$$

For optical wavelengths, the slit separation d in a demonstration of the Young experiment is typically on the order of a fraction of a millimeter, while D is about a meter.

• When Thomas Young published his results in 1802 he encountered a great deal of criticism from the proponents of Newton's particle theory of light. One objection was that the interference experiment was inconsistent with the law of energy conservation (Problem 9.4). Although it was eventually recognized by nearly all scientists that Young's experiment had put an end to Newton's particle theory, Young was discouraged by the criticism of his work, and gave up his research in optics for other endeavors. His accomplishments were remarkable. For instance, he made a major contribution to Egyptology by deciphering the Rosetta stone. His theory of color vision is widely cited even today, and the same is true of his work on elasticity. He also made pioneering contributions to the studies of sound, tides, and the human voice. His work on the wave theory of light was done shortly after he began a medical practice in London. •

Imagine now that the Young experiment is performed with a single photon of light. Can we obtain Young's interference pattern in this case? According to quantum theory, the intensity interference pattern indicated in Figure 9.7 represents only the relative *probability distribution* for detecting the photon somewhere on the observation screen. That is, a single photon by itself does not produce an interference pattern. Instead, there is a relatively high probability of detecting the photon at points satisfying the constructive interference condition (9.4.1a), and the photon will never be found at points satisfying (9.4.1b) because at such points the probability is exactly zero.

In other words, the quantum theory of radiation says that the classical interference pattern is correct, but not in the sense of classical theory. The strictly classical (wave) interpretation of the interference pattern is that the *entire* pattern is observed regardless of the intensity of the light incident on the screen containing the slits. However, the quantum interpretation is that, with light so dim that only a single photon is involved, we do not in fact observe a pattern but only a single photon at some point on the classically calculated pattern. To summarize: *the wave and particle aspects of two-slit interference are reconciled by associating a particle (photon) probability distribution function with the classical (wave) intensity pattern.*

At the risk of belaboring the point, we emphasize that the interference pattern *applies* to a single photon (in the probabilistic sense) but would not be *revealed* by a single photon. When the Young experiment is performed with a light beam containing a very large number of photons, even points of low probability (or, classically speaking, low intensity) receive some photons because of the large numbers involved. In this case we observe the complete interference pattern, just as if we perform the single-photon experiment a very large number of times. In nearly all interference experiments, of course, an extremely large number of photons is involved, as in all of classical optics (Problem 9.1). Under such circumstances we observe smooth and entire intensity patterns, exactly as predicted by classical wave theory. This is why rather delicate experiments are necessary to see departures from classical wave theory. Because a succession of identical independent single-photon experiments eventually builds up the same pattern associated with a many-photons-at-once experiment, the entire pattern must be "known" to each single photon. It is conventional, therefore, to say, following P. A. M. Dirac, that a photon interferes only with itself.

In quantum mechanics both radiation and massive particles (electrons, protons, neutrons, etc.) display a wave–particle duality. A particle with linear momentum of magnitude p has a de Broglie wavelength $\lambda = h/p$ associated with it. For a photon the de Broglie wavelength is $\lambda = h/(h\nu/c) = c/\nu = \lambda$, just the wavelength of a wave of frequency ν. The wave fields in this case satisfy Maxwell's equations. For material particles the wave fields satisfy Schrödinger's equation, at least in the nonrelativistic regime.

The association of a particle probability function with a wave interference pattern correctly describes many phenomena that cannot be explained with wave or particle concepts alone. Wave and particle attributes are interwoven in the microscopic world, both being oversimplifications derived from our experience with macroscopic phenomena.

9.5 PHOTON COUNTING

Imagine an ideal photocathode surface in which every incident photon ejects a photoelectron. By counting the number of photoelectrons we are in effect counting the number of incident photons. Of course real experiments are much more involved than this naive idealization would suggest, but let us suppose nevertheless that we can count photons in this manner.

Consider the following experiment. We have a shutter that we can open and close instantaneously. We place the shutter between a photon detector and an incident quasimonochromatic cw beam of light. We open the shutter at some time t and close it at a time $t + T$, and record the number of photons that were counted while the shutter was open. Next we open the shutter for another time interval T, and again take note of the number of photons counted in that "run." We repeat this procedure several thousand times and make a histogram of the number of photons counted in a time interval T. Our histogram might look like that shown in Figure 9.8.

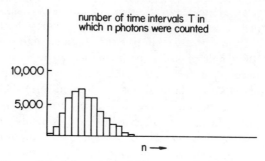

number of time intervals T in
which n photons were counted

Figure 9.8 A hypothetical histogram of photon counts.

What we would like is a theory of such a photon count distribution. Based on the scatter in the number of photon counts from one run to the next, we base our theory on probabilities. That is, we only try to calculate the probability $P_n(T)$ of counting n photons in a given time interval T.

Consider a very short sub-interval Δt within the time interval T. We assume that $p(t)\,\Delta t$, the probability of ejecting one electron in the time interval Δt, is given by

$$p(t)\,\Delta t = \alpha I(t)\,\Delta t, \tag{9.5.1}$$

where $I(t)$ is the incident intensity, averaged over a few optical periods, and α is a constant depending on the details of the photon detector (density of atoms, size of exposed surface, etc.). The time interval Δt is taken to be so short that the probability of ejecting more than one electron during this time is completely negligible. Δt is not an intrinsic parameter of our experiment, but merely a theoretical construct, and so we can make it as small as we please.

Let $P_n(t)$ be the probability of counting n photons during a time t ($0 \leq t \leq T$) during which the shutter is open. It is convenient to consider $P_n(t + \Delta t)$, where Δt is defined above. There are two mutually exclusive ways of getting n photons in the time interval $t + \Delta t$: we can get $n - 1$ photons in the time interval t and 1 more in the interval from t to $t + \Delta t$, or n photons in the time t and none in the interval from t to $t + \Delta t$. The probability of the first way is

$$P_{n-1}(t)\,p(t)\,\Delta t = (\text{probability of } n - 1 \text{ photons in time } t)$$

$$\times\ (\text{probability of 1 photon in } \Delta t) \tag{9.5.2}$$

These probabilities multiply for the same reason that the probability of getting two heads in two successive flips of a coin is given by the product $(\frac{1}{2})\,(\frac{1}{2}) = \frac{1}{4}$. Similarly the probability of the second alternative is

$$P_n(t) \left[1 - p(t) \, \Delta t \right] = \text{(probability of } n \text{ photons in time } t)$$

$$\times \text{ (probability of no photon in } \Delta t) \quad (9.5.3)$$

Since each alternative leads to the same end result—n photons counted in a time interval $t + \Delta t$—we have

$$P_n(t + \Delta t) = P_{n-1}(t) \, p(t) \, \Delta t + P_n(t) \left[1 - p(t) \, \Delta t \right] \quad (9.5.4)$$

Here we add the probabilities for the two possible alternatives because they are mutually exclusive. The probability of getting one head and one tail in two flips of a coin, for instance, is the probability $(\frac{1}{2})(\frac{1}{2})$ of getting a head and then a tail, plus the probability $(\frac{1}{2})(\frac{1}{2})$ of getting a tail followed by a head.

Rearranging (9.5.4), we have

$$\frac{P_n(t + \Delta t) - P_n(t)}{\Delta t} = \left[P_{n-1}(t) - P_n(t) \right] p(t) \quad (9.5.5)$$

Since Δt is at our disposal, let us make it so small that the left side becomes the derivative of $P_n(t)$:

$$\frac{P_n(t + \Delta t) - P_n(t)}{\Delta t} \approx \lim_{\Delta t \to 0} \frac{P_n(t + \Delta t) - P_n(t)}{\Delta t} = \frac{dP_n}{dt} \quad (9.5.6)$$

Therefore (9.5.5) becomes a differential equation for $P_n(t)$:

$$\frac{dP_n}{dt} = p(t) \left[P_{n-1}(t) - P_n(t) \right] \quad (9.5.7)$$

or, using (9.5.1),

$$\frac{dP_n}{dt} = \alpha I(t) \left[P_{n-1}(t) - P_n(t) \right] \quad (9.5.8)$$

To compare the theory leading to (9.5.8) with the results of our photon counting experiment, we must solve (9.5.8) for the probability $P_n(T)$ of counting n photons in any one of our counting intervals of duration T. It is shown below that the desired solution is

$$P_n(T) = \frac{\left[X(T) \right]^n}{n!} \, e^{-X(T)} \quad (9.5.9)$$

where

$$X(T) = \alpha \int_0^T I(t)\, dt \tag{9.5.10}$$

Equation (9.5.9) is correct as far as it goes, but there is a modification of our theory to be made before we can meaningfully compare it to experiment. In an actual experiment we are dealing with a large number of time intervals T, each starting at a different time t. For a time interval from t to $t + T$ rather than from 0 to T we must replace (9.5.10) by

$$X(t, T) = \alpha \int_t^{t+T} I(t')\, dt' \tag{9.5.11}$$

It is convenient to write this as

$$X(t, T) = \alpha T \bar{I}(t, T) \tag{9.5.12}$$

where

$$\bar{I}(t, T) = \frac{1}{T} \int_t^{t+T} I(t')\, dt' \tag{9.5.13}$$

is the average incident intensity during the time interval from t to $t + T$. Then (9.5.9) becomes

$$P_n(t, T) = \frac{1}{n!} [X(t, T)]^n \exp[-X(t, T)] \tag{9.5.14}$$

Finally, to compare with experiment we must average (9.5.14) over all the "starting times" t. That is, the theoretical photon counting probability distribution is

$$P_n(T) = \langle P_n(t, T) \rangle$$

$$= \left\langle \frac{1}{n!} [\alpha T \bar{I}(t, T)]^n \exp[-\alpha T \bar{I}(t, T)] \right\rangle \tag{9.5.15}$$

where $\langle \cdots \rangle$ denotes an average over t. Equation (9.5.15), which was first derived by L. Mandel in 1958, is used frequently in the analysis of photon counting experiments. In the following section we will consider an especially important example of the use of this formula.

• To verify (9.5.9), consider the function

$$P_n(t) = \frac{1}{n!} X(t)^n e^{-X(t)} \qquad (9.5.16)$$

The derivative of this function is

$$
\begin{aligned}
\frac{dP_n}{dt} &= \frac{1}{n!} X(t)^n \frac{d}{dt} e^{-X(t)} + \frac{1}{n!} e^{-X(t)} \frac{d}{dt} X(t)^n \\
&= \frac{1}{n!} X(t)^n \left(-\frac{dX}{dt} e^{-X(t)} \right) + \frac{1}{n!} e^{-X(t)} \left(nX(t)^{n-1} \frac{dX}{dt} \right) \\
&= \frac{dX}{dt} \left(\frac{[X(t)]^{n-1}}{(n-1)!} e^{-X(t)} - \frac{[X(t)]^n}{n!} e^{-X(t)} \right) \\
&= \frac{dX}{dt} \left[P_{n-1}(t) - P_n(t) \right]
\end{aligned}
$$

$$\qquad (9.5.17)$$

From (9.5.10) it follows that

$$\frac{dX(t)}{dt} = \alpha \frac{d}{dt} \int_0^t I(t') \, dt' = \alpha I(t) \qquad (9.5.18)$$

and therefore (9.5.17) is just (9.5.8). In other words, the function (9.5.16) is a solution of the differential equation (9.5.8).

We want a solution of (9.5.8) satisfying $P_n(0) = 0$ because the probability of counting any photons at exactly the time $t = 0$, when the shutter has suddenly been opened, is zero. The function (9.5.16) satisfies this condition because $X(0) = 0$. Therefore (9.5.16) with $t = T$, i.e., (9.5.9), is the desired solution for $P_n(t = T)$. •

9.6 THE POISSON DISTRIBUTION

Suppose the intensity $I(t)$ of the incident beam in our photon counting experiment is constant, i.e.,

$$I(t) = I = \text{constant} \qquad (9.6.1)$$

Then (9.5.13) becomes simply

$$\bar{I}(t, T) = \frac{1}{T} \int_t^{t+T} I(t') \, dt' = \frac{1}{T} I \int_t^{t+T} dt' = I \qquad (9.6.2)$$

In this case the starting time average in (9.5.15) is superfluous. Using (9.6.2) in (9.5.15), we obtain the photon-counting probability distribution

$$P_n(T) = \frac{(\alpha I T)^n}{n!} e^{-\alpha I T} \tag{9.6.3}$$

or

$$P_n(T) = \frac{(\bar{n})^n}{n!} e^{-\bar{n}} \tag{9.6.4}$$

where

$$\bar{n} = \alpha I T \tag{9.6.5}$$

The probability distribution (9.6.4) is called the *Poisson distribution*. Note that it is properly normalized to unity, as any valid probability distribution must be. That is, the sum of the probabilities of all possible outcomes is equal to one:

$$\sum_{n=0}^{\infty} P_n = \sum_{n=0}^{\infty} \frac{(\bar{n})^n}{n!} e^{-\bar{n}} = e^{-\bar{n}} \sum_{n=0}^{\infty} \frac{(\bar{n})^n}{n!} = 1 \tag{9.6.6}$$

since the sum in parentheses is just the power series for the function $e^{\bar{n}}$.

One important quantity we can calculate, given $P_n(T)$, is the average number of photons counted in a time interval T, denoted $\langle n(T) \rangle$:

$$\langle n(T) \rangle = \sum_{n=1}^{\infty} n P_n(T) = \sum_{n=1}^{\infty} n \frac{(\bar{n})^n}{n!} e^{-\bar{n}}$$

$$= \bar{n} e^{-\bar{n}} \sum_{n=1}^{\infty} \frac{(\bar{n})^{n-1}}{(n-1)!}$$

$$= \bar{n} e^{-\bar{n}} \sum_{n=0}^{\infty} \frac{(\bar{n})^n}{n!} = \bar{n} e^{-\bar{n}} e^{\bar{n}}$$

$$= \bar{n} = \alpha I T \tag{9.6.7}$$

This is a reasonable result: the average number of photons counted in a time T is equal to the rate of ejection of photoelectrons, αI, times the time T.

We can also calculate the average of the square of the number of photons counted in a time interval T (Problem 9.5):

$$\langle n(T)^2 \rangle = \sum_{n=0}^{\infty} n^2 P_n(T) = \bar{n}^2 + \bar{n} \qquad (9.6.8)$$

The quantity

$$\langle \Delta n(T)^2 \rangle = \left\langle \left[n(T) - \langle n(T) \rangle \right]^2 \right\rangle \qquad (9.6.9)$$

i.e., the average of the square of the deviation $n(T) - \langle n(T) \rangle$ of $n(T)$ from its average value $\langle n(T) \rangle$, is the mean square deviation of $n(T)$ from its average. It gives a measure of the "spread" of $n(T)$ values about the average. From the definition (9.6.9) it follows that

$$\langle \Delta n(T)^2 \rangle = \left\langle n(T)^2 - 2n(T) \langle n(T) \rangle + \langle n(T) \rangle^2 \right\rangle$$
$$= \langle n(T)^2 \rangle - 2\langle n(T) \rangle^2 + \langle n(T) \rangle^2$$
$$= \langle n(T)^2 \rangle - \langle n(T) \rangle^2 \qquad (9.6.10)$$

since $\langle\langle \cdots \rangle\rangle$ is the same as $\langle \ldots \rangle$. From (9.6.7) and (9.6.8), therefore,

$$\langle \Delta n(T)^2 \rangle = \bar{n} \quad \text{(Poisson distribution)} \qquad (9.6.11)$$

for the Poisson distribution. Similarly the root-mean-square (rms) deviation, $\Delta n(T)_{\text{rms}}$, for the Poisson distribution is

$$\Delta n(T)_{\text{rms}} = \sqrt{\bar{n}} \quad \text{(Poisson distribution)} \qquad (9.6.12)$$

In Figure 9.9 we plot the Poisson distribution for various values of $\bar{n}$. As $\bar{n}$ increases, the rms deviation in n increases as $\sqrt{\bar{n}}$, but the relative rms deviation, $\Delta n_{\text{rms}}/\bar{n} = 1/\sqrt{\bar{n}}$, decreases.

Actually (9.5.15) reduces to a Poisson distribution not only when $I(t)$ is constant, but more generally whenever the *mean* intensity $\bar{I}(t, T)$ given by (9.5.13) is independent of t. In this more general case the average over t in (9.5.15) is again

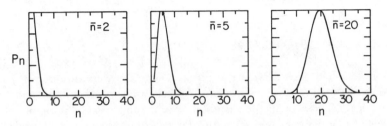

Figure 9.9 The Poisson distribution (9.6.4) for $\bar{n} = 2, \bar{n} = 5, \bar{n} = 20$.

superfluous, because everything to be averaged over t is in fact independent of t. We obtain the Poisson distribution (9.6.4) with the average photon count

$$\bar{n} = \alpha \int_t^{t+T} I(t') \, dt' = \alpha \int_0^T I(t') \, dt' \tag{9.6.13}$$

in a time interval T.

One important example of a field having Poisson photon counting statistics is the idealized monochromatic plane wave:

$$E(z, t) = E_0 \cos \omega \, (t - z/c) \tag{9.6.14}$$

in which case $I(t) = (c\epsilon_0/2) \, E_0^2$ is obviously independent of t [and therefore so is $\bar{I}(t, T)$]. In Chapter 15 we will see that laser light and ordinary light give rise to different photon statistics, the Poisson distribution applying to laser radiation.

• The Poisson distribution appears in many contexts in science and engineering. One famous application is to radioactive decay: if a sample emits alpha particles, say, at a rate of r particles per second, then the average number of alpha particles emitted in a time interval T is $\bar{n} = rT$, and the probability that n particles are counted in any time interval T is given by (9.6.4). Numerous other applications are discussed in textbooks on probability and statistics. •

9.7 PHOTONS AND LASERS

The photon concept is an integral part of the modern description of light. In this chapter we have described some effects that cannot be understood without knowing that there are indivisible units (photons) of radiation energy and momentum.

However, the wave concept is an equally important part of the modern theory of light. For practical applications it is sometimes more useful to think in terms of light waves and sometimes in terms of photons. We will be able to describe all of the practically interesting aspects of lasers using classical wave theory. This is because we will be dealing mainly with phenomena involving enormously large numbers of photons, where the classical wave theory is quite adequate.

APPENDIX 9.A PHOTON DETECTORS

Radiation detectors are of two basic types: There are *thermal detectors*, which respond to the heating of some part of the detector by incident light, and *photon* (or *quantum*) *detectors*, which measure directly the rate of absorption of photons. Examples of thermal detectors are the *Golay cell*, which detects the expansion of a gas heated by the absorption of light, and the *bolometer*, which responds to a change in electrical resistance of a material that is heated by absorption. In this appendix we will describe in general terms some aspects of photon detectors, which essentially count photons without first converting their energy to heat. Photon detectors are used in photon counting experiments of the type considered in Section 9.5.

A *phototube* is a photon detector whose principle of operation is based on the photoelectric effect. Its simplest form is the *vacuum photodiode*, an evacuated tube containing a photoemissive surface called the photocathode, and an anode that collects electrons (because of its higher voltage) emitted from the photocathode (Figure 9.10). The photoelectric ejection of electrons by incident radiation thus gives rise to an electric current in a circuit containing the phototube. It is the electric current that is directly measured, but since this current is proportional to the number of incident photons, the phototube responds to the rate of absorption of incident photons, i.e., it is a "photon detector."

Phototubes are most effective in the visible and near-visible portions of the electromagnetic spectrum. At lower frequencies the incident photons are not energetic enough to eject electrons from photoemissive surfaces (i.e., to overcome the "work function" of a light-sensitive surface). At higher frequencies there is another technical difficulty, namely the absorption of the incident radiation by the window of the phototube; this difficulty may be alleviated by coating the window with a phosphor that emits at a lower frequency than it absorbs.

The sensitivity of a phototube may be increased by filling it with a low-pressure gas. An electron ejected from the photocathode can ionize the atoms of the gas, producing more electrons (called "secondary electrons") which can themselves collide with atoms to produce more electrons. This avalanche process results in a greater current, and thus makes the phototube more sensitive to low-intensity radiation. A vacuum photodiode may have a response time, determined by the transit

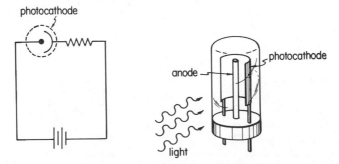

Figure 9.10 A vacuum photodiode acts as a current source in an electric circuit. The measured current in the circuit is proportional to the rate of absorption of incident photons.

time of the photoelectrons to the anode, as low as 10 nsec; the addition of gas to increase the sensitivity, however, tends to increase this response time by orders of magnitude.

Fast response times and high sensitivity are best achieved with *photomultipliers*. A photomultiplier is basically a phototube with a series of additional anodes called dynodes, each at a higher voltage (typically 100 V) than the preceding one. The dynodes are coated with a material that loses secondary electrons when it is struck by incident electrons. An electron ejected from the photocathode by incident radiation is electrostatically focused to the first dynode, generating secondary electrons. These are focused to the next dynode, producing more electrons, and the process continues with a total of perhaps ten dynodes in all. There is thus a multiplication of the number of electrons produced at the photocathode by the photoelectric effect, or equivalently an amplification of the current. This results in high sensitivity combined with short response times. A photomultiplier is thus capable of a prompt response to a single photon, but not all incident photons eject electrons into the dynode chain, so photomultipliers operate with a detection efficiency below 50%.

Photon detectors can also operate on the principle of the *internal photoelectric effect*. In this case, an incident photon is not sufficiently energetic to free an electron from the surface of a material, but nevertheless gives the electron more energy inside the material. An important example is *photoconductivity*, in which an electron is promoted to the conduction band of a semiconductor by the absorption of a photon. That is, shining light on a photoconductive material changes its electrical conductivity. (See also Sec. 16.6 for a discussion of avalanche and PIN photodiode detectors.)

Common photoconductive materials for photon detectors are lead sulfide and cadmium sulfide. The latter is used in the visible, whereas the former has a good response for wavelengths as high as 3 or 4 μm in the infrared. Photoconductive detectors have longer response times than vacuum photodiodes or photomultipliers, typical values being fractions of a microsecond at best. For wavelengths extending into the infrared it is frequently necessary to cool photoconductors down to liquid-nitrogen temperatures for good sensitivity.

For technical details of various detectors the reader is referred to the specialized literature.[2]

• The human eye is a surprisingly good photon detector. The detector units on the retina are of two types, called rods and cones because of their shapes. The response of the rods and cones is different at different light wavelengths—the cones are responsible for color vision and have a peak sensitivity at about 5600 A, while the rods are most sensitive at shorter wavelengths around 5100 A.

2. See, for instance, R. J. Keyes and R. H. Kingston, "A Look at Photon Detectors," *Physics Today*, March 1972, p. 48. A good introduction to optical radiation detection is R. W. Boyd, *Radiometry and the Detection of Optical Radiation* (Wiley, New York, 1983).

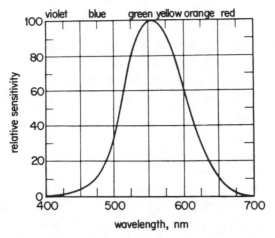

Figure 9.11 Relative sensitivity of the human eye as a function of wavelength for a "standard observer." From D. Halliday and R. Resnick, *Physics* (Wiley, New York, 1978).

The rods and cones also have different saturation properties. The rods saturate more easily and become relatively ineffective at high light levels, but are more sensitive at very low light levels. A comparatively high light level (and thus a strong contribution from the cone response) is necessary for color vision. The cones have a peak sensitivity at about 5600 Å and are more sensitive to longer wavelengths than the rods, as evidenced by the difficulty we have seeing red through dark-adapted eyes (Figure 9.11). Dark-adapted eyes operate almost exclusively by rod response, and, as noted in Section 2.8, are almost color-blind.

The sensitivity of the dark-adapted eye was studied in the classic work of M. H. Pirenne and his colleagues.[3] They found that the typical human retina can respond to as few as about ten photons around 5100 A, with an efficiency of about 60%. The overall detection efficiency of the eye is actually much less than this, because 80–90% of the light incident upon the eye is lost before it can be absorbed by rod cells. [Of course many animals have enormously more efficient and sensitive eyes than humans do.]

The time constant of the human eye is about 0.1 sec, and this is associated with the "persistence of vision" that makes movies possible: we do not notice the changing of frames if the rate is faster than about 15 frames per second. Moreover, the eye does not "integrate" a signal much beyond about 0.1 sec. The experimental results of Pirenne and collaborators cited above, for instance, were obtained with millisecond pulses of light. If the same total electromagnetic energy were incident upon the eye over a longer time, say 1 sec, there would be no response, because too few photons would be available over the 0.1-sec integration period. •

3. See S. Hecht, S. Shlaer, and M. H. Pirenne, "Energy, Quanta, and Vision," *Journal of General Physiology* **25**, 819 (1942) and A. Rose, "Quantum Effects in Human Vision," *Biological and Medical Physics*, vol. 5, 211 (1957).

PROBLEMS

9.1 **(a)** Estimate the average number of photons per second per square centimeter reaching the earth's surface from the sun.

(b) A certain source of radiation puts out 1 mW at a wavelength of 6000 Å. How many 6000-Å photons are emitted per second?

9.2 **(a)** Two Polaroid sheets are placed one over the other with their polarization axes orthogonal. Is there a nonzero probability that any photon will pass through both sheets?

(b) A third Polaroid sheet is inserted between the two sheets in part (a), so that its axis makes an angle of $45°$ with respect to each of the other two. Is it possible for any photon to pass through all three sheets?

9.3 A certain atom has a transition frequency $\nu_0 = (E_2 - E_1)/h$, where 1 and 2 refer to the ground level and first excited level, respectively. It is moving with velocity v away from a source of radiation of frequency ν. Using conservation of energy and linear momentum, show that the atom will absorb a photon of frequency ν provided that

$$\nu_0 = \nu \left(1 - \frac{v}{c} \right).$$

Assume that $v \ll c$ and $h\nu \ll mc^2$, so that relativistic effects may be ignored. Also ignore any broadening of the atomic transition due to collisions, spontaneous emission, etc.

9.4 At points of constructive interference in the Young two-slit experiment, the intensity is twice the intensity calculated by adding the intensities associating with each individual slit. This does not violate the principle of conservation of energy. Indicate why not (a) qualitatively and (b) semiquantitatively.

9.5* Note that the Poisson distribution (9.6.4) has the property

$$\bar{n} \frac{\partial}{\partial \bar{n}} \left[e^{\bar{n}} P_n(T) \right] = n P_n(T) \, e^{\bar{n}}.$$

(a) Therefore, show that $\langle n \rangle = \Sigma \, n P_n(T) = \bar{n} e^{-\bar{n}} \, \partial (e^{\bar{n}})/\partial \bar{n}$

(b) Find a similar formula for $n^k P_n(T) \, e^{\bar{n}}$, and use it to verify (9.6.8) for $\langle n^2 \rangle$ and to compute the analogous result for $\langle n^3 \rangle$.

10 LASER OSCILLATION: GAIN AND THRESHOLD

10.1 INTRODUCTION

In our superficial analysis of the laser in Chapter 1 we introduced certain concepts such as gain, threshold, and feedback, and indicated their importance in our understanding of lasers. We also introduced certain coefficients (a, b, f, p), which we did not derive or explain very carefully. In the intervening chapters we have laid the foundation necessary for a deeper understanding of these concepts, and in this chapter we will begin a detailed description of laser oscillation.

The physical system we consider is a collection of atoms (or molecules) between two mirrors. By some pumping process, such as absorption of light from a flashlamp or electron-impact excitation in a gaseous discharge, some of these atoms are promoted to excited states. The excited atoms begin radiating spontaneously, as in an ordinary fluorescent lamp. A spontaneously emitted photon can induce an excited atom to emit another photon of the same frequency and direction as the first. The more such photons are produced by stimulated emission, the faster is the production of still more photons, because the stimulated emission rate is proportional to the flux of photons already in the stimulating field [recall the rate formula (7.2.8)].

The mirrors of the laser keep photons from escaping completely, so that they can be redirected into the active laser medium to stimulate the emission of more photons. By making the mirrors partially transmitting, some of the photons are allowed to escape. They constitute the output laser beam. The intensity of the output laser beam is determined by the rate of production of excited atoms, the reflectivities of the mirrors, and certain properties of the active atoms. We will see, in this chapter and the next, exactly how the laser output depends on these quantities.

10.2 GAIN

In Chapter 1 the growth rate of the number of laser photons in the cavity was described by an amplification coefficient a. It is closely related to the conventional "gain coefficient" that was derived in Section 8.5.

We now consider the propagation of narrowband radiation in a medium of atoms which have a transition frequency equal, or nearly equal, to the frequency

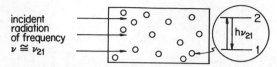

Figure 10.1 The propagation of radiation of frequency $\nu \approx \nu_{21}$ in a medium of atoms with a transition frequency ν_{21}.

of the radiation (Figure 10.1). If more atoms are in the upper level of the transition than the lower, we can expect that there will be more stimulated emission than absorption, and that the radiation will be amplified as it propagates. In such a case we say there is *gain* at the resonant frequency.

To begin with, we will not specialize our analysis to a cavity bounded at two ends by mirrors. Instead we want first to set up some equations describing the amplification of light.

We recall from Eq. (2.6.12) that the intensity $I(z)$ of a light wave is equal to the energy density $u(z)$ times the wave propagation velocity. Consider a quasi-monochromatic plane wave of frequency ν propagating in the $+z$ direction. The rate at which electromagnetic energy passes through a plane of cross-sectional area A at z is $I_\nu(z)\,A$. At an adjacent plane at $z + \Delta z$ this rate is $I_\nu(z + \Delta z)\,A$, and the difference is

$$\left[I_\nu(z + \Delta z) - I_\nu(z)\right] A = \frac{\partial}{\partial z}\left(I_\nu A\right)\Delta z \tag{10.2.1}$$

in the limit in which Δz is very small. Equation (10.2.1) gives the rate at which electromagnetic energy leaves the volume $A\,\Delta z$, i.e.,

$$-\frac{\partial}{\partial t}\left(u_\nu A\,\Delta z\right) = \frac{\partial}{\partial z}\left(I_\nu A\right)\Delta z \tag{10.2.2}$$

Since A and Δz are constant, and $u_\nu = I_\nu/c$, we may write this equation in the form

$$\frac{1}{c}\frac{\partial}{\partial t}I_\nu + \frac{\partial}{\partial z}I_\nu = 0 \tag{10.2.3}$$

the so-called *equation of continuity*. It is an example of Poynting's theorem, in one space dimension, and is applicable to a plane wave propagating in vacuum.

If the wave propagates in a medium, however, we must replace the zero on the right side of (10.2.3) by the rate per unit volume at which electromagnetic energy changes because of the medium. We can compute this from the rate of change of upper-state (or lower-state) population density due to both absorption and stimulated emission. The last term of (7.3.4a), for example, gives the rate (per unit volume) at which the ground-state population increases due to these two processes. Thus $h\nu$ times this term is the rate at which energy (per unit volume) is added to the field. Therefore, in the laser medium (10.2.3) becomes

$$\left(\frac{1}{c}\frac{\partial}{\partial t} + \frac{\partial}{\partial z}\right) I_\nu = \sigma(\nu) I_\nu (N_2 - N_1) \tag{10.2.4}$$

where we have used $h\nu\Phi = I_\nu$. The reader should convince himself that the two sides of (10.2.4) are dimensionally consistent.

Recalling (7.4.19), $\sigma(\nu) = (h\nu/c) BS(\nu)$, as well as Table 7.1, including the possibility of level degeneracies, we may rewrite (10.2.4) in several other forms, involving the Einstein A or B coefficient and the atomic lineshape factor $S(\nu)$:

$$\left(\frac{1}{c}\frac{\partial}{\partial t} + \frac{\partial}{\partial z}\right) I_\nu = \frac{h\nu}{c} B(N_2 - N_1) S(\nu) I_\nu \quad (\text{no degeneracies})$$

$$= \frac{h\nu}{c} B\left(N_2 - \frac{g_2}{g_1} N_1\right) S(\nu) I_\nu \quad (\text{degeneracies included})$$

$$= \frac{\lambda^2 A}{8\pi}\left(N_2 - \frac{g_2}{g_1} N_1\right) S(\nu) I_\nu$$

$$= \sigma(\nu)\left[N_2 - \frac{g_2}{g_1} N_1\right] I_\nu \tag{10.2.5}$$

Here and below we will use the symbols A and B as shorthand for the Einstein coefficients $A(2, 1)$ and $B(2, 1)$ defined in Section 7.A. It is conventional to group all of the factors multiplying I_ν on the right side of (10.2.5) into a single *gain coefficient* $g(\nu)$:

$$\frac{\partial I_\nu}{\partial z} + \frac{1}{c}\frac{\partial I_\nu}{\partial t} = g(\nu) I_\nu \tag{10.2.6}$$

where

$$g(\nu) = \frac{\lambda^2 A}{8\pi}\left(N_2 - \frac{g_2}{g_1} N_1\right) S(\nu)$$

$$= \sigma(\nu)\left(N_2 - \frac{g_2}{g_1} N_1\right) \tag{10.2.7}$$

In the absence of degeneracies this reduces to the gain coefficient derived in

(8.5.14) after a much more complicated derivation from the coupled Maxwell-Bloch equations.

The selection of the term "gain coefficient" for $g(\nu)$ can easily be understood by considering the temporal steady state. In this case I_ν is independent of time, and varies only with the distance z of propagation. Then (10.2.6) simplifies to

$$\frac{dI_\nu}{dz} = g(\nu) I_\nu \tag{10.2.8a}$$

If furthermore the gain coefficient is independent of I_ν and z, it follows from (10.2.8a) that I_ν grows exponentially at the rate $g(\nu)$:

$$I_\nu(z) = I_\nu(0) e^{g(\nu) z} \tag{10.2.8b}$$

In general, however, this exponential gain is valid only for low intensities. We will see later in the chapter that g actually "saturates" with increasing field intensity, so that (10.2.8) overestimates $I_\nu(z)$ except at low intensities.

We note furthermore that g is positive only if $N_2 > (g_2/g_1) N_1$, that is, only if there is a *population inversion*: more atoms available to amplify the field by stimulated emission than to attenuate it by absorption. The opposite is the case if practically all the atoms are in the lower level of the transition. In this case, then,

$$N_2 - \frac{g_2}{g_1} N_1 \approx -\frac{g_2}{g_1} N_1 \approx -\frac{g_2}{g_1} N$$

and

$$g(\nu) \approx -\frac{\lambda^2 A}{8\pi} \frac{g_2}{g_1} NS(\nu)$$

$$= -N \frac{g_2}{g_1} \sigma(\nu) = -a(\nu) \tag{10.2.9}$$

Thus, for an unexcited resonant medium, *the gain coefficient is identical, except for its sign, to the absorption coefficient.*

The relation to the corresponding classical formula can be clarified by extracting the oscillator strength $f = f(1, 2)$ from the factors in (10.2.9) to get

$$a(\nu) = \frac{1}{4\pi\epsilon_0} \frac{\pi e^2 f}{mc} NS(\nu) \tag{10.2.10}$$

This is in fact just the expression given in Table 3.2 for the classical absorption coefficient.

The formula (10.2.7) for the gain coefficient shows that the same atomic lineshape function applies for both absorption and stimulated emission. Our discussion of lineshape functions in Chapter 3 is therefore relevant to laser media as well as absorbing media. The only thing that distinguishes amplifying media from absorbing media is the sign of the population inversion $N_2 - (g_2/g_1) N_1$. In Chapter 11 we will discuss some methods for achieving (positive) population inversions.

• The expression (10.2.7) for the gain coefficient may be generalized to include the effect of the refractive index of the host medium. This modification is significant in solid-state lasers, where n may be appreciably different from unity. For our purposes it is sufficient merely to write the result, which may be obtained by replacing λ by λ/n in (10.2.7):

$$g(\nu) = \frac{\lambda^2 A}{8\pi n^2} \left(N_2 - \frac{g_2}{g_1} N_1 \right) S(\nu) \qquad (10.2.11)$$

The corresponding modification of the cross section is given by

$$\sigma(\nu) = \frac{\lambda^2 A}{8\pi n^2} S(\nu) \qquad (10.2.12) \ \bullet$$

10.3 FEEDBACK

Equation (10.2.8) provides an overly optimistic estimate of the growth of intensity in an amplifying ($g > 0$) medium, for it assumes that the gain is independent of intensity. As mentioned earlier, this is a valid assumption only for low intensities. In Section 7.5 we explained what it means to have a "high" intensity, and in Section 10.11 we will discuss the implication for $g(\nu)$. For the present, however, let us accept the prediction of exponential growth as the first approximation to the actual behavior of light in an amplifier.

It is reasonable to expect that a laser can be built in the form of a pencil-shaped container of atoms for which $g > 0$ (Figure 10.2). The consequences of such a

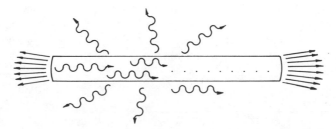

Figure 10.2 A mirrorless "laser." Photons emitted spontaneously along the axis of the tube of excited atoms are multiplied by stimulated emission, resulting in a burst of radiation.

geometry are easy to predict. Some photons are emitted along the axis of the container, where they can encounter other atoms and so induce the emission of more photons propagating in the same direction with the same frequency, by stimulated emission. As the number of such photons grows, the stimulated emission rate grows proportionately, so that we expect a burst of radiation to emerge from either end of the container. The direction and cross-sectional area of the beam of light so produced are determined by the container of the excited atoms.

As an example, suppose we have an amplifying medium with a gain coefficient $g = 0.01$ cm^{-1}, an achievable gain in many laser media. With a length $L = 1$ m for the gain cell, a spontaneously emitted photon at one end of the cell leads, according to (10.2.8), to a total of

$$e^{(0.01 \text{ cm}^{-1})(100 \text{ cm})} = e^1 \approx 2.72 \qquad (10.3.1)$$

photons emerging at the other end. The output of such a "laser" is obviously not very impressive.

The way to increase the photon yield from such a device is to catch photons emerging from one end and repeatedly feed them back for more amplification. In this way we can, in effect, increase the length of the gain cell. The practical way to achieve this *feedback*, of course, is to have mirrors at the ends of the container.

• It is possible, in media with very high gain, to build mirrorless (sometimes called "superradiant") lasers. The light from such a device resembles that from a conventional laser insofar as it is bright, is quasimonochromatic, and forms a small spot when shined on a screen. However, it does not have the same degree of temporal and spatial coherence usually associated with lasers. We discuss these coherence properties in Chapter 15. •

10.4 THRESHOLD

In a laser there is not only an increase in the number of cavity photons because of stimulated emission, but also a decrease because of loss effects. These include scattering and absorption of radiation at the mirrors, as well as the "output coupling" of radiation in the form of the usable laser beam. In order to sustain laser oscillation the stimulated amplification must be sufficient to overcome these losses. This sets a lower limit on the gain coefficient $g(\nu)$, below which laser oscillation does not occur.

One thing we can do now is to predict, given the various losses that tend to diminish the intensity of radiation within the cavity, what minimum gain is necessary to achieve laser oscillation. The condition that the gain coefficient is greater than or equal to this lower limit is called the *threshold condition* for laser oscillation.

Ordinarily the scattering and absorption of radiation within the gain medium of active atoms is quite small compared with the loss occurring at the mirrors of the laser. We will therefore consider in detail only the losses associated with the mirrors. Figure 10.3 shows a stylized version of a laser resonator, i.e., a laser cavity bounded on two sides by highly reflecting mirrors. A beam of intensity I incident upon one of these mirrors is transformed into a reflected beam of intensity rI, where r is the *reflection coefficient* of the mirror. A beam of intensity tI, where t is the *transmission coefficient*, passes through the mirror. We might expect from the law of conservation of energy that

$$r + t = 1,$$

that is, the fraction of power reflected plus the fraction transmitted should be unity. Actually, however, some of the incident beam power may be absorbed by the mirror, tending to raise its temperature. Or some of the incident beam may be scattered away because the mirror surface is not perfectly smooth. Thus the law of conservation of energy takes the form

$$r + t + s = 1, \tag{10.4.1}$$

where s represents the fraction of the incident beam power that is absorbed or scattered by the mirror.

Each of the mirrors of Figure 10.3 is characterized by a set of coefficients r, t, and s. At the mirror at $z = L$ we have (Figure 10.3)

$$I_\nu^{(-)}(L) = r_2 I_\nu^{(+)}(L) \tag{10.4.2a}$$

and similarly

$$I_\nu^{(+)}(0) = r_1 I_\nu^{(-)}(0) \tag{10.4.2b}$$

for the mirror at $z = 0$. Equations (10.4.2) are boundary conditions that must be

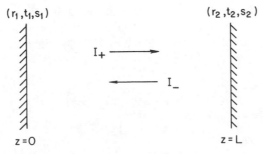

Figure 10.3 The two oppositely propagating beams in a laser cavity.

satisfied by the solution of the equations describing the propagation of intensity inside the laser cavity.

What are these equations? We are now interested only in steady-state, or continuous-wave (cw), laser oscillation. Near the threshold of laser oscillation the intracavity intensity is very small, and therefore Eq. (10.2.8) is applicable. For light propagating in the positive z direction, therefore, we have

$$\frac{dI_\nu^{(+)}}{dz} = g(\nu) I_\nu^{(+)} \tag{10.4.3a}$$

near the threshold, where g may be taken to be constant. Light propagating in the negative z direction also sees the same gain medium and so satisfies a similar equation (see Problem 10.1):

$$\frac{dI_\nu^{(-)}}{dz} = -g(\nu) I_\nu^{(-)} \tag{10.4.3b}$$

The solutions of these equations are

$$I_\nu^{(+)}(z) = I_\nu^{(+)}(0) e^{g(\nu) z} \tag{10.4.4a}$$

and

$$I_\nu^{(-)}(z) = I_\nu^{(-)}(L) \exp\left[g(\nu)(L - z)\right] \tag{10.4.4b}$$

From (10.4.4) we see that

$$I_\nu^{(+)}(L) = I_\nu^{(+)}(0) e^{g(\nu) L} \tag{10.4.5a}$$

at the right mirror ($z = L$), and the left-going beam has intensity

$$I_\nu^{(-)}(0) = I_\nu^{(-)}(L) e^{g(\nu) L} \tag{10.4.5b}$$

at the left mirror ($z = 0$). In steady state the left-going beam has a fraction r_1 of itself reflected at the left mirror (at $z = 0$), and this fraction is just the right-going beam at $z = 0$. A similar consideration applies at the right mirror. Thus we have

$$
\begin{aligned}
I_\nu^{(+)}(0) &= r_1 I_\nu^{(-)}(0) \\
&= r_1\left[e^{g(\nu) L} I_\nu^{(-)}(L)\right] \\
&= r_1 e^{g(\nu) L}\left[r_2 I_\nu^{(+)}(L)\right] \\
&= r_1 r_2 e^{g(\nu) L}\left[I_\nu^{(+)}(0) e^{g(\nu) L}\right] \\
&= \left[r_1 r_2\, e^{2g(\nu) L}\right] I_\nu^{(+)}(0) \tag{10.4.6}
\end{aligned}
$$

Similar manipulations, applied to any of the quantities $I_\nu^{(+)}(L)$, $I_\nu^{(-)}(L)$, and $I_\nu^{(-)}(0)$, lead to the same result. Therefore, if $I_\nu^{(+)}(0)$ is not zero, we must have, at steady state,

$$r_1 r_2 e^{2gL} = 1 \qquad (10.4.7)$$

The steady-state value of gain that allows (10.4.7) to be satisfied is also the value at which laser action begins. For smaller values there is net attenuation of I_ν in the cavity. Thus the value of g that satisfies (10.4.7) is labeled g_t and called the *threshold gain*:

$$g_t = \frac{1}{2L} \ln\left(\frac{1}{r_1 r_2}\right) = -\frac{1}{2L} \ln\left(r_1 r_2\right) \qquad (10.4.8)$$

This expression can be rewritten usefully in the common case that $r_1 r_2 \approx 1$. Then we define $r_1 r_2 = 1 - x$, or $x = 1 - r_1 r_2$, and use the first term in the Taylor series expansion $\ln(1 - x) \approx -x$, valid when $x \ll 1$, to obtain

$$g_t = \frac{1}{2L}(1 - r_1 r_2) \qquad \text{(high reflectivities)} \qquad (10.4.9)$$

which is a satisfactory approximation to (10.4.8) if $r_1 r_2 > 0.90$.

Note that if we are given the mirror reflectivities r_1 and r_2, and their separation L, and therefore the threshold gain, we can determine the population inversion necessary to achieve laser action from (10.2.7) and the atomic cross section.

Our derivation of (10.4.7) assumes that the gain medium fills the entire distance L between the mirrors. This assumption is valid for many solid-state lasers in which the ends of the gain medium are polished and coated with reflecting material. In gas and liquid lasers, however, the gain medium is usually contained in a cell of length $l < L$ that is not joined to the mirrors (Figure 10.4). In this case the threshold condition is

$$g_t = -\frac{1}{2l} \ln\left(r_1 r_2\right)$$

$$\approx \frac{1}{2l}(1 - r_1 r_2) \qquad \text{(high reflectivities)} \qquad (10.4.10)$$

The threshold condition (10.4.10) [or (10.4.8)] assumes that "loss" occurs only at the mirrors. This loss is associated with transmission through the mirrors,

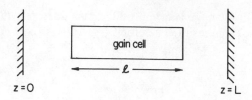

Figure 10.4 A laser in which the gain medium does not fill the entire distance L between the mirrors.

absorption by the mirrors, and scattering off the mirrors into nonlasing modes. Absorption and scattering are minimized as much as possible by using mirrors of high optical quality. Transmission, of course, is necessary if there is to be any output from the laser.

Other losses might arise from scattering and absorption within the gain medium (from nearly resonant but nonlasing transitions). Such losses are usually small, but they are not difficult to account for in the threshold condition. If a is the effective loss per unit length associated with these additional losses, then the threshold condition (10.4.10) is modified as follows:

$$g_t = -\frac{1}{2l} \ln (r_1 r_2) + a \qquad (10.4.11)$$

For our purposes these "distributed losses" (i.e., losses not associated with the mirrors) may usually be ignored.

It is instructive at this point to consider an example. A typical 6328-Å He–Ne laser might have a gain cell of length $l = 50$ cm and mirrors with reflectivities $r_1 = 0.998$ and $r_2 = 0.980$. Thus

$$g_t = \frac{-1}{2(50)} \ln \left[(0.998)(0.980) \right] \text{ cm}^{-1}$$

$$= 2.2 \times 10^{-4} \text{ cm}^{-1} \qquad (10.4.12)$$

is the threshold gain.

Using this value for g_t and Eq. (10.2.11), we may calculate the *threshold population inversion* necessary to achieve lasing:

$$\Delta N_t = \left(N_2 - \frac{g_2}{g_1} N_1 \right)_t = \frac{8\pi n^2 g_t}{\lambda^2 A S(\nu)} = \frac{g_t}{\sigma(\nu)} \qquad (10.4.13)$$

The A coefficient for the 6328-Å transition in Ne is

$$A \approx 1.4 \times 10^6 \text{ sec}^{-1} \qquad (10.4.14)$$

For $T \approx 400$ K and the Ne atomic weight $M \approx 20$ g, we obtain from Table 3.3 the Doppler width $\delta\nu_D \approx 1500$ MHz. Thus

$$S(\nu) \approx 6.3 \times 10^{-10} \text{ sec}$$

and

$$\Delta N_t \approx \frac{(8\pi)(2.2 \times 10^{-4} \text{ cm}^{-1})}{(6328 \times 10^{-8} \text{ cm})^2(1.4 \times 10^6 \text{ sec}^{-1})(6.3 \times 10^{-10} \text{ sec})}$$
$$= 1.6 \times 10^9 \text{ atoms/cm}^3 \qquad (10.4.15)$$

This is a lot of atoms, but it is nevertheless quite a small number compared with the total number of Ne atoms. For a (typical) Ne partial pressure of 0.2 Torr, the total number of Ne atoms per cm^3 is [Eq. (3.10.8)] about 4.8×10^{15}. Thus the ratio of the threshold population inversion to the total density of atoms of the lasing species is only

$$\frac{\Delta N_t}{N} = \frac{1.6 \times 10^9}{4.8 \times 10^{15}} = \tfrac{1}{3} \times 10^{-6} \qquad (10.4.16)$$

Sometimes the quantity e^{gl} is called the gain, and expressed in decibels, i.e.,

$$G_{dB} = 10 \log_{10}(e^{gl}) = 10 \log_{10}(10^{0.434gl})$$
$$= 4.34gl \qquad (10.4.17)$$

The threshold gain in our example is thus

$$(G_{dB})_t = (4.34)(2.2 \times 10^{-4} \text{ cm}^{-1})(50 \text{ cm})$$
$$= 0.048 \text{ dB}$$

In the laser research literature gain is usually expressed in cm^{-1} and this is the unit we will use throughout this book.

In Table 10.1 we collect the formulas and terms we have used in discussing gain and threshold.

TABLE 10.1 Quantities and Formulas Related to Gain and Threshold

---

The Gain Coefficient

$$g(\nu) = \frac{\lambda^2 A}{8\pi n^2} \left(N_2 - \frac{g_2}{g_1} N_1 \right) S(\nu)$$

$$= \sigma(\nu) \left(N_2 - \frac{g_2}{g_1} N_1 \right)$$

$\lambda = c/\nu = $ wavelength of radiation

$A = $ Einstein A coefficient for spontaneous emission on the $2 \rightarrow 1$ transition

$n = $ refractive index at wavelength λ

$N_2, N_1 = $ number of atoms per unit volume in levels 2 and 1

$g_2, g_1 = $ degeneracies of levels 2 and 1

$S(\nu) = $ lineshape function (Table 3.3)

Threshold Gain

$$g_t = \frac{-1}{2l} \ln (r_1 r_2) + a \approx \frac{1}{2l} (1 - r_1 r_2) + a$$

$l = $ length of gain medium

$r_1, r_2 = $ mirror reflectivities

$a = $ distributed loss per unit length

10.5 RATE EQUATIONS FOR PHOTONS AND POPULATIONS

To describe time-dependent phenomena, such as pulsed laser operation or the startup of continuous-wave lasing, we must include the time derivative $\partial I_\nu / \partial t$ in the propagation equation (10.2.6). For the right- and left-going waves in the laser resonator (Figure 10.3), we write

$$\frac{\partial I_\nu^{(+)}}{\partial z} + \frac{1}{c} \frac{\partial I_\nu^{(+)}}{\partial t} = g(\nu) I_\nu^{(+)} \tag{10.5.1}$$

and

$$-\frac{\partial I_\nu^{(-)}}{\partial z} + \frac{1}{c} \frac{\partial I_\nu^{(-)}}{\partial t} = g(\nu) I_\nu^{(-)} \tag{10.5.2}$$

respectively. Addition of these equations gives

$$\frac{\partial}{\partial z}(I_\nu^{(+)} - I_\nu^{(-)}) + \frac{1}{c}\frac{\partial}{\partial t}(I_\nu^{(+)} + I_\nu^{(-)}) = g(\nu)(I_\nu^{(+)} + I_\nu^{(-)}) \quad (10.5.3)$$

We will see in Chapter 11 that in many lasers there is very little gross variation of either $I_\nu^{(+)}$ or $I_\nu^{(-)}$ with z. Assuming this result, we neglect any z variation of intensity. Thus the first term in (10.5.3) can be dropped, and we obtain the *ordinary* differential equation

$$\frac{d}{dt}(I_\nu^{(+)} + I_\nu^{(-)}) = cg(\nu)(I_\nu^{(+)} + I_\nu^{(-)}) \quad (10.5.4a)$$

Note that here we are assuming spatial uniformity, just as we assumed temporal uniformity (steady state) in the preceding section. In Section 11.5 we will discuss the temporal steady-state rate equation that results from a more detailed consideration of the spatial boundary conditions.

If the gain medium does not completely fill the resonator (Figure 10.4), then $g(\nu) = 0$ outside it. If we integrate both sides of (10.5.4a) over z in the region $0 < z < L$, then the left side, which is independent of z, is simply multiplied by L. However, the right side is multiplied by l, which is less than L, since $g(\nu)$ is different from zero only inside the gain medium. Thus

$$\frac{d}{dt}(I_\nu^{(+)} + I_\nu^{(-)}) = \frac{cl}{L}g(\nu)(I_\nu^{(+)} + I_\nu^{(-)}) \quad (10.5.4b)$$

is the generalization of (10.5.4a) to the case $l < L$. Since the number of photons inside the cavity is proportional to the total intensity, we may also write

$$\frac{dq_\nu}{dt} = \frac{cl}{L}g(\nu)\,q_\nu \quad (10.5.5)$$

where q_ν is the number of cavity photons associated with the frequency ν.

Equation (10.5.5) describes the growth in time of the number of cavity photons as a result of the absorption and induced emission of photons by the gain medium. The factor $cg(\nu)\,l/L$ is the growth rate. Of course we must also consider the loss of cavity photons due to output coupling, absorption and scattering at the mirrors, etc. We can take systematic account of the loss associated with the output coupling of laser radiation from the cavity, which frequently is the most important loss mechanism, as follows.

Radiation reflected from the mirror at $z = L$ (Figure 10.3) has an intensity that is r_2 times the incident intensity. After it is reflected from the mirror at $z = 0$, therefore, it has an intensity $r_1 r_2$ times its intensity before the round trip inside the resonator. In other words, a fraction $1 - r_1 r_2$ of intensity is lost. Since the time it takes to make a round trip is $2L/c$, the *rate* at which intensity is lost due to the imperfect reflectivity of the mirrors is $c(1 - r_1 r_2)/2L$. In terms of photons, this loss rate is

$$\left(\frac{dq_\nu}{dt}\right)_{\text{output coupling}} = -\frac{c}{2L}(1 - r_1 r_2) q_\nu \qquad (10.5.6)$$

The total rate at which the number of cavity photons changes is therefore

$$\frac{dq_\nu}{dt} = \left(\frac{dq_\nu}{dt}\right)_{\text{gain processes}} + \left(\frac{dq_\nu}{dt}\right)_{\text{output coupling}}$$

$$= \frac{cl}{L} g(\nu) q_\nu - \frac{c}{2L}(1 - r_1 r_2) q_\nu$$

$$= \frac{cl}{L} g(\nu) q_\nu - \frac{cl}{L} g_t q_\nu \qquad (10.5.7)$$

If there are significant losses besides those occurring at the mirrors, they may be accounted for in a similar fashion. We will assume that these other losses are negligible, in which case (10.5.7) gives the rate of change with time of the number of cavity photons. Equivalently, we may write the rate equation

$$\frac{dI_\nu}{dt} = \frac{cl}{L} g(\nu) I_\nu - \frac{c}{2L}(1 - r_1 r_2) I_\nu \qquad (10.5.8)$$

for the total intensity $I_\nu = I_\nu^{(+)} + I_\nu^{(-)}$ inside the cavity. The gain coefficient is given by (10.2.7). If we assume equal upper and lower level degeneracy, $g_1 = g_2$, for simplicity, then we may write (10.5.8) as

$$\frac{dI_\nu}{dt} = \frac{cl}{L}\frac{\lambda^2 A}{8\pi}(N_2 - N_1) S(\nu) I_\nu - \frac{c}{2L}(1 - r_1 r_2) I_\nu$$

$$= \frac{cl}{L}\sigma(\nu)(N_2 - N_1) I_\nu - \frac{c}{2L}(1 - r_1 r_2) I_\nu \qquad (10.5.9)$$

• The rate of change of the photon number q_ν due to the combined effects of stimulated emission and cavity loss is given by Eq. (10.5.7). If q_ν is to increase, we must obviously have $dq_\nu/dt > 0$, or

$$g(\nu) > \frac{1}{2l}(1 - r_1 r_2) \qquad (10.5.10)$$

Thus the threshold gain that must be exceeded in order to have $dq_\nu/dt > 0$ is

$$g_t = \frac{1}{2l}\left(1 - r_1 r_2\right) \tag{10.5.11}$$

This agrees with our previous expression for the threshold gain, given in (10.4.10), only if $1 - r_1 r_2 \ll 1$, for then

$$-\ln\left(r_1 r_2\right) = -\ln\left[1 - \left(1 - r_1 r_2\right)\right] \approx 1 - r_1 r_2 \tag{10.5.12}$$

which follows from the Taylor series

$$\ln\left(1 - x\right) = -x - \tfrac{1}{2}x^2 - \tfrac{1}{3}x^3 - \cdots \tag{10.5.13}$$

The limit $1 - r_1 r_2 \ll 1$ is appropriate in most lasers. The difference between our derivations of g_t in this section and in Section 10.4 can be traced to our assumption here that the intracavity field is spatially uniform. We will see in the next chapter that spatial uniformity is a good approximation when $1 - r_1 r_2$ is small, i.e., when the mirrors are highly reflecting. When this is not the case, our approximation leading to (10.5.11) is inaccurate. In many cases, however, such as in the example at the end of Section 10.4, (10.5.11) is a satisfactory expression of the threshold gain. •

The population densities N_1 and N_2 also change in time, of course, due to stimulated emission, absorption, collisions, and spontaneous emission. The net rate of change in N_1 and N_2 due to these processes is given by Eqs. (7.3.2).

However, these are not the only processes contributing to the rate of change of the population densities N_2 and N_1 of the lasing transition. In particular, of course, we must account for the pumping process that produces the population inversion $(N_2 - N_1)$. To account for the pumping process most simply we add a term K to the population equations and call it the pumping rate into the upper level. There are several methods of arranging pumping of this kind, as discussed in Chap. 13. For the time being we simply insert a term K. With this minor modification of the population equations (7.3.2), we have the following set of coupled equations for the light and the atoms in the laser cavity:

$$\frac{dN_1}{dt} = -\Gamma_1 N_1 + A N_2 + g(\nu)\,\Phi_\nu \tag{10.5.14a}$$

$$\frac{dN_2}{dt} = -\left(\Gamma_2 + A\right)N_2 - g(\nu)\,\Phi_\nu + K \tag{10.5.14b}$$

$$\frac{d\Phi_\nu}{dt} = \frac{cl}{L}\,g(\nu)\,\Phi_\nu - \frac{c}{2L}\left(1 - r_1 r_2\right)\Phi_\nu \tag{10.5.14c}$$

We have used $I_\nu = h\nu\Phi_\nu$ in rewriting (10.5.8) as (10.5.14c).

For some purposes it is useful to rewrite Eqs. (10.5.14) so that they refer to absolute numbers, rather than densities, of atoms and photons. This is easy to do. The total number of atoms in level 2 is $n_2 = N_2 V_g$, where V_g is the volume of the gain medium. Likewise the total number of atoms in the lower level of the laser transition is $n_1 = N_1 V_g$. The electromagnetic energy density in the cavity is related to intensity and photon flux by

$$u_\nu = I_\nu/c = (h\nu/c)\Phi_\nu \qquad (10.5.15a)$$

and it is related to photon number q_ν by

$$u_\nu = h\nu q_\nu/V \qquad (10.5.15b)$$

where V is the cavity volume. This relation assumes a uniform distribution of intensity within the cavity. Thus we have

$$\Phi_\nu = cq_\nu/V \qquad (10.5.15c)$$

and (10.5.14) may be rewritten in the form

$$\frac{dn_1}{dt} = -\Gamma_1 n_1 + An_2 + \frac{cl}{L} g(\nu) q_\nu \qquad (10.5.16a)$$

$$\frac{dn_2}{dt} = -\Gamma_2 n_2 - An_2 - \frac{cl}{L} g(\nu) q_\nu + p \qquad (10.5.16b)$$

$$\frac{dq_\nu}{dt} = \frac{cl}{L} g(\nu) q_\nu - \frac{c}{2L} (1 - r_1 r_2) q_\nu \qquad (10.5.16c)$$

where we have used the relations

$$\frac{V_g}{V} = \frac{l}{L} \quad \text{and} \quad KV_g = p \qquad (10.5.17)$$

Eqs. (10.5.16b, c) imply that

$$\frac{d}{dt}(n_2 + q_\nu) = -(\Gamma_2 + A) n_2 + p - \frac{c}{2L}(1 - r_1 r_2) q_\nu \qquad (10.5.18)$$

This equation has an obvious interpretation. The left-hand side is the rate of change of the total number of *excitations*, i.e., the number of atoms in the upper level 2 of the lasing transition plus the number of photons. The first term on the right is the rate of decrease in the number of these excitations as a result of inelastic collisions and spontaneous emission from level 2. The second term is the rate of change associated with pumping of level 2. The last term is the rate at which excitation in the form of photons is lost from the cavity. Note that contributions from stimulated emission (or absorption) do not appear in (10.5.18), because they have canceled out; an increase in q_ν is always accompanied by an equal decrease in n_2. Further features of (10.5.18) are pointed out in Problem 10.2.

10.6 COMPARISON WITH THE THEORY OF CHAPTER 1

In Chapter 1, Sec. 1.5, we developed an intuitive quantum theory of the laser, introducing various rate constants in a largely *ad hoc* fashion. Now that we have developed rate equations for level populations and photons, it is interesting to return to this intuitive model and examine its validity. We have also done this in Section 8.5 in connection with semiclassical laser theory.

First recall that $g(\nu) = \sigma(\nu)(N_2 - N_1)$ if the level degeneracies are equal. Thus we have

$$g(\nu) \approx \sigma(\nu) N_2 = \sigma(\nu) n_2 / V_g \qquad (10.6.1)$$

if there is negligible occupation of level 1: $N_2 \gg N_1$. Now define two constant coefficients a and b:

$$a = \frac{c\sigma(\nu) l/L}{V_g} = \frac{c\sigma(\nu)}{V} \qquad (10.6.2)$$

and

$$b = \frac{c}{2L}(1 - r_1 r_2) \qquad (10.6.3)$$

Then Eq. (10.5.16c) for the photon number q_ν can be written in the compact form

$$\frac{dq_\nu}{dt} = an_2 q_\nu - bq_\nu \qquad (10.6.4)$$

which is exactly Eq. (1.5.1). Recall that in Chapter 1 we identified n as the number of atoms in level 2, here denoted n_2.

The equation for n_2 is easily obtained from (10.5.16b). We again invoke the assumption $n_2 \gg n_1$ to get

$$\frac{dn_2}{dt} = -an_2 q_\nu - fn_2 + p \qquad (10.6.5)$$

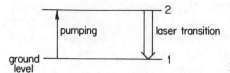

Figure 10.5 A two-level laser scheme.

where

$$f = \Gamma_2 + A \qquad (10.6.6)$$

We see that Eq. (10.6.5) is the same as Eq. (1.5.2).

Thus if the population inversion is large enough that N_1 is negligible compared with N_2, the theory developed in Chapter 1 agrees with our coupled photon-population rate equations. If N_2 is not much larger than N_1, the theory of Chapter 1 requires some minor modifications. What we were not able to do in Chapter 1, however, was to identify the constants a, b, f, and p in terms of fundamental atomic parameters like the Einstein A coefficient, the inelastic collision rate, the atomic absorption cross section, and mirror reflectivities. That has now been accomplished and the results agree with those found in Section 8.5.

10.7 THREE-LEVEL LASER SCHEME

Thus far we have not specified where levels 1 and 2 appear in the overall energy-level scheme of the lasing atoms. We might imagine that level 1 is the ground level and level 2 the first excited level of an atom. (Fig. 10.5) When we attempt to achieve continuous laser oscillation using the two-level scheme of Figure 10.5, however, we encounter a serious difficulty: the mechanism we use to excite atoms to level 2 may also deexcite them. For example, if we try to pump atoms from level 1 to level 2 by irradiating the medium, the radiation will induce both upward transitions $1 \rightarrow 2$ (absorption) and downward transitions $2 \rightarrow 1$ (stimulated emission).[1]

The *best* we can do by this process of cw optical pumping is to produce the same number of atoms in level 2 as in level 1 by saturating the $1 \leftrightarrow 2$ transition (recall Section 7.3); we cannot obtain a positive steady-state population inversion using only two atomic levels in the pumping process. This is clear from the solution (7.3.12) for the upper-level population of a two-state atom. In this case N_2 is always less than $\frac{1}{2}N$, regardless of the intensity of the pumping field.

One resolution of this difficulty is to make use of a third level, as in the *three-*

---

1. In quantum mechanics the forward and reverse rates are related by the principle of detailed balance, a manifestation of microscopic reversibility. Recall equations (7.2.3), and see also Appendix 7.A.

level laser inversion scheme of Figure 10.6. In such a laser, some pumping process acts between level 1 and level 3. An atom in level 3 cannot stay there forever. As a result of the pumping process it may return to level 1, but for other reasons such as spontaneous emission or a collision with another particle, the atom may drop to a different level of lower energy. In the case of spontaneous emission the energy lost by the atom appears as radiation. In the case of collisional deexcitation, the energy lost by the atom may appear as internal excitation in a collision partner, or as an increase in the kinetic energy of the collision partners, or both. The key to the three-level inversion scheme of Figure 10.6 is to have atoms in the pumping level 3 drop very rapidly to the upper laser level 2. This accomplishes two purposes. First, the pumping from level 1 is, in effect, directly from level 1 to the upper laser level 2, because every atom finding itself in level 3 converts quickly to an atom in level 2. Second, the rapid depletion of level 3 does not give the pumping process much of a chance to act in reverse and repopulate the ground level 1.

We will characterize the pumping process by a rate P, so that PN_1 is the number of atoms per cubic centimeter per second that are taken from the ground level 1 to the level 3. Thus the rate of change of the population N_1 of atoms per cubic centimeter in level 1 is

$$\left(\frac{dN_1}{dt}\right)_{\text{pumping}} = -PN_1 \qquad (10.7.1)$$

as a result of the pumping process. Since the pumping takes atoms from level 1 to level 3, and level 3 is assumed to decay very rapidly into level 2, we may also write (see Problem 10.3)

$$\left(\frac{dN_2}{dt}\right)_{\text{pumping}} \approx \left(\frac{dN_3}{dt}\right)_{\text{pumping}} = -\left(\frac{dN_1}{dt}\right)_{\text{pumping}}$$

$$= PN_1 \qquad (10.7.2)$$

for the rate of change of population of level 2 due to pumping.

Atoms in level 2 can decay, by spontaneous emission or via collisions, as indicated in the population equations (7.3.2) or (10.5.14). For simplicity we will

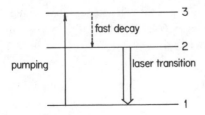

Figure 10.6 A three-level laser. Level 1 is the ground level, and laser oscillation occurs on the $2 \rightarrow 1$ transition.

now assume that level 2 decays only into level 1 by these processes, and we will denote the rate by Γ_{21}. That is, we assume

$$\left(\frac{dN_2}{dt}\right)_{decay} = -\Gamma_{21}N_2 \qquad (10.7.3a)$$

$$\left(\frac{dN_1}{dt}\right)_{decay} = \Gamma_{21}N_2 \qquad (10.7.3b)$$

for the population changes associated with the decay of level 2. The total rates of change of the population of levels 1 and 2 are therefore

$$\frac{dN_1}{dt} = -PN_1 + \Gamma_{21}N_2 + \sigma\Phi_\nu(N_2 - N_1) \qquad (10.7.4a)$$

$$\frac{dN_2}{dt} = PN_1 - \Gamma_{21}N_2 - \sigma\Phi_\nu(N_2 - N_1) \qquad (10.7.4b)$$

Equations (10.7.4) imply the conservation law

$$\frac{d}{dt}(N_1 + N_2) = 0$$

or

$$N_1 + N_2 = \text{constant} = N_T \qquad (10.7.5)$$

By ignoring any other atomic energy levels, and assuming that level 3 decays practically instantaneously into level 2, we are assuming that each active atom of the gain medium must be either in level 1 or level 2. Therefore the conserved quantity N_T is simply the total number of active atoms per unit volume.

We can now draw some important conclusions about the "threshold region" of steady-state (cw) laser oscillation. Near threshold the number of cavity photons is small enough that stimulated emission may be omitted from Eqs. (10.7.4). In particular, we can determine from these equations the threshold pumping rate necessary to achieve a population inversion, together with the threshold power expended in the process.

In the steady state N_1 and N_2 are not changing in time. The steady-state values $\overline{N}_1$ and $\overline{N}_2$ therefore satisfy Eqs. (10.7.4) with $\dot{N}_1 = \dot{N}_2 = 0$. Thus, if Φ is so small that the last terms in (10.7.4) are negligible, we find

$$\overline{N}_2 = \frac{P}{\Gamma_{21}}\overline{N}_1 \qquad (10.7.6)$$

in the steady state. Since (10.7.5) must hold for all possible values of N_1 and N_2 including the steady-state values $\overline{N}_1$ and $\overline{N}_2$, we also have

$$\overline{N}_1 + \overline{N}_2 = N_T \tag{10.7.7}$$

Equations (10.7.6) and (10.7.7) may be solved for $\overline{N}_1$ and $\overline{N}_2$ to obtain

$$\overline{N}_1 = \frac{\Gamma_{21}}{P + \Gamma_{21}} N_T \tag{10.7.8a}$$

and

$$\overline{N}_2 = \frac{P}{P + \Gamma_{21}} N_T \tag{10.7.8b}$$

The steady-state threshold-region population inversion is therefore

$$\overline{N}_2 - \overline{N}_1 = \frac{P - \Gamma_{21}}{P + \Gamma_{21}} N_T \tag{10.7.9}$$

In order to have a positive steady-state population inversion, and therefore a positive gain, we must obviously have

$$P > \Gamma_{21} \tag{10.7.10}$$

which simply says that the pumping rate into the upper laser level must exceed the decay rate. The greater the pumping rate with respect to the decay rate, the greater the population inversion and gain.

The pumping of an atom from level 1 to level 3 requires an energy

$$E_3 - E_1 = h\nu_{31} \tag{10.7.11}$$

The power per unit volume delivered to the active atoms in the pumping process is therefore

$$\frac{\text{Pwr}}{V} = h\nu_{31} P \overline{N}_1 \tag{10.7.12}$$

in the steady state. Using (10.7.8), we may write this as

$$\frac{\text{Pwr}}{V} = \frac{h\nu_{31} P \Gamma_{21}}{P + \Gamma_{21}} N_T \tag{10.7.13}$$

Now from (10.7.10) we may regard

$$P_{\min} = \Gamma_{21} \qquad (10.7.14)$$

as the minimum pumping rate necessary to reach positive gain. Substituting $P_{\min}$ for P in (10.7.13), we obtain

$$\left(\frac{\text{Pwr}}{V}\right)_{\min} = \tfrac{1}{2} \Gamma_{21} N_T h\nu_{31} \qquad (10.7.15)$$

as the minimum power per unit volume that must be exceeded to produce a positive gain. With this amount of pumping power delivered to the active medium, we see from (10.7.8) (with $P = P_{\min} = \Gamma_{21}$) that half the active atoms are in the lower level of the laser transition and half are in the upper level. A pumping power density greater than (10.7.15) makes $\overline{N}_2 > \overline{N}_1$.

10.8   FOUR-LEVEL LASER SCHEME

Another useful model for achieving population inversion is the *four-level laser* scheme shown in Figure 10.7. Pumping proceeds from the ground level 0 to the level 3 which, as in the three-level laser, decays rapidly into the upper laser level 2. In this model the lower laser level 1 is not the ground level, but an excited level which can itself decay into the ground level. This represents an advantage over the three-level laser, for the depletion of the lower laser level obviously enhances the population inversion on the laser transition. That is, a decrease in N_1 results in an increase in $N_2 - N_1$.

As in a three-level laser the decay from level 3 to level 2 is ideally instantaneous, i.e., extremely rapid compared with any other rates in the population rate

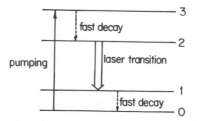

Figure 10.7 A four-level laser. The lower laser level 1 decays into the ground level 0.

equations. Then we may take $N_3 \approx 0$, and the population rate equations for the four-level laser take the form

$$\frac{dN_0}{dt} = -PN_0 + \Gamma_{10}N_1 \qquad (10.8.1a)$$

$$\frac{dN_1}{dt} = -\Gamma_{10}N_1 + \Gamma_{21}N_2 + \sigma(\nu)\,(N_2 - N_1)\,\Phi_\nu \qquad (10.8.1b)$$

$$\frac{dN_2}{dt} = PN_0 - \Gamma_{21}N_2 - \sigma(\nu)\,(N_2 - N_1)\,\Phi_\nu \qquad (10.8.1c)$$

where P is again the pumping rate out of the ground level 0, and PN_0 is the upper level pumping rate denoted K in (10.5.14b). Γ_{21} and Γ_{10} are the rates for the decay processes $2 \rightarrow 1$ and $1 \rightarrow 0$, respectively, and we have made the same approximation (10.7.2) for the pumping process as in the three-level case. Note that Eqs. (10.8.1) imply the conservation law

$$N_0 + N_1 + N_2 = \text{constant} = N_T \qquad (10.8.2)$$

If the stimulated emission rate is very small compared with the pumping and decay rates, Eqs. (10.8.1) give the steady-state populations

$$\bar{N}_0 = \frac{\Gamma_{10}\Gamma_{21}}{\Gamma_{10}\Gamma_{21} + \Gamma_{10}P + \Gamma_{21}P}\,N_T \qquad (10.8.3a)$$

$$\bar{N}_1 = \frac{\Gamma_{21}P}{\Gamma_{10}\Gamma_{21} + \Gamma_{10}P + \Gamma_{21}P}\,N_T \qquad (10.8.3b)$$

$$\bar{N}_2 = \frac{\Gamma_{10}P}{\Gamma_{10}\Gamma_{21} + \Gamma_{10}P + \Gamma_{21}P}\,N_T \qquad (10.8.3c)$$

The steady-state population inversion of the laser transition is therefore (Problem 10.4)

$$\bar{N}_2 - \bar{N}_1 = \frac{P(\Gamma_{10} - \Gamma_{21})\,N_T}{\Gamma_{10}\Gamma_{21} + \Gamma_{10}P + \Gamma_{21}P} \qquad (10.8.4)$$

Thus the pumped ($P \neq 0$) four-level system will always have a steady-state population inversion when

$$\Gamma_{10} > \Gamma_{21} \tag{10.8.5}$$

that is, when the lower laser level decays more rapidly than the upper laser level. When we have

$$\Gamma_{10} \gg \Gamma_{21}, P \tag{10.8.6}$$

then $N_0 \approx N_T$, $N_1 \approx 0$ and Eq. (10.8.4) reduces to

$$\overline{N}_2 - \overline{N}_1 \approx N_2 \approx \frac{P}{P + \Gamma_{21}} N_T \tag{10.8.7}$$

10.9 COMPARISON OF PUMPING REQUIREMENTS FOR THREE- AND FOUR-LEVEL LASERS

It is interesting to compare the pumping rates necessary for laser oscillation in the three- and four-level lasers. To achieve laser oscillation, we must produce a gain larger than the threshold value g_t, and therefore a population inversion greater than ΔN_t, where

$$g_t \equiv \sigma(\nu) \, \Delta N_t \tag{10.9.1}$$

is the threshold gain for frequency ν. Setting $\overline{N}_2 - \overline{N}_1$ equal to the threshold inversion in (10.7.9), we obtain

$$(P_t)_{\text{three-level laser}} = \frac{N_T + \Delta N_t}{N_T - \Delta N_t} \Gamma_{21} \tag{10.9.2}$$

for the threshold pumping rate for laser oscillation in the three-level laser. From (10.8.7), on the other hand, the pumping rate necessary for a four-level laser satisfying (10.8.6) is

$$(P_t)_{\text{four-level laser}} = \frac{\Delta N_t}{N_T - \Delta N_t} \Gamma_{21} \tag{10.9.3}$$

The ratio of (10.9.3) to (10.9.2) is

$$\frac{(P_t)_{\text{four-level laser}}}{(P_t)_{\text{three-level laser}}} = \frac{\Delta N_t}{N_T - \Delta N_t} \tag{10.9.4}$$

Ordinarily the population inversion (at threshold or otherwise) is very small compared with the total number of active atoms; recall our example at the end of Section 10.4. From (10.9.4), therefore, we see that

$$\left(P_t\right)_{\text{four-level laser}} \ll \left(P_t\right)_{\text{three-level laser}} \tag{10.9.5}$$

Thus a much larger pumping rate is necessary to achieve laser oscillation in a three-level laser than in a four-level laser.

In the three-level laser the pumping power density necessary to establish the threshold inversion ΔN_t is

$$\left(\frac{(\text{Pwr})_t}{V}\right)_{\text{three-level laser}} = h\nu_{31}\,(N_1)_t(P_t)_{\text{three-level laser}} \tag{10.9.6}$$

where $(N_1)_t$ is given by (10.7.8a) with $P = (P_t)_{\text{three-level laser}}$. Assuming $\Delta N_t \ll N_T$, we conclude from (10.9.2) that

$$\left(P_t\right)_{\text{three-level laser}} \approx \Gamma_{21} \tag{10.9.7}$$

and from (10.7.8a) that

$$(N_1)_t \approx N_T/2 \tag{10.9.8}$$

Equation (10.9.6) therefore becomes

$$\left(\frac{(\text{Pwr})_t}{V}\right)_{\text{three-level laser}} \approx \tfrac{1}{2}\,h\nu_{31}N_T\Gamma_{21} \tag{10.9.9}$$

which, because of the approximation $\Delta N_t \ll N_T$, is the same as (10.7.15). Thus when $\Delta N_t \ll N_T$ the minimum pumping power density necessary to achieve positive gain is approximately the same as that necessary to reach threshold and laser oscillation. In the four-level case, on the other hand, we obtain

$$\left(\frac{(\text{Pwr})_t}{V}\right)_{\text{four-level laser}} \approx h\nu_{30}\,\Delta N_t\Gamma_{21} \tag{10.9.10}$$

for $\Delta N_t \ll N_T$. The ratio of (10.9.10) to (10.9.9) is

$$\frac{\left((\text{Pwr})_t/V\right)_{\text{four-level laser}}}{\left((\text{Pwr})_t/V\right)_{\text{three-level laser}}} \approx \frac{2\nu_{30}\,\Delta N_t}{\nu_{31}N_T} \tag{10.9.11}$$

if we take the upper-laser-level decay rate Γ_{21} to be the same in the two cases. This shows that, other things being roughly commensurate, much less power will be necessary to achieve laser oscillation in the four-level case.

10.10 EXAMPLES OF THREE- AND FOUR-LEVEL LASERS

Real lasers seldom fit very neatly into our three- and four-level categories. However, these idealizations sometimes provide a useful framework for rough estimates of pump power requirements. We will illustrate this for cw ruby and Nd:YAG solid-state lasers, which are approximately three-level and four-level systems, respectively.

It was mentioned in Section 3.1 that ruby is the crystal Al_2O_3 with chromium ions (Cr^{+3}) replacing some of the aluminum ions (Al^{+3}); the concentration of chromium is only about 0.05% by weight. The relevant energy levels for the ruby laser are those of the Cr^{+3} ion in the host crystal lattice. The laser is *optically pumped*, i.e., a population inversion is obtained by the absorption of radiation from a lamp, typically a high-pressure Xe or Hg flashlamp. It is approximately a three-level laser, although the third "level" really consists of two broad bands of energy, both decaying rapidly (rate $\approx 10^7$ sec^{-1}) into the upper laser level 2 (See Figure 13.3). At room temperature the decay rate of the upper laser level is

$$\Gamma_{21} \approx \tfrac{1}{2} \times 10^3 \text{ sec}^{-1} \qquad (10.10.1)$$

The density of "active atoms" (i.e., Cr^{+3} ions) is (Problem 10.5)

$$N_T \approx 1.6 \times 10^{19} \text{ cm}^{-3} \qquad (10.10.2)$$

The energy required from the pump, the difference between the ground level and the lower pump band, is about 3.6×10^{-12} erg, corresponding to a wavelength of about 5500 Å (green). From (10.9.9), therefore, the minimum pumping power density necessary to achieve nonnegative gain [and also, according to (10.7.15), approximately the pumping power density necessary for laser oscillation] is

$$\left(\frac{\text{Pwr}}{V}\right)_{\text{min}} \approx \tfrac{1}{2}(\tfrac{1}{2} \times 10^3 \text{ sec}^{-1})(1.6 \times 10^{19} \text{ cm}^{-3})(3.6 \times 10^{-12} \text{ erg})$$

$$\approx 1 \text{ kW/cm}^3 \qquad (10.10.3)$$

For a 5-cm-long ruby rod of radius 2 mm, the required pump power is

$$\text{Pwr} \approx (1 \text{ kW/cm}^3)\pi(0.2 \text{ cm})^2 (5 \text{ cm})$$

$$= 600 \text{ W} \qquad (10.10.4)$$

This is a rather large amount of power. In fact much *more* power is actually required to operate a cw ruby laser, because (10.10.4) only gives the power that must actually be absorbed by the Cr^{+3} ions. In reality only a small fraction (typically $\approx 0.1\%$) of the electrical power delivered to the lamp is converted to useful laser radiation.

It is also interesting to estimate the population inversion necessary for threshold gain in a typical ruby laser. At room temperature the 6943-Å laser transition has a Lorentzian lineshape of width (HWHM)

$$\delta\nu_0 \approx 170 \text{ GHz} \qquad (10.10.5)$$

and the A coefficient for spontaneous emission is

$$A \approx 230 \text{ sec}^{-1} \qquad (10.10.6)$$

The line-center cross section for stimulated emission is therefore

$$\sigma = \frac{\lambda^2 A}{8\pi n^2} \frac{1}{\pi \delta\nu_0}$$

$$\approx 2.7 \times 10^{-20} \text{ cm}^2 \qquad (10.10.7)$$

where we have used the value $n = 1.76$ for the refractive index of ruby. If we assume a resonator with mirror reflectivities $r_1 = 1.0$ and $r_2 = 0.96$, and a scattering loss of 3% per round-trip pass through the gain cell, then the threshold gain for laser oscillation is [Eq. (10.4.11)]

$$g_t = \frac{-1}{2(5 \text{ cm})} \ln (0.96) + \frac{.03}{2(5 \text{ cm})} = 7.1 \times 10^{-3} \text{ cm}^{-1} \qquad (10.10.8)$$

for a ruby rod 5 cm long. From (10.4.8), (10.10.7), and (10.10.8), therefore, we calculate a population inversion threshold

$$\Delta N_t \approx \frac{7.1 \times 10^{-3} \text{ cm}^{-1}}{2.7 \times 10^{-20} \text{ cm}^2}$$

$$= 2.6 \times 10^{17} \text{ cm}^{-3} \qquad (10.10.9)$$

This is much larger than the sort of population inversion necessary for a typical He–Ne laser [Eq. (10.4.15)]. The difference stems from the much larger stimulated-emission cross section of the 6328-Å laser transition of Ne, which in turn results from the much larger A coefficient and much smaller linewidth than in ruby. This illustrates an important point: gas lasers obviously have a much smaller density of atoms than solid-state lasers, but this does not necessarily mean that they have smaller gains. In fact, many of the most powerful lasers are gas lasers. The reasons for this are discussed in Chapter 13.

In the 1.06-μm (1064 nm) neodymium YAG laser the active atoms are also impurities in a crystal lattice, in this case Nd^{+3} ions in yttrium aluminum garnet ($Y_3Al_5O_{12}$, called YAG). The Nd:YAG laser is approximately a four-level system, with upper-level decay rate

$$\Gamma_{21} \approx 1800 \text{ sec}^{-1} \tag{10.10.10}$$

and stimulated-emission cross section

$$\sigma \approx 9 \times 10^{-19} \text{ cm}^2 \tag{10.10.11}$$

at room temperature. If we assume the same threshold gain (10.10.8) as in our calculation for the ruby laser, we obtain a population inversion threshold

$$\Delta N_t \approx \frac{7.1 \times 10^{-3} \text{ cm}^{-1}}{9 \times 10^{-19} \text{ cm}^2} \approx 8 \times 10^{15} \text{ cm}^{-3} \tag{10.10.12}$$

which, because of the relatively large stimulated emission cross section for Nd^{+3}, is considerably smaller than the value (10.10.9) for ruby.

The pump "level 3" for the Nd:YAG laser is actually a series of energy bands located between about 2.6×10^{-12} and 5.0×10^{-12} erg above the ground level. If we take the energy difference $E_3 - E_0$ in our four-level model (Figure 10.7) to be the average value, 3.8×10^{-12} erg, we obtain from (10.9.10) the pumping power density for threshold:

$$\frac{(\text{Pwr})_{\text{min}}}{V} \approx 5 \text{ W/cm}^3 \tag{10.10.13}$$

For a 5-cm Nd:YAG rod of radius 2 mm, therefore, the threshold pump power is

$$\text{Pwr} \approx 3 \text{ W} \tag{10.10.14}$$

—much smaller than the estimate (10.10.4) for ruby.

10.11 SMALL-SIGNAL GAIN AND GAIN SATURATION

Equation (10.7.9) gives the steady-state population inversion for a three-level laser when the stimulated emission rate is negligible. In general, of course, the stimulated emission rate is not negligible, and here we consider the steady-state population inversion in the more general case. For this we require the steady-state solutions of Eqs. (10.7.4). These may be obtained by noting that the following replacements for P and Γ_{21} (in the equations *without* stimulated emission):

$$P \to P + \sigma\Phi_\nu \tag{10.11.1a}$$

$$\Gamma_{21} \to \Gamma_{21} + \sigma\Phi_\nu \tag{10.11.1b}$$

are sufficient to reinstate all stimulated-emission terms. Here Φ_ν is the steady-state (i.e., time-independent) cavity photon flux. Thus the steady-state solutions of (10.7.4) may be obtained by making the same replacements in the solutions (10.7.8). Likewise the steady-state population inversion $\overline{N}_2 - \overline{N}_1$ in the general case follows when the replacements (10.11.1) are made in (10.7.9):

$$\overline{N}_2 - \overline{N}_1 = \frac{(P - \Gamma_{21})N_T}{P + \Gamma_{21} + 2\sigma\Phi_\nu} \tag{10.11.2}$$

This is the generalization of (10.7.9) to the case in which the stimulated emission rate is not negligible compared with P and Γ_{21}. Note that it is also the generalization (to the case of nonzero pumping rate P) of the power-broadened steady-state inversion implied by (7.3.12) (Problem 10.6).

The steady-state gain coefficient for a three-level laser follows from (10.2.7) and (10.11.2). Assuming $g_1 = g_2$, we have

$$g(\nu) = \frac{\sigma(\nu)(P - \Gamma_{21})\, N_T}{P + \Gamma_{21} + 2\sigma(\nu)\, \Phi_\nu} \tag{10.11.3}$$

The steady-state gain is thus a decreasing function of the cavity photon flux Φ_ν. Physically, the decrease of the gain with increasing Φ_ν is due to the fact that a large cavity photon number, and therefore a large stimulated emission (and absorption) rate, tends to equalize the populations $\overline{N}_1$ and $\overline{N}_2$. In this case the gain is said to be *saturated*. We will see that gain saturation plays a very important role in determining the output power of a laser.

It is useful to recognize what is going on microscopically that gives rise to gain saturation. Both spontaneous and stimulated emission apply to an atom in the upper laser level and cause it to drop to the lower level and emit a photon. Once it is in the lower level, it can decay further or it can absorb a photon back from the field. Recall that the lower level's absorption rate is exactly equal to the upper level's stimulated emission rate. When the field is so strong that these rates are much greater than the levels' decay rates, the atom jumps so rapidly between the upper and lower levels that it has effectively the same probability of being in one or the other of these levels. Then the atom is equally often an absorber and an emitter, and in this extreme limit the gain is zero. Thus in general the gain coefficient of the medium must be reduced as the cavity photon number is increased. Equation

(10.11.3) is a quantitative expression of this decrease of gain with increasing photon number.

The expression (10.11.3) can be cast in several different forms by trivial rearrangements. These rearrangements illustrate the role of *saturation flux* Φ_ν^{sat} in determining the gain. Let us factor $P + \Gamma_{21}$ in the denominator to get

$$g(\nu) = \frac{\sigma(\nu)(P - \Gamma_{21}) N_T}{P + \Gamma_{21}} \frac{1}{1 + [2\sigma(\nu)\Phi_\nu/(P + \Gamma_{21})]}$$

$$= \frac{g_0(\nu)}{1 + \Phi_\nu/\Phi_\nu^{sat}} \tag{10.11.4}$$

Here we have defined the *small-signal gain*

$$g_0(\nu) = \frac{\sigma(\nu)(P - \Gamma_{21}) N_T}{P + \Gamma_{21}} \tag{10.11.5}$$

and the *saturation flux*

$$\Phi_\nu^{sat} = \frac{P + \Gamma_{21}}{2\sigma(\nu)} \tag{10.11.6}$$

The corresponding saturation expressions for intensity and photon number are

$$I_\nu^{sat} = \frac{h\nu(P + \Gamma_{21})}{2\sigma(\nu)} \tag{10.11.7}$$

$$q_\nu^{sat} = \frac{(P + \Gamma_{21})}{2c\sigma(\nu)} V \tag{10.11.8}$$

Equation (10.11.4) shows that

$$g(\nu) \to g_0(\nu) \quad \text{as} \quad \Phi_\nu \to 0 \tag{10.11.9}$$

This is the reason $g_0(\nu)$ is termed the small-signal gain. When the absorption lineshape is Lorentzian we have [recall (7.4.10) and see Problem 10.8]

$$\sigma(\nu) = \frac{e^2 f}{4\pi\epsilon_0 mc} \frac{\delta\nu_{21}}{(\nu - \nu_{21})^2 + (\delta\nu_{21})^2}$$

$$= \sigma(\nu_{21}) \frac{1}{(\nu_{21} - \nu)^2/(\delta\nu_{21})^2 + 1} \tag{10.11.10}$$

From this formula and the definition (10.11.5) it is clear that the gain is appreciable only within $\delta\nu_{21}$ of line center ($\nu = \nu_{21}$). Thus $\delta\nu_{21}$ can be called the small-signal *gain bandwidth*: $\Delta\nu_g = \delta\nu_{21}$.

With the Lorentzian formula for $\sigma(\nu)$ we can derive a particularly transparent form for $g(\nu)$ from (10.11.3) by multiplying numerator and denominator by $\sigma(\nu_{21})/\sigma(\nu)$ and then using (10.11.10) to get:

$$g(\nu) = g_0(\nu_{21}) \frac{1}{(\nu_{21} - \nu)^2/(\delta\nu_{21})^2 + 1 + (\Phi_\nu/\Phi_{\nu_{21}}^{sat})} \qquad (10.11.11)$$

Here $g_0(\nu_{21})$ is the *line-center small-signal gain* [i.e., $g_0(\nu)$ at exact resonance]:

$$g_0(\nu_{21}) = \frac{\sigma(\nu_{21})(P - \Gamma_{21}) N_T}{P + \Gamma_{21}} \qquad (10.11.12)$$

and $\Phi_{\nu_{21}}^{sat}$ is given by (10.11.6) when $\nu = \nu_{21}$.

It is obvious in (10.11.11) that $g_0(\nu_{21})$ is the maximum gain, and that it is achieved at line center ($\nu = \nu_{21}$) for weak cavity excitation ($\Phi_\nu \to 0$). In Figure 10.8 we plot $g(\nu)$ vs. ν as given in (10.11.11) for several values of Φ_ν, and $g(\nu)$ vs. Φ_ν for several values of ν. Clearly it is harder to saturate $g(\nu)$ away from line center. Alternatively, for higher fluxes the half width of $g(\nu)$ is greater. This is exactly the same as the *power broadening* discussed in Section 7.3. The power-broadened gain bandwidth is the half width implied by (10.11.11), namely

$$\Delta\nu_g = \delta\nu_{21}\left(1 + \frac{\Phi_\nu}{\Phi_{\nu_{21}}^{sat}}\right)^{1/2} \qquad (10.11.13)$$

It is greater than the small-signal gain bandwidth and is seen to be exactly $1/2\pi$ times the angular frequency half width $\Delta_{1/2}$ given in (7.3.14), as it should be, if we put $P = 0$ in the expression for Φ_ν^{sat}.

In the case of a four-level laser we can obtain in a similar fashion a gain with the flux dependence (10.11.4), but with a saturation flux different from (10.11.6) (Problem 10.9). The same basic flux dependence is also obtained when we include the degeneracy and refractive-index corrections to the gain coefficient, as described in Section 10.2.

Three- and four-level lasers are idealizations that are seldom fully realized in practice. The gain-saturation formulas (10.11.4) and (10.11.11) are, however, applicable to a wide variety of actual lasers. That is, although (10.11.4) may be derived from simple models, it often applies outside the range of validity of these models. It is the most commonly assumed formula for the intensity dependence of the gain on a homogeneously broadened laser transition. In Section 11.7 we will consider the case of an inhomogeneously broadened transition.

In Eq. (10.11.6) the rates P and Γ_{21} are the "decay rates" of the lower and

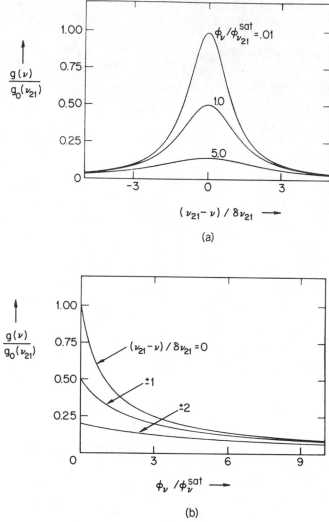

Figure 10.8 Saturated gain curves, according to Eq. (10.11.11).

upper laser levels, respectively [cf. Eqs. (10.7.4)]. The larger the decay rates, the larger the saturation flux. This makes good sense physically, for the larger the decay rates, the larger must be the stimulated emission rate necessary to saturate the transition, i.e., to equalize the population densities $\overline{N}_1$ and $\overline{N}_2$. In fact the saturation flux (10.11.6) for a three-level laser is just the intensity for which the stimulated emission rate is the average of the upper- and lower-level decay rates (Problem 10.7). In general, the larger these decay rates, the larger the saturation flux. Equation (10.11.6) is an example of this general result.

The relation $\sigma(\nu) = (h\nu/c)BS(\nu)$ [recall (7.4.19)] determines that the saturation flux (10.11.6) is inversely proportional to the Einstein B coefficient for the transition. This is another general result, and is hardly surprising, because the smaller B is, the greater the intensity necessary to achieve a given stimulated emission rate. For a Lorentzian lineshape function, (10.11.6) also predicts that the line-center saturation flux is directly proportional to the transition linewidth $\delta\nu_{21}$:

$$\Phi_{\nu_{21}}^{sat} = \frac{P + \Gamma_{21}}{2\sigma(\nu_{21})} = \frac{4\pi^2 \, \delta\nu_{21}}{\lambda^2 A}(P + \Gamma_{21}) \qquad (10.11.14)$$

This too is a general result that is applicable beyond the three- and four-level models.

In practically all lasers the saturation flux is independent of the rate at which the gain medium is pumped, in contrast to the result given in (10.11.6) for the idealized three-level laser. However, even though equations (10.11.5) and (10.11.6) are specific just to the three-level model, the results (10.11.4) and (10.11.11) are applicable to a wide variety of real lasers under conditions of homogeneous line broadening. It will usually be difficult to *calculate* g_0 and I_ν^{sat}, but we can often be confident nevertheless that the *form* of the intensity dependence of the gain described by (10.11.4) or (10.11.11) is correct.

Equations (10.11.4) and (10.11.11) are applicable regardless of whether g_0 is positive (gain) or negative (absorption). That is, a medium may be saturated regardless of whether it is amplifying or absorbing. Thus the absorption coefficient $a(\nu)$ of an absorbing medium will decrease as the intensity of the radiation is raised. When the intensity is much larger than the line-center saturation flux $\Phi_{\nu_{21}}^{sat}$ characteristic of the medium, the absorption coefficient is very small [$a(\nu) \approx 0$], which means that the medium is practically transparent to high-intensity radiation. In this case the medium is sometimes said to be "bleached," because it no longer absorbs radiation that is resonant with one of its transition frequencies. What is happening in the case of such strong saturation is that the stimulated emission (and absorption) rate has become much greater than the decay rate of the *upper* level of the transition. An atom that has absorbed a photon will then be quickly induced to return to the lower level and give the photon back to the field by stimulated emission. This occurs, with high probability, before the absorbed energy can be dissipated as heat or fluorescence. Thus no energy is lost by the incident field; the medium has been made effectively transparent ("bleached") by virtue of the high intensity of the field.

10.12 SPATIAL HOLE BURNING

In this section we will consider more carefully the meaning of the "intensity." Intensity refers to the electromagnetic energy flow per unit area per unit time, but in most lasers we have *standing* waves rather than traveling waves. The gain-saturation formulas (10.11.4) and (10.11.11) are often written with Φ_ν assumed to be the sum of the fluxes of the two traveling waves (Figure 10.3):

$$\Phi_\nu \rightarrow \Phi_\nu = \Phi_\nu^{(+)} + \Phi_\nu^{(-)} \tag{10.12.1}$$

However, this is not quite correct, for it ignores the interference of the two traveling waves. The electromagnetic energy density $(h\nu/c)\Phi_\nu$ is proportional to the square of the electric field:[2]

$$\frac{h\nu}{c}\,\Phi_\nu = \epsilon_0 E^2(\mathbf{r}, t)$$

$$= \epsilon_0 E_0^2 \cos^2\omega t \, \sin^2 kz \tag{10.12.2}$$

We replace $\cos^2\omega t$ by its average value $(\frac{1}{2})$ over times long compared with an optical period $2\pi/\omega \approx 10^{-14}$ sec, and write

$$\frac{h\nu}{c}\,\Phi_\nu = \frac{\epsilon_0}{2}\,E_0^2 \sin^2 kz \tag{10.12.3}$$

Now a cavity standing-wave field is the sum of two oppositely propagating traveling-wave fields:

$$E(z, t) = E_0 \cos\omega t \, \sin kz$$

$$= \tfrac{1}{2} E_0 \left[\sin(kz - \omega t) + \sin(kz + \omega t) \right]$$

$$= E_+(z, t) + E_-(z, t) \tag{10.12.4}$$

where the two electric waves

$$E_\pm(z, t) = \tfrac{1}{2} E_0 \sin(kz \mp \omega t) \tag{10.12.5}$$

propagate in the positive $(+)$ and negative $(-)$ z directions. The time-averaged square of the electric field (10.12.5) gives a field energy density

$$\frac{h\nu}{c}\,\Phi_\nu^{(\pm)} = \frac{\epsilon_0}{8}\,E_0^2 \tag{10.12.6}$$

---

2.   For the sake of simplicity we assume in this section that the refractive index $n \approx 1$.

From (10.12.3) and (10.12.6), therefore, it follows that

$$\Phi_\nu = 2\left[\Phi_\nu^{(+)} + \Phi_\nu^{(-)}\right] \sin^2 kz \qquad (10.12.7)$$

or, in terms of the intensity $I_\nu = h\nu\Phi_\nu$,

$$I_\nu = 2\left[I_\nu^{(+)} + I_\nu^{(-)}\right] \sin^2 kz \qquad (10.12.8)$$

Thus it is not correct to use (10.12.1) as the flux in the gain-saturation formulas (10.11.4) and (10.11.11). We should use (10.12.7), which accounts properly for the interference of the two traveling wave fields. Then the gain saturation formula for a homogeneously broadened transition is

$$g(\nu) = \frac{g_0(\nu)}{1 + 2\left[(\Phi_\nu^{(+)} + \Phi_\nu^{(-)})/\Phi_\nu^{\text{sat}}\right] \sin^2 kz} \qquad (10.12.9)$$

This saturation formula replaces (10.11.4) when the standing-wave nature of the cavity field is properly accounted for.

The $\sin^2 kz$ term in Eq. (10.12.9) gives rise to what is called *spatial hole burning* in the gain coefficient $g(\nu)$. At points z for which $\sin^2 kz = 0$, $g(\nu)$ takes on its maximum value, namely the small-signal value $g_0(\nu)$. Where $\sin^2 kz = 1$, however, $g(\nu)$ has its minimum value, i.e., it is most strongly saturated; a "hole" is "burned" in the curve of $g(\nu)$ vs. z (Figure 10.9). The holes in this curve are separated by $\Delta z = \pi/k = \lambda/2$. Thus $g(\nu)$ varies with z on the scale of the laser wavelength.

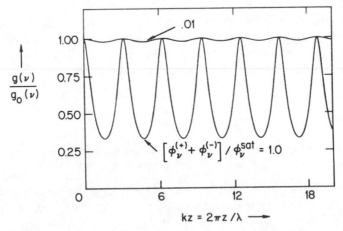

Figure 10.9 Spatial hole burning in the gain curve, according to Eq. (10.12.9).

This rapid variation of $g(\nu)$ with z suggests the approximation of replacing $\sin^2 kz$ by its spatial average, $\frac{1}{2}$, in Eq. (10.12.9). In this approximation we take

$$g(\nu) = \frac{g_0(\nu)}{1 + (\Phi_\nu^{(+)} + \Phi_\nu^{(-)})/\Phi_\nu^{\text{sat}}} \tag{10.12.10}$$

which is the result obtained by using (10.12.1) in the gain-saturation formula (10.11.4). This approximation ignores the spatial dependence of the intracavity field, and therefore also the spatial hole burning of the gain coefficient. It is called the *uniform-field approximation* or the *mean-field approximation*.

10.13 SUMMARY

In Chapter 1 we gave a very crude description of laser action and introduced some fundamental laser concepts such as gain and threshold. In the intervening chapters we have gone more deeply into the theory of the interaction of light and matter, and in the present chapter we have begun to apply what we have learned.

The most important theoretical tools for our understanding of lasers are the rate equations for atomic (or molecular) level populations and field intensities. These equations generally include effects of pumping, collisions, absorption, spontaneous and stimulated emission, field gain and loss, and other processes that may be pertinent for a particular laser. We have used such equations to discuss three- and four-level lasers, and in fact the use of rate equations will be a dominant theme of the following chapters.

The rate equations for emission and absorption are based on familiar transition rates and cross sections, effectively on the Einstein A and B coefficients. We should not forget, of course, the approximations employed in deriving the rate coefficients in Chapter 6. Rate equations do not apply in every situation and are significantly wrong in some cases. Fortunately, however, the most important features of lasers do not require attention to the limits of rate-equation validity, and our approach will be entirely adequate for our purpose in this book.

PROBLEMS

10.1 Show how (10.2.3) and (10.4.3a) are modified if the light propagates toward $-z$ rather than $+z$, and thus derive (10.4.3b).

10.2 (a) Solve (10.5.18) as a function of time for the (unusual) case that the equation's loss parameters satisfy $\Gamma_2 + A_{21} = (c/2L)(1 - r_1 r_2)$. Give the steady-state value of $n_2 + q_\nu$.

   (b) Find the steady-state solution for q_ν in terms of p and n_2 for arbitrary loss parameters.

10.3 (a) Write the full set of equations for a three-level laser by modifying (10.7.4b) and including this equation for the third level (as shown in Figure 10.6):

$$\dot{N}_3 = PN_1 - \Gamma_{32}N_3$$

and show that the full set of equations satisfies $N_1 + N_2 + N_3 = $ constant $= N_T$.

(b) Determine the steady-state values of the three level populations.

(c) Find the condition under which it is satisfactory to neglect level 3 [$N_3 \approx 0$] and to use Eqs. (10.7.2) and (10.7.4b) as written in the text.

10.4 Solve equations (10.8.1) for the steady-state value of $\overline{N}_2 - \overline{N}_1$, and show that (10.8.4) gives the limiting value as the stimulated emission rate decreases to zero.

10.5 Estimate the density of chromium ions in ruby, assuming that the concentration of chromium in ruby is about 0.05% by weight.

10.6 Use (10.11.2) and $\overline{N}_1 + \overline{N}_2 = N_T$ to derive a solution for $\overline{N}_2$ alone. What is the relation of $\overline{N}_2$ to the solution for N_2 given in (7.3.12)?

10.7 We showed in equation (10.11.7) that the saturation intensity I^{sat} of a three-level laser is the intensity for which the stimulated emission rate is the average of the upper- and lower-level decay rates of the laser transition. Find the corresponding expression that follows from laser equations (10.5.14).

10.8 Derive Eqs. (10.11.10)–(10.11.12).

10.9 (a) Derive the formula equivalent to (10.11.3) for a four-level laser.

(b) Derive an expression for the saturation intensity of a four-level laser.

11 LASER OSCILLATION: POWER AND FREQUENCY

11.1 INTRODUCTION

In this chapter we will consider the output power and frequency of a continuous-wave (cw) laser, i.e., a laser in the steady-state, time-independent mode of operation. Such a laser emits a steady, continuous beam of radiation. The other common mode of operation, in which the laser output is in the form of single or repeated pulses of radiation, will be discussed in the following chapter.

We will also restrict ourselves to the case of laser oscillation on a single resonator mode, postponing until the next chapter the case of multimode lasing. The assumption of single-mode oscillation allows us to focus on the essential ideas without undue complication.

11.2 OUTPUT INTENSITY: UNIFORM-FIELD APPROXIMATION

We will now use the gain-saturation formula (10.12.10) to derive an expression for the output intensity of a cw laser oscillating on a single cavity mode. We continue to assume for the present that the lasing transition is homogeneously broadened. In Section 11.4 we will consider the effect of spatial hole burning, but our discussion in the present section will be restricted to the mean-field (uniform-field) approximation.

In cw laser oscillation the cavity photon number is constant in time. This means that the field amplification due to stimulated emission exactly balances the attenuation due to the output coupling, scattering, and other cavity loss processes. That is, the growth rate of the cavity photon number equals the decay rate.

We will assume all loss processes to be independent of the cavity intensity. This implies that the field attenuation rate in steady-state oscillation is no different from that at the threshold of oscillation, where the cavity intensity is practically zero. Thus the growth rate of cavity photons in steady-state oscillation must also be the same as its threshold value. In other words, *in cw oscillation the gain is precisely equal to its threshold value g_t*. The gain is sometimes said to be "clamped" at its threshold value. Since the steady-state gain equals g_t, this clamping should determine the cavity intensity of the laser. We will now justify these assertions and examine their implications.

From the photon rate equation (10.5.8) [which applies when $g(\nu)$ is independent of z, consistent with the uniform-field approximation] we infer that

$$\frac{cl}{L} g(\nu) I_\nu = \frac{c}{2L} (1 - r_1 r_2) I_\nu \qquad (11.2.1)$$

in steady-state oscillation ($dI_\nu/dt = 0$). Dividing through by I_ν, therefore, we have the condition

$$g(\nu) = \frac{1}{2l} (1 - r_1 r_2) \qquad (11.2.2)$$

for steady-state oscillation. But from (10.5.11) we recognize the right side of (11.2.2) as the threshold gain g_t necessary for oscillation. Equation (11.2.2), we recall, is applicable when $r_1 r_2 \approx 1$; our discussion in Section 10.5 implies that this case is consistent with the uniform-field approximation. Thus the condition for steady-state oscillation in the uniform-field approximation is simply

$$g(\nu) = g_t \qquad \text{(steady-state oscillation)} \qquad (11.2.3)$$

as asserted above. Given the threshold gain g_t, which may be calculated from the length of the gain medium, the mirror reflectivities, and any significant scattering or other loss coefficients, we therefore have also the "clamped" gain for cw oscillation.

Now from the gain saturation formula (10.12.10), and the cw oscillation condition (11.2.3), we must evidently have (remembering that $I_\nu = h\nu\Phi_\nu$)

$$g(\nu) = \frac{g_0(\nu)}{1 + (I_\nu^{(+)} + I_\nu^{(-)})/I_\nu^{\text{sat}}} = g_t \qquad (11.2.4)$$

or

$$I_\nu^{(+)} + I_\nu^{(-)} = I_\nu^{\text{sat}} \left(\frac{g_0(\nu)}{g_t} - 1 \right) \qquad (11.2.5)$$

This simple formula gives the total cw cavity intensity in terms of the saturation intensity I_ν^{sat}, the small-signal gain $g_0(\nu)$, and the threshold gain g_t.

Of course the intensity on the left side of (11.2.5) is not the laser output intensity, but the intracavity intensity. The output intensity is

$$I_\nu^{\text{out}} = t_1 I_\nu^{(-)} + t_2 I_\nu^{(+)} \qquad (11.2.6)$$

where t_1 and t_2 are the transmission coefficients of the mirrors (Figure 11.1a), which we assume to be independent of the frequency ν. Usually only one of the mirrors is transmitting (Figure 11.1b), in which case we write

$$I_\nu^{\text{out}} = t I_\nu^{(+)} \qquad (11.2.7)$$

Figure 11.1 (a) A laser cavity with mirror transmission coefficients t_1 and t_2. (b) A laser cavity with one output mirror.

instead of (11.2.6). Here t is the transmission coefficient of the output mirror, and $I_\nu^{(+)}$ is the intensity of the traveling wave propagating towards the output mirror. For the cavity mode defined by (10.12.2) we have

$$I_\nu^{(+)} = I_\nu^{(-)} \tag{11.2.8}$$

The equality of $I_\nu^{(+)}$ and $I_\nu^{(-)}$ holds whenever the cavity mirrors of Figure 11.1 are perfectly reflecting, and so we might expect (11.2.8) to be a good approximation if the mirrors are highly reflecting, as is the case in most lasers. This expectation will be borne out in Section 11.5.

Combining the results (11.2.5), (11.2.7), and (11.2.8), we obtain the equation

$$I_\nu^{\text{out}} = \tfrac{1}{2} t I_\nu^{\text{sat}} \left(\frac{g_0(\nu)}{g_t} - 1 \right) \tag{11.2.9}$$

for the output intensity. Now according to (10.5.11) and (10.4.1) we have

$$g_t = \frac{1}{2l}(1 - r) = \frac{1}{2l}(t + s) \tag{11.2.10}$$

so that

$$I_\nu^{\text{out}} = \tfrac{1}{2} t I_\nu^{\text{sat}} \left(\frac{2 g_0(\nu)\, l}{t + s} - 1 \right) \tag{11.2.11}$$

for "small" (and typical) output couplings, for which (10.5.11) describes the threshold gain.

It may be worthwhile to summarize the assumptions made in deriving the output intensity (11.2.11) of a single-mode, homogeneously broadened, cw laser. First of all, we have ignored any spatial variation of the intensity transverse to the cavity axis (the z direction). Furthermore we have assumed that the intensity is also constant along the cavity axis; this is the uniform-field approximation, and it implies that the gain coefficient is likewise independent of z. Thus we have assumed that the gain and intensity are constant *throughout* the laser cavity. In the derivation leading to (11.2.11) we have also assumed that one of the mirrors is essentially perfectly reflecting, while the output mirror's reflectivity is high enough that $I_\nu^{(+)} \approx I_\nu^{(-)}$.

Equation (11.2.11) gives the output intensity in terms of the small-signal gain $g_0(\nu)$, the saturation intensity I_ν^{sat}, and the length l of the gain medium, plus the coefficients t and s of the output mirror. Thus a given gain medium, characterized by g_0, I_ν^{sat}, and l, can yield different laser beam intensities, depending on how the laser cavity (i.e., t and s) is chosen. We will next determine the largest possible output intensity that can be obtained from a given gain medium.

11.3 OPTIMAL OUTPUT COUPLING

It is a simple exercise in calculus to determine the transmission coefficient t_{opt} of the output mirror that optimizes the output intensity (11.2.11). We obtain (Problem 11.1)

$$t_{\text{opt}} = \sqrt{2g_0(\nu)\, ls} - s \tag{11.3.1}$$

and therefore the maximum possible output intensity is

$$[I_\nu^{\text{out}}]_{\text{max}} = I_\nu^{\text{sat}} \left[\sqrt{g_0(\nu)\, l} - \sqrt{s/2}\right]^2 \tag{11.3.2}$$

When $t = t_{\text{opt}}$, the threshold gain, and therefore the gain in steady-state oscillation, is

$$\begin{aligned}
(g_t)_{\text{opt}} &= \frac{1}{2l}(1 - r)_{\text{opt}} \\[2mm]
&= \frac{1}{2l}(t_{\text{opt}} + s) \\[2mm]
&= \frac{1}{2l}\sqrt{2g_0(\nu)\, ls} \\[2mm]
&= \sqrt{\frac{g_0(\nu)\, s}{2l}}
\end{aligned} \tag{11.3.3}$$

The small-signal gain $g_0(\nu)$ must be greater than $s/2l$ in a laser, or else the threshold condition (11.2.10) could not be satisfied. If the scattering and absorption losses are small enough that $g_0(\nu) \gg s/2l$, then from (11.3.2) we have

$$[I_\nu^{\text{out}}]_{\text{max}} \approx g_0(\nu) I_\nu^{\text{sat}}\, l \qquad [g_0(\nu) \gg s/2l] \tag{11.3.4}$$

for the maximum possible intensity of radiation at frequency ν that can be extracted from the medium. That is, if the small-signal gain is much greater than the scattering loss coefficient $s/2l$, and we design the resonator to have the optimal output coupling (11.3.1), we will extract the maximum intensity (11.3.4) at frequency ν. Since the small-signal gain is generally greatest at line center ($\nu = \nu_0$), the maximum output intensity extractable from the medium is

$$[I_{\nu_{21}}^{\text{out}}]_{\text{max}} \approx g_0(\nu_{21}) I^{\text{sat}} l \qquad (11.3.5a)$$

where $I^{\text{sat}} \equiv I_{\nu_{21}}^{\text{sat}}$ is given by (10.11.7) in the special case of a three-level laser.

The result (11.3.4) may perhaps be better appreciated by deriving it in a different way, using population rate equations. For this purpose we consider again the specific example of the ideal three-level laser, the population rate equations for which are given by (10.7.4). Since the ground level may be taken to have zero energy, the rate of change, due to stimulated emission, of the energy per unit volume stored in the atoms is

$$h\nu \left(\frac{dN_2}{dt}\right)_{\text{stimulated emission}} = -\sigma(\nu)\,(N_2 - N_1)\,I_\nu \qquad (11.3.5b)$$

in steady-state oscillation. Using Eq. (10.12.8) in the uniform-field approximation (replacing $\sin^2 kz$ by its average value $\frac{1}{2}$), we may write (11.3.5b) as

$$h\nu \left(\frac{dN_2}{dt}\right)_{\text{stimulated emission}} = -g(\nu)\,(I_\nu^{(+)} + I_\nu^{(-)}) \qquad (11.3.6)$$

This is the power per unit volume extracted from the active atoms by stimulated emission.

It follows that $g(I_\nu^{(+)} + I_\nu^{(-)})$ is the growth rate per unit volume of laser field energy in the active medium. But in steady-state oscillation this must equal the loss rate due to output coupling plus the loss rate associated with scattering and absorption processes; these are characterized by t and s, respectively. If scattering and absorption losses are small compared with output coupling, therefore, then $g(I_\nu^{(+)} + I_\nu^{(-)})$ is approximately the power per unit volume lost by the active medium in the form of *output* laser radiation. From (11.2.5),

$$g(\nu)\,(I_\nu^{(+)} + I_\nu^{(-)}) = g(\nu) I_\nu^{\text{sat}} \left(\frac{g_0(\nu)}{g_t} - 1\right)$$

$$= g_t I_\nu^{\text{sat}} \left(\frac{g_0(\nu)}{g_t} - 1\right)$$

$$= I_\nu^{\text{sat}} \left[g_0(\nu) - g_t\right] \qquad (11.3.7)$$

The maximum value of the power per unit volume that can be extracted as output laser radiation of frequency ν is therefore

$$\left(\frac{\text{power}}{\text{volume}} \to \text{laser radiation}\right)_{\text{max}} = g_0(\nu) I_\nu^{\text{sat}} \qquad (11.3.8)$$

and this is obtained when the small-signal gain g_0 is much larger than the threshold gain g_t. Thus we can interpret the maximum possible intensity (11.3.4) that can be extracted from the gain medium as simply the maximum possible power per unit volume $g_0 I^{\text{sat}}$ that can be extracted from the medium, multiplied by the length l of the medium.

The theoretical upper limit (11.3.8) to the extracted power per unit volume of the gain medium is useful because it depends only on the properties of the active atoms and the pumping process. Consider as an example the ideal three-level laser. The input power per unit volume to the gain cell in steady-state oscillation is given by Eq. (10.7.12). The theoretical upper limit of the input-to-output *power conversion efficiency* is therefore

$$e_{\text{max}} = \frac{g_0(\nu_{21}) I^{\text{sat}}}{h\nu_{31} P\overline{N}_1} \qquad (11.3.9)$$

for line-center operation. $g_0(\nu_{21})$ and I^{sat} for the three-level laser are given by (10.11.12) and (10.11.7), respectively, and some simple algebra yields

$$e_{\text{max}} = \frac{(P - \Gamma_{21}) N_T h\nu_{21}}{2h\nu_{31} P\overline{N}_1} \qquad (11.3.10)$$

Since we are considering the case of optimal output coupling and $g_0 \gg s/2l$ for the purpose of obtaining a theoretical upper limit to the power conversion efficiency, we may take $I_{\nu_{21}}^{(+)} + I_{\nu_{21}}^{(-)} \gg I^{\text{sat}}$, for from (11.2.5) and (11.3.3),

$$\frac{I_{\nu_{21}}^{(+)} + I_{\nu_{21}}^{(+)}}{I^{\text{sat}}} = \frac{g_0(\nu_{21})}{g_t} - 1$$

$$= \frac{g_0(\nu_{21})}{\sqrt{g_0(\nu_{21}) s/2l}} - 1$$

$$= \sqrt{\frac{2g_0(\nu_{21}) l}{s}} - 1 \gg 1 \qquad (11.3.11)$$

In this case the laser transition is strongly saturated, i.e., $\overline{N}_1 \approx \overline{N}_2 \approx \frac{1}{2} N_T$, and therefore

$$e_{\text{max}} \approx \frac{P - \Gamma_{21}}{P} \frac{\nu_{21}}{\nu_{31}} \qquad (11.3.12)$$

Finally we assume $P \gg \Gamma_{21}$ in order to have a large small-signal gain [Eq. (10.11.5)]. Thus

$$e_{\text{max}} \approx \frac{\nu_{21}}{\nu_{31}} = \frac{E_2 - E_1}{E_3 - E_1} \qquad (11.3.13)$$

This is the theoretical upper limit to the power conversion efficiency. It is just the ratio of the quantum of energy $h\nu_{21}$ associated with the laser transition, to the quantum of energy $h\nu_{31}$ associated with the pump transition of the three-level laser (Figure 10.6). This ratio is called the *quantum efficiency* of the three-level laser. It is a property only of the energy-level structure of the active atoms. Similarly ν_{21}/ν_{30} is the quantum efficiency of the ideal four-level laser. In the ruby and Nd:YAG lasers the pump level 3 is not a single, sharply defined level. Viewing them as approximately three- and four-level lasers, and using the numbers given in Section 10.10, we calculate that the ruby and Nd:YAG lasers have quantum efficiencies $\leq 80\%$ and $\leq 50\%$, respectively (Problem 11.2).

Needless to say, the quantum efficiency is seldom approached in real lasers. First of all, the input-to-output power conversion efficiency, of which the quantum efficiency is the theoretical upper limit, does not give the actual overall efficiency of operation of the laser. It only gives the fraction of the power *actually delivered* to the active medium that is converted to laser output power. There is no account of the efficiency with which the pump power is generated and delivered.

In a carefully designed cw ruby laser, for example, about 25% of the electric power used by the lamp is actually converted to radiation with frequencies lying within the pump bands of the chromium ion, and of course not all of this radiation is actually incident on the ruby rod. The fraction of the incident radiation actually absorbed by the ruby is about 4%, and of this only the fraction equal to the quantum efficiency may be used for lasing. All things considered, the actual operating efficiency of a cw ruby laser system is on the order of a tenth of a per cent. Although much higher efficiencies are available with other types of lasers, the point is that the quantum efficiency defined by (11.3.13) usually has little bearing on the actual operating efficiency of the complete laser system consisting of the pump, the gain cell, and the laser resonator.

11.4 EFFECT OF SPATIAL HOLE BURNING

The gain saturates according to the formula (10.12.9), i.e.,

$$g(\nu) = \frac{g_0(\nu)}{1 + (2I_\nu^{(+)}/I_\nu^{\text{sat}}) \sin^2 kz} \qquad (11.4.1)$$

where we have used Eq. (11.2.8) to write

$$I_\nu = I_\nu^{(+)} + I_\nu^{(-)} = 2I_\nu^{(+)} \qquad (11.4.2)$$

Our result (11.2.11) for the laser output intensity is based on the approximation of replacing $\sin^2 kz$ by its average value, i.e., by ignoring the spatial dependence of the gain arising from the interference of the two traveling waves. We will now consider the effect of retaining the spatial variation (11.4.1) of the gain coefficient. In other words, we will now improve upon the uniform-field approximation by including the effect of spatial hole burning.

Equation (11.3.6) was written in the uniform-field approximation. Without this approximation we arrive at the expression

$$-h\nu \left(\frac{dN_2}{dt}\right)_{\text{stimulated emission}} = 2g(\nu)I_\nu \sin^2 kz$$

$$= \frac{2g_0(\nu)I_\nu \sin^2 kz}{1 + (2I_\nu/I_\nu^{\text{sat}})\sin^2 kz} \qquad (11.4.3)$$

This is the power (at frequency ν) per unit volume, *at the point z*, extracted from the gain medium by stimulated emission. Equation (11.3.6) follows when $\sin^2 kz$ is replaced by $\frac{1}{2}$, its average value over distances large compared with a wavelength.

The gain "clamping" condition (11.2.3) does not apply in the "exact" theory in which the gain and intensity vary with z. In other words, if g is a function of z, we can no longer say that the gain and loss coefficients at *every point* in the gain medium are equal in steady-state oscillation. It must still be true, however, that the rate at which the field gains energy equals the rate at which it loses energy. The former follows from the generalization (11.4.3) of (11.3.6):

$$\int_0^l gI\,dz = 2g_0 I \int_0^l \frac{\sin^2 kz\,dz}{1 + (2I/I^{\text{sat}})\sin^2 kz} \qquad (11.4.4)$$

The rate of field intensity loss from the cavity is just

$$(t + s)I^{(+)} = \tfrac{1}{2}(t + s)I = g_t lI \qquad (11.4.5)$$

Note that the one-way intensity $I^{(+)} = I/2$ in the direction of the output mirror is independent of z. The right-hand sides of (11.4.4) and (11.4.5) must be equal in cw oscillation, and this equality determines I. From a table of integrals we find that, for $kl \gg 1$,

$$\int_0^l \frac{\sin^2 kz\,dz}{1 + (2I/I^{\text{sat}})\sin^2 kz} \simeq \frac{lI^{\text{sat}}}{2I}\left(1 - \frac{1}{\sqrt{1 + 2I/I^{\text{sat}}}}\right) \qquad (11.4.6)$$

and therefore, from the equality of (11.4.4) and (11.4.5),

$$1 - \frac{1}{\sqrt{1 + 2I/I^{\text{sat}}}} = \frac{g_t I}{g_0 I^{\text{sat}}} \tag{11.4.7}$$

This expression can be written more simply:

$$\sqrt{x} = \frac{2g_0}{g_t} - x \tag{11.4.8}$$

where

$$x = 1 + \frac{2I}{I^{\text{sat}}} \tag{11.4.9}$$

Squaring both sides of (11.4.8), we obtain a quadratic equation for x, with the two solutions

$$x = 1 + \frac{2I}{I^{\text{sat}}} = \frac{2g_0}{g_t} + \frac{1}{2} \pm \sqrt{\frac{2g_0}{g_t} + \frac{1}{4}} \tag{11.4.10}$$

Since x should be equal to 1 ($I = 0$) when $g_0/g_t = 1$, the desired solution is the one with the minus sign in the last term on the right:

$$1 + \frac{2I}{I^{\text{sat}}} = \frac{2g_0}{g_t} + \frac{1}{2} - \sqrt{\frac{2g_0}{g_t} + \frac{1}{4}}$$

or

$$I = I^{\text{sat}} \left(\frac{g_0}{g_t} - \frac{1}{4} - \sqrt{\frac{g_0}{2g_t} + \frac{1}{16}} \right) \tag{11.4.11}$$

The output intensity is $I_\nu^{\text{out}} = t I_\nu^{(+)} = (t/2) I_\nu$, exactly as in the uniform-field approximation in which spatial hole burning is not included. This is because I_ν^{out} is determined directly by the *one-way* intensity $I_\nu^{(+)}$ (Figure 11.1), and there are no interference terms to worry about. Thus

$$I_\nu^{\text{out}} = \frac{t}{2} I_\nu^{\text{sat}} \left(\frac{g_0(\nu)}{g_t} - \frac{1}{4} - \sqrt{\frac{g_0(\nu)}{2g_t} + \frac{1}{16}} \right) \tag{11.4.12}$$

which is different from the result (11.2.11) obtained when spatial hole burning is neglected.

Figure 11.2 shows the curve of output intensity vs. output coupling predicted by (11.2.11) for the example $g_0 l = 0.10$ and $s = 0.034$. Also shown is the curve predicted by the formula (11.4.12). The two predictions are seen to differ significantly, typically by about 30%. The effect of spatial hole burning is to *reduce* the output intensity.

11.5 LARGE OUTPUT COUPLING

Our analysis of output power thus far has assumed that the output coupling is small and that the two traveling waves have equal intensities, $I^{(+)} = I^{(-)}$. We have also assumed that the time-averaged intensities $I^{(+)}$ and $I^{(-)}$ are independent of the axial coordinate z. We will now allow arbitrary output coupling and therefore allow the possibility that $I^{(+)}$ and $I^{(-)}$ may vary with z. We assume, however, that the variation of interest is much more gradual than the $\sin^2 kz$ variation due to spatial hole burning, and replace $\sin^2 kz$ by its average value $\frac{1}{2}$.

Thus we work with the gain-saturation formula (11.2.4), which we now write in the form

$$g(z) = \frac{g_0}{1 + [I^{(+)}(z) + I^{(-)}(z)]/I^{\text{sat}}} \qquad (11.5.1)$$

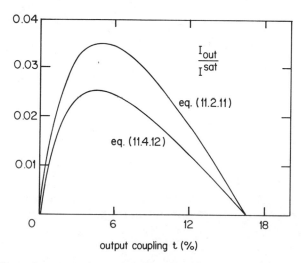

Figure 11.2 Effect of spatial hole burning on output intensity, assuming $g_0 l = 0.10$ and $s = 0.034$.

For notational simplicity we have suppressed the ν dependence of the various terms in this equation, but we indicate explicitly the z dependence. In principle, g_0 could also depend on z, for example if the pumping rate P depended on z, but here we assume that it does not. In steady-state oscillation the intensities $I^{(+)}$ and $I^{(-)}$ both satisfy Eq. (10.4.3):

$$\frac{dI^{(+)}}{dz} = g(z)\, I^{(+)}(z) \tag{11.5.2a}$$

$$\frac{dI^{(-)}}{dz} = -g(z)\, I^{(-)}(z) \tag{11.5.2b}$$

We will assume that all cavity loss processes (output coupling, scattering, absorption) occur at the mirrors. Thus we will not include terms accounting for "distributed" loss within the cavity.

Equations (11.5.2) imply that

$$\frac{d}{dz}\left[I^{(+)}I^{(-)}\right] = I^{(+)}\frac{dI^{(-)}}{dz} + I^{(-)}\frac{dI^{(+)}}{dz} = 0 \tag{11.5.3}$$

or

$$I^{(+)}(z)\, I^{(-)}(z) = \text{constant} = C \tag{11.5.4}$$

where the constant C can be evaluated at either mirror:

$$I^{(+)}(0)\, I^{(-)}(0) = I^{(+)}(L)\, I^{(-)}(L) = C \tag{11.5.5}$$

Let us now make use of (11.5.4) and (11.5.1) in (11.5.2):

$$\frac{1}{I^{(+)}}\frac{dI^{(+)}}{dz} = \frac{g_0}{1 + \dfrac{I^{(+)} + C/I^{(+)}}{I^{\text{sat}}}} \tag{11.5.6a}$$

and

$$\frac{1}{I^{(-)}}\frac{dI^{(-)}}{dz} = \frac{-g_0}{1 + \dfrac{I^{(-)} + C/I^{(-)}}{I^{\text{sat}}}} \tag{11.5.6b}$$

In this form the equations for $I^{(+)}(z)$ and $I^{(-)}(z)$ are uncoupled and can be solved separately.

Consider first the equation for $I^{(+)}(z)$. Writing (11.5.6a) in the form

$$
\begin{aligned}
g_0 \, dz &= \frac{1}{I^{(+)}} \left[1 + \frac{1}{I^{\mathrm{sat}}} \left(I^{(+)} + \frac{C}{I^{(+)}} \right) \right] dI^{(+)} \\
&= \frac{dI^{(+)}}{I^{(+)}} + \frac{1}{I^{\mathrm{sat}}} \, dI^{(+)} + \frac{C}{I^{\mathrm{sat}}} \frac{dI^{(+)}}{I^{(+)2}}
\end{aligned}
\tag{11.5.7}
$$

and integrating from $z = 0$ to $z = L$ (Figure 10.4), we have

$$
\begin{aligned}
\int_0^L g_0 \, dz &= \int_{I^{(+)}(0)}^{I^{(+)}(L)} \frac{dI^{(+)}}{I^{(+)}} + \frac{1}{I^{\mathrm{sat}}} \int_{I^{(+)}(0)}^{I^{(+)}(L)} dI^{(+)} \\
&\quad + \frac{C}{I^{\mathrm{sat}}} \int_{I^{(+)}(0)}^{I^{(+)}(L)} \frac{dI^{(+)}}{I^{(+)2}}
\end{aligned}
$$

or

$$
\begin{aligned}
g_0 l &= \ln \frac{I^{(+)}(L)}{I^{(+)}(0)} + \frac{1}{I^{\mathrm{sat}}} \left[I^{(+)}(L) - I^{(+)}(0) \right] \\
&\quad - \frac{C}{I^{\mathrm{sat}}} \left(\frac{1}{I^{(+)}(L)} - \frac{1}{I^{(+)}(0)} \right)
\end{aligned}
\tag{11.5.8a}
$$

Considering the equation (11.5.6b), we obtain similarly

$$
\begin{aligned}
g_0 l &= \ln \frac{I^{(-)}(0)}{I^{(-)}(L)} + \frac{1}{I^{\mathrm{sat}}} \left[I^{(-)}(0) - I^{(-)}(L) \right] \\
&\quad - \frac{C}{I^{\mathrm{sat}}} \left(\frac{1}{I^{(-)}(0)} - \frac{1}{I^{(-)}(L)} \right)
\end{aligned}
\tag{11.5.8b}
$$

Since the small-signal gain g_0 is assumed to be constant inside the gain cell of length l (Figure 10.4), but vanishes outside the gain cell, we have used, in both of Eqs. (11.5.8),

$$
\int_0^L g_0 \, dz = g_0 l
\tag{11.5.9}
$$

The boundary conditions (10.4.2), together with (11.5.4), may be used to express $I^{(-)}(0)$, $I^{(-)}(L)$, and $I^{(+)}(0)$ in terms of $I^{(+)}(L)$. Thus

$$
I^{(-)}(L) = \frac{C}{I^{(+)}(L)} = r_2 I^{(+)}(L)
\tag{11.5.10}
$$

so that

$$C = r_2 [I^{(+)}(L)]^2 \tag{11.5.11}$$

Furthermore

$$I^{(+)}(0) = \frac{C}{I^{(-)}(0)} = r_1 I^{(-)}(0) \tag{11.5.12}$$

which means we may also write

$$I^{(-)}(0) = \sqrt{r_2/r_1} \, I^{(+)}(L) \tag{11.5.13}$$

and from (11.5.12),

$$I^{(+)}(0) = \sqrt{r_1 r_2} \, I^{(+)}(L) \tag{11.5.14}$$

We can use (11.5.11), (11.5.12), and (11.5.14) to express the right sides of (11.5.8) in terms of $I^{(+)}(L)$. From (11.5.8a),

$$g_0 l = \ln \frac{1}{\sqrt{r_1 r_2}} + \frac{I^{(+)}(L)}{I^{sat}} (1 - \sqrt{r_1 r_2})$$

$$- \frac{I^{(+)}(L)}{I^{sat}} \left(r_2 - \sqrt{\frac{r_2}{r_1}} \right) \tag{11.5.15a}$$

while from (11.5.8b),

$$g_0 l = \ln \frac{1}{\sqrt{r_1 r_2}} + \frac{I^{(+)}(L)}{I^{sat}} \left(\sqrt{\frac{r_2}{r_1}} - r_2 \right)$$

$$- \frac{I^{(+)}(L)}{I^{sat}} (\sqrt{r_1 r_2} - 1) \tag{11.5.15b}$$

These equations are identical. We now solve for $I^{(+)}(L)$:

$$I^{(+)}(L) = \frac{I^{sat}(g_0 l + \ln \sqrt{r_1 r_2})}{1 + \sqrt{r_2/r_1} - r_2 - \sqrt{r_1 r_2}}$$

$$= \frac{\sqrt{r_1} \, I^{sat}}{(\sqrt{r_1} + \sqrt{r_2})(1 - \sqrt{r_1 r_2})} (g_0 l + \ln \sqrt{r_1 r_2}) \tag{11.5.16}$$

Finally we can use (11.5.13) and (11.5.16) to obtain the output intensity:

$$I_{\text{out}} = t_1 I^{(-)}(0) + t_2 I^{(+)}(L)$$

$$= I^{\text{sat}} \left(t_2 + \sqrt{\frac{r_2}{r_1}}\, t_1 \right) \frac{\sqrt{r_1}}{(\sqrt{r_1} + \sqrt{r_2})(1 - \sqrt{r_1 r_2})}$$

$$\times (g_0 l + \ln \sqrt{r_1 r_2}) \qquad (11.5.17)$$

These results generalize our previous ones in that they apply to arbitrary values of output coupling. The principal assumptions in our derivation of (11.5.17) have been that (1) the gain medium is homogeneously broadened; (2) the gain saturates according to the formula (11.5.1); (3) the small-signal gain g_0 and the saturation intensity I^{sat} are constant throughout the gain medium; (4) spatial variations of the cavity intensity transverse to the resonator axis can be neglected as a first approximation; (5) loss occurs only at the mirrors; and (6) spatial hole burning is averaged.

The analysis leading to (11.2.11) assumed all these things, and also that the output coupling (or other losses) is small. The analysis just given should therefore reproduce (11.2.11) in the limit of high mirror reflectivities.

To see that this is so, suppose that one of the mirrors is perfectly reflecting ($r_1 = 1$, $t_1 = s_1 = 0$). Then (11.5.17) becomes

$$I_{\text{out}} = t\, I^{\text{sat}} \frac{1}{(1 + \sqrt{r})(1 - \sqrt{r})}(g_0 l + \ln \sqrt{r})$$

$$= \frac{t}{1 - r} I^{\text{sat}} \left(g_0 l - \tfrac{1}{2} \ln \frac{1}{r} \right)$$

$$= \frac{t}{t + s} I^{\text{sat}} \left[g_0 l - \tfrac{1}{2} \ln (1 - t - s)^{-1} \right]$$

$$= \frac{(\tfrac{1}{2} t) \ln (1 - t - s)^{-1}}{t + s} I^{\text{sat}} \left(\frac{2 g_0 l}{\ln (1 - t - s)^{-1}} - 1 \right) \qquad (11.5.18)$$

where $r = r_2$, $t = t_2$, $s = s_2$. This is a generalization of (11.2.11) (Problem 11.3). When $t + s \ll 1$ we have

$$I^{\text{out}} \approx \tfrac{1}{2} t\, I^{\text{sat}} \left(\frac{2 g_0 l}{t + s} - 1 \right) \qquad (11.5.19)$$

which is just (11.2.11). In going from (11.5.18) to (11.5.19) we have used the approximation

$$-\ln (1 - t - s) \approx t + s \qquad (t + s \ll 1) \qquad (11.5.20)$$

One interesting result of our analysis is that the total two-way intensity is relatively uniform in the z direction, even for moderately large output couplings. A simple way to see this is to use the boundary conditions (11.5.10), (11.5.13), and (11.5.14) to calculate the ratio of the total intensity at the output mirror to that at the other mirror, assumed again to be perfectly reflecting. With $r_1 = 1$ and $r_2 = r$, we obtain

$$\frac{I^{(+)}(L) + I^{(-)}(L)}{I^{(+)}(0) + I^{(-)}(0)} = \frac{1 + r}{2 \sqrt{r}} \tag{11.5.21}$$

This is plotted in Figure 11.3. It is seen that the total intensities are comparable at the two mirrors for reflectivities as low as 50%. This conclusion is consistent with a more detailed analysis in which $I^{(+)}(z)$ and $I^{(-)}(z)$ are calculated for $0 \leq z \leq L$; Eqs. (11.5.6) are the starting points for such an analysis. We can conclude that for most (but by no means all) lasers, I^+ and I^- vary only mildly with z. This justifies the approximate theory of Section 11.4.

These results for arbitrarily large output couplings were first obtained by W. W. Rigrod.[1] The Rigrod analysis predicts an optimal output coupling that reduces to (11.3.1) when $t + s \ll 1$. We will not bother to show this. The reader is referred to Rigrod's paper for graphs of optimal output coupling and output intensity as a function of small-signal gain g_0 and scattering loss coefficient s. Spatial hole burning in the case of arbitrary output couplings has subsequently been included in numerical computations. The effect is to reduce the output intensity calculated without spatial hole burning by about the same relative magnitude as in the case of small output coupling (cf. Figure 11.2).

1. W. W. Rigrod, *Journal of Applied Physics* **36**, 27 (1965).

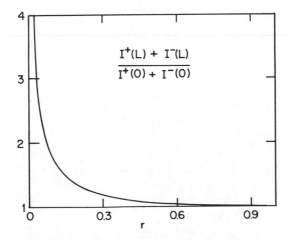

Figure 11.3 Ratio of total intensities at the two mirrors [Eq. (11.5.21)].

11.6 MEASURING SMALL-SIGNAL GAIN AND OPTIMAL OUTPUT COUPLING

Equation (11.2.11) for output intensity, or its generalization (11.5.18), has been shown experimentally to be quite accurate, and it is used extensively in laser design. This is so despite the fact that our formulas for laser output intensity were derived without taking atomic motion into account. Atomic motion tends to smear out the effect of spatial hole burning: an atom at a field nodal point does not stay there forever, as assumed in our simple analysis. The result of rapid atomic motion will be to average the field spatial variations, and so spatial hole burning is usually assumed to be negligible in gas lasers.

Figure 11.4 shows experimental results for the output power of a He–Ne laser as a function of mirror transmission. The three curves drawn through the experimental points are based on (11.2.11), and are seen to fit the data very nicely.

In general the small-signal gain g_0 and the saturation intensity I^{sat} are difficult to calculate accurately, because the pumping and decay rates of the relevant atomic (molecular) levels may not be well known. One way to measure g_0 is the *maximal-loss method*. In this method the cavity loss is increased until the laser oscillation ceases. Since laser oscillation requires $g_0 > g_t$, and the cavity loss determines g_t, the maximal loss allowing laser oscillation is in fact just g_0. As illustrated in Figure 11.5, the cavity loss may be varied by inserting a reflecting knife-edge into the

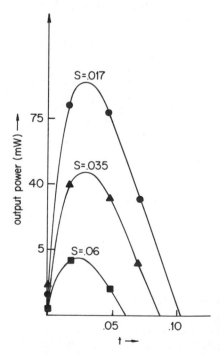

Figure 11.4 Experimental data on the output power of a 6328-Å He–Ne laser. The solid curves are based on Eq. (11.2.11) for output intensity vs. output coupling. The three curves correspond to $s = 0.06$, 0.035, and 0.017. [P. Laures, *Physics Letters* **10**, 61 (1964).]

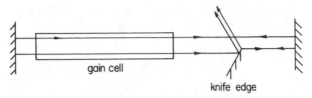

Figure 11.5 A knife-edge may be used to determine the small-signal gain by the maximal-loss method. By a micrometer adjustment the knife-edge is made to occlude an increasing fraction of the intracavity beam, until lasing ceases because the total loss exceeds the gain. This determines the maximal loss, and therefore the small-signal gain for laser oscillation.

cavity. A micrometer adjustment determines the fraction of the intracavity intensity that is occluded by the knife-edge, and thus the cavity loss is varied by turning the micrometer screw.

A variant of this method[2] may sometimes be used to determine not only the small-signal gain, but also the optimal output coupling t_{opt} and the output power obtainable with $t = t_{opt}$. This method may be understood with reference to Figure 11.6. The knife-edge may be a small cube that has been coated on two adjacent faces with a reflecting material. The knife-edge deflects part of the intracavity field out of the cavity and onto a power meter D; the amount of power deflected to D, P_{in}, may be varied by a micrometer adjustment. M_F is an extracavity folding mirror that directs the output power P_{out} onto the same detector D.

With the knife-edge inserted in the cavity there is a scattering coefficient

$$s = P_{in}/P^+ \tag{11.6.1}$$

so that the effective output coupling is $t + s$, where t is the transmission coefficient

2. T. F. Johnston, *IEEE Journal of Quantum Electronics* **12**, 310 (1976).

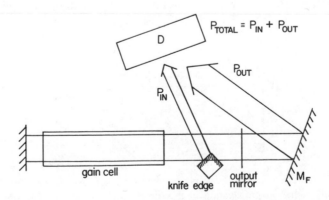

Figure 11.6 An experimental setup for determining the small-signal gain, the optimal output coupling, and the maximum possible output power.

of the output mirror M_0; here $P^+ = I^+A$ is the power associated with the wave traveling toward M_0. Since $t = P_{out}/P^+$, we have

$$s = \frac{P_{in}}{P_{out}} t \qquad (11.6.2)$$

We are assuming that the knife-edge represents a small perturbation, so that I^+ is the intensity at both M_0 and the knife-edge. Now the sum

$$P_{total} = P_{in} + P_{out} \qquad (11.6.3)$$

represents the total output power of the laser. The micrometer setting can be varied until a value s_{opt} is obtained for which P_{total} is a maximum. This gives the optimal output coupling as (if we assume $t < t_{opt}$)

$$t_{opt} = s_{opt} + t \qquad (11.6.4)$$

Furthermore the output power at this optimal output coupling is just the maximum value obtained for P_{total}. The small-signal gain is determined by the value of s at which laser oscillation stops. In practice this must be determined by extrapolation, since both P_{in} and P_{out} go to zero as the laser threshold is approached, and (11.6.2) becomes indeterminate.

In lasers with liquid or solid gain media the spatial hole burning is not smoothed out by atomic motion to the same extent as in gas lasers, although other processes may tend to weaken its effect. Convincing evidence for spatial hole burning has been obtained with liquid dye lasers. The output power of a single-mode dye laser is found to increase significantly when operated as a ring laser instead of the two-mirror standing-wave configuration. In the ideal traveling-wave ring laser, sketched in Figure 11.7, there is no standing-wave interference pattern and therefore no spatial hole burning.

We will see in the following chapter that spatial hole burning plays a crucial role in the multimode behavior of many lasers.

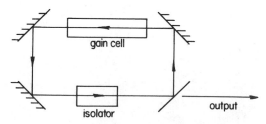

Figure 11.7 A unidirectional (traveling-wave) ring laser. An optical "isolator," which transmits only radiation propagating in the direction shown, is used to obtain traveling-wave rather than standing-wave laser oscillation. Such an isolator might use polarizers and a material that rotates the polarization of a light wave (e.g., via the Faraday effect).

11.7 INHOMOGENEOUSLY BROADENED LASER MEDIA

In an inhomogeneously broadened gain medium the different active atoms have different central transition frequencies ν_{21}. This may be due to their different velocities and the Doppler effect (recall Section 3.11), spatially nonuniform electric and magnetic fields,[3] the presence of different isotopes having slightly different energy levels, or a host of other effects.

In the case of inhomogeneous broadening the theory of laser oscillation can become enormously complex. For one thing, it becomes much more difficult to justify our assumption of single-mode oscillation, as we will see in Section 11.12. Here we will simply restrict ourselves to a few remarks.

Just as in the case of homogeneous broadening, it is possible to define a small-signal gain and a saturation intensity for an inhomogeneously broadened gain medium. The expressions we obtained for the gain coefficient in Chapter 10 are, of course, always applicable; the small-signal gain is simply the gain when the field intensity is so small that it does not affect the population difference $N_2 - N_1$. In the case of Doppler broadening, for example, the small-signal gain is given by

$$
\begin{aligned}
g_0(\nu) &= \frac{\lambda^2 A}{8\pi}(\Delta N)_0 S(\nu) \\
&= \frac{\lambda^2 A}{8\pi}(\Delta N)_0 \frac{1}{\delta\nu_D}\left(\frac{4\ln 2}{\pi}\right)^{1/2} \exp\left[-4(\nu - \nu_{21})^2 (\ln 2)/\delta\nu_D^2\right]
\end{aligned}
$$

$$(11.7.1)$$

where $(\Delta N)_0 = (N_2 - N_1)_0$, the small-signal (unsaturated) population difference, and where we have used Eq. (3.11.10) to write the Doppler lineshape function. Thus

$$
g_0(\nu) = g_0(\nu_{21}) \exp\left[-4(\nu - \nu_{21})^2 (\ln 2)/\delta\nu_D^2\right] \qquad (11.7.2)
$$

which replaces the formula (10.11.11) in the small-signal limit when the gain profile has the Doppler lineshape.

Real complications arise when we consider the saturation characteristics of an inhomogeneously broadened gain medium. Atoms with central transition frequency ν_{21} will be saturated according to (10.11.11), but there is a *distribution* of resonance frequencies ν_{21}. The gain coefficient is obtained by integrating the contributions from the different frequency components, each of which saturates to a different degree depending on its detuning from the cavity mode frequency ν. For

---

3. The energy levels of atoms and molecules may be shifted by electric or magnetic fields. The energy-level shifts produced by electric and magnetic fields are called the Stark and Zeeman effects, respectively. The relative shift is usually very small.

many inhomogeneously broadened laser media (in particular, low-pressure gas lasers) this integration results in a gain-saturation formula of the form

$$g(\nu) = \frac{g_0(\nu)}{\sqrt{1 + I/I_{\text{sat}}}} \tag{11.7.3}$$

if spatial hole burning and power broadening are ignored. Thus Eq. (11.2.9) is replaced by

$$I_\nu^{\text{out}} = \frac{t}{2} I^{\text{sat}} \left[\left(\frac{g_0(\nu)}{g_t} \right)^2 - 1 \right] \tag{11.7.4}$$

in the case of inhomogeneous broadening (and small output coupling). However, these results apply to single-mode operation, and we will see in the next chapter that single-mode operation is seldom achieved with inhomogeneously broadened gain media.

In general the gain profile is determined by both homogeneous and inhomogeneous broadening mechanisms. In a gaseous medium, for example, there is inhomogeneous broadening due to atomic motion. The lineshape function $S(\nu)$ entering the small-signal gain formula is then the Voigt profile, exactly as in the classical theory of absorption. Thus the assumptions of pure homogeneous broadening or pure inhomogeneous broadening are, in general, *approximations* that apply when one type of broadening is dominant.

11.8 SPECTRAL HOLE BURNING AND THE LAMB DIP

We have noted that in the case of inhomogeneous broadening the gains associated with different ν_{21} frequency components, or *spectral packets*, are saturated to different degrees, depending on their detuning from the field frequency ν. Equation (10.11.11) implies that the gain for spectral packets with frequency $\nu_{21} \approx \nu$ is saturated more strongly than others; spectral packets with frequencies detuned from ν by much more than the homogeneous linewidth, i.e., $|\nu_{21} - \nu| \gg \delta\nu_{21}$, are hardly saturated at all. This is illustrated in Figure 11.8. This selective saturation

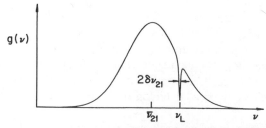

Figure 11.8 Spectral hole burning in an inhomogeneously broadened gain profile. Radiation of frequency ν_L saturates only a spectral packet of atoms with frequency $\nu \approx \nu_L$.

leads to *spectral hole burning* in the gain curve. The width of a hole is just the homogeneous linewidth $\delta\nu_{21}$, while the depth is determined by the field intensity. When the intensity is very large the hole "touches down," i.e., the gain at the center of the hole is fully saturated.

Spectral hole burning is especially interesting in the case of a purely Doppler-broadened gain medium. Suppose the cavity mode frequency ν is different from the center frequency of the Doppler gain profile. Consider, for example, the traveling-wave field propagating in the positive z direction. This wave will strongly saturate the spectral packet of atoms with Doppler-shifted frequencies $\nu'_{21} = \nu$; the Doppler effect has brought these atoms into resonance with the wave. Therefore those atoms have the z component of velocity given by (cf. Section 3.11)

$$\nu = \bar{\nu}\left(1 + \frac{v}{c}\right)$$

or

$$\frac{v}{c} = \frac{\nu - \bar{\nu}}{\bar{\nu}} \tag{11.8.1}$$

where $\bar{\nu}$ is the resonance frequency of a stationary atom. If $\nu > \bar{\nu}$ then v is positive. This means that the atoms that have been Doppler-shifted into resonance are moving in the positive z direction, the same direction in which the traveling wave is propagating. If $\nu < \bar{\nu}$, on the other hand, v/c is negative, or in other words the atom must be moving in the negative z direction, opposite to the traveling wave, in order to be shifted into resonance.

In the same way the traveling wave propagating to the *left* will strongly saturate those atoms with the z component of velocity given by

$$\frac{v}{c} = -\frac{\nu - \bar{\nu}}{\bar{\nu}} \tag{11.8.2}$$

Therefore the *standing wave* cavity fields will burn *two* holes in the Doppler line profile, as shown in Figure 11.9.

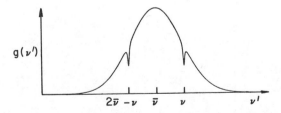

Figure 11.9 A standing-wave field burns two holes in the gain curve. These are centered at velocities $v = \pm c(\nu - \bar{\nu})/\bar{\nu}$, where $\bar{\nu}$ is the Bohr frequency of a stationary atom. This results in the burning of two holes in the Doppler-broadened gain profile.

If the laser is operating with its cavity mode frequency detuned from the resonance frequency $\bar{\nu}$ of a stationary active atom (i.e., from the center of the Doppler line), it will burn a hole on either side of line center (Figure 11.9). In other words, the cavity mode is "feeding" off two spectral packets. When the mode frequency is exactly at the center of the Doppler line, however, the two holes merge together, because the field can now strongly saturate only those atoms having no z component of velocity. In this case the cavity mode is feeding off only a single spectral packet, and we would thus expect to find a lower output power exactly at resonance than slightly off resonance. This dip in output power at line center was predicted in 1963, and is called the *Lamb dip*.

Figure 11.10 shows very early experimental results confirming the prediction of the Lamb dip. The data also show, as expected, that the dip becomes more pronounced at higher power levels, where the degree of selective saturation (hole burning) is greatest.

The observation of the Lamb dip requires that the cavity mode frequency be swept across the gain profile. Since the cavity mode frequencies are given (approximately) by $\bar{\nu} = mc/2L$, this may be done by slowly varying the cavity length L, for example by mounting one of the cavity mirrors on a piezoelectric crystal, which expands or contracts in certain directions when an electric field is applied.

11.9 CAVITY FREQUENCY AND FREQUENCY PULLING

We have mostly ignored the effect of the refractive index of the gain medium on laser oscillation, except insofar as it enters into the equations of Table 10.1 for gain and threshold. However, it turns out that the refractive index of the gain

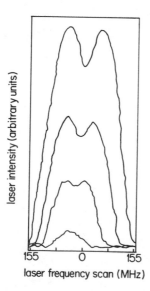

Figure 11.10 Experimental results on the Lamb dip in the output power of a 1.15-μm He–Ne laser. From A. Szöke and A. Javan, *Physical Review Letters* **10**, 512 (1963).

medium actually determines to some extent the laser oscillation frequency. We will now examine how this occurs.

A laser will oscillate at a frequency ν such that the optical length of the cavity is an integral number of half wavelengths. That is, $L = m\lambda/2$, or

$$\nu = \frac{mc}{2L} = \nu_m.$$

(11.9.1)

This applies to the *bare-cavity* case in which the gain and refractive index of the active medium are not taken into account.

In general, however, the effective *optical* length of a medium is not just its physical length, but rather the product of its physical length and its refractive index $n(\nu)$. To account for the index of refraction of the active medium, therefore, we divide the cavity length into two parts $L = l + (L - l)$, where l is the length of the gain cell and the remainder is empty cavity, as in Figure 10.4. The *optical* length of the gain cell is $n(\nu)l$. Thus (11.9.1) should be replaced by

$$\nu = \frac{mc/2}{n(\nu)l + (L - l)}$$

(11.9.2a)

or

$$\frac{l}{L}[n(\nu) - 1]\nu = \nu_m - \nu$$

(11.9.2b)

where $\nu_m = mc/2L$ is a bare-cavity mode frequency. Equation (11.9.2) shows that *the laser oscillation frequency ν will be different from a bare-cavity mode frequency ν_m if $n(\nu) \neq 1$.*

Equations (11.9.2) determining the laser oscillation frequency may be written in a different form. To do this, let us assume that $n(\nu)$ is determined primarily by the single nearly resonant, lasing atomic transition. In other words, we will assume that $n(\nu)$ is essentially the resonant (or "anomalous") refractive index (Section 3.4). Since other transitions contributing to the refractive index will usually be off resonance by many transition linewidths, this will often be an excellent approximation.

In the case of an absorbing medium the resonant refractive index is simply related to the absorption coefficient [recall (3.4.11) and Problem 3.6]. The same relation applies to an amplifying (gain) medium, simply by replacing the absorption coefficient by the negative of the gain coefficient. Thus, for a homogeneously broadened gain medium, we have

$$n(\nu) - 1 = -\frac{\lambda_{21}}{4\pi} \frac{\nu_{21} - \nu}{\delta\nu_{21}} g(\nu)$$

(11.9.3)

where as usual $\delta\nu_{21}$ is the homogeneous linewidth (HWHM). From (11.9.2) and (11.9.3), therefore, we obtain, using the fact that $\nu \approx c/\lambda_{21}$,

$$\nu = \frac{\nu_{21}[cg(\nu)l/4\pi L] + \nu_m\delta\nu_{21}}{[cg(\nu)l/4\pi L] + \delta\nu_{21}} \tag{11.9.4}$$

The quantity

$$\delta\nu_c = \frac{cg(\nu)l}{4\pi L} \tag{11.9.5}$$

is related to the mirror reflectivities through the gain-clamping condition (11.2.3), and has the dimensions of frequency. It is called the *cavity bandwidth*. Thus Eq. (11.9.4), the equation for the laser oscillation frequency ν, is usually written in the form

$$\nu = \frac{\nu_{21}\delta\nu_c + \nu_m\delta\nu_{21}}{\delta\nu_c + \delta\nu_{21}} \tag{11.9.6a}$$

or alternatively

$$\delta\nu_c(\nu - \nu_{21}) = \delta\nu_{21}(\nu_m - \nu) \tag{11.9.6b}$$

Note that the second of these two expressions establishes that the lasing frequency ν lies *between* ν_{21} and ν_m, no matter which of them is larger. The actual frequency of laser radiation is therefore "pulled" toward the center of the gain profile and away from the bare-cavity frequency. This effect is called *frequency pulling* and was already mentioned in Sections 3.5 and 8.5.

We will discuss the cavity bandwidth in some detail at the end of this section. For now it suffices to mention that $\delta\nu_{21} \gg \delta\nu_c$ in most lasers. In most lasers, therefore,

$$\nu = \frac{\nu_{21}\delta\nu_c/\delta\nu_{21} + \nu_m}{1 + \delta\nu_c/\delta\nu_{21}} \approx \left(\nu_{21}\frac{\delta\nu_c}{\delta\nu_{21}} + \nu_m\right)\left(1 - \frac{\delta\nu_c}{\delta\nu_{21}}\right)$$

$$\approx \nu_m + (\nu_{21} - \nu_m)\frac{\delta\nu_c}{\delta\nu_{21}} \quad \text{(homogeneous broadening)} \tag{11.9.7a}$$

Now the starting point of our analysis leading to (11.9.7a) was the relation (11.9.3) between the resonant refractive index and the gain. Although we have assumed a homogeneously broadened line, it is in fact *always* possible to relate the index and the gain (or absorption). (See Problem 3.7.) In the case of a Doppler-broadened medium, for example, we are led by analogous manipulations (for $|\nu_{21} - \nu| \ll \delta\nu_D$) to the formula

$$\nu \approx \nu_m + (\nu_{21} - \nu_m) \frac{\delta \nu_c}{\delta \nu_D} \sqrt{4 \ln 2/\pi}$$

$$\approx \nu_m + 1.88 (\nu_{21} - \nu_m) \frac{\delta \nu_c}{\delta \nu_D} \qquad \text{(inhomogeneous broadening)} \qquad (11.9.7b)$$

If we use the gain-clamping condition (11.2.3), then (11.9.5) gives

$$\delta \nu_c = \frac{cl}{4\pi L} g_t \tag{11.9.8}$$

and therefore

$$\frac{\delta \nu_c}{\delta \nu_{21}} = \frac{cl}{4\pi L} \frac{g_t}{\delta \nu_{21}} \tag{11.9.9a}$$

$$\frac{\delta \nu_c}{\delta \nu_D} = \frac{cl}{4\pi L} \frac{g_t}{\delta \nu_D} \tag{11.9.9b}$$

The larger the threshold gain g_t, the greater the frequency pulling for a fixed gain linewidth ($\delta \nu_{21}$ or $\delta \nu_D$). Since the gain of the medium must exceed g_t for laser action to occur, these results indicate that frequency pulling will be more readily observed for high-gain media with narrow gain profiles. This prediction is in fact borne out experimentally.

For a 6328-Å He–Ne laser with a threshold gain of 0.001 cm^{-1} and a Doppler width of 1500 MHz, for example, we have

$$\frac{\delta \nu_c}{\delta \nu_D} \approx 0.0016 \qquad (l \approx L) \tag{11.9.10}$$

The 3.39-μm He–Ne laser, on the other hand, has a smaller Doppler width of about 280 MHz owing to the larger value of the transition wavelength. Furthermore, the gain at the 3.39-μm transition is typically much larger than at the 6328-Å transition, so that lasing can be achieved with $g_t \approx 0.03$ cm^{-1}. In this case

$$\frac{\delta \nu_c}{\delta \nu_D} \approx 0.26 \qquad (l \approx L) \tag{11.9.11}$$

Frequency pulling is therefore readily observed with a 3.39-μm He–Ne laser.

Frequency pulling is especially pronounced in the low-pressure He–Xe laser operating on the 3.51-μm transition of Xe. In this case the Doppler width is only about 100 MHz owing to the relatively large mass of Xe. Furthermore the gain may be as high as 1 cm^{-1}. With $g_t = 0.5$ cm^{-1} we obtain for this laser the ratio

$$\frac{\delta\nu_c}{\delta\nu_D} \approx 12 \quad (l \approx L) \tag{11.9.12}$$

In this case the frequency pulling is so pronounced that the approximation (11.9.7) is inapplicable.

We can write (11.9.6) as

$$\nu^{(m)} = \frac{\nu_{21}\,\delta\nu_c + \nu_m\,\delta\nu_{21}}{\delta\nu_c + \delta\nu_{21}} \tag{11.9.13}$$

where $\nu^{(m)}$ denotes the "frequency-pulled" value of the bare-cavity mode frequency ν_m. The bare-cavity mode spacing $\nu_{m+1} - \nu_m = c/2L$ given in (11.9.1) is therefore "renormalized" by frequency pulling to the value

$$\nu^{(m+1)} - \nu^{(m)} = \frac{(\nu_{m+1} - \nu_m)\,\delta\nu_{21}}{\delta\nu_c + \delta\nu_{21}}$$

$$= \frac{c}{2L}\,\frac{1}{1 + \delta\nu_c/\delta\nu_{21}} \tag{11.9.14}$$

An analogous expression applies for an inhomogenously broadened gain medium.

If a laser is oscillating on several modes, therefore, the effect of frequency pulling is to reduce the mode spacing from $c/2L$ to the value (11.9.14), or the analogous expression for an inhomogeneous line. This effect of frequency pulling can be observed by looking at the beat (difference) frequency of the laser output with a sufficiently fast photodetector. In this way a reduction on the order of 2 has been observed in the mode spacing of a He–Xe laser.

When spectral hole burning is present in the case of inhomogeneous broadening, the analysis of frequency pulling becomes rather complicated, and our simple theory gives only a crude approximation to the actual situation. Since frequency pulling in many lasers is only a small effect, we will not discuss the complications due to hole burning.

It is not difficult to see why $\delta\nu_c$ is called the cavity bandwidth. According to Eq. (10.5.8), in the absence of an active medium in the cavity the cavity intensity satisfies the rate equation

$$\frac{dI_\nu}{dt} = -\frac{c}{2L}\,(1 - r_1 r_2)\,I_\nu$$

$$= -\left(c\frac{l}{L}\,g_t\right) I_\nu \tag{11.9.15}$$

We have used Eq. (10.5.11) for the threshold gain. In a bare cavity, therefore, the cavity intensity decays exponentially:

$$I_\nu(t) = I_\nu(0) \exp\left[-c(l/L) \, g_t \, t\right] \tag{11.9.16}$$

Since the intensity is proportional to the square of the electric field amplitude E_ν, we may write

$$E_\nu(t) = E_\nu(0) \exp\left[-c(l/2L) \, g_t \, t\right] e^{-2\pi i\nu t} \tag{11.9.17}$$

for the decay of the electric field in a bare cavity.

Now the fact that the field decays in time implies that it cannot be truly monochromatic. In fact the frequency spectrum associated with the time-dependent field (11.9.17) is Lorentzian:

$$s(\nu') = \frac{\delta\nu_c/\pi}{(\nu' - \nu)^2 + (\delta\nu_c)^2} \tag{11.9.18}$$

where

$$\delta\nu_c = \frac{1}{2\pi} \frac{clg_t}{2L} \tag{11.9.19}$$

which of course is the same as (11.9.5) when the gain-clamping condition $g(\nu) = g_t$ is used. The step from (11.9.17) to (11.9.18) is a standard result of Fourier-transform theory: the frequency spectrum of a quantity that decays exponentially in time is a Lorentzian function of frequency. We have already seen an example of this in Chapter 3, where we found that the exponential decay of the atomic dipole moments due to collisions led to a Lorentzian lineshape function.

Note that $\delta\nu_c$ depends only on properties of the bare cavity:

$$\delta\nu_c = \frac{1}{2\pi} \frac{cl}{2L} \left[\frac{1}{2l} \ln\left(\frac{1}{r_1 r_2}\right) \right]$$

$$= \frac{1}{4\pi} \left[\frac{c}{2L} \ln\left(\frac{1}{r_1 r_2}\right) \right] \tag{11.9.20}$$

Frequently a laser cavity is characterized by the dimensionless *quality factor*

$$Q = \frac{1}{2} \frac{\nu}{\delta\nu_c} \tag{11.9.21}$$

A high-Q cavity is one with low loss, whereas a low-Q cavity has a high power loss rate. These results are easily generalized to include loss effects other than output coupling.

11.10 THE LASER BANDWIDTH

The bandwidth of an operating laser is in general different from the bare-cavity bandwidth. In steady-state laser oscillation the decay of the field amplitude described by (11.9.17) is exactly balanced by amplification due to stimulated emission. That is, the field amplitude does not actually decay. One might therefore expect that laser radiation should have a very narrow spectral width, and no natural minimum value for the width is readily apparent. Nevertheless, laser radiation, even from a cw single-mode laser, can never be made perfectly monochromatic.

The fundamental reason for this is spontaneous emission. An excited atom in the gain medium can drop spontaneously to the lower laser level, rather than be stimulated to do so by the cavity field. Whereas stimulated emission adds *coherently* to the stimulating field, i.e., with a definite phase relationship, the spontaneously emitted radiation bears no phase relation to the cavity field. It adds *incoherently* to the cavity field. And the spontaneously emitted radiation has an inherent, Lorentzian distribution of frequencies (Section 7.6). Spontaneous emission therefore sets a fundamental lower limit on the laser linewidth.

We mentioned in Section 7.6 that a proper treatment of spontaneous emission requires the quantum theory of radiation. Therefore the problem of determining the fundamental lower limit to the spectral width of a laser can be solved truly rigorously only by using quantum electromagnetic theory. However, it is possible to give a heuristic argument that leads to the same answer given by the quantum theory of radiation.

The heuristic argument is based on the energy–time uncertainty relation

$$\Delta E \, \Delta t > \hbar \tag{11.10.1}$$

which is analogous to the position–momentum uncertainty relation. In (11.10.1) ΔE is the uncertainty in the energy measured in a time interval Δt. If an energy measurement process requires a time Δt, there is a fundamental lower limit $\Delta E = \hbar/\Delta t$ to the precision with which the energy can be determined.

In terms of the number of cavity photons q_ν, the energy E in the cavity mode is $E = h\nu q_\nu$, so that E may have uncertainties arising from both ν and q_ν:

$$\Delta E = h\nu \, \Delta q_\nu + h q_\nu \, \Delta \nu \tag{11.10.2}$$

Now the condition that the gain balances the loss in steady-state oscillation suggests that the amplitude of the cavity field should have a small relative uncertainty compared with the frequency, i.e.,

$$\frac{|\Delta q_\nu|}{q_\nu} \ll \frac{|\Delta \nu|}{\nu} \tag{11.10.3}$$

In other words, we suspect that the field amplitude should be relatively stable compared with the frequency; any small fluctuation of q_ν from its steady-state value $\bar{q}_\nu$ should quickly relax to zero. Therefore we assume

$$\Delta E \approx h\bar{q}_\nu \, \Delta\nu, \qquad (11.10.4)$$

or, using (11.10.1),

$$\Delta\nu \geq \frac{1}{\bar{q}_\nu \, \Delta t} \qquad (11.10.5)$$

The cavity-mode energy is determined both by stimulated emission and spontaneous emission into the cavity mode. The uncertainty ΔE, we argue, is due to spontaneous emission, which has not been accounted for in our laser intensity equations up to now [recall (10.5.14c) or (10.6.4)]. It is reasonable to assume in this heuristic approach to linewidth that any sort of measurement of E requires a time interval Δt no larger than the spontaneous-emission lifetime, i.e., $1/\Delta t$ must be at least as large as the spontaneous emission rate into the lasing cavity mode. Thus we write (11.10.5) as

$$\Delta\nu \geq C/\bar{q}_\nu \qquad (11.10.6)$$

where C is the rate at which photons are emitted spontaneously into the single cavity mode.

Now we use the fact (recall Section 7.8) that the rate of spontaneous emission into any one field mode is equal to the rate of stimulated emission when there is one photon already in that mode. Therefore

$$C = (\text{stimulated emission rate for one atom when } q_\nu = 1)$$

$$\times \, (\text{number of atoms in upper level}) \qquad (11.10.7)$$

The first factor in parentheses is just $c\sigma(\nu)/V$ [recall (10.6.2) and (10.6.5) with $n_2 = q_\nu = 1$]. The second factor is $N_2 V_g$. Therefore (11.10.7) is equivalent to

$$C = c\sigma(\nu) \, N_2 \frac{V_g}{V} = c\sigma(\nu) \, \frac{l}{L} \, \bar{N}_2 \qquad (11.10.8)$$

in steady-state oscillation, whence it follows from (11.10.6) that

$$\Delta\nu \geq \frac{c\sigma(\nu) \, \bar{N}_2 \, l/L}{\bar{q}_\nu} \qquad (11.10.9)$$

This is our expression for the minimum possible linewidth of laser radiation.

It is convenient to rewrite this result in standard form:

$$c\sigma(\nu) = \frac{cg(\nu)}{N_2 - N_1} = \frac{cg_t}{\Delta N_t} \qquad (11.10.10)$$

where the second equality follows from the steady-state gain-clamping condition. As in Section 10.4, ΔN_t represents the threshold population inversion for laser oscillation. Furthermore $\bar{q}_\nu$ can be related to the output power P_{out} of the laser, which is just $h\nu\bar{q}_\nu$ times the rate $cg_t l/L$ [cf. Eq. (10.5.6)] at which photons are removed from the cavity:

$$\bar{q}_\nu = \frac{P_{out}}{h\nu cg_t} \frac{L}{l} \qquad (11.10.11)$$

Using (11.10.10) and (11.10.11) in (11.10.9), therefore, we have

$$\Delta\nu \geq \frac{h\nu\bar{N}_2}{\Delta N_t} \left(\frac{cg_t l}{L}\right)^2 \frac{1}{P_{out}} \qquad (11.10.12)$$

Finally we use (11.9.8) to write cg_t in terms of the cavity bandwidth $\delta\nu_c$:

$$\Delta\nu \geq \frac{\bar{N}_2}{\Delta N_t} \frac{h\nu(4\pi \, \delta\nu_c)^2}{P_{out}} \qquad (11.10.13)$$

A similar result was first obtained by Schawlow and Townes[4] in 1958.

Equation (11.10.13) gives the theoretical lower limit set by spontaneous emission for the spectral bandwidth of laser radiation. The inverse power dependence is reasonable, since higher power levels increase stimulated emission relative to spontaneous emission. Since $\delta\nu_c$ is proportional to g_t and therefore ΔN_t, $\Delta\nu$ increases with $\bar{N}_2$. This behavior is also reasonable, because a higher degree of excited-state population increases the number of spontaneously emitted photons.

The theoretical lower limit (11.10.13) for the bandwidth has never been reached. Even in highly stabilized lasers there are slight mechanical vibrations and mirror jitters that change the mirror separation L and therefore shift the oscillation frequency ν uncontrollably. Laboratory lasers with bandwidths as small as several hundred hertz have been built, but even such a small width is still considerably larger than the quantum-mechanical limit.

Consider, for example, a typical 6328-Å He–Ne laser with mirror reflectivities $r_1 \approx 1.0$, $r_2 = 0.97$, a mirror separation of 30 cm, and an output power of 1 mW.

---

4. A. L. Schawlow and C. H. Townes, *Physical Review* **112**, 1940 (1958). This paper is widely recognized as containing the first analysis of the conditions necessary to achieve laser oscillation at optical frequencies.

Assuming that $\overline{N}_2/\Delta N_t$ is on the order of unity, we obtain a theoretical lower limit of ≈ 0.07 Hz for the bandwidth of the laser. Lasers can *in principle* produce such extremely narrow-line radiation, but practical problems have so far always broadened the emission spectrum far beyond the theoretical lower limit.

11.11 LASER POWER AT THRESHOLD

We have mostly ignored the threshold regime except to identify it as the point beyond which useful laser output is obtained. We will now consider the threshold of laser oscillation more carefully. For this purpose we use the model of a three-level laser. In steady-state oscillation the population inversion for a three-level laser is given by the formula (10.11.2). When $P \gg \Gamma_{21}$ we have

$$\overline{N}_2 - \overline{N}_1 \approx \overline{N}_2 \approx \frac{PN_T}{P + 2\sigma(\nu)\,\Phi_\nu}$$

$$= \frac{N_T}{1 + \overline{q}_\nu/q^{\text{sat}}} \tag{11.11.1}$$

where q^{sat} is defined by Eq. (10.11.8).

Near threshold spontaneous emission may not be negligible compared with stimulated emission, contrary to what we have assumed in writing the photon rate equation (10.5.7). Therefore let us amend (10.5.7) to include a term giving the growth of q_ν due to spontaneous emission. For this purpose we use again the fact that the rate of spontaneous emission into a single field mode is just the rate of stimulated emission with one incident photon [recall Sec. 7.8]. Thus we may simply replace q_ν by $q_\nu + 1$ in the first term on the right side of (10.5.7):

$$\frac{dq_\nu}{dt} = c\sigma(\nu)\frac{l}{L}\overline{N}_2\,(q_\nu + 1) - \frac{l}{L}\,cg_t q_\nu \tag{11.11.2}$$

The steady-state solution of this equation is obtained by setting the right-hand side to zero, which gives

$$\overline{q}_\nu = \frac{\sigma(\nu)\,\overline{N}_2}{g_t - \sigma(\nu)\,\overline{N}_2} \tag{11.11.3}$$

Now

$$cg_t = c\sigma(\nu)\,\Delta N_t \tag{11.11.4}$$

and if we define two convenient dimensionless parameters

$$x = \frac{\overline{N}_2}{\Delta N_t} \quad \text{and} \quad y = \frac{N_T}{\Delta N_t} \tag{11.11.5}$$

we may write (11.11.3) in the simplified form

$$\bar{q}_\nu = \frac{x}{1 - x} \tag{11.11.6}$$

and we may similarly write (11.11.1) as

$$x = \frac{N_T/\Delta N_t}{1 + \bar{q}_\nu/q^{\text{sat}}} = \frac{y}{1 + \bar{q}_\nu/q^{\text{sat}}} \tag{11.11.7}$$

Equations (11.11.6) and (11.11.7) may be solved simultaneously for $\bar{q}_\nu$. Some simple algebra yields the quadratic equation

$$\bar{q}_\nu^2 + q^{\text{sat}}(1 - y)\bar{q}_\nu - q^{\text{sat}}y = 0 \tag{11.11.8}$$

which has the solution

$$\frac{\bar{q}_\nu}{q^{\text{sat}}} = \tfrac{1}{2}(y - 1) + \frac{1}{2}\sqrt{(y - 1)^2 + \frac{4y}{q^{\text{sat}}}} \tag{11.11.9}$$

In order to obtain a nonnegative value for the cavity photon number $\bar{q}_\nu$ we have taken in (11.11.9) the positive-definite root of (11.11.8).

Equation (11.11.1) shows that, for an ideal three-level laser, $\overline{N}_2 \approx N_T$ when $\bar{q}_\nu \to 0$. That is, N_T is equal to the small-signal population inversion. It follows that y is equal to the small-signal gain divided by the threshold gain:

$$y = g_0/g_t \tag{11.11.10}$$

If $y < 1$ the device is below the threshold for laser oscillation; if $y > 1$ it is above threshold. Far above threshold ($y \gg 1$) we have from (11.11.9) the result

$$\bar{q}_\nu \approx q^{\text{sat}}\left(\frac{g_0}{g_t} - 1\right) \tag{11.11.11}$$

which may be used to obtain our previous expression (11.2.9) for the laser output intensity.

In order to study the threshold region $y \approx 1$, we need an estimate of the number q^{sat} appearing in the equation (11.11.9) for the cavity photon number. q^{sat} is defined by (10.11.8). Assuming again that $P \gg \Gamma_{21}$, we have

$$q^{\text{sat}} \approx \left(2\pi \frac{\epsilon_0 m}{e^2}\right)\left(\frac{1}{f}\right) \delta\nu_{21} \, (PV) \qquad (11.11.12)$$

where f is the oscillator strength of the laser transition. In many lasers PV is roughly on the order of 10^3 sec^{-1}. Taking $f \approx 1$ and $\delta\nu_{21} \approx 10$ GHz, therefore, we have the reasonable estimate $q^{\text{sat}} \approx 10^{10}$ for the saturation photon number. We will therefore study the properties of (11.11.9) near the threshold region $y \approx 1$ by *defining* $q^{\text{sat}} = 10^{10}$ as a reasonable value for typical lasers.

Exactly at threshold the cavity photon number is given by (11.11.9) with $y = 1$:

$$\left(\bar{q}_\nu\right)_{\text{threshold}} = \tfrac{1}{2} q^{\text{sat}} \sqrt{4/q^{\text{sat}}} = \sqrt{q^{\text{sat}}} = 10^5 \qquad (11.11.13)$$

This is, to be sure, a small number of photons compared with what (11.11.11) predicts well above threshold. However, it is much larger than the average photon number per mode of a thermal field,

$$\left(q_\nu\right)_{\text{thermal}} = \left(e^{h\nu/kT} - 1\right)^{-1} \qquad (11.11.14)$$

for any realistic value of temperature. For ν corresponding to the He–Ne 6328-Å line, for example,

$$\left(q_\nu\right)_{\text{thermal}} \approx 0.023 \qquad (11.11.15)$$

for the solar temperature $T \approx 6000$ K. The conclusion is that a laser exactly at threshold channels many more photons into the "lasing" mode than it would if it were an ordinary thermal source. We will see that the same thing happens even fairly far *below* threshold.

It is interesting to consider the rate of change of $\bar{q}_\nu$ with y in the vicinity of threshold. From (11.11.9) we calculate

$$\frac{d\bar{q}_\nu}{dy} = \tfrac{1}{2} q_{\text{sat}}\left(1 + \frac{y - 1 \, 2/q_{\text{sat}}}{\sqrt{(y-1)^2 + 4y/q^{\text{sat}}}}\right) \qquad (11.11.16)$$

so that exactly at threshold ($y = 1$) we have

$$\left(\frac{d\bar{q}_\nu}{dy}\right)_{\text{threshold}} \approx \tfrac{1}{2} q^{\text{sat}} = \tfrac{1}{2} \times 10^{10} \qquad (11.11.17)$$

Thus the curve of $\bar{q}_\nu$ vs. y has an extremely large, positive slope at $y = 1$. This is evident in Figure 11.11, which plots $\bar{q}_\nu$ versus y from equation (11.11.9) for $q^{\text{sat}} = 10^{10}$. We conclude that there is an extremely rapid rise in the cavity photon

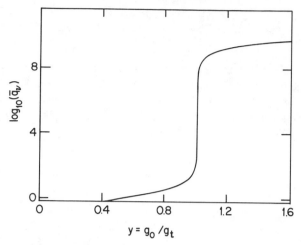

Figure 11.11 The remarkable jump in the cavity photon number at threshold.

number at the point where the medium is pumped just above threshold. We note also that $\bar{q}_\nu$ is much larger than the typical "thermal" value given in Eq. (11.11.15) even for $y = 0.1$.

The sudden transition at threshold that occurs in a laser is so abrupt that it is not usually observed unless an experiment is designed specifically for its observation.

• An even more profound transition occurs at threshold in the *photon-statistical* properties of the laser output radiation. The subject of photon statistics was treated briefly in Chapter 9 in connection with photon counting. A detailed discussion of photon statistics is beyond our scope, but in Chapter 15 we will summarize the photon-statistical properties of laser radiation. •

11.12   OBTAINING SINGLE-MODE OSCILLATION

In our discussion of laser theory we have thus far ignored any variations of the cavity intensity transverse to the cavity axis (in the x–y plane). We will extend the theory beyond this crude approximation in Chapter 14, where we discuss the cavity modes of actual laser resonators.

Our assumption of single-mode oscillation in this chapter, therefore, has really been the assumption of a single *longitudinal* mode, whose frequency is given by (11.9.1): $\nu_m = mc/2L$. For many applications (e.g., holography) single-mode oscillation is highly desirable, and it is appropriate to close this chapter by considering some ways of obtaining it.

The number of possible lasing modes can be estimated by counting the number of cavity modes, separated in frequency by $\Delta\nu = c/2L$, that lie within the gain bandwidth $\Delta\nu_g$. This number is on the order of $\Delta\nu_g/\Delta\nu$ for $\Delta\nu_g > \Delta\nu$. If $\Delta\nu >$

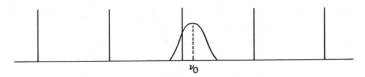

Figure 11.12 A case in which the cavity mode frequency spacing $c/2L$ is larger than the width of the gain profile. Only a single longitudinal mode can lase.

$\Delta \nu_g$, however, the cavity mode frequencies are separated by more than the width of the gain profile, and at most only a single cavity frequency can have gain and lase (Figure 11.12). In other words, *we can obtain single-mode oscillation by making the cavity short enough.* For a He–Ne laser with gain bandwidth $\Delta \nu_g = \delta \nu_D \approx 1500$ MHz, for example, we require

$$\frac{c}{2L} = \frac{3 \times 10^{10} \text{ cm sec}^{-1}}{2L} > 1.5 \times 10^9 \text{ sec}^{-1}$$

or $L < 10$ cm. A disadvantage of this way of achieving single-mode oscillation is thus evident: it requires a small gain cell and therefore results in low output power, typically on the order of 1 μW.

The linewidths of liquid- and solid-laser transitions are usually much larger than in gases, 100 GHz being a typical order of magnitude. In these lasers the short-cavity approach to single-mode oscillation is not practical.

Actually the gain-clamping condition (11.2.3) leads to a surprising conclusion: all lasers operating on a homogeneously broadened transition should oscillate on only a single mode, that for which the small-signal gain is greatest. This is illustrated in Figure 11.13. In Figure 11.13*a* we show a small-signal gain profile broad enough to allow five cavity modes to be above threshold. The *saturated* gain in steady-state oscillation must equal the threshold gain g_t (i.e., gain equals loss)

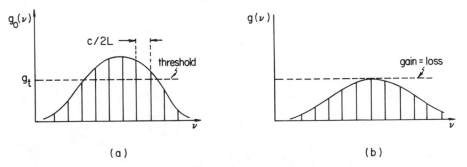

(a)   (b)

Figure 11.13 (*a*) A case in which five cavity modes have a small-signal gain g_0 larger than the threshold g_t for laser oscillation. (*b*) If the gain saturates homogeneously, only the mode with the largest small-signal gain is expected to lase. The others are saturated below the gain g_t necessary for laser oscillation.

according to the gain-clamping condition. The saturated-gain profile is shown in Figure 11.13b. The saturated gain of one of the cavity modes—that with the largest small-signal gain—equals the threshold gain g_t. But the laser oscillation on this mode has saturated the gain to the extent that all other modes have gains *below* the threshold value; these other modes therefore cannot lase. Only the one mode can lase (Problem 11.7).

Many lasers with homogeneously broadened gain media do indeed oscillate on a single longitudinal mode. However, our argument for this assumes the validity of the gain-clamping condition (11.2.3) and ignores spatial hole burning. When spatial hole burning is present the atoms of the gain medium are saturated to different degrees, depending on where they are within the standing-wave field, as indicated by Eq. (11.4.1). In this case the argument for single-mode oscillation illustrated in Figure 11.13 does not apply. Multimode oscillation in a homogeneously broadened medium is therefore permitted. Thus the ruby laser, which operates on a homogeneously broadened transition, will oscillate in a multimode fashion unless something is done to produce single-mode oscillation.

In gas lasers for which the pressure is large enough to make the gain medium predominantly homogeneously (collision-) broadened, however, the output tends to be single-mode. In this case the effect of spatial hole burning is largely mitigated by atomic motion.

The argument illustrated in Figure 11.13 is also inapplicable to inhomogeneously broadened laser media. In this case the presence of *spectral* hole burning means that different atoms saturate differently. The gain does not saturate to the same degree (i.e., homogeneously) for different spectral packets of atoms, in contrast to the homogeneous case, where there is in fact only a single spectral packet. One cavity mode might burn a deep hole in a particular spectral packet, without at all saturating other spectral packets on which other modes may lase. The He–Ne laser, for example, generally oscillates multimode.

Spatial and spectral hole burning thus invalidate the argument for single-mode oscillation illustrated in Figure 11.13. If spatial hole burning is negligible, single-mode oscillation can be expected for a homogeneously broadened medium. Since spectral hole burning will always be present to some degree in an inhomogeneously broadened medium, however, we generally expect multimode oscillation in this case. These expectations are borne out experimentally.

This leads us to consider methods other than the short-cavity approach for achieving single-mode oscillation. These other methods have one feature in common: an additional loss mechanism is introduced to discriminate against all possible laser modes but one. That is, a situation is created in which all modes but one have a gain less than their loss. Note that this is just an extended application of the open-cavity principle which is basic for optical-wave oscillation. An open cavity is extremely lossy except for axial waves. What we want now is a way to make the losses very large for most of these axial waves as well.

One important way to do this is to use a *Fabry–Perot etalon*. Consider the situation illustrated in Figure 11.14a. An incident plane monochromatic wave is normally incident from a medium of refractive index n' onto a slab of material of

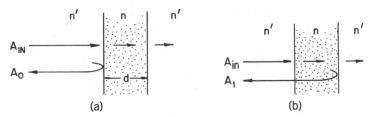

Figure 11.14 (a) A plane wave incident normally on a Fabry–Perot etalon and the zeroth contribution to the reflected wave. (b) Another contribution (A_1) to the reflected wave in a Fabry–Perot etalon.

index n and thickness d. Reflection and transmission occur at both interfaces. If $n > n'$, the reflected wave at the first interface is shifted in phase by π radians from the incident wave. This is a simple consequence of the Fresnel formulas for reflection and refraction.

Now the Fabry–Perot etalon can also produce a reflected wave as a result of reflection off the back face of the etalon and transmission again across the first interface (Figure 11.14b). If $n > n'$, there is no phase change associated with reflection from the internal back face of the etalon. The phase of this wave reflected from the etalon is thus shifted from that of the incident wave only because of propagation, i.e., because it makes a round trip inside the etalon. This phase change is simply 2π times the optical length of propagation in wavelengths:

$$2\pi \frac{n(d + d)}{\lambda} = \frac{4\pi nd}{\lambda} = \frac{4\pi nd}{c} \nu \tag{11.12.1}$$

The phase difference between the two reflected waves shown in Figure 11.14b is thus

$$\frac{4\pi nd}{c} \nu - \pi = 2\pi \left(\frac{2\nu nd}{c} - \frac{1}{2} \right) \tag{11.12.2}$$

These two reflected waves will interfere destructively if their phase difference is equal to an odd integral multiple of π radians, i.e., if

$$2\pi \left(\frac{2\nu nd}{c} - \frac{1}{2} \right) = (2m + 1)\pi = 2\pi(m + \tfrac{1}{2}), \quad m = 0, 1, 2, \ldots$$

$$\tag{11.12.3}$$

In other words, destructive interference occurs for frequencies

$$\nu_m = m \frac{c}{2nd}, \quad m = 1, 2, 3, \ldots \tag{11.12.4}$$

These are the "resonance frequencies" of the Fabry–Perot etalon in the sense that they undergo minimal reflection and therefore maximal transmission (Problem 11.8).

Actually we should include the effect of multiple reflections within the etalon. As shown in Appendix 11.A, this leads to the same resonance frequencies (11.12.4) for maximal transmission. Furthermore we can easily generalize to the case of an arbitrary angle of incidence (Problem 11.8), with the result that the resonance frequencies become

$$\nu_m = m \left(\frac{c}{2nd \cos \theta} \right), \quad m = 1, 2, 3, \ldots \tag{11.12.5}$$

If d is small enough, the spacing

$$\Delta \nu = \nu_{m+1} - \nu_m = \frac{c}{2nd \cos \theta} \tag{11.12.6}$$

between adjacent resonance frequencies of the etalon will be large compared with the width $\Delta \nu_g$ of the gain profile. By adjusting θ, a resonance frequency can be brought near the center of the gain profile, while the next resonance frequency lies outside the gain profile.[5]

The Fabry–Perot etalon is also widely used in spectroscopy. Because of its general importance in laser technology, we devote Appendix 11.A to a more detailed discussion of its properties.

In referring to cavity modes in this section we have ignored polarization. A very common and convenient way of obtaining a linearly polarized output from a laser is to use *Brewster windows*. To understand this technique, it may be worthwhile to review briefly some results of electromagnetic theory.

For a plane wave incident upon a plane interface between two dielectric media, we define the plane of incidence as the plane formed by the propagation direction and a line perpendicular to the interface. *For polarization parallel to this plane of incidence*, it turns out that there is a particular angle of incidence θ_B, called *Brewster's angle*, for which there is no reflected wave. If the wave is incident from a medium of index $n' \approx 1$ onto a medium of index n, the Brewster angle is given by

$$\theta_B = \tan^{-1} n \tag{11.12.7}$$

For an air-to-glass interface, θ_B is about 56°.

Now if a plane wave of mixed polarization is incident at the angle θ_B, we can imagine it to be composed of components parallel and perpendicular to the plane of incidence. The components parallel to the plane of incidence will not be re-

---

5. In practice the tilt angle θ must not be too small, or else the etalon modifies the cavity frequencies and can no longer be regarded as a simple frequency filter.

flected. If a wave of arbitrary polarization is incident at Brewster's angle, the reflected field will therefore be completely polarized perpendicular to the plane of incidence. This is in fact a way, albeit inefficient, of producing polarized light. Furthermore the reflected wave will be partially polarized even if the angle is close but not quite equal to θ_B. This result explains the success of Polaroid sunglasses in reducing the glare of reflected sunlight.

Now we can understand the use of *Brewster-angle windows* for obtaining linearly polarized laser radiation. Figure 11.15 illustrates a laser in which the ends of the gain cell are cut at the Brewster angle with respect to the cavity axis. The plane of incidence associated with the cavity field is obviously just the plane of the figure. Laser radiation that is linearly polarized in this plane will not suffer any reflection off the ends of the gain cell. Radiation polarized perpendicular to the plane of the figure, however, will have a greater loss coefficient because it is reflected at the windows. Lasing is therefore more favorable to linear polarization in the plane of incidence, as indicated in Figure 11.15.

APPENDIX 11.A THE FABRY–PEROT ETALON

Consider again the case of a monochromatic plane wave incident normally upon a Fabry–Perot etalon (Figure 11.14). With the first interface ($n' \rightarrow n$) between the two media we associate reflection and transmission coefficients r and t, respectively. We similarly denote by r' and t' the reflection and transmission coefficients for the second interface ($n \rightarrow n'$). The coefficients r, r', t, and t' are given in terms of n and n' by the Fresnel formulas.

If A_{in} is the (complex) amplitude of the incident wave, then

$$A_0 = \sqrt{r}\, A_{\text{in}} e^{i\pi} = -\sqrt{r}\, A_{\text{in}} \tag{11.A.1}$$

is the amplitude of the wave reflected from the first interface. The phase shift of π introduced in (11.A.1) is again a consequence of our assumption that $n > n'$. (The reflected *intensity* is r times the incident intensity, and is independent of whether $n > n'$ or $n < n'$.)

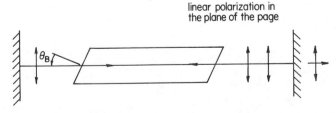

linear polarization in the plane of the page

Figure 11.15 A laser with the gain-cell windows cut at the Brewster angle. The indicated polarization (parallel to the plane of incidence, which is the plane of the figure) will suffer no reflective loss at the windows, and therefore will lase preferentially. The orthogonal polarization will have a greater loss due to reflection at the windows.

Now the etalon can also produce a reflected wave as a result of transmission at the first interface ($n' \to n$), reflection at the second interface ($n' \to n$), and finally transmission across the first interface ($n \to n'$). The amplitude of the reflected wave is $\sqrt{tr't'}\, A_0$, and it also has a phase shift

$$\frac{2\pi}{\lambda}(2dn) = \frac{4\pi\nu nd}{c} = \Phi \tag{11.A.2}$$

with respect to the incident wave. Because $n > n'$, there are no π phase shifts associated with reflections off the inner faces of the etalon. Thus we write

$$A_1 = (\sqrt{r'tt'}\, e^{i\Phi})A_{\text{in}} \tag{11.A.3}$$

A contribution to the total reflected wave also arises from transmission across the first interface, *two* round-trip passes through the etalon as a result of three reflections, and finally transmission across the first interface. For this contribution we have

$$A_2 = \sqrt{tr'r'r't'}\, e^{2i\Phi}A_{\text{in}} \tag{11.A.4}$$

Continuing in this manner, it is easy to see that the total reflected field due to multiple reflections inside the etalon has an amplitude

$$\begin{aligned}
A_R &= A_0 + A_1 + A_2 + A_3 + \cdots \\
&= \left[-\sqrt{r} + \sqrt{r'}\,\sqrt{tt'}\, e^{i\Phi}\left(1 + \sqrt{r'r'}\, e^{i\Phi} + \sqrt{r'r'r'r'}\, e^{2i\Phi} + \cdots\right)\right]A_{\text{in}}
\end{aligned}$$
$$\tag{11.A.5}$$

From the Fresnel formulas we recall that for normal incidence

$$r = \left(\frac{n - n'}{n + n'}\right)^2 = r' \tag{11.A.6}$$

It is convenient to define

$$R = r = r' \tag{11.A.7a}$$
$$T = \sqrt{tt'} \tag{11.A.7b}$$

and it follows from energy conservation (and the Fresnel formulas) that

$$R + T = 1 \tag{11.A.8}$$

Then we may write (11.A.5) as

$$\frac{A_R}{A_{\text{in}}} = -\sqrt{R}\left[1 - Te^{i\Phi}(1 + Re^{i\Phi} + R^2 e^{2i\Phi} + \cdots)\right] \tag{11.A.9}$$

The infinite series can be summed easily:

$$1 + x + x^2 + x^3 + \cdots = \frac{1}{1-x} \qquad (|x| < 1) \qquad (11.A.10)$$

and so we have

$$1 + Re^{i\Phi} + R^2 e^{2i\Phi} + \ldots = \frac{1}{1 - Re^{i\Phi}} \qquad (11.A.11)$$

in Eq. (11.A.9). Thus

$$\frac{A_R}{A_{in}} = -\sqrt{R}\left(1 - \frac{Te^{i\Phi}}{1 - Re^{i\Phi}}\right)$$

$$= -\sqrt{R}\,\frac{1 - (R+T)e^{i\Phi}}{1 - Re^{i\Phi}}$$

$$= -\frac{1 - e^{i\Phi}}{1 - Re^{i\Phi}}\sqrt{R} \qquad (11.A.12)$$

where the last step follows from (11.A.8). The fraction of the incident *intensity* reflected by the etalon is therefore given by the Airy formula

$$\frac{I_R}{I_{in}} = \left|\frac{A_R}{A_{in}}\right|^2 = \left|\frac{1 - e^{i\Phi}}{1 - Re^{i\Phi}}\right|^2 R$$

$$= \frac{4R\sin^2(\Phi/2)}{(1-R)^2 + 4R\sin^2(\Phi/2)} \qquad (11.A.13)$$

Similarly the fraction of the transmitted intensity is

$$\frac{I_T}{I_{in}} = 1 - \frac{I_R}{I_{in}} = \frac{(1-R)^2}{(1-R)^2 + 4R\sin^2(\Phi/2)} \qquad (11.A.14)$$

It may be shown that for nonnormal incidence we simply replace (11.A.2) by

$$\Phi = \frac{4\pi\nu nd}{c}\cos\theta \qquad (11.A.15)$$

From Eqs. (11.A.13) and (11.A.14) it follows that the Fabry–Perot etalon is perfectly transmitting for ν and θ such that

$$\Phi/2 = m\pi, \qquad m = 1, 2, 3, \ldots \qquad (11.A.16)$$

This is precisely equivalent to the condition (11.12.5) obtained by considering only a single reflection inside the etalon.

The multiple-reflection analysis leading to (11.A.11) and (11.A.14) allows us to investigate the bandpass characteristics of the Fabry–Perot etalon. That is, we can determine how the transmission through the etalon falls off as the frequency ν is displaced from a resonance frequency (11.12.5). In Figure 11.16 we plot the transmission function (11.A.14) versus Φ for several values of the parameter

$$F = \frac{4R}{(1 - R)^2} \qquad (11.A.17)$$

In terms of this parameter, the transmitted fraction is

$$\frac{I_T}{I_{in}} = \frac{1}{1 + F \sin^2 (\Phi/2)} \qquad (11.A.18)$$

For small values of F there is considerable transmission of *all* frequencies. For large values of F ($R \to 1$), however, only a narrow band of frequencies centered at each resonance frequency (11.12.5) is transmitted. In spectroscopic applications this results in a tradeoff between the "throughput," or amount of intensity transmitted, and the "resolution," or narrowness of the bandwidths of transmitted frequencies (Figure 11.17).

The resonance frequency spacing (11.12.6) is called the *free spectral range* of the Fabry–Perot. The ratio of the free spectral range to the half-width of the frequency band centered on a resonance frequency may be shown to be

$$\mathfrak{F} = \frac{\pi}{2} \sqrt{F} = \frac{\pi \sqrt{R}}{1 - R} \qquad (11.A.19)$$

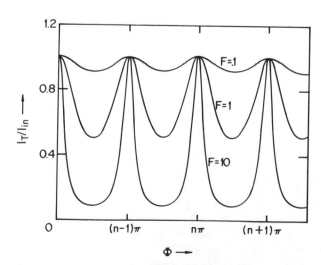

Figure 11.16 Transmission function of the Fabry–Perot etalon for three values of $\mathfrak{F}$.

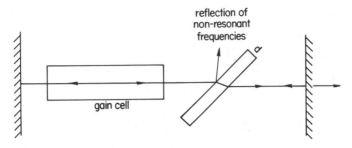

Figure 11.17 An intracavity Fabry–Perot etalon can be used to filter out all cavity mode frequencies except those satisfying the condition (11.A.16) for transmission without reflection. If the "free spectral range" $c/2nd \cos \theta$ is large compared with the width of the gain profile, only a single longitudinal mode can lase.

$\mathcal{F}$ is called the *finesse* of the Fabry–Perot. The greater the finesse, the sharper the bands of transmitted frequencies relative to their separation. Practical difficulties, such as surface roughness of the Fabry–Perot plates, place a normal upper limit of around 30 on the finesse, though values in the range $\mathcal{F} \approx 100$ can be achieved under special circumstances.

PROBLEMS

11.1 Show that the output coupling that maximizes the output intensity (11.2.11) is given by (11.3.1), and determine the output maximum.

11.2 Using numerical values given in Section 10.10, estimate the quantum efficiencies of the ruby and Nd:YAG lasers.

11.3 Assuming $s = 0.04$, plot Eq. (11.2.11) for the output intensity as a function of the output coupling t for the three cases $g_0 l = 0.1, 1.0, 30.0$. Compare these results with those based on Eq. (11.5.18).

11.4 A high-power CO_2 laser has a small-signal gain $g_0(\nu_0) \approx 0.005 \text{ cm}^{-1}$ at line center. The laser transition is homogeneously broadened with a Lorentzian linewidth (HWHM) $\delta\nu_{21} \approx 1$ GHz. The gain medium fills nearly the entire 50 cm between the cavity mirrors. One of the mirrors is nominally perfectly reflecting, while the output mirror is characterized by a scattering–absorption coefficient $s = 2\%$.

   (a) Determine the output mirror transmission coefficient t that will produce the greatest amount of output power from this laser.

   (b) The saturation intensity I^{sat} for this laser is estimated to be about 100 kW/cm^2. What is the output intensity if the cavity is designed to have the optimal output power?

   (c) Estimate the intracavity intensity. Why might such a laser be designed to have water-cooled mirrors?

11.5 A cavity for a 6328-Å He–Ne laser is 50 cm long with reflection coefficient $r_1 = 1.0$ for one mirror and $r_2 = 0.98$ for the other. Losses other than output coupling are very small and may be ignored. The output power of the laser is measured to be about 10-mW on a single mode.

   (a) What is the cavity photon number and what is the output rate of photons?

   (b) It has been written that "if the light from a thousand suns were to shine in the sky, that would be the glory of the Mighty One." Assume that the sun is an ideal blackbody radiator at $T = 6000$ K. Estimate the flux of photons in a frequency band of width $\delta\nu \approx 10$ MHz centered at 6328 Å that can be obtained from 1000 suns. How does this compare with the photon flux that can be obtained from He-Ne or other lasers?

11.6 Should the Lamb dip occur with any inhomogeneously broadened gain medium, or only the specific case of Doppler broadening?

11.7 Do you think that most lasers have a cavity bandwidth much larger or smaller than the linewidth of the gain profile? Is any implicit assumption about this made in our discussion related to Figure 11.13?

11.8 Derive Eq. (11.12.5) for the resonance frequencies of a Fabry–Perot etalon for an arbitrary angle of incidence.

12 MULTIMODE AND TRANSIENT OSCILLATION

12.1 INTRODUCTION

Thus far we have restricted our study of the laser to the case of continuous-wave, single-mode operation. In this chapter we will consider time-dependent, transient effects, including relaxation oscillations and Q switching. We will also extend our single-mode theory somewhat to the case in which several or many cavity modes can oscillate simultaneously. This allows us in particular to understand the important technique called mode locking, a way to obtain ultrashort pulses of light.

12.2 RATE EQUATIONS FOR INTENSITIES AND POPULATIONS

In the preceding two chapters we have found it convenient and instructive to describe the strength of the cavity field either in terms of intensity I_ν or photon number q_ν. In the present chapter it will be convenient to use the intensity description. We will therefore begin with a brief review of the appropriate equations coupling the intensity and the laser level population densities N_2 and N_1.

In general the cavity intensity will vary both in time and space. We will continue in this chapter to make the plane-wave approximation in which the intensity is assumed to be uniform in any plane perpendicular to the cavity axis. Furthermore we showed in Section 11.5 that, for the most common situation in which the mirror reflectivities are large (say, $> 50\%$), the cavity intensity is approximately uniform along the cavity axis if we ignore the rapidly varying $\sin^2 kz$ interference term. So it is useful again to make the uniform-field approximation, but now to include the time dependence of the cavity intensity. First we recall equation (10.5.8):

$$\frac{dI_\nu}{dt} = \frac{cl}{L}\left(g(\nu)I_\nu - \frac{1}{2l}(1 - r_1 r_2)I_\nu\right)$$

$$= \frac{cl}{L}[g(\nu) - g_t]I_\nu \tag{12.2.1}$$

For simplicity we will assume that the gain cell fills the entire space between the mirrors. Then $l = L$ and

$$\frac{dI_\nu}{dt} = c[g(\nu) - g_t]I_\nu \tag{12.2.2}$$

Recall that $I_\nu = I_\nu^{(+)} + I_\nu^{(-)}$ is the sum of the two traveling-wave intensities; in the case of high mirror reflectivities, the two are approximately equal.

In terms of the cavity intensity we can write the population rate equations [cf. (10.5.14) with $\Gamma_1 = \Gamma_2 = 0$ and $A \to \Gamma_{21}$]

$$\frac{dN_2}{dt} = -\frac{g(\nu)I_\nu}{h\nu} - \Gamma_{21}N_2 + K_2 \qquad (12.2.3a)$$

$$\frac{dN_1}{dt} = \frac{g(\nu)I_\nu}{h\nu} + \Gamma_{21}N_2 + K_1 \qquad (12.2.3b)$$

where the rates Γ_{21}, K_2, and K_1 are, again, level decay and pumping rates. Since N_2 and N_1 are populations per unit volume, the pumping rates have units of (volume)$^{-1}$ (time)$^{-1}$. Equations (12.2.2) and (12.2.3) are coupled rate equations for I_ν, N_2, and N_1. The coupling is through the gain coefficient

$$g(\nu) = \frac{\lambda^2 A}{8\pi}(N_2 - N_1)S(\nu)$$

$$= \sigma(\nu)(N_2 - N_1) \qquad (12.2.4)$$

where we assume for simplicity that $g_2/g_1 = 1$.

The population rate equations (12.2.3) are easily modified to suit a particular laser medium. We have already described such modifications in the case of the stylized three- and four-level models. Further modifications are described in Chapter 13, where we consider specific population inversion mechanisms. Since we will be describing in this chapter some rather general phenomena that transcend specific inversion schemes, it will be adequate to use the simple rate equations (12.2.3) for the laser level population densities.

For many purposes the rate equations (12.2.2) and (12.2.3) may be simplified somewhat. One simplifying assumption is that $N_1 \ll N_2$, i.e., that the lower laser level population is negligible compared with the upper laser level population. This would be the case in a four-level laser, where the lower level decays very rapidly compared with the stimulated emission (absorption) rate. Then $g(\nu) = \sigma(\nu)N_2$, and (12.2.2) and (12.2.3a) become

$$\frac{dI_\nu}{dt} = c\sigma(\nu)N_2 I_\nu - cg_t I_\nu \qquad (12.2.5a)$$

$$\frac{dN_2}{dt} = -\frac{\sigma(\nu)}{h\nu}N_2 I_\nu - \Gamma_{21}N_2 + K_2 \qquad (12.2.5b)$$

12.3   RELAXATION OSCILLATIONS

The coupled equations (12.2.5) for I_ν and N_2 are simple in appearance, but they have no known general solution. However, it is easy to find the *steady-state* solutions which we denote $\bar{I}_\nu$ and $\bar{N}_2$. These are obtained simply by replacing the left sides of (12.2.5) by zero and solving the resulting algebraic equations, with the result

$$\bar{I}_\nu = h\nu \left(\frac{K_2}{g_t} - \frac{\Gamma_{21}}{\sigma(\nu)} \right) \tag{12.3.1a}$$

$$\bar{N}_2 = \frac{g_t}{\sigma(\nu)} \tag{12.3.1b}$$

These solutions may also be written in a different form to show explicitly how $\bar{N}_2$ saturates with increasing $\bar{I}_\nu$ (Problem 12.1).

It is possible to solve these equations approximately if the laser is operating very near to steady state. In this case we write

$$I_\nu = \bar{I}_\nu + \epsilon \tag{12.3.2a}$$

$$N_2 = \bar{N}_2 + \eta \tag{12.3.2b}$$

and assume

$$|\epsilon| \ll \bar{I}_\nu \tag{12.3.3a}$$

$$|\eta| \ll \bar{N}_2 \tag{12.3.3b}$$

This approximation allows the equations (12.2.5) to be linearized and solved, as follows.

Using (12.3.2) in (12.2.5a), we have

$$\frac{d}{dt} (\bar{I}_\nu + \epsilon) = c\sigma(\bar{N}_2 + \eta)(\bar{I}_\nu + \epsilon) - cg_t(\bar{I}_\nu + \epsilon) \tag{12.3.4}$$

which is the same (since $d\bar{I}_\nu/dt = 0$) as

$$\frac{d\epsilon}{dt} = c\sigma(\bar{N}_2 + \eta)(\bar{I}_\nu + \epsilon) - cg_t(\bar{I}_\nu + \epsilon)$$

$$= c\sigma(\bar{N}_2\bar{I}_\nu + \bar{N}_2\epsilon + \eta\bar{I}_\nu + \eta\epsilon) - cg_t(\bar{I}_\nu + \epsilon) \tag{12.3.5}$$

Now $\bar{I}_\nu$ and $\bar{N}_2$ are such as to make the right sides of (12.2.5) vanish. In particular,

$$c\sigma\bar{N}_2\bar{I}_\nu - cg_t\bar{I}_\nu = 0 \tag{12.3.6}$$

Using this relation in (12.3.5), we obtain the much simpler equation

$$\frac{d\epsilon}{dt} = c\sigma\eta\,\bar{I}_\nu + c\sigma\eta\epsilon \tag{12.3.7}$$

This is still nonlinear (because of the term $\eta\epsilon$), but now the nonlinearity is very small because it involves the product of the small quantities η and ϵ. Near enough to steady state [recall (12.3.3)], such second-order-small terms can be dropped altogether without significant error. Thus we obtain the following linear equation for the time dependence of the departure of the cavity intensity from its steady-state value:

$$\frac{d\epsilon}{dt} = (c\sigma\bar{I}_\nu)\,\eta \tag{12.3.8}$$

where the factor in parentheses is constant in time.

The same procedure can be applied to (12.2.5b). Again the product $\eta\epsilon$ is very small and can be dropped, and again the definitions of $\bar{I}_\nu$ and $\bar{N}_2$ can be used to cancel some terms. The result is

$$\frac{d\eta}{dt} = -\frac{g_t}{h\nu}\,\epsilon - \frac{\sigma K_2}{g_t}\,\eta \tag{12.3.9}$$

Equations (12.3.8) and (12.3.9) are still coupled to each other, but they are now linear and easily solved. We use (12.3.8) to replace η in (12.3.9) by $(c\sigma\bar{I}_\nu)^{-1}\,d\epsilon/dt$ to get

$$\frac{d^2\epsilon}{dt^2} + \gamma\frac{d\epsilon}{dt} + \omega_0^2\epsilon = 0 \tag{12.3.10}$$

where we define

$$\gamma = \sigma K_2/g_t \tag{12.3.11}$$

and

$$\omega_0^2 = \frac{c\sigma g_t}{h\nu}\bar{I}_\nu \tag{12.3.12}$$

The solution to (12.3.10) is easily found to be

$$\epsilon(t) = A\,e^{-\gamma t/2}\cos(\omega t + \phi) \tag{12.3.13}$$

where A and ϕ are the initial amplitude and phase of $\epsilon(t)$, and the frequency of oscillation is given by

$$\omega = \sqrt{\omega_0^2 - \frac{\gamma^2}{4}} \tag{12.3.14}$$

For definiteness we assume $\omega_0 > \gamma/2$, making ω real. Thus, near to the steady state, the cavity intensity oscillates about the steady-state value $\bar{I}_\nu$, and gradually approaches $\bar{I}_\nu$ at the (exponential) rate $\gamma/2$:

$$I_\nu = \bar{I}_\nu + A\,e^{-\gamma t/2}\cos{(\omega t + \phi)} \qquad (12.3.15)$$

This is called a *relaxation oscillation*. Similar behavior is observed in a wide variety of nonlinear systems.

Although the relaxation-oscillation solution (12.3.13) is valid only if $|\epsilon| \ll \bar{I}_\nu$ [recall (12.3.3)], the nature of the solution is of general importance. The critical feature of the solution is that γ is positive. This guarantees that the steady-state solution $\bar{I}_\nu$ is a *stable* solution. That is, if some outside agent slightly disturbs the laser while it is running in steady state, the effect of the disturbance decays to zero, thus returning the laser to steady state again. If γ were negative, a small disturbance would grow, and the steady state would therefore be unstable, and thus of very little practical significance.

We may write the period T_r and lifetime τ_r of the relaxation oscillations as (Problem 12.1)

$$T_r = \frac{2\pi}{\omega_0} = \frac{2\pi}{\sqrt{(c/\tau_{21})(g_0 - g_t)}} \qquad (12.3.16)$$

and

$$\tau_r = \frac{1}{\gamma} = \frac{g_t}{g_0}\,\tau_{21} \qquad (12.3.17)$$

where g_0 is the small-signal gain and $\tau_{21} = \Gamma_{21}^{-1}$ is the lifetime of the upper laser level. From (12.3.17) or (12.3.11) we see that the duration of the relaxation oscillations decreases with increasing pumping rate K_2 of the gain medium. Likewise the period T_r of the relaxation oscillations should decrease with increased g_0. These predicted trends are consistent with many experimental observations.

It is possible to observe relaxation oscillations in the output intensity of a laser after it is turned on and approaches a steady-state operation. Perturbations in the pumping power can also cause relaxation oscillations to appear spontaneously. In some cases, especially in solid-state lasers, the relaxation time τ_r may be relatively large, making relaxation oscillations readily apparent on an oscilloscope trace of the laser output intensity.

As an example, consider a ruby laser with mirror reflectivities $r_1 \approx 1.0$, $r_2 \approx 0.94$, and a ruby rod of length $l = 5.0$ cm. For such a laser $g_t \approx (1/2l)(1 - r_2) = 0.006$ cm^{-1}, so that $cg_t \approx 1.8 \times 10^8$ sec^{-1}. For ruby the upper level lifetime $\tau_{21} \approx 2 \times 10^{-3}$ sec. Assuming a pumping level such that $g_0/g_t = 2.0$, we compute from (12.3.16) and (12.3.17) the period and lifetime of relaxation oscillations:

$$T_r \approx 21\ \mu\text{sec} \qquad (12.3.18)$$

$$\tau_r \approx 2\ \text{msec} \qquad (12.3.19)$$

Relaxation-oscillation periods are often in the microsecond range, as in this example. The damping time τ_r is particularly large in ruby because of its unusually long upper-level lifetime τ_{21}. Relaxation oscillations are therefore particularly pronounced in ruby. The output of a continuously pumped ruby laser typically consists of a series of irregular spikes, and this spiking behavior is usually attributed to relaxation oscillations being continuously excited by various mechanical and thermal perturbations.

12.4 Q SWITCHING

Q switching is a way of obtaining short, powerful pulses of laser radiation. Q refers to the quality factor of the laser resonator, as discussed in Section 11.9; recall that a high-Q cavity is one with low loss, whereas a "lossy" cavity will have a low Q. The term Q switching therefore refers to an abrupt change in the cavity loss. Specifically, it is a sudden switching of the cavity Q from a low value to a high value, i.e., a sudden lowering of the cavity loss. In this section we will describe how Q switching works, and in the following section how it is achieved in actual lasers.

Suppose we pump a laser medium inside a very lossy cavity. Because the loss is so large, laser action is precluded even if the upper level population N_2 is pumped to a very high value. No field builds up by stimulated emission in the gain cell. Obviously this means that the gain cannot be saturated, and if pumping is very strong it can grow to a large, small-signal value. Suddenly we lower the loss to a value permitting laser oscillation. We now have a small-signal gain much larger than the threshold gain for oscillation.

What happens in this situation, of course, is that there is a rapid growth of intensity inside the cavity. The intensity builds up quickly to a large value, resulting in a large stimulated emission rate and therefore a rapid extraction of energy from the gain cell. The result of the Q switching is therefore a short, intense pulse of laser radiation, sometimes called a *giant pulse*. Pulses as short as 10^{-7}–10^{-8} sec are routinely obtained by Q switching.

This qualitative explanation of Q switching may be substantiated by solving the rate equations (12.2.5). For this purpose it is convenient to define the dimensionless quantities

$$x = \frac{I_\nu}{ch\nu \, N_t} \qquad (12.4.1)$$

$$y = \frac{N_2}{N_t} \qquad (12.4.2)$$

where N_t is the threshold population inversion density. The threshold gain is $g_t = \sigma(\nu) N_t$. Clearly y is the ratio of the population inversion to the threshold inversion, or, equivalently, the ratio of gain g to threshold gain g_t; and x is easily shown to be the ratio of the cavity photon density to the threshold population inversion density (Problem 12.2).

In terms of x, y, and the dimensionless time variable

$$\tau = cg_t t \tag{12.4.3}$$

the equation (12.2.5a) for the cavity intensity takes the form (Problem 12.2)

$$\frac{dx}{d\tau} = (y - 1)x \tag{12.4.4a}$$

We will assume that the duration of the "giant pulse" is short enough that pumping and spontaneous decay of N_2 during this interval is negligible, and only stimulated decay due to the intense pulse occurs. This assumption allows us to ignore the second and third terms on the right side of the rate equation (12.2.5b), and to write the simpler equation (Problem 12.2)

$$\frac{dy}{d\tau} = -xy \tag{12.4.4b}$$

The validity of this assumption can always be checked after a solution of equations (12.4.4) has been obtained.

The result of a numerical integration of equations (12.4.4) is shown in Figure 12.1. The pumping level prior to Q switching is assumed to be such that $y(0) = 2$. We observe that the normalized intensity x grows until the population inversion drops below threshold, at which point the intensity begins to decrease.

As a specific example, consider the case of a 6943-Å ruby laser, in which $\sigma \approx 2.7 \times 10^{-20}$ cm^2. Suppose one of the mirrors is highly reflecting ($r_1 \approx 1.0$), the other has a reflectivity $r \approx 0.90$, and the ruby rod has a length $l = 5$ cm. Then

$$g_t \approx \frac{1}{2l}(1 - r) = 0.01 \text{ cm}^{-1} \tag{12.4.5}$$

and $cg_t \approx 3 \times 10^8$ sec^{-1}. From (12.4.3), therefore,

$$\tau = 3 \times 10^8 t \quad (t \text{ measured in seconds}) \tag{12.4.6}$$

The Q-switched pulse of Figure 12.1 has a width of about $\tau = 4$, corresponding to an actual pulse duration of

$$t_p = (3 \times 10^8)^{-1}(4) = 13 \text{ nsec} \tag{12.4.7}$$

The variable x in Figure 12.1 has a peak value of about 0.3, corresponding to a peak intensity of [Eq. (12.4.1)]

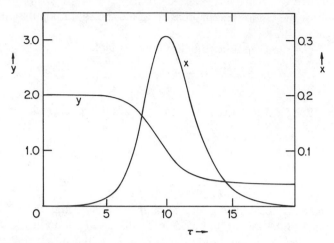

Figure 12.1 Solution of Eqs. (12.4.4) for $y(0) = 2$ and $x(0) \approx 0$.

$$I_{\text{peak}} = (ch\nu) N_t (0.3) = \left(\frac{c^2 h}{\lambda}\right) \left(\frac{g_t}{\sigma}\right) (0.3)$$

$$\approx 10^9 \text{ W/cm}^2 \qquad (12.4.8)$$

as the reader may easily verify. This is a very large amount of power—much larger than would be obtainable if the same laser were operated as a continuous-wave device (Problem 12.3). For a beam cross-sectional area of 0.1 cm^2 the total energy in the Q-switched pulse is

$$\text{energy} \approx (I_{\text{peak}}) (t_p) (0.1 \text{ cm}^2)$$

$$\approx 1 \text{ J} \qquad (12.4.9)$$

• Equations (12.4.4) imply that, if $x(0) = 0$, then x and y remain fixed at their initial values. Physically, this is incorrect, and occurs only because in writing (12.4.4) we left out the effect of spontaneous emission. Spontaneous emission has the effect of giving x a small but nonzero initial value, allowing it to grow from this initial value. In other words, spontaneous emission provides the first few "seed" photons needed to initiate the growth of laser intensity by stimulated emission.

In obtaining the numerical results shown in Figure 12.1 a fourth-order Runge–Kutta integration algorithm was used, with a step size $\Delta\tau = 0.01$. An initial value of 10^{-4} was assumed for $x(0)$. The numerical results for the pulse shape, duration, and peak value are insensitive to the (small) initial value assumed for x. This is because x grows to values large compared with its initial value. Then the number of cavity photons becomes so large that spontaneous emission is negligible compared with stimulated emission.

The value of τ at which the pulse intensity reaches its peak, however, does depend upon the choice of $x(0)$. If this aspect of the problem is of concern, therefore, one should include properly the effect of spontaneous emission in the rate equations (12.4.4) as well as various details of the *Q* switching.

The reader may wish to experiment with Eqs. (12.4.4), or various other differential equations that appear in this book and in the laser research literature. We include in Appendix 12.A both BASIC and FORTRAN listings of the Runge–Kutta algorithm used to obtain the results in Figure 12.1. •

12.5 METHODS OF *Q* SWITCHING

There are various ways to *Q*-switch a laser. The most popular ones switch the cavity *Q* factor within a time interval that is short compared with the photon lifetime $(cg_t)^{-1}$, allowing the gain to build up to a large value before the onset of laser oscillation. We will discuss three common methods of *Q* switching.

Rotating Mirrors

One way to *Q*-switch is to have one of the cavity mirrors rotating about an axis perpendicular to the cavity axis (Figure 12.2). The loss is then very large except during the brief period when the mirrors are nearly parallel. A typical angular velocity of the rotating mirror is about 10,000 revolutions per minute (rpm).

A similar mechanical method of *Q* switching involves a rotating chopper wheel. In this method, however, the *Q* switching is effected relatively slowly, even for a wheel velocity of 10,000 rpm. This is because lasing can begin before the shutter fully exposes the gain cell to the cavity mirrors.

Electro-optical Switches

Electro-optical shutters can be used to control the cavity *Q* by means of an applied voltage. To understand the operation of these switches, we must first describe the *electro-optical effect*. In Section 2.9 we identified birefringence as a difference in refractive indices for light of different linear polarizations. The electro-optical effect refers to birefringence that occurs in certain media when a voltage is applied. One example is the Kerr effect, in which the degree of birefringence is proportional to the square of the applied voltage. Another is the Pockels effect, in which the birefringence is linearly proportional to the voltage. Kerr cells typically require voltages in the 10-20 kilovolt range and Pockels cells somewhat less.

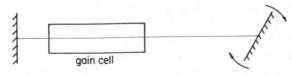

gain cell

Figure 12.2 A laser cavity with a rotating mirror for *Q* switching.

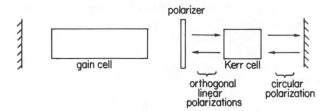

Figure 12.3 Q switching with a Kerr cell. With the Kerr cell on the cavity loss is large, but is suddenly lowered when the Kerr cell is switched off.

Q switching can be effected by using a polarizer and an electro-optical cell, as illustrated in Figure 12.3. The voltage and orientation of the Kerr cell are such that the (linearly polarized) light passing through the polarizer is converted to circularly polarized light. After reflection off the cavity mirror this circularly polarized light is converted by the Kerr cell to light linearly polarized *orthogonally* to the polarizer axis. The presence of the Kerr cell thus prevents feedback, and the cavity is in effect a very lossy one. If the voltage across the Kerr cell is switched off, however, the cell is no longer birefringent. Then the cavity Q has suddenly been increased, and a giant pulse develops.

Saturable Absorbers

Another way of Q switching is to place in the laser cavity a "shutter" consisting of a cell of absorbing material whose absorption coefficient can be saturated (or "bleached") by the laser radiation. Saturation can occur in both absorbing and amplifying media. The absorption (or gain) coefficient decreases with increasing intensity of resonant radiation, becoming nearly zero when the intensity is much larger than a characteristic saturation intensity I^{sat} of the medium (recall Fig. 10.8).

When the gain medium is first pumped, the gain threshold is very high. The cavity loss is too large, because of the absorbing cell, to allow laser oscillation. The medium can therefore be pumped to a high gain without generating significant light intensity. Once the gain is high enough to overcome the loss, however, the cavity intensity grows rapidly. This in turn rapidly saturates the absorption cell, and the effective cavity loss drops abruptly. The whole process leads to a giant-pulse output in a manner similar to that with a mechanical Q switch (Problem 12.4).

The use of a saturable absorber for Q switching is often called *passive Q* switching, in contrast to the *active Q* switching achieved mechanically or electro-optically as described above.

The passive Q switch is obviously simpler in terms of the necessary auxiliary equipment than the two active Q switches we have described. It enjoys an additional advantage: a passive Q switch will often give an output pulse concentrated mostly in a single mode. The reason for this is that it takes a finite time for the saturable absorber to become highly saturated and thus to raise the cavity Q. In the meantime the cavity intensity originating from spontaneous-emission noise builds up on different modes, and it grows to a greater degree on those modes with the lowest loss per pass. Since a photon can typically make several thousand round trips between the cavity mirrors before the absorber saturates, even small differences in the losses of different modes become significant. The result is that only the lowest-loss mode (or modes) appear in the Q-switched pulse.

In the active Q switches, the switch to high Q is much more rapid, typically occurring during only several tens of photon round trips in the cavity. Small differences in mode losses per pass may then not be sufficient to discriminate among different modes. The output frequency spectrum of a Q-switched ruby laser has long been known to be narrower if a passive Q switch is used instead of a rotating mirror or a Kerr (or Pockels) cell.

12.6 MULTIMODE LASER OSCILLATION

In Section 11.12 we noted that a laser with a homogeneously broadened gain medium tends to oscillate on a single longitudinal mode if the effect of spatial hole burning is small. This expectation is borne out in collision-broadened gas lasers, where atomic motion tends to smear out the effect of spatial hole burning. A similar effect can occur in solid-state lasers in which there is a diffusion of excitation among the atoms. In general, however, oscillation will occur on many longitudinal modes, especially when the gain medium is pumped far above threshold, allowing many modes under the gain curve to meet the threshold condition.

Ruby lasers, for example, are predominantly homogeneously broadened. Due to spatial hole burning, however, they generally oscillate multimode, especially when strongly pumped. As the pumping rate is increased, furthermore, the power associated with any particular mode tends to rise at a slower rate than the total output power on all lasing modes.

Single-longitudinal-mode oscillation is generally precluded by spectral hole burning in inhomogeneously broadened lasers, unless the laser cavity is very short (Figure 11.12) or the pumping rate permits only one mode to reach threshold (Figure 12.4). He–Ne lasers, for example, usually oscillate on several longitudinal

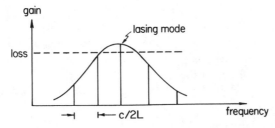

Figure 12.4 A case in which several modes lie under the gain curve, but only one can lase.

(a)   (b)

Figure 12.5 Typical output spectra of a 1-m-long He–Ne laser for low (a) and high (b) pumping (discharge current) levels.

modes in the absence of any mode selection mechanism (e.g., an etalon). Figure 12.5 shows the output spectrum of a typical, low-pressure, 6328-Å He–Ne laser having a mirror separation $L = 1$ m. Figure 12.5a is the result obtained at a relatively low pumping level. Only one longitudinal mode is above threshold. As the pumping level is raised by increasing the discharge current, however, several modes under the 1700-MHz Doppler profile can oscillate (Figure 12.5b), and their frequency spacing is near $c/2L = 150$ MHz, as expected.

A rigorous theory of laser oscillation must therefore describe the case in which several or many modes oscillate simultaneously. In this case we cannot formulate the theory in terms of a single cavity photon number or intensity. Instead we must specify the photon number or intensity *for each mode*. The analysis is especially complicated by spectral hole burning in the case of inhomogeneous broadening.

The rate equations (10.5.14), or their simplified version (12.2.5b), describe the rate of change of the total number of atoms per unit volume in a particular atomic level. They do not take account of the fact that different atoms may have different line-center frequencies and therefore different stimulated-emission cross sections $\sigma(\nu)$ for radiation of frequency ν. That is, there is no account of inhomogeneous line broadening. If there is inhomogeneous broadening, we must write separate rate equations for different "spectral packets" of atoms. Different spectral packets will then saturate to different degrees (spectral hole burning), and the complications can be enormous in the multimode case. A proper description of this case would, for our purposes, be inordinately lengthy.

In spite of these complexities, there are situations where the gain of a multimode, inhomogeneously broadened laser saturates homogeneously in the sense that every spectral packet saturates in the same manner. In this case a saturation formula like (10.11.4) is applicable, and the total output power on all modes is well described by the Rigrod-type analysis discussed in the preceding chapter. One situation in which this is realized approximately is when the longitudinal mode spacing $c/2L$ is small compared with the homogeneous linewidth $\delta\nu_0$, i.e., when there are many longitudinal modes lying within the frequency interval $\delta\nu_0$. Evidence for the validity of this approximation may be found in the results of experiments with a low-pressure 3.51-μm He–Xe laser, which is highly inhomogeneously (Doppler) broadened.[1] The cavity mirrors were separated by over 10 m in

1. L. W. Casperson, *IEEE Journal of Quantum Electronics* **QE-9**, 250 (1973).

order to permit the oscillation of a large number of longitudinal modes. The total output power was described quite well by the Rigrod theory for a *homogeneously* broadened laser. (Section 11.5)

There are other effects tending to "homogenize" the gain saturation. In a gas laser, for instance, collisions will change the z component of an atom's velocity and tend to "fill in" a spectral hole. In other words, collisions act to preserve the Maxwell–Boltzmann velocity distribution, and therefore the Doppler gain profile. Collisions thus act in opposition to the spectral hole-burning effect of the field. At high intensity levels the effective homogeneous linewidth is also increased due to power broadening (Section 10.11).

12.7 PHASE-LOCKED OSCILLATORS

In a Q-switched laser the light pulse must make several passes through the gain medium after the cavity Q is switched. Feedback is necessary in order to build up a large field amplitude by stimulated emission. For some applications it is desirable to have pulses of light even shorter than can be achieved by Q switching. Such powerful, ultrashort pulses of light can be obtained by a technique known as *mode locking*.

Whereas Q switching may involve either a single mode or many modes, mode locking is a fundamentally multimode phenomenon. Specifically, mode locking involves the "locking" together of the phases of many cavity longitudinal modes. The purpose of this section is to consider a simple analog of a mode-locked laser. We will consider the problem of adding the displacements of N harmonic oscillators with equally spaced frequencies. That is, we consider the sum of

$$x_n(t) = x_0 \sin(\omega_n t + \phi_0) \tag{12.7.1}$$

where

$$\omega_n = \omega_0 + n\Delta,$$

$$n = -\frac{N-1}{2}, \; -\frac{N-1}{2} + 1, \; -\frac{N-1}{2} + 2, \ldots, \; \frac{N-1}{2} \tag{12.7.2}$$

In other words, the amplitudes x_0 and phases ϕ_0 of the oscillators are identical, and their frequencies ω_n are equally spaced by Δ and centered at ω_0, as shown in Figure 12.6. The sum of the displacements (12.7.1) is

$$X(t) = \sum_n x_n(t) = \sum_{-(N-1)/2}^{(N-1)/2} x_0 \sin(\omega_n t + \phi_0) \tag{12.7.3}$$

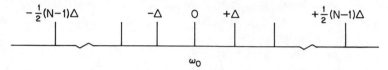

Figure 12.6 A collection of N frequencies running from $\omega_0 - \frac{1}{2}(N-1)\Delta$ to $\omega_0 + \frac{1}{2}(N-1)\Delta$ as in Eq. (12.7.2).

Since $\sin x$ is the imaginary part of e^{ix}, we may write this as

$$X(t) = x_0 \, \text{Im} \left(\sum_n e^{i(\omega_0 t + \phi_0 + n\Delta t)} \right)$$

$$= x_0 \, \text{Im} \left(e^{i(\omega_0 t + \phi_0)} \sum_n e^{in\Delta t} \right) \tag{12.7.4}$$

The general identity

$$\sum_{-(N-1)/2}^{(N-1)/2} e^{iny} = \frac{\sin(Ny/2)}{\sin(y/2)} \tag{12.7.5}$$

proved below allows us to write (12.7.4) as

$$X(t) = x_0 \, \text{Im} \left(e^{i(\omega_0 t + \phi_0)} \frac{\sin(N\Delta t/2)}{\sin(\Delta t/2)} \right)$$

$$= x_0 \sin(\omega_0 t + \phi_0) \left(\frac{\sin(N\Delta t/2)}{\sin(\Delta t/2)} \right)$$

$$= A_N(t) \, x_0 \sin(\omega_0 t + \phi_0) \tag{12.7.6}$$

The function $A_N(t)$ is plotted in Figure 12.7 for $N = 3$ and $N = 7$. In general $A_N(t)$ has equal maxima

$$A_N(t)_{max} = N \tag{12.7.7}$$

at values of t given by

$$t_m = m \left(\frac{2\pi}{\Delta} \right) \equiv mT, \quad m = 0, \pm 1, \pm 2, \ldots \tag{12.7.8}$$

As N increases, the maxima of $A_N(t)$ become larger. They also become more

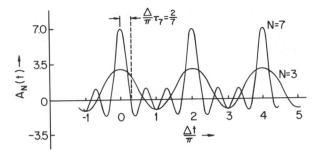

Figure 12.7 The function $A_N(t) = (\sin \frac{1}{2} N \, \Delta t)/(\sin \frac{1}{2} \Delta t)$ vs. $\Delta t/\pi$.

sharply peaked. A measure of their width is the time interval τ_N indicated in Figure 12.7 for $N = 7$:

$$\tau_N = \frac{2\pi}{N\Delta} = \frac{T}{N} \tag{12.7.9}$$

We have thus shown that the addition of N oscillators of equal amplitudes and phases, and equally spaced frequencies (12.7.2), gives maximum total oscillation amplitudes equal to N times the amplitude of a single oscillator. These maximum amplitudes occur at intervals of time T [Eq. (12.7.8)]. For large N we have, loosely speaking, a series of large-amplitude "spikes." The smaller the frequency spacing Δ between the individual oscillators, the larger the time interval $T = 2\pi/\Delta$ between spikes, and conversely. The temporal duration of each spike is $\tau_N = T/N$, so the spikes get sharper as N is increased.

We have assumed for simplicity that each oscillator has the same phase ϕ_0 [Eq. (12.7.1)]. A more general kind of phase locking occurs when the phase differences of the oscillators are constant but not necessarily zero:

$$\phi_n = \phi_0 + n\alpha \tag{12.7.10a}$$

or

$$\phi_{n+1} - \phi_n = \alpha \tag{12.7.10b}$$

In this case the sum of the oscillator displacements (12.7.3) is replaced by

$$X(t) = \sum_n x_0 \sin x_0(\omega_n t + \phi_n)$$

$$= x_0 \, \mathrm{Im} \left(e^{i(\omega_0 t + \phi_0)} \sum_{-(1/2)(N-1)}^{(N-1)/2} e^{in(\Delta t + \alpha)} \right) \tag{12.7.11}$$

and this may be evaluated to give the total displacement

$$X(t) = x_0 \sin\left(\omega_0 t + \phi_0\right) \left(\frac{\sin N(\Delta t + \alpha)/2}{\sin\left(\Delta t + \alpha\right)/2}\right) \qquad (12.7.12)$$

having basically the same properties as (12.7.6) obtained with $\alpha = 0$. (See also Problem 12.5.)

• We prove (12.7.5) as follows. Let the sum be denoted S_N. For convenience we will first evaluate S_{N+1}.

$$S_{N+1} = \sum_{n=-N/2}^{+N/2} e^{iny} \qquad (12.7.13)$$

The first step is to shift the summation label by introducing

$$m = n + N/2 \qquad (12.7.14)$$

so that

$$S_{N+1} = \sum_{m=0}^{N} e^{i(m-N/2)y}$$

$$= e^{-iNy/2} \sum_{m=0}^{N} e^{imy}$$

$$= e^{-iNy/2} \sum_{m=0}^{N} \left(e^{iy}\right)^m \qquad (12.7.15)$$

The second step is to make use of the identity

$$\sum_{m=0}^{N} x^m = \frac{1 - x^{N+1}}{1 - x} \qquad (12.7.16)$$

Then we can write

$$S_{N+1} = e^{-iNy/2} \frac{1 - e^{i(N+1)y}}{1 - e^{iy}}$$

$$= e^{-iNy/2} \frac{e^{i(N+1)y/2}}{e^{iy/2}} \frac{e^{-i(N+1)y/2} - e^{i(N+1)y/2}}{e^{-iy/2} - e^{iy/2}}$$

$$= \frac{\sin\left(N+1\right)y/2}{\sin y/2} \qquad (12.7.17)$$

Thus we have proved

$$S_N = \frac{\sin Ny/2}{\sin y/2} \qquad (12.7.18)$$

as claimed in (12.7.5). •

12.8 MODE LOCKING

What we have in the example of the preceding section is a simple model of a mode-locked laser. The individual oscillators in the model play the role of individual longitudinal-mode fields, while their frequency spacing Δ represents the mode (angular) frequency separation $2\pi(c/2L) = \pi c/L$. The assumption of equal oscillator phase differences α ("phase locking") in the model corresponds to the locking together of the phases of the different cavity modes.

Our oscillator model suggests that, if we can somehow manage to lock together the phases of N longitudinal modes of a laser, then the light coming out of the laser will consist of a *train of pulses* separated in time by $T = 2\pi/\Delta = 2L/c$. The temporal duration of each pulse in the train will be $\tau_N = T/N = 2L/cN$. The larger the number N of phase-locked modes, the greater the amplitude, and the shorter the duration, of each individual pulse in the train. As we will see, this is indeed the essence of the mode-locking technique for obtaining ultrashort, powerful laser pulses.

The number of longitudinal modes that can simultaneously lase is determined by the gain linewidth (FWHM) $\Delta\nu_g$ and the frequency separation $c/2L$ between modes (cf. Figure 12.8). Under sufficiently strong pumping of the gain medium we expect that approximately

$$M = \frac{\Delta\nu_g}{c/2L} = \frac{2L}{c}\Delta\nu_g \qquad (12.8.1)$$

longitudinal modes can oscillate simultaneously. The shortest pulse length we expect to achieve by mode locking is therefore

$$\tau_{\min} = \tau_M = \frac{2L}{cM} = \frac{1}{\Delta\nu_g} \qquad (12.8.2)$$

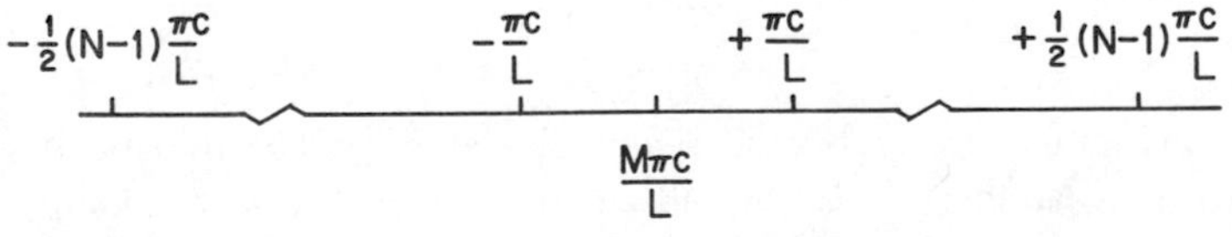

Figure 12.8 The distribution of N cavity mode frequencies as given by Eq. (12.8.7). The situation is exactly the same as in Figure 12.7 for the case of N phase-locked oscillators.

That is, *the shortest pulse duration we can achieve by mode locking is (approximately) the reciprocal of the gain linewidth.*

As an example, consider the 6328-Å He–Ne laser with a gain linewidth $\Delta \nu_g = \delta \nu_D = 1700$ MHz. For such a laser the shortest pulses obtainable by mode locking are of duration

$$\frac{1}{\delta \nu_D} = \frac{1}{1700 \times 10^6 \text{ sec}^{-1}} \approx 1 \text{ nsec} \qquad (12.8.3)$$

In other words, for this laser, mode locking is not much of an improvement over Q switching for the production of short pulses. This is often true of gas lasers. Their gain linewidths are so narrow that ultrashort (say, picosecond, 10^{-12} sec duration) pulses cannot be obtained by mode locking.

On the other hand, consider a 6934-Å ruby laser with $\Delta \nu_g \approx 10^{11} \text{ sec}^{-1}$. For this laser mode-locked pulses of 10^{-11} sec may be obtained.

Liquid dye lasers typically have broad gain profiles, with $\Delta \nu_g \approx 10^{12} \text{ sec}^{-1}$ or more. With such lasers mode-locked pulses in the picosecond range are routinely obtained.

A basic understanding of mode-locked laser oscillation may be reached by extending only slightly our analysis of phase-locked oscillators. We associate with the mth longitudinal mode an electric field

$$\mathbf{E}_m(z, t) = \hat{\mathbf{\varepsilon}}_m \, \mathcal{E}_m(z) \sin(\omega_m t + \phi_m)$$

$$= \hat{\mathbf{\varepsilon}}_m \, \mathcal{E}_m \sin k_m z \, \sin(\omega_m t + \phi_m) \qquad (12.8.4)$$

where

$$k_m = m \frac{\pi}{L}, \qquad m = 1, 2, 3, \dots \qquad (12.8.5a)$$

and

$$\omega_m = k_m c = m \frac{\pi c}{L}, \qquad m = 1, 2, 3, \dots \qquad (12.8.5b)$$

For simplicity let us assume that the mode fields all have the same magnitude ($\mathcal{E}_0$) and polarization, so that we can do our calculations below with scalar quantities. Furthermore let us consider, without much loss of generality, the simplest example of phase locking, in which all $\phi_m = 0$. Then the total electric field in the cavity is

$$E(z, t) = \sum_m E_m(z, t)$$

$$= \mathcal{E}_0 \sum_m \sin k_m z \sin \omega_m t \qquad (12.8.6)$$

where the summation is over all oscillating modes.

For a cavity 1 m long, $\pi c/L \approx 9 \times 10^8$ Hz. For near-optical frequencies, of course, the lasing frequencies ω_m will be much larger; at a wavelength of 6000 Å, $\omega = 2\pi c/\lambda \approx 3 \times 10^{15}$ Hz. The integer m in (12.8.5) will therefore typically be in the millions. So let us write (12.8.5) as

$$k_m = (M + n)\, \pi/L \qquad (12.8.7a)$$

$$\omega_m = (M + n)\, \pi c/L \qquad (12.8.7b)$$

where M is a very large positive integer ($M \approx 10^6$) and n runs from $-\frac{1}{2}(N - 1)$ to $+\frac{1}{2}(N - 1)$, corresponding to a total of $N\ (\ll M)$ modes centered at the frequency $M\pi c/L$ (Figure 12.8). Then (12.8.6) becomes

$$E(z, t) = \mathcal{E}_0 \sum_{-(N-1)/2}^{(N-1)/2} \sin \frac{(M + n)\pi z}{L} \sin \frac{(M + n)\pi ct}{L}$$

$$= \tfrac{1}{2} \mathcal{E}_0 \sum_n \left(\cos \frac{(M + n)\pi(z - ct)}{L} - \cos \frac{(M + n)\pi(z + ct)}{L} \right)$$

$$(12.8.8)$$

for the total electric field in the laser cavity.

Now we proceed as in the preceding section. The sum

$$\sum_{-(N-1)/2}^{(N-1)/2} \cos \frac{(M + n)\, \pi(z - ct)}{L}$$

$$= \mathrm{Re} \sum_{n=-(N-1)/2}^{(N-1)/2} e^{i(M+n)\pi(z-ct)/L}$$

$$= \mathrm{Re} \left(e^{iM\pi(z-ct)/L}\, \frac{\sin \pi N(z - ct)/2L}{\sin \pi(z - ct)/2L} \right)$$

$$= \left(\cos \frac{M\pi(z - ct)}{L} \right) \frac{\sin \pi N(z - ct)/2L}{\sin \pi(z - ct)/2L} \qquad (12.8.9)$$

where we have again used the identity (12.7.5). Similarly

$$\sum_n \cos \frac{(M + n)\, \pi(z + ct)}{L} = \left(\cos \frac{M\pi(z + ct)}{L} \right) \frac{\sin \pi N(z + ct)/2L}{\sin \pi(z + ct)/2L}$$

$$(12.8.10)$$

From (12.8.8), then,

$$E(z, t) = \frac{\mathcal{E}_0}{2} \left(\cos k_0(z - ct) \frac{\sin N\pi(z - ct)/2L}{\sin \pi(z - ct)/2L} \right.$$

$$\left. - \cos k_0(z + ct) \frac{\sin N\pi(z + ct)/2L}{\sin \pi(z + ct)/2L} \right) \qquad (12.8.11)$$

where $k_0 = \pi M/L$.

The functions

$$A_N^{(\pm)}(z, t) = \frac{\sin N\pi(z \pm ct)/2L}{\sin \pi(z \pm ct)/2L} \qquad (12.8.12)$$

appearing in (12.8.11) have basically the same form—and effect—as the function $A_N(t)$ appearing in Eq. (12.7.6) for the phase-locked oscillator model. In particular, $A_N^{(\pm)}(z, t)$ has maxima occurring at

$$z \pm ct = m(2L), \qquad m = 0, \pm 1, \pm 2, \dots \qquad (12.8.13)$$

If we put our attention on a fixed value of z inside the cavity, for instance, there are pulses of peak amplitude $N\mathcal{E}_0/2$ appearing at time intervals of $2L/c$, each pulse having a duration T/N (Figure 12.9). If we fix our attention on the spatial distribution of $E(z, t)$ at a fixed time t, we find pulses of amplitude $N\mathcal{E}_0/2$ with spatial separation $2L$, each pulse having a spatial extent of $2L/N$ (Figure 12.10).

In other words, the field (12.8.11) represents two trains of pulses, one moving in the positive z direction and the other in the negative z direction. In the usual situation in which output is obtained through one of the cavity mirrors, the laser radiation appears as a single train of pulses of temporal separation and duration $2L/c$ and $2L/cN$, respectively. All this confirms our conclusions deduced from the phase-locked oscillator model.

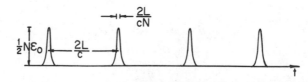

Figure 12.9 A mode-locked pulse train as a function of time, observed at a fixed position z.

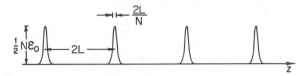

Figure 12.10 A mode-locked pulse train as a function of coordinate z, observed at a fixed instant of time.

The fact that the pulses of a mode-locked train are separated in time by the round-trip cavity transit time $2L/c$ suggests a "bouncing-ball" picture of a mode-locked laser: we can regard the mode locking as generating a pulse of duration $2L/cN$, and this pulse keeps bouncing back and forth between the cavity mirrors. Focusing our attention on a particular plane of constant z in the resonator, we observe a train of identical pulses moving in either direction.

In most lasers the phases ϕ_n of the different modes will undergo random and uncorrelated variations in time. In this case the total intensity is the sum of the individual mode intensities. In mode-locked lasers, however, the mode phases are correlated and the total intensity is not simply the sum of the individual mode intensities. In fact the individual pulses in the mode-locked train have an intensity N times larger than the sum of the individual mode intensities. The average power, however, is essentially unaltered by mode-locking the laser (Problem 12.6).

• Before discussing how mode locking can be accomplished, it is worth noting that "phase locking" or "synchronization" phenomena occur in many *nonlinear* oscillatory systems besides lasers, and indeed these phenomena have been known for a very long time. C. Huygens (1629–1695), for instance, observed that two pendulum clocks hung a few feet apart on a thin wall tend to have their periods synchronized as a result of their small coupling via the vibrations of the wall. Near the end of the nineteenth century Lord Rayleigh found that two organ pipes of slightly different resonance frequencies will vibrate at the same frequency when they are sufficiently close together.

The contractive pulsations of the heart's muscle cells become phase-locked during the development of the fetus. Fibrillation of the heart occurs when they get out of phase for some reason, and results in death unless the heart can be shocked back into the normal condition of cell synchronization. There are other biological examples of phase locking, but detailed theoretical analyses are obviously extremely difficult or impossible for such complex systems. Modern applications of synchronization principles are made in high-precision motors and control systems. •

12.9 AM MODE LOCKING

The process by which phase or mode locking is forced upon a laser is fundamentally a nonlinear one, and a rigorous analysis of it is complicated. We will therefore rely largely on semiquantitative explanations.

Consider again the scalar electric field

$$E_m(z, t) = \mathcal{E}_m \sin k_m z \sin (\omega_m t + \phi_m) \qquad (12.9.1)$$

associated with a longitudinal mode. Suppose that the amplitude $\mathcal{E}_m$ is not constant but rather is modulated periodically in time according to the formula

$$\mathcal{E}_m = \mathcal{E}_0 (1 + \epsilon \cos \Omega t) \qquad (12.9.2)$$

where Ω is the modulation frequency and $\mathcal{E}_0$ and ϵ are constants. Thus we have an amplitude-modulated field

$$E_m(z, t) = \mathcal{E}_0 (1 + \epsilon \cos \Omega t) \sin (\omega_m t + \phi_m) \sin k_m z \qquad (12.9.3)$$

Since

$$\cos \Omega t \sin (\omega_m t + \phi_m) = \tfrac{1}{2} \sin (\omega_m t + \phi_m + \Omega t)$$

$$+ \tfrac{1}{2} \sin (\omega_m t + \phi_m - \Omega t) \qquad (12.9.4)$$

we can write the field (12.9.3) as a sum of harmonically varying parts:

$$E_m(z, t) = \mathcal{E}_0 \Big\{ \sin (\omega_m t + \phi_m) + \frac{\epsilon}{2} \sin \big[(\omega_m + \Omega)t + \phi_m \big]$$

$$+ \frac{\epsilon}{2} \sin \big[(\omega_m - \Omega)t + \phi_m \big] \Big\} \sin k_m z \qquad (12.9.5)$$

The frequency spectrum of the field (12.9.5) is shown in Figure 12.11. The amplitude modulation of the field (12.9.1) of frequency ω_m has generated *sidebands* of frequency $\omega_m \pm \Omega$. These sidebands are displaced from the *carrier frequency* ω_m by precisely the modulation frequency Ω. Sideband generation is a well-known consequence of amplitude modulation.

In a laser the mode amplitudes $\mathcal{E}_m$ are determined by the condition that the gain equals the loss. If the loss (or gain) is periodically modulated at a frequency Ω, we expect the fields $E_m(z, t)$ associated with the various modes to be amplitude-modulated (AM) with this frequency. In other words, we expect sidebands to be generated about each mode frequency ω_m, as in (12.9.5). In particular, if the modulation frequency Ω is equal to the mode frequency spacing

$$\Delta = \omega_{m+1} - \omega_m = \pi c / L \qquad (12.9.6)$$

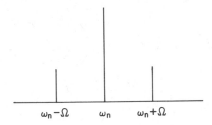

Figure 12.11 Frequency spectrum of the amplitude modulated field (12.9.5). The sidebands at $\omega_n \pm \Omega$ have amplitudes $\epsilon/2$ times as large as the carrier amplitude at ω_n. In this case $\epsilon/2 < 1$.

the sidebands associated with each mode match exactly the frequencies of the two adjacent modes (Figure 12.12). In this case each mode becomes strongly coupled to its nearest-neighbor modes, and it turns out that there is a tendency for the modes to lock together in phase. Loss or gain modulation at the mode separation frequency Δ is therefore one way of mode locking. Borrowing terminology from radio engineering, we call this *AM mode locking*.

The dimensionless factor ϵ appearing in (12.9.2) is called the modulation index. It is usually small, but it must be large enough to couple the different modes sufficiently strongly. This is analogous to the synchronization phenomenon observed in the 17th century by Huygens with pendulum clocks. Their frequencies were locked together when the clocks were mounted just a meter or so apart, but larger separations weakened their coupling and destroyed the locking effect. If ϵ is too large, on the other hand, the locking effect is also weakened. This is analogous to the distortion arising in AM radio electronic systems when the carrier wave is "overmodulated," i.e., when $\epsilon > 1$. (See also Problem 12.7.)

A heuristic way to understand why AM mode locking occurs in lasers is first to suppose that lasing can occur only in brief intervals when the periodically modulated loss is at a minimum. These minima occur in time intervals of $T = 2\pi/\Delta = 2L/c$ if the modulation frequency $\Omega = \Delta$. Between these times of minimum loss the loss is too large for laser oscillation. Thus we can have laser oscillation only if it is possible to generate a train of short pulses separated in time by T. This is possible if the modes lock together and act in unison, for then we generate a mode-locked train of pulses separated by time T. Thus mode locking has been described as a kind of "survival of the fittest" phenomenon.

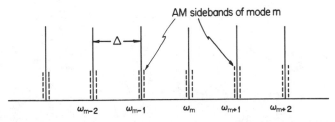

AM sidebands of mode m

Figure 12.12 Longitudinal modes amplitude-modulated at the frequency Δ equal to their spacing. For clarity the AM sidebands are indicated as dashed lines slightly dispaced from the mode frequencies ω_m.

12.10   FM MODE LOCKING

We will now consider the case where the *phase* of the field (12.9.1) is periodically modulated rather than the amplitude:

$$E_m(z, t) = \mathcal{E}_m \sin k_m z \sin (\omega_m t + \phi_m + \delta \cos \Omega t) \qquad (12.10.1)$$

The dimensionless constant δ gives the amplitude of the modulation of frequency Ω. As in the case of amplitude modulation, this phase modulation gives rise to sideband frequencies about the carrier frequency ω_m. As we will now see, however, the phase modulation produces a whole series of sidebands.

The time-dependent part of (12.10.1) may be written as

$$\sin (\omega_m t + \phi_m + \delta \cos \Omega t) = \sin (\omega_m t + \phi_m) \cos (\delta \cos \Omega t)$$

$$+ \cos (\omega_m t + \phi_m) \sin (\delta \cos \Omega t) \qquad (12.10.2)$$

Now we make use of two mathematical identities:[2]

$$\cos (x \cos \theta) = J_0(x) + 2 \sum_{k=1}^{\infty} (-1)^k J_{2k}(x) \cos (2k\theta) \qquad (12.10.3a)$$

and

$$\sin (x \cos \theta) = 2 \sum_{k=0}^{\infty} (-1)^k J_{2k+1}(x) \cos [(2k + 1) \theta] \qquad (12.10.3b)$$

where $J_n(x)$ is the Bessel function of the first kind of order n. The first few lowest-order Bessel functions are plotted in Figure 12.13. These plots are all we will need

2. See, for example M. Abramowitz and I. A. Stegun, *Handbook of Mathematical Functions* (Dover, New York, 1971), formulas 9.1.44 and 9.1.45.

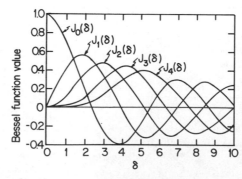

Figure 12.13 The first few lowest-order Bessel functions of the first kind, $J_n(\delta)$.

to know about them. The functions (12.10.3) appear in (12.10.2) with $x = \delta$ and $\theta = \Omega t$. Thus

$$\sin(\omega_m t + \phi_m + \delta \cos \Omega t)$$

$$= \sin(\omega_m t + \phi_m)\left(J_0(\delta) + 2 \sum_{k=1}^{\infty} (-1)^k J_{2k}(\delta) \cos(2k\Omega t)\right)$$

$$+ 2 \cos(\omega_m t + \phi_m) \sum_{k=0}^{\infty} (-1)^k J_{2k+1}(\delta) \cos[(2k+1)\Omega t]$$

$$= \sin(\omega_m t + \phi_m)[J_0(\delta) - 2J_2(\delta) \cos 2\Omega t$$

$$+ J_4(\delta) \cos 4\Omega t - 2J_6(\delta) \cos 6\Omega t + \cdots]$$

$$+ 2 \cos(\omega_m t + \phi_m)[J_1(\delta) \cos \Omega t - J_3(\delta) \cos 3\Omega t$$

$$+ J_5(\delta) \cos 5\Omega t - \ldots] \tag{12.10.4}$$

Using the identities

$$\sin x \cos y = \tfrac{1}{2}[\sin(x+y) + \sin(x-y)]$$

$$\cos x \cos y = \tfrac{1}{2}[\cos(x+y) + \cos(x-y)]$$

therefore, we have

$$\sin(\omega_m t + \phi_m + \delta \cos \Omega t)$$

$$= J_0(\delta) \sin(\omega_m t + \phi_m)$$

$$+ J_1(\delta)\{\cos[(\omega_m + \Omega)t + \phi_m] + \cos[(\omega_m - \Omega)t + \phi_m]\}$$

$$- J_2(\delta)\{\sin[(\omega_m + 2\Omega)t + \phi_m] + \sin[(\omega_m - 2\Omega)t + \phi_m]\}$$

$$- J_3(\delta)\{\cos[(\omega_m + 3\Omega)t + \phi_m] + \cos[(\omega_m - 3\Omega)t + \phi_m]\}$$

$$+ J_4(\delta)\{\sin[(\omega_m + 4\Omega)t + \phi_m] + \sin[(\omega_m - 4\Omega)t + \phi_m]\}$$

$$+ J_5(\delta)\{\cos[(\omega_m + 5\Omega)t + \phi_m] + \cos[(\omega_m - 5\Omega)t + \phi_m]\}$$

$$- \ldots \tag{12.10.5}$$

after a simple rearrangement of terms in (12.10.4).

Whereas amplitude modulation produces one sideband on either side of the carrier frequency ω_m, phase modulation in general produces a whole series of pairs of sidebands. If the "modulation index" δ is somewhat less than unity, however, we observe from (12.10.5) and Figure 12.13 that the first pair of sidebands

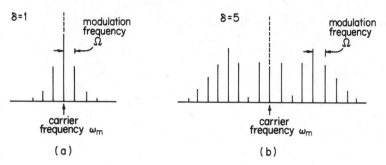

Figure 12.14 Frequency spectrum of the function (12.10.5) for the modulation index (a) $\delta = 1$ and (b) $\delta = 5$.

at $\omega_m \pm \Omega$ is strongest. As the strength of the modulation increases, i.e., as δ increases, more sideband pairs become important. Figure 12.14 shows the frequency spectrum of the function (12.10.5) for $\delta = 1$ and $\delta = 5$.

Again borrowing the terminology of radio engineering, we refer to this type of modulation as *frequency modulation* (FM). As in the AM case, frequency modulation at the mode separation frequency $\Omega = \Delta = \pi c/L$ causes the sidebands associated with each mode to be in resonance with the carrier frequencies of other modes. This results in a strong coupling of these modes and a tendency for them to lock together and produce a mode-locked train of pulses. This is called *FM mode locking*.

• Information cannot be transmitted with a purely monochromatic wave. The basic idea of radio communication is to modulate a monochromatic (carrier) wave in some way (AM or FM), transmit it, then demodulate it at a receiver to recover the information contained in the original modulation. In the AM case the sidebands imposed on the carrier wave are displaced from the carrier by an amount equal to the modulation frequency, independently of the modulation index ϵ. In the FM case, on the other hand, the "width" of the modulation about the carrier is directly proportional to the corresponding index δ, approximately independently of the modulation frequency Ω. This makes FM transmission less susceptible to interference from extraneous sources (lightning, electric power generators, etc.) than AM if its modulation index is large. At the same time, there is a disadvantage to FM in that the amplifiers in the transmitter and receiver must have large bandwidths in order to pick up a good portion of the sideband spectrum. A large bandwidth is most easily obtained at higher carrier frequencies; this is analogous to the fact that the bandwidth of a laser cavity increases with frequency if the cavity Q is held constant [Eq. (11.9.21)], and explains why FM radio stations broadcast at higher frequencies than AM stations (Problem 2.9). •

12.11   METHODS OF MODE LOCKING

Lasers can be mode-locked in a variety of ways. We will focus our attention on three common and illustrative techniques.

Acoustic Loss Modulation

This method is based on the diffraction of light by sound waves, i.e., on Brillouin scattering. A sound wave is basically a wave of density variation—and therefore refractive index variation (recall Section 2.4)—in a material medium. As discussed in Appendix 12.B, a sound wave can therefore act as a "diffraction grating" for light. A sound wave of wavelength λ_s diffracts light of wavelength λ with diffraction angle θ (Figure 12.19) satisfying (Eq. 12.B.3)

$$\sin \theta = \frac{\lambda}{2 n \lambda_s} \qquad (12.11.1)$$

where n is the refractive index of the medium.

A standing sound wave in a medium may be represented by a refractive index variation of the form

$$\Delta n(x, t) = a \sin(\omega_s t + \theta) \sin k_s x \qquad (12.11.2)$$

The periodic spatial modulation $\sin k_s x$ of the refractive index gives rise to diffraction at the angle θ given by (12.11.1) with $\lambda_s = 2\pi/k_s$. The temporal oscillation at frequency ω_s means that the diffraction is most effective at times t such that $\sin(\omega_s t + \theta) = \pm 1$, for at these times the "diffraction grating" represented by $\sin k_s x$ has its largest amplitude ($\pm a$). Thus the diffracting strength of the standing acoustic wave varies harmonically in time with frequency $2\omega_s$.

We can now understand how the diffraction of light by sound can be used to periodically modulate the cavity loss in a laser, and thereby to achieve AM mode locking. If a block of material having a standing acoustic wave is inside the cavity, the diffractive loss associated with it will oscillate with frequency $2\omega_s$. If $2\omega_s = \Delta = \pi c/L$, the cavity loss is modulated at the mode frequency separation, as desired for mode locking. Since audible sound waves have frequencies roughly from 20 Hz to 2×10^4 Hz, while the mode separations in a laser are typically much larger, it is clear that ultrasonic acoustic modulation is required for mode locking. This may be done by driving a block of quartz with a piezoelectric crystal.

Electro-optical Phase Modulation

This method is based on the electro-optical effect. Consider a linearly polarized monochromatic wave propagating in the z direction in a medium with refractive index n:

$$\mathbf{E}(z, t) = \hat{\mathbf{x}} \mathcal{E}_0 \cos(\omega t - kz) = \hat{\mathbf{x}} \mathcal{E}_0 \cos \omega \left(t - \frac{n}{c} z \right) \qquad (12.11.3)$$

Suppose we have a Pockels-type electro-optical medium in which the refractive index for light polarized in the x direction is linearly proportional to an applied electric field E_a:

$$n = n_0 + \beta E_a \tag{12.11.4}$$

Therefore the electric field (12.11.3) in such an electro-optical medium in which an external field E_a is applied is

$$\mathbf{E}(z, t) = \hat{\mathbf{x}} \mathcal{E}_0 \cos\left(\omega t - \frac{n_0 \omega}{c} z - \frac{\beta \omega}{c} E_a z\right)$$

$$= \hat{\mathbf{x}} \mathcal{E}_0 \cos\left[\omega\left(t - \frac{n_0}{c} z\right) - \phi\right] \tag{12.11.5}$$

where

$$\phi = \frac{\beta \omega}{c} E_a z \tag{12.11.6}$$

After a distance l of propagation in the medium the field has the phase

$$\phi = \frac{\beta \omega}{c} V \tag{12.11.7}$$

where $V = E_a l$ is the potential difference due to the field E_a. Thus if an electro-optical cell is inserted in a laser cavity, the laser can be FM mode-locked by varying the applied voltage V sinusoidally at the mode separation frequency Δ.

In general a linearly polarized electric field entering an electro-optical medium can be decomposed into two orthogonally polarized components, each of which has a different refractive index. The two orthogonal polarization directions are determined by the orientation of the cell and the applied field E_a. In deriving (12.11.7) we have assumed that the incident field is linearly polarized along one of these directions. In the general case the field will have components in both directions, and in a Pockels cell the two components will have different values of β. This results in a phase difference between the two field components. If the cell produces a total phase change of $90°$, for example, the incident linearly polarized field will be converted to a circularly polarized field, as in the case illustrated in Figure 12.3 for a Kerr cell. That is, a cell containing an electro-optical material can act as a quarter-wave plate. The advantage of using electro-optical media rather than naturally birefringent materials, of course, is the switching and control capabilities one has through the adjustment of the bias voltage.

Saturable Absorbers

As in the case of Q switching, a "passive" AM mode locking may be achieved through the use of a saturable absorber. Assume for simplicity that the absorption coefficient a of the absorption cell saturates according to the formula

$$a = \frac{a_0}{1 + I/I^{\text{sat}}} \tag{12.11.8}$$

for a homogeneously broadened line. It is also convenient (but not necessary) to assume that the saturation intensity I^{sat} of the absorption line is very large compared with the laser intensity I. Then (12.11.8) is approximated by

$$a \approx a_0 - a_0 I/I^{\text{sat}} \tag{12.11.9}$$

Suppose first that there are two oscillating cavity modes, so that the total cavity electric field is

$$E(z, t) = \mathcal{E}_1 \sin k_1 z \sin(\omega_1 t + \phi_1) + \mathcal{E}_2 \sin k_2 z \sin(\omega_2 t + \phi_2) \tag{12.11.10}$$

and the cavity intensity is

$$\begin{aligned}
I(z, t) &= c\epsilon_0 E(z, t)^2 \\
&= c\epsilon_0 \{ \mathcal{E}_1^2 \sin^2 k_1 z \sin^2(\omega_1 t + \phi_1) \\
&\quad + \mathcal{E}_2^2 \sin^2 k_2 z \sin^2(\omega_2 t + \phi_2) \\
&\quad + 2\mathcal{E}_1 \mathcal{E}_2 \sin k_1 z \sin k_2 z \sin(\omega_1 t + \phi_1) \sin(\omega_2 t + \phi_2) \}
\end{aligned} \tag{12.11.11}$$

Now the last term can be rewritten using the identity

$$\begin{aligned}
2 \sin(\omega_1 t + \phi_1) \sin(\omega_2 t + \phi_2) &= \cos\left[(\omega_1 - \omega_2)t + \phi_1 - \phi_2\right] \\
&\quad - \cos\left[(\omega_1 + \omega_2)t + \phi_1 + \phi_2\right]
\end{aligned} \tag{12.11.12}$$

The frequencies ω_1, ω_2, and $\omega_1 + \omega_2$ are very large compared with the mode separation frequency $\omega_1 - \omega_2 = \Delta$. If we average the intensity (12.11.11) over a few optical periods, therefore, we obtain

$$\begin{aligned}
\bar{I}(z, t) = \frac{c\epsilon_0}{2} [&\mathcal{E}_1^2 \sin^2 k_1 z + \mathcal{E}_2^2 \sin^2 k_2 z \\
&+ 2\mathcal{E}_1 \mathcal{E}_2 \sin k_1 z \sin k_2 z \cos(\Delta t + \phi_1 - \phi_2)]
\end{aligned} \tag{12.11.13}$$

The intensity $\bar{I}$ has a time dependence that is simply a sinusoidal oscillation at the mode beat frequency Δ. The absorption coefficient (12.11.9) averaged over a few optical periods will therefore have this same time dependence. In other words, if a saturable absorber described by (12.11.8) is placed inside the laser cavity, it results in a cavity loss modulated at the mode separation frequency, and therefore acts to mode-lock the laser. The argument may be extended to the case of N cavity modes, and we conclude that mode-locked operation may be achieved by placing a cell containing a saturable absorber inside the cavity.

This technique is commonly used in mode-locked solid-state and dye lasers, which, as discussed in Section 12.8, are especially attractive in this regard.

• Although both Q switching and mode locking may be accomplished by inserting a cell containing a saturable absorber into the laser cavity, there are somewhat different requirements for the absorber in the two cases.

In the case of mode locking the absorber should respond very quickly to any changes in the cavity intensity. This was implied in our discussion above, where it was assumed that the saturation behavior of the absorber is fixed according to (12.11.8); there are no transient terms showing how a changes from $a_0(1 + I_1/I^{sat})^{-1}$ to $a_0(1 + I_2/I^{sat})^{-1}$ as I changes from I_1 to I_2. Rather, it was assumed that a reacts instantaneously to variations in I, or at least with a response time shorter than $2L/c$. This requires the absorber to have a short relaxation time, whereas a longer one would be tolerable for Q switching.

Similarly, it is desirable for Q switching that the saturation intensity of the absorber be considerably smaller than that of the laser gain medium. This ensures a large, unsaturated gain after the loss associated with the absorption cell is fully saturated, and allows the giant pulse to build up. In the case of mode locking, however, a relatively large absorber saturation intensity can still give rise to the required modulation. Our assumption $I \ll I^{sat}$ in (12.11.9) was made for convenience, not necessity.

Thus it is possible to Q-switch a laser with one absorption cell and mode-lock it with another. Frequently, however, both effects are present with a saturable absorber, and a Q-switched laser will show signs of mode locking. The output of such a Q-switched, mode-locked laser is indicated in Figure 12.15. •

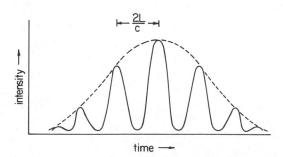

Figure 12.15 Output intensity vs. time of a Q-switched, mode-locked laser. The dashed curve is the envelope of the mode-locked pulse train. Each contributing mode viewed individually has the time dependence of the Q-switched envelope, but because the modes are locked the total output is in the form of a group of pulses separated in time by $2L/c$.

12.12 AMPLIFICATION OF SHORT OPTICAL PULSES

In many applications it is necessary to amplify laser radiation by passing it through a medium with a population inversion on a resonant transition. The amplifier is often made of the same material, and pumped in the same way, as the gain cell of the laser. The most important difference between the laser and the amplifier is simply that the amplifier does not have a resonator with mirrors for feedback. Radiation incident on the amplifier undergoes amplification by stimulated emission and emerges at the other end with greater energy. A series of amplifiers may be employed in tandem, and mirrors may be used to allow the beam to make several passes through a single amplifier. In this section we will consider a pulse of radiation making a single pass through an amplifier.

We will assume that the duration of the laser pulse is short compared with any pumping or relaxation times, so that the changes in level populations in the amplifier are due mainly to stimulated emission and absorption. The rate equations for the level population densities of the amplifying transition are then simply

$$\frac{\partial N_2}{\partial t} = -\frac{\sigma}{h\nu}(N_2 - N_1)I \tag{12.12.1a}$$

$$\frac{\partial N_1}{\partial t} = \frac{\sigma}{h\nu}(N_2 - N_1)I \tag{12.12.1b}$$

since pumping and relaxation processes do not affect N_2 and N_1 significantly during the pulse; this condition for a "short" pulse typically requires pulse lengths shorter than about a nanosecond. Equations (12.12.1) may be combined to form a single equation for the population difference $N \equiv N_2 - N_1$:

$$\frac{\partial N}{\partial t} = -\frac{2\sigma}{h\nu}NI \tag{12.12.2}$$

We also write the (plane-wave) equation for the variation in space and time of the intensity:

$$\frac{\partial I}{\partial z} + \frac{1}{c}\frac{\partial I}{\partial t} = \sigma NI \tag{12.12.3}$$

Let us integrate both sides of (12.12.3) over time:

$$\int_{-\infty}^{\infty}\left(\frac{\partial I}{\partial z} + \frac{1}{c}\frac{\partial I}{\partial t}\right)dt = \int_{-\infty}^{\infty}\sigma NI\,dt \tag{12.12.4}$$

Here $t = -\infty$ and $t = +\infty$ denote times long before and after the pulse has "turned on" at z, so that $I(z, t = -\infty) = I(z, t = +\infty) = 0$. Thus

$$\int_{-\infty}^{\infty} \frac{\partial I}{\partial t} \, dt = I(z, t=+\infty) - I(z, t=-\infty) = 0 \qquad (12.12.5)$$

and so (12.12.4) becomes

$$\frac{d\phi}{dz} = \sigma \int_{-\infty}^{\infty} N(z, t) \, I(z, t) \, dt \qquad (12.12.6)$$

where

$$\phi(z) \equiv \int_{-\infty}^{\infty} I(z, t) \, dt \qquad (12.12.7)$$

is called the *fluence*, and is a measure of the total energy content of the pulse. Note that the fluence has units of energy per unit area, and should not be confused with the photon flux Φ (number of photons per unit area and time) that was introduced in Chapter 7 and used extensively in Chapter 10.

Equation (12.12.6) may be simplified by solving (12.12.2) for $N(z, t)$:

$$N(z, t) = N(z, -\infty) \exp\left(-\frac{2\sigma}{h\nu} \int_{-\infty}^{t} I(z, t') \, dt'\right) \qquad (12.12.8)$$

where $N(z, -\infty)$ is the population inversion at z before the pulse has arrived. Thus we find

$$
\begin{aligned}
\frac{d\phi}{dz} &= \sigma N(z, -\infty) \int_{-\infty}^{\infty} I(z, t) \exp\left(-\frac{2\sigma}{h\nu} \int_{-\infty}^{t} I(z, t') \, dt'\right) dt \\
&= -\sigma N(z, -\infty) \int_{-\infty}^{\infty} \frac{h\nu}{2\sigma} \frac{\partial}{\partial t} \left[\exp\left(-\frac{2\sigma}{h\nu} \int_{-\infty}^{t} I(z, t') \, dt'\right) \right] dt \\
&= -\frac{h\nu}{2} N(z, -\infty) \left[\exp\left(-\frac{2\sigma}{h\nu} \int_{-\infty}^{\infty} I(z, t) \, dt\right) - 1 \right] \qquad (12.12.9)
\end{aligned}
$$

Then use of (12.12.7) allows us to write

$$
\begin{aligned}
\frac{d\phi}{dz} &= \frac{h\nu}{2} N(z, -\infty) \left[1 - \exp\left(-\frac{2\sigma\phi(z)}{h\nu}\right) \right] \\
&= \frac{h\nu}{2} N(z, -\infty) \left[1 - \exp\left(-\frac{\phi(z)}{\phi_{\text{sat}}}\right) \right] \qquad (12.12.10)
\end{aligned}
$$

where we define the *saturation fluence*

$$\phi_{\text{sat}} \equiv h\nu/2\sigma \qquad (12.12.11)$$

This is just the photon energy divided by twice the stimulated-emission cross section.

The differential equation (12.12.10) may be written in terms of the ratio $\theta(z) = \phi(z)/\phi_{sat}$:

$$\frac{d\theta}{dz} = g_0(1 - e^{-\theta}) \qquad (12.12.12)$$

where we have used the fact that the small-signal gain coefficient $g_0(z, -\infty) = \sigma N(z, -\infty)$. In many cases of interest the spatial variations of g_0 are small, and we can take g_0 to be a constant in the differential equation (12.12.12) for the fluence. Then this equation has the solution

$$\theta(z) = \ln\left[1 + e^{g_0 z}(e^{\theta(0)} - 1)\right] \qquad (12.12.13)$$

or

$$\phi_{out} = \phi_{sat} \ln\left[1 + G_0(\exp(\phi_{in}/\phi_{sat}) - 1)\right] \qquad (12.12.14)$$

where $\phi_{out} \equiv \phi(L)$ is the output fluence of an amplifier of length L with small-signal total gain $G_0 = e^{g_0 L}$, given the input fluence $\phi_{in} = \phi(0)$ to the amplifier. We can also write (12.12.14) in terms of the total gain $G \equiv \phi_{out}/\phi_{in}$:

$$G = \frac{\phi_{out}}{\phi_{in}} = X^{-1}\ln\left[1 + G_0(e^X - 1)\right], \qquad X \equiv \frac{\phi_{in}}{\phi_{sat}} \qquad (12.12.15)$$

It is important to note that this solution for the output fluence is independent of the shape of the pulse as a function of time. As long as the pulse is confined, to a good approximation, to a finite duration, and this duration is short compared with any pumping and relaxation times, Eq. (12.12.15) gives us the output fluence as a function of the small-signal gain, length, and saturation fluence of the amplifier. In Figure 12.16 we plot G as a function of X, assuming a small-signal total gain factor $G_0 = 5000$.

If $X = \phi_{in}/\phi_{sat} \ll 1$, then (12.12.15) becomes

$$G \approx X^{-1}\ln(1 + G_0 X) \qquad (12.12.16)$$

If furthermore $G_0 X \ll 1$, then $\ln(1 + G_0 X) \approx G_0 X$ and we have the small-signal limit on total gain

$$G \approx G_0 = e^{g_0 L} \qquad (12.12.17)$$

If $e^X = \exp(\phi_{in}/\phi_{sat}) \gg 1$, on the other hand, then

$$G \approx X^{-1}\ln(G_0 e^X) = X^{-1}\ln(\exp(g_0 L + X))$$

$$= X^{-1}(g_0 L + X) = 1 + g_0 L/X \qquad (12.12.18a)$$

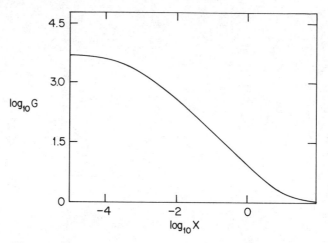

Figure 12.16 Gain $G(X)$ [Eq. (12.12.15)] for $G_0 = 5000$.

or

$$\phi_{\text{out}} \approx \phi_{\text{in}} + (g_0 \phi_{\text{sat}}) L \qquad (12.12.18b)$$

This result identifies $g_0 \phi_{\text{sat}}$ as the largest energy per unit volume that can be extracted from the amplifier when ϕ_{in} is large compared with ϕ_{sat}. This is analogous to the result (11.3.8) for a cw laser, where $g_0 I_{\text{sat}}$ is the largest possible rate of energy extraction per unit volume. Using the fact that $g_0 = \sigma N$ and $\phi_{\text{sat}} = h\nu/2\sigma$, we can write (12.12.18) in the form

$$\phi_{\text{out}} \approx \phi_{\text{in}} + \frac{Nh\nu}{2} L \qquad (12.12.19)$$

This says that *the largest extractable energy density corresponds to taking half a photon, on average, from each excited atom of the amplifier*. The reason for the factor $\frac{1}{2}$ is simply that in the limit of large $\phi_{\text{in}}/\phi_{\text{sat}}$ under consideration, the amplifier is well saturated, with the upper- and lower-level populations having equal probabilities.

This theory of short-pulse amplification is often referred to as the *Frantz–Nodvik model*,[3] and is useful in the design and interpretation of short-pulse amplification experiments. It is worth noting, however, that the model, which is based on the rate-equation approximation, does not account for coherent effects like 2π-pulse formation (Section 8.2 and Problem 8.4).

3.   L. M. Frantz and J. S. Nodvik, *Journal of Applied Physics* **24**, 2346 (1963).

12.13 AMPLIFIED SPONTANEOUS EMISSION

The theory of pulse propagation presented in the preceding section is used frequently in the analysis of laser amplifiers. However, for high-gain systems this theory ignores an important phenomenon: the amplifier can amplify not only the input field from a laser oscillator (or another amplifier), but also the spontaneous radiation emitted by the excited molecules of the amplifier itself. It is easy to see that spontaneously emitted photons at one end of an amplifier, which happen to be directed along the amplifier axis, or close to that direction, can *stimulate* the emission of more photons and lead to substantial output radiation at the other end of the amplifier. This radiation, which appears regardless of whether there is any input radiation, is called *amplified spontaneous emission* (ASE).

It is clear that ASE will have at least some properties resembling laser radiation. In particular, it will be narrow-band in frequency and it will also be highly directional, simply because the amplifier is long and thin. For these reasons high-gain systems emitting ASE are often referred to as "mirrorless lasers." Such mirrorless lasing, also called "superradiance," is well known in the 3.39 μm He-Ne laser, and in high-gain excimer, dye, and semiconductor laser media.

For a simple quantitative description of ASE, let us consider the steady-state equation for the propagation of intensity in an amplifying medium characterized by the gain coefficient g, namely

$$\frac{dI}{dz} = gI \qquad (12.13.1)$$

If $I(0) = 0$, then this equation predicts that $I(z) = 0$ for all z. In other words, this equation does not account for ASE. To include the possibility of ASE we add to (12.13.1) the effect of spontaneous emission:

$$\frac{dI}{dz} = gI + (A_{21}N_2h\nu)\,(\Omega/4\pi) \qquad (12.13.2)$$

The added term is the contribution to dI/dz from spontaneous emission of photons of energy $h\nu$ by N_2 excited molecules per unit volume with spontaneous emission rate A_{21}. Since spontaneous emission is (statistically) isotropic, we have included a factor $\Omega/4\pi$, where Ω is an appropriate solid angle; this factor accounts for the fact that only a fraction $\Omega/4\pi$ of spontaneously emitted photons are emitted in directions for which amplification can occur. In the simplest approximation Ω is taken to be A/L^2, where A is the cross-sectional area of the amplifier and L is its length.

In the small-signal regime in which g and N_2 are independent of I, we have the following solution of equation (12.13.2):

$$I(z) = \left(\frac{A_{21}h\nu\Omega}{4\pi}\right)\left(\frac{N_2}{g}\right)(e^{gz} - 1)$$

$$\approx \left(\frac{A_{21}h\nu\Omega}{4\pi}\right)\left(\frac{N_2}{g}\right)e^{gz} \tag{12.13.3}$$

for $\exp(gz) \gg 1$. For simplicity we will assume that the lower-level population of the amplifying transition is negligible, so that $g \cong \sigma N_2$, where σ is the stimulated emission cross section. For a homogeneously broadened transition having a Lorentzian lineshape of full width at half-maximum $\Delta\nu$ we have

$$\sigma = \left(\frac{\lambda^2 A_{21}}{8\pi}\right)\left(\frac{2}{\pi\Delta\nu}\right) = \frac{\lambda^4 A_{21}}{4\pi^2 c\Delta\lambda} \tag{12.13.4}$$

where $\Delta\lambda = (c/\nu^2)\Delta\nu$ is the width in the emission wavelength $\lambda = c/\nu$. In this case (12.13.3) becomes

$$I(z) \approx \left(\frac{\pi h c^2 \Omega}{\lambda^5}\right)\Delta\lambda e^{gz} \tag{12.13.5}$$

for the growth of intensity with propagation in the amplifier.

The result (12.13.5) allows us to estimate the importance of ASE in high-gain amplifiers. [In careful design analyses, of course, it is necessary to include gain saturation, which has not been done in the derivation of (12.13.5).] ASE can be detrimental to the performance of amplifier systems for two reasons. First, the ASE can significantly deplete the upper-level population, thus diminishing the gain available to the input signal to be amplified. Second, the ASE radiation can irradiate a target before the arrival of the amplified signal, and thus modify an experiment in unintended ways.

ASE can be expected to deplete seriously the upper-level population of the amplifying transition if it becomes comparable in magnitude to the saturation intensity, I^{sat}, of the transition. Setting $I(L)$ given by (12.13.5) equal to I^{sat}, we have

$$gL = \log\left[\frac{\lambda^5 I^{\text{sat}}}{\pi h c^2 \Omega\Delta\lambda}\right] \tag{12.13.6}$$

for the gain-length product of the amplifier at which gain depletion due to ASE may be an important consideration. It is clear from this formula (and (12.13.5)) that large-bore amplifiers, subtending large solid angles, are most susceptible to ASE problems.

• Amplified spontaneous emission is sometimes described from the standpoint of an "effective noise input." This approach is based on (12.13.1) rather than (12.13.2). It begins by noting that, since equation (12.13.1) gives $I(z) = 0$ for all z if $I(0) = 0$, some effective input $I(0) = I_{eff} \neq 0$ is necessary in order to obtain a nonvanishing $I(z)$. An expression for I_{eff} is obtained by recalling that in a cavity of dimensions large compared with a wavelength there are $(8\pi\nu^2/c^3)\Delta\nu$ modes of the field per unit volume in the frequency interval $[\nu, \nu + \Delta\nu]$. (Section 1.3) And according to the quantum theory of radiation each of these modes has a zero-point energy $\frac{1}{2}h\nu$. There is therefore a zero-point field energy per unit volume

$$\rho_o(\nu) = \left(\frac{8\pi\nu^2}{c^3}\right)\frac{1}{2}h\nu\Delta\nu = \left(\frac{4\pi h\nu^3}{c^3}\right)\frac{c\Delta\lambda}{\lambda^2} \tag{12.13.7}$$

in the wavelength interval $[\lambda, \lambda + \Delta\lambda]$. Clearly we should take $\Delta\lambda$ to be on the order of the spectral width of the laser transition. For our purposes we replace $\Delta\lambda$ in (12.13.7) by $\pi\Delta\lambda$, where $\Delta\lambda$ is the transition linewidth (FWHM). Then the "quantum noise" intensity at the laser transition wavelength λ is given by

$$I_{eff} = c\rho_o(\nu)\Omega/4\pi = \left(\frac{\pi hc^2\Omega}{\lambda^5}\right)\Delta\lambda \tag{12.13.8}$$

where we have inserted the factor $\Omega/4\pi$ to account for the fact that only those modes within the solid angle Ω appropriate to the amplifier can act as effective noise sources. Thus, from (12.13.1),

$$I(z) = I_{eff}e^{gz} = \left(\frac{\pi hc^2\Omega}{\lambda^5}\right)\Delta\lambda e^{gz} \tag{12.13.9}$$

which reproduces (12.13.5), and thus validates the approach to ASE based on an effective noise input. •

ASE radiation can have spatial coherence comparable to true laser radiation, but it generally lacks the same degree of temporal coherence. (See Chapter 15 for a discussion of spatial and temporal coherence.) The latter property is understandable from the fact that ASE is basically amplified "noise." The bandwidth of ASE is typically a few times smaller than the gain linewidth $\Delta\nu$.

12.14   ULTRASHORT LIGHT PULSES

With mode-locked lasers it is possible to produce ultrashort, extremely intense pulses of radiation. Mode-locked lasers using saturable absorbers are used to produce, rather routinely, picosecond pulses with peak powers in some cases exceeding 10^{11} W (100 GW). There are techniques for producing even shorter light pulses.

There are many scientific and technological applications of these ultrashort light pulses. For instance, they can be used to study extremely fast photoprocesses in molecules and semiconductors. Especially promising applications may be possible in biological systems, where it has already been determined that certain fundamental chemical reactions occur on picosecond time scales. Intense laser pulses have also been of interest in connection with laser isotope separation and controlled thermonuclear fusion, and it seems safe to say that many new applications will be developed in the future. Not surprisingly, therefore, an entire field of research has grown up around the generation of ultrashort light pulses. Although it is well beyond our scope in this book to discuss in detail these developments, we will mention briefly some techniques in use.

One of these is the *colliding-pulse laser*. This is a mode-locked, three-mirror ring laser with two counter-circulating pulse trains (Figure 12.17). Pulses in each direction pass through a very thin (about 10 μm) jet of a saturable absorbing dye. The absorption coefficient of the absorber is smallest when the intensity is largest. Therefore the cavity loss is least when the counter-circulating pulses collide and overlap within the thin dye jet. The thinness of the absorber forces the pulses to overlap within a very short distance and thus over a very short time interval ($\tau \sim 10\ \mu\text{m}/c \sim 3 \times 10^{-14}$ sec).

Another technique involves *chirping*, by which an ultrashort pulse from a laser can be further compressed. A *chirped* pulse is one in which the carrier frequency ω has a small time dependence. In particular, it has a linear time dependence of the form $\omega = \omega_0 + \beta t$. A pulse can be deliberately chirped by passing it through a medium with a nonlinear refractive index, i.e., a medium in which the refractive index depends upon the electric field (Chapter 17). The chirping results in a spectral broadening of the pulse, i.e., it extends the range of frequency components contained in the pulse. A chirped pulse can be compressed by passing it through a dispersive (i.e., frequency-dependent) *delay line*. If the higher-frequency components of the pulse travel more slowly than the lower-frequency components, the delay line is designed to make them catch up with the lower frequencies on the leading temporal edge of the pulse. By broadening the spectral width of the pulse by chirping, therefore, it is possible to narrow it in time.

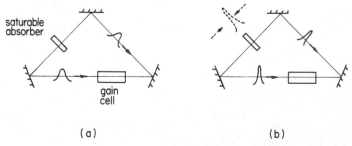

(a)   (b)

Figure 12.17 (*a*) A colliding-pulse ring laser with countercirculating pulses. (*b*) The lowest-loss condition is for the colliding pulses to synchronize and overlap inside the thin saturable absorber.

Optical pulses shorter than 10 fs (1 femtosecond $= 10^{-15}$ sec) have been achieved by compressing in this way ultrashort pulses from a colliding-pulse dye laser.

APPENDIX 12.A   RUNGE–KUTTA FORTRAN AND BASIC PROGRAMS

The Runge–Kutta algorithm is an easily implemented and frequently used method for the numerical integration of ordinary differential equations. For the first-order differential equation

$$\frac{dy}{dx} = F(x, y) \tag{12.A.1}$$

the fourth-order Runge–Kutta algorithm for $y(x + h)$ is

$$y(x + h) = y(x) + \tfrac{1}{6}(k_1 + 2k_2 + 2k_3 + k_4) \tag{12.A.2}$$

where

$$k_1 = hF(x, y)$$

$$k_2 = hF\left(x + \frac{h}{2}, y + \frac{k_1}{2}\right)$$

$$k_3 = hF\left(x + \frac{h}{2}, y + \frac{k_2}{2}\right)$$

$$k_4 = hF(x + h, y + k_3) \tag{12.A.3}$$

The form of (12.A.2) and (12.A.3) is such as to duplicate the Taylor series for $y(x + h)$ up to fourth order in the step size h. The method is easily extended to systems of equations, as in the example below.

The Runge–Kutta method is described in detail in many textbooks on mathematical methods and numerical analysis. For the reader's convenience we list here a FORTRAN program for the Runge–Kutta integration of the two coupled equations (12.4.4):

```
PROGRAM MAIN
DIMENSION Y(2), DY(2), W(2,5)
W IS A WORK ARRAY USED IN SUBROUTINE RUNG
T=0.
Y(1) = 1.E−3
Y(2) = 2.0
DT = .01
NSTEP = 1000
```

```
         DO 1 N = 1, NSTEP
         CALL RUNG(2,Y,DY,T,DT,W)
1        CONTINUE
         PRINT 100,T,Y(1),Y(2)
100      FORMAT(3E16.7)
         STOP
         END
         SUBROUTINE DERIV (T,Y,DY)
         DIMENSION Y(2),DY(2)
         DY(1)=(Y(2)-1.)*Y(1)
         DY(2)=-Y(1)*Y(2)
         RETURN
         END
         SUBROUTINE RUNG(N,Y,DY,T,DT,W)
C        N IS THE NUMBER OF EQUATIONS
         DIMENSION Y(N),DY(N),W(N,5)
         DO 10 J= 1,N
10       W(J,1)=Y(J)
         CALL DERIV(T,Y,DY)
         DO 20 J=1,N
20       W(J,2)=DY(J)*DT
         Z=T+.5*DT
         DO 40 I=2,3
         DO 30 J=1,N
30       Y(J)=W(J,1)+W(J,I)/2.
         CALL DERIV(Z,Y,DY)
         DO 40 J=1,N
40       W(J,I+1)=DY(J)*DT
         Z=T+DT
         DO 50 J=1,N
50       Y(J)=W(J,1)+W(J,4)
         CALL DERIV(Z,Y,DY)
         DO 60 J=1,N
60       W(J,5)=DY(J)*DT
         DO 70 J=1,N
70       Y(J)=W(J,1)+(W(J,2)+2.*(W(J,3)+W(J,4))+W(J,5))/6.
         T=T+DT
         RETURN
         END
```

The dependent variables are stored in the array Y. In this case [Eqs. (12.4.4)] $Y(1) = x$ and $Y(2) = y$. T is the independent variable (τ), and the step size DT [denoted by h in (12.A.3)] is taken to be 0.01. In general DT should be taken small enough to give an accurate solution of the equations, but not so small that the program is unnecessarily time-consuming. The accuracy of the solution can always be checked by halving the step size and noting whether there is a significant change in the computed solution.

NSTEP is the number of integration steps. Each call to RUNG moves time forward by DT. In the way our program is set up, therefore, the final values of Y after the last call to RUNG are $Y(1) = x(t_{max})$ and $Y(2) = y(t_{max})$, where $t_{max} =$ NSTEP*DT. In general the intermediate values of Y can be stored in arrays for plotting purposes.

Subroutine DERIV simply defines the derivatives. In our example $DY(1) = dx/d\tau$ and $DY(2) = dy/d\tau$. Subroutine RUNG does a fourth-order Runge-Kutta integration, using the derivatives defined in DERIV. N is the number of (first-order) simultaneous differential equations to be solved; in our example $N = 2$. W is a work array that must be dimensioned N by 5. Only the MAIN and DERIV routines in our example depend on the specific problem, the RUNG subroutine being a "canned" routine, usable as it stands in every problem.

It may be worth noting that RUNG can also be used to solve higher-order differential equations of mixed order. For example, to solve the system

$$\frac{d^2x}{dt^2} + x = 0 \qquad x(0) = \left(\frac{dx}{dt}\right)_{t=0} = 1 \qquad (12.A.4)$$

using RUNG, we let $Y(1) = x$, $Y(2) = dx/dt$. Then (12.A.4) is equivalent to the two first-order equations

$$DY(1) = Y(2), \qquad DY(2) = -Y(1) \qquad (12.A.5)$$

with initial values $Y(1) = Y(2) = 1$.

RUNG can be used as it stands to solve complex as well as real systems of equations. One simply writes the equations of DERIV in complex form and declares the appropriate variables COMPLEX.

It is easy to convert the FORTRAN program above to a BASIC program. We list below a BASIC program suitable for use on many personal computers. The program is set up specifically to solve an autonomous set of equations of the type shown in Eqs. (12.4.4), where the right sides do not depend explicitly on the independent variable, but it can easily be modified to solve nonautonomous systems.

```
10      DIM Y(2), DY(2), W(2,5)
20      T=0
30      N=2
40      REM N IS THE NUMBER OF EQUATIONS
50      Y(1)=.001
60      Y(2)=2
70      DT=.01
80      NSTEP=1000
90      FOR K=1 TO NSTEP
100     GOSUB 1000
110     NEXT K
120     PRINT T, Y(1), Y(2)
500     REM DERIV
510     DY(1)=(Y(2)-1)*Y(1)
```

```
520      DY(2)=-Y(1)*Y(2)
530      RETURN
1000     REM INTEG
1010     FOR J=1 TO N
1020     W(J,1)=Y(J)
1030     NEXT J
1040     GOSUB 500
1050     FOR J=1 TO N
1060     W(J,2)=DT*DY(J)
1070     NEXT J
1080     FOR I=2 TO 3
1090     FOR J=1 TO N
1100     Y(J)=W(J,1)+W(J,I)/2
1110     NEXT J
1120     GOSUB 500
1130     FOR J=1 TO N
1140     W(J,I+1)=DT*DY(J)
1150     NEXT J
1160     NEXT I
1170     FOR J=1 TO N
1180     Y(J)=W(J,1)+W(J,4)
1190     NEXT J
1200     GOSUB 500
1210     FOR J=1 TO N
1220     W(J,5)=DT*DY(J)
1230     NEXT J
1240     FOR J=1 TO N
1250     Y(J)=W(J,1)+(W(J,2)+2*(W(J,3)+W(J,4))+W(J,5))/6
1260     NEXT J
1270     T=T+DT
1280     RETURN
```

APPENDIX 12.B DIFFRACTION OF LIGHT BY SOUND

The diffraction of light by sound waves can be understood by analogy with the diffraction of X-rays by crystals. The atoms of a crystal are spaced in a regular pattern, and consequently they scatter radiation cooperatively, with well-defined phase relations between the fields scattered by different atoms. This results in scattering only in certain well-defined directions, and the process is usually called "diffraction" instead of "scattering." Figure 12.18 shows a wave incident upon a stack of crystal planes separated by a distance d. The allowed diffraction angles are determined by the condition of constructive interference of the fields "reflected" from different planes. As shown in the figure, these diffraction angles satisfy the Bragg diffraction formula

$$2d \sin \theta = m\lambda, \qquad m = 1, 2, 3, \ldots \tag{12.B.1}$$

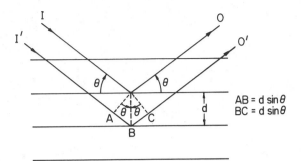

Figure 12.18 Diffraction of a plane wave by a stack of crystal planes. Constructive interference of the two rays IO and $I'O'$ occurs when their path difference AB + BC is equal to an integral multiple m of the wavelength λ. This gives the Bragg diffraction formula $2d \sin \theta = m\lambda$ for the allowed diffraction angles θ. A more complete analysis shows that these angles give the *only* directions in which scattering occurs.

where λ is the wavelength of radiation and d is the separation distance between adjacent crystal planes.

• Since d is on the order of 1 Å in actual crystals, only wavelengths in the X-ray region can satisfy (12.B.1) and the requirement $|\sin \theta| \leq 1$. The measurement of X-ray diffraction angles thus provides information about crystal structure. Indeed the use of crystals as "diffraction gratings" for X-rays has been one of the most important techniques of modern science for probing the structure of matter. This technique was originally suggested by Max von Laue. The idea arose in connection with the question of whether X-rays were particles or waves. L. Brillouin predicted the diffraction of radiation by sound waves in 1922, and it was first observed ten years later by P. W. Debye and F.W. Sears. •

The refractive-index variation associated with a sound wave of wavelength λ_s has a spatial dependence of the form

$$\Delta n(x) = a \sin k_s x \qquad (12.B.2)$$

where $k_s = 2\pi/\lambda_s$ and a depends on material constants of the medium and the intensity of the sound wave. Equation (12.B.2) arises from the fact that a sound wave is basically a wave of density variation. Figure 12.19 shows an intuitive way of understanding the diffraction of light by a sound wave. We regard the planes of constant x where (12.B.2) is a maximum as "crystal" planes which, because of their regular spacing λ_s, will diffract light only in certain well-defined directions. Indeed it turns out that the diffraction angles are given by the Bragg formula (12.B.1) with $d = \lambda_s$ and $m = 1$:

$$2\lambda_s \sin \theta = \lambda/n \qquad (12.B.3)$$

We have included the effect of the refractive index n of the medium.

The important difference between (12.B.1) and (12.B.3) is that there are no higher-order diffraction angles corresponding to $m > 1$ in (12.B.3). This differ-

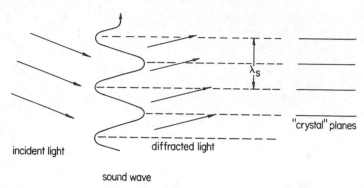

Figure 12.19 Intuitive picture of diffraction of light by sound as diffraction from a fictitious set of "crystals" planes defined by the intensity maxima of the sound.

ence arises from the fact that the "diffraction grating" associated with the sound wave indicated in Figure 12.19 is really a spatially continuous one, not a discrete set of crystal planes with nothing in between.

Equation (12.B.3) gives only the diffraction angle. It does not tell us the strength with which the sound wave diffracts light, i.e., the fraction of light intensity diffracted after a given distance of propagation. This is determined by a and the wavelength of the light.[4]

PROBLEMS

12.1   (a) Write the steady-state solutions of equations (12.2.5) in such a way as to show the saturation of $\overline{N}_2$ with increasing $\overline{I}_\nu$.

  (b) Verify the equations (12.3.16) and (12.3.17) for the period and lifetime of relaxation oscillations.

12.2   (a) Show that the quantity x defined in Eq. (12.4.1) is the cavity photon number density divided by the threshold population inversion density.

  (b) Show that Eqs. (12.2.5) may be written as (12.4.4) when the change of variables (12.4.1)–(12.4.3) is made.

12.3   (a) Why is it that so much more power can be obtained from a Q-switched laser than in ordinary continuous-wave operation?

  (b) Suppose a Q-switched laser using a rotating mirror or a saturable absorber is pumped continuously. How do you expect the laser to behave?

12.4   Set up the equations for x and y [Eqs. (12.4.1) and (12.4.2)] in the case

---

4. See, for example, R. Adler, "The Interaction between Light and Sound," *IEEE Spectrum*, May 1967, p. 42.

where Q switching is done with a saturable absorber with absorption coefficient $a = a_0 (1 + I/I^{sat})^{-1}$. If you have access to a computer, solve these equations numerically using, for example, the Runge–Kutta algorithm of Appendix 12.A. You will have to assume values of the small-signal absorption coefficient a_0 and the saturation intensity I^{sat} of the absorber. Determine how x and y depend on the choice of a_0 and I^{sat}.

12.5 (a) Suppose we have N oscillators whose frequencies are given by (12.7.2) and whose phases ϕ_m are fixed but not "locked" according to (12.7.10). Discuss the properties of the sum $X(t)$ of the oscillator displacements in this case. Can the maximum value of $X(t)$ be as large as in the phase-locked case?

   (b)*Suppose that the phases ϕ_n are randomly chosen from an ensemble and are completely uncorrelated, so that $\langle e^{i(\phi_m - \phi_n)} \rangle = \delta_{mn}$, where $\langle \cdots \rangle$ indicates an ensemble average. Then compute $\langle X(t) \rangle$ and $\langle X^2(t) \rangle$.

12.6 (a) Show that each pulse of a mode-locked pulse train has an intensity N times larger than the sum of the intensities of the individual modes constituting it.

   (b) Show that the average intensity of a mode-locked pulse train is equal to the sum of the intensities of the individual modes constituting it.

12.7 Make a plot of the time-dependent factor in the brackets in Eq. (12.9.5), choosing $\phi_m = 0$, $\Omega = \omega_m/10$, and (a) $\epsilon = \frac{1}{2}$, (b) $\epsilon = 1$, (c) $\epsilon = 5$.

12.8 Consider the 6328-Å He–Ne laser.

   (a) Estimate the shortest pulse that can be obtained by mode-locking such a laser.

   (b) What is the duration of each pulse of the mode-locked train if the gain tube has length $l = 10$ cm and the mirror separation $L = 40$ cm?

   (c) What is the separation between the mode-locked pulses in part (b)?

   (d) Why do liquid dye and solid-state lasers produce much shorter mode-locked pulses than typical gas lasers?

12.9 (a) Estimate the average power, in watts, expended by a normal human adult. Assume that a "normal human adult" consumes 2500 dietitian's calories (2500 × 4185 J) per day, and that his output energy just balances his input energy.

   (b) Estimate the intensity at a distance of 1 m from a 60-W light bulb.

   (c) Estimate the average electrical power used to operate a typical house in your area.

12.10 It is possible to "dump" the cavity of a pulsed laser by making the reflectivity of the output mirror effectively zero at the moment of peak intensity. What is the advantage of "cavity dumping" in this way?

13 SPECIFIC LASERS AND PUMPING MECHANISMS

13.1 INTRODUCTION

In thermodynamic equilibrium the ratio of (nondegenerate) upper- and lower-state populations of an atomic or molecular transition is

$$\frac{N_2}{N_1} = \frac{e^{-E_2/kT}}{e^{-E_1/kT}}$$
$$= e^{-(E_2 - E_1)/kT} = e^{-h\nu/kT} \tag{13.1.1}$$

This ratio is always less than one. This means that a medium in complete thermal equilibrium is always an absorber rather than an amplifier of radiation, regardless of how hot it is.

In order to have a population inversion on a transition, therefore, the level populations N_2 and N_1 must have a nonthermal distribution. If we insist on thinking in terms of a temperature, we see from (13.1.1) that a population inversion ($N_2 > N_1$) is associated with an unphysical "negative (absolute) temperature." The concept of negative absolute temperature was sometimes used in the early days of maser and laser research, but it can be misleading because it applies only to the lasing levels, while the rest of the atom exists at an entirely different and generally positive temperature.

The amplification of radiation by stimulated emission requires a population inversion. In order to have a population inversion we must "pump" the medium to overcome its natural tendency to reach a thermal equilibrium. The suitability of a material as a laser medium thus depends, among other things, on how readily we can force it away from thermal equilibrium and establish a population inversion.

Lasers are often classified according to the method used to obtain a population inversion and gain. Thus we speak of optically pumped lasers, electric-discharge lasers, chemical lasers, gas-dynamic lasers, etc. We will now discuss, in a semi-quantitative fashion, some of the principal lasers in relation to the methods of obtaining population inversions in them. Each of these methods draws on principles associated with different, broad areas of research and development in chemistry, physics and engineering.

13.2 SOLID-STATE RARE-EARTH ION LASERS AND OPTICAL PUMPING

A laser is said to be *optically pumped* when the population inversion is obtained by the absorption of radiation illuminating the laser medium. An example is the ruby laser, for which the active element is the Cr^{3+} ion in an Al_2O_3 (corundum) lattice. A typical design for the optical pumping of a ruby rod is shown in Figure 13.1. The flashlamp is similar in some ways to a photographic flashlamp. Typically the lamp contains Xe at a pressure of about a hundred or several hundred Torr. An electric discharge in the Xe gas produces an intense burst of spontaneous emission lying in a visible spectral range that is absorbed by the ruby. The flashlamp has a helical shape and wraps around the ruby rod (typically about the size of a cigarette) in order to expose as much of the rod as possible to the lamp's output radiation.

Usually the lamp is not on continuously but is pulsed ("flashed"), a typical pulse duration being on the order of a millisecond. The flashlamp is excited by the discharge of a capacitor bank with a capacitance of several hundred microfarads and charged to several kilovolts. Considerable heat develops in the ruby rod when the lamp is on, and the rod must be cool before it is exposed to another flash of radiation.

Another flashlamp pumping configuration is shown in Figure 13.2. In this case the flashlamp has the same (linear) cylindrical shape as the rod, and the lamp and rod are placed along the focal axes of an elliptical reflecting tube. This permits the focusing of a large portion of the lamp's output onto the laser rod.

Laser action was first obtained in 1960 by T. H. Maiman, using a ruby rod in the helical-lamp configuration of Figure 13.1. We will consider the example of ruby in some detail, as it illustrates some of the concepts involved in the optical pumping of other solid-state lasers.

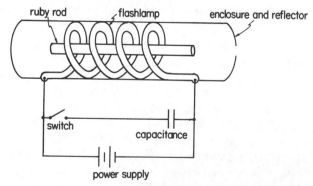

Figure 13.1 Helical flashlamp arrangement for the optical pumping of a ruby laser. The mirrors may be external to the ruby rod, or may be silver coatings applied directly onto the ends of the rod.

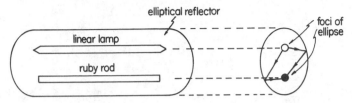

Figure 13.2 Linear flashlamp arrangement for the optical pumping of a ruby laser.

Figure 13.3 shows the relevant energy levels of the Cr^{+3} ion in ruby. The levels labeled $^4F_1$ and $^4F_2$ (a standard spectroscopic notation) are broad *bands* of energy, as indicated in the figure. The level labeled ^2E actually consists of two separate levels, labeled $2\bar{A}$ and $\bar{E}$ in conventional notation. These two levels are separated by about 29 cm^{-1}, or 8.7×10^{11} Hz, and it is the lower one, $\bar{E}$, that serves as the upper laser level in the ruby laser. The ground level is labeled $^4A_2$, and it serves as the lower laser level of the 6943-Å laser transition.

The levels $^4F_1$ and $^4F_2$ each decay very rapidly into the ^2E level, the decay rate being about 10^7 sec^{-1}. The decay process is not spontaneous emission, but rather a *nonradiative decay*, in which the energy lost by the chromium ion is converted to thermal energy (heat) of the crystal lattice. The upper laser level, on the other hand, is metastable. That is, it has a long lifetime, about 3 msec, for (spontaneous-emission) decay to the ground level. This decay is accompanied by the emission of 6943-Å photons.

The absorption of radiation of wavelength around 4000 or 5700 Å will populate the $^4F_1$ or $^4F_2$ levels, respectively. Figure 13.4 shows the absorption coefficient versus wavelength in the visible region for ruby. Flashlamps used to pump ruby lasers should obviously emit radiation at those wavelengths where ruby is strongly

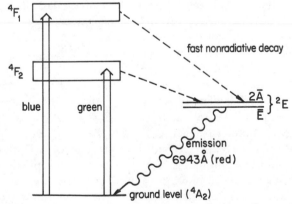

Figure 13.3 Simplified energy-level diagram of the Cr^{+3} ion in ruby.

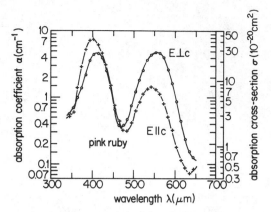

Figure 13.4 Absorption coefficient of ruby. From D. C. Cronemeyer, *Journal of the Optical Society of America* **56**, 1703 (1966).

absorbing. The rapid decay of the pumped bands into the upper laser level, with the ground level itself acting as the lower laser level, means that ruby is approximately a three-level laser system.

In Section 10.10 we estimated a pumping power density of about $1 \text{ kW}/\text{cm}^3$ for laser oscillation in ruby. For a 1-msec excitation pulse this translates to a pumping energy density on the order of $1 \text{ J}/\text{cm}^3$. Since not all of the electrical energy delivered to the flashlamp is converted to radiation, and not all of the radiation is actually absorbed by the ruby rod, an electrical power on the order of megawatts must be delivered to the flashlamp in typical ruby lasers.

The very broad absorption coefficient of ruby (Figure 13.4) and other solid-state or liquid laser materials is, of course, advantageous. It is well matched by the broad emission bandwidth of a "conventional" (i.e., nonlaser) light source such as a flashlamp, which is much like a blackbody radiator. A medium with very sharp absorption lines would be much more difficult to pump optically with a conventional broadband light source. In a Doppler-broadened gaseous absorber, for instance, an absorption linewidth is approximately v/c times the transition frequency, where v is the average atomic velocity and c is the speed of light (Problem 13.1). This v/c ratio is typically about 10^{-5} or 10^{-6} which explains why optical pumping is seldom used for gas lasers. In solid-state or liquid media, however, the absorption linewidth is more like 10^{-2}–10^{-4} times the transition frequency.

Another common solid-state system is the 1.06-μm Nd:YAG (yttrium aluminum garnet) laser. The relevant energy levels of the Nd^{+3} ions in the YAG crystal lattice are shown in Figure 13.5. Again we label the levels according to a conventional spectroscopic notation. In this case we have a good approximation to a four-level pumping scheme. As in ruby, the pump bands are broad and decay rapidly into the upper laser level. Because it is a four-level system, however, and furthermore has a much larger stimulated emission cross section than ruby, the YAG laser has much lower pumping requirements (Section 10.10).

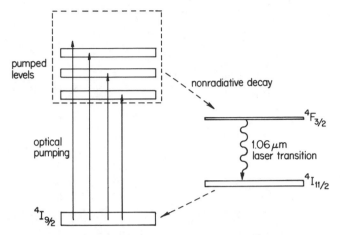

Figure 13.5 Simplified energy-level diagram for the Nd^{+3} ions in a YAG crystal.

The Nd : glass laser is similar to Nd : YAG, except that the Nd^{+3} ions are present as impurities in glass rather than YAG. Unlike Nd : YAG, it is almost always used in the pulsed mode of excitation, because the low thermal conductivity of glass makes it too difficult to cool efficiently under continuous excitation. The Nd : glass laser is especially useful for mode-locked operation because of its large gain linewidth ($\approx 3 \times 10^{12}$ Hz).

13.3   DYE LASERS AND OPTICAL PUMPING

Liquid dye lasers provide especially interesting examples of population inversion by optical pumping. The active molecules in these lasers are large organic dye molecules in a solvent such as an alcohol or water.[1] The most useful aspect of dye lasers is their tunability: by means of some adjustment, a dye laser can be made to oscillate over a wide range of optical wavelengths. This tunability is extremely useful in atomic spectroscopy.

Figure 13.6 is an energy-level diagram typical of dye molecules. With each electronic level of the molecule is associated a set of vibrational and rotational energy levels, which are spaced very closely compared with electronic energy-level spacings. The vibrational energy levels are typically separated by 1200–1700 cm^{-1}, whereas the rotational level spacings are roughly two orders of magnitude smaller. The symbols S_0, S_1, S_2, T_1, T_2 in Figure 13.6 label electronic energy levels and the associated manifold of vibrational and rotational levels. S and T stand for singlet and triplet electronic levels, respectively. In a singlet the total electron spin

---

1.   Many dyes have been made to lase. Lists of laser dyes have been compiled by K. H. Drexhage in *Dye Lasers*, edited by F. P. Schafer (Springer, Berlin, 1973), and L. G. Nair, *Progress in Quantum Electronics* **7**, 153 (1982).

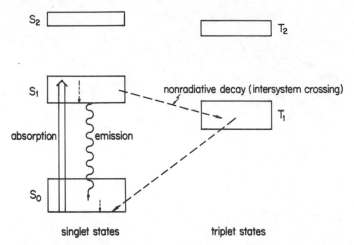

Figure 13.6 Typical energy-level diagram of a dye molecule.

quantum number is zero ($S = 0$), whereas in a triplet level $S = 1$. A level having a total spin quantum number S is ($2S + 1$)-fold degenerate, whence the names singlet ($2S + 1 = 1$) and triplet ($2S + 1 = 3$). For our purposes only one thing about these S and T levels is important: *$S \leftrightarrow S$ and $T \leftrightarrow T$ transitions are radiatively allowed, whereas $S \leftrightarrow T$ transitions are radiatively forbidden.* That is, the oscillator strengths for $S \leftrightarrow T$ transitions are zero. This is a consequence of the dipole selection rule $\Delta S = 0$ for electron spin.

As shown in Figure 13.6, the ground level S_0 is a singlet state. In thermal equilibrium practically all the dye molecules are in the S_0 level. Because $S \rightarrow T$ radiative transitions are forbidden, optical pumping can promote the molecule from the ground level S_0 to the higher-energy singlet states (S_1, S_2, etc.) but not to any of the triplet states. To begin with, therefore, let us focus our attention only on the allowed singlet–singlet transitions.

Optical pumping by external radiation takes the molecule from one of the vibrational–rotational levels of the ground electronic state (S_0) to one of the vibrational–rotational levels of the first excited singlet electronic state (S_1). The $S_0 \rightarrow S_1$ transition covers a broad frequency range because there is a broad range of vibrational–rotational levels associated with both S_0 and S_1. The transition frequencies typically lie in the visible or near-visible portion of the electromagnetic spectrum, and so the dye gives the solution a certain color because of selective absorption. Furthermore the oscillator strengths for allowed transitions in dye molecules are usually quite large. This can be understood from the large size of organic dye molecules. In the hydrogen atom, for instance, the transition electric dipole moments are on the order of ea_0, where $a_0 \approx 0.53$ Å is the Bohr radius, roughly the "size" of a hydrogen atom. Dye molecules are much larger, and consequently have much larger dipole moments and oscillator strengths. Dye molecules therefore have strong absorption bands.

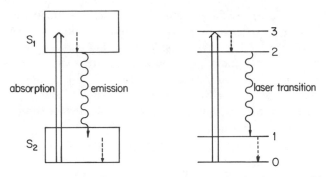

Figure 13.7 Approximate four-level picture of a dye laser transition.

An excited dye molecule tends to decay very quickly to the lowest-lying vibra-tional level of a given electronic state. The decay process is nonradiative, and typical lifetimes are in the picosecond range. A dye molecule in electronic state S_1, for example, will quickly decay to the "bottom" of the S_1 manifold, as indi-cated in Figure 13.6. A crude description of the pumping of a dye laser therefore follows the four-level scheme of Section 10.8 (Figure 13.7). Absorption of radia-tion takes the molecule from the bottom level of S_0 to one of the S_1 levels, where nonradiative decay quickly brings it to the bottom level of S_1. The latter serves as the upper laser level, the lower laser level being one of the vibrational–rotational levels of S_0.

Unfortunately things are not as simple as the four-level picture suggests. For although $S \leftrightarrow T$ transitions are radiatively forbidden, they may occur nonradia-tively in collisions between molecules. This is called *intersystem crossing*[2]. The nonradiative, intersystem decay of S_1 into T_1, and T_1 into S_0, is indicated in Figure 13.6.

Intersystem crossing has some undesirable ramifications for the pumping of dye lasers. The decay $S_1 \rightarrow T_1$, for instance, obviously reduces the number of mole-cules in the upper laser level, and in effect reduces the number of laser-active molecules. Since triplet–triplet transitions are strongly allowed, furthermore, the optical pumping process $S_0 \rightarrow S_1$ also induces absorptive transitions $T_1 \rightarrow T_2$, which have frequencies in the same region as $S_0 \rightarrow S_1$ transitions.

However, it is possible to get around the problems raised by intersystem cross-ing. By dissolving oxygen in the dye solution, for example, the decay rate for the intersystem crossing $T_1 \rightarrow S_0$ can be enhanced by orders of magnitude (to about $10^7 \ sec^{-1}$). Continuous-wave dye-laser oscillation can then be achieved if the $T_1 \rightarrow S_0$ decay is faster than the $S_1 \rightarrow T_1$ decay. Otherwise the laser can only be operated in a pulsed mode in which the excitation pulse duration is short compared with the $S_1 \rightarrow T_1$ decay. In the latter case, loosely speaking, lasing occurs before S_1 has a chance to be significantly depleted by intersystem crossing.

2.   There are in fact also radiative contributions to intersystem crossing because forbidden transitions are not *strictly* forbidden, they are only much less likely than allowed transitions.

Because of their large oscillator strengths, the $S_0 \rightarrow S_1$ transitions have small spontaneous-emission lifetimes, typically in the nanosecond range. The upper laser level therefore decays quickly, and a flashlamp used to optically pump a dye laser must have fast-rising, high-intensity output in order to produce a substantial population inversion. In this regard the requirements on the flashlamps are more stringent than in the case of ruby or Nd : YAG and Nd : glass lasers.

Dye lasers are also frequently optically pumped with the radiation from another laser. Pulsed N_2 lasers with output in the ultraviolet (3370 Å) are particularly useful for pumping pulsed dye lasers. Continuous-wave dye lasers are frequently pumped by the blue–green radiation of an argon ion laser.

• One very important characteristic of laser dye molecules is that their emission spectra are shifted in wavelength from their absorption spectra. This fortunate circumstance prevents the laser radiation from being strongly absorbed by the dye itself. We can understand this characteristic based on the so-called *Franck–Condon principle* and the fact that the vibrational relaxation associated with any electronic state is very rapid.

The Franck–Condon principle is basically just the statement that electron motion in molecules is very rapid compared with the vibrations of the individual atoms. This means that electronic transitions occur very quickly, essentially without adjustment of the interatomic coordinate R. We indicate in Figure 13.8 an absorptive transition between the two lowest electronic states of a molecule. The transition is shown to proceed vertically (without

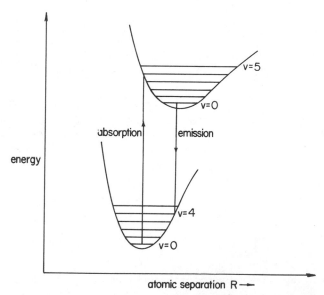

Figure 13.8 An illustration of the Franck–Condon principle. After absorption R changes due to the molecular vibration, so that the emission occurs at a different R, and therefore a different wavelength, than the absorption.

change of R). It starts near the bottom of the potential curve of the lower electronic state, because in thermal equilibrium most molecules will be in the lower vibrational states of the ground electronic level. Following the absorption, the molecule is in a vibrational state of the upper electronic level. The vibrational motion of the molecule then changes R. The vibrational state also relaxes quickly to $v = 0$, and so the eventual downward electronic transition (due to spontaneous emission) proceeds along a different vertical line. The emission wavelength is thus longer than the absorption wavelength. In Figure 13.9 we show the singlet-state absorption and emission spectra of a solution of rhodamine 6G, the most commonly used laser dye molecule. The emission spectrum is shifted to longer wavelength as expected from the Franck–Condon principle. The shift between the peaks of the two curves is called the Stokes shift. •

As noted earlier, the most important aspect of dye lasers is their tunability. This tunability is a consequence of the broad emission curve of a dye molecule (Figure 13.9), which allows dye laser radiation to extend over a broad band of wavelengths, typically 10–60 Å. Tuning within this band is accomplished by discriminating against most of the frequencies, i.e., by making the cavity loss larger than the gain for most frequencies. The most common way of doing this makes use of the "Littrow arrangement" sketched in Figure 13.10. In the arrangement shown one of the cavity mirrors is replaced by a diffraction grating, which reflects radiation of wavelength λ only in those directions satisfying the Bragg condition

$$2d \sin \theta = m\lambda, \qquad m = 1, 2, \ldots \qquad (13.3.1)$$

where d is the spacing between lines of the grating. Wavelengths not satisfying (13.3.1) are not fed back along the cavity axis and consequently have large losses. Thus the bandwidth of the laser radiation is greatly reduced (typically to around 1 Å or less), and tuning is accomplished by rotation of the grating.

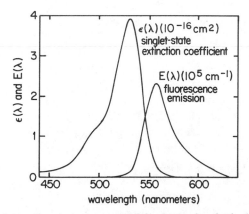

Figure 13.9 Absorption and emission spectra of the dye molecule rhodamine 6G in ethanol (10^{-4} molar solution). From B. B. Snavely, *Proceedings of the IEEE* **57**, 1374 (1969).

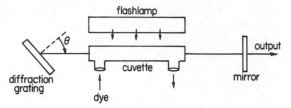

Figure 13.10 Littrow arrangement for wavelength tuning of a dye laser. Tuning is achieved by rotation of the grating.

13.4   ELECTRON IMPACT EXCITATION

We have mentioned that optical pumping is generally not a very efficient way of pumping gas lasers. The most common means of pumping gas lasers is with an electric discharge.

An electric discharge may be produced in a gas contained inside a glass tube by applying a high voltage to electrodes on either side of the tube. Electrons are ejected from the negative electrode (the cathode) and drift toward the positive electrode (the anode). When an electron collides with an atom (or molecule) there is some quantum-mechanical probability of raising it to some higher-energy state. This process of *electron impact excitation* occurs in neon lamps, in which neon atoms excited by collisions with electrons undergo spontaneous emission and emit the deep red light so familiar from neon-sign advertisements. Electron impact excitation is also the microscopic basis of fluorescent lamps. In this case the electrons excite mercury atoms, which emit strongly in the ultraviolet, and the tube is coated with a material that absorbs in the ultraviolet and emits in the visible.

We will discuss in Sections 13.6 and 13.9 the electrical excitation of He–Ne and CO_2 lasers. First, however, it will be worthwhile to discuss some general aspects of electron–atom (–molecule) collisions.

The simplest electron–atom process is an elastic collision, in which the kinetic energy of the electron–atom system is conserved in the collision. That is, none of the initial kinetic energy in an elastic collision is converted to internal energy of the atom. Furthermore there is relatively little exchange of kinetic energy between the electron and the atom in such an elastic collision. This is a consequence of the much smaller mass of the electron than the atom (Problem 13.2).

In an inelastic collision with an atom, the kinetic energies of the electron before and after the collision are different. In an inelastic *collision of the first kind*, the electron loses kinetic energy. Energy lost by the electron is converted to internal excitation energy of the atom so that, of course, the total energies (kinetic plus internal) before and after the collisions are the same. These collisions of the first kind are what we normally refer to as electron impact excitation processes. In electron impact excitation of molecules, the internal energy added to the molecule may be in the form of vibrational and rotational energy as well as electronic energy.

If an electron collides with an already excited atom or molecule, it can cause the atom or molecule to drop to a lower level of excitation, the energy difference now going into an increase in the kinetic energy of the system. This type of inelastic collision is called a *collision of the second kind* (or a "superelastic" collision).

A gas laser may be pumped *directly* by electron impact excitation, in the sense that collisions of the active atoms with electrons are the sole source of the population inversion. In this case the rates for the various excitation (collisions of the first kind) and deexcitation (collisions of the second kind) processes enter into the population rate equations as pumping and decay rates, respectively. Frequently, however, electron impact excitation produces a population inversion *indirectly*, in the sense that it sets the stage for another process that acts more directly to produce a positive gain. The most important of these other processes is excitation transfer from one atom (molecule) to another, to which we turn our attention in the following section.

The rates at which electrons excite atoms and molecules in collisions of the first kind are determined by the collision cross sections σ. These in turn depend upon the relative velocity of the electron–atom (–molecule) pair. Thus, if a monoenergetic beam of N_e electrons per unit volume is incident upon an atom, the rate at which the atom is raised from level i to level f ($E_f > E_i$) is

$$R_{if} = N_e \sigma_{if}(v)v \tag{13.4.1}$$

where v is the relative velocity prior to collision. The cross section $\sigma_{if}(v)$ for the $i \rightarrow f$ atomic transition has the dimensions of area, just as do the cross sections for the scattering or absorption of radiation (Sections 2.6 and 3.6), or the collision of two atoms or molecules (Section 3.10). It is easy to see, therefore, that R_{if} has the dimension of a rate. The determination of the electron impact excitation cross sections $\sigma_{if}(v)$ for electron–atom, electron–molecule, and electron–ion collisions is an old and active branch of atomic and molecular physics.[3] These cross sections are important not only for the understanding of electric-discharge lasers, but for a great many other phenomena, including such things as lightning and the aurora borealis.

In an electric discharge the electrons do not all have the same kinetic energy. Rather, there is a distribution $f(E)$ of electron energies, defined such that $f(E)\,dE$ is the fraction of electrons with energy in the interval $[E, E + dE]$. In this case R_{if} is obtained by averaging over the electron energy distribution:

3. See, for example, H. S. W. Massey and E. H. S. Burhop, *Electronic and Ionic Impact Phenomena* (Oxford University Press, London, 1969), Volumes I and II.

$$R_{if} = N_e \left(\frac{2}{m}\right)^{1/2} \int_0^\infty \sigma_{if}(E)\, E^{1/2}\, f(E)\, dE \qquad (13.4.2)$$

where we have used $E = \frac{1}{2} m v^2$ to relate the electron energy and velocity. Collisions with electrons can also ionize atoms and molecules, and break apart (dissociate) molecules. Equation (13.4.2) applies to any of these processes, each characterized by a cross section $\sigma(E)$. The rate for a collision of the second kind may also be expressed in the form (13.4.2).[4]

Electron energy distribution functions $f(E)$ in electric-discharge lasers are often not well described by a Boltzmann distribution. They are frequently approximated by such a distribution, however, in which case the "electron temperature" is much greater than the temperature of the atomic (or molecular) gas in the discharge, typically being measured in thousands of degrees. The distribution $f(E)$ is sometimes calculated by direct numerical solution of the Boltzmann transport equation. In this case the cross sections for all important electron collision processes are essential data for the computer program.

13.5 EXCITATION TRANSFER

One way for an excited atom to transfer energy to another atom is by photon transfer: the photon spontaneously emitted by one atom is absorbed by the other. In this way the first atom drops to a lower level and the second atom is raised to a higher level, i.e., there is an excitation transfer between the two atoms. This process has a negligible probability of occurrence unless the photon emitted by the donor atom is within the absorption linewidth of the acceptor atom, i.e., there must be a resonance (or near-resonance) of the atomic transitions.

Actually the process of excitation transfer via spontaneous emission is quite negligible compared with other transfer processes that result from a direct nonradiative (e.g., collisional) interaction between two atoms. The calculation of excitation transfer rates between atoms (and molecules) is usually very complicated, and experimental determinations of transfer rates are essential. Indeed, such studies form an entire field of research.

It would take us too far afield to discuss experimental and theoretical techniques used to obtain excitation transfer cross sections and rates. We will only describe some salient results of such studies. In the following sections we discuss the essential role of excitation transfer in two important lasers, He–Ne and CO_2.

The most important fact about excitation transfer is that the transfer cross section is large when the corresponding atomic or molecular transition frequencies are approximately equal. However, excitation transfer can occur between two spe-

---

4. The cross section $\sigma_{fi}(E)$ for a collision of the second kind may be related to the cross section $\sigma_{if}(E)$ for the reverse process (collision of the first kind) by using the principle of detailed balance.

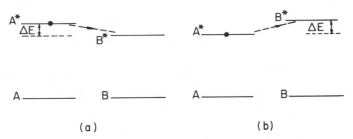

Figure 13.11 Exothermic (*a*) and endothermic (*b*) excitation transfer from atom A to atom B.

cies *A* and *B* even if the transitions are not precisely resonant. The *energy defect* ΔE (Figure 13.11) can be made up from translational degrees of freedom, so there is no contradiction with the law of conservation of energy.

In the case of a positive energy defect, for instance, the "extra" energy ΔE after the excitation transfer appears as additional kinetic energy of *A* and *B* after the transfer. In other words, *A* and *B* have more translational energy following the transfer than before. Since the temperature of a gas is a measure of its total translational energy content, this means that an *exothermic* (ΔE positive) process raises the temperature of the *A–B* system.

Similarly, in an *endothermic* (ΔE negative) process the "defect" in energy is made up for at the expense of the kinetic energy of the collision partners. The kinetic energy of *A* and *B* after the excitation transfer is less than that before the transfer. Therefore endothermic excitation transfer processes tend to lower the temperature of the system. It may be shown from the so-called principle of detailed balance that the rate for an exothermic process is a factor $e^{\Delta E/kT}$ greater than the reverse endothermic process. Thus, if *R* is the rate for an exothermic process, then $R\,e^{-\Delta E/kT}$ is the rate for the reverse process.

It is conventional to designate an excited atom or molecule by an asterisk. The exothermic process indicated in Figure 13.11*a* is written out as a reaction as follows:

$$A* + B \rightarrow A + B* + \Delta E \tag{13.5.1a}$$

Likewise the endothermic process of Figure 13.11*b* is written symbolically as

$$A* + B \rightarrow A + B* - \Delta E \tag{13.5.1b}$$

We can write the following rate equation for the number density N_{A*} of excited atoms (or molecules) of species *A* due to the process (13.5.1a):

$$\frac{dN_{A*}}{dt} = -R\left(N_{A*}N_B - e^{-\Delta E/kT}N_A N_{B*}\right) \tag{13.5.2}$$

The first term on the right is associated with the process (13.5.1a), described by the rate constant R. Since this process occurs only if an A atom is excited *and* a B atom is not, the rate of decrease of excited A atoms is proportional to the product[5] of the number densities of excited A atoms (N_{A*}) and unexcited B atoms (N_B). Similarly the second term, associated with the process of Figure 13.11b, is proportional to N_{B*} times N_A because the process occurs only when a B atom is excited and an A atom is unexcited. The principle of detailed balance, which is discussed below, has been used to relate the forward and reverse rates in (13.5.2). In addition to (13.5.2) we may write rate equations for N_A, N_{B*}, and N_B for the processes indicated in Figure 13.11 (Problem 13.3).

• In one form the principle of detailed balance is the requirement that in thermodynamic equilibrium the rate of *any* process must be exactly balanced by the rate associated with the reverse of that process. In thermodynamic equilibrium, therefore, the right side of (13.5.2) must vanish. This means that

$$\frac{N_{A*}}{N_A} = e^{-\Delta E/kT}\frac{N_{B*}}{N_B} \tag{13.5.3}$$

for thermal equilibrium. But since

$$\Delta E = (E_{B*} - E_B) - (E_{A*} - E_A) \tag{13.5.4}$$

(recall Fig. 13.11a) we can write (13.5.3) as

$$\frac{N_{A*}}{N_A} = e^{-(E_{A*} - E_A)/kT}\left(\frac{N_{B*}}{N_B}e^{(E_{B*} - E_B)/kT}\right) \tag{13.5.5}$$

However, it is always true in thermal equilibrium that

$$\frac{N_{A*}}{N_A} = e^{-(E_{A*} - E_A)/kT} \tag{13.5.6a}$$

and

$$\frac{N_{B*}}{N_B} = e^{-(E_{B*} - E_B)kT} \tag{13.5.6b}$$

5. The product is a consequence of the fact that the *probability* of A being excited and B being unexcited is the probability of A being excited times the probability of B being unexcited. That is, the two "events"—A excited and B unexcited—are assumed to be statistically independent.

according to the Boltzmann law. Therefore, we see that (13.5.2) is consistent with the steady-state Boltzmann law (for which $dN_{A*}/dt = 0$). However, it is important to recognize that (13.5.2) applies regardless of whether the level populations are actually in thermal equilibrium, as long as the *translational* degrees of freedom of the gas are in thermal equilibrium at the temperature T.

Thus the principle of detailed balance relates the forward and reverse rates for a process in just such a way as to be consistent with the equilibrium distribution of states that the process would establish were it acting alone. In the present example of collisional excitation transfer in a gas whose translational degrees of freedom are characterized by a temperature T, the appropriate distribution is that of Boltzmann. In the case of electron impact excitation, in which the electrons may not be well described by a Boltzmann distribution, the principle of detailed balance relates the forward and reverse rates in a way that would bring the atomic states into equilibrium with whatever the electron energy distribution happens to be. If both electron impact excitation and collisional excitation transfer are occurring, as in many gaseous laser media, the steady-state distribution of the active atomic levels will not in general be a Boltzmann distribution. That is, the active levels will not be populated according to the statistical distribution (13.5.6). Obviously, if they were, collisional excitation would be unattractive as a laser pumping mechanism. •

It is important to recognize that these energy transfer processes are not the only ones that can occur in a collision involving an excited atom or molecule. In a molecular gas, for example, collisions between molecules result in vibration-to-vibration (VV) energy transfer between the molecules. But there is also a probability that in a collision a vibrationally excited molecule will jump to a lower vibrational level, with the difference in energy between the two levels appearing as an increase in the translational kinetic energy of the colliding molecules. The latter process is called vibration-to-translation (VT) energy transfer.

13.6   He–Ne LASERS

The He–Ne electric-discharge laser, with its red output beam at 6328 Å, is the most common and best-known laser. He–Ne lasers can be made to operate at many other (mainly infrared) wavelengths. Whatever the operating wavelength, the active lasing species in He–Ne lasers is the Ne atom. It is excited by the transfer of excitation from He atoms, which in turn are excited by collisions with electrons. The population-inversion mechanism in He–Ne lasers thus involves a combination of electron impact excitation (of He) and excitation transfer (from He to Ne).

Figure 13.12 shows simplified energy-level diagrams for the neon and helium atoms. The 3.39-μm, 1.15-μm, and 6328-Å lines of neon, which are the strongest lasing transitions in He–Ne lasers, are indicated. The common upper level of the 3.39-μm and 6328-Å transitions, designated $3s_2$, is populated by excitation transfer from nearly resonant He atoms excited by electron impact to the 2^1S level. The upper level of the 1.15-μm transition is nearly resonant with the 2^2S level of He, and is populated by excitation transfer from He atoms in that excited state. Ac-

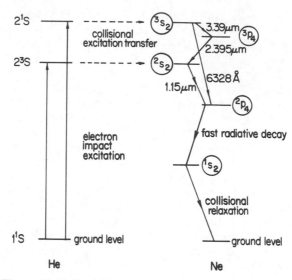

Figure 13.12 Partial energy-level diagrams for He and Ne.

tually Ne is also pumped directly into excited states by electron impacts, but the excitation transfer from He is the dominant pumping mechanism. The excited levels 2^1S_0 and 2^3S_1 of He, in addition to being nearly resonant with levels of Ne (and therefore allowing strong collisional excitation transfer) have the advantage of being forbidden by a selection rule from spontaneous emission. This allows these levels to "hold" energy for delivery to Ne during collisions. (The total decay rates of the He 2^1S and 2^3S levels due to collisions with Ne atoms are about 2×10^5 sec^{-1} and 10^4 sec^{-1}, respectively, per Torr of Ne.)

The partial pressures of Ne and He in typical He–Ne lasers are roughly 0.1 and 1 Torr, respectively. At these low pressures the upper-state lifetimes are determined predominantly by spontaneous emission rather than collisional deexcitation. At very low pressures, the $3s_2$ and $2s_2$ levels[6] of Ne have short radiative (i.e., spontaneous-emission) lifetimes, roughly 10–20 nsec, due to the strong allowed ultraviolet transitions to the ground state. For Ne pressures typical of He–Ne lasers, however, these radiative lifetimes are actually about 10^{-7} sec because of *radiative trapping*. This occurs when the spontaneously emitted photons are reabsorbed by atoms in the ground state, thereby effectively increasing the lifetime of the emitting level. Since the ground state is generally the most highly populated level even under conditions of population inversion [recall the numerical estimates of Section 10.4], radiative trapping is significant only for levels connected to the ground level by an allowed transition. Thus the Ne $3p_4$ and $2p_4$ levels, which are

---

6.   We follow here the Paschen notation for the Ne levels. A more modern and systematic notation, based on so-called Racah symbols, is not often encountered in the laser literature.

forbidden by a selection rule from decaying spontaneously to the ground level, are not radiatively trapped, and have lifetimes of about 10^{-8} sec, roughly 10 times shorter than the $3s_2$ and $2s_2$ levels.

This means that the $s \rightarrow p$ transitions indicated in Figure 13.12 have favorable lifetime ratios for lasing, i.e., their lower (p) levels decay more quickly than their upper (s) levels, making it easier to establish a population inversion. The integrated absorption coefficients of the 6328-Å and 3.39-μm lines have roughly the same magnitude, but the 3.39-μm line has a Doppler width about 5.4 times smaller than the 6328-Å line, and consequently a considerably larger line-center gain. Without some mechanism for suppressing oscillation on the 3.39-μm line, therefore, the familiar 6328-Å line would not lase (Problem 13.4).

The 6328-Å and 1.15-μm transitions have a common lower level, $2p_4$, which decays rapidly into the $1s$ level. The latter is forbidden by a selection rule from decaying radiatively into the ground level, and is therefore relatively long-lived. This is bad for laser oscillation on the 6328-Å and 1.15-μm lines, because electron impact excitation can pump Ne atoms from $1s$ to $2p_4$, thereby reducing the population inversion on these lines. However, Ne atoms in the $1s$ level can decay to the ground level when they collide with the walls of the gain tube. In fact it is found that the gain on the 6328-Å and 1.15-μm lines increases when the tube diameter is decreased; this is attributed to an increase in the atom–wall collision rate with decreasing tube diameter.

The first gas laser, which was also the first cw laser, was a 1.15-μm He–Ne laser constructed in 1960 by A. Javan, W. R. Bennett, Jr., and D. R. Herriott. Inexpensive He–Ne lasers have long since been available commercially, and have many practical applications. In most of these applications a *visible* laser beam is desired, and it is therefore necessary to suppress the infrared lines in order to obtain oscillation at 6328 Å. This is done by discriminating against the infrared lines by using a cavity in which these lines have greater loss than the 6328-Å line, and therefore a higher threshold for oscillation. Typically this is accomplished by coating the cavity mirrors with dielectric materials that reflect in the visible but transmit in the infrared.

• The preparation of laser mirrors is a delicate manufacturing process. The mirrors must be of much finer optical quality, (i.e., smoothness) than the lenses and mirrors used in many other optical instruments. Special surfacing machines polish the mirrors for many hours, even days, and the mirrors must then be carefully cleaned and prepared for the coating process, which is done in a vacuum chamber. The coating may be done in several tens of layers, each a fraction of a wavelength thick, while a technician monitors the mirror reflectance.

Figure 13.13 indicates the general structure of a commercially available He–Ne laser. For reasons discussed in the next chapter, one of the mirrors is flat whereas the other is curved. The gain tube, which is typically 10–30 cm long, is shown with Brewster-angle windows to give a linearly polarized output (Section 11.12). The mirrors can be attached to the glass tube with an epoxy resin. However, epoxy eventually deteriorates and gas leaks out of the tube. ''Hard-seal mirrors'' have been developed to increase the life of the gain

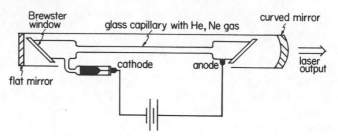

Figure 13.13 Basic structure of a He–Ne laser.

tube; the mirrors are bonded directly to the metal housing of the laser to form a tight seal without any epoxy. These lasers are low-power devices, producing outputs measured in milliwatts. •

13.7   RATE-EQUATION MODEL OF POPULATION INVERSION IN He–Ne LASERS

We will consider a rate-equation model of population inversion in a 3.39-μm He–Ne laser. Our goal is to account for some observed results on the variation of small-signal gain with the electric current i through the gain tube.

Figure 13.14 summarizes our notation for the pertinent energy-level population densities. N_2 and N_1 denote the population densities (atoms/cm^3) of the upper and lower levels of the 3.39-μm laser transition. The ground-level population densities of He and Ne are denoted by $\tilde{N}_0$ and N_0 respectively, and $\tilde{N}_2$ denotes the population density of the excited level 2^1S of He. In our simplified model we will ignore all other levels of the He and Ne atoms. In other words, we assume that other levels are required only to explain some finer details that are not presently of interest to us.

The rate of change of N_2 due to excitation transfer with He is given by

$$\left(\frac{dN_2}{dt}\right)_{\substack{\text{He–Ne excita-}\\\text{tion transfer}}} = R\tilde{N}_2 N_0 - Re^{-\Delta E/kT}\tilde{N}_0 N_2 \qquad (13.7.1)$$

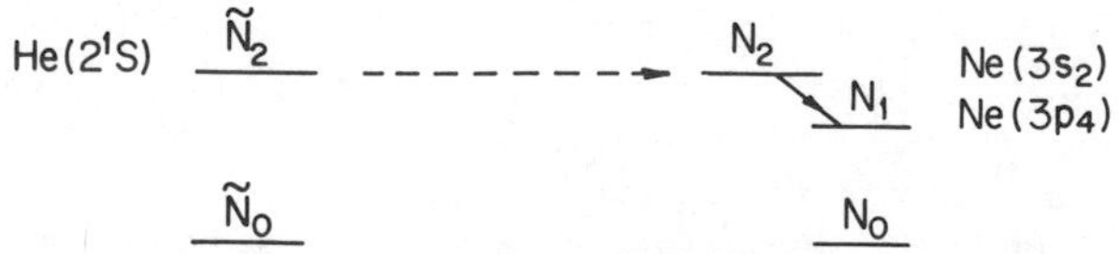

He(2^{1}S) $\underline{\quad\tilde{N}_2\quad}$ $\;-\;-\;-\;-\;\longrightarrow\;$ $\underline{\quad N_2\quad}$ Ne(3s$_2$)
$\qquad\qquad\qquad\qquad\qquad\qquad\qquad\quad\diagdown\;\underline{N_1}\quad$ Ne(3p$_4$)

$\qquad\qquad\quad\underline{\quad\tilde{N}_0\quad}$ $\qquad\qquad\qquad\qquad\underline{\quad N_0\quad}$

Figure 13.14 Simplified version of Figure 13.12 for a model of population inversion on the 3.39-μm transition of Ne.

where R is the rate constant for the excitation-transfer collisions and

$$\Delta E = E[\text{He}(2^1S)] - E[\text{Ne}(3s_2)] \qquad (13.7.2)$$

is the energy defect of the $\text{He}(2\,^1S) \rightarrow \text{Ne}(3s_2)$ inelastic collision. This energy defect is quite small, about 0.04 eV, and so we will take $\exp(-\Delta E/kT) \sim 1$ as an approximation in (13.7.1) (Problem 13.5). We therefore write

$$\frac{dN_2}{dt} = \left(\frac{dN_2}{dt}\right)_{\substack{\text{He-Ne excita-} \\ \text{tion transfer}}} - R_2 N_2$$

$$\approx R\tilde{N}_2 N_0 - R\tilde{N}_0 N_2 - R_2 N_2 \qquad (13.7.3)$$

where the decay rate R_2 is the rate of decrease of N_2 due to spontaneous emission and other deexcitation processes.

Similarly we have

$$\frac{d\tilde{N}_2}{dt} = \left(\frac{d\tilde{N}_2}{dt}\right)_{\substack{\text{He-Ne excita-} \\ \text{tion transfer}}} + \left(\frac{d\tilde{N}_2}{dt}\right)_{\substack{\text{electron} \\ \text{impact}}}$$

$$+ \left(\frac{d\tilde{N}_2}{dt}\right)_{\substack{\text{decay} \\ \text{processes}}} \qquad (13.7.4)$$

The first term is simply the negative of (13.7.1) (Problem 13.3). The second term is the rate of change of $\tilde{N}_2$ due to electron impact excitation (and deexcitation) of He 2^1S, which we write as

$$\left(\frac{d\tilde{N}_2}{dt}\right)_{\substack{\text{electron} \\ \text{impact}}} = K_1 \tilde{N}_0 - K_2 \tilde{N}_2 \qquad (13.7.5)$$

K_1 is the rate for the electron impact excitation of He 2^1S from the ground level of He, i.e., for the process

$$\text{He}(1^1S) + e \rightarrow \text{He}(2^1S) + e \qquad (13.7.6)$$

whereas K_2 is the rate for the reverse, "superelastic" collision. Finally, the third term on the right side of (13.7.4) is the rate of decrease of $\tilde{N}_2$ due to other deexcitation processes, and is assumed to be characterized by some constant $\tilde{R}_2$. Thus

$$\frac{d\tilde{N}_2}{dt} = -R\tilde{N}_2 N_0 + RN_2\tilde{N}_0 + K_1\tilde{N}_0 - K_2\tilde{N}_2 - \tilde{R}_2\tilde{N}_2 \qquad (13.7.7)$$

In a similar fashion we can write rate equations for $\tilde{N}_0$ and N_0. However, as noted in the preceding section, these ground-state populations are very large compared with excited-state populations, and remain relatively unchanged by the pumping process. We will therefore make the approximation that $\tilde{N}_0$ and N_0 are constants. This reduces (13.7.3) and (13.7.7) to linear differential equations with constant coefficients. In particular the steady-state values of N_2 and $\tilde{N}_2$, denoted $\bar{N}_2$ and $\bar{\tilde{N}}_2$, are easily obtained by setting the derivatives in these equations to zero:

$$\bar{\tilde{N}}_2 = \frac{K_1 A \tilde{N}_0}{1 + K_2 A} \tag{13.7.8a}$$

$$\bar{N}_2 = \frac{RN_0}{R\tilde{N}_0 + R_2} \frac{K_1 A \tilde{N}_0}{1 + K_2 A} \tag{13.7.8b}$$

where A is a constant in our simple-minded model:

$$A = \frac{R_2 + R\tilde{N}_0}{\tilde{R}_2(R_2 + R\tilde{N}_0) + RR_2 N_0} \tag{13.7.9}$$

Now the electron-impact rates K_1 and K_2 are directly proportional to the number density (N_e) of electrons and therefore to the electric current i in the discharge tube, as is evident from Eq. (13.4.2). We can therefore write equations (13.7.8) in terms of the current:

$$\bar{N}_2 = \frac{ai}{1 + bi} \tag{13.7.10a}$$

$$\bar{\tilde{N}}_2 = \frac{ci}{1 + bi} \tag{13.7.10b}$$

where a, b, and c are constants.

In order to express the gain in terms of the current, we need an expression for $\bar{N}_1$, the (steady-state) population density of the lower laser level. We will assume that N_1 varies according to the rate equation

$$\frac{dN_1}{dt} = K_3 N_0 - R_1 N_1 \tag{13.7.11}$$

where K_3 is the rate at which electron impacts pump ground-state Ne atoms up to the $3p_4$ level, and R_1 is the total decay rate of that level.[7] The steady-state solution of this equation is obtained as usual by setting $dN_1/dt = 0$, and is given by

---

7. It is not known precisely how the lower levels, such as $3p_4$, are populated in He–Ne lasers. There is apparently some electron impact excitation from ground-state Ne atoms, as evidence by the fact that pure Ne can be made to lase (weakly). The p levels may also be populated via the decay of higher-lying Ne levels that are populated by electron impact excitation from ground-state or excited-state Ne atoms. In neglecting the deexcitation of $3p_4$ by electron impact, we are assuming that the corresponding rate is much smaller than the decay rate R_1 due to spontaneous emission, etc.

$$\overline{N}_1 = K_3 N_0 / R_1 \propto \text{current} \qquad (13.7.12)$$

From the expression for the gain coefficient g in Table 10.1, therefore, we obtain from (13.7.10) and (13.7.12) the gain–current relation

$$g = \frac{\alpha i}{1 + bi} - \beta i \qquad (13.7.13)$$

where α and β are constants.

Note that, since no stimulated-emission terms were included in our simple rate-equation model, Eq. (13.7.13) is actually a prediction about the variation of *small-signal* gain with current. This variation was studied in the early days of He–Ne laser research and, as indicated in Figure 13.15, the functional form (13.7.13), with certain values of the constants α, β, and b, can be used to fit very well the measured data on small-signal gain vs. current.

One can also deduce relative values of various level population densities by measuring the intensities of fluorescent "sidelight" radiation at different wavelengths. That is, by measuring the strength of the spontaneous emission from a side of the laser tube, one can estimate the relative population of the upper level from which this emission proceeds (Problem 13.6). Such results are shown in Figure 13.16, and provide evidence for the approximate proportionality of the He 2^1S and Ne $3s_2$ populations, as predicted by (13.7.10). They show furthermore that the Ne $3p_4$ (and $2p_4$) population density is very nearly proportional to the current, as predicted by (13.7.12).

For small values of the current the small-signal gain increases linearly with current. As the current is raised, however, two effects work against a further in-

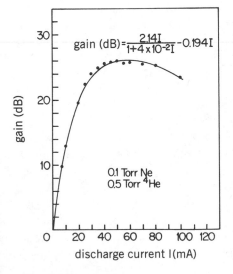

Figure 13.15 Measured variation of gain with current in a He–Ne laser. From A. D. White and E. I. Gordon, *Applied Physics Letters* **3**, 197 (1963).

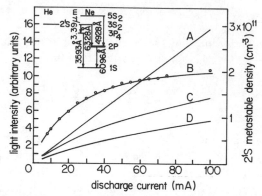

Figure 13.16 Relative populations in a He–Ne laser measured by White and Gordon. The four curves refer to: A—4928 Å line, B—6328 Å line, C—6096 Å line, D—3593 Å line, and the circles on curve B are measurements of the 2^1S metastable population density.

crease of the gain. First there is the deexcitation of He 2^1S by electron impact; this is described by the rate constant $K_2 \propto b$, and is thus associated with the denominator $1 + bi$ in (13.7.13). Because of this denominator, the first term in (13.7.13) does not increase linearly with i, but rather saturates to the constant value α/b for large current values ($bi \gg 1$). Second, there is the proportionality of the lower-laser-level population Ne $3p_4$ to the current, which is associated with the term βi in (13.7.13). Beyond a certain range of current values, therefore, the output power of a He–Ne laser will decrease with increasing current, and lasing will eventually cease altogether.

It should be emphasized that we have not formulated a first-principles theory of population inversion on the 3.39-μm line of a He–Ne laser. For instance, we have not specified the values of the electron-atom collision rates K_1, K_2, and K_3, although we have mentioned in Section 13.4 how they might be calculated. This approach to understanding population inversion processes—that is, trying to understand general trends rather than obtaining detailed quantitative predictions—is more often the rule than the exception in laser theory and design. This is partly because many of the rates for the processes determining population inversion are often not well known and, because of the complexity of the processes, theoretical analyses are highly involved but not highly reliable. Furthermore, in a given device various quantities, such as number densities of atoms and electrons, may be hard to specify accurately, so that even a trustworthy theory might be of only limited value. In spite of these disclaimers, however, some level of theoretical analysis is indispensable to the building of new lasers and the improvement of old ones.

13.8   RADIAL GAIN VARIATION IN He–Ne LASER TUBES

In this book, and in most of laser theory, it is usually assumed that the small-signal gain g_0 is approximately constant within the gain tube. For many purposes this is an adequate approximation, but it is generally not strictly true. In this section we will describe how the small-signal gain in a He–Ne laser varies radially with distance r from the axis of the tube, which is assumed to have a circular cross section.

We will show below that the (free-) electron number density N_e varies with r according to the formula

$$N_e(r) = N_e(0) \, J_0(2.405 \, r/R) \tag{13.8.1}$$

where R is the tube radius, J_0 is the zeroth-order Bessel function,[8] and $N_e(0)$ is the electron number density at the tube axis, where $r = 0$ $[J_0(x) = 1$ for $x = 0]$. Since $J_0(2.405 \ldots) = 0$, it follows that N_e is zero on the tube wall, as we might expect.

Assuming that the current density j is directly proportional to N_e, we have

$$j(r) = j(0) \, J_0(2.405 \, r/R) \tag{13.8.2}$$

The current i is the integral of j over the cross-sectional area of the tube:

$$i = 2\pi \int_0^R j(r) \, r \, dr$$

$$= 2\pi j(0) \int_0^R J_0\left(2.405 \, \frac{r}{R}\right) r \, dr$$

$$\approx 1.36 \, j(0) \, R^2 \tag{13.8.3}$$

where we have used the properties $\int_0^z J_0(x) \, x \, dx = zJ_1(z)$ and $J_1(2.405) \approx 0.519$.

In the preceding section we derived the formula (13.7.13) for the current dependence of the gain, neglecting any r variation of the electron number density. In other words, we took N_e and j as simply proportional to the current i, with no r dependence. Let us assume that (13.7.13) actually gives the small-signal gain at the *center* of the tube, i.e., that the left side of (13.7.13) is really $g(r = 0)$. Then, since N_e (and j) varies radially as $J_0(2.405 \, r/R)$, we have for $g(r \neq 0)$ the formula (13.7.13) with i replaced by $iJ_0(2.405 \, r/R)$:

$$g(r) = \frac{\alpha i J_0(2.405 \, r/R)}{1 + bi J_0(2.405 \, r/R)} - \beta i J_0(2.405 \, r/R) \tag{13.8.4}$$

A graph of $g_0(r)$ is given in Figure 13.17 for several values of i, using the values of Figure 13.15 for the parameters α, b, and β. For small currents $g(r)$ goes as $J_0(2.405 \, r/R)$, having its maximum on-axis and falling off to zero at $r = R$. As the current is raised $g(r)$ becomes flatter near the axis, and with higher currents

8. A graph of the function $J_0(x)$ is given in Figure 12.13.

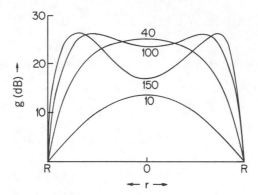

Figure 13.17 Radial dependence of small-signal gain, obtained by plotting the function (13.8.4) for the parameters α, b, β given in Figure 13.15. The four curves correspond to currents $i = 10$, 40, 100, and 150 mA, as indicated.

has a dip on axis. That is, as the current is raised, a stage is reached where the small-signal gain has a local minimum along the axis of the tube.

Similar results for the radial variation of the small-signal gain are obtained whenever the on-axis gain-vs.-current curve has a maximum and then "turns over" with increasing current, as in Figure 13.15. Therefore radial variations of the form shown in Figure 13.17 are expected not only in He–Ne lasers, but in a wide variety of other electric-discharge lasers. Although these lasers employ somewhat different population-inversion processes, they all follow the general trends predicted in Figure 13.17. Experimental results for He–Ne, He–Xe, and CO_2 lasers are shown in Figure 13.18. We refer the reader to the papers cited in Figure 13.18 for details of the different measurements.

• The basic starting point of the analysis above is the formula (13.8.1) for the electron number density in the (cylindrical) discharge tube. The origin of this formula may be understood from the following argument.

If the density of electrons has some nonvanishing gradient, the electrons will tend to redistribute, or *diffuse*, just as a gradient in the temperature of a gas gives rise to a flow of heat. We will assume that the diffusion of electrons follows the *Fick law*, i.e., that the rate of diffusion of electrons across a given area is directly proportional, but opposite in direction, to the gradient. In other words, the flux **f** (particles per second per unit area) of electrons is given by

$$\mathbf{f} = -D\nabla N_e \tag{13.8.5}$$

where the constant of proportionality D is called the *diffusion coefficient*. If the number density of electrons is increasing in some direction (positive gradient), Equation (13.8.5) says there will be a compensating diffusion of electrons in the opposite direction. We will not attempt to justify (13.8.5), but it is worth mentioning that this Fick law accurately

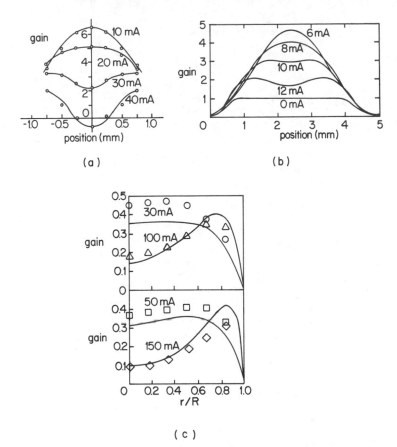

Figure 13.18 Measured radial variation of gain in (*a*) a He–Ne laser, from I. P. Mazanko, M. I. Molchanov, N.-D. D. Ogurok, and M. V. Sviridov, *Optics and Spectroscopy* **30**, 495 (1971); (*b*) a He–Xe laser, from P. A. Wolff, N. B. Abraham, and S. R. Smith, *IEEE Journal of Quantum Electronics* **QE-13**, 400 (1977); (*c*) a CO_2 laser, from W. J. Wiegand, M. C. Fowler, and J.A. Benda, *Applied Physics Letters* **18**, 365 (1971).

describes many diffusion processes, such as the diffusion of gas particles due to a density or thermal gradient, or the diffusion of neutrons in a nuclear reactor.

The rate at which electrons diffuse out of some closed surface S is given by the surface integral of the flux:

$$\text{(rate of diffusion out of volume } V) = \int_S \mathbf{f} \cdot \hat{\mathbf{n}} \, dA \qquad (13.8.6)$$

where V is the volume enclosed by S, and $\hat{\mathbf{n}}$ is the unit vector outwardly normal to S. From the divergence theorem (i.e., Gauss's law) and (13.8.5) we have

$$\text{(rate of diffusion out of volume } V) = \int_V \nabla \cdot \mathbf{f} \, dV$$

$$= -D \int_V \nabla^2 N_e \, dV \qquad (13.8.7)$$

Now in a steady-state discharge the rate of diffusion of electrons out of V must be exactly balanced by the rate at which free electrons are produced inside V, in order that the total number of electrons within V be constant. Free electrons are produced when an electron collides with an atom (or ion) and ionizes it, leaving another free electron plus a positive ion. This electron impact ionization is the electron production process that must balance the diffusive loss (13.8.7). Letting Q_i be the ionization rate, we have

$$\text{(rate of production of free electrons inside } V) = \int_V Q_i N_e \, dV \qquad (13.8.8)$$

Equating (13.8.7) and (13.8.8), we have

$$\nabla^2 N_e + \frac{Q_i}{D} N_e = 0 \qquad (13.8.9)$$

This equation, subject to whatever boundary conditions are to be imposed, determines the electron number density N_e.

For a cylindrical discharge tube it is convenient to write out the Laplacian ∇^2 in terms of the cylindrical coordinates r, θ, z:

$$\nabla^2 N_e = \left(\frac{\partial^2 N_e}{\partial r^2} + \frac{1}{r} \frac{\partial N_e}{\partial r} + \frac{1}{r^2} \frac{\partial^2 N_e}{\partial \theta^2} + \frac{\partial^2 N_e}{\partial z^2} \right) \qquad (13.8.10)$$

We will assume circular symmetry (no θ dependence of N_e), and that the z dependence of N_e can be ignored to a good approximation. Then the last two terms in $\nabla^2 N_e$ above may be dropped, and (13.8.9) becomes

$$\frac{d^2 N_e}{dr^2} + \frac{1}{r} \frac{dN_e}{dr} + \frac{Q_i}{D} N_e = 0 \qquad (13.8.11)$$

This differential equation has the solution

$$N_e(r) = N_e(0) \, J_0(r\sqrt{Q_i/D}) \qquad (13.8.12)$$

where the constant $N_e(0)$ is the value of N_e at the tube axis, $r = 0$.

To satisfy the boundary condition that the electron number density vanishes on the wall of the tube, i.e., that $N_e(R) = 0$, we require that

$$J_0(R\sqrt{Q_i/D}) = 0 \qquad (13.8.13)$$

In other words, $R\sqrt{Q_i/D}$ must be a zero of the zeroth-order Bessel function J_0. In order to ensure that $N_e(r)$ given by (13.8.12) is positive-definite for all values of $r \leq R$, further-more, we require $R\sqrt{Q_i/D}$ to be the *first* zero of J_0, which is about 2.405. Thus

$$\sqrt{Q_i/D} \approx 2.405/R \qquad (13.8.14)$$

which, together with (13.8.12), gives (13.8.1).

The electrons in the discharge, because of their much higher average velocity, might be expected to diffuse to the walls much more quickly than the positive ions, producing an excess of negative charge near the walls. However, an electric field is set up by the charges (a "space-charge" field) in such a way as to retard the diffusion of electrons and effectively "drag along" the positive ions. In this *ambipolar diffusion* both the positive and negative charge carriers have the same diffusion constant. Our simple derivation of (13.8.1) assumes ambipolar diffusion, and D in our analysis is in fact the ambipolar diffusion coefficient •

13.9 CO$_2$ ELECTRIC-DISCHARGE LASERS

The electric-discharge carbon dioxide laser has a population inversion mechanism similar in some respects to the He–Ne laser: the upper CO$_2$ laser level is pumped by excitation transfer from the nitrogen molecule, with N$_2$ itself excited by electron impact.

The relevant energy levels of the CO$_2$ and N$_2$ molecules are vibrational–rotational levels of their electronic ground states. We discussed the vibrational–rotational characteristics of the CO$_2$ molecule in Section 4.7, and indicated in Figure 4.9 the relative energy scales of the three normal modes of vibration, the so-called symmetric stretch, bending, and asymmetric stretch modes (Figure 4.8). N$_2$, like all diatomic molecules, has a single "ladder" of vibrational levels corresponding to a single mode of vibration (Figure 4.6). In Figure 13.19 we show the CO$_2$ and N$_2$ vibrational energy level diagrams side by side.

Figure 13.19 shows that the first excited vibrational level ($v = 1$) of the N$_2$ molecule lies close to the level (001) of CO$_2$. Because of this near-resonance there is a rapid excitation transfer between N$_2$ ($v = 1$) and CO$_2$ (001), the upper laser level. N$_2$ ($v = 1$) is itself a long-lived (metastable) level, so it effectively stores energy for eventual transfer to CO$_2$ (001); it is also efficiently pumped by electron impact excitation. As in the case of the He–Ne laser, therefore, advantage is taken of a fortuitous near-resonance between an excited state of the lasing species and an excited, long-lived collision partner.

Laser action in CO$_2$ lasers occurs on the vibrational transition (001) → (100) of CO$_2$. This transition has a wave number around (2349 − 1388) cm^{-1} = 961 cm^{-1} (Figure 13.19), or a wavelength around (961 cm^{-1})$^{-1}$ = 10.4 μm in the infrared. The laser wavelength depends also on the rotational quantum numbers of the upper and lower laser levels. For the case in which the upper and lower levels are characterized by $J = 19$ and 20, respectively, the wavelength is about 10.6 μm. This is perhaps the most common CO$_2$ laser wavelength.

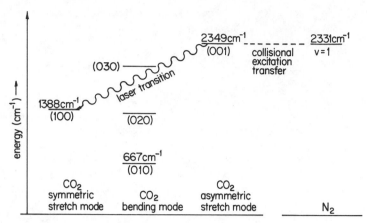

Figure 13.19 Vibrational energy levels of CO_2 and N_2. The energies are given in cm^{-1}, a unit corresponding to a frequency $(c)(1 \text{ cm}^{-1}) \approx 3.0 \times 10^{10}$ Hz, or an energy $h\nu \approx 1.2 \times 10^{-4}$ eV.

The (100) and (020) vibrational levels of CO_2 are essentially resonant. This "accidental degeneracy" results in a strong quantum-mechanical coupling in which states in effect lose their separate identities.[9] Furthermore the (010) and (020) levels undergo a very rapid vibration-to-vibration (VV) energy transfer:

$$CO_2(020) + CO_2(000) \rightarrow CO_2(010) + CO_2(010) \qquad (13.9.1)$$

For practical purposes, then, the stimulated emission on the (001) → (100) vibrational band takes CO_2 molecules from (001) to (010). The (010) level thus acts in effect like a lower laser level that must be rapidly "knocked out" in order to avoid a bottleneck in the population inversion.

Fortunately it is relatively easy to deexcite the (010) level by vibration-to-translation (VT) processes:

$$CO_2(010) + A \rightarrow CO_2(000) + A \qquad (13.9.2)$$

where A represents some collision partner. The VT deexcitation of (010) effectively depopulates the lower laser level, and also puts CO_2 molecules in the ground level, where they can be pumped into the upper laser level by the VV excitation transfer process

$$N_2(v = 1) + CO_2(000) \rightarrow N_2(v = 0) + CO_2(001) \qquad (13.9.3)$$

9. For a discussion of this *Fermi resonance* effect see, for instance, G. Herzberg, *Infrared and Raman Spectra of Polyatomic Molecules* (Van Nostrand, New York, 1949).

In high-power CO_2 lasers the lifetime of the $CO_2(010)$ level may be on the order of 1 μsec due to *VT* collisions of CO_2 with He, N_2, and CO_2 itself. Of course the *VT* process (13.9.2) is exothermic, and results in a heating up of the laser medium; some of the other *VT* and *VV* processes in the CO_2 laser have the same effect. This heating of the laser medium is a very serious problem in high-power lasers. In the next section we will see how it may be overcome.

Electron impacts excite CO_2 as well as N_2 vibrationally. Furthermore there are various other processes that have to be accounted for in an accurately predictive rate-equation model of a CO_2 laser. Because of the many applications of high-power CO_2 lasers, such models are available, and are often quite accurate. These models are computer programs that numerically integrate rate equations for the various level populations and the intensity. They also compute the electron energy distribution function, and from this the electron impact excitation rates (Section 13.4). Our discussion captures only the bare essence of the population inversion process, but it is sufficient for a qualitative understanding of CO_2 lasers.

In the laser research literature one finds expressions for the gain coefficients of CO_2 and other infrared molecular-vibration lasers; at first glance these expressions often do not resemble the "standard" formula for the gain coefficient given in Table 10.1. In Appendix 13.A we carry out the steps leading from the formula in Table 10.1 to an expression that appears frequently in the literature. As an example of the use of this expression, we estimate the absorption coefficient for 10.6-μm CO_2 laser radiation propagating in air at sea level.

13.10 GAS-DYNAMIC LASERS

We mentioned in Section 10.10 that gas lasers have generally been the best sources of very high-power, nearly monochromatic, coherent radiation. Although solid laser media have much higher molecular number densities than gases, and perhaps therefore a potential for higher gains, they have serious drawbacks, especially for high-power cw lasers. One problem is that the pumping and lasing of the gain medium produces heat that can damage it or, at the least, induce distortions in it that degrade the spatial coherence of the laser radiation. The major damage problem for gaseous media, however, is photoionization, and this is usually not a concern except at extremely high intensities, perhaps 10^{10}–10^{14} W/cm^2, depending on the circumstances.

The pumping and lasing of a gaseous medium also generates waste heat, and this is deleterious to the scaling of a gas laser up to very high powers. Various factors, such as an increase in collisional deexcitation rates, contrive to reduce the power and coherence properties of the laser radiation when the gas temperature gets too high. By the late 1960s the highest-power cw lasers were CO_2 lasers generating several kilowatts of power. This certainly represents a good deal of radiation intensity when it is concentrated in a narrow laser beam; such a beam

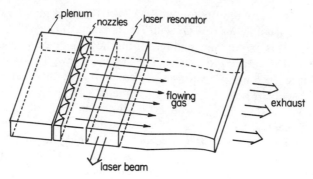

Figure 13.20 Schematic picture of a flowing-gas arrangement used in some high-power gas lasers.

can drill a hole in quarter-inch steel in a matter of seconds. But these lasers typically had "folded" resonator designs that made the gain medium effectively several hundred feet long in some cases. Because of the heat generated as an unavoidable by-product of lasing, they seemed at the time to be approaching a sort of practical upper limit in the quest for higher and higher powers.

In 1968, however, a major breakthrough was achieved.[10] A CO_2 laser was developed that produced more than 60 kW of continuous-wave output power—an improvement of approximately a factor of 10. The new idea was to remove the waste heat by using a laser medium consisting of a gas *flowing* through the laser cavity (Figure 13.20). In this way the hot gas is expelled while fresh, cooler gas is continually flowing in and lasing. The high-power lasers operated in this manner are called *gas-dynamic lasers*, and the gas flow velocity is supersonic.

It should be evident by now that laser research involves many diverse facets of atomic, molecular, electronic, and optical physics. The gas-dynamic laser triggered a whole new branch of laser research connected with aerodynamic effects. We will not be able to discuss any technical aspects of this interplay of radiation physics and aerodynamics; instead we will describe qualitatively the population-inversion mechanism in gas-dynamic lasers.

In the CO_2 gas-dynamic laser a gas mixture containing CO_2 and N_2 is heated in a high-pressure container, or plenum. The temperature in the plenum may be 1500–2000 K, with a pressure on the order of several tens of atmospheres. The translational, rotational, and vibrational degrees of freedom of the CO_2 and N_2 molecules in the plenum are in thermal equilibrium, and their population densities satisfy a (high-temperature) Boltzmann distribution.

The gas is then suddenly allowed to leave the plenum through an array of nozzles. A supersonic expansion results in which the gas (translational) temperature and pressure are drastically reduced, to about 300–400 K and 50 Torr, respec-

---

10. This advance, which was made at the Avco-Everett Research Laboratories, was not declassified by the U.S. government until 1970.

tively. The rotational degrees of freedom of the molecules also relax quickly to this new, much cooler thermal equilibrium. The key point, however, is that the *vibrational* relaxation rates (VV and VT) are much lower; the vibrational degrees of freedom are thus temporarily "frozen" near the original, high-temperature Boltzmann distribution. On the other hand, the gas temperature itself as measured by the average translational energy $\frac{3}{2} kT$ of the molecules, is greatly decreased by the expansion. We then have a *nonequilibrium flow*, in which various degrees of freedom in the gas are characterized by very different temperatures.

Population inversion in the nonequilibrium flow results for two reasons. First, the VT collision rates at the gas translational temperature (300–400 K) are such that the decay of the lower level of a CO_2 vibrational–rotational transition near 10.6 μm is rapid, giving a favorable lifetime ratio of upper and lower levels. Second, the N_2 vibrational decay rates (due to spontaneous emission and VT collisions) are very low, so that the N_2 molecules store energy for excitation transfer to CO_2 (001).

Note that the population inversion achieved in this manner does not require any "external" process such as optical pumping or an electric discharge. In this sense the gas-dynamic laser is thermodynamically similar to classical power generators (e.g., steam engines) in that, using hot and cold reservoirs, it converts thermal energy into a more useful form of energy.

13.11 CHEMICAL LASERS

Chemical reactions frequently produce excited-state species. So-called *chemical lasers* are those in which a population inversion is established as a result of a chemical reaction of two or more species. In other words, chemical energy of molecular bonding is converted into electromagnetic energy in the form of laser radiation.

We will consider briefly one example of a chemical laser, the hydrogen fluoride (HF) laser. HF lasers operate on several (sometimes many) HF vibrational–rotational transitions in the neighborhood of 2.6–2.8 μm. Vibrationally excited HF molecules are produced as a result of two exothermic chemical reactions:

$$F + H_2 \rightarrow HF^* + H + \Delta H_1 \qquad (13.11.1)$$

$$H + F_2 \rightarrow HF^* + F + \Delta H_2 \qquad (13.11.2)$$

The heats of reaction of these two processes are $\Delta H_1 \approx 31.6$ kcal/mole and $\Delta H_2 \approx 98.0$ kcal/mole, and they are therefore referred to as the "cold" and "hot" reactions, respectively. HF* denotes a vibrationally excited HF molecule. The cold reaction (13.11.1) typically produces HF molecules in excited vibrational levels up to about $v = 3$, whereas for the hot reaction (13.11.2) it is known that levels up to $v = 10$ are significantly populated.

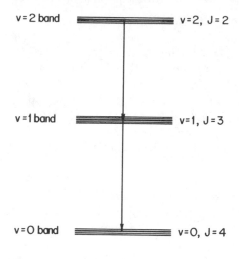

v = 2 band   v=2, J = 2

v =1 band   v=1, J=3

v=0 band   v=0, J=4

Figure 13.21 HF and other molecular lasers can lase simultaneously on two or more coupled transitions, as indicated above for a $(v = 2, J = 2) \rightarrow (v = 1, J = 3) \rightarrow (v = 0, J = 4)$ cascade.

The overall result of the cold and hot reactions is summarized by writing

$$H_2 + F_2 \rightarrow 2HF^* + \Delta H_1 + \Delta H_2 \tag{13.11.3}$$

However, this disguises the role of atomic fluorine (F) in the cold reaction (13.11.1); without F, the production rate of HF* is too slow. Since F atoms bond to form F_2, the F_2 molecules must somehow be dissociated into two F atoms. There are several ways of doing this. One method utilizes the collisional dissociation of F_2 by a collision partner A, i.e.,

$$F_2 + A \rightarrow 2F + A \tag{13.11.4}$$

in a high-temperature chamber. Another method involves the use of radiation to free F atoms from chemical bonding; this photochemical process is called *photolysis*, or, if pulsed radiation is used, *flash photolysis*.[11] Another common means of getting F atoms for HF chemical lasers is by electron impact dissociation of F_2 or another molecule bonding F atoms, such as sulfur hexafluoride:

$$SF_6 + e \rightarrow SF_5 + F + e \tag{13.11.5}$$

HF typically lases on low-lying vibrational–rotational transitions, such as $(v = 1, J = 3) \rightarrow (v = 0, J = 4)$. Lasing tends to occur on two or more vibrational–rotational transitions simultaneously, as indicated in Figure 13.21. The various chemical and VT processes that occur in HF lasers result in a considerable heating

---

11.   An important example of photolysis occurs in the upper atmosphere of the earth. O_2 molecules are dissociated by ultraviolet solar radiation, and the freed O atoms react with O_2 to produce ozone, O_3. The "ozone layer" absorbs far-ultraviolet solar radiation that is harmful to living organisms on earth.

of the gain medium. High-power HF chemical lasers are therefore frequently of the flowing-gas type.

13.12 EXCIMER LASERS

There are some molecules that can exist only in excited electronic levels, the ground level being dissociative. In such a molecule the potential energy curve for the ground level has no local minimum, and so there is no stable ground level (Figure 13.22). A molecule of this sort is called an *excimer*, a contraction for "excited dimer." In the transition indicated in Figure 13.22 the lower level very quickly dissociates into two unbound atoms. The dissociation time is on the order of a vibrational period, around 10^{-13} sec. This effective absence of any lower-level population is the most significant feature of an excimer laser operating on such a bound–free transition. Obviously such a laser has a very favorable lifetime ratio of upper and lower levels.

Another attractive feature of excimer lasers is their wavelength, which extends from the visible to the ultraviolet, depending on the particular excimer. Moreover the bound–free nature of the laser transition allows for tunability over a considerable range of wavelengths (≈ 50 Å), since well-defined vibrational–rotational transitions do not occur. Such a tunable source of coherent ultraviolet radiation is very useful for a variety of applications, such as high-resolution studies of molecular electronic spectra.

The KrF excimer laser at around 248 nm is one of the most efficient high-power ultraviolet lasers. The population inversion process for this and other rare-gas monohalide lasers (e.g., ArF, XeF, XeCl) involves a relatively large number of reactions; computer models of such lasers sometimes include about 100 rate equations for different processes. These lasers are pumped either by an electric discharge or by an electron beam. Of the processes leading directly to excited XY molecules,

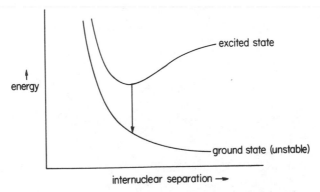

Figure 13.22 An excimer molecule has no stable ground state, because the potential energy curve has no local minimum.

where X and Y refer to a rare-gas atom and a halogen, respectively (e.g., $X = Kr$, $Y = F$), two are especially important. One is the ion-ion recombination process in which ions X^+ and Y^- combine in the presence of a third body to produce an excited XY molecule:

$$X^+ + Y^- \rightarrow XY^* \tag{13.12.1}$$

The other is the "harpooning reaction"

$$X^* + YR \rightarrow XY^* + R \tag{13.12.2}$$

where R represents some radical attached to the halogen Y. The rate for such a process tends to be largest when $R = Y$, as in the reaction

$$Kr^* + F_2 \rightarrow KrF^* + F \tag{13.12.3}$$

in the KrF laser. Ion–ion recombination processes like

$$Kr^+ + F^- \rightarrow KrF^* \tag{13.12.4}$$

are very important in the monofluoride excimer lasers because of the rapid production of F^- ions by the dissociative electron attachment reaction

$$F_2 + e \rightarrow F^- + F \tag{13.12.5}$$

High-power KrF lasers typically contain a gas mix of around 90% Ar, less than 10% Kr, and about 0.5% F_2. An electron beam produces electron–ion pairs, and the "secondary" electrons so generated take part in processes such as (13.12.5). For pressures less than about 1 atm, Ar^+ and F^- undergo ion–ion recombination to form ArF*, which reacts with Kr to form KrF*. At higher pressures the charge-transfer reactions

$$Ar^+ + 2Ar \rightarrow Ar_2^+ + Ar \tag{13.12.6}$$

and

$$Ar_2^+ + Kr \rightarrow Kr^+ + 2Ar \tag{13.12.7}$$

provide Kr^+ ions for the reaction (13.12.4). Kr_2^+ ions can also be formed by reactions such as

$$Kr^+ + 2Kr \rightarrow Kr_2^+ + Kr \tag{13.12.8}$$

and the molecular krypton ions can then react with F^- in an ion–ion recombination reaction to form KrF*. The reader is referred to the literature for details concerning the molecular kinetics of KrF and other excimer lasers.

13.13 FREE-ELECTRON LASERS

Early sources of coherent radiation for radar applications used electron beams at velocities $v \ll c$ to generate radiation of wavelength ≈ 1 cm in closed microwave cavities. Despite numerous designs, it was not possible to generate much shorter wavelengths with these "free-electron" devices. The free-electron "laser" generates much shorter wavelengths than these devices by employing relativistic electrons ($v \approx c$) and an open cavity design.

The free-electron laser (often abbreviated FEL) differs in several ways from other kinds of laser. Although like all lasers it is based on the principle of "light amplification by stimulated emission of radiation," the stimulated emission does not occur on a transition between bound electron states. In fact, the process of stimulated emission and gain in the free-electron laser is fairly subtle, and most of our discussion will focus on this process of stimulated emission by free electrons.

The basic scheme for a free electron laser is illustrated in Figure 13.23. The "wiggler" field shown is a static magnetic field that varies periodically in space. We denote the spatial period by λ_w. The arrows in Figure 13.23 indicate the direction of the wiggler field

$$\mathbf{B}_w = \hat{\mathbf{x}} B_w \cos \frac{2\pi z}{\lambda_w} \qquad (13.13.1)$$

at points spaced by $\lambda_w/2$ along the z direction. Instead of such a linear transverse wiggler we might have a helical wiggler:

$$\mathbf{B}_w = B_w \left(\hat{\mathbf{x}} \cos \frac{2\pi z}{\lambda_w} + \hat{\mathbf{y}} \sin \frac{2\pi z}{\lambda_w} \right) \qquad (13.13.2)$$

The wiggler exerts on an electron of velocity $\mathbf{v}$ a force

$$\mathbf{F}_w = e\mathbf{v} \times \mathbf{B}_w \qquad (13.13.3)$$

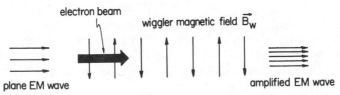

electron beam

wiggler magnetic field $\vec{B}_w$

plane EM wave

amplified EM wave

Figure 13.23 Basic "wiggler" design for a free-electron laser.

Since $\mathbf{F}_w \cdot \mathbf{v} = 0$ is identically zero, whatever the form of $\mathbf{B}_w$, the wiggler does no work on the electron. However, it does make the electron oscillate ("wiggle") transverse to the direction z. [The reader should have no difficulty convincing himself of this, using (13.3.3) and (13.13.1) or (13.13.2).] Such an oscillating electron produces synchrotron radiation. Moreover, the oscillating electric field of the applied plane electromagnetic wave indicated in Figure 13.23 can exchange energy with the transversely oscillating electron, since $e\mathbf{E} \cdot \mathbf{v} \neq 0$. It is such an energy exchange that can be used to amplify an electromagnetic wave propagating along the direction of an electron beam.

Let us consider in a little more detail the forces acting on an electron with velocity

$$\mathbf{v} = v_x\hat{\mathbf{x}} + v_y\hat{\mathbf{y}} + v_z\hat{\mathbf{z}} \tag{13.13.4}$$

where the longitudinal velocity v_z is nearly as large as c, the velocity of light. Using (13.13.3), we obtain for a helical wiggler (Problem 13.9) the following Newton equations of motion for the transverse components of the electron velocity:

$$\dot{v}_x = -\frac{e}{m} B_w c \sin \frac{2\pi ct}{\lambda_w} \tag{13.13.5a}$$

$$\dot{v}_y = \frac{e}{m} B_w c \cos \frac{2\pi ct}{\lambda_w} \tag{13.13.5b}$$

where we have used the approximation $v_z \approx c$ and written $z = v_z t \approx ct$ for the z coordinate of the moving electron. Integrating (13.13.5), we have

$$v_x = Kc \cos \frac{2\pi z}{\lambda_w} \tag{13.13.6a}$$

$$v_y = Kc \sin \frac{2\pi z}{\lambda_w} \tag{13.13.6b}$$

where we have gone back to writing z for ct and have defined the dimensionless wiggler parameter

$$K = eB_w\lambda_w/2\pi mc \tag{13.13.7}$$

Obviously we have chosen a particular initial condition in the integration of (13.13.5), but this choice will not turn out to be of any real consequence.

Actually (13.13.5) and (13.13.6) are incorrect for relativistic electrons, because the Newton equations of motion do not correctly describe the behavior of particles moving at velocities near the velocity of light. For one thing, the mass of a moving particle depends, according to the theory of relativity, on the velocity of the particle. For a particle of velocity $\mathbf{v}$ the mass becomes

$$M = \frac{m}{\sqrt{1 - v^2/c^2}} \equiv \gamma m \qquad (13.13.8)$$

where m, the mass when $\mathbf{v} = 0$, is the rest mass of the particle. We can rectify (13.13.6) by replacing m with M, i.e., by replacing K with K/γ:

$$v_x = \frac{Kc}{\gamma} \cos \frac{2\pi z}{\lambda_w} \qquad (13.13.9a)$$

$$v_y = \frac{Kc}{\gamma} \sin \frac{2\pi z}{\lambda_w} \qquad (13.13.9b)$$

Using the definition of γ in (13.13.8), we have

$$\begin{aligned}
v_z^2 &= v^2 - v_x^2 - v_y^2 \\
&= c^2 \frac{\gamma^2 - 1}{\gamma^2} - v_x^2 - v_y^2 \\
&= c^2 \frac{\gamma^2 - 1}{\gamma^2} - \frac{K^2 c^2}{\gamma^2} \\
&= c^2 \left(1 - \frac{1 + K^2}{\gamma^2} \right)
\end{aligned} \qquad (13.13.10)$$

where we have used (13.13.9) to replace $v_x^2 + v_y^2$ by $K^2 c^2/\gamma^2$. For $v \approx c$, γ is large (perhaps $\approx 10^3$) and

$$\begin{aligned}
v_z &= c \left(1 - \frac{1 + K^2}{\gamma^2} \right)^{1/2} \\
&\approx c \left(1 - \frac{1 + K^2}{2\gamma^2} \right)
\end{aligned} \qquad (13.13.11)$$

provided that K is not large. This approximation is consistent with our assumption that $v_z \approx c$ and therefore that v_x and v_y, according to Eq. (13.13.9), are both small compared with c.

We suppose that the wiggler field is strong enough to determine the electron trajectory, in contrast to the applied electromagnetic plane wave which, at least in the small-signal regime of the laser, is negligible by comparison. The wiggler field does not exchange energy with the electron but, in determining its trajectory, facilitates the exchange of energy between the electron and the plane wave.

We write the electric field of this plane wave in the form

$$\mathbf{E} = E_0 \left[\hat{\mathbf{x}} \sin \left(\frac{2\pi z}{\lambda} - \omega t + \phi_0 \right) + \hat{\mathbf{y}} \cos \left(\frac{2\pi z}{\lambda} - \omega t + \phi_0 \right) \right] \quad (13.13.12)$$

The form (13.13.12) is convenient because it allows us to combine sines and cosines as in (13.13.13), given the helical form (13.13.2) assumed for the wiggler. The form of the wiggler determines the polarization of the free electron laser radiation, with linear transverse and helical wigglers leading to linear and circular polarization, respectively. The rate at which the plane wave does work on the electron with velocity components (13.13.9) is then

$$\dot{W} = e\mathbf{E} \cdot \mathbf{v}$$

$$= eE_0 \frac{Kc}{\gamma} \left[\cos \frac{2\pi z}{\lambda_w} \sin \left(\frac{2\pi z}{\lambda} - \omega t + \phi_0 \right) \right.$$

$$+ \sin \frac{2\pi z}{\lambda_w} \cos \left(\frac{2\pi z}{\lambda} - \omega t + \phi_0 \right) \Bigg]$$

$$= \frac{eE_0 Kc}{\gamma} \sin \left[2\pi \left(\frac{1}{\lambda} + \frac{1}{\lambda_w} \right) z - \omega t + \phi_0 \right]$$

$$= \frac{eE_0 Kc}{\gamma} \sin \phi \quad (13.13.13)$$

where

$$\phi \equiv 2\pi \left(\frac{1}{\lambda} + \frac{1}{\lambda_w} \right) z - \omega t + \phi_0 \quad (13.13.14)$$

According to Einstein's mass–energy formula $E = Mc^2 = \gamma mc^2$, we may write (13.13.13) as

$$\dot{W} = \dot{\gamma} mc^2 = \frac{eE_0 Kc}{\gamma} \sin \phi$$

or

$$\dot{\gamma} = \frac{eE_0 K}{\gamma mc} \sin \phi \quad (13.13.15)$$

Using $\dot{z} = v_z$, with v_z given by (13.13.11), we can also write an equation for the rate of change of ϕ seen by the moving electron:

$$\dot{\phi} = 2\pi \left(\frac{1}{\lambda} + \frac{1}{\lambda_w} \right) v_z - \omega$$

$$= 2\pi c \left(\frac{1}{\lambda} + \frac{1}{\lambda_w} \right) \left(1 - \frac{1 + K^2}{2\gamma^2} \right) - \frac{2\pi c}{\lambda}$$

$$= \frac{2\pi c}{\lambda_w} \left[1 - \left(1 + \frac{\lambda_w}{\lambda} \right) \frac{1 + K^2}{2\gamma^2} \right] \qquad (13.13.16)$$

Now in a free-electron laser the wiggler period λ_w is on the order of centimeters, whereas the laser wavelength λ is much shorter. That is, $\lambda_w/\lambda \gg 1$ and so

$$\dot{\phi} \approx \frac{2\pi c}{\lambda_w} \left(1 - \frac{\lambda_w}{\lambda} \frac{1 + K^2}{2\gamma^2} \right) \qquad (13.13.17)$$

We note that $\dot{\phi} = 0$ for that value of the electron energy (γmc^2) such that

$$\gamma^2 = \gamma_R^2 \equiv \frac{\lambda_w}{2\lambda} (1 + K^2) \qquad (13.13.18)$$

γ_R defines the *resonant electron energy*. We can understand this terminology as follows. As an electron moves along the z axis a distance Δz in a time $\Delta t = \Delta z/v_z$, it sees a change in the phase of the field (13.13.12). The magnitude of this phase change is

$$\Delta\theta = \omega \left(\Delta t - \frac{\Delta z}{c} \right) = \frac{2\pi c}{\lambda} \Delta z \left(\frac{1}{v_z} - \frac{1}{c} \right)$$

$$= \frac{2\pi \Delta z}{\lambda} \left(\frac{c}{v_z} - 1 \right)$$

$$\approx \frac{2\pi \Delta z}{\lambda} \frac{1 + K^2}{2\gamma^2} \qquad (13.13.19)$$

where we have used (13.13.11) for v_z ($\approx c$). In particular, for $\Delta z = \lambda_w$ and $\gamma^2 = \gamma_R^2$ we have $\Delta\theta = 2\pi$. That is, an electron with the resonant energy sees the period of the electromagnetic field and the period of the wiggler to be equal. According to (13.13.18), this resonant electron energy depends on the wiggler period λ_w and also on the strength of the wiggler magnet B_w.

We now have the coupled equations (13.13.15) and (13.13.17) for γ and ϕ:

$$\dot{\gamma} = \frac{eE_0K}{\gamma mc} \sin \phi \qquad (13.13.20a)$$

$$\dot{\phi} = \frac{2\pi c}{\lambda_w} \left(1 - \frac{\gamma_R^2}{\gamma^2} \right) \qquad (13.13.20b)$$

where we have used the definition (13.13.18) of γ_R^2 in (13.13.17). These equations describe a single electron in the wiggler field and the applied monochromatic plane wave. It is clear from these equations that an electron can gain energy ($\dot{\gamma} > 0$) or lose energy ($\dot{\gamma} < 0$) to the plane wave, depending on the value of the phase term ϕ. This gain or loss of electron energy corresponds to absorption or stimulated emission, respectively, of radiation of wavelength λ. However, a pulse of electrons injected into the wiggler can be expected to have an approximately uniform distribution of ϕ values, giving an average $\sin \phi$ of zero and therefore an average $\dot{\gamma}$ of zero. In other words, we might expect to have neither absorption nor stimulated emission of radiation for a current made up of many electrons.

But this is not the whole story. We note from Eqs. (13.13.11) and (13.13.18) that

$$v_z \approx c \left(1 - \frac{\lambda \gamma_R^2}{\lambda_w \gamma^2} \right) \qquad (13.13.21a)$$

and therefore

$$\dot{v}_z \approx \frac{2c\lambda \gamma_R^2}{\lambda_w \gamma^3} \dot{\gamma} \qquad (13.13.21b)$$

That is, an electron is accelerated or decelerated longitudinally depending on whether it is gaining ($\dot{\gamma} > 0$) or losing ($\dot{\gamma} < 0$) energy to the plane wave. Thus the exchange of energy with the plane wave causes not only changes in the electron energies, but also a redistribution of electron positions along the z axis, with the faster electrons catching up to the slower ones. On a macroscopic scale the distribution of electrons remains uniform, but the "bunching" effect on a microscopic scale can lead to net gain rather than absorption.

We show in Appendix 13.B that the gain depends on the distribution of electron energies, and in particular that electrons at the resonant energy do not contribute to the gain. Electrons of energy greater than the resonant energy give rise to gain, whereas electrons of energy less than the resonant energy give rise to absorption. We may summarize the main results of the calculation in Appendix 13.B as follows:

   i. There is gain if the electron energy is such that $\gamma > \gamma_R$, but loss if $\gamma <$ γ_R.

  ii. The maximum gain occurs at the electron energy $\gamma \approx (1 + 0.2/N_w)\gamma_R$, where N_w is the number of wiggler periods.

 iii. The net gain is very small unless the electron beam has a narrow distribution of electron energies.

The stimulated emission in a free-electron laser is a classical process: we can understand the free-electron laser without recourse to quantum theory. This is because the electron orbits are macroscopic and quantum effects are negligible. Thus we have shown that nonlinear modifications of purely classical equations can be sufficient to produce stimulated emission, but not in the sense of Einstein, which manifestly involves quantum states. On this basis it is questionable whether the FEL should be called a laser.

The fact that the lasing process in a free-electron laser is not confined to a particular transition of an atom or molecule is very important, for it means that a free-electron laser is tunable over a very wide range of wavelengths. The relation (13.13.18) defines for any wavelength λ a resonant electron energy, near which there is gain at the wavelength λ. Since the resonant electron energy also depends on the characteristics of the wiggler field (λ_w and B_w), tuning to a particular wavelength region can be accomplished by varying either the wiggler parameters or the electron beam energy. It is possible, at least in principle, that free-electron lasers can be made to operate at wavelengths ranging from millimeters to a few angstroms. Free-electron lasers appear to have the potential for generating very high powers in either the cw or pulsed modes of operation over a range of wavelengths unavailable to any other type of laser.

Electron beam sources for free-electron lasers must deliver peak currents of typically a few amperes in order to obtain substantial gains. Because the gain coefficient is antisymmetric about the resonant electron energy, a broad distribution of electron energies will result in a zero or very small gain. The spread in electron energies must be small enough that the condition (Problem 13.11)

$$\Delta\gamma \lesssim \gamma_R/2N_w \qquad (13.13.22)$$

is realized. In addition the angular spread of the electrons must be small. Such considerations are important in the design of electron accelerators and storage rings for free-electron lasers.

The bandwidth of a free-electron laser may be estimated as follows. Suppose the electrons are injected in bunches of length l_e. Then since there is gain only where the electron density is nonzero, the laser radiation is also produced in pulses of length $\sim l_e$, and so we expect a frequency bandwidth $\delta\nu$ given by $\delta\nu \sim c/l_e$, or a relative bandwidth of

$$\frac{\delta \nu}{\nu} \sim \frac{\lambda}{l_e} \qquad\qquad (13.13.23)$$

As in all lasers, the cavity intensity in a free-electron laser builds up initially from spontaneous radiation, and in steady state is determined from the condition that the saturated gain just balances the total cavity loss. In our discussion of the energy exchange between the radiation and the electrons, we have implicitly assumed an electron density small enough that many-electron collective effects are negligible. This is the so-called Compton regime of the free-electron laser, as opposed to what is called the Raman regime. In the latter the exchange of energy occurs via collective wave motions of many electrons. Furthermore the wiggler need not be a magnetic field but may be, for instance, a periodic electrostatic field.

13.14   SEMICONDUCTOR LASERS

In Section 4.10 we discussed light-emitting diodes (LEDs), in which radiation is produced by the (radiative) recombination of electrons and holes at a *pn* junction. If a large density of electrons and holes is produced by forward biasing a junction of heavily doped *p* and *n* materials, this radiation can *stimulate* the recombination of electrons and holes, and laser action can be realized if the amplification of radiation exceeds the loss. This is the basis of semiconductor diode lasers. The most distinctive features of these lasers are their small size and the degree to which their output can be modulated by varying the current. These properties of laser diodes are important in optical communication technology, as discussed in Chapter 16. This section is devoted to a brief introduction to the principles of diode lasers.

To get some idea of the sort of parameters involved in diode lasers, we consider first the geometry shown in Figure 13.24. The gain medium consists of a *pn* junc-

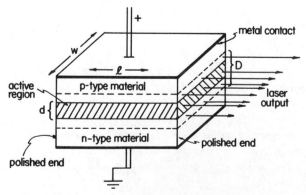

Figure 13.24 A laser diode (homojunction design).

tion with an active region of width d. In this active region there are sufficient numbers of electrons and holes under forward biasing to produce positive gain. The width d may be estimated from the diffusion lengths of the charge carriers, and is very small, typically on the order of 1 μm. The width D, which is discussed further below, is used to indicate the size of the mode volume of the radiation field. Whereas in most lasers $D < d$ (Chapter 14), the reverse is true in diode lasers.

The threshold gain for a diode laser may be expressed in terms of a threshold current flowing through the diode. If the density of holes in the active region is large, we may take $N_1 \approx 0$ in the gain expression (10.2.11). This means that the density of electrons in the lower level (band) is small enough that the absorption of a photon, and the simultaneous promotion of an electron to the conduction band, is negligible. Then

$$g(\nu) \approx \frac{\lambda^2 A}{8\pi n^2} N_2 S(\nu) \tag{13.14.1}$$

where N_2 is the density of electrons injected (by forward biasing) into the active region from the conduction band of the n-type material. We assume a homogeneously broadened transition with Lorentzian linewidth (HWHM) $\delta\nu_0$, so that

$$g(\nu_0) \approx \frac{\lambda^2 A N_2}{8\pi^2 n^2 \, \delta\nu_0} \tag{13.14.2}$$

at line center. Here A is the spontaneous emission rate for radiative recombination. The threshold gain coefficient is given by equation (10.4.11):

$$g_t = a - \frac{1}{2l} \ln r_1 r_2 \tag{13.14.3}$$

where l is the length of the gain medium, r_1 and r_2 are the mirror reflectivities, and a is the loss per unit length due to effects other than imperfect reflections at the mirrors. The threshold condition $g(\nu_0) = g_t$ gives

$$(N_2)_t = \frac{8\pi^2 n^2 \delta\nu_0}{\lambda^2 A} \left(a - \frac{1}{2l} \ln r_1 r_2 \right) \tag{13.14.4}$$

for the threshold value of N_2.

Diode lasers usually do not employ mirrors for feedback. This is because the refractive index n is large enough to give considerable reflection at the semiconductor/air interfaces. From the Fresnel formulas for normal incidence we have the reflection coefficient

$$r \approx \frac{(n-1)^2}{(n+1)^2} \qquad (13.14.5)$$

if we approximate the refractive index of air by unity. For GaAs $n \approx 3.6$ and therefore

$$r \approx (2.6/4.6)^2 \approx 0.32 \qquad (13.14.6)$$

for the reflection coefficient of the "mirrors." By polishing two opposite ends of the diode, and leaving the remaining sides rough (so that they are poor specular reflectors), laser oscillation is favored along the axis joining the polished ends (Figure 13.24). Sometimes the polished ends are given coatings to increase their reflectivities.

It is convenient to express $(N_2)_t$ in (13.14.4) in terms of the current density through the diode. Let J be the current density, i.e., the flow of charge per unit area per unit time. In terms of J the rate per unit volume at which electrons are injected into the active region is J/ed. Electrons are also being lost at a total rate R_e due to both radiative and nonradiative recombination processes, and in steady state the injection rate is equal to the loss rate. Then $J/ed = R_e N_2$, or

$$J = (edR_e) N_2 \qquad (13.14.7)$$

and we may rewrite (13.14.4) in terms of a threshold current density J_t necessary for laser oscillation:

$$J_t = (edR_e)(N_2)_t$$
$$= \frac{8\pi^2 n^2 \, \delta\nu_0}{\lambda^2} (ed) \frac{R_e}{A} \left(a - \frac{1}{2l} \ln r_1 r_2 \right) \qquad (13.14.8)$$

The expression (13.14.8) does not account for the fact that the mode volume is wider than the width d of the active region. The greater density of charge carriers in the active region results in a greater refractive index than in the surrounding medium, and therefore some confinement of radiation to the active region. However, this "waveguiding" effect is very weak, and the radiation therefore has a mode volume of width $D > d$. The effective gain coefficient is then smaller than (13.14.2) by a factor $\approx d/D$, and so the actual threshold current density is a factor D/d larger than (13.14.8):

$$J_t = \frac{8\pi n^2 \, \delta\nu_0}{\lambda^2} (eD) \frac{R_e}{A} \left(a - \frac{1}{2l} \ln r_1 r_2 \right) \qquad (13.14.9)$$

An important diode laser material is GaAs, for which $\lambda \approx 8400$ Å, $n \approx 3.6$, and $\delta\nu_0 \sim 10^{13}$ Hz. The ratio A/R_e, the so-called internal quantum efficiency, represents the fraction of injected electrons undergoing *radiative* recombination, and is close to unity. If we assume in addition the reasonable values $D = 2$ μm, $l = 500$ μm, and $a = 10$ cm^{-1}, then from (13.14.9) and (13.14.6) we estimate

$$J_t \approx 500 \text{ amp/cm}^2 \qquad (13.14.10)$$

For a junction area $lw = 500 \times 250$ μm^2 (Figure 13.24) this implies a threshold current $J_t lw \approx 1$ amp.

Several points emerge from this numerical example, for which we have used typical values for the various parameters appearing in the formula (13.14.9). First, the gain cell, which is essentially the entire laser, is less than 1 mm across in any direction in our example. As noted above, this small size of diode lasers is very important for certain applications. Furthermore the fact that the internal quantum efficiency is close to unity suggests that diode lasers are potentially very efficient; in fact they are among the most efficient of all lasers. GaAs lasers cooled to liquid-nitrogen temperatures (77 K) give cw output powers as high as a few watts, with overall efficiencies (laser output power divided by input electrical power) of around 30%.

The operating temperature of a diode laser is an important consideration. Our estimate of 500 amp/cm^2 for the threshold current density of a GaAs laser is in order-of-magnitude agreement with measurements for low-temperature GaAs lasers. However, our considerations leading up to this result were rather simplistic because we did not take account of the fact that there are distributions of electron and hole states associated with the conduction and valence *bands*. When such effects are taken into account it turns out that, for all but the lowest operating temperatures, our simple-minded theory predicts much too small a value for the threshold current density. At room temperature, current densities more like 50,000 amp/cm^2 are needed to reach threshold.

The most useful applications of diode lasers call for room-temperature, and often cw, operation. The large current densities required for the *homojunction* design of Figure 13.24 then pose serious problems. This difficulty was solved in the early 1970s with the development of *heterojunction* diode lasers. Figure 13.25 illustrates a *double* heterojunction design, which employs not only n- and p-type GaAs layers, but also n- and p-type layers of an AlAs–GaAs alloy, denoted AlGaAs. The AlGaAs material has a larger energy band gap than GaAs, and also a smaller refractive index. Because of the band-gap differences at the two GaAs–AlGaAs junctions, there is a greater confinement of electrons and holes in the active region (p-type GaAs in the example shown). This comes about because the band-gap differences act in effect as potential-energy barriers for the electrons and holes, preventing them from diffusing out of the active GaAs layer into which they are injected by forward biasing. Furthermore the greater refractive index of GaAs than that of the AlGaAs compound acts to confine the radiation to the active region, thus reducing the width D appearing in Eq. (13.14.9). In addition, the loss

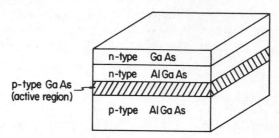

Figure 13.25 Heterojunction laser diode.

coefficient a will be smaller than in the homojunction case, simply because any radiation that "spills over" into the AlGaAs layers finds itself in a nonresonant, nonabsorbing medium (because of the large band gap compared with the radiation frequency). Largely for these reasons, cw heterojunction lasers can operate at room temperature with current densities typically ≤ 4000 amp/cm^2.

Threshold currents are reduced still further by confining the current across the active region to a narrow stripe, as shown in Figure 13.26. This may be done in a variety of ways, such as by building high-resistance regions into the diode (Figure 13.27). Stripe-geometry heterojunction devices give output powers ~ 10 mW with currents less than 1 amp. One advantage of the stripe geometry is that the cross-sectional area of the output radiation is reduced, making it easier to couple the radiation into an optical fiber (Chapter 16). Stripe lasers also tend to have a more stable output.

Diode lasers can be made about as small as the eye of a needle. Because the radiation is emitted from such a tiny area, it has a large angular divergence (Chapter 15), and for some purposes the laser may be treated as a point source of a spherical wave.

It has been said that diode lasers bear a relation to gas lasers analogous to the relation of transistors to vacuum-tube amplifiers. The small size, reliability, low power consumption, and low cost of diode lasers are responsible for whole new technologies not only in fiber-optic communications but also for things like compact audio digital discs, and this is reminiscent of the new technologies spawned, for similar reasons, by the transistor. Diode lasers have become really useful not only because of heterojunction and stripe designs, but also because of materials

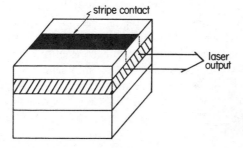

Figure 13.26 Laser diode with stripe contact to confine the current to a small part of the active layer.

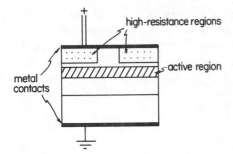

high-resistance regions

active region

metal contacts

+

Figure 13.27 Stripe design achieved by building high-resistance regions into a laser diode.

research and the evolution of various epitaxial technologies necessary for precise layer control. It seems likely that further important progress will be made, given the great commercial interest in diode lasers.

13.15 REMARKS

It should be clear that different lasers are very similar in their basic operating principles; the concepts of gain, threshold, and feedback are central to an understanding of any laser, regardless of the physical (or chemical) processes by which gain is established.

However, lasers can differ greatly in various ways. Compare, for instance a tiny diode laser with a high-power CO_2 or free-electron laser, which may occupy a large room or an entire building. Or consider the vastly different technologies, involving supersonic flows or miniscule junctions or strong wiggler magnets, necessary for the operation of these lasers. Obviously there is a correspondingly wide diversity in the basic physical principles necessary for an understanding of different lasers.

The development of various lasers has been a largely evolutionary process involving the efforts of large numbers of people throughout the world. All indications point to a continuation of this evolution, and for this reason we have not attempted to give any account of the latest ideas and possibilities. The reader interested in the latest advances in specific areas of laser technology and electro-optics is best referred to various technical and trade periodicals.

APPENDIX 13.A THE GAIN COEFFICIENT FOR MOLECULAR-VIBRATION LASERS

From Table 10.1 we have the gain coefficient

$$g(\nu) = \frac{\lambda^2 A}{8\pi} \left(N_2 - \frac{g_2}{g_1} N_1 \right) S(\nu) \qquad (13.A.1)$$

if we take the refractive index $n \approx 1$. Let $N_2 = N(v_2, J_2)$ be the number density of molecules in vibrational level v_2 and rotational level J_2, the upper level of a molecular vibrational–rotational laser transition. Similarly let $N_1 = N(v_1, J_1)$ be the number density of the lower vibrational–rotational level. Since the degeneracy of rotational level J is $2J + 1$, we have

$$g(\nu) = \frac{\lambda^2 A}{8\pi} \left(N(v_2, J_2) - \frac{2J_2 + 1}{2J_1 + 1} N(v_1, J_1) \right) S(\nu) \qquad (13.\text{A}.2)$$

As noted in Chapter 4, rotational level spacings in molecules are much smaller than vibrational spacings. In many molecules the separation of adjacent rotational energy levels is small compared with kT. This means that the spacing between rotational levels is small compared with the average kinetic energy of a molecule. In this case we might expect that, as a result of collisions, the rotational motion of the molecules will be in thermal equilibrium at the gas (translational) temperature T. In other words, we expect the rotational levels to be distributed according to the Boltzmann distribution:

$$N(v_2, J_2) = N(v_2) \frac{g_{J_2}}{Z} \exp\left(-\frac{E_{J_2}}{kT} \right) \qquad (13.\text{A}.3\text{a})$$

$$N(v_1, J_1) = N(v_1) \frac{g_{J_1}}{Z} \exp\left(-\frac{E_{J_1}}{kT} \right) \qquad (13.\text{A}.3\text{b})$$

Here $N(v_2)$ and $N(v_1)$ are the total vibrational population densities regardless of rotation, i.e.,

$$N(v_{2,1}) = \sum_{J=0}^{\infty} N(v_{2,1}, J) \qquad (13.\text{A}.4)$$

$g_{J_2} = 2J_2 + 1$ and $g_{J_1} = 2J_1 + 1$ are the rotational degeneracies, and Z is the rotational partition function

$$Z = \sum_{J=0}^{\infty} g_J \, e^{-E_J/kT} \qquad (13.\text{A}.5)$$

where

$$E_J \approx hc \, B_e J(J + 1) \qquad (13.\text{A}.6)$$

For CO_2, $B_e \approx 0.39 \text{ cm}^{-1}$, and it is easily checked that $hc B_e \ll kT$, and therefore $E_J \ll kT$ except for very low temperatures or states with very large J. Partly for this reason we can expect (13.A.3), the assumption of *rotational thermal equilibrium*, to be an excellent approximation for CO_2 and many other molecules.

Using (13.A.6) in (13.A.5), we have

$$Z = \sum_{J=0}^{\infty} (2J + 1)\, e^{-(hcB_e/kT)J(J+1)}$$

$$= \sum_{J=0}^{\infty} (2J + 1)\, e^{-xJ(J+1)}$$

$$\approx \int_0^{\infty} e^{-xy}\, dy = \frac{1}{x} = \frac{kT}{hc\,B_e} \qquad (x \ll 1) \qquad (13.A.7)$$

The replacement of the sum by the integral is a good approximation if $x \ll 1$; then we may replace $J(J + 1)$ by the continuous variable y. Note that $dy/dJ = 2J + 1$. The reader may readily check the validity of (13.A.7) using a programmable pocket calculator. Thus, using (13.A.3)–(13.A.7) in (13.A.2), we have the gain coefficient

$$g(\nu) = \frac{\lambda^2 A}{8\pi} \frac{\overline{B}_e}{T} (2J_2 + 1)\left[N(v_2)\, e^{-\overline{B}_e J_2(J_2 + 1)/T} - N(v_1)e^{-\overline{B}_e J_1(J_1 + 1)/T}\right] S(\nu)$$

$$(13.A.8)$$

where $\overline{B}_e = hc\,B_e/k$ is the rotational constant expressed in kelvins. For CO_2, $\overline{B}_e \approx 0.565$ K.

Recall the selection rule $\Delta J = 0, \pm 1$ for the change in the rotational quantum number in a molecular transition. Let us consider for definiteness $J_1 = J_2 + 1$, a so-called P-branch transition. In this case (13.A.8) becomes

$$g(\nu) = \frac{\lambda^2 A}{8\pi} \frac{\overline{B}_e}{T} (2J_2 + 1)\, e^{-\overline{B}_e J_2(J_2 + 1)/T}$$

$$\times \left[N(v_2) - N(v_1)\, e^{-2\overline{B}_e(J_2 + 1)/T}\right] S(\nu) \qquad (13.A.9)$$

for the gain on the vibrational–rotational transition $(v_2, J_2) \rightarrow (v_1, J_2 + 1)$. Expressions like this are found frequently in the laser research literature. The advantage of such an expression is that only the vibrational level densities (13.A.4) are required for its evaluation. The rotational level densities are "frozen," so to speak, at the Boltzmann distribution (13.A.3). In a rate-equation model of the laser, therefore, rate equations may be written only for the vibrational populations rather than the myriad vibrational–rotational populations. The assumption of rotational thermal equilibrium thus represents an enormous simplification.

Equation (13.A.9) shows that we can have a positive gain even if $N(v_2) < N(v_1)$, provided that

$$N(v_2) > N(v_1)\, e^{-2\overline{B}_e\,(J_2 + 1)/T} \qquad (13.A.10)$$

In other words, a population inversion can be achieved on a (*P*-branch) vibrational–rotational transition without having an inversion on the total vibrational populations. In this case the gain is sometimes said to be due to a *partial population inversion*.

In the case of CO_2, Eq. (13.A.9) requires a slight modification. It turns out, for reasons we will not go into, that the (001) and (100) vibrational levels have associated with them only odd and even rotational quantum numbers, respectively. The rotational partition function (13.A.7) is therefore effectively halved, whereas the expression (13.A.9) for the gain coefficient is multiplied by 2. For the (001) $\to$ (100) vibrational band of CO_2, $A \approx \frac{1}{3}$ sec^{-1}. Using $\lambda = 10.6$ μm, $\overline{B}_e = 0.565$ K, $T = 293$ K, and $J_2 = 19$, therefore, we obtain

$$g(10.6 \ \mu\text{m}) = 6.4 \times 10^{-10} \left[N(001) - 0.93 \ N(100) \right] S(\nu) \ \text{cm}^2\text{-sec}^{-1}$$

$$(13.\text{A}.11)$$

where the vibrational population densities $N(001)$ and $N(100)$ have units of cm^{-3}, and the lineshape function $S(\nu)$ is expressed in seconds. These quantities will depend on various factors determined by the laser medium, such as the pressure, temperature, and gas mix.

For definiteness, let us focus our attention on a situation in which the parameters are known reasonably well, namely, CO_2 in the earth's atmosphere. For $T = 293$ K and $P = 1$ atm, the total molecular number density found from (3.10.8) is 2.5×10^{19} cm^{-3}. Assuming a concentration of 0.033 % CO_2 in the earth's atmosphere, therefore, we estimate about $8.3 \times 10^{15}/\text{cm}^3$ at sea level. Assuming a Boltzmann distribution of the CO_2 vibrational levels, and using the vibrational energies given in Figure 13.19, we estimate that the fractions of CO_2 molecules in the (001) and (100) levels at 293 K are 9.7×10^{-6} and 1.1×10^{-3}, respectively, so that

$$N(001) = (9.7 \times 10^{-6}) (8.3 \times 10^{15} \ \text{cm}^{-3}) = 8.1 \times 10^{10} \ \text{cm}^{-3} \quad (13.\text{A}.12a)$$

and

$$N(100) = (1.1 \times 10^{-3}) (8.3 \times 10^{15} \ \text{cm}^{-3}) = 9.1 \times 10^{12} \ \text{cm}^{-3} \quad (13.\text{A}.12b)$$

From (13.A.11), therefore, we have the absorption coefficient

$$a(10.6 \ \mu\text{m}) = -g(10.6 \ \mu\text{m})$$

$$\approx 5.4 \times 10^3 \ S(\nu) \ \text{cm}^{-1}\text{-sec}^{-1} \quad (13.\text{A}.13)$$

This gives the absorption coefficient due to CO_2 in the earth's atmosphere at sea level. Before a numerical result can be obtained, it remains to estimate the lineshape factor $S(\nu)$.

At atmospheric pressures the 10.6 μm absorption line of CO_2 is collision broadened. At line center, therefore, $S(\nu)$ is given by (3.6.20), with the collision linewidth $\delta\nu_0 = 2\pi(1/\tau)$ given by (3.10.5). Since N_2 and O_2 are the most frequent collision partners of CO_2 in the earth's atmosphere, we can approximate $\delta\nu_0$ by including only contributions from N_2 and O_2. We will assume the following collision cross sections:

$$\sigma(CO_2, O_2) = 95 \text{ Å}^2 \qquad (13.A.14a)$$

$$\sigma(CO_2, N_2) = 120 \text{ Å}^2 \qquad (13.A.14b)$$

so that

$$\delta\nu_0 = 2.7 \times 10^9 \text{ sec}^{-1} \qquad (13.A.14c)$$

for the atmospheric concentrations ≈ 0.78 and 0.21 of N_2 and O_2, respectively (Problem 13.8). Thus

$$S(\nu) = \frac{1}{\pi \, \delta\nu_0} = 1.2 \times 10^{-10} \text{ sec} \qquad (13.A.15)$$

and from (13.A.13) it follows that

$$a(\nu) = 6.5 \times 10^{-7} \text{ cm}^{-1} \qquad (13.A.16)$$

This result is in good agreement with published estimates of the absorption coefficient (at 10.6 μm) associated with CO_2 in the atmosphere. The actual absorption coefficient for 10.6-μm radiation in the earth's atmosphere has a nonnegligible contribution from water vapor, and particulate matter ("aerosols") must also be taken into account for practical purposes. Our purpose in deriving (13.A.16) has only been to provide an example of the use of (13.A.2).

APPENDIX 13.B TRAJECTORY AND ENERGY ANALYSIS OF THE FREE-ELECTRON LASER

From the discussion leading up to (13.13.21) it is evident that the redistribution of electrons along the z axis is crucial to the free-electron laser. Since each electron is characterized, according to (13.13.20), by a "trajectory" in the two-dimensional space (γ, ϕ), we can approach the many-electron dynamics through the use of an electron distribution function in this space. First, however, we simplify things by defining the dimensionless energy parameter

$$\eta = \frac{\gamma - \gamma_R}{\gamma_R} \qquad (13.B.1)$$

and assuming $|\eta|$ is small compared with unity. This assumption allows us to write (13.13.20) in the approximate form (Problem 13.10)

$$\dot{\eta} = \omega_0 (\Omega/\omega_0)^2 \sin\phi \qquad (13.B.2a)$$

$$\dot{\phi} = 2\omega_0 \eta \qquad (13.B.2b)$$

where we have defined the frequencies

$$\omega_0 = 2\pi c/\lambda_w \qquad (13.B.3a)$$

$$\Omega = (eE_0 K\omega_0/\gamma_R^2 mc)^{1/2}$$

$$= (e^2 E_0 B_w/\gamma_R^2 m^2 c)^{1/2} \qquad (13.B.3b)$$

We define an electron distribution function $f(\eta, \phi, t)$ such that $f\,d\eta\,d\phi$ is the fraction of electrons with η, ϕ values in the intervals $[\eta, \eta + dn]$ and $[\phi, \phi + d\phi]$ at time t. Since the rate of change of the number of electrons in an "area" element $d\eta\,d\phi$ is equal to the rate at which electrons enter or leave this area element, the total time derivative of $f(\eta, \phi, t)$ must be zero:

$$\frac{df}{dt} = \frac{\partial f}{\partial t} + \frac{\dot{\eta}\partial f}{\partial \eta} + \frac{\dot{\phi}\partial f}{\partial \phi} = 0 \qquad (13.B.4)$$

This partial differential equation is called the Vlasov equation (or the collisionless Boltzmann equation). Using the single-electron equations of motion (13.B.2) in the Vlasov equation, we have

$$\frac{\partial f}{\partial t} + \omega_0 \left(\frac{\Omega}{\omega_0}\right)^2 \sin\phi\, \frac{\partial f}{\partial \eta} + 2\omega_0 \eta\, \frac{\partial f}{\partial \phi} = 0 \qquad (13.B.5)$$

Once we solve this equation, we can calculate quantities like the average electron energy:

$$\langle \eta(t) \rangle = \int_{-\infty}^{\infty} \int_0^{2\pi} \eta f(\eta, \phi, t)\, d\eta\, d\phi \qquad (13.B.6)$$

We now outline a perturbative approach to the solution of (13.B.5), omitting details that may be checked by the reader.

For typical parameter values the ratio Ω/ω_0 is small, and so we seek a solution of (13.B.5) of the form

$$f(\eta, \phi, t) = f_0(\eta, \phi, t) + \left(\frac{\Omega}{\omega_0}\right)^2 f_1(\eta, \phi, t) + \left(\frac{\Omega}{\omega_0}\right)^4 f_2(\eta, \phi, t) + \cdots$$

$$(13.B.7)$$

where the terms on the right are successively smaller corrections. Using this series in (13.B.5), and equating coefficients of equal powers of Ω/ω_0, we obtain coupled equations for the $f_n(\eta, \phi, t)$, the first three of which are

$$\frac{\partial f_0}{\partial t} + 2\omega_0\eta \frac{\partial f_0}{\partial \phi} = 0 \tag{13.B.8a}$$

$$\frac{\partial f_1}{\partial t} + 2\omega_0\eta \frac{\partial f_1}{\partial \phi} = -\omega_0 \sin\phi \frac{\partial f_0}{\partial \eta} \tag{13.B.8b}$$

$$\frac{\partial f_2}{\partial t} + 2\omega_0\eta \frac{\partial f_2}{\partial \phi} = -\omega_0 \sin\phi \frac{\partial f_1}{\partial \eta} \tag{13.B.8c}$$

Since f_0 is the distribution function when $\Omega = 0$, i.e., when there is no wiggler or plane wave acting on the electrons, we assume f_0 describes a steady-state electron distribution. Then $\partial f_0/\partial\phi = 0$, so that f_0 is a function only of the electron energy:

$$f_0(\eta, \phi, t) = F(\eta) \tag{13.B.9}$$

Equation (13.B.8b) then becomes

$$\frac{\partial f_1}{\partial t} + 2\omega_0\eta \frac{\partial f_1}{\partial \phi} = -\omega_0 \sin\phi \frac{dF}{d\eta} \tag{13.B.10}$$

with solution

$$f_1(\eta, \phi, t) = \frac{1}{2\eta}\frac{dF}{d\eta}\left[\cos\phi - \cos(\phi - 2\omega_0\eta t)\right] \tag{13.B.11}$$

The fact that $f_1(\eta, \phi, t)$ is a function of ϕ, and therefore of z according to Eq. (13.13.14), means that to order $(\Omega/\omega_0)^2$ the electron distribution $f(\eta, \phi, t)$ is not a uniform distribution in z defined by the zeroth-order distribution function $f_0 = F(\eta)$. To order $(\Omega/\omega_0)^2$ we have from (13.B.6) the average electron energy

$$\langle \eta(t) \rangle = \int_{-\infty}^{\infty} \int_0^{2\pi} \eta\left[f_0 + \left(\frac{\Omega}{\omega_0}\right)^2 f_1\right] d\eta \, d\phi$$

$$= 2\pi \int_{-\infty}^{\infty} \eta F(\eta) \, d\eta \equiv \langle \eta \rangle_0 \tag{13.B.12}$$

since the integral over ϕ of (13.B.11) gives zero. Thus, although to order $(\Omega/\omega_0)^2$ there is a redistribution of the electrons along the z axis, there is no change in the average electron energy, and therefore no absorption or gain for the electromagnetic plane wave.

The next correction to $f(\eta, \phi, t)$ involves f_2, the equation for which follows from (13.B.8c) and (13.B.11):

$$\frac{\partial f_2}{\partial t} + 2\omega_0\eta \frac{\partial f_2}{\partial \phi} = -\frac{\omega_0}{4}\frac{\partial}{\partial \eta}\left(\frac{1}{\eta}\frac{dF}{d\eta}(\sin 2\phi - \sin 2\phi \cos 2\omega_0\eta t\right.$$

$$\left. - \sin 2\omega_0\eta t + \cos 2\phi \sin 2\omega_0\eta t)\right) \qquad (13.B.13)$$

Inspection of the form of the solution of this equation indicates that only the part of the solution that is independent of ϕ will make a nonvanishing contribution to the integral (13.B.6). This ϕ-independent solution, which we denote by f_{20}, satisfies the equation

$$\frac{\partial f_{20}}{\partial t} = \frac{\omega_0}{4}\frac{\partial}{\partial \eta}\left(\frac{1}{\eta}\frac{dF}{d\eta}\sin 2\omega_0\eta t\right) \qquad (13.B.14)$$

Thus

$$f_{20}(\eta, t) = \frac{\omega_0}{4}\frac{\partial}{\partial \eta}\left(\frac{1}{\eta}\frac{dF}{d\eta}\int_0^t \sin 2\omega_0\eta t' \, dt'\right)$$

$$= \frac{1}{4}\frac{\partial}{\partial \eta}\left[\left(\frac{\sin \omega_0\eta t}{\eta}\right)^2 \frac{dF}{d\eta}\right] \qquad (13.B.15)$$

and, up to fourth order in Ω/ω_0,

$$\langle \eta(t)\rangle = \langle \eta\rangle_0 + 2\pi\left(\frac{\Omega}{\omega_0}\right)^4 \int_{-\infty}^{\infty} \eta f_{20}(\eta, t) \, d\eta$$

$$= \langle \eta\rangle_0 + \frac{\pi}{2}\left(\frac{\Omega}{\omega_0}\right)^4 \int_{-\infty}^{\infty} \eta\frac{\partial}{\partial \eta}\left[\left(\frac{\sin \omega_0\eta t}{\eta}\right)^2 \frac{dF}{d\eta}\right] d\eta$$

$$= \langle \eta\rangle_0 + \frac{\pi}{2}\left(\frac{\Omega}{\omega_0}\right)^4 \int_{-\infty}^{\infty} F(\eta)\frac{\partial}{\partial \eta}\left(\frac{\sin \omega_0\eta t}{\eta}\right)^2 d\eta \qquad (13.B.16)$$

where the last equality is the result of two partial integrations. Performing the differentiation in the integrand, and rearranging terms, we have finally

$$\langle \eta(t)\rangle = \langle \eta\rangle_0 + \frac{\pi}{2}\left(\frac{\Omega}{\omega_0}\right)^4 (2\omega_0 t)^3 \int_{-\infty}^{\infty} F(\eta) \, G(2\omega_0\eta t) \, d\eta \qquad (13.B.17)$$

where

$$G(x) \equiv x^{-3} \left(\cos x - 1 + \frac{x}{2} \sin x \right) \qquad (13.B.18)$$

The function $G(x)$ is plotted in Figure 13.28.

If the spread of electron energies about some energy η_0 is very narrow, so that $F(\eta)$ is negligible except at $\eta \approx \eta_0$, then (13.B.17) becomes

$$\langle \eta(t) \rangle \approx \langle \eta \rangle_0 + \frac{\pi}{2} \left(\frac{\Omega}{\omega_0} \right)^4 (2\omega_0 t)^3 F(\eta_0) \, G(2\omega_0 \eta_0 t) \quad (13.B.19)$$

Since it takes a time $t_e = N_w \lambda_w / v_z \approx N_w \lambda_w / c$ for the electrons to cross a total of N_w wiggler periods, and

$$2\omega_0 \eta_0 t_e \approx \frac{4\pi c}{\lambda_w} \eta_0 \frac{N_w \lambda_w}{c} = 4\pi \eta_0 N_w \qquad (13.B.20)$$

we have

$$\Delta \eta = \langle \eta(t_e) \rangle - \langle \eta \rangle_0 \approx 32\pi^4 \left(\frac{\Omega}{\omega_0} \right)^4 N_w^3 F(\eta_0) \, G(4\pi \eta_0 N_w) \quad (13.B.21)$$

for the average change in energy as an electron traverses the entire wiggler field.

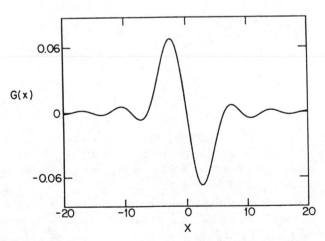

Figure 13.28 The function $G(x)$ defined by Eq. (13.B.18).

Since the actual electron energy in terms of the dimensionless parameter η is $W = \gamma mc^2 = \gamma_R(\eta + 1) mc^2$, (13.B.21) implies (Problem 13.10)

$$\Delta W = \frac{16\pi^2\sqrt{2}\ e^2 E_0^2}{mc^2} \lambda^{3/2} \lambda_w^{1/2}\ K^2 \left(1 + K^2\right)^{-3/2}$$

$$\times\ N_w^3\ F(\eta_0)\ G(4\pi\eta_0 N_w) \tag{13.B.22}$$

for the net change in energy.

We have thus obtained an approximate expression for the average change in energy of electrons passing through the wiggler in the presence of an electromagnetic plane wave. Since we have assumed the trajectory of each electron to be determined by the wiggler, and neglected effects of the plane wave, our analysis applies only to the small-signal regime of a free-electron laser. We can go further and derive an expression for the gain coefficient, but the result (13.B.22) for the average energy change of the electrons will suffice for our purposes.

From Figure 13.28 we see that $\Delta W = 0$ when $\eta_0 = 0$. According to (13.B.1), this occurs when the electrons have energy equal to the resonant value defined by γ_R. In this case the electrons do not exchange energy with the plane wave, and so there is no gain or loss. When $\eta_0 > 0$, however, so that the electron energy is greater than the resonant energy $\gamma_R mc^2$, we see from (13.B.22) and Figure 13.28 that $\Delta W < 0$. In this case the electrons *lose* energy to the plane wave, and so there is positive gain and amplification of the field. When $\eta_0 < 0$, on the other hand, $\Delta W > 0$ and so the plane wave loses energy to the electrons; this absorption occurs for electron energies less than the resonant energy.

Since the function $G(x)$ has its minimum value at $x \approx 2.6$ (Figure 13.28), the largest electron energy loss—and therefore the greatest field gain—occurs when $4\pi\eta_0 N_w \approx 2.6$, or $\eta_0 \approx 0.2/N_w$. This corresponds to an electron energy given by

$$\gamma \approx \left(1 + \frac{0.2}{N_w}\right) \gamma_R \tag{13.B.23}$$

PROBLEMS

13.1   (a) In Section 13.1 we showed that a transition between two non-degenerate states is always absorbing in thermal equilibrium, i.e., the gain coefficient is negative. Show that the same conclusion applies regardless of the level degeneracies.

(b) Show that the absorption linewidth of a Doppler-broadened transition is typically about 10^{-5}–10^{-6} times the transition frequency.

(c) Use numerical estimates to show that optical pumping of a gaseous laser medium with a blackbody source of radiation will usually not be very practical.

13.2 Consider an elastic collision between an electron and an atom. Show that the kinetic energy exchanged is relatively small.

13.3 For the excitation transfer process indicated in Figure 13.11*b*, write rate equations like (13.5.2) for the populations N_A, N_B, and N_{B^*}.

13.4 According to our discussion in Section 13.6, the higher gain of the 3.39-μm transition in the He–Ne laser would ordinarily preclude lasing at 6328 Å, unless the 3.39-μm line is deliberately suppressed. Why is this so?

13.5 In Section 13.7 we replaced the Boltzmann factor $e^{-\Delta E/kT}$ by unity, where ΔE is the energy difference between $He(2\,^1S)$ and $Ne(3s_2)$. Was this a reasonable approximation for our purposes?

13.6 In Section 13.7 we mentioned that White and Gordon deduced relative level populations by monitoring intensities of spontaneously emitted "sidelight." Show that a knowledge of the Einstein A-coefficient of a transition, combined with a frequency filter and an absolute intensity measurement, allows an absolute determination of the upper-level population. [For instance, White and Gordon reported a $He(2\,^1S)$ population of about 2.5×10^{11} cm^{-3} under lasing conditions.] Does radiative trapping afffect such a measurement?

13.7 Why is only a small concentration of a dye typically required to "dye" a solution?

13.8 Using the collision cross sections (13.A.14), estimate the absorption linewidth of the 10.6-μm transition of CO_2 in the earth's atmosphere. For a discussion of the cross sections (13.A.14) see T. W. Meyer, C. K. Rhodes, and H. A. Haus, *Physical Review A* **12,** 1993 (1975).

13.9 Verify Eqs. (13.13.5).

13.10 (a) Show that (13.B.2) follows if the dimensionless energy parameter η is small.

(b) Show that (13.B.22) follows from (13.B.21).

13.11 Show that the gain in a free-electron laser will be very small unless the spread in electron energies satisfies (13.13.22).

14 LASER RESONATORS

14.1 INTRODUCTION

Until now we have supposed a laser resonator to consist of two highly reflecting, flat, parallel mirrors separated by some distance L. The only important property of such a resonator for our purposes thus far is that it has "longitudinal" modes separated in frequency by $c/2L$. We have not concerned ourselves with how the field inside the resonator varies in directions transverse to the line joining the centers of the mirrors. In fact we have assumed the field to be uniform in any plane perpendicular to this so-called optical axis. In this chapter we will consider laser resonators more realistically. We will consider some of the important characteristics of actual laser resonators, beginning with a rather simple approach based on geometrical optics, and gradually working our way up to a description based on Maxwell's equations.

Most of our treatment of laser resonators will assume that the laser medium is *passive*. That is, the electromagnetic modes of the laser resonator will be assumed to be the same as the modes of an *empty* resonator having no gain medium. This is a good approximation if the gain coefficient and refractive index of the medium are fairly uniform throughout the medium. This is obviously a useful approximation, because it allows us to consider laser resonators independently of the laser medium. Fortunately it is often an accurate approximation.

Figure 14.1a shows a light ray normal to the mirrors of a resonator with flat, parallel mirrors. The ray keeps retracing its path on successive reflections from the mirrors. If the mirrors are not perfectly parallel, however, the ray will eventually escape from the resonator, as indicated in Figure 14.1b. The misaligned resonator of Figure 14.1b requires greater gain for laser oscillation than the resonator of Figure 14.1a. We might find, for instance, that a laser with flat mirrors turns off

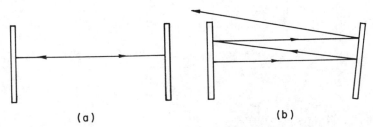

(a)   (b)

Figure 14.1 A laser resonator with flat, parallel mirrors. A light ray parallel to the optical axis remains inside the resonator if the mirrors are perfectly parallel (*a*). Otherwise it eventually escapes (*b*).

Figure 14.2 A laser resonator with mirrors that are spherical surfaces with radii of curvature R_1 and R_2.

(i.e., laser action ceases) at the slightest misalignment of the mirrors. Obviously this is undesirable if we wish to construct a practical and durable laser. Figure 14.2 shows a much more commonly used type of laser resonator, consisting of mirrors with spherical surfaces. This is the type of resonator used in most commercially available lasers. In Section 14.3 we will see why.

14.2   THE RAY MATRIX

In geometrical optics light propagation is described in terms of rays. We may define a ray at each point on a wave as an arrow drawn normal to the wave front. (Figure 14.3). We will assume that the direction of a ray is the direction of energy flow. There is no physical significance to the "length" of a ray; a ray merely represents a direction of propagation at a given point. When we adopt this ray picture we are ignoring the polarization of the light waves. Our ray picture is a crude but useful representation of the actual physical situation.

In this section we will develop a convenient formalism for ray propagation. This formalism will turn out to be appropriate for the description of *Gaussian* laser beams, which are discussed in Section 14.5.

In situations of practical interest we are dealing with light waves traveling more or less in a single direction, which we will call the z direction. The rays we envision point almost parallel to the z axis. At any point on the wave we imagine a ray having a lateral displacement $r(z)$, measured from the z axis, and a slope (Figure 14.4)

$$r'(z) = \frac{dr}{dz} \qquad (14.2.1)$$

Because of our assumption of nearly unidirectional propagation along z, the slope $r'(z)$ of a ray will be very small, so that (Figure 14.4)

$$r'(z) = \tan \theta \approx \sin \theta \approx \theta \qquad (14.2.2)$$

Figure 14.3 Rays drawn on a wave represent the direction of propagation.

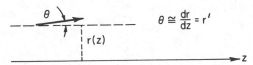

Figure 14.4 A ray is characterized by its displacement r and slope r' measured from some z axis.

Such rays are called *paraxial rays*. We will assume, as is implicit in our definition of the ray displacement r and slope r', that we have cylindrical symmetry about the z axis. The slope of a ray is taken to be positive or negative depending whether the displacement r is increasing or decreasing in the direction of propagation.

We would like to relate the displacement and slope of a ray at a point z to the displacement and slope at a point z'. Consider, for example, the simple case of vacuum propagation from z_1 to z_2. In vacuum there is nothing to change the direction of a ray, so we have (Figure 14.5):

$$r(z_2) = r(z_1) + r'(z_1)(z_2 - z_1) \tag{14.2.3}$$

and

$$r'(z_2) = r'(z_1) \tag{14.2.4}$$

In matrix notation we may write Eqs. (14.2.3) and (14.2.4) as

$$\begin{bmatrix} r(z_2) \\ r'(z_2) \end{bmatrix} = \begin{bmatrix} 1 & z_2 - z_1 \\ 0 & 1 \end{bmatrix} \begin{bmatrix} r(z_1) \\ r'(z_1) \end{bmatrix} \tag{14.2.5}$$

A ray is completely characterized by the 2×1 matrix, or *column vector*,

$$\begin{bmatrix} r \\ r' \end{bmatrix}$$

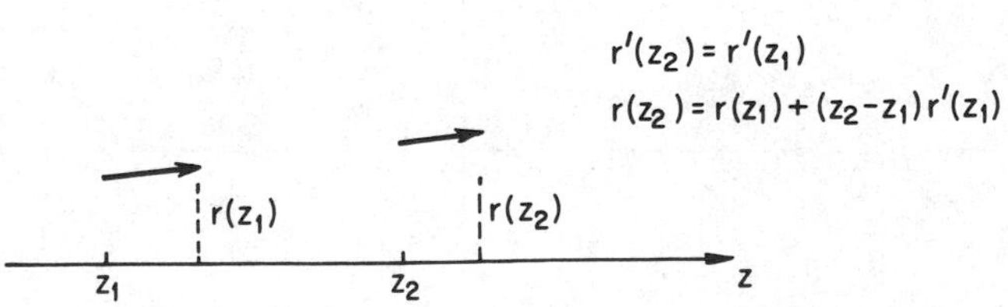

$$r'(z_2) = r'(z_1)$$
$$r(z_2) = r(z_1) + (z_2 - z_1)r'(z_1)$$

Figure 14.5 The transformation of a ray as a result of free propagation over a distance $z_2 - z_1$.

and Eq. (14.2.5) relates the *final ray*

$$\begin{bmatrix} r_f \\ r_f' \end{bmatrix} = \begin{bmatrix} r(z_2) \\ r'(z_2) \end{bmatrix} \qquad (14.2.6)$$

to the *initial ray*

$$\begin{bmatrix} r_i \\ r_i' \end{bmatrix} = \begin{bmatrix} r(z_1) \\ r'(z_1) \end{bmatrix} \qquad (14.2.7)$$

Thus, according to Eq. (14.2.5), the vacuum propagation of a ray through a distance $d = z_2 - z_1$ is described by the matrix equation

$$\begin{bmatrix} r_f \\ r_f' \end{bmatrix} = \begin{bmatrix} 1 & d \\ 0 & 1 \end{bmatrix} \begin{bmatrix} r_i \\ r_i' \end{bmatrix} \qquad (14.2.8)$$

Given the initial ray with displacement r_i and slope r_i', this equation tells us how that ray is modified by propagation through a distance d.

Consider next the more interesting example of the transformation of a (paraxial) ray by a thin lens of focal length f (Figure 14.6). Immediately to the right of the lens the ray's lateral displacement r_f is the same as the initial displacement r_i immediately to the left:

$$r_f = r_i \qquad (14.2.9)$$

The slope of the ray, however, is changed by the lens. From the thin lens equation relating the object and image distances with the focal length of the lens, we obtain (Figure 14.6)

$$r_f' = r_i' - \frac{r_i}{f} \qquad (14.2.10)$$

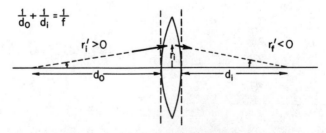

Figure 14.6 Ray transformation by a thin lens.

In matrix notation, Eqs. (14.2.9) and (14.2.10) take the form

$$\begin{bmatrix} r_f \\ r_f' \end{bmatrix} = \begin{bmatrix} 1 & 0 \\ -1/f & 1 \end{bmatrix} \begin{bmatrix} r_i \\ r_i' \end{bmatrix} \qquad (14.2.11)$$

One more example will be of interest to us, namely, the case of a spherical mirror with radius of curvature R. The displacement of the ray is the same immediately before and after reflection from the mirror, i.e., $r_f = r_i$. The slope of the ray after reflection, however, is (Figure 14.7)

$$r_f' = r_i' - \frac{2r_i}{R} \qquad (14.2.12)$$

In matrix notation, therefore, the ray transformation by the spherical mirror is given by the equation

$$\begin{bmatrix} r_f \\ r_f' \end{bmatrix} = \begin{bmatrix} 1 & 0 \\ -2/R & 1 \end{bmatrix} \begin{bmatrix} r_i \\ r_i' \end{bmatrix} \qquad (14.2.13)$$

There is a sign convention for r', namely, $r' > 0$ if r is increasing with propagation, $r' < 0$ otherwise. With this in mind, our sign convention for the radius of curvature R of a spherical mirror is easily checked: R is positive for a concave mirror (Figure 14.7) and negative for a convex mirror. Similarly the focal length f of a lens is positive for a converging lens (Figure 14.6) and negative for a diverging lens. These statements may be verified by making sketches like those in Figures 14.6 and 14.7. Thus (14.2.11) and (14.2.13) apply also to diverging lenses

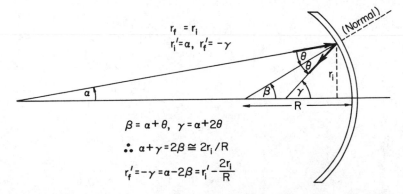

Figure 14.7 Paraxial ray transformation by a spherical mirror surface with radius of curvature R. The relation between r_f' and r_i' is obtained by applying the trigonometric theorem that an exterior angle of a triangle equals the sum of the two opposite interior angles, and the approximation $\beta \approx 2r_i/R$ that holds for paraxial rays.

and convex mirrors, respectively, provided f and R are taken to be negative in those cases.

We have considered thus far the transformation of a ray by three different "optical elements"—empty space of length d, a thin lens of focal length f, and a spherical mirror of radius of curvature R. In general an optical element will transform a ray according to the matrix equation

$$
\begin{bmatrix} r_f \\ r_f' \end{bmatrix} = \begin{bmatrix} A & B \\ C & D \end{bmatrix} \begin{bmatrix} r_i \\ r_i' \end{bmatrix}
\tag{14.2.14}
$$

The 2×2 matrix on the right-hand side of this equation is called the *ray matrix*, or *ABCD* matrix, for the optical element. Equations (14.2.8), (14.2.11), and (14.2.13) give the ray matrices for a straight section of length d, a thin lens of focal length f, and a spherical mirror of radius of curvature R, respectively.

Let us consider the effect on a ray of an open path section of length d followed by a thin lens of focal length f. If a ray has displacement r_i and slope r_i' initially, then after the open section of propagation it has displacement r and slope r' given by Eq. (14.2.8):

$$
\begin{bmatrix} r \\ r' \end{bmatrix} = \begin{bmatrix} 1 & d \\ 0 & 1 \end{bmatrix} \begin{bmatrix} r_i \\ r_i' \end{bmatrix}
\tag{14.2.15}
$$

This gives the "initial" ray displacement and slope immediately before passage through the lens. The "final" ray displacement and slope are therefore given by Eq. (14.2.11):

$$
\begin{aligned}
\begin{bmatrix} r_f \\ r_f' \end{bmatrix} &= \begin{bmatrix} 1 & 0 \\ -1/f & 1 \end{bmatrix} \begin{bmatrix} r \\ r' \end{bmatrix} \\[6pt]
&= \begin{bmatrix} 1 & 0 \\ -1/f & 1 \end{bmatrix} \begin{bmatrix} 1 & d \\ 0 & 1 \end{bmatrix} \begin{bmatrix} r_i \\ r_i' \end{bmatrix} \\[6pt]
&= \begin{bmatrix} 1 & d \\ -1/f & 1 - d/f \end{bmatrix} \begin{bmatrix} r_i \\ r_i' \end{bmatrix}
\end{aligned}
\tag{14.2.16}
$$

The matrix

$$
\begin{bmatrix} 1 & d \\ -1/f & 1 - d/f \end{bmatrix} = \begin{bmatrix} 1 & 0 \\ -1/f & 1 \end{bmatrix} \begin{bmatrix} 1 & d \\ 0 & 1 \end{bmatrix}
\tag{14.2.17}
$$

is therefore the ray matrix for the combined optical system consisting of an open section of length d followed by a thin lens of focal length f. It is the product of the ray matrices for an open section and a lens. It follows that if we have any number of optical elements in some sequence, then the ray matrix for the system comprising all these elements is the matrix product of the ray matrices of the individual elements. Since the matrix product $M_1 M_2$ is in general not the same as $M_2 M_1$, the order of the matrices in the product is important. Thus the system ray matrix is the ray matrix of the first optical element encountered, multiplied *on the left* by the ray matrix of the second optical element, multiplied *on the left* by the ray matrix of the third element, etc. The reader may easily show, for instance, that the ray matrix for the system consisting of an open section followed by a thin lens is different from the ray matrix for a thin lens followed by an open section (Problem 14.1). This means, of course, that the effects of the two systems on a ray are different.

14.3 RESONATOR STABILITY

One of the simplest but most important questions concerning a laser resonator is whether it is *stable*. To see what this means, consider an arbitrary (paraxial) ray bouncing back and forth between the mirrors of a resonator. If the ray remains within the resonator, the resonator is said to be stable. If, however, the ray escapes from the resonator after a sufficiently large number of reflections, the resonator is unstable. Figure 14.1b, for example, shows that a misaligned flat-mirror resonator is unstable. In general a stability criterion for a laser resonator can be expressed in terms of the radii of curvature of the mirrors and the distance separating the mirrors. We will now derive this stability criterion with the aid of the *ABCD* matrix.

Consider the resonator sketched in Figure 14.2, consisting of mirrors of radii of curvature R_1 and R_2, separated by a distance L. As drawn, the mirrors are concave. Our analysis, however, will apply also to the case of convex mirrors if we recall that a convex mirror by convention has a negative radius of curvature. We note also that a flat mirror may be regarded as a spherical mirror surface with an infinite radius of curvature.

Imagine a ray starting at the left mirror of Figure 14.2. After a round trip through the resonator, this ray will have been transformed by a straight section of length L, a spherical mirror of radius of curvature R_2, another straight section of length L, and finally a spherical mirror of radius of curvature R_1. The ray matrix describing the ray transformation by a round trip through the resonator is

$$
\begin{bmatrix} A & B \\ C & D \end{bmatrix} = \begin{bmatrix} 1 & 0 \\ -2/R_1 & 1 \end{bmatrix} \begin{bmatrix} 1 & L \\ 0 & 1 \end{bmatrix} \begin{bmatrix} 1 & 0 \\ -2/R_2 & 1 \end{bmatrix} \begin{bmatrix} 1 & L \\ 0 & 1 \end{bmatrix}
$$

$$
= \begin{bmatrix} 1 - \dfrac{2L}{R_2} & 2L - \dfrac{2L^2}{R_2} \\[2ex] \dfrac{4L}{R_1 R_2} - \dfrac{2}{R_1} - \dfrac{2}{R_2} & 1 - \dfrac{2L}{R_2} - \dfrac{4L}{R_1} + \dfrac{4L^2}{R_1 R_2} \end{bmatrix} \qquad (14.3.1)
$$

After N round trips through the resonator, therefore, the initial ray with displacement r_i and slope r_i' is transformed to the ray with displacement r_N and slope r_N' given by

$$\begin{bmatrix} r_N \\ r_N' \end{bmatrix} = \begin{bmatrix} A & B \\ C & D \end{bmatrix}^N \begin{bmatrix} r_i \\ r_i' \end{bmatrix} \tag{14.3.2}$$

where the ray ($ABCD$) matrix is defined by (14.3.1). The ray matrix (14.3.1) has determinant[1]

$$AD - BC = 1 \tag{14.3.3}$$

Using this fact, and defining an angle θ by

$$\cos \theta = \tfrac{1}{2}(A + D) \tag{14.3.4}$$

it may be shown (see below) that

$$\begin{bmatrix} A & B \\ C & D \end{bmatrix}^N = \frac{1}{\sin \theta} \begin{bmatrix} A \sin N\theta - \sin(N-1)\theta & B \sin N\theta \\ C \sin N\theta & D \sin N\theta - \sin(N-1)\theta \end{bmatrix}$$

$$\tag{14.3.5}$$

• The result (14.3.5) for a 2×2 matrix satisfying (14.3.3) is sometimes called "Sylvester's theorem." It may be proved by induction: it obviously holds for the case $N = 1$, and so we try to show that if it holds for a single given (but arbitrary) N it must hold also for $N + 1$. If we can show this, Sylvester's theorem is proved.

Thus let us assume that (14.3.5) holds, so that

$$\begin{bmatrix} A & B \\ C & D \end{bmatrix}^{N+1} = \begin{bmatrix} A & B \\ C & D \end{bmatrix} \begin{bmatrix} A & B \\ C & D \end{bmatrix}^N$$

$$= \frac{1}{\sin \theta} \begin{bmatrix} A & B \\ C & D \end{bmatrix} \begin{bmatrix} A \sin N\theta - \sin(N-1)\theta & B \sin N\theta \\ C \sin N\theta & D \sin N\theta - \sin(N-1)\theta \end{bmatrix}$$

$$= \frac{1}{\sin \theta} \begin{bmatrix} (A^2 + BC) \sin N\theta - A \sin(N-1)\theta & B(A+D) \sin N\theta - B \sin(N-1)\theta \\ C(A+D) \sin N\theta - C \sin(N-1)\theta & (BC + D^2) \sin N\theta - D \sin(N-1)\theta \end{bmatrix}$$

$$\tag{14.3.6}$$

---

1. The simplest way to check this is to note that the ray matrix (14.3.1) is a product of four matrices, each having determinant equal to one. Since the determinant of the product of matrices is equal to the product of the determinants, (14.3.3) follows.

Using (14.3.3) and (14.3.4), we see that the $(1, 1)$ element of this matrix is

$(A^2 + BC) \sin N\theta - A \sin (N - 1)\theta$

$= (A^2 + AD - 1) \sin N\theta - A \sin (N - 1)\theta$

$= A(A + D) \sin N\theta - \sin N\theta - A \sin (N - 1)\theta$

$= 2A \sin N\theta \cos \theta - \sin N\theta - A \sin (N - 1)\theta$

$= 2A[\frac{1}{2} \sin (N + 1)\theta + \frac{1}{2} \sin (N - 1)\theta] - \sin N\theta - A \sin (N - 1)\theta$

$= A \sin (N + 1)\theta - \sin N\theta$ \hfill (14.3.7)

The remaining three matrix elements of (14.3.6) may be evaluated similarly. We obtain

$$\begin{bmatrix} A & B \\ C & D \end{bmatrix}^{N+1} = \frac{1}{\sin \theta} \begin{bmatrix} A \sin (N + 1)\theta - \sin N\theta & B \sin (N + 1)\theta \\ C \sin (N + 1)\theta & D \sin (N + 1)\theta - \sin N\theta \end{bmatrix}$$

$$(14.3.8)$$

But this is just Eq. (14.3.5) with N replaced by $N + 1$. Thus (14.3.5) is true for $N = 1$, and we have just shown that if it is true for any N, then it must be true also for $N + 1$. This proves Sylvester's theorem. •

It now follows from Eq. (14.3.2) that

$$\begin{bmatrix} r_N \\ r_N' \end{bmatrix} = \frac{1}{\sin \theta} \begin{bmatrix} A \sin N\theta - \sin (N - 1)\theta & B \sin N\theta \\ C \sin N\theta & D \sin N\theta - \sin (N - 1)\theta \end{bmatrix} \begin{bmatrix} r_i \\ r_i' \end{bmatrix}$$

$$(14.3.9)$$

where, from (14.3.4) and (14.3.1),

$$\cos \theta = \frac{1}{2}\left(1 - \frac{2L}{R_2} + 1 - \frac{2L}{R_2} - \frac{4L}{R_1} + \frac{4L^2}{R_1 R_2}\right)$$

$$= 1 - \frac{2L}{R_1} - \frac{2L}{R_2} + \frac{2L^2}{R_1 R_2} \qquad (14.3.10)$$

Equation (14.3.9) gives the ray displacement and slope after N round trips through the resonator. We observe that r_N (and r_N') stays finite as long as θ is real. If θ is a complex number, however, then $\sin N\theta = (e^{iN\theta} - e^{-iN\theta})/2i$ can be very large for large N, and in fact diverges as $N \to \infty$. In other words, if θ is not purely real, r_N itself will diverge, i.e., the ray will escape from the confines of the resonator. Thus the condition for resonator stability is for θ to be real, which means that $|\cos \theta| \leq 1$, or, from (14.3.10),

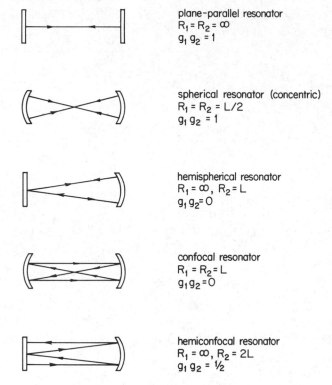

plane-parallel resonator
$R_1 = R_2 = \infty$
$g_1 g_2 = 1$

spherical resonator (concentric)
$R_1 = R_2 = L/2$
$g_1 g_2 = 1$

hemispherical resonator
$R_1 = \infty, \ R_2 = L$
$g_1 g_2 = 0$

confocal resonator
$R_1 = R_2 = L$
$g_1 g_2 = 0$

hemiconfocal resonator
$R_1 = \infty, \ R_2 = 2L$
$g_1 g_2 = \tfrac{1}{2}$

Figure 14.8 Examples of stable resonators.

$$-1 \leqq 1 - \frac{2L}{R_1} - \frac{2L}{R_2} + \frac{2L^2}{R_1 R_2} \leqq 1$$

$$-2 \leqq -\frac{2L}{R_1} - \frac{2L}{R_2} + \frac{2L^2}{R_1 R_2} \leqq 0$$

$$0 \leqq 1 - \frac{L}{R_1} - \frac{L}{R_2} + \frac{L^2}{R_1 R_2} \leqq 1 \tag{14.3.11}$$

This stability condition is usually written in the laser literature as

$$0 \leqq g_1 g_2 \leqq 1 \tag{14.3.12}$$

where

$$g_1 = 1 - \frac{L}{R_1} \tag{14.3.13a}$$

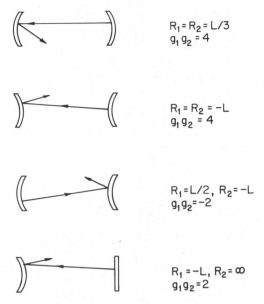

$R_1 = R_2 = L/3$
$g_1 g_2 = 4$

$R_1 = R_2 = -L$
$g_1 g_2 = 4$

$R_1 = L/2, R_2 = -L$
$g_1 g_2 = -2$

$R_1 = -L, R_2 = \infty$
$g_1 g_2 = 2$

Figure 14.9 Examples of unstable resonators.

and

$$g_2 = 1 - \frac{L}{R_2} \qquad (14.3.13b)$$

These are called the *g parameters* of the resonator. If the *g* parameters are such that (14.3.12) is satisfied, the resonator is stable. If $g_1 g_2 < 0$ or $g_1 g_2 > 1$, however, the resonator is unstable.

The ray-matrix approach allows us to check immediately whether a given resonator is stable, without having to perform a ray trace such as that shown in Figure 14.1*b*. Whether a given resonator is stable or unstable depends only on the radii of curvature of the mirrors and the distance separating them. Figures 14.8 and 14.9 show examples of stable and unstable resonators, respectively. The reader may easily check in each case whether the resonator is stable or unstable (Problem 14.2).

Our stability analysis has assumed perfect mirror reflectivities. In reality, of course, some energy will be taken from the intraresonator laser field because of imperfect mirror reflectivities. We have already noted (Chapter 10) that transmissive output coupling through one (or both) of the mirrors is one such loss mechanism. In addition to such loss mechanisms as output coupling, scattering, or absorption, a laser with an unstable resonator will have a large loss associated with the escape of radiation *past* the mirrors, as indicated by ray tracing as in Figures 14.1*b* and 14.9. Because of this additional loss factor, unstable resonators typically require media with higher gain to sustain laser oscillation. This is not to say that unstable resonators should always be avoided. On the contrary, unstable resonators offer several advantages for certain high-power lasers (Section 14.13). In more familiar devices such as commercial He–Ne lasers, however, stable resonators are usually employed.

The plane-parallel resonator of Figure 14.1 is not used for practical lasers because it becomes unstable with only slight misalignment of the mirrors. The resonators of most lasers have at least one spherical mirror surface. The hemispherical resonator of Figure 14.8, for instance, is perhaps the most commonly used design for He–Ne lasers.

14.4 THE PARAXIAL WAVE EQUATION

Many important properties of laser resonators are consequences of the wave nature of light. A complete understanding of laser resonators therefore demands a treatment based on Maxwell's equations rather than geometrical rays. In this section we will examine an approximate solution of the Maxwell wave equation that turns out to be very important for laser resonators.

Let us first recall the wave equation (see Section 2.1) for the electric field in vacuum:

$$\nabla^2 E(\mathbf{r},\, t) - \frac{1}{c^2} \frac{\partial^2}{\partial t^2} E(\mathbf{r},\, t) = 0 \qquad (14.4.1)$$

We have written the scalar wave equation instead of the full vector equation. Our treatment will therefore account for diffraction and interference of the radiation inside a resonator, but not for polarization effects. A fully vectorial treatment of laser resonators is very complicated, so fortunately the scalar theory is quite adequate for our purposes. We will be interested in solutions of (14.4.1) of the form

$$E(\mathbf{r},\, t) = \mathcal{E}(\mathbf{r})\, e^{-i\omega t} \qquad (14.4.2)$$

i.e., monochromatic fields. When this expression is used in the wave equation (14.4.1), we obtain the *Helmholtz equation* for $\mathcal{E}(\mathbf{r})$:

$$\nabla^2 \mathcal{E}(\mathbf{r}) + k^2 \mathcal{E}(\mathbf{r}) = 0 \qquad (14.4.3)$$

where

$$k^2 = \omega^2/c^2 \qquad (14.4.4)$$

A solution of the Helmholtz equation for $\mathcal{E}(\mathbf{r})$ will provide a monochromatic solution (14.4.2) of the wave equation.

One solution of (14.4.3) is

$$\mathcal{E}(\mathbf{r}) = \mathcal{E}_0 \, e^{i\mathbf{k} \cdot \mathbf{r}} \tag{14.4.5}$$

where $\mathcal{E}_0$ is a constant and $\mathbf{k}$ is a vector whose squared magnitude is given by (14.4.4). Such a plane-wave solution was discussed in Chapter 2. It has the same value for all points in any plane normal to $\mathbf{k}$. If we take $\mathbf{k}$ to point in the z direction, for instance, the solution (14.4.5) has the same value ($\mathcal{E}_0 e^{ikz}$) in any plane defined by a constant value of z. Another solution of (14.4.3), valid for all $r \neq 0$, is (Problem 14.3)

$$\mathcal{E}(\mathbf{r}) = \frac{A}{r} \, e^{ikr} \tag{14.4.6}$$

This solution has a constant value on any sphere centered at the origin, and is therefore called a *spherical wave*. This is the form of solution we would associate with a point source at the origin. It represents a (spherical) wave emanating from the origin, with the intensity of the wave (square of $|\mathcal{E}|$) decreasing with distance r according to the inverse square law.

Consider the spherical-wave solution (14.4.6) in the plane ($z = R$) (Figure 14.10). In this plane

$$r = (x^2 + y^2 + R^2)^{1/2} = R \left(1 + \frac{x^2 + y^2}{R^2} \right)^{1/2} \tag{14.4.7}$$

If we restrict ourselves to a small "patch" of observation about the point ($x = 0$, $y = 0$, $z = R$), so that $x^2 + y^2$ is small compared with R^2, then

$$\left(1 + \frac{x^2 + y^2}{R^2} \right)^{1/2} = 1 + \frac{x^2 + y^2}{2R^2} + \cdots \tag{14.4.8}$$

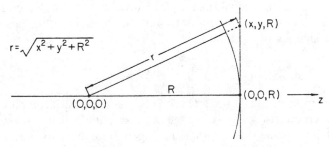

Figure 14.10 Geometry for Eq. (14.4.7).

according to the binomial expansion. Thus we use the approximation

$$kr \approx kR + \frac{k(x^2 + y^2)}{2R} \qquad (14.4.9)$$

The field on the plane $(z = R)$ in the vicinity of $x \approx 0$, $y \approx 0$ is therefore

$$\mathcal{E}(\mathbf{r}) = \frac{A}{R} e^{ikR} e^{ik(x^2 + y^2)/2R} \qquad (14.4.10)$$

at points for which (14.4.8) is satisfied. Note that we have simply replaced r by R for the approximate evaluation of the factor A/r in (14.4.6). In the factor e^{ikr}, however, we have retained in (14.4.10) both terms on the right-hand side of (14.4.9). This is necessary because, although $(x^2 + y^2)/2R$ may be very small compared with R, it need not be very small compared with a wavelength ($\lambda = 2\pi/k$), and so the second term in (14.4.9) cannot be neglected in e^{ikr}. In order for (14.4.10) to be a good approximation the next term in the binomial expansion of r must be very small compared with a wavelength. This condition is (Problem 14.3)

$$\frac{a^2}{\lambda R} \ll \left(\frac{R}{a}\right)^2 \qquad (14.4.11)$$

where

$$a^2 = x^2 + y^2 \qquad (14.4.12)$$

The field (14.4.10) is an accurate approximation to the spherical wave (14.4.6) when (14.4.8) or (14.4.11) is satisfied, i.e., when we consider a small enough radius a of observation about the point $(0, 0, R)$ in the plane $z = R$. This approximation is used frequently in physical optics.

Now a laser beam propagates as a nearly unidirectional wave with some *finite* cross-sectional area. Plane and spherical waves are obviously not beams. A spherical wave is not unidirectional, and a plane wave has an infinite cross-sectional area. We therefore seek solutions of (14.4.3) that look more like beams. To this end we try a solution of the form

$$\mathcal{E}(\mathbf{r}) = \mathcal{E}_0(\mathbf{r}) e^{ikz} \qquad (14.4.13)$$

which differs from the plane wave (14.4.5) by the fact that its amplitude is not a constant. In writing (14.4.13) we are seeking a solution which has nearly the unidirectionality of a plane wave, without having an infinite beam cross section. We assume that the variations of $\mathcal{E}_0(r)$ and $\partial \mathcal{E}_0(r)/\partial z$ within a distance of the order of a wavelength in the z direction are negligible, i.e.,

$$\lambda \left| \frac{\partial \mathcal{E}_0}{\partial z} \right| \ll |\mathcal{E}_0|, \qquad \lambda \left| \frac{\partial^2 \mathcal{E}_0}{\partial z^2} \right| \ll \left| \frac{\partial \mathcal{E}_0}{\partial z} \right| \qquad (14.4.14)$$

or, since $k = 2\pi / \lambda$,

$$\left| \frac{\partial \mathcal{E}_0}{\partial z} \right| \ll k |\mathcal{E}_0|, \qquad \left| \frac{\partial^2 \mathcal{E}_0}{\partial z^2} \right| \ll k \left| \frac{\partial \mathcal{E}_0}{\partial z} \right| \qquad (14.4.15)$$

Thus we are assuming that $\mathcal{E}(r)$ varies approximately as e^{ikz} over distances z on the order of several wavelengths.

The field (14.4.13) must satisfy the Helmholtz equation:

$$\left(\frac{\partial^2}{\partial x^2} + \frac{\partial^2}{\partial y^2} + \frac{\partial^2}{\partial z^2} \right) \mathcal{E}_0(\mathbf{r}) \, e^{ikz} + k^2 \, \mathcal{E}_0(\mathbf{r}) \, e^{ikz} = 0 \qquad (14.4.16)$$

Now

$$\frac{\partial^2}{\partial z^2} \mathcal{E}_0(\mathbf{r}) \, e^{ikz} = \left(\frac{\partial^2 \mathcal{E}_0}{\partial z^2} + 2ik \frac{\partial \mathcal{E}_0}{\partial z} - k^2 \mathcal{E}_0 \right) e^{ikz}$$

$$\approx \left(2ik \frac{\partial \mathcal{E}_0}{\partial z} - k^2 \mathcal{E}_0 \right) e^{ikz} \qquad (14.4.17)$$

in the approximation (14.4.15), and thus from (14.4.16) we have

$$\left(\frac{\partial^2}{\partial x^2} + \frac{\partial^2}{\partial y^2} + 2ik \frac{\partial}{\partial z} \right) \mathcal{E}_0(\mathbf{r}) \cong 0 \qquad (14.4.18)$$

We therefore seek solutions to this equation, i.e., fields of the form (14.4.13) satisfying (14.4.15). The equation

$$\nabla_T^2 \mathcal{E}_0 + 2ik \frac{\partial \mathcal{E}_0}{\partial z} = 0 \qquad (14.4.19)$$

is called the *paraxial wave equation*. Here the *transverse Laplacian* is

$$\nabla_T^2 = \frac{\partial^2}{\partial x^2} + \frac{\partial^2}{\partial y^2} \qquad (14.4.20)$$

We will now consider important solutions of this partial differential equation, which has a clear similarity to Schrödinger's wave equation [recall (5.2.6)] for a free particle in two space variables (x and y), with t replaced by z.

14.5 GAUSSIAN BEAMS

The intensity of a "beamlike" wave propagating in the z direction is negligible at points sufficiently far from the z axis. A Gaussian beam intensity profile, for example, has the form

$$I(x, y, z) \sim \left| \mathcal{E}_0(\mathbf{r}) \right|^2 e^{-2(x^2 + y^2)/w^2} \qquad (14.5.1)$$

in a plane normal to the direction (z) of propagation. At a lateral distance w from the z axis, the intensity is a factor e^2 ($=7.389$) smaller than its value on axis. If the beam were projected onto a screen we would see a spot of radius $\sim w$, and so w is called the *spot size* of the Gaussian beam.

Laser beams are frequently observed to have an intensity profile like (14.5.1). With this in mind, we try to construct a solution of (14.4.19) having the form

$$\mathcal{E}_0(\mathbf{r}) = A \, e^{ik(x^2 + y^2)/2q(z)} \, e^{ip(z)} \qquad (14.5.2)$$

where A is a constant and $q(z)$ and $p(z)$ are to be determined. Note that if

$$\frac{1}{q} = \frac{2i}{kw^2} = \frac{i\lambda}{\pi w^2} \qquad (14.5.3)$$

then (14.5.2) gives the Gaussian intensity profile (14.5.1). In assuming a solution of the form (14.5.2), in which q depends upon z, we are allowing for the possibility that the spot size of a Gaussian beam can vary with distance of propagation, which is in fact known to occur in laser beams.

For the function (14.5.2) we have

$$\frac{\partial \mathcal{E}_0}{\partial z} = iA \left(\frac{dp}{dz} - \frac{k}{2} (x^2 + y^2) \frac{1}{q^2} \frac{dq}{dz} \right) e^{ik(x^2 + y^2)/2q(z)} \, e^{ip(z)} \qquad (14.5.4)$$

and

$$\nabla_T^2 \mathcal{E}_0 = A \left[\frac{2ik}{q} - \frac{k^2}{q^2} (x^2 + y^2) \right] e^{ik(x^2 + y^2)/2q(z)} \, e^{ip(z)} \qquad (14.5.5)$$

so that

$$\nabla_T^2 \mathcal{E}_0 + 2ik \frac{\partial \mathcal{E}_0}{\partial z} = A \left[\frac{k^2}{q^2} (x^2 + y^2) \left(\frac{dq}{dz} - 1 \right) \right.$$

$$\left. - 2k \left(\frac{dp}{dz} - \frac{i}{q} \right) \right] e^{ik(x^2 + y^2)/2q(z)} \, e^{ip(z)} \qquad (14.5.6)$$

Therefore the form (14.5.2) is indeed a solution to Eq. (14.4.19) if $p(z)$ and $q(z)$ satisfy

$$\frac{dq}{dz} = 1 \tag{14.5.7}$$

and

$$\frac{dp}{dz} = \frac{i}{q} \tag{14.5.8}$$

These equations have the solutions

$$q(z) = q_0 + z \tag{14.5.9}$$

and

$$p(z) = i \ln \frac{q_0 + z}{q_0} \tag{14.5.10}$$

where $q_0 = q(0)$ and we assume $p(0) = 0$.

Since q may be complex, we write

$$\frac{1}{q(z)} = \frac{1}{R(z)} + \frac{i\lambda}{\pi w^2(z)} \tag{14.5.11}$$

with R and w real. This way of writing $1/q$ is suggested by (14.5.3), to which (14.5.11) reduces when $R \to \infty$, i.e., when q is purely imaginary. With $1/q$ written this way, we have

$$e^{ik(x^2+y^2)/2q(z)} = e^{ik(x^2+y^2)/2R(z)} \, e^{-(x^2+y^2)/w^2(z)} \tag{14.5.12}$$

Using (14.5.10) and (14.5.11), we also have

$$e^{ip(z)} = \exp\left(-\ln \frac{q_0 + z}{q_0} \right)$$

$$= \frac{q_0}{q_0 + z}$$

$$= \frac{1}{1 + z/q_0}$$

$$= \frac{1}{1 + z/R_0 + i\lambda z/\pi w_0^2} \tag{14.5.13}$$

where R_0 and w_0 denote the values of R and w at $z = 0$.

If R_0 and w_0 are known, Eqs. (14.5.9) and (14.5.11) give $R(z)$ and $w(z)$ for all values of z. Since the designation $z = 0$ is arbitrary, let us choose the plane $z = 0$ to be that for which R_0 is infinitely large, i.e.,

$$R_0 = \infty \tag{14.5.14}$$

and

$$\frac{1}{q_0} = \frac{i\lambda}{\pi w_0^2} \tag{14.5.15}$$

It then follows from (14.5.9) that

$$
\begin{aligned}
\frac{1}{q(z)} &= \frac{1}{q_0 + z} = \frac{1/q_0}{1 + z(1/q_0)} \\
&= \frac{i\lambda/\pi w_0^2}{1 + iz\lambda/\pi w_0^2} \\
&= \frac{i\lambda/\pi w_0^2 + 1/z \, (\lambda z/\pi w_0^2)^2}{1 + (\lambda z/\pi w_0^2)^2} \\
&= \frac{1}{R(z)} + \frac{i\lambda}{\pi w^2(z)}
\end{aligned}
\tag{14.5.16}
$$

Equating separately the real and imaginary parts, we have

$$R(z) = z + \frac{z_0^2}{z} \tag{14.5.17}$$

$$w(z) = w_0\sqrt{1 + z^2/z_0^2} \tag{14.5.18}$$

where we have defined z_0 by

$$z_0 = \frac{\pi w_0^2}{\lambda} \tag{14.5.19}$$

This new parameter is known as the *Rayleigh range*, and is discussed below. The alternative term *confocal parameter* (exactly twice the Rayleigh range) is also used to characterize Gaussian beams.

Finally let us note that (14.5.14) allows us to write (14.5.13) as

$$
\begin{aligned}
e^{ip(z)} &= \frac{1}{1 + iz/z_0} \\
&= \frac{1}{\sqrt{1 + z^2/z_0^2}} \, e^{-i\phi(z)}
\end{aligned}
\tag{14.5.20}
$$

where

$$\phi(z) = \tan^{-1}(z/z_0). \tag{14.5.21}$$

With this result and Eq. (14.5.12), we can write our solution (14.5.2) of the paraxial wave equation in the form

$$\mathcal{E}_0(\mathbf{r}) = \frac{A\, e^{-i\phi(z)}}{\sqrt{1 + z^2/z_0^2}}\, e^{ik(x^2+y^2)/2R(z)}\, e^{-(x^2+y^2)/w^2(z)} \tag{14.5.22}$$

with A a constant and $R(z)$, $w(z)$, and z_0 satisfying (14.5.17)–(14.5.19).

By multiplying (14.5.22) by e^{ikz} we obtain the full expression (14.4.13) for a "beamlike" solution to the wave equation, or at least one that is valid within the approximation (14.4.15). The solution (14.5.22) has the Gaussian intensity profile (14.5.1) in any plane $z = $ constant. The spot size $w(z)$ has a minimum value w_0 in some plane $z = 0$, and grows with distance from this plane according to the formula (14.5.18). This behavior is sketched in Figure 14.11. Note that we do not have an expression for the minimum spot size w_0. The solution (14.5.22) is characterized, except for the trivial constant A, by w_0 and the wavelength λ. For obvious reasons (Figure 14.11) the plane $z = 0$ is called the *beam waist*.

The distance z_0 defined by (14.5.19) is such that

$$w(z_0) = w_0\sqrt{2} \tag{14.5.23}$$

The *Rayleigh range* z_0 is thus a measure of the length of the waist region, where the spot size is smallest. The smaller the spot size w_0 at the beam waist, the smaller the Rayleigh range, and thus the greater the rate of growth with z of the spot size from the waist. This result is similar to what happens when a plane wave is diffracted by a circular aperture in an opaque screen. The smaller the aperture di-

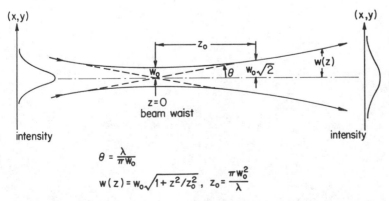

Figure 14.11 Variation of the spot size $w(z)$ of a Gaussian beam.

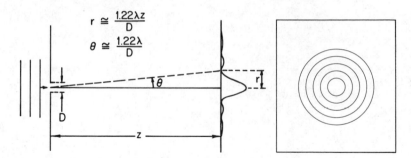

Figure 14.12 Diffraction of a monochromatic plane wave of wavelength λ by a circular aperture of diameter D. The diffraction intensity pattern in the fair field consists of a central bright spot of radius r surrounded by faint concentric rings, each of width $\Delta r \approx r$.

ameter D, the greater the diffraction. In fact the far-field divergence angle of the diffracted beam (Figure 14.12) may be defined by

$$\theta = 1.22 \frac{\lambda}{D} \qquad (14.5.24)$$

This result is derived in Section 14.10. We may define the divergence angle of our Gaussian beam similarly, as (Figure 14.11)

$$\theta \approx \frac{w(z)}{z} \approx \frac{w_0}{z_0} = \frac{\lambda}{\pi w_0}, \qquad z \gg z_0 \qquad (14.5.25)$$

Thus the divergence angle of a Gaussian beam is of the same order as that associated with the diffraction of a plane wave by an aperture of diameter $D \sim w_0$ located at the beam waist.

The intensity of the field (14.5.22) averaged over an optical period is (Problem 14.3)

$$I(x, y, z) = \frac{c\epsilon_0}{2} \left| \mathcal{E}(x, y, z) \right|^2 = \frac{(c\epsilon_0/2)\left| A \right|^2}{1 + z^2/z_0^2} e^{-2(x^2+y^2)/w^2(z)} \qquad (14.5.26)$$

The rate at which energy crosses any plane defined by a constant value of z is therefore

$$\int_{-\infty}^{\infty} \int_{-\infty}^{\infty} dx \, dy \, I(x, y, z) = \frac{(c\epsilon_0/2)\left| A \right|^2}{1 + z^2/z_0^2} \int_{-\infty}^{\infty} \int_{-\infty}^{\infty} dx \, dy \, e^{-2(x^2+y^2)/w^2(z)}$$

$$= \frac{c\epsilon_0}{4} \left| A \right|^2 (\pi w_0^2) \qquad (14.5.27)$$

The fact that this expression is independent of z is, of course, consistent with the conservation of energy. For $z \gg z_0$ the beam intensity (14.5.26) has an inverse-square dependence on the distance z from the waist:

$$I(x, y, z) \approx \frac{I_0}{z^2} z_0^2 \, e^{-2(x^2+y^2)/w^2(z)}, \qquad z \gg z_0 \qquad (14.5.28)$$

where

$$I_0 = \frac{c\epsilon_0}{2} |A|^2 \qquad (14.5.29)$$

The spot size at large distances from the beam waist is

$$w(z) \approx \frac{w_0 z}{z_0} = \frac{\lambda z}{\pi w_0}, \qquad z \gg z_0 \qquad (14.5.30)$$

and is seen to grow linearly with distance z from the beam waist.

It is also interesting to consider the electric field at large radial distances from the beam waist:

$$\mathcal{E}(\mathbf{r}) \approx \frac{A z_0}{z} \, e^{i(kz-\pi/2)} \, e^{ik(x^2+y^2)/2R(z)} \, e^{-(x^2+y^2)/w^2(z)} \qquad (14.5.31)$$

for $z \gg z_0$. In this limit Eq. (14.5.17) gives

$$R(z) \approx z, \qquad z \gg z_0 \qquad (14.5.32)$$

so that

$$\mathcal{E}(\mathbf{r}) \approx -iAz_0 \left[\frac{1}{z} \, e^{ikz} \, e^{ik(x^2+y^2)/2z} \right] e^{-(x^2+y^2)/w^2(z)} \qquad (14.5.33)$$

The factor in brackets has exactly the form (14.4.10) of a spherical wave with its center of curvature located at the beam waist ($z = 0$) (Figure 14.13). The field

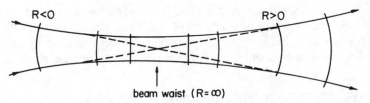

Figure 14.13 Variation of the radius of curvature of a Gaussian beam. At large distances from the beam waist the surfaces of constant phase are spheres centered on the waist.

TABLE 14.1   Gaussian Beam Solution of the Paraxial Wave Equation

$E(\mathbf{r}) = \mathcal{E}(\mathbf{r}) \, e^{-i\omega t}$	(electric field)

$$\mathcal{E}(\mathbf{r}) = A \frac{w_0}{w(z)} \, e^{i[kz - \tan^{-1}(z/z_0)]} \, e^{ik(x^2+y^2)/2R(z)} \, e^{-(x^2+y^2)/w^2(z)}$$

$$I(\mathbf{r}) = \frac{c\epsilon_0}{2} |A|^2 \left(\frac{w_0}{w(z)}\right)^2 e^{-2(x^2+y^2)/w^2(z)} \qquad \text{(intensity)}$$

$$w(z) = w_0 \sqrt{1 + \frac{z^2}{z_0^2}} \qquad \text{(spot size)}$$

$$R(z) = z + \frac{z_0^2}{z} \qquad \text{(radius of curvature)}$$

$$z_0 = \pi w_0^2/\lambda \qquad \text{(Rayleigh range)}$$

$$\theta = \lambda/\pi w_0 \qquad \text{(divergence angle)}$$

(14.5.33) in fact has *exactly* the form of a spherical wave for points close enough to the beam axis that

$$e^{-(x^2+y^2)/w^2(z)} \approx 1 \qquad\qquad (14.5.34)$$

The beamlike fields of the type (14.5.22) are sometimes called *Gaussian spherical waves*.

The properties of the Gaussian beam field (14.5.22) are collected in Table 14.1. These properties are illustrated in Figures 14.11 and 14.13. Note that to the left of the beam waist the radius of curvature is negative, corresponding to concave surfaces of constant phase in the direction of propagation (Figure 14.13). The field (14.5.22), except for the ''strength'' constant A, is completely specified by the wavelength λ and the spot size w_0 at the beam waist. Given λ and w_0, we can determine the spot size and radius of curvature everywhere. A monochromatic Gaussian beam is therefore fully characterized by three parameters: w_0, the location of its waist, and the field amplitude.

Actually there is one restriction on w_0, namely, w_0 must be large compared with the wavelength λ. This restriction is found by requiring the field (14.5.22) to satisfy (14.4.15), which was assumed in the derivation of the paraxial wave equation (14.4.19). In other words, our solution (14.5.22) must be consistent with the approximations used in obtaining it. This restriction implies that the beam divergence angle (14.5.25) must be small, i.e., that our Gaussian beams are paraxial.

14.6   THE *ABCD* LAW FOR GAUSSIAN BEAMS

We have found that a Gaussian beam remains a Gaussian beam as it propagates in vacuum. The beam spot size and radius of curvature change with propagation, but the basic Gaussian spherical wave form (14.5.22) is always maintained. The equations (14.5.17) and (14.5.18) for $R(z)$ and $w(z)$ follow from the simple propagation law (14.5.9) for the q parameter of a Gaussian beam. This propagation law describes the effect on a Gaussian beam of propagation in empty space, i.e., propagation through an "optical element" consisting of a straight section of empty space of length z. If the q parameter has the value q_i in the plane $z = z_i$, then its value in the plane $z = z_f$ is

$$q_f = q_i + d \qquad (14.6.1)$$

where $d = z_f - z_i$. This result, which follows trivially from (14.5.7), is the generalization of (14.5.9) to the case in which z_i is not necessarily zero. We will now consider the effect on a Gaussian beam of other optical elements, such as lenses and spherical mirrors.

Let us consider first the effect of a thin lens of focal length f on a Gaussian beam. Suppose the radius of curvature and spot size immediately before the lens are R_1 and w_1 respectively. Immediately to the right of the lens the spot size should also be w_1, since the lens is not expected to alter the transverse intensity distribution of the beam. The beam curvature, however, will be changed by the lens. We can determine the beam curvature immediately to the right of the lens using the thin-lens equation (Figure 14.14). The spherical wavefront immediately to the left of the lens has the same phase distribution as it would have if there were a point source on the lens axis at a distance $d_0 = R_1$ to the left of the lens. Such a point object will focus to a point image at a distance d_i to the right of the lens. The phase distribution of the beam immediately to the right of the lens should therefore correspond to a spherical wave of radius of curvature R_2 converging to a point at d_i (Figure 14.14). In our sign convention, a spherical wavefront propagating in the positive z direction has a positive curvature if it is convex when viewed from $z = \infty$, and negative curvature if it is concave. This is consistent with our convention that concave spherical mirrors have positive curvature and convex mirrors have negative curvature. If a plane wave is incident on a concave mirror, for example, the reflected wave will have a negative radius of curvature, the same as that of the mirror. Thus R_1 is positive and R_2 negative for the case in

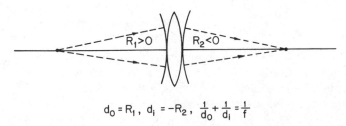

$$d_0 = R_1, \ d_i = -R_2, \ \frac{1}{d_0} + \frac{1}{d_i} = \frac{1}{f}$$

Figure 14.14 The transformation of a diverging spherical wavefront of curvature R_1 to a converging spherical wavefront of curvature R_2 by a thin lens of focal length f.

Figure 14.14, or $d_0 = R_1$, $d_i = -R_2$, with d_0 and d_i both positive for the converging lens shown. The thin-lens equation, i.e.,

$$\frac{1}{d_0} + \frac{1}{d_i} = \frac{1}{f} \tag{14.6.2}$$

then gives

$$\frac{1}{R_2} = \frac{1}{R_1} - \frac{1}{f} \tag{14.6.3}$$

which is the desired relation between the radii of curvature of the spherical wavefront immediately before and after transformation by the lens. Equation (14.6.3) shows that R_2 must be *positive* if $0 < R_1 < f$. In this case the lens is too weak to focus the spherical wave to a point behind the lens. Equation (14.6.3) also applies to the case of a diverging lens, for which $f < 0$.

Using (14.6.3), we may relate the *final* q parameter q_f of a Gaussian spherical wave, just after passage through the lens, to the *inital* value q_i just before the lens. From the definitions

$$\frac{1}{q_i} = \frac{1}{R_1} + \frac{i\lambda}{\pi w_1^2} \tag{14.6.4}$$

and

$$\frac{1}{q_f} = \frac{1}{R_2} + \frac{i\lambda}{\pi w_2^2} \tag{14.6.5}$$

we have, since $w_1 = w_2$,

$$\frac{1}{q_f} = \frac{1}{q_i} - \frac{1}{f} \tag{14.6.6}$$

or

$$q_f = \frac{1}{1/q_i - 1/f} = \frac{q_i}{-q_i/f + 1} \tag{14.6.7}$$

Thus, whereas (14.6.1) gives the transformation of the q parameter of a Gaussian beam arising from free propagation through a distance d, (14.6.7) is the transformation effected by a thin lens of focal length f.

By similar reasoning we can determine the transformation of the q parameter effected by a straight section of propagation of length d, followed by a lens of focal length f. The q transformation due to this "optical system" is found to be

$$q_f = \frac{q_i + d}{-q_i/f + 1 - d/f} \tag{14.6.8}$$

In each of these three examples, the q parameter of a Gaussian beam is transformed according to an equation of the form

$$q_f = \frac{Aq_i + B}{Cq_i + D} \tag{14.6.9}$$

The coefficients A, B, C, and D in the three cases (14.6.1), (14.6.7), and (14.6.8) may be read off from the matrices

$$\begin{bmatrix} A & B \\ C & D \end{bmatrix} = \begin{bmatrix} 1 & d \\ 0 & 1 \end{bmatrix} \tag{14.6.10}$$

$$\begin{bmatrix} A & B \\ C & D \end{bmatrix} = \begin{bmatrix} 1 & 0 \\ -1/f & 1 \end{bmatrix} \tag{14.6.11}$$

and

$$\begin{bmatrix} A & B \\ C & D \end{bmatrix} = \begin{bmatrix} 1 & d \\ -1/f & 1 - d/f \end{bmatrix} \tag{14.6.12}$$

Now we recall from Section 14.2 that (14.6.10) is just the ray matrix associated with a straight section of propagation of length d [Eq. (14.2.8)]. Similarly (14.6.11) is the ray matrix for a thin lens of focal length f [Eq. (14.2.11)]. And finally (14.6.12) is precisely the ray matrix for a straight section of length d followed by a thin lens of focal length f [Eq. (14.2.17)], i.e., it is the matrix (14.6.10) multiplied from the left by the matrix (14.6.11).

These results are special cases of the *ABCD law for Gaussian beams*: The transformation of a Gaussian beam by an optical system may be obtained from the equation (14.6.9), where the coefficients A, B, C, and D are given by the ray matrix of the optical system.

The advantage of the *ABCD* law is that it allows us to evaluate the transformation of a Gaussian beam using the ray matrix of *geometrical* optics. The transformation by an optical system of the Gaussian beam (14.5.22), which is a solution of the paraxial wave equation, may be inferred from the way the system transforms geometrical, paraxial rays. If we know the ray matrix for the optical system, we can predict how it modifies the q parameter, and therefore the spot size and radius of curvature, of a Gaussian beam. This remarkable property of Gaussian beams proves very useful in tracing the behavior of a (Gaussian) laser beam through various optical systems of lenses and mirrors.

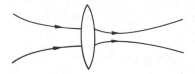

Figure 14.15 A Gaussian beam incident on a lens at its waist is focused to a new waist.

As an example of the application of the *ABCD* law, consider the arrangement shown in Figure 14.15. A Gaussian beam is incident upon a lens located at the waist of the Gaussian beam, where $R = \infty$ and the q parameter is

$$q_i = -\frac{i\pi w_0^2}{\lambda} \tag{14.6.13}$$

with w_0 again the spot size at the waist. The beam passes through the lens, and we wish to determine the distance d behind the lens at which the transmitted beam has its waist. First we calculate the ray matrix for the optical system consisting of a thin lens of focal length f followed by a straight section of propagation of length d:

$$\begin{bmatrix} A & B \\ C & D \end{bmatrix} = \begin{bmatrix} 1 & d \\ 0 & 1 \end{bmatrix} \begin{bmatrix} 1 & 0 \\ -1/f & 1 \end{bmatrix} = \begin{bmatrix} 1 - d/f & d \\ -1/f & 1 \end{bmatrix} \tag{14.6.14}$$

The q parameter of the Gaussian beam is transformed by this optical system to

$$q_f = \left[\frac{1}{R(d)} + \frac{i\lambda}{\pi w^2(d)} \right]^{-1} = \frac{-(i\pi w_0^2/\lambda)(1 - d/f) + d}{(i\pi w_0^2/\lambda f) + 1} \tag{14.6.15}$$

where we have used the *ABCD* law (14.6.9) with the ray matrix (14.6.14) and the initial q parameter (14.6.13). By equating the real and imaginary parts of both sides of (14.6.15), we obtain after some straightforward algebra the expressions

$$R(d) = \frac{(d/z_0)^2 + (1 - d/f)^2}{d/z_0^2 - (1/f)(1 - d/f)} \tag{14.6.16}$$

and

$$w^2(d) = w_0^2 \left(1 - \frac{d}{f}\right)^2 + w_0^2 \left(\frac{d}{z_0}\right)^2 \tag{14.6.17}$$

for the radius of curvature $R(d)$ and the spot size $w(d)$ of the beam at any distance d behind the lens. Here z_0 is the Rayleigh range of the beam incident upon the lens.

The waist of the transmitted beam occurs at the distance d behind the lens for which $R(d) = \infty$. From (14.6.16), therefore, we see that d must satisfy the equation

$$\frac{d}{z_0^2} - \frac{1}{f}\left(1 - \frac{d}{f}\right) = 0 \qquad (14.6.18)$$

so that

$$d = \frac{f}{1 + f^2/z_0^2} \qquad (14.6.19)$$

is the distance behind the lens where the new waist is located. Using this value of d in Eq. (14.6.17), we obtain the spot size w_0' at the new waist:

$$w_0' = \frac{\lambda f}{\pi w_0} \frac{1}{\sqrt{1 + f^2/z_0^2}} \qquad (14.6.20)$$

Equation (14.6.20) is an important result, for it indicates that *a Gaussian beam can be focused to a very small spot.* This property of Gaussian laser beams is important in many applications in which it is necessary to focus radiation onto a small area.

In the following section we will see how to compute the spot size of a laser beam emerging from a given resonator. For a He–Ne laser the spot size w_0 at the waist is typically on the order of 1 mm. From (14.6.20) with $\lambda = 6328$ Å, therefore, we obtain (for $f \ll z_0$)

$$w_0' \approx 2 \times 10^{-4} f \qquad (14.6.21)$$

For lens focal lengths f on the order of centimeters, therefore, we see that the new spot size at the (new) waist is considerably smaller than the unfocused value, $w_0 \sim 1$ mm.

14.7 GAUSSIAN BEAM MODES OF LASER RESONATORS

The mirrors of a laser force radiation to pass through the gain cell repeatedly, thereby enhancing its amplification by stimulated emission. The time taken by light to traverse the distance L (measured in cm) between the mirrors is

$$T = \frac{L}{c} \approx \tfrac{1}{3} \times 10^{-10} L \text{ sec} \qquad (14.7.1)$$

If we measure the transverse intensity profile of the radiation from a laser, we

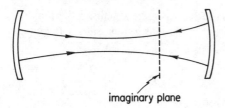

Figure 14.16 Laser resonator with imaginery plane drawn through the optical axis.

normally find a steady profile that does not change in a time corresponding to successive reflections off the mirrors. Such a steady spatial pattern of intensity implies a steady spatial pattern of the field inside the resonator too. This is called a *mode* of the resonator.

To bring out more clearly the idea of a resonator mode, consider a plane normal to the optical axis of the resonator (Figure 14.16). Radiation passing through this plane from the left will propagate to the mirror on the right, be reflected, and then pass through the plane from the right. After reflection from the mirror on the left, it propagates back to our imaginary plane, thus completing a round trip through the resonator. The key point is that, if the field is a mode of the resonator, it must have exactly the same value on the imaginary plane after a round trip as before. And this must be true regardless of where the plane is chosen inside the resonator.

If the Gaussian field of Table 14.1 is to be a mode of a resonator, then evidently its q parameter must not change in a round trip through the resonator. This condition may be examined using the *ABCD* law (14.6.9) for Gaussian beams. The condition for a Gaussian beam mode of a resonator is thus

$$q(z) = \frac{Aq(z) + B}{Cq(z) + D} \qquad (14.7.2)$$

where A, B, C, and D are the elements of the ray matrix for the "optical system" defined by a round trip through the resonator; these matrix elements will generally depend upon z. In order to have a Gaussian beam mode, Eq. (14.7.2) must hold for all values of z between the mirrors.

The condition (14.7.2) for a Gaussian beam mode is general. It can be applied to more complicated resonators than we are considering. For instance, it can be used for the case in which a lens is placed between the mirrors. For our purposes, however, the algebra is simpler if we follow a more intuitive approach. In this approach we consider the propagation of the field from one mirror to the other, as sketched in Figure 14.17.

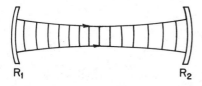

R_1 R_2

Figure 14.17 A Gaussian beam with radii of curvature at the mirrors equal in magnitude to those of the mirrors. A mode of the resonator is formed by the right-going Gaussian beam and the reflected, left-going beam.

If a Gaussian beam is to be a mode of a resonator with spherical mirrors, *its radius of curvature at each mirror must be equal in magnitude to that of the mirror*. If this were not true, the mirror would change the magnitude of the beam radius upon reflection, and we would therefore not have a mode of the resonator. This physically reasonable result may be proved formally using the *ABCD* law for Gaussian beams (Problem 14.4). The radius of curvature of the Gaussian beam at the left-hand mirror of Figure 14.17 is [recall Eq. (14.5.17)]

$$R(z_1) = z_1 + \frac{z_0^2}{z_1} \tag{14.7.3}$$

where z_0 is the Rayleigh range of the beam and z_1 gives the location of the mirror as measured from the beam waist. The beam at z_1 in Figure 14.17 is a converging Gaussian spherical wave, and therefore has a negative radius of curvature according to our sign convention. On the other hand, the (concave) mirror at z_1 has a positive radius of curvature, R_1. Therefore

$$R(z_1) = z_1 + \frac{z_0^2}{z_1} = -R_1 \tag{14.7.4}$$

if the radius of curvature of the Gaussian beam is to have the same magnitude as that of the mirror. Similarly,

$$R(z_2) = z_2 + \frac{z_0^2}{z_2} = R_2 \tag{14.7.5}$$

where R_2 is the radius of curvature of the right-hand mirror of Figure 14.17. The Gaussian beam in this case is diverging and thus has a positive radius of curvature, equal in magnitude and sign to that of the (concave) mirror drawn. It is easily checked that (14.7.4) and (14.7.5) apply regardless of whether the mirrors are concave or convex.

If the mirror separation is L, then we may also write the equation

$$z_2 - z_1 = L \tag{14.7.6}$$

Equations (14.7.4)–(14.7.6) are three equations for the three "unknowns" z_1, z_2, z_0^2. Their solution gives us z_1, z_2, and z_0^2 in terms of the mirror radii (R_1 and R_2) and mirror separation L. After straightforward algebra we obtain

$$z_1 = \frac{-L(R_2 - L)}{R_1 + R_2 - 2L} \tag{14.7.7}$$

$$z_2 = \frac{L(R_1 - L)}{R_1 + R_2 - 2L} \tag{14.7.8}$$

$$z_0^2 = \frac{L(R_1 - L)(R_2 - L)(R_1 + R_2 - L)}{(R_1 + R_2 - 2L)^2} \qquad (14.7.9)$$

In Section 14.4, where we discussed Gaussian beams as solutions of the paraxial wave equation in free space, the location of the beam waist was arbitrary. There was nothing to tell us its location. Equation (14.7.7) [or (14.7.8)], however, locates the beam waist of a Gaussian beam mode with respect to a mirror of the resonator. Similarly, the spot size w_0 at the waist was essentially a free parameter in Section 14.4, whereas (14.7.9) allows us to express w_0 explicitly in terms of the wavelength, the mirror radii of curvature, and their separation. Specifically, we obtain from (14.5.19) and (14.7.9) the minimum beam spot size

$$w_0 = \frac{(\lambda/\pi)^{1/2} \left[L(R_1 - L)(R_2 - L)(R_1 + R_2 - L) \right]^{1/4}}{(R_1 + R_2 - 2L)^{1/2}} \qquad (14.7.10)$$

In Table 14.2 we express z_1, z_2, and w_0 in terms of the resonator g parameters (14.3.10). From these expressions we obtain the beam spot sizes at the mirrors using (14.5.18):

$$w_1 = w_0 \sqrt{1 + \frac{z_1^2}{z_0^2}} \qquad (14.7.11a)$$

and

$$w_2 = w_0 \sqrt{1 + \frac{z_2^2}{z_0^2}} \qquad (14.7.11b)$$

which are given in terms of g_1 and g_2 in Table 14.2. The spot size for $z_1 < z < z_2$ may also be obtained from the propagation law (14.5.18). Similarly the beam radius of curvature for any z may be determined using (14.5.17).

The results listed in Table 14.2 follow from the requirement that the radii of curvature of the Gaussian beam at the mirrors coincide in magnitude with those of the mirrors. The field reflected from the right-hand mirror of Figure 14.17 is also a Gaussian beam whose radii of curvature at the mirrors match (in magnitude) those of the mirrors. The standing wave formed by the left- and right-going Gaussian beams thus form a mode of the resonator: They are not changed on successive reflections, and consequently the standing-wave pattern stays fixed.

The phase of our Gaussian-mode field along the optical axis ($r = 0$) of the resonator is (Table 14.1)

$$\theta(z) = kz - \tan^{-1}(z/z_0) \qquad (14.7.12)$$

(It is convenient to consider the phase change along the optical axis because there the mirror surfaces are separated by exactly L.) The condition for a mode is that

TABLE 14.2 Properties of Lowest-Order Gaussian Modes (TEM$_{00}$) of Stable Resonators

---

Definitions and Conventions

$$g_i = 1 - L/R_i$$

$R_i > 0$ for concave mirrors, $R_i < 0$ for convex mirrors.

Location of Mirrors with Respect to Beam Waist

$$z_1 = \frac{-Lg_2(1 - g_1)}{g_1 + g_2 - 2g_1g_2}, \qquad z_2 = z_1 + L$$

Spot Sizes at Mirrors

$$w_1 = \left(\frac{\lambda L}{\pi}\right)^{1/2} \left(\frac{g_2}{g_1(1 - g_1g_2)}\right)^{1/4}$$

$$w_2 = \left(\frac{\lambda L}{\pi}\right)^{1/2} \left(\frac{g_1}{g_2(1 - g_1g_2)}\right)^{1/4}$$

$$w_0 = \left(\frac{\lambda L}{\pi}\right)^{1/2} \left(\frac{g_1g_2(1 - g_1g_2)}{(g_1 + g_2 - 2g_2)^2}\right)^{1/4}$$

= spot size at beam waist.

Resonance Frequencies

$$\nu_q = \frac{c}{2L}\left(q + \frac{1}{\pi}\cos^{-1}\sqrt{g_1g_2}\right)$$

---

the field does not change in a round trip through the resonator. This means that the phase change of the field in a round trip should be an integral multiple of 2π, or that the one-way phase change is an integral multiple of π:

$$\theta(z_2) - \theta(z_1) = k(z_2 - z_1) - \left[\tan^{-1}\frac{z_2}{z_0} - \tan^{-1}\frac{z_1}{z_0}\right]$$

$$= q\pi, \qquad q = 0, 1, 2, \ldots \tag{14.7.13}$$

This expression gives the allowed values of k, and therefore the resonance frequencies $\nu = kc/2\pi$. Using Eqs. (14.7.6)–(14.7.8), we find after some algebra (Problem 14.5) the allowed mode frequencies

$$\nu_q = \frac{c}{2L}\left(q + \frac{1}{\pi}\cos^{-1}\sqrt{g_1g_2}\right) \tag{14.7.14}$$

where the sign of the square root is understood to be the same as the sign of g_1

(equal to the sign of g_2 for a stable resonator). In labeling the mode frequencies by the integer q we are following a convention of the laser literature. This integer should not be confused with the (complex) q parameter of a Gaussian beam.

A resonator with plane parallel mirrors has $g_1 g_2 = 1$, and so (14.7.14) gives back the result found earlier for this case [recall (11.9.1)]:

$$\nu_q = q \frac{c}{2L} \tag{14.7.15}$$

In the case of spherical but nearly flat mirrors we obtain (Problem 14.5)

$$\nu_q \approx \frac{c}{2L} \left[q + \frac{1}{\pi} \left(\frac{L}{R_1} + \frac{L}{R_2} \right)^{1/2} \right], \quad L \ll R_1, R_2 \tag{14.7.16}$$

In all cases for which our Gaussian mode applies we have the frequency spacing

$$\Delta \nu = \nu_q - \nu_{q-1} = c/2L \tag{14.7.17}$$

between different modes, *exactly as for plane-parallel resonator mirrors*.

We note from Table 14.2 that all the properties of our Gaussian modes—the radius of curvature and spot size as a function of z inside the resonator, the location of the beam waist in terms of the mirror locations, and the allowed mode frequencies—follow from the resonator g parameters, the wavelength λ, and the mirror separation L. Actually the mode with the properties listed in Table 14.2 is a special type of *zero-order* Gaussian mode. We will understand this when we consider higher-order Gaussian modes in the following section. We must first, however, add three caveats.

First we note from Table 14.2 that w_0^4 is negative if

$$g_1 g_2 (1 - g_1 g_2) < 0 \tag{14.7.18}$$

When this occurs we do not have a beamlike solution with a finite cross section. Thus our Gaussian mode can only be valid for values of $g_1 g_2$ not satisfying (14.7.18), i.e., for

$$0 \leq g_1 g_2 \leq 1 \tag{14.7.19}$$

This is precisely the condition (14.3.9) for resonator stability. Therefore *Gaussian beam modes apply only to stable resonators*. In fact our analysis also generally breaks down if either $g_1 g_2 = 1$ or $g_1 g_2 = 0$, for then the spot size becomes infinite on at least one of the mirrors (Table 14.2). The cases $g_1 g_2 = 0$ and $g_1 g_2 = 1$ are the boundaries between stability and instability, the region of *marginal stability*.

The second restriction on the validity of our Gaussian beam modes is that the mirrors must be large enough to intercept the beam without any spillover. Otherwise the beam is not simply reflected at a mirror, and a more complicated, diffraction analysis is required. The transverse dimensions of the mirrors when viewed along the optical axis may be characterized by some effective radius a if the x and y dimensions are not too disparate. In the case of identical, flat, circular mirrors, a will be just the mirror radius. In order for the mirror to reflect a Gaussian beam without appreciable spillover, we require that

$$a \gg w_1, w_2 \tag{14.7.20}$$

where the spot sizes w_1 and w_2 at the mirrors are given in Table 14.2. These expressions for w_1 and w_2 are of the form

$$w_1 = (\lambda L/\pi)^{1/2} F_1(g_1, g_2) \tag{14.7.21a}$$

$$w_2 = (\lambda L/\pi)^{1/2} F_2(g_1, g_2) \tag{14.7.21b}$$

where $F_1(g_1, g_2)$ and $F_2(g_1, g_2)$ are typically of order unity. Thus (14.7.20) requires in this case

$$a \gg (\lambda L/\pi)^{1/2}$$

or

$$N_F = a^2/\lambda L \gg 1 \tag{14.7.22}$$

where N_F is called the *Fresnel number* of the resonator.

The condition (14.7.22) is normally not difficult to meet. Consider, for example, a He–Ne laser wavelength of 6328 Å and a mirror separation L of 50 cm. The condition (14.7.22) in this example becomes

$$a \gg 0.56 \text{ mm} \tag{14.7.23}$$

which will obviously be satisfied for reasonably designed mirrors. For a 10.6-μm CO_2 laser with the same mirror separation we require

$$a \gg 2.3 \text{ mm} \tag{14.7.24}$$

which again is a reasonable condition.

The third restriction on the validity of the Gaussian-mode analysis is that

$$N_F \ll (L/a)^2 \tag{14.7.25}$$

must be satisfied. This condition may be understood from essentially the same condition (14.4.11) for the accurate approximation of the spherical wave (14.4.6) by (14.4.10). In the present context (14.7.25) ensures that the Gaussian beam field (14.5.22) has approximately spherical wave fronts which match the spherical mirror surfaces. The condition (14.7.25) is well satisfied in most practical resonators.

14.8 HIGHER-ORDER GAUSSIAN BEAM MODES

The Gaussian beam of Table 14.1 results from the solution (14.5.22) of the paraxial wave equation (14.4.19). However, this is not the only solution. In this section we will consider a more general type of Gaussian-beam solution of the paraxial wave equation. The Gaussian beam of Table 14.1 will emerge as a special case of this more general solution.

 We arrived at the solution (14.5.22) of the paraxial wave equation by guessing a solution of the form (14.5.2). In attempting to obtain other solutions we will proceed in a similar fashion, assuming a solution of the form

$$\mathcal{E}_0(\mathbf{r}) = Ag\left(\frac{x}{w(z)}\right)h\left(\frac{y}{w(z)}\right)e^{iP(z)}e^{ik(x^2+y^2)/2q(z)} \qquad (14.8.1)$$

We assume $w(z)$ and $q(z)$ are the same as before, i.e., that the spot size and radius of curvature of our more general Gaussian beam are given in Table 14.1. In fact if

$$P(z) = p(z)$$

and

$$g\left(\frac{x}{w(z)}\right) = h\left(\frac{y}{w(z)}\right) = 1$$

then the trial solution (14.8.1) reduces exactly to (14.5.2). The fact that g and h are functions of $x/w(z)$ and $y/w(z)$, respectively, means that the intensity pattern associated with (14.8.1) will scale according to the spot size $w(z)$. This intensity pattern will be a function of $x/w(z)$ and $y/w(z)$, as is the intensity pattern given in Table 14.1. Our task is to find g, h, and P such that (14.8.1) satisfies the paraxial wave equation.

 Using our trial solution (14.8.1) in the paraxial wave equation (14.4.19), we obtain differential equations for g, h, and P. Since the algebra is straightforward but rather tedious, we will omit the details of the derivation and give only the main steps. First we use the fact that g and h are functions of the independent variables

$$\xi = \frac{x}{w(z)} \qquad (14.8.2a)$$

and

$$\eta = \frac{y}{w(z)} \qquad (14.8.2b)$$

respectively, to write

$$\frac{\partial g}{\partial x} = \frac{dg}{d\xi} \frac{\partial \xi}{\partial x} = \frac{1}{w(z)} \frac{dg}{d\xi} \tag{14.8.3a}$$

$$\frac{\partial^2 g}{\partial x^2} = \frac{1}{w^2(z)} \frac{d^2 g}{d\xi^2} \tag{14.8.3b}$$

$$\frac{\partial g}{\partial z} = \frac{dg}{d\xi} \frac{\partial \xi}{\partial z} = -\frac{1}{w^2(z)} \frac{dw}{dz} \frac{dg}{d\xi} \tag{14.8.3c}$$

with analogous results for the partial derivatives of h. We then use these results, together with Eqs. (14.5.11), (14.5.17), and (14.5.18), in the paraxial wave equation (14.4.19). We obtain

$$\frac{1}{g(\xi)} \left(\frac{d^2 g}{d\xi^2} - 4\xi \frac{dg}{d\xi} \right) + \frac{1}{h(\eta)} \left(\frac{d^2 h}{d\eta^2} - 4\eta \frac{dh}{d\eta} \right)$$
$$+ \left(\frac{2ik}{q(z)} - 2k \frac{dP}{dz} \right) w^2(z) = 0 \tag{14.8.4}$$

after division by $g(\xi) h(\eta)$. The functions $g(\xi)$, $h(\eta)$, and $P(z)$ must satisfy this equation in order for (14.8.1) to satisfy the paraxial wave equation.

Now the first term on the left-hand side of (14.8.4) is a function only of the independent variable ξ, the second term is a function only of the independent variable η, and the third term is a function only of the independent variable z. Thus Eq. (14.8.4) cannot hold for all values of the *independent* variables ξ, η, and z unless each of these terms is separately constant. Therefore we write

$$\frac{1}{g(\xi)} \left(\frac{d^2 g}{d\xi^2} - 4\xi \frac{dg}{d\xi} \right) = -a_1 \tag{14.8.5}$$

$$\frac{1}{h(\eta)} \left(\frac{d^2 h}{d\eta^2} - 4\eta \frac{dh}{d\eta} \right) = -a_2 \tag{14.8.6}$$

and

$$\left(\frac{2ik}{q(z)} - 2k \frac{dP}{dz} \right) w^2(z) = a_1 + a_2 \tag{14.8.7}$$

where a_1 and a_2 are constants. Thus we have reduced the problem of solving the partial differential equation (14.4.19) in three independent variables to the problem of solving the three ordinary differential equations (14.8.5)–(14.8.7). This is another example of the method of "separation of variables," which was also used in Chapter 2.

It is convenient to write (14.8.5) in a slightly different form by defining the new variable

$$u = \sqrt{2}\, \xi \tag{14.8.8}$$

Since

$$\frac{dg}{d\xi} = \frac{dg}{du}\frac{du}{d\xi} = \sqrt{2}\,\frac{dg}{du} \tag{14.8.9a}$$

and

$$\frac{d^2 g}{d\xi^2} = 2\frac{d^2 g}{du^2} \tag{14.8.9b}$$

we have

$$\frac{d^2 g}{du^2} - 2u\frac{dg}{du} + \frac{a_1}{2}g = 0 \tag{14.8.10}$$

The reason we have chosen to write (14.8.5) in this form is that the equation (14.8.10) arises in many different problems. It appears, for example, in the quantum mechanics of the harmonic oscillator, as described in Problem 5.8. A solution of (14.8.10) stays finite as $u \to \infty$ only if the constant a_1 satisfies

$$\frac{a_1}{2} = 2m, \qquad m = 0, 1, 2, \ldots \tag{14.8.11}$$

The allowed (finite) solutions of (14.8.10) are the Hermite polynomials, the first few of which are listed in Appendix 5.B. The allowed solutions for the function g in our trial solution (14.8.1) are thus

$$g\left(\frac{x}{w(z)}\right) = H_m\left(\sqrt{2}\,\frac{x}{w(z)}\right), \qquad m = 0, 1, 2, \ldots \tag{14.8.12}$$

In a similar fashion we obtain the allowed solutions

$$h\left(\frac{y}{w(z)}\right) = H_n\left(\sqrt{2}\,\frac{y}{w(z)}\right), \qquad n = 0, 1, 2, \ldots \tag{14.8.13}$$

for the function h.

It remains to determine $P(z)$. Using Eqs. (14.5.11), (14.5.17), and (14.5.18), we obtain from (14.8.7) the differential equation

$$\frac{dP}{dz} = \frac{iz}{z^2 + z_0^2} - \frac{z_0(m + n + 1)}{z^2 + z_0^2} \tag{14.8.14}$$

which may be integrated to give

$$P(z) = i \ln \sqrt{1 + \frac{z^2}{z_0^2}} - (m + n + 1)\, \phi(z) \tag{14.8.15}$$

or

$$e^{iP(z)} = \frac{e^{-i(m+n+1)\,\phi(z)}}{\sqrt{1 + z^2/z_0^2}} = \frac{w_0}{w(z)}\, e^{-i(m+n+1)\,\phi(z)} \tag{14.8.16}$$

where $\phi(z) = \tan^{-1}(z/z_0)$, as in (14.5.21). Collecting the results (14.8.12), (14.8.13), and (14.8.16), we have a solution (14.8.1) to the paraxial wave equation. The electric field is thus

$$\mathcal{E}_{mn}(x, y, z) = \frac{Aw_0}{w(z)} H_m\left(\sqrt{2}\,\frac{x}{w(z)}\right) H_n\left(\sqrt{2}\,\frac{y}{w(z)}\right)$$
$$\times\, e^{i[kz - (m+n+1)\tan^{-1}z/z_0]}$$
$$\times\, e^{ik(x^2+y^2)/2R(z)}\, e^{-(x^2+y^2)/w^2(z)} \tag{14.8.17}$$

Note that when $m = n = 0$ we recover the solution (14.5.22), so our previous Gaussian-beam solution is therefore the "lowest-order" or "zero-order" case of (14.8.17). Another important point is that $R(z)$ and $w(z)$ are independent of m and n; all higher-order Gaussian beams are characterized by the same values of $R(z)$ and $w(z)$ as the lowest-order one. Furthermore, all higher-order Gaussian beams satisfy the same *ABCD* law as the lowest-order one: a Gaussian beam of order (m, n) remains a Gaussian beam of the same order after propagation in free space or transformation by a thin lens or a spherical mirror, but its q parameter is changed according to the *ABCD* law.

Because they have the same spot size and radius of curvature as the lowest-order beam, the higher-order Gaussian beams also form modes of stable resonators satisfying (14.7.20). All the properties of Table 14.2 apply as well to such higher-order modes, *except* for the resonance frequencies. Following exactly the same approach that led to (14.7.14), we obtain for a Gaussian mode of order (m, n) the allowed mode frequencies

$$\nu_{qmn} = \frac{c}{2L}\left(q + \frac{1}{\pi}(m + n + 1)\cos^{-1}\sqrt{g_1 g_2}\right) \tag{14.8.18}$$

with q a positive integer or zero, and the sign convention of (14.7.14).

Gaussian modes characterized by different values of m and n are said to be different *transverse modes*, because their intensity patterns transverse to the optical axis are different. Modes associated with different values of q are said to be different *longitudinal modes*. Thus a given transverse mode (m, n) may be associated with different longitudinal modes (q), and vice versa. A Gaussian mode is specified by the three integers (q, m, n), i.e., by its longitudinal and transverse mode character. The transverse character of a Gaussian mode is conventionally designated TEM_{mn}, meaning "transverse electromagnetic of order (m, n)."

There is a wealth of experimental evidence to corroborate our Gaussian-mode analysis for stable resonators satisfying the conditions (14.7.20) and (14.7.25). One way to record the intensity pattern of a laser beam is to move a photometer (which could be a specially designed "laser power meter") slowly across the pattern, with the photometer output hooked up to a strip-chart recorder. Another way is illustrated in Figure 14.18, while Figure 14.19 shows a "power-in-the-bucket" method of measuring the spot size of a lowest-order (TEM_{00}) Gaussian beam.

Figure 14.20 shows actual intensity patterns recorded with a He–Ne laser operating at 6328 Å. The reader may easily show that these intensity patterns may be associated with the various low-order transverse Gaussian modes indicated (Problem 14.7). It should be noted that the intensity patterns tend to be larger for the higher-order modes. For such modes the condition (14.7.22) may not hold as well as for lower-order modes. Consequently the higher-order modes tend to suffer greater loss due to diffractive spillover at the mirrors.

In general a laser will oscillate on a number of transverse (and longitudinal) modes. In order to achieve oscillation on a single transverse mode, as in the intensity patterns of Figure 14.20, it is necessary to have some sort of *mode discrimination*. That is, it is necessary to have high losses for all transverse modes except one. A laser in which the gain is concentrated near the optical axis, for instance, will tend to oscillate on lower-order modes. In many applications a TEM_{00} Gaussian mode is desired. To discriminate against the higher-order modes in this

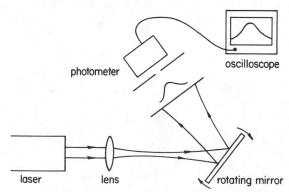

Figure 14.18 One way of recording the intensity pattern of a laser beam. The lens is used to expand the laser beam, which normally has a very small diameter (~ 1 mm). The mirror rotates at an angular velocity of a few revolutions per second.

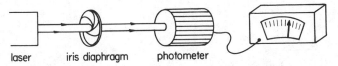

laser   iris diaphragm   photometer

Figure 14.19 Determining the spot size of a TEM_{00} Gaussian beam. The power is first measured with the iris diaphragm fully open. The iris aperture is then reduced until the measured power is 86.5% of the initial value. Then the aperture radius is equal to the spot size. (See Problem 14.6.)

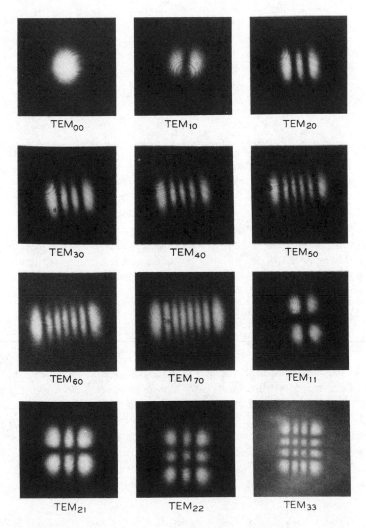

Figure 14.20 Mode patterns obtained with a He–Ne laser [H. Kogelnik and W. W. Rigrod, *Proceedings of the IRE (Correspondence)* **50**, 220 (1962)].

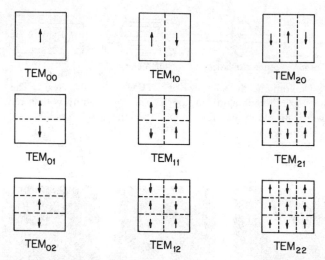

Figure 14.21 Variation of linear polarization across some low-order Gaussian mode patterns.

case, a circular aperture may be inserted into the resonator to produce high losses on all but the lowest-order mode.

The Gaussian modes we have found have rectangular symmetry, and thus would appear to be inapplicable in the case of mirrors with a circular cross section. In practice, however, slight mirror misalignments or other "perturbations" will result in rectangularly symmetric Gaussian modes rather than circularly symmetric ones.[2] One such perturbation is the use of Brewster-angle windows. In this case our scalar electric field may be assumed to be the field component having the favored linear polarization. The directions of the electric field vectors for our Gaussian modes in this case are shown in Figure 14.21.

14.9 RESONATORS FOR He–Ne LASERS

Many factors are involved in the design of laser resonators, including, of course, the intended applications of the laser. Such things as mechanical stability and thermal expansion coefficients of Brewster windows and mirrors must also be considered. A detailed technical discussion of resonator construction is inappropriate here. Instead we will briefly apply some of the results obtained in the preceding sections to the design of resonators for commercially available He–Ne lasers.

He–Ne lasers usually have hemispherical resonators (Figure 14.8). This type of resonator has low sensitivity to mirror misalignments and is easily adjusted.

2. Circularly symmetric modes involve Laguerre rather than Hermite polynomials. In mathematical terms, both the rectangularly symmetric and circularly symmetric Gaussian modes form a complete set of modes, linear combinations of which produce any possible resonator mode.

Unlike the confocal resonator, for example, it is not difficult to obtain TEM_{00} oscillation with a hemispherical resonator. This mode is desirable for applications such as alignment and holography. Higher-order Gaussian modes do not offer the same low beam divergence or the same ability to be focused down to a tiny spot.

For the hemispherical resonator we have $g_1 = 1$, $g_2 = 0$, and $g_1 g_2 = 0$. It is therefore on the border between stability and instability, as indicated by the fact that the spot size w_2 in Table 14.2 is infinite, whereas w_1 vanishes. Actual "hemispherical" resonators are used with the mirror separation L slightly less than the radius of curvature R_2, so that $g_2 > 0$ and $g_1 g_2 > 0$ (Figure 14.22). The resonator is then stable.

Taking $R_1 = \infty$ (flat mirror), $R_2 = R$, and

$$L = R - \Delta L, \qquad R \gg \Delta L > 0, \qquad (14.9.1)$$

we can use the results of Table 14.2 to obtain approximate expressions for the spot sizes in terms of L and ΔL. At the flat mirror we have the spot size (Problem 14.8)

$$w_1 \approx \left(\frac{\lambda^2 L\, \Delta L}{\pi^2} \right)^{1/4} \qquad (14.9.2)$$

and at the spherical mirror

$$w_2 \approx \left(\frac{\lambda^2 L^3}{\pi^2\, \Delta L} \right)^{1/4} \qquad (14.9.3)$$

Thus the Gaussian mode spot size is much larger at the spherical mirror than at the flat mirror, as indicated in Figure 14.22. In fact it follows from Table 14.2 that the beam waist occurs at the flat mirror.

The spot sizes of Gaussian beam modes are usually very small. This is inefficient in the sense that the laser beam intersects only a small fraction of the total volume of the gain medium. Equation (14.9.3) indicates that w_2 can be made large by making ΔL smaller. By using a micrometer adjustment screw to vary ΔL, w_2 can be made to cover a significant portion of the output coupling mirror of Figure 14.22. In this way the size of the output beam can be varied while the laser is on (Problem 14.8). In practice, however, the spot sizes of commercial He–Ne lasers are not adjustable. In fact, many such lasers are of the "hard-seal" type, in which the mirror spacing is permanently fixed (Section 13.6).

Many commercial gas lasers have *collimating mirrors* at the output port. Figure

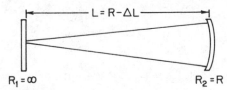

$L = R - \Delta L$

$R_1 = \infty$ $R_2 = R$

Figure 14.22 Quasihemispherical resonator used in many commercial lasers.

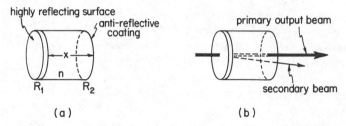

Figure 14.23 (*a*) Design of collimating mirror and (*b*) formation of a secondary beam.

14.23 illustrates a design of such a mirror. The surface with radius of curvature R_1 is highly reflecting, while the bulk of the "mirror" consists of quartz or glass of refractive index n. R_1, R_2, and n are chosen in such a way that the output beam is collimated, i.e., there is a new waist near the out-coupling mirror (Problem 14.9). On the outer surface with radius of curvature R_2 in Figure 14.23 is an antireflective coating to minimize the power in any secondary beam. (For an additional cost of $10, one manufacturer of He–Ne lasers has guaranteed that the secondary beam will be no brighter than $\frac{1}{500}$ of the primary beam.)

Equation (14.5.25) shows that the larger the spot size at the waist of a Gaussian beam, the smaller the beam divergence angle. In many applications it is important to minimize the spot size over large distances of propagation. This can be accomplished by mounting a "beam expander" to the output port of the laser. A beam expander is basically a telescope in reverse. Figure 14.24 shows the basic principle of operation of Galilean and Keplerian beam-expanding telescopes. A typical commercially available beam expander magnifies the beam waist by a factor of ten. Figure 14.25 shows the spot size as a function of distance for a He–Ne laser (λ = 6328 Å) with and without such a beam expander. Figure 14.26 shows a way of measuring the divergence angle of the beam (see Problem 14.11).

The reader interested in the construction details of laser resonators is advised to consult the brochures and handbooks provided by various manufacturers. Such publications often include useful summaries of the properties of Gaussian beam modes. The material in this chapter should provide the reader with sufficient background to read these publications as well as the research literature on resonators.

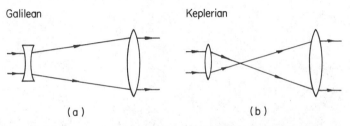

Figure 14.24 Beam expanders of the (*a*) Galilean and (*b*) Keplerian type.

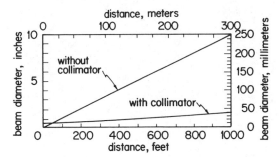

Figure 14.25 Spot size as a function of distance for a He–Ne laser beam with and without a beam collimator. (From Hughes Aircraft Company catalogue.)

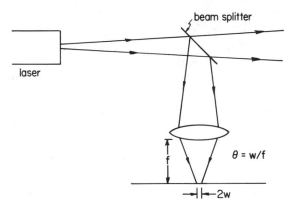

Figure 14.26 Setup for measuring the divergence angle θ of a Gaussian beam.

14.10 DIFFRACTION

We define diffraction, in broad terms, as the bending of light around some obstacle. As such, diffraction is a distinctly wavelike phenomenon, inexplicable from the viewpoint of geometrical rays of light. Some appreciation of diffraction is necessary for a more complete understanding of laser resonators. We will also require some aspects of diffraction theory in the following chapter on the coherence properties of radiation fields.

Let us start by mentioning *Huygens' principle*. Huygens' principle says that *we can imagine every point on a wave front to be a point source for a spherical wave* (Figure 14.27). This way of thinking about waves allows us to make accurate estimates about how they propagate, either in free space or in a medium or around obstacles. The first thing we will do is to express Huygens' principle in quantitative terms, so that we will be able not only to understand why diffraction occurs, but also to calculate diffraction patterns for different obstacles. As in our discussion of Gaussian beams, we will confine ourselves to a scalar-wave approach, ignoring for simplicity the polarization of electromagnetic fields.

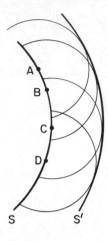

A B C D S S'

Figure 14.27 Huygens' principle says we can imagine each point (such as A, B, C, D) on a wave front S to be a source of spherical waves. The superposition of all these spherical waves gives the wave front (S') elsewhere in space.

Consider the situation indicated in Figure 14.28. We have a monochromatic wave propagating in the z direction, and we are given the field values $\mathcal{E}(x, y, 0)$ in the xy plane. We want to know what the field is on some observation plane a distance z away. In other words, what is $\mathcal{E}(x, y, z)$, given $\mathcal{E}(x, y, 0)$? Huygens' principle tells us how to calculate the field on the observation plane: we do it by regarding every point in the plane $(x, y, z = 0)$ as a point source of spherical waves of the type (14.4.6). Every point $(x', y', 0)$ in the "source plane" acts as a source for a spherical wave that contributes to the point (x, y, z) on the observation plane a field

$$\Delta\mathcal{E}(x, y, z) = -\frac{i}{\lambda}\frac{e^{ikr}}{r}\,\mathcal{E}(x', y', 0)\,\Delta a' \qquad (14.10.1)$$

where

$$r = \left[(x - x')^2 + (y - y')^2 + z^2\right]^{1/2}, \qquad (14.10.2)$$

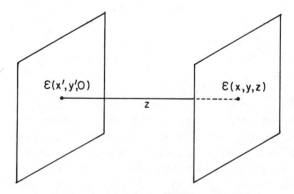

Figure 14.28 Given the field distribution of a monochromatic wave in the plane $z = 0$, what is the field in a plane $z > 0$? We can answer this question using Huygens' principle.

$k = 2\pi/\lambda$, and $\Delta a' = dx'\,dy'$ is a tiny element of area surrounding the point $(x',$ $y', 0)$ on the source plane, where the field has the value $\mathcal{E}(x', y', 0)$. We will not attempt to explain the origin of the factor $-i/\lambda$ in (14.10.1); it arises in a slightly more mathematical version of Huygen's principle. The factor i will not be very important to us anyway, because it will not affect the calculation of measureable intensities. The appearance of the wavelength λ in (14.10.1) ensures that the two sides of the equation are dimensionally consistent.

Following exactly the same approach as that leading from (14.4.6) to (14.4.10), we replace the Huygens spherical wave (14.10.1) by

$$\Delta\mathcal{E}(x, y, z) \approx -\frac{i}{\lambda z} e^{ikz} e^{ik[(x-x')^2 + (y-y')^2]/2z} \mathcal{E}(x', y', 0)\,\Delta a' \quad (14.10.3)$$

The theory of diffraction based on the approximation (14.10.3) is called the Fresnel approximation, or simply *Fresnel diffraction*. Finally we calculate the complete field at the point (x, y, z) on the observation plane by integrating over all the point sources on the source plane, i.e., by summing up the contributions from all the Huygens spherical waves:

$$\mathcal{E}(x, y, z) \approx -\frac{ie^{ikz}}{\lambda z} \int\int \mathcal{E}(x', y', 0)\, e^{ik[(x-x')^2 + (y-y')^2]/2z}\,dx'\,dy' \quad (14.10.4)$$

The exponential in the integrand of (14.10.4) can be rewritten

$$e^{ik(x^2+y^2)/2z}\, e^{ik(x'^2+y'^2)/2z}\, e^{-ik(xx'+yy')/z} \quad (14.10.5)$$

The first factor on the right is independent of the integration variables x', y', and may be pulled outside the integral in (14.10.4). The important special case of *Fraunhofer diffraction* occurs if the inequality

$$z \gg k(x'^2 + y'^2) \quad (14.10.6)$$

holds for all points (x', y') on the source plane aperture. In this case,

$$e^{ik(x'^2+y'^2)/2z} \approx 1$$

and

$$\mathcal{E}(x, y, z) \approx -\frac{ie^{ikz}}{\lambda z} e^{ik(x^2+y^2)/2z} \int\int \mathcal{E}(x', y', 0)\, e^{-ik(xx'+yy')/z}dx'\,dy'$$

$$(14.10.7)$$

The approximate expression given in (14.10.7) can be called the *Fraunhofer dif-*

fraction integral. It is basically the two-dimensional Fourier transform of the aperture distribution $\mathcal{E}(x', y', 0)$.

• The Fresnel approximation retains only the first two terms of the series expansion

$$kr = k[(x - x')^2 + (y - y')^2 + z^2]^{1/2}$$

$$= kz + \frac{k}{2z}[(x - x')^2 + (y - y')^2] - \frac{k}{8z^3}[(x - x')^2 + (y - y')^2]^2 + \cdots$$

$$(14.10.8)$$

in e^{ikr}. As such it is a valid approximation if the third term is small, i.e., if

$$z^3 \gg \frac{\pi}{4\lambda}[(x - x')^2 + (y - y')^2]^2_{\max} \qquad \text{(Fresnel approximation)} \qquad (14.10.9)$$

where $[\cdots]_{\max}$ denotes the largest value of interest of the quantity in brackets. The Fraunhofer approximation, on the other hand, assumes that

$$z \gg \frac{\pi}{\lambda}[x'^2 + y'^2]_{\max} \qquad \text{(Fraunhofer approximation)} \qquad (14.10.10)$$

which is less often satisfied in the sense that larger values of z are required than in the Fresnel approximation. An example of (14.10.10) will be given in the following section. •

14.11   DIFFRACTION BY AN APERTURE

In this section we will use the Fraunhofer diffraction formula to treat the diffraction of light by an aperture. This example is useful for a qualitative understanding of other diffraction problems, as we will see. The Fraunhofer diffraction formula is very important, and it is generally much easier to work with than the Fresnel formula (i.e., the integrals are easier to work out). In the following sections we will relate the Fresnel diffraction formula (14.10.4) to the paraxial wave equation, Gaussian beams, and more general types of laser resonators.

We will consider a circular aperture of diameter $D = 2a$, upon which is incident a uniform monochromatic plane wave. We have already indicated in Figure 14.12 the solution of this diffraction problem, and now we will derive the solution using the Fraunhofer diffraction formula (14.10.7).

Since we are considering a uniform plane wave incident upon the aperture, we have $\mathcal{E}(x', y', 0) = \mathcal{E}_0$ everywhere on the aperture and $\mathcal{E}(x', y', 0) = 0$ everywhere on the opaque screen. That is, the aperture is uniformly illuminated with a field of amplitude $\mathcal{E}_0$, whereas the screen is perfectly absorbing and as such is not a source of any Huygens wavelets. The integral in (14.10.7) therefore becomes an integral over the $x'y'$ coordinates of the aperture alone:

$$\iint \mathcal{E}(x', y', 0) \, e^{-ik(xx'+yy')/z} \, dx' \, dy' = \mathcal{E}_0 \iint e^{-ik(xx'+yy')/z} \, dx' \, dy'$$

$$(14.11.1)$$

Because the aperture is circular it is convenient to use circular coordinates for both the source and observation planes:

$$x = r \cos \theta, \qquad y = r \sin \theta$$
$$x' = r' \cos \theta', \qquad y' = r' \sin \theta'$$
$$(14.11.2a)$$

$$xx' + yy' = rr' \, (\cos \theta \cos \theta' + \sin \theta \sin \theta') \qquad (14.11.2b)$$
$$= rr' \cos (\theta' - \theta)$$

$$dx' \, dy' = r' \, dr' \, d\theta' \qquad \begin{array}{c} (14.11.2c) \\ (14.11.2d) \end{array}$$

In terms of these coordinates we have

$$\iint\limits_{\text{aperture}} e^{-ik(xx'+yy')/z} \, dx' \, dy'$$

$$= \int_0^a r' \, dr' \int_0^{2\pi} e^{-ikrr' \cos (\theta' - \theta)/z} \, d\theta'$$

$$= 2\pi \int_0^a J_0 \left(\frac{krr'}{z} \right) r' \, dr' \qquad (14.11.3)$$

where we have written the integral over θ' as a zeroth-order Bessel function of the first kind, J_0. This θ' integral representation of J_0 is a standard one that may be found in integral tables. Furthermore

$$\int_0^a J_0 \left(\frac{krr'}{z} \right) r' \, dr' = \frac{a^2}{2} \frac{2J_1(kar/z)}{kar/z} \qquad (14.11.4)$$

where J_1 is the first-order Bessel function of the first kind, a graph of which is given in Figure 12.13. In Figure 14.29 we plot the function $2J_1(x)/x$ appearing in Eq. (14.11.4).

Using these results in (14.10.7), we have

$$\mathcal{E}(r, z) = (-ie^{ikz} \, e^{ikr^2/2z}) \frac{\pi a^2}{\lambda z} \mathcal{E}_0 \frac{2J_1(kar/z)}{(kar/z)} \qquad (14.11.5)$$

for the field at the observation plane. Note that $\mathcal{E}$ is independent of θ, compatible

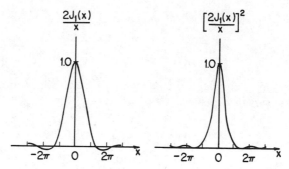

Figure 14.29 The functions $2J_1(x)/x$ and $[2J_1(x)/x]^2$.

with the circular symmetry in our example. The intensity distribution over the observation plane is determined by $|\mathcal{E}|^2$, and is therefore independent of the first factor in (14.11.5), since it has unit modulus. Thus, after writing $2\pi/\lambda$ for k everywhere, the intensity distribution corresponding to (14.11.5) is given by

$$I(r, z) = I_0 \left(\frac{\pi a^2}{\lambda z}\right)^2 \frac{\left[2J_1(2\pi ar/\lambda z)\right]^2}{(2\pi ar/\lambda z)^2} \tag{14.11.6}$$

where I_0 is the intensity of the incident (uniform) plane wave. This is the Fraunhofer diffraction pattern for a circular aperture. Its general properties may be inferred from Figure 14.29.

For a given distance z of the observation plane from the plane containing the aperture, the intensity has its maximum at $r = 0$, i.e., "on axis," where the intensity is

$$I(0, z) = (\pi a^2/\lambda z)^2 I_0 \tag{14.11.7}$$

We may write (14.11.6) as

$$I(r, z) = I(0, z) \frac{\left[2J_1(2\pi ar/\lambda z)\right]^2}{(2\pi ar/\lambda z)^2} \tag{14.11.8}$$

The factor multiplying $I(0, z)$ determines the variation of intensity with r in the observation plane. The intensity distribution (14.11.8) is called the *Airy pattern*, after the British astronomer G. B. Airy,[3] who derived it in 1835.

Away from $r = 0$ in the observation plane the intensity decreases, reaching

---

3. Airy is believed to be the first person (in 1827) to correct astigmatism in the eye (his own) by using cylindrical eyeglass lenses.

zero at the value of r satisfying $J_1(2\pi ar/\lambda z) = 0$. Since $J_1(x) = 0$ at $x \approx 1.22\pi$ (Figure 12.13), there is a zero in the Airy pattern at

$$r_0 \cong 1.22\pi \frac{\lambda z}{2\pi a} = 1.22 \frac{\lambda z}{D} \qquad (14.11.9)$$

As indicated in Figure 14.12, this radius of the central bright spot, or *Airy disk*, increases linearly with z. We may define the divergence angle

$$\theta = \tan^{-1}\frac{r_0}{z} \approx \frac{r_0}{z} = 1.22 \frac{\lambda}{D} \qquad (14.11.10)$$

which is Eq. (14.5.24). The small-angle approximation in (14.11.10) is made because we are considering apertures large compared with a wavelength, i.e., λ/D is small.

As we increase r beyond r_0, the intensity rises above zero again, but it is much smaller than in the central bright region. There is a pattern of wider and successively dimmer concentric rings, as indicated in Figure 14.12. The intensity in the observation plane does not cut off sharply at any radius r, and the light spreads out beyond the geometrical size of the aperture $r = a$. Of course this is just what we mean by "diffraction."

Another feature of the Airy pattern is that the radius r_0 of the central bright spot is directly proportional to the wavelength and inversely proportional to the diameter of the aperture. This feature is characteristic of the diffraction of light by apertures of arbitrary shapes. That is, *the larger the wavelength and the smaller the aperture, the more pronounced will be the diffractive "spreading" of the field.*

• To check the condition (14.10.10) for the validity of the Fraunhofer approximation, consider the following example. Let $\lambda = 6238$ Å and $a = 1$ cm. In this case $(x'^2 + y'^2)_{max} = a^2$ and (14.10.10) gives the condition $z \gg 496$ m for the validity of the Fraunhofer approximation. [The reader may easily convince himself that the condition (14.10.9) for the validity of the Fresnel approximation can be satisfied with much smaller values of z]. This makes Fraunhofer diffraction seem of doubtful relevance in the laboratory. However, it turns out that the use of lenses restores the relevance of Fraunhofer diffraction even for small values of z. Indeed, a whole branch of optics called *Fourier optics* is based essentially on this property of lenses.[4]

It may be shown, for instance, that the image of a point source in the focal plane of a circular lens is not a point, as predicated by geometrical ray optics, but an Airy pattern. Thus the image of a star formed in a telescope is an Airy pattern. If D is the diameter of

---

4. See, for example, J. W. Goodman, *Introduction to Fourier Optics* (McGraw-Hill, New York, 1968). Fourier optics is so named because it deals extensively with Fourier transforms such as, for instance, Eq. (14.10.7).

the lens, and f its focal length, this Airy pattern is given by (14.11.8) with $a = D/2$ and $z = f$. The diameter of the Airy disk is

$$d_0 = 2r_0 = 2.44 \frac{f\lambda}{D} \tag{14.11.11}$$

The fact that the image of a point is an Airy pattern rather than a point is essential to the understanding of *resolution limits* of optical instruments such as the human eye. If two distant objects subtend a very small angle, it is hard to distinguish (or "resolve") them if their diffraction patterns in the focal plane of a lens overlap. A useful measure of resolution is provided by the *Rayleigh criterion*: we say that two points are "just resolved" if the central maximum of the Airy pattern of the image of one coincides with the first zero in the Airy pattern of the other (Figure 14.30). If the point sources were brought closer together we would say (somewhat arbitrarily) that they were no longer resolvable. Thus two point sources are just resolved if their angular separation ($\Delta\theta$) is equal to the angular radius of the Airy disk of the image of either source (Figure 14.30):

$$(\Delta\theta)_{min} = \frac{r_0}{f} = 1.22 \frac{\lambda}{D} \tag{14.11.12}$$

As an example, consider the human eye. Assuming a lens aperture (i.e., pupil) diameter $D = 2.5$ mm, and a mean optical wavelength $\lambda = 5000$ Å, we have a resolution limit of

$$(\Delta\theta)_{min} = 2.7 \times 10^{-4} \text{ rad} \tag{14.11.13}$$

If we are to distinguish the headlights of a car as two separate sources their angular separation subtended at the eye must, according to the Rayleigh criterion, be no smaller than $(\Delta\theta)_{min}$, i.e.,

$$h/d \geq (\Delta\theta)_{min} \tag{14.11.14}$$

where h is the distance between the headlights and d is the distance from the observer to

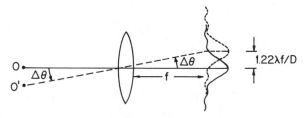

Figure 14.30 According to the Rayleigh criterion, two points O and O' are just resolved in the focal plane of a lens when the first minimum of the image of one coincides with the central maximum of the image of the other.

the car. Thus the two headlights are resolvable if the car is not larger than a distance $d_{max} = h/(\Delta\theta)_{min}$ away. Taking $h = 1.2$ m, this distance is

$$d_{max} = \frac{1.2 \text{ m}}{2.7 \times 10^{-4}} = 4.4 \text{ km} \tag{14.11.15}$$

If the car is farther away, the headlights will be blurred together and not resolvable. (Actually we are ignoring the refractive index of the vitreous humor, and the fact that the pupil diameter of the dark-adapted eye is dilated.)

Equation (14.11.11) for the diameter of the Airy disk associated with a point source is frequently written as

$$d_0 = 2.44\lambda f\# \tag{14.11.16}$$

where the dimensionless number $f\# = f/D$ is called the *f-number*. The intensity of light in the focal plane of the lens may be shown to be inversely proportional to the square of the *f*-number; this is probably intuitively reasonable to anyone who has ever tried to burn things using sunlight and a magnifying glass.

The "stops" marked on cameras reflect this dependence of focal-plane intensity (or, optically speaking, "illumination") on *f*-number. Sequences of numbers like 2, 2.8, 4, 5.6, 8, 11, 16, 22 on the diaphragm (aperture) setting for the camera lens are *f*-numbers such that the square of each *f*# is approximately twice the square of the preceding one. A shift from one *f*# to the next one requires that the exposure time be doubled in order to get the same film exposure ("exposure" equals illumination times exposure time). An $f/2.8$ setting with an exposure time of $\frac{1}{250}$ sec, for instance, will give the same exposure as an $f/4$ setting with a $\frac{1}{125}$-sec exposure time. The minimum *f*# of a given lens is obtained with the diaphragm wide open. The smaller this minimum *f*#, the "faster" the lens is said to be (and usually the more expensive). •

14.12 DIFFRACTION THEORY OF LASER RESONATORS

The Fresnel diffraction formula (14.10.4) is intimately related to our theory of Gaussian beams based on the paraxial wave equation. If $\mathcal{E}_0(x, y, z)$ satisfies (14.4.19), and is specified in the plane $(x, y, z = 0)$, then

$$\mathcal{E}_0(x, y, z) = -\frac{i}{\lambda z} \iint \mathcal{E}_0(x', y', 0)$$
$$\cdot\, e^{ik[(x-x')^2 + (y-y')^2]/2z}\, dx'\, dy', \tag{14.12.1}$$

and therefore [recall Eq. (14.4.13)]

$$\mathcal{E}(x, y, z) = -\frac{ie^{ikz}}{\lambda z} \iint \mathcal{E}(x', y', 0)$$
$$\cdot\, e^{ik[(x-x')^2 + (y-y')^2]/2z}\, dx'\, dy' \tag{14.12.2}$$

which is precisely (14.10.4). In other words, *the Fresnel diffraction formula* (14.10.4) *is the solution of the paraxial wave equation* when the field is specified in a plane $(x, y, z = 0)$ transverse to the propagation direction.

• We will not take the time to derive (14.12.1) from (14.4.19). For the reader familiar with such things, we note that the function

$$K(x, y; x', y'; z) = -\frac{ie^{ikz}}{\lambda z} e^{ik[(x-x')^2 + (y-y')^2]/2z} \qquad (14.12.3)$$

is a Green function, or *propagator*, of the paraxial wave equation.

Our Gaussian beam solutions of the paraxial wave equation therefore satisfy (14.12.2). This is most easily seen in the Fraunhofer limit (14.10.7). In this case $\mathcal{E}(x, y, z)$ and $\mathcal{E}(x', y', 0)$ are related by a Fourier transform, and since the Fourier transform of a Gaussian function is again a Gaussian function, it follows that a Gaussian beam remains a Gaussian beam as it propagates through space. •

Suppose we have a laser resonator and we know the field $\mathcal{E}_0(x, y, 0)$ on the mirror at $z = 0$. Then from (14.12.1) we know that the field at the mirror at $z = L$ is

$$\mathcal{E}_0(x, y, L) = -\frac{i}{\lambda L} \iint \mathcal{E}_0(x', y', 0)$$

$$\cdot\ e^{ik[(x-x')^2 + (y-y')^2]/2L}\, dx'\, dy'$$

$$= \iint K(x, y; x', y')\, \mathcal{E}_0(x', y', 0)\, dx'\, dy' \qquad (14.12.4)$$

Similarly we can use (14.12.4) in (14.12.1) to obtain the field on the first mirror *after one round trip through the resonator*:

$$\tilde{\mathcal{E}}_0(x, y, 0) = \iint K(x, y; x', y')\, \mathcal{E}_0(x', y', L)\, dx'\, dy'$$

$$= \iint K(x, y; x', y')\, dx'\, dy'$$

$$\times \iint K(x', y', x'', y'')\, \mathcal{E}_0(x'', y'', 0)\, dx''\, dy''$$

$$= \iint\left\{\iint K(x, y, x', y')\, K(x', y'; x'', y'')\, dx'\, dy'\right\}$$

$$\times\ \mathcal{E}_0(x'', y'', 0)\, dx''\, dy''$$

$$= \iint \tilde{K}(x, y; x'', y'')\, \mathcal{E}_0(x'', y'', 0)\, dx''\, dy'' \qquad (14.12.5)$$

where

$$\tilde{K}(x, y; x'', y'') = \int\int K(x, y; x', y') \, K(x', y'; x'', y'') \, dx' \, dy' \quad (14.12.6)$$

Continuing in this manner, we obtain the field at each mirror after any arbitrary number of round trips (or "bounces") through the resonator. Now according to our discussion in Section 14.7, a *mode* of the resonator is defined as a field distribution that does not change on successive bounces inside the resonator. Because we are dealing with the modes of an empty resonator with no gain medium, the mode amplitudes will decrease on successive bounces due to diffraction at the mirrors, but the field distribution, or spatial pattern, will not. That is, a mode of the field is such that $\tilde{\mathcal{E}}_0(x, y, 0)$ is simply $\mathcal{E}_0(x, y, 0)$ times some (complex) number:

$$\tilde{\mathcal{E}}_0(x, y) = \gamma \mathcal{E}_0(x, y) \quad (14.12.7a)$$

or

$$\gamma \mathcal{E}_0(x, y) = \int\int \tilde{K}(x, y; x', y') \, \mathcal{E}_0(x', y') \, dx' \, dy' \quad (14.12.7b)$$

where for simplicity we drop the explicit reference to the z dependence.

Since diffractive losses at the mirrors can only diminish the total field energy inside the resonator, we must have $|\gamma| < 1$. In an actual "loaded" laser resonator, stimulated emission in the gain medium will compensate for this diffractive loss and all other losses. Equation (14.12.7) is simply a condition that the empty-cavity field distribution must satisfy if steady-state mode patterns are to exist.

For resonators with Gaussian beam modes, the integral equation (14.12.7) does not give any new results. However, the formulation of the mode problem based on (14.12.7) is useful when the spot sizes w_1, w_2 given in Table 14.2 are comparable to or larger than the mirror radii. As discussed in Section 14.7, our Gaussian-beam-mode analysis in this case is inapplicable. From Table 14.2 it is seen that this happens when $g_1 g_2$ is close to 0 or 1, or is such that the stability condition (14.7.19) is violated.

When the Gaussian-beam-mode analysis breaks down, the resonator modes may be found by solving (14.12.7) numerically. According to the method developed by Fox and Li in 1961, this is usually done by iteration: One starts by *assuming* some field $\mathcal{E}_0(x, y)$ on a mirror, usually just $\mathcal{E}_0(x, y)$ = constant. The field is then "propagated" to the other mirror by doing the integral (14.12.4) numerically. The field obtained on the second mirror is then propagated back to the first mirror by another numerical computation based on (14.12.4). This procedure is then iterated until the field on the mirrors is unchanged (within some prescribed numerical error) on successive iterations, except for a constant multiple γ. The field so

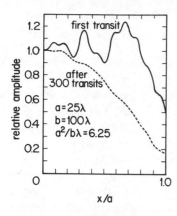

Figure 14.31 Field amplitude distribution computed by Fox and Li for a resonator with flat rectangular mirrors. The mirrors have length 20 in one direction, and are effectively infinite in the other direction. The parameters are such that the Fresnel number $a^2/b\lambda = 6.25$. The iteration process was begun assuming a uniform plane-wave field on one of the mirrors.

obtained is then a solution of (14.12.7), i.e., it is a mode of the resonator. In practice this method will yield straightforwardly only one mode, that of lowest round-trip loss, but certain numerical "tricks" can be employed to obtain higher-loss modes.

Figures 14.31 and 14.32 are reproduced from the original Fox–Li paper. For the example shown, about 300 iterations were necessary for the iterative procedure to converge on a mode of the resonator.

- Equation (14.12.7) may be written in the *operator form*

$$\Gamma \mathcal{E}_0 = \gamma \mathcal{E}_0 \tag{14.12.8}$$

where Γ is the operator corresponding to a round trip through the resonator, i.e., Γ is *defined* by its effect on functions $f(x, y)$:

$$\Gamma f(x, y) \equiv \iint \tilde{K}(x, y; x', y') f(x', y') \, dx' \, dy' \tag{14.12.9}$$

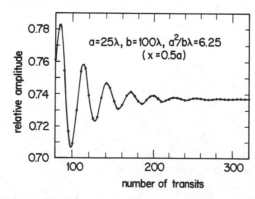

Figure 14.32 Dependence on iteration number of the amplitude at a fixed point on a mirror, as computed by Fox and Li for the case of Figure 14.31.

According to (14.12.8), the modes of a resonator are the eigenfunctions of the operator Γ for the resonator. The number γ is the eigenvalue corresponding to the eigenfunction $\mathcal{E}_0(x, y)$. This sounds a bit like the mathematics of quantum mechanics. Here Γ is generally not a Hermitian operator; for instance, the eigenvalues γ are complex, whereas the eigenvalues of a Hermitian operator are always real. •

14.13   UNSTABLE RESONATORS FOR HIGH-POWER LASERS

Our emphasis on stable laser resonators should not be taken to imply that *unstable* resonators have no practical applications. On the contrary, unstable resonators enjoy certain advantages, and they are essential to the design of many important high-power lasers.

Stable resonators have some drawbacks if one wants to build a high-power device. A major disadvantage is that the modes of stable resonators tend to be concentrated in very thin, needlelike regions within the resonator. Therefore they do not overlap a very large portion of the gain medium, and this obviously presents a problem if high power extraction from the medium is desired. A Gaussian beam mode of a stable resonator, for instance, has a spot size on the order of $(\lambda L / \pi)^{1/2}$ [see Eq. (14.7.21)]. For a CO_2 laser with $\lambda = 10.6 \ \mu$m and $L = 1$ m,

$$(\lambda L / \pi)^{1/2} = 1.8 \text{ mm} \tag{14.13.1}$$

a typical sort of beam "size" for Gaussian beam modes of stable resonators.

Unstable resonators, however, typically have much larger mode volumes, and can therefore make better use of the available gain region. Figure 14.33 shows an important practical example of an unstable resonator, the so-called positive-branch (because $g_1 g_2 > 1$) confocal resonator. As indicated, the intracavity field fills a large portion of the cavity, and can be made larger simply by using larger mirrors. The "magnification" M is a function only of the g parameters of the mirrors.

Iterative computations of the Fox–Li type reveal that the modes of unstable resonators like that shown in Figure 14.33 are distinctly non-Gaussian. To a first approximation the lowest-loss mode has a nearly uniform intensity profile on the mirrors. The output beam for the resonator shown in Figure 14.33 is a collimated annular (doughnut-shaped) beam in the *near field* close to the resonator. In the far

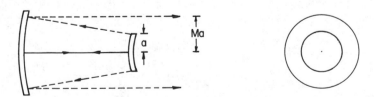

Figure 14.33 A positive-branch ($g_1 g_2 > 1$) confocal unstable resonator. The near-field output is a collimated, annular beam.

field this output beam has a central bright spot on axis. In the limit of large magnification this far field approaches an Airy pattern, with most of the intensity concentrated in the central bright spot.

Unstable resonators offer other advantages in addition to their large mode volumes. For instance, they tend to yield higher output powers when operating on the lowest-loss transverse mode rather than on several (or many) modes. This property is not generally shared by stable-resonator lasers, and it is an important advantage in many applications. In addition, unstable-resonator lasers use *all-reflective optics*. That is, the output does not pass through any mirrors but simply spills around the mirror edges. At high power levels, where mirror damage is an important consideration, the mirrors can often be water-cooled without much difficulty. Obviously the problem of mirror damage and thermal distortion is not so easily surmountable in stable laser resonators employing transmissive output coupling.

The theory of unstable-resonator lasers does not differ in any fundamental way from that of stable-resonator lasers. For this reason, and because stable resonators are more common, we will not consider in any detail the mode characteristics of unstable resonators.

14.14   BESSEL BEAM MODES

In discussing Gaussian beams in Section 14.4 we introduced both the paraxial factorization (14.4.13)

$$\mathcal{E}(\mathbf{r}) = \mathcal{E}_0(\mathbf{r}) \, e^{ikz} \tag{14.14.1}$$

and several approximations [recall Eqs. (14.4.14) and (14.4.15)] based on the presumed slow variation of the plane-wave envelope $\mathcal{E}_0(\mathbf{r})$. These are natural steps to take when dealing with beams that spread very little. Surprisingly, they do not lead to a description of the beams that spread the least of all. There is a set of ideally nonspreading beams that are described by Bessel functions rather than by Gaussian–Hermite or Gaussian–Laguerre functions. Although Bessel beam modes have not played any role in laser development so far, we include a description of them here for completeness.

In contrast to the "weak" paraxial factorization (14.14.1), there is also a "strong" factorization. It is introduced by requiring the envelope function $\mathcal{E}_0(\mathbf{r})$ to be completely independent of z. That is, in place of (14.14.1) we write

$$\mathcal{E}(\mathbf{r}) = \mathcal{E}_0(x, y) \, e^{i\beta z} \tag{14.14.2}$$

where we have indicated explicitly the absence of z dependence in the envelope function. In this case, in place of (14.4.19) we find the equation:

$$[\nabla_T^2 + \beta^2] \, \mathcal{E}_0(x, y) = 0 \tag{14.14.3}$$

where ∇_T^2 is the transverse Laplacian defined in (14.4.20). It is easy to check that Eq. (14.14.3), which is the Helmholtz equation in two dimensions rather than three, is an *exact* consequence of the strong paraxial factorization (14.14.2). There are no leftover terms required to be negligible, as there were in the transition from (14.4.16) to (14.4.18).

The solution to the two-dimensional Helmholtz equation was known to be given in terms of Bessel functions at least 50 years before the time of Helmholtz. The solutions are most conveniently expressed in cylindrical coordinates ρ and ϕ:

$$x = \rho \cos \phi \quad \text{and} \quad y = \rho \sin \phi \qquad (14.14.4)$$

The solution that is nonsingular at the origin is given by

$$\mathcal{E}_0 \rightarrow \mathcal{E}_m(x, y) = A J_m(\alpha\rho) \, e^{im\phi} \qquad (14.14.5)$$

where A is a constant and $J_m(x)$ is the mth Bessel function, the same functions introduced in our discussion of FM mode locking in Section 12.10 and shown in Figure 12.13. In order for (14.14.5) to satisfy the two-dimensional Helmholtz equation, and for (14.14.2) to satisfy the full three-dimensional Helmholtz equation (14.4.3), it is only necessary that α and β be connected by the frequency of the light:

$$\alpha^2 + \beta^2 = (\omega/c)^2 \qquad (14.14.6)$$

It is clear from (14.14.5) that only the lowest-order solution, the one with $m = 0$, is cylindrically symmetric (independent of ϕ). This is analogous to the situation found earlier with Gaussian–Hermite modes in Section 14.8. Inspection of Figure 12.13 shows that the lowest-order mode also gives the most intense beam near the axis (near $\rho = 0$).

The most remarkable feature of the Bessel mode solutions described here is that they are, in the ideal case, *completely nondiffracting*. To show precisely what this statement means, let us compute the intensity of radiation associated with the general Bessel solution (14.14.5). The physical electric field is the real part of $\mathcal{E}(\mathbf{r}) \, e^{-i\omega t}$, which in this case is given by

$$E_m(\mathbf{r}, t) = A J_m(\alpha\rho) \cos (\omega t - kz + m\phi) \qquad (14.14.7)$$

The cycle-averaged power flow of the mth Bessel beam mode is then easily seen to be given by

$$I_m(\mathbf{r}, t) = \frac{c\epsilon_0}{2} A^2 J_m^2(\alpha\rho) \qquad (14.14.8)$$

and this function has the property that it is completely independent of z.

That is, at every value of z (every distance of propagation) the intensity of an ideal Bessel beam has exactly the same x, y dependence. By contrast, Gaussian beams are characterized by their waist function $w(z)$, which grows from a minimum value w_0 in the course of propagation, and it is the growth of the beam waist that determines a Gaussian beam's far-field divergence angle (recall Figure 14.11). By contrast, Bessel beams are characterized by the same transverse distribution of intensity (the same "waist") at every value of z and so do not diverge at all. In this sense they can be said to constitute nondiffracting beams.

One may recall that the intensity distribution of light propagated through a circular aperture, graphed in Figure 14.29, is expressed in (14.11.6) in terms of the Bessel function J_1. The argument of the Bessel function in that application, however, depends explicitly on z and so is quite different from that derived here for a Bessel beam. This comparison of (14.11.6) and (14.14.5) shows that nonspreading Bessel beams will not be created by transmitting a plane wave through a circular aperture. That raises the question how Bessel beams can be realized in practice. One might think that any practical realization would involve apertures that would inevitably prevent the development of the strong paraxial character expressed in (14.14.2), and thus lead back to a Gaussian beam. However, this is not the case, and the remarkable properties of Bessel beams can actually be realized over substantial regions of space [see the cover photo].

To explain one method for creating a Bessel beam in practice we will concentrate on the most important case, the cylindrically symmetric lowest-order Bessel beam. The zero-order Bessel function $J_0(\alpha\rho)$ can be represented by an integral

$$J_0(\alpha\rho) = \frac{1}{2\pi} \int_0^{2\pi} e^{i\alpha(x\cos\phi + y\sin\phi)} \, d\phi \qquad (14.14.9)$$

The full x, y, z dependence of the solution is then obtained by combining (14.14.2) with (14.14.5) for $m = 0$, and using the integral expression for J_0 to get

$$\mathcal{E}_0(\mathbf{r}) = \frac{A}{2\pi} \int_0^{2\pi} e^{i[\alpha(x\cos\phi + y\sin\phi) + \beta z]} \, d\phi \qquad (14.14.10)$$

This can be rewritten and interpreted directly in physical terms by defining a "wave vector" $\mathbf{q}$ whose components are given by

$$\mathbf{q} = (\alpha \cos \phi, \, \alpha \sin \phi, \, \beta) \qquad (14.14.11)$$

Then we have

$$\mathcal{E}_0(\mathbf{r}) = \frac{A}{2\pi} \int_0^{2\pi} e^{i\mathbf{q} \cdot \mathbf{r}} \, d\phi \qquad (14.14.12)$$

The advantage of this form is that it can be interpreted physically. It says that a

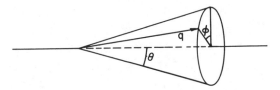

Figure 14.34 The cone of wave vectors making up the Bessel beam defined by (14.14.11) and (14.14.12).

zero-order Bessel mode consists of all possible plane waves with wave vectors $\mathbf{q}$ whose length is restricted by $\alpha^2 + \beta^2 = (\omega/c)^2$, whose polar angle of inclination to the z axis is fixed at the value $\tan\theta = \sqrt{q_x^2 + q_y^2}/q_z = \alpha/\beta$, and whose azimuthal angles ϕ are completely unrestricted. In other words, these are all wave vectors of length $q = \omega/c$ lying on the surface of a cone with opening angle θ, as shown in Figure 14.34.

This mathematical description suggests a method of creating a Bessel beam in practice. A very narrow annular aperture normal to the z axis can be illuminated from one side. If the aperture slit is narrow enough it acts as a circular line source of light. A lens placed with the aperture in its focal plane will then transmit a cone of light just as (14.14.12) requires. In the space beyond the lens the optical field will be given by $J_0(\alpha\rho)$. Of course the inevitably finite size of the circular aperture and of the lens will tend to destroy some of the ideal features of the Bessel beam. In particular, as Figure 14.35 shows, the cone of light will be only finitely wide and its elements will not overlap on the z axis beyond a certain point.

We denote by $Z_{\max}$ the point on the z axis where the geometrical rays shown in the figure no longer overlap. It is found by simple geometrical arguments to be given by

$$Z_{\max} = r/\tan\theta \qquad (14.14.13)$$

where $\tan\theta = \alpha/\beta$ and r is the radius of the lens. However, the Bessel solution is more than just a geometrical ray property of the light field, and experiments have shown that the nondiffracting nature of the Bessel mode persists beyond the point $z = Z_{\max}$, although the intensity is much diminished.

As an example of some of the striking differences between Bessel and Gaussian beams, let us consider the propagation of a very narrow Bessel beam of wavelength $\lambda = 5000$ Å in a region empty of lenses or other media or boundaries. Suppose that at $z = 0$ the aperture (lens diameter in Figure 14.35) has area $\pi r^2 = (5 \text{ mm})^2$ and that the central Bessel spot is rather small with area $\pi a^2 = (50 \ \mu\text{m})^2$. It can easily be determined geometrically that $(Z_{\max})^2 = (\pi r^2/\lambda)(\pi a^2/\lambda)$, or $Z_{\max} = \pi r a/\lambda$, which in this example means $Z_{\max} \approx 1.5$ m. For comparison, a Gaussian beam with its waist in the aperture plane and as small as the central Bessel spot, [namely, a beam area of $(50 \ \mu\text{m})^2$] will have a Rayleigh range [recall Eq. (14.5.19)] given by $z_0 \approx 1.5$ cm, two orders of magnitude smaller than $Z_{\max}$.

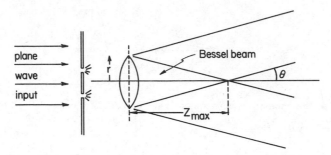

Figure 14.35 Cross section of an arrangement for creating the Bessel beam (14.14.10). Illumination of a narrow annular slit from the left by a plane wave creates a circle of coherent point sources. These give rise to the required cone of rays on the right side of the lens. Here $Z_{max} = r/\tan \theta = r\beta/\alpha$. [From J. Durnin, J. J. Miceli, Jr., and J. H. Eberly, Physical Review Letters **58**, 1499 (1987).]

This illustration is designed to be "favorable" for a Bessel beam, to show that it can propagate without spreading and with significant intensity much farther than a Gaussian beam with an equally small central spot. However, the Gaussian beam would not have to have its waist in the aperture plane, as we assumed here. With a sufficiently large lens at $z = 0$ we could arrange for the Gaussian waist (focal spot) to occur at any distance down the z axis. However, even then, wherever the $(50 \ \mu m)^2$ Gaussian focal spot is placed, it will have a depth of field of only 1.5 cm. The significant difference between Bessel and Gaussian beams is therefore the much greater depth of field of the Bessel beam's "focal region."

Bessel mode beams are of course not ideal in every respect, and the lack of beam divergence is achieved at some cost, both theoretically and practically. In principle the Bessel solution is valid only in infinite transverse x-y space. This is true also of the Gaussian beam solution. However, the Gaussian beam solution falls to zero so rapidly with increasing x and y (away from the z axis) that a Gaussian beam does not "notice" that infinite x-y space is actually not available to it in any real laboratory. This is less true of the Bessel beam, which falls to zero in the transverse directions rather slowly:

$$I_m(\alpha\rho) = \frac{c\epsilon_0}{2} A^2 J_m^2(\alpha\rho)$$

$$\approx \frac{1}{\alpha\rho} \cos^2 (\alpha\rho + \delta) \qquad (14.14.14)$$

for large values of x and y. An important consequence of the cosine-squared character of (14.14.14) is that the energy of a Bessel beam is contained in concentric rings of width given by $\alpha\rho = \pi$, and the intensity is approximately equal in each ring:

$$\int_{\text{one ring}} I_m(\alpha\rho)\, \rho\, d\rho\, d\phi \approx \frac{2\pi}{\alpha} \int \cos^2(\alpha\rho + \delta)\, d\rho = \frac{\pi}{\alpha^2} \quad (14.14.15)$$

where the transverse integration is carried out over one period of J_m. Thus we see that an ideal Bessel-mode beam must carry an infinite amount of energy in its transverse skirt. This is in strong contrast to a Gaussian beam, for which only a very small fraction of energy will be in the skirt outside the central beam spot.

Finally, we should mention that Eq. (14.14.14) explains the apparent conflict between the existence of nondiffracting beams and the familiar Fourier principle (or, in quantum theory, the Heisenberg uncertainty principle), which requires

$$\Delta k_\perp \Delta\rho > (2\pi)^2 \quad (14.14.16)$$

where $\Delta k_\perp$ is the dispersion of the transverse component of the beam wave vector ($\hbar k_\perp$ is the transverse photon momentum in quantum theory) and $\Delta\rho$ is the transverse dispersion in the beam wave solution. As it happens, because of the slow falloff of the intensity of the Bessel beam's skirt, as given in (14.14.14), the dispersion $\Delta\rho$ actually diverges. Because $\Delta\rho$ is infinite, $\Delta k_\perp$ can be zero (no diffraction at all) without violating (14.14.16).

PROBLEMS

14.1 (a) Prove that the ray matrix for the optical system consisting of a straight section followed by a thin lens is different from that for a thin lens followed by a straight section.

  (b) Verify Eq. (14.3.1).

  (c) Using ray-matrix multiplication, show that two thin lenses of focal lengths f_1 and f_2 placed in contact are equivalent to a single thin lens of focal length

$$f = \frac{f_1 f_2}{f_1 + f_2}, \quad \text{or} \quad \frac{1}{f} = \frac{1}{f_1} + \frac{1}{f_2}$$

14.2 Show that the resonators sketched in Figures 14.8 and 14.9 are stable and unstable, respectively.

14.3 (a) Prove that the spherical wave (14.4.6) satisfies the Helmholtz equation (14.4.3).

(b) Verify the condition (14.4.11) for the validity of (14.4.10).

(c) Verify the equation (14.5.26) for the intensity of a Gaussian beam.

(d) Show that the intensity (14.5.26) of a Gaussian beam may be written in the form

$$I(r, z) = \frac{2P}{\pi w^2(z)} e^{-2r^2/w^2(z)}$$

where P is the total beam power, i.e.,

$$P = 2\pi \int_0^\infty I(r, z)\, r\, dr$$

14.4 Show that the magnitude of the radius of curvature of a Gaussian beam is changed upon reflection from a spherical mirror, unless (a) the mirror has infinite radius of curvature (flat mirror), or (b) the radius of curvature of the mirror equals that of the Gaussian beam.

14.5 Verify equations (14.7.14) and (14.7.16) for the resonance frequencies of TEM_{00} Gaussian modes of resonators with nearly flat mirrors.

14.6 Consider the arrangement sketched in Figure 14.19 for the measurement of the spot size of TEM_{00} Gaussian beam. Show that the aperture passes about 86.5% of the total beam power when its radius equals the spot size.

14.7 (a) Show that the intensity patterns of Figure 14.20 may be associated with the transverse Gaussian modes indicated.

(b) Show that the low-order, linearly polarized Gaussian modes have the polarization patterns indicated in Figure 14.21.

14.8 (a) Derive the expressions (14.9.2) and (14.9.3) for the Gaussian-mode spot sizes on the mirror of a quasihemispherical resonator.

(b) A 6328-Å He–Ne laser with a mirror separation of 50 cm has a micrometer adjustment to vary the separation between a flat mirror and an output mirror of radius of curvature $R = 50$ cm. If spot sizes between 1 and 2 mm are desired for the output beam, over what range must the mirror separation vary? Over what range will the spot size at the flat mirror vary?

14.9 Consider the optical element sketched in Figure 14.23. Take the refractive indices inside and outside the element to be n and 1.0, respectively.

(a) Show that the ray matrices associated with the first and second spherical interfaces are

$$\begin{bmatrix} 1 & 0 \\ \dfrac{n-1}{R_1} & \dfrac{1}{n} \end{bmatrix} \quad \text{and} \quad \begin{bmatrix} 1 & 0 \\ \dfrac{1-n}{R_2} & n \end{bmatrix}$$

respectively.

(b) Determine the ray matrix for the optical element in terms of n, R_1, R_2, and the thickness x.

(c) A particular He–Ne laser (Spectra-Physics Model 124) employed a collimating mirror with $R_1 = 2$ m, $R_2 = 64$ cm, and $x = 4$ mm. The waist of the output beam occurred at a distance of 28 cm from the collimating mirror. Determine the refractive index n.

14.10 Figure 14.25 plots the output beam spot size vs. distance for a 6328-Å He–Ne laser, both with and without a beam collimator with a magnification of 10. Derive the equations satisfied by these two performance curves.

14.11 Figure 14.26 shows a possible experimental setup for measuring the divergence angle θ of a Gaussian beam. Show that

$$\theta = w/f,$$

where w is the beam spot size in the focal plane of the lens of focal length f. Does it matter where we intercept the beam from the laser?

15 OPTICAL COHERENCE AND LASERS

15.1 INTRODUCTION

Laser radiation is different from ordinary radiation like that from the sun or a fluorescent lamp. There are the obvious differences, such as the very bright and nearly monochromatic nature of laser light, and its propagation as directed beams, but there are also subtle differences that distinguish laser radiation in other ways. For instance, if the light from the sun were filtered in such a way that only a single, quasimonochromatic and unidirectional component of it remained, it could still be distinguished from laser radiation.

Of course, the obvious differences are very important. In many applications, for instance, it is only the *brightness* of laser radiation that is needed. For this reason we will begin in the following section with a discussion of this aspect of laser radiation. The remainder of the chapter deals with *coherence* properties of radiation. After a careful consideration of the concept of coherence we can begin to appreciate the fundamental differences between lasers and ordinary radiation sources.

15.2 BRIGHTNESS

Consider a blackbody source of radiation. The radiation inside a blackbody cavity has a spectral energy density $\rho(\nu)$ given by (7.7.5). If we divide the frequency band into small finite elements $\delta\nu$, the intensity of radiation "at" frequency ν emitted by a blackbody is $I_\nu = \int_{\delta\nu} I(\nu)\,d\nu \approx I(\nu)\delta\nu$, or

$$I_\nu = \tfrac{1}{4}\,c\rho(\nu)\,\delta\nu$$

$$= \frac{(2\pi h\nu^3/c^2)\,\delta\nu}{e^{h\nu/kT} - 1} \tag{15.2.1}$$

There are two reasons for the factor $\frac{1}{4}$ [recall Section 7.7], both stemming from the fact that an ideal blackbody emits radiation isotropically. First, a factor of $\frac{1}{2}$ arises because there are photons propagating in all directions inside a blackbody cavity, whereas we are only interested in photons propagating *outward* from the blackbody. And another factor of $\frac{1}{2}$ arises because we are interested in photons

533

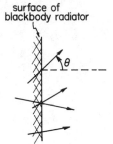

surface of
blackbody radiator

Figure 15.1 Photons emitted by a blackbody radiator have average outward component of velocity equal to the average of $c \cos \theta$ over solid angles of emission.

propagating in a particular direction, whereas the blackbody emits isotropically: the average outward component of photon velocity from the emitting surface is (Figure 15.1)

$$\frac{c \int_0^{2\pi} d\phi \int_0^{\pi/2} \cos \theta \sin \theta \, d\theta}{\int_0^{2\pi} d\phi \int_0^{\pi/2} \sin \theta \, d\theta} = \frac{c}{2} \qquad (15.2.2)$$

As an example of a blackbody radiator, consider the sun. For wavelengths between about 1000 Å and 1 cm, the solar spectrum is approximately that of a blackbody at $T \approx 6000$ K (Section 7.7). At the He–Ne laser wavelength $\lambda = 6328$ Å the intensity given by (15.2.1) for such a blackbody is

$$I_\nu = (1.14 \times 10^{-11} \text{ J/cm}^2) \, \delta\nu \qquad (15.2.3)$$

If we take $\delta\nu = 100$ MHz, then $I_\nu = 1.14$ mW/cm^2.

Now a 6328-Å He–Ne laser might have an output power $P = 1$ mW and a Gaussian-beam spot size $w_0 = 1$ mm. The peak intensity at the waist of the (lowest-order) Gaussian beam in this case is (Problem 14.3) $I_{max} = 2P/\pi w_0^2 = 64$ mW/cm^2. The spectral width of the laser may be larger than the value $\delta\nu = 100$ MHz used in our example above, or it could be much smaller. The point is that even the low output powers of He–Ne lasers give higher radiation intensities, within a narrow spectral range, than the sun at its surface.

The total intensity radiated by a blackbody is obtained by integrating (15.2.1) over all frequencies. The result of the integration is the well-known Stefan law:

$$I_{total} = \int_0^\infty \frac{c}{4} \rho(\nu) \, d\nu = \sigma T^4 \qquad (15.2.4)$$

where $\sigma = 5.67 \times 10^{-5}$ erg-cm^{-2}-deg^{-4}-sec^{-1} is the Stefan–Boltzmann constant. The total intensity of radiation from the sun is given by (15.2.4) with $T \approx 5800$ K (Problem 15.1):

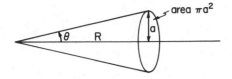

Figure 15.2 The solid angle Ω associated with a divergence angle θ is $\Omega \equiv \pi a^2/R^2 = \pi\theta^2$ if $\theta = \tan^{-1}(a/R) \approx a/R$, i.e., if θ is small.

$$I = 6.4 \times 10^{10} \text{ erg/cm}^2\text{-sec} = 6.4 \text{ kW/cm}^2 \quad (\text{sun}) \quad (15.2.5)$$

This intensity is beyond the range of (unfocused) He–Ne lasers, but it is easily exceeded by higher-power lasers, both cw and pulsed. Furthermore the laser radiation is all concentrated within a narrow frequency range, whereas (15.2.5) is a sum over all frequencies, distributed according to the Planck law.

• One convenient way to measure these differences in brightness between laser radiation and thermal radiation is to recall [Eq. (7.7.20)] that the average number of photons per mode in a thermal (blackbody) field is $(e^{h\nu/kT} - 1)^{-1}$. For the solar temperature $T = 5800$ K, this number (defined as the photon "degeneracy factor" in Section 15.12) is about 0.02. For a laser, on the other hand, this number can be enormously greater, as we will see in Section 15.12. The very first gas laser constructed, for instance, had a photon degeneracy factor of about 10^{12}. •

For many purposes, power and intensity are not adequate measures of "brightness." Instead we define the *brightness* (or *radiance*) of a source as the emitted power per unit area per unit solid angle. This concept of brightness is useful in practical applications, especially when the radiation from a source is to be focused by a lens to increase its intensity. A fundamental theorem in optics states that *the brightness of a source is an invariant quantity*, unchangeable by a lens or any other passive optical system.[1] That is, the intensity of a light beam can be increased by focusing, but the brightness cannot.

An important aspect of brightness is that it is inversely proportional to the solid angle. The solid angle subtended by a beam is proportional to the square of the divergence angle θ (Figure 15.2). For a Gaussian beam, for instance, the solid angle is

$$\Omega = \pi\theta^2 = \lambda^2/\pi w_0^2 \quad (15.2.6)$$

and is thus inversely proportional to the beam area (πw_0^2). Since brightness is power per unit area per unit solid angle, it is clear that the brightness of a Gaussian beam does not change as it propagates. Furthermore, since a Gaussian beam remains Gaussian under focusing by a lens, it is also clear that the brightness of a Gaussian beam cannot be changed by focusing it down to a smaller, more intense spot.

---

1. This is true provided the refractive indices of the object and image spaces are the same—a minor technical point that will not concern us.

Laser beams have a high degree of directionality, i.e., their divergence angles are small. Therefore their solid angles of divergence are also small, and consequently the brightness of a laser beam is very high. To see this, consider again the peak intensity of a (lowest-order) Gaussian beam at its waist (Problem 14.3):

$$I_{max} = 2P/\pi w_0^2 \tag{15.2.7}$$

where P is the total power transported by the beam. From (15.2.6) it follows that the corresponding brightness is

$$B = \frac{I_{max}}{\Omega} = \frac{2P}{\lambda^2} \tag{15.2.8}$$

For a He–Ne laser with $P = 1$ mW and $\lambda = 6328$ Å this brightness is 5×10^5 W/cm^2-sr. This is a modest brightness for lasers. A Q-switched ruby laser, for instance, might have a brightness of 10^{12} W/cm^2-sr, and brightnesses many orders of magnitude larger than this are possible with other lasers. Conventional light sources, even very powerful ones, have much lower brightnesses because their radiation lacks the directionality of laser beams. For example, the sun has a brightness of about 130 W/cm^2-sr at its surface; this is hundreds of times smaller than the brightness of He–Ne lasers, and billions of times smaller than the brightnesses possible with high-power lasers.

In many applications laser radiation is focused to produce an intense spot. Brightness is extremely important in this respect, because *the intensity that can be obtained in the focal plane of a lens is proportional to the brightness of the beam.*

Consider the focusing of a Gaussian beam. Without focusing, the peak intensity at the beam waist is given by (15.2.7). With a lens of focal length f we can focus the beam to a spot size (Problem 15.2)

$$w_f = \lambda f/\pi w_0 \tag{15.2.9}$$

in the focal plane of the lens. The focused beam is still Gaussian, and so its peak intensity is given by (15.2.7) with w_0 replaced by w_f:

$$I_{max}(f) = \frac{2P}{\pi w_f^2} = \frac{2P}{f^2} \frac{\pi w_0^2}{\lambda^2}$$

$$= \frac{2P}{f^2 \Omega} \tag{15.2.10}$$

where Ω [Eq. (15.2.6)] is the solid-angle divergence of the unfocused beam. This result indicates that, for a beam of given power and area, the intensity that can be obtained by focusing is directly proportional to the beam brightness. It shows explicitly why the small divergence (i.e., directionality) of laser beams is so important for obtaining high intensities by beam focusing.

Consider again the example of a He–Ne laser with $P = 1$ mW, $\lambda = 6328$ Å, $w_0 = 1$ mm. From Table 14.1 we compute a divergence angle $\theta = 2 \times 10^{-4}$ rad, and therefore $\Omega = \pi\theta^2 = 1.3 \times 10^{-7}$ sr. Equation (15.2.10) gives $I_{max}(f) = 15$ kW/cm^2 for the peak intensity of a beam focused with a lens of focal length $f = 1$ cm. A laser with the same λ and w_0, but a power of 1 W, gives $I(f) = 15$ MW/cm^2. For reasons discussed below, these estimates are somewhat high, but they serve to indicate the sorts of intensities available with lasers of quite modest output powers. These large intensities are a consequence of the low divergence angles (high brightness) of laser beams.

By contrast, an ordinary lamp emits in all directions. If it delivers a power P over an ideal lens, then the intensity is

$$I \approx P/f^2 \qquad (15.2.11)$$

in the focal plane. This differs from (15.2.10) by the absence of Ω in the denominator, which is very small for a laser beam. Consequently, lamps would have to emit tens or hundreds of thousands of watts to match the intensities achievable by focusing low-power milliwatt He–Ne lasers.

The brightness of laser beams, which makes high intensities achievable by focusing, makes lasers useful in applications such as drilling and welding. It is even easy to vaporize metal surfaces with laser radiation, so that in laser welding special care must be taken to avoid vaporization.

For safety's sake, it must be remembered that the lens of the eye focuses radiation onto the retina, so that direct viewing of even a low-power laser beam, even with "laser goggles," can result in severe retinal damage (Problem 15.3). On the other hand, the use of lasers in *repairing* detached retinas, and in other surgical procedures, has become practically routine.

• Our estimates of focal-spot intensities above apply to the ideal case. We are ignoring lens imperfections (aberrations) and assuming that the laser is oscillating on a single mode, the transverse mode pattern corresponding to a TEM$_{00}$ Gaussian beam. For reasons discussed in Section 15.9, the divergence angles of laser beams can in practice be considerably larger than those in the ideal limit we have assumed. In Table 15.1 we list typical divergence angles of several types of laser. •

TABLE 15.1 Typical Beam Divergence Angles

Laser	Divergence Angle θ (mrad)	Solid Angle Ω (sr)
He–Ne	0.2–1	$(0.1–3) \times 10^{-6}$
CO_2	1–10	$(3–300) \times 10^{-6}$
Ruby	1–10	$(3–300) \times 10^{-6}$
Nd:YAG	1–20	$(3–1300) \times 10^{-6}$
Nd:glass	0.5–10	$(1–300) \times 10^{-6}$

15.3 THE COHERENCE OF LIGHT

The essential features of light coherence are displayed in the Young two-slit interference experiment as shown in Figure 15.3. Light from a source S is incident upon a screen containing two narrow slits, S_1 and S_2. At a second screen, a distance L away from the first screen, we observe the intensity distribution of the light emerging from the two slits. For some sources we see interference fringes on the second screen, i.e., the intensity is not simply the sum of the intensities associated with each slit. In such cases we say that the radiation has a certain measure of *coherence*.

   An elementary approach to the explanation of the interference fringes is to assume both that the source emits monochromatic radiation, and that the slit separation d is much smaller than the screen separation L.

   At a point P on the observation screen the path difference from the two slits is (Figure 15.4)

$$l = d \sin \theta \approx d \frac{Y}{L} \qquad (15.3.1)$$

if the two slits subtend a small angle θ at P. Since $\omega/c = 2\pi/\lambda$, the intensity has its maxima at points P for which this optical path difference is an integral number of wavelengths, i.e., $\omega(l_2 - l_1)/c = n \times 2\pi$. This condition gives

$$Y^{(\text{max})} = n \frac{\lambda L}{d}, \qquad n = 0, 1, 2, \ldots \qquad (15.3.2)$$

for the location of the intensity maxima on the observation screen. There is destructive interference of the light from the two slits if the path difference is an odd

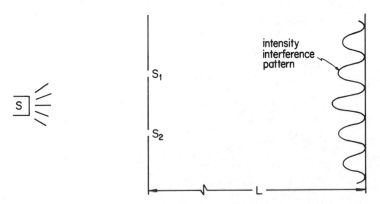

Figure 15.3 The Young two-slit interference experiment.

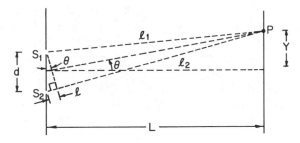

Figure 15.4 If the angle θ subtended by S_1 and S_2 at P is small, the path difference l is approximately dY/L.

integral number of half wavelengths, i.e., $\omega(l_2 - l_1)/c = (n + \frac{1}{2}) \times 2\pi$. At the points where y has one of the values

$$Y^{(\min)} = (n + \tfrac{1}{2}) \frac{\lambda L}{d}, \qquad n = 0, 1, 2, \ldots \qquad (15.3.3)$$

therefore, there are minima in the intensity pattern on the observation screen. For optical wavelengths the separation

$$\Delta = Y^{(\max)} - Y^{(\max)} = \lambda L/d \qquad (15.3.4)$$

between intensity maxima (minima) on the observation screen is quite small even for very closely spaced slits. For example, for $\lambda = 6000 \ \text{Å}$, $d = 0.1$ mm, and $L = 20$ cm, we compute $\Delta = 1.2$ mm.

A real source of radiation will not be perfectly monochromatic. If the source emits radiation with a spread $\delta\lambda$ in wavelengths about λ, interference maxima of one wavelength may coincide with minima of another, causing the interference pattern to be washed out. The difference in fringe separation for two wavelengths separated by $\delta\lambda$ is, from (15.3.4), $\delta\Delta = L\,\delta\lambda/d$. The washing out of the interference pattern when the source is not perfectly monochromatic is therefore minimal if

$$\frac{\delta\Delta}{\Delta} = \frac{L\,\delta\lambda/d}{L\lambda/d} = \frac{\delta\lambda}{\lambda} = \frac{\delta\nu}{\nu} \ll 1 \qquad (15.3.5)$$

Radiation with bandwidth $\delta\nu \ll \nu$ is said to be *quasimonochromatic*.

Suppose the two-slit experiment is performed with an ordinary source of light, such as a mercury arc lamp, or a sodium lamp, or some other nonlaser source of light that may be frequency-filtered in some way to make it quasimonochromatic. We can summarize the results of experiment as follows, without trying to explain them at this point. If the slit separation d is small enough, we observe an intensity interference pattern, and the maxima and minima are accurately located by (15.3.2) and (15.3.3). However, as we increase d we find the interference pattern becoming less sharp, and beyond a certain value of d the interference fringes disappear altogether. A similar effect is observed when we hold d fixed and bring the slits closer to the observation screen. As the interference fringes fade out the intensity at any point P on the observation screen becomes simply the sum of the intensities associated with the two slits individually.

It is convenient to characterize the sharpness of the interference fringes by defining a quantity

$$V = \frac{I_{max} - I_{min}}{I_{max} + I_{min}} \tag{15.3.6}$$

called the *visibility*. I_{max} and I_{min} are respectively the maximum and minimum intensities on the observation screen. If at any point there is complete destructive interference, so that $I_{min} = 0$, it follows from (15.3.6) that the visibility $V = 1$, its maximum possible value. If, on the other hand, there is no interference pattern, in which case the fields from the slits add *incoherently*, then $I_{min} = I_{max}$ and therefore $V = 0$. The visibility thus provides a quantitative measure of the ability of light to produce interference fringes; in other words, it is a measure of the *coherence* of light.

Our example indicates that coherence is not an intrinsic characteristic of a source of light, but depends on the experimental situation (e.g., on the slit separation d, or the distance L). If the light source in our experiment were a laser instead of an ordinary thermal source, we would typically find high fringe visibilities for considerably larger slit separations. We say, somewhat loosely, that the laser radiation is much more coherent than thermal radiation, even though high fringe visibility is also obtainable with a thermal source if d is small enough.

In order to understand the results of the two-slit experiment for different slit separations and different sources, we must next consider more carefully what it is that the experiment actually measures. This will also lead us to a deeper appreciation of the concept of coherence.

15.4 THE MUTUAL COHERENCE FUNCTION

Suppose a quasimonochromatic field is incident upon the screen containing the two slits in Figure 15.4, and that the field is uniform over each slit. Suppose that the width of each slit is a. In order to have any interference pattern on the observation screen, the slits must cause enough diffraction that the field from each slit has a transverse spread much larger than Δ [Eq. (15.3.4)]. Since a diffraction angle $\theta \sim \lambda/a$ is associated with a slit of width a, we require that

$$\theta L = \frac{\lambda L}{a} \gg \Delta = \frac{\lambda L}{d} \tag{15.4.1}$$

or $a \ll d$. Provided the slits are much narrower than their separation, therefore, the fields from the slits will be diffracted enough to produce interference fringes on the screen. We will assume this condition is satisfied.

Since we are now considering a quasimonochromatic field rather than a perfectly monochromatic one, it is no longer appropriate to write

$$E(\mathbf{r}, t) = \mathcal{E}(\mathbf{r}) \, e^{-i\omega t} \qquad (15.4.2)$$

for the complex electric field [cf. (14.4.2)]. If we have a field with N distinct frequency components, we might write instead

$$E(\mathbf{r}, t) = \sum_m \mathcal{E}_m(\mathbf{r}) \, e^{-i\omega_m t} \qquad (15.4.3)$$

In the case $N = 1$, we recover (15.4.2) with $\mathcal{E}(\mathbf{r}) = \mathcal{E}_1(\mathbf{r})$. More generally, we may be dealing with a field having a continuous distribution of frequencies, in which case we replace (15.4.3) by the so-called *analytic signal*[2]

$$E(\mathbf{r}, t) = \int_0^\infty \tilde{\mathcal{E}}(\mathbf{r}, \omega) \, e^{-i\omega t} \, d\omega \qquad (15.4.4)$$

where $\tilde{\mathcal{E}}(r, \omega)$ is the Fourier transform of the real physical field:

$$\tilde{\mathcal{E}}(r, \omega) = \int_{-\infty}^\infty \left[E(\mathbf{r}, t) + E^*(\mathbf{r}, t) \right] e^{+i\omega t} \, dt/2\pi \qquad (15.4.5)$$

In practice, instead of (15.4.3) or (15.4.4) it is often more useful to write

$$E(\mathbf{r}, t) = \mathcal{E}(\mathbf{r}, t) \, e^{-i\omega t} \qquad (15.4.6)$$

where the complex amplitude $\mathcal{E}(\mathbf{r}, t)$ varies very slowly in time. [Recall that such a representation was used in Chapter 8].

All of formulas (15.4.3)–(15.4.6) indicate possible generalizations of (15.4.2) and they all indicate that the complex field E is to be associated with the *positive frequency part* of the real field. This is seen quite clearly in (15.4.4), where the frequency integral is over the range 0 to ∞. [It is unfortunately firmly conventional that the *positive* frequency part of the field is the one that goes with the exponential $e^{-i\omega t}$ having the *negative* sign.]

We will consistently use (15.4.6) to denote a quasimonochromatic field. Given the identity

$$\text{Re}[E(\mathbf{r}, t)] = (\tfrac{1}{2}) \left[E(\mathbf{r}, t) + E^*(\mathbf{r}, t) \right] \qquad (15.4.7)$$

2. In optical coherence theory, $1/2 \, E(\mathbf{r}, t)$ is called the *analytic signal* associated with the field $\text{Re} \, E(\mathbf{r}, t)$. It is usually denoted by $V(\mathbf{r}, t)$.

we can then write the intensity in terms of the positive and negative frequency parts of the field as follows:

$$I(\mathbf{r}, t) = c\epsilon_0 \left[\text{Re } E(\mathbf{r}, t)\right]^2$$

$$= \frac{c\epsilon_0}{4} \left[E^2(\mathbf{r}, t) + E^{*2}(\mathbf{r}, t) + 2E^*(\mathbf{r}, t) E(\mathbf{r}, t)\right] \quad (15.4.8)$$

In the special case of a purely monochromatic field, for instance, $E(\mathbf{r}, t)$ is given by (15.4.2), so that

$$I(\mathbf{r}, t) = \frac{c\epsilon_0}{4} \left[\mathcal{E}^2(\mathbf{r}) e^{-2i\omega t} + \mathcal{E}^{*2}(\mathbf{r}) e^{2i\omega t} + 2\left|\mathcal{E}(\mathbf{r})\right|^2\right] \quad (15.4.9)$$

For optical fields ω is on the order of 10^{15} sec^{-1}, which means that the first two terms in (15.4.9) oscillate sinusoidally with a period far shorter than the resolving time of any available detector. That is, no available detector will be able to follow the rapid oscillations at frequency 2ω in (15.4.9). What is measured is an average of (15.4.9) over many optical periods. Since the first two terms of (15.4.9) average to zero over an optical period, we ignore them and write the measurable intensity as

$$I(\mathbf{r}, t) = \frac{c\epsilon_0}{2} \left|\mathcal{E}(\mathbf{r})\right|^2 \quad (15.4.10)$$

In the quasimonochromatic case, the first two terms in (15.4.9) are replaced by almost purely sinusoidal oscillations which likewise average to zero over any realistic measurement time. In other words, *in general* we can write the measurable intensity as

$$I(\mathbf{r}, t) = \frac{c\epsilon_0}{2} \left|E(\mathbf{r}, t)\right|^2 \quad (15.4.11)$$

Note that the transition from (15.4.9) to (15.4.10) is exactly in the character of the rotating-wave approximation used for atomic wave-function amplitudes in Chapter 6 [recall the transition from (6.3.13) to (6.3.14)].

In the two-slit interference experiment, the field $E(P, t)$ at the point P on the observation screen is the sum of the fields diffracted by the slits. The field $E(S_1, t)$ at slit S_1 will give rise to the field

$$E(P, t) = K_1 E\left(S_1, t - \frac{l_1}{c}\right) \quad (15.4.12)$$

at the point P. The *retardation time* l_1/c is just the time it takes for light to propagate from S_1 to P (Figure 15.4). K_1 is a function of the distance l_1 and of other geometrical details of the particular experimental arrangement. It may be derived from diffraction theory, but for our purposes its precise form is unnecessary; it will be convenient, however, to know that K_1 (and K_2 below) is a pure dimensionless imaginary number, i.e., $K_1 = -K_1^*$. This property of K_1 and K_2 is related to the factor of i used in our statement of Huygens' principle, Eq. (14.10.1). In any case (15.4.12) has an intuitively reasonable form. It says simply that the field $E_1(P, t)$ at time t is due to diffraction of the field that was incident on S_1 at the earlier time $t - l_1/c$.

The field $E(P, t)$ at P due to both slits is then (recall the simplified discussion in Section 15.3):

$$
\begin{aligned}
E(P, t) &= E_1(P, t) + E_2(P, t) \\
&= K_1 E\left(S_1, t - \frac{l_1}{c}\right) + K_2 E\left(S_2, t - \frac{l_2}{c}\right)
\end{aligned}
\tag{15.4.13}
$$

Using this result in (15.4.11), we have

$$
\begin{aligned}
I(P, t) = &\frac{c\epsilon_0}{2}\left|K_1 E\left(S_1, t - \frac{l_1}{c}\right)\right|^2 + \frac{c\epsilon_0}{2}\left|K_2 E\left(S_2, t - \frac{l_2}{c}\right)\right|^2 \\
&+ c\epsilon_0 \mid K_1 K_2 \mid \operatorname{Re}\left[E^*\left(S_1, t - \frac{l_1}{c}\right) E\left(S_2, t - \frac{l_2}{c}\right)\right]
\end{aligned}
\tag{15.4.14}
$$

for the intensity at point P on the observation screen, averaged over a few optical periods.

In addition to the fluctuations of intensity due to the regular but very rapid time variation of the factors $e^{\pm 2i\omega t}$, every light field is subject to small irregular fluctuations arising from a variety of causes. One fundamental source of such fluctuations is the (necessarily random) spontaneous emission component of every light beam. Other less fundamental but generally much more important sources include fluctuations in the atmosphere and mechanical vibrations of optical elements in the path of the beam.

To account for the influence of these fluctuations in a detailed way would be impossible. Fortunately it is satisfactory to assume that these fluctuations can be treated in an average sense. Recall that we adopted such a point of view in treating the effect of collisions on a Lorentz oscillator [Section 3.9] when we assumed that, on average, an oscillator has zero displacement and velocity immediately after every collision. On this basis we developed equations of motion for the average oscillator rather than trying to account for the details of the collisional history of an individual atom.

For the same reasons we will now assume that the average light field is representative of the collection of all possible light fields compatible with the fluctuations mentioned. This collection of light fields can be termed a statistical *ensemble* of light fields, and we will associate observable properties of light fields with averages over this collection, so-called ensemble averages. We will denote an ensemble average by angular brackets, so the average intensity will be written $\langle I \rangle$. Thus, following (15.4.11), we have

$$\langle I(P, t) \rangle = \frac{c\epsilon_0}{2} \langle E(P, t) E^*(P, t) \rangle \qquad (15.4.15)$$

and so on. Note that the ensemble average is not the same as a time average. In particular, the ensemble average may be time-dependent, and a time average by definition could not be. Of course there can arise situations in which a light signal is detected (by counting photons, say) over a long time period $2T$, not instantaneously at time t. In this case the time-dependent ensemble average must be further time-averaged in order to correspond to the measuring process and we can add a bar above the symbol to denote this:

$$\langle \bar{I}(P, t) \rangle = \frac{1}{2T} \int_{-T}^{+T} \langle I(P, t + t') \rangle \, dt' \qquad (15.4.16)$$

Upon averaging both sides of (15.4.14) over the field fluctuations we have

$$\begin{aligned}
\langle I(P, t) \rangle = &\frac{c\epsilon_0}{2} |K_1|^2 \left\langle \left| E\left(S_1, t - \frac{l_1}{c}\right) \right|^2 \right\rangle \\
&+ \frac{c\epsilon_0}{2} |K_2|^2 \left\langle \left| E\left(S_2, t - \frac{l_2}{c}\right) \right|^2 \right\rangle \\
&+ c\epsilon_0 |K_1 K_2| \, \text{Re} \left\langle E^*\left(S_1, t - \frac{l_1}{c}\right) E\left(S_2, t - \frac{l_2}{c}\right) \right\rangle \qquad (15.4.17)
\end{aligned}$$

The function

$$\Gamma(\mathbf{r}_1, t_1; \mathbf{r}_2, t_2) = \langle E^*(\mathbf{r}_1, t_1) E(\mathbf{r}_2, t_2) \rangle \qquad (15.4.18)$$

appearing in (15.4.17) is called the *mutual coherence function* of the fields at $\mathbf{r}_1$, t_1 and $\mathbf{r}_2$, t_2. It is also called the two-point function or auto-correlation function of the electric field. In terms of the mutual coherence function, we may write (15.4.17) as

$$\begin{aligned}
\langle I(P, t) \rangle = &\langle I_1(P, t) \rangle + \langle I_2(P, t) \rangle \\
&+ c\epsilon_0 |K_1 K_2| \, \text{Re} \, \Gamma\left(S_1, t - \frac{l_1}{c}; S_2, t - \frac{l_2}{c}\right) \qquad (15.4.19)
\end{aligned}$$

where

$$\langle I_i(P, t) \rangle = \frac{c\epsilon_0}{2} |K_i|^2 \left\langle \left| E\left(S_i, t - \frac{l_i}{c}\right) \right|^2 \right\rangle$$

$$= |K_i|^2 \left\langle I\left(S_i, t - \frac{l_i}{c}\right) \right\rangle, \quad i = 1, 2 \qquad (15.4.20)$$

is the intensity that would be measured at P if slit S_i were acting alone, i.e., if the other slit were closed. The intensity (15.4.19) is not just the sum of the intensities I_1 and I_2 associated with each slit alone, unless the mutual coherence function vanishes. We see, therefore, that the mutual coherence function is intimately con- nected with the ability of the fields to produce interference fringes, i.e., to act coherently.

The definition of the mutual coherence function is the principal result of this section. In the following sections we will use this important quantity to discuss the concepts of spatial coherence and temporal coherence.

15.5 STATIONARY FIELDS

We will consider sources of radiation that have reached a more or less "steady- state" operation after they have been turned on. The radiation from such sources may be assumed in most instances to have a property called *stationarity*.

A stationary field has the property that the mutual coherence function $\Gamma(\mathbf{r}_1, t_1; \mathbf{r}_2, t_2)$ depends on t_1 and t_2 only through the difference $t_2 - t_1$. This is sometimes expressed by saying that the mutual coherence function for a stationary field is independent of the origin of time. For instance, for a stationary field, we have for any value s

$$\Gamma(\mathbf{r}_1, t_1; \mathbf{r}_2, t_2) = \Gamma(\mathbf{r}_1, t_1 + s; \mathbf{r}_2, t_2 + s) \qquad (15.5.1)$$

and so we can shorten the notation and write

$$\Gamma(\mathbf{r}_1, t_1; \mathbf{r}_2, t_2) = \Gamma(\mathbf{r}_1, \mathbf{r}_2, t_2 - t_1) = \Gamma(\mathbf{r}_1, \mathbf{r}_2, \tau) \qquad (15.5.2a)$$

for stationary fields, where

$$\tau = t_2 - t_1 \qquad (15.5.2b)$$

Note that, since

$$\langle I(P, t) \rangle = \frac{c\epsilon_0}{2} \langle E^*(P, t) E(P, t) \rangle = \frac{c\epsilon_0}{2} \Gamma(P, t; P, t) \qquad (15.5.3)$$

the time independence of the measured intensity of a stationary field follows from the property (15.5.1) of the mutual coherence function, i.e.,

$$\langle I(P, t)\rangle = \langle I(P)\rangle = \frac{c\epsilon_0}{2}\Gamma(P, P, 0) \qquad (15.5.4)$$

in the notation (15.5.2). Equation (15.5.1) may therefore be considered as the defining characteristic of a stationary field.

In many cases it is useful to have a specific model of Γ that exhibits the commonly observed property that Γ decreases as the separation $|\tau| = |t_2 - t_1|$ increases. One such model for a quasimonochromatic laser field with frequency ω_L is given by

$$\Gamma = \langle \mathcal{E}_0^*(\mathbf{r}_1)\, \mathcal{E}_0(\mathbf{r}_2)\rangle\, e^{-|\tau/\tau_c|} e^{-i\omega_L t} \qquad (15.5.5)$$

where τ_c is called the field's correlation time. Obviously the field is poorly correlated with itself (Γ is small) over time displacements greater than $|t_2 - t_1| \approx \tau_c$. According to the Wiener–Khintchine theorem, the field's spectrum $S(\omega)$ can be defined as the Fourier transform of its autocorrelation function, so that in this case we can easily determine

$$S(\omega) = 2\pi\, \langle \mathcal{E}_0^*(\mathbf{r}_1)\, \mathcal{E}_0(\mathbf{r}_2)\rangle\, \frac{1/\pi\tau_c}{\left(\omega - \omega_L\right)^2 + \left(1/\tau_c\right)^2} \qquad (15.5.6)$$

The laser spectrum is thus predicted to be a smoothly peaked function centered at ω_L, with a halfwidth given by $1/\tau_c$. The spectral lineshape of most lasers is not Lorentzian, but the other features of this model are satisfactory, especially the identification of the spectral linewidth with the inverse coherence time of the light field [recall Section 1.1].

A trivial example of a stationary field is a perfectly monochromatic field. To see this, use (15.4.2) in (15.4.18):

$$\Gamma(\mathbf{r}_1, t_1; \mathbf{r}_2, t_2) = \langle E^*(\mathbf{r}_1, t_1)\, E(\mathbf{r}_2, t_2)\rangle$$

$$= \langle \mathcal{E}^*(\mathbf{r}_1)\, \mathcal{E}(\mathbf{r}_2)\, e^{-i\omega(t_2 - t_1)}\rangle$$

$$= \langle \mathcal{E}^*(\mathbf{r}_1)\, \mathcal{E}(\mathbf{r}_2)\rangle\, e^{-i\omega\tau} \qquad (15.5.7)$$

Thus the mutual coherence function depends on t_1 and t_2 only through the difference $t_2 - t_1 = \tau$, and so a monochromatic field is stationary.

In the two-slit experiment the intensity (15.4.19) becomes

$$\langle I(P, t)\rangle = I_1 + I_2 + c\epsilon_0|K_1 K_2|\, \mathrm{Re}\, \Gamma\!\left(S_1, S_2, \frac{l}{c}\right), \qquad l = l_2 - l_1 \qquad (15.5.8)$$

when a stationary field is incident on the slits. Here

$$I_i = \langle I_i(P) \rangle = |K_i|^2 \langle I(S_i) \rangle \tag{15.5.9}$$

is the intensity associated with slit S_i alone, and is independent of time because the source is stationary. The intensity (15.5.8) is also time-independent.

In the case of a monochromatic field the mutual coherence function is given by (15.5.7), and so

$$\langle I(P) \rangle = I_1 + I_2 + c\epsilon_0 |K_1 K_2| \, \text{Re} \left[\mathcal{E}^*(S_1) \, \mathcal{E}(S_2) \, e^{-i\omega l/c} \right] \tag{15.5.10}$$

Suppose that $\mathcal{E}(S_1) = \mathcal{E}_0$ and $\mathcal{E}(S_2) = \mathcal{E}_0 e^{-i\Phi}$. This means that the fields at the slits differ only by a constant phase term. In this case

$$c\epsilon_0 |K_1 K_2| \, \text{Re} \left[\mathcal{E}^*(S_1) \, \mathcal{E}(S_2) \, e^{-i\omega l/c} \right]$$

$$= c\epsilon_0 |K_1 K_2 \mathcal{E}_0^2| \cos \left(\frac{\omega l}{c} + \Phi \right)$$

$$= 2 \left(\frac{c\epsilon_0}{2} |K_1 \mathcal{E}_0|^2 \frac{c\epsilon_0}{2} |K_2 \mathcal{E}_0|^2 \right)^{1/2} \cos \left(\frac{2\pi l}{\lambda} + \Phi \right)$$

$$= 2 \left(I_1 I_2 \right)^{1/2} \cos \left(\frac{2\pi l}{\lambda} + \Phi \right) \tag{15.5.11}$$

and therefore

$$\langle I(P) \rangle = I_1 + I_2 + 2 (I_1 I_2)^{1/2} \cos \left(\frac{2\pi l}{\lambda} + \Phi \right) \tag{15.5.12}$$

If $\Phi = 0$, there is constructive interference at point P if $\cos(2\pi l/\lambda) = 1$, i.e., if the path difference l in Figure 15.4 is an integral number of wavelengths. Similarly, there is destructive interference at points on the observation screen where l is an odd integral number of half wavelengths. We have merely justified our assumptions leading to (15.3.2) and (15.3.3).

If $\Phi \neq 0$, (15.5.12) implies there is constructive interference at points P where $2\pi l/\lambda + \Phi = 2\pi n$, or

$$l = n\lambda + \lambda(\Phi/2\pi), \quad n = 1, 2, \ldots \tag{15.5.13}$$

If the two slits subtend a small angle at P, then (15.5.13) leads to

$$Y_n^{(\text{max})} = n \frac{\lambda L}{d} + \frac{\Phi}{2\pi} \frac{\lambda L}{d}, \quad n = 0, 1, 2, \ldots \tag{15.5.14a}$$

the generalization of (15.3.2) when $\Phi \neq 0$. Similarly we obtain

$$Y_n^{(\min)} = \left(n + \tfrac{1}{2}\right)\frac{\lambda L}{d} + \frac{\Phi}{2\pi}\frac{\lambda L}{d}, \qquad n = 0, 1, 2, \ldots \quad (15.5.14b)$$

in place of (15.3.3). Equations (15.5.14) indicate that a phase difference Φ has the effect of shifting the positions of the intensity maxima and minima by the amount $\Phi \lambda L / 2\pi d$. The overall interference pattern, though shifted upwards or downwards (depending on the sign of Φ), is otherwise basically unchanged. In particular, the separation Δ between intensity maxima (and minima) is the same as the value (15.3.3) for the case $\Phi = 0$.

15.6   FRINGE VISIBILITY

The intensity on the observation screen in the two-slit experiment is given by (15.5.8) whenever the field is stationary. It is convenient to write this as

$$\langle I(P)\rangle = I_1 + I_2 + 2\sqrt{I_1 I_2}\ \mathrm{Re}\ \gamma(S_1, S_2, l/c) \qquad (15.6.1)$$

where the dimensionless complex number γ is given by

$$
\begin{aligned}
\gamma(S_1, S_2, l/c) &= \frac{(c\epsilon_0/2)\,|K_1 K_2|\,\Gamma(S_1, S_2, l/c)}{\sqrt{I_1 I_2}} \\[2mm]
&= \frac{(c\epsilon_0/2)\,|K_1 K_2|\,\Gamma(S_1, S_2, l/c)}{\sqrt{|K_1|^2\,\langle I(S_1)\rangle \cdot |K_2|^2\,\langle I(S_2)\rangle}} \\[2mm]
&= \frac{(c\epsilon_0/2)\,\Gamma(S_1, S_2, l/c)}{\sqrt{\langle I(S_1)\rangle\,\langle I(S_2)\rangle}}
\end{aligned}
\qquad (15.6.2)
$$

or in general

$$\gamma(\mathbf{r}_1, \mathbf{r}_2, \tau) = \frac{\Gamma(\mathbf{r}_1, \mathbf{r}_2, \tau)}{\sqrt{\Gamma(\mathbf{r}_1, \mathbf{r}_1, 0)\,\Gamma(\mathbf{r}_2, \mathbf{r}_2, 0)}} \qquad (15.6.3)$$

$\gamma(\mathbf{r}_1, \mathbf{r}_2, \tau)$ is called the *complex degree of coherence*. As we will now show, it is intimately related to the fringe visibility V defined by Eq. (15.3.6).

In the case of a purely monochromatic field, where $\Gamma(\mathbf{r}_1, \mathbf{r}_2, \tau)$ is given by (15.5.7), the complex degree of coherence is simply

$$\gamma(\mathbf{r}_1, \mathbf{r}_2, \tau) = e^{-i\omega\tau} \qquad (15.6.4)$$

For quasimonochromatic light we may assume that

$$\gamma(\mathbf{r}_1, \mathbf{r}_2, \tau) = \left|\gamma(\mathbf{r}_1, \mathbf{r}_2, \tau)\right| e^{-i\omega\tau} \qquad (15.6.5)$$

where ω is the central frequency and $|\gamma(\mathbf{r}_1, \mathbf{r}_2, \tau)|$ is a slowly varying function of τ compared with $e^{-i\omega\tau}$. In this case (15.6.1) becomes

$$\langle I(P)\rangle = I_1 + I_2 + 2\sqrt{I_1 I_2}\left|\gamma\left(S_1, S_2, \frac{l}{c}\right)\right| \cos\frac{2\pi l}{\lambda} \qquad (15.6.6)$$

Consider a region around P much larger than a wavelength. Over this region the factor $\cos(2\pi l/\lambda)$ varies rapidly between -1 and $+1$ as l is varied, whereas the (slowly varying) factor $|\gamma(S_1, S_2, l/c)|$ is practically unchanged. Thus the maximum and minimum intensities in the neighborhood of P are

$$I_{max} = I_1 + I_2 + 2\sqrt{I_1 I_2}\,|\gamma| \qquad (15.6.7a)$$

and

$$I_{min} = I_1 + I_2 - 2\sqrt{I_1 I_2}\,|\gamma| \qquad (15.6.7b)$$

and so the fringe visibility (15.3.6) is

$$V = \frac{2\sqrt{I_1 I_2}\,|\gamma|}{I_1 + I_2} \qquad (15.6.8)$$

The modulus $|\gamma|$ of the complex degree of coherence is thus a direct measure of the fringe visibility.

In the special case $I_1 = I_2$, the visibility and the modulus of the complex degree of coherence are identical:

$$V = |\gamma| \qquad (15.6.9)$$

From the definition of the visibility, it is clear that $0 \le V \le 1$ in general, and therefore also[3]

$$0 \le |\gamma| \le 1 \qquad (15.6.10)$$

When $|\gamma| = 1$ or 0 we have complete coherence or incoherence, respectively. When $0 < |\gamma| < 1$ the light is said to be *partially coherent*.

---

3. This general property of γ may be derived from the definition (15.6.2), using the Schwarz inequality.

It is important to note that for a monochromatic field

$$|\gamma| = 1 \qquad \text{(monochromatic field)} \qquad (15.6.11)$$

which follows trivially from (15.6.4). Therefore the idealized monochromatic field always gives the maximum possible fringe visibility. However, nonmonochromatic radiation can also satisfy the condition $|\gamma| = 1$ for complete coherence.

The mutual coherence function $\Gamma(\mathbf{r}_1, \mathbf{r}_2, \tau)$ is the average of the product of $E^*(\mathbf{r}_1, t)$ and $E(\mathbf{r}_2, t + \tau)$. During the measurement time over which the average is taken, the fields at $\mathbf{r}_1, t$ and $\mathbf{r}_2, t + \tau$ may undergo rapid fluctuations. This is to be expected, because the fields are due to a large number of individual radiators (atoms and molecules) that themselves fluctuate due to collisions, thermal motion, etc. The functions Γ and γ characterize the degree to which the fields at $\mathbf{r}_1, t$ and $\mathbf{r}_2, t + \tau$ are correlated, or able to produce interference fringes, in spite of these fluctuations.

The total electric field is always the sum of the fields from all the individual sources. The intensity is proportional to the square of the total field, and therefore has contributions arising from the interference of the fields from different sources. If we had a hypothetical detector that could respond *instantaneously* to the field fluctuations, we would always measure interference fringes. Usually the interference terms fluctuate too rapidly to be observed in a realistic measurement time.

In other words, whether the fields at $\mathbf{r}_1, t$ and $\mathbf{r}_2, t + \tau$ exhibit any mutual coherence depends not only on the intrinsic properties of the fields, or their sources, but also on what we measure. The situation here is quite analogous to our discussion of unpolarized light in Section 2.7.

15.7 SPATIAL COHERENCE OF THE LIGHT FROM ORDINARY SOURCES

The mutual coherence function determines the mutual coherence of fields at two different spatial points ($\mathbf{r}_1$ and $\mathbf{r}_2$) *and* at two different times (t_1 and t_2). To isolate the spatial characteristics of the mutual coherence function, we take $t_1 = t_2 = t$; then $\Gamma(\mathbf{r}_1, t; \mathbf{r}_2, t)$ determines the mutual coherence of the fields at two different points in space at the same time. In this case we speak of *spatial coherence*. For stationary sources, spatial coherence is characterized by the mutual coherence function $\Gamma(\mathbf{r}_1, \mathbf{r}_2, 0)$ and the complex degree of coherence $\gamma(\mathbf{r}_1, \mathbf{r}_2, 0)$. This is the case of most practical interest, and so we will henceforth always assume stationarity.

To get a physical picture of spatial coherence, consider a beam of constant intensity propagating in the z direction, and consider any two points P_1 and P_2 on the beam in a plane perpendicular to the z axis. Then the beam may be considered to be fully spatially coherent if the phase difference measured at the (arbitrary) points P_1 and P_2 remains constant in time. That is, the phase at any point may be fluctuating randomly, but if the beam is spatially coherent the phase *difference* between any two points remains fixed.

To test for the spatial coherence of our beam, we can intercept it with a screen containing pinholes at P_1 and P_2. We then measure the fringe visibility of the light from the pinholes falling on a second, observation screen. If the phase difference at P_1 and P_2 is constant, we will observe an interference pattern, just as in the two-slit experiment with a phase difference Φ between the fields at the slits (Equations (15.5.14). If the field is *not* spatially coherent, the phase difference in the light from P_1 and P_2 will fluctuate and the interference fringes will shift around. No consistent interference pattern will be observed. In the latter case the phases are not sticking together, or cohering, to produce any interference.

If we imagine an ordinary light source to consist of myriads of individual, independent radiators, it may seem surprising at first that such a source can have any spatial coherence at all. However, ordinary sources can (and frequently do) radiate spatially coherent fields, fields which produce interference fringes in the two-slit experiment, for instance. We will now explain how an ordinary incoherent source can emit spatially coherent radiation. As a result of our discussion, we will be able to understand the experimentally observed dependence of the fringe visibility on slit separation in the Young experiment, as described in Section 15.3.

It is worth emphasizing that for monochromatic radiation $|\gamma(\mathbf{r}_1, \mathbf{r}_2, 0)| = 1$ [recall Eq. (15.6.4)]. Needless to say, monochromaticity is an idealization that cannot be attained in the real world. However, we can produce *quasi*monochromatic radiation quite readily, and so we will focus our attention on this more realistic case. We will denote by λ the central wavelength of our quasimonochromatic field. The spread in wavelengths, $\delta\lambda$, is very small compared with λ for such a field [Eq. (15.3.5)].

We note first that a *point source* of radiation, one with dimensions much smaller than a wavelength, will always produce spatially coherent radiation. For even though the radiated field may vary quite erratically in its amplitude and phase, as a result of fluctuations in the source, every point on the wavefront has the *same* variation, that dictated by the single point source (Figure 15.5). Thus the variations are perfectly correlated across any wavefront, and the emitted field is spatially coherent. The real question, therefore, is how an actual *extended* source, com-

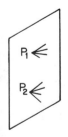

Figure 15.5 Two points P_1 and P_2 on a spherical wave emitted by a point source have exactly the same amplitude and phase variations, these being determined only by the fluctuations in the source.

prising many independently fluctuating point sources, can produce spatially coherent radiation.

Consider the case of two independent point sources, one on the axis in a two-slit experiment and the other a distance ρ off axis (Figure 15.6). The source on the axis is equidistant from the two slits and, because it is a point source, produces spatially coherent radiation. It therefore produces fringes on the observation screen with perfect visibility ($V = 1$); the positions of intensity maxima and minima are given by (15.3.2) and (15.3.3) for $L \gg d$.

The second point source is not equidistant from the two slits. The difference in path length is $D = \rho d/R$ (Figure 15.6), corresponding to a phase difference

$$\Phi = \frac{2\pi D}{\lambda} = \frac{2\pi(\rho d)}{\lambda R} \tag{15.7.1}$$

This point source therefore produces an interference pattern with intensity maxima and minima given by (15.5.14). In other words, the interference pattern associated with the second source is shifted a distance

$$\frac{\Phi}{2\pi}\frac{\lambda L}{d} = \frac{\rho L}{R} \tag{15.7.2}$$

from the interference pattern of the source on axis.

The two point sources are assumed to be completely independent. Thus their fields fluctuate independently, and do not interfere for any measurable time interval. However, this does not necessarily mean that an interference pattern cannot be observed, for the fringes associated with one (point) source may practically coincide with the fringes of the other. This happens if the displacement (15.7.2) of their interference patterns is small compared with the fringe spacing Δ (equation

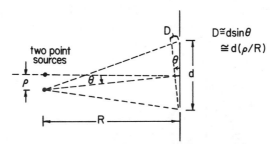

Figure 15.6 Two point sources in a two-slit experiment, one equidistant from the slits, the other displaced by ρ from the axis of equidistance. For the second source the distances to the slits differ by D. (The two angles labeled θ are equal because their sides are perpendicular.)

(15.3.4)) of the interference pattern associated with each individual source, i.e., if

$$\frac{\rho L}{R} < \Delta = \frac{\lambda L}{d} \qquad (15.7.3)$$

Equation (15.7.3) says that, using the *two* point sources of Figure 15.6, there will be interference fringes in the two-slit experiment if the slit separation d is small enough, namely if

$$d < \frac{\lambda}{\rho} R \qquad (15.7.4)$$

The factor λ/ρ is approximately the diffraction angle for light of wavelength λ incident upon an aperture of radius ρ (Section 14.11). This connection with diffraction theory is the essence of the *van Cittert–Zernike theorem*, which relates the mutual coherence function of the field from an ordinary source to the diffraction pattern for an aperture of the same dimensions as the source. We will discuss one important example.

Consider a plane circular disk source of radius ρ. The van Cittert–Zernike theorem gives a simple expression in this case for the degree of spatial coherence $|\gamma(\mathbf{r}_1, \mathbf{r}_2, 0)|$ in a plane parallel to the source and a distance R from it (Figure 15.7) in terms of the first-order Bessel function $J_1(x)$:

$$\left|\gamma(\mathbf{r}_1, \mathbf{r}_2, 0)\right| = \left|\frac{2J_1(x)}{x}\right| \qquad (15.7.5)$$

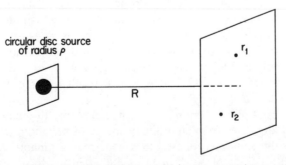

circular disc source
of radius ρ

R

r_1

r_2

Figure 15.7 The van Cittert–Zernike theorem gives $|\gamma(\mathbf{r}_1, \mathbf{r}_2, 0)|$ in a plane a distance R from an ordinary (nonlaser) source of quasimonochromatic radiation. We consider the case of a circular disk source of radius ρ, for which $|\gamma(\mathbf{r}_1, \mathbf{r}_2, 0)|$ is given by (15.7.5).

where

$$x = \frac{2\pi\rho}{\lambda R}\left|\mathbf{r}_1 - \mathbf{r}_2\right| = \frac{2\pi\rho d}{\lambda R} \tag{15.7.6}$$

and it is assumed that ρ and d are much smaller than R. Comparing this with our results in Section 14.11, we see that $|\gamma(\mathbf{r}_1, \mathbf{r}_2, 0)|^2$ is the Airy pattern associated with Fraunhofer diffraction by a uniformly illuminated circular *aperture* of radius ρ. Equation (15.7.5) gives the degree of coherence of the radiation at a distance R from the source. Since $2J_1(x)/x \approx 0.88$ for $x = 1$, the radiation has a degree of spatial coherence $|\gamma(\mathbf{r}_1, \mathbf{r}_2, 0| \geq 88\%$ if $x \leq 1$, i.e., if

$$\frac{2\pi\rho d}{\lambda R} \leq 1 \tag{15.7.7}$$

or $d \leq (1/2\pi)(\lambda R/\rho)$. In other words, the radiation has a high degree of spatial coherence over a circular area of diameter

$$d_{\text{coh}} = \frac{1}{2\pi}\frac{\lambda R}{\rho} \approx 0.16\frac{\lambda R}{\rho} \tag{15.7.8}$$

This result of the van Cittert–Zernike theorem supports our intuitive argument leading to (15.7.4).

Now we can understand the results of the two-slit experiment when an ordinary source is used (Section 15.3). The light from such a source has spatial coherence over a limited area ($\approx \pi d_{\text{coh}}^2/4$) on the screen containing the slits. As the slit separation is increased, the degree of coherence of the fields at the two slits decreases. For slit separations large compared with d_{coh}, the fringe visibility approaches zero. The fringe visibility also decreases if the slit separation is kept constant and R is decreased, i.e., if the source is brought closer to the slits. Actually, if the slit separation is increased so that the fringe visibility falls from near unity to zero, a continued increase in the slit separation causes the visibility to increase and then decrease again repeatedly, but with very small secondary maxima. This is simply a reflection of the oscillatory behavior of the Bessel function $J_1(x)$ in (15.7.5) (cf. Figure 12.13). A similar result is obtained if the slit separation is fixed while R is decreased.

Using a lens-and-pinhole arrangement, it is possible to obtain a spatially coherent beam of large area from an ordinary source of radiation. (Figure 15.8) The radiation from the source is focused to a spot near a pinhole of radius a. The pinhole, which acts as a plane circular source, is in the focal plane of a second lens of focal length f. The beam emerging from the second lens is spatially coherent over a circle of diameter given by (15.7.8) with $\rho = a$ and $R = f$:

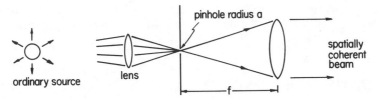

Figure 15.8 A way of obtaining a spatially coherent beam of large area from an ordinary source of radiation.

$$d_{coh} = 0.16 \frac{\lambda f}{a} \qquad (15.7.9)$$

A small pinhole therefore gives rise to a large spatially coherent beam.

• Consider as an example the light from the sun. As an approximation let us treat the sun as a disk source of mean wavelength 5500 Å. Over what linear dimensions on earth is the light from the sun spatially coherent? The answer is given by (15.7.8) with $\rho = 6.96 \times 10^{10}$ cm, the radius of the sun, and $R = 1.5 \times 10^{13}$ cm, the mean distance of the earth from the sun:

$$d_{coh} = \frac{(0.16)(5500 \times 10^{-8} \text{ cm})(1.5 \times 10^{13} \text{ cm})}{6.96 \times 10^{10} \text{ cm}}$$

$$= 0.02 \text{ mm} \qquad (15.7.10)$$

We could use the arrangement of Figure 15.8 to achieve a larger region of spatial coherence, but the larger beam area would result in a diminution of intensity. Lasers, on the other hand, can give large coherence areas *and* high intensity.

We can write (15.7.8) in the form

$$d_{coh} = 0.16 \frac{\lambda}{\theta} \qquad (15.7.11)$$

where $\theta = \rho/R$ is the angle subtended by the source at the observation plane. The sun, for instance, subtends an angle $\theta \approx 4.6 \times 10^{-3}$ rad at the earth.

The smaller the angle θ subtended by the source, the greater the diameter d_{coh} over which its radiation is spatially coherent. The star Betelguese, for instance, subtends an angle $\approx 2 \times 10^{-7}$ rad at the Earth. For $\lambda = 5500$ Å, therefore, (15.7.8) gives $d_{coh} \approx 80$ cm. In other words, stellar radiation is spatially coherent over fairly large areas at the Earth's surface. There are techniques that take advantage of this to measure the angular diameters of stars.

The planets in our solar system typically subtend angles several orders of magnitude larger than those of stars. Their radiation (i.e., the solar radiation they scatter to the earth) is therefore spatially coherent over much smaller distances at the earth. This is partly responsible for the fact that stars twinkle, whereas planets normally do not. The twinkling is an interference effect arising from refractive-index fluctuations in the Earth's atmosphere.

Similarly, distant streetlights appear to twinkle, whereas closer ones do not. Such effects were noted by Aristotle (384–322 B.C.), even before streetlights were common in Athens. •

15.8 SPATIAL COHERENCE OF LASER RADIATION

Laser radiation can have a high degree of spatial coherence. This in itself is not remarkable, for we have seen that spatially coherent beams of light can be obtained from the radiation of ordinary sources. What *is* unique about lasers is that they can combine spatial coherence with high intensity, or at least a high enough intensity to be useful in applications. It is this property of lasers that makes them so useful for holography.

A laser oscillating on a single transverse mode has perfect spatial coherence.[4] A two-slit experiment will therefore show interference fringes of high visibility. For a Gaussian mode, for instance, sharp fringes are observed even if the slit separation is considerably larger than the spot size w of the laser beam.

It is worth noting that this spatial coherence has nothing to do with stimulated emission *per se*. The spatial coherence of a laser oscillating on a single transverse mode is a consequence of the fact that the field is a mode of a resonator. As a result, the field values at any two points across the wavefront are perfectly correlated, i.e., in step with one another. For instance, even an emitter operating below the threshold for laser oscillation exhibits perfect spatial coherence if the radiation is associated with a single transverse mode.

However, *a laser operating on more than one transverse mode does not have perfect spatial coherence*. In particular, a laser operating on many transverse modes has spatial coherence properties much like those of ordinary sources of radiation, where the van Cittert–Zernike theorem is applicable. This is why single-mode operation is so important in holography, for instance.

To get an intuitive picture of why oscillation on more than one transverse mode reduces the spatial coherence, recall that different transverse modes have different field distributions, as in the case of a lowest-order Gaussian beam compared with a higher-order one. It can thus be imagined that the different modes are being excited by quite different groups of active atoms, and are therefore associated with completely independent sources. This brings us close to our picture of an ordinary source of radiation.

Figure 15.9 shows results of an experiment to determine the spatial coherence of a 6328-Å He–Ne laser. The degree of spatial coherence was determined by a two-slit arrangement with aperture spacings from 2 to 20 mm. When the laser was oscillating on a single transverse mode, the result was $|\gamma(\mathbf{r}_1, \mathbf{r}_2, 0)| \approx 1$. However, when the resonator was adjusted so that two transverse modes oscillated, $|\gamma(\mathbf{r}_1, \mathbf{r}_2, 0)|$ dropped dramatically with increasing $|\mathbf{r}_1 - \mathbf{r}_2|$. In fact, $|\gamma|$ for

---

4.   This is true even if the laser is oscillating on more than one longitudinal mode. See Section 15.11.

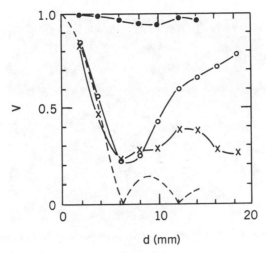

Figure 15.9 $V = |\gamma(\mathbf{r}_1, \mathbf{r}_2, 0)|$ for a 6328-Å He–Ne laser, as a function of $d = |\mathbf{r}_1 - \mathbf{r}_2|$. Observations of single mode (closed circles), double mode (open circles), and multimode (crosses) types of operation are shown along with the calculated visibility function of a thermal source (dashed curve). From M. Young and P. L. Drewes, Optics Communications **2**, 253 (1970).

the case of only two transverse modes already approaches that for an ordinary thermal source. As the number of oscillating transverse modes increases further, $|\gamma(\mathbf{r}_1, \mathbf{r}_2, 0)|$ comes close to the functional form (15.7.5) for a thermal source.

These experimental results, and others like them, show how crucial is oscillation on a single transverse mode for the spatial coherence of laser radiation. Unfortunately, the restriction to a single transverse mode often reduces the total output power of the laser. For when several modes oscillate, each with their different field distributions, the overall mode volume covers a greater portion of the available gain medium.

• One result of the spatial coherence of a laser beam is the *speckle effect* that is observed when an expanded laser beam shines on a surface with fine-scale irregularities (e.g., a "diffuse" surface like a wall). The reflected light has a speckled appearance, consisting of irregularly shaped but sharply defined bright and dark areas. The bright and dark areas are associated respectively with constructive and destructive interference of the light from the various surface scattering elements (Figure 15.10). Because the surface has more or less random irregularities, the speckle pattern itself appears random and irregular. Laser speckle is a consequence of spatial coherence. If the radiation incident on the scattering surface were not spatially coherent, the uncorrelated fluctuations in the field at nearby points on the surface would wash out the interference pattern.

When we view an object illuminated by laser light, we often find it difficult to focus on the object. This is because our eyes involuntarily try to focus on the speckle. This cannot

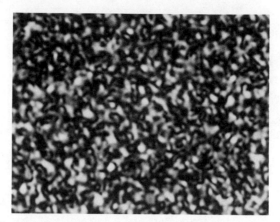

Figure 15.10 A speckle pattern [Courtesy of G.M. Morris].

be done, because the speckle pattern is not "on" the object or any other plane in space. Indeed, we see the interference pattern even if we focus our eyes on a plane *between* the object and ourselves.

If a near-sighted observer moves his head from side to side, the speckle pattern appears to move in the opposite direction, whereas a far-sighted person will see it moving in the same direction he is moving his head. This effect has a simple explanation (Problem 15.4). •

15.9 DIRECTIONALITY OF LASER RADIATION

The high degree of directionality of laser beams, and therefore their high brightness, is intimately related to their spatial coherence.

Our treatment of diffraction in Chapter 14 assumed perfect spatial coherence: we dealt with time-independent field amplitudes and phases, so there were no fluctuations of these quantities to be averaged. For a spatially coherent beam propagating in free space the divergence angle obeys the relation

$$\theta \sim \lambda/D \tag{15.9.1}$$

where D is the beam diameter. The precise value of the divergence angle depends on the intensity distribution across the beam. For a Gaussian beam $\theta = 2/\pi(\lambda/D)$, where $D = 2w_0$.

If the beam has only partial spatial coherence, the Huygens wavelets from different points on the beam do not all add up coherently. Imagine, for instance, that the beam is spatially coherent only over distances $d < D$ across the beam. In this case the divergence angle is

$$\theta \sim \frac{\lambda}{d} > \frac{\lambda}{D} \qquad (15.9.2)$$

In other words, if the beam is not spatially coherent, the divergence angle is greater than in the spatially coherent case with the same intensity distribution. In particular, the divergence angle associated with a laser operating on more than one transverse mode will generally be greater than that for the single-mode case.

The divergence angle of a laser beam can be reduced simply by increasing the beam diameter. This may be done, for instance, by letting the beam pass backwards through a Keplerian telescope (Figure 14.24). The divergence angle is inversely proportional to the beam diameter, and so the angle θ_f after passage through the telescope is related to the initial angle θ_i by

$$\frac{\theta_f}{\theta_i} = \frac{D_i}{D_f} = \frac{1}{M} \qquad (15.9.3)$$

where M is the magnification of the telescope. Low divergence angles are obviously essential in such applications as alignment or surveying, where a laser beam is used as a straight line.

Because of diffraction, a laser beam can never have a divergence angle of zero, just as it cannot be focused with a lens to the geometrical point predicted by ray optics. The beam divergence is minimized, however, if the beam is spatially coherent. In this case, because diffraction sets the ultimate lower limit on the beam spread, we say we have reached the *diffraction limit*. Realization of the diffraction limit in practice requires that aberrations and other defects in components such as lenses and mirrors are negligible. Unfortunately the term "diffraction limit" is used without general agreement on its precise meaning. Frequently its intended meaning has to be understood from the context in which it is used.

15.10 COHERENCE AND THE MICHELSON INTERFEROMETER

If we take $\mathbf{r}_1 = \mathbf{r}_2 = \mathbf{r}$ in the definition (15.4.18), then $\Gamma(\mathbf{r}, t_1; \mathbf{r}, t_2)$ determines the mutual coherence of the fields at the same point in space but at two different times. In this case we speak of *temporal coherence*. We are interested in stationary fields, for which $\Gamma(\mathbf{r}, t_1; \mathbf{r}, t_2) = \Gamma(\mathbf{r}, \mathbf{r}, \tau)$ depends on t_1 and t_2 only through the difference $\tau = t_2 - t_1$.

The significance of temporal coherence can be illustrated by considering as an example the Michelson interferometer shown in Figure 15.11. The incident beam is split by a 50:50 beam splitter (BS) into two beams of equal intensity. One of these beams is reflected off mirror M_1 and makes its way to BS again, where part of it is transmitted. Similarly the other beam reflects off mirror M_2 and propagates back to BS, where part of it is reflected. At a point such as P in Figure 15.11 there

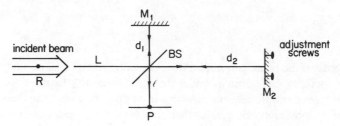

Figure 15.11 Basic setup for a Michelson interferometer.

is thus a superposition of two fields. Because each of these fields is twice incident on BS, it has an intensity one-fourth the intensity (at R, say) originally incident upon the interferometer; the field amplitude, therefore, has been cut in half. So the total field at P at time t is

$$E(P, t) = \tfrac{1}{2} E\left(R, t - \frac{l_1}{c}\right) + \tfrac{1}{2} E\left(R, t - \frac{l_2}{c}\right) \qquad (15.10.1)$$

where

$$l_1 = L + 2d_1 + l \qquad (15.10.2a)$$

$$l_2 = L + 2d_2 + l \qquad (15.10.2b)$$

In writing (15.10.1) we are assuming that the field amplitudes are reduced only by the beam splitter. The first term is the field at P resulting from propagation via the upper arm of the interferometer in Figure 15.11. Except for the factor $\tfrac{1}{2}$, this field is the same as that at R at the earlier time $t - l_1/c$, where l_1/c is the time it takes light to propagate from R to P via the upper arm. The second term has the same interpretation, except that it arises because of the second arm of the interferometer.

The intensity measured at P is

$$\langle I(P, t)\rangle = \frac{c\epsilon_0}{2}\left\langle\left|E(P, t)\right|^2\right\rangle$$

$$= \frac{c\epsilon_0}{2}\left[\tfrac{1}{4}\;\left|E\left(R, t - \frac{l_1}{c}\right)\right|^2 + \tfrac{1}{4}\;\left|E\left(R, t - \frac{l_2}{c}\right)\right|^2\right.$$

$$\left. + \tfrac{1}{2}\,\mathrm{Re}\;\;E^*\left(R, t - \frac{l_1}{c}\right) E\left(R, t - \frac{l_2}{c}\right)\;\right] \qquad (15.10.3)$$

For stationary fields, every term in this equation is independent of t, and furthermore the mutual coherence function appearing on the right depends only on the time difference

$$\left(t - \frac{l_2}{c}\right) - \left(t - \frac{l_1}{c}\right) = \frac{l_1 - l_2}{c} = 2\frac{d_1 - d_2}{c} = \tau \qquad (15.10.4)$$

The measured intensity at P for a stationary field is thus

$$\langle I(P)\rangle = \tfrac{1}{4}\left[\langle I(R)\rangle + \langle I(R)\rangle + c\epsilon_0 \text{ Re } \Gamma(R, R, \tau)\right]$$

$$= \tfrac{1}{2}\langle I(R)\rangle\left[1 + \text{Re } \gamma(R, R, \tau)\right] \qquad (15.10.5)$$

where the complex degree of coherence is defined by (15.6.3):

$$\gamma(R, R, \tau) = \frac{(c\epsilon_0/2)\,\Gamma(R, R, \tau)}{\sqrt{\langle I(R)\rangle\,\langle I(R)\rangle}}$$

$$= \frac{(c\epsilon_0/2)\,\Gamma(R, R, \tau)}{\langle I(R)\rangle} \qquad (15.10.6)$$

In the case of perfectly monochromatic light, for which γ is given by (15.6.4), we have from (15.10.5)

$$\langle I(P)\rangle = \tfrac{1}{2}\langle I(R)\rangle\,(1 + \cos \omega\tau)$$

$$= \langle I(R)\rangle \cos^2 \tfrac{1}{2}\omega\tau$$

$$= \langle I(R)\rangle \cos^2\left(\frac{\omega}{c}(d_1 - d_2)\right)$$

$$= \langle I(R)\rangle \cos^2\left(\frac{2\pi}{\lambda}(d_1 - d_2)\right) \qquad (15.10.7)$$

There is therefore constructive interference at P when

$$|d_1 - d_2| = n\lambda, \qquad n = 0, 1, 2, \ldots \qquad (15.10.8a)$$

and destructive interference when

$$|d_1 - d_2| = (n + \tfrac{1}{2})\,\lambda, \qquad n = 0, 1, 2, \ldots \qquad (15.10.8b)$$

just as we should have expected. As the arm separation $|d_1 - d_2|$ is varied, there is a sequence of alternately bright and dark spots at P.

For quasimonochromatic light, where γ is given by (15.6.5), we obtain from (15.10.5) the intensity

$$\langle I(P)\rangle = \tfrac{1}{2}\langle I(R)\rangle \left[1 + |\gamma(R, R, \tau)| \cos\left(\frac{2\pi}{\lambda}(d_1 - d_2)\right) \right] \quad (15.10.9)$$

The visibility in this case may be defined as

$$V = \frac{\langle I(P)\rangle_{max} - \langle I(P)\rangle_{min}}{\langle I(P)\rangle_{max} + \langle I(P)\rangle_{min}} = |\gamma(R, R, \tau)| \quad (15.10.10)$$

The Michelson interferometer thus provides a way of measuring temporal coherence, just as the Young two-slit experiment may be used to measure spatial coherence (Problem 15.5).

• The Michelson interferometer was invented by Albert A. Michelson, who began his study of optics as a student at the U.S. Naval Academy. There he was considered below average in seamanship, but he excelled in science. His best-known work involved the use of his interferometer to test for the motion of the Earth through the "ether." The null result of the Michelson–Morley experiment in 1887 led eventually to the abandonment of the ether concept.

The Michelson interferometer can be used to determine the wavelength of quasimonochromatic radiation (Problem 15.6), and in this role has for a long time been a very useful spectroscopic tool. •

15.11 TEMPORAL COHERENCE

Experimentally it is found that the visibility (15.10.10) decreases with increasing τ. Furthermore the visibility decreases more rapidly for larger bandwidths $\delta\nu$ of the (quasimonochromatic) radiation. In other words, the more nearly monochromatic the radiation, the greater its temporal coherence.

To understand this, suppose we have radiation of spectral width $\delta\lambda$ incident upon a Michelson interferometer. Then the total intensity at P is the sum of contributions like (15.10.7) if we add intensities of different frequency components (Problem 15.7). Since each wavelength component of the incident radiation is associated with a different pattern of bright and dark spots as $|d_1 - d_2|$ is varied, the pattern will be smeared out if $\delta\lambda$ is large enough. We will assume for simplicity that the intensity is constant for wavelengths between $\lambda - \tfrac{1}{2}\delta\lambda$ and $\lambda + \tfrac{1}{2}\delta\lambda$, and zero outside this range, as shown in Fig. 15.12. Then we expect that the interference pattern is smeared out if $|d_1 - d_2|$ is large enough that the largest

intensity

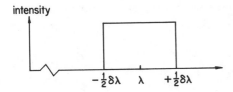

Figure 15.12 Hypothetical intensity distribution as a function of wavelength.

wavelength $\lambda + \frac{1}{2}\delta\lambda$ corresponds to an intensity maximum whereas the smallest wavelength $\lambda - \frac{1}{2}\delta\lambda$ corresponds to an intensity minimum (or vice versa). From (15.10.8) we have therefore the two conditions

$$|d_1 - d_2| = n(\lambda + \tfrac{1}{2}\delta\lambda) \qquad (15.11.1a)$$

$$|d_1 - d_2| = (n + \tfrac{1}{2})(\lambda - \tfrac{1}{2}\delta\lambda) \qquad (15.11.1b)$$

in this case, or

$$\frac{|d_1 - d_2|}{\lambda + \tfrac{1}{2}\delta\lambda} = n \qquad (15.11.2a)$$

$$\frac{|d_1 - d_2|}{\lambda - \tfrac{1}{2}\delta\lambda} = n + \tfrac{1}{2} \qquad (15.11.2b)$$

where n is an integer. Subtraction of (15.11.2a) from (15.11.2b) yields

$$|d_1 - d_2| \left(\frac{1}{\lambda - \tfrac{1}{2}\delta\lambda} - \frac{1}{\lambda + \tfrac{1}{2}\delta\lambda} \right) = \frac{1}{2} \qquad (15.11.3)$$

Since $\delta\lambda \ll \lambda$, we can combine the two fractions to obtain

$$|d_1 - d_2| \frac{\delta\lambda}{\lambda^2} = \frac{1}{2} \qquad (15.11.4)$$

where we have dropped $(\delta\lambda)^2/4$ in the denominator compared with λ^2. Thus we find

$$|d_1 - d_2| = c\tau = \frac{\lambda^2}{2\,\delta\lambda} \qquad (15.11.5)$$

Since $\lambda = c/\nu$, it follows that differential increments of wavelength and frequency are related by

$$\left| \frac{d\lambda}{d\nu} \right| = \frac{c}{\nu^2} = \frac{\lambda}{\nu} \qquad (15.11.6)$$

and so we have, for sufficiently small finite positive increments,

$$\frac{\delta\lambda}{\delta\nu} = \frac{\lambda}{\nu} \qquad (15.11.7)$$

or

$$\frac{\delta\lambda}{\lambda} = \frac{\delta\nu}{\nu} \qquad (15.11.8)$$

Using this relation in (15.11.5), we obtain

$$c\tau = \frac{\lambda}{2} \frac{\lambda}{\delta\lambda} = \frac{\lambda\nu}{2\,\delta\nu} = \frac{c}{2\,\delta\nu},$$

that is,

$$\tau = \frac{|d_1 - d_2|}{c} = \frac{1}{2\,\delta\nu} \qquad (15.11.9)$$

Equation (15.11.9) gives the value of the interferometer path separation $|d_1 - d_2|$, or time difference τ, at which we expect the interference pattern to be smeared out. For separations larger than τ the visibility should be very small or zero. In agreement with experiment, τ decreases with increasing bandwidth $\delta\nu$. The intensity distribution shown in Fig. 15.12 was chosen for convenience, so the relation (15.11.9) derived from it cannot be regarded as fundamental. Also, as in the case of spatial coherence, there is some arbitrariness to the boundary between coherence and incoherence. Instead of (15.11.9) it is conventional to define

$$\tau_{\mathrm{coh}} = \frac{1}{2\pi\,\delta\nu} \qquad (15.11.10)$$

as the *coherence time* of quasimonochromatic radiation of bandwidth $\delta\nu$. The distance $c\tau_{\mathrm{coh}}$ is called the *coherence length*. If a beam is divided into two parts, the coherence length is the path difference beyond which there will be very little interference (or fringe visibility) when the two fields are superposed. Note that the coherence length arises from temporal coherence, and is thus unrelated to the coherence area of Section 15.7, which is a measure of spatial coherence.

A good non-laser source of "monochromatic" radiation might have a bandwidth $\delta\nu = 100$ MHz. This translates into a coherence time

$$\tau_{\text{coh}} = \frac{1}{(2\pi)(10^8 \text{ sec}^{-1})} = 1.6 \text{ nsec} \qquad (15.11.11)$$

and a coherence length

$$c\tau_{\text{coh}} = 48 \text{ cm} \qquad (15.11.12)$$

More typical of such sources are coherence times and lengths on the order of 10^{-10} sec and a few centimeters, respectively (Problem 15.8). With such sources the path separation (in a Michelson interferometer, for instance) must be less than a centimeter or two if interference fringes are to be observed.

A laser operating on a single transverse mode will have perfect spatial coherence, whereas its temporal coherence will be determined by the bandwidth of the output radiation. If it is operating on a single longitudinal mode, $\delta\nu$ is often so small that the coherence length is practically infinite for many purposes. As discussed in Section 11.10, the theoretical lower limit to $\delta\nu$ for a laser has never been achieved. Effects like mirror vibrations and thermal expansion of the resonator lead to bandwidths much larger than the theoretical lower limit set by the Heisenberg uncertainty principle. Nevertheless, various stabilization methods can be used to make $\delta\nu$ very small, say 300 Hz; for this bandwidth the coherence length is about 160 km, or 100 miles. In fact sophisticated stabilization techniques have been used to obtain coherence lengths thousands of times greater even than this. However, this ultrastabilization cannot usually be maintained for more than a few minutes.

A laser operating on more than one longitudinal mode, however, will have a much larger bandwidth, and therefore a much smaller coherence length, than in the single-mode case. Many He–Ne lasers, for instance, operate on two longitudinal modes separated in frequency by $c/2L$. In this case $\delta\nu \sim c/2L$, and therefore $\tau_{\text{coh}} \sim L/\pi c$, so $c\tau_{\text{coh}} \sim L/\pi$. The coherence length of the laser radiation in this case is less than the length of the laser itself.

If many modes are lasing, the laser may emit radiation over virtually the entire gain bandwidth. In this case the temporal coherence of the laser radiation is about the same as a "monochromatic" thermal source that derives its radiation from the spontaneous emission of a single atomic transition. That is, the coherence length in the multimode case may be only on the order of a centimeter.

A laser operating on several longitudinal and transverse modes will therefore resemble a thermal source in both its temporal and spatial coherence properties. It remains true, of course, that the laser can emit more power than one can ever hope to obtain from a conventional source of radiation.

• The quantity $|\gamma(\mathbf{r}, \mathbf{r}, \tau)|$ is related to the Fourier transform of the spectral lineshape function. For a Lorentzian lineshape of HWHM $\delta\nu$, for instance, it may be shown that

$$\left|\gamma(\mathbf{r}, \mathbf{r}, \tau)\right| = e^{-2\pi\delta\nu\tau} = e^{-\tau/\tau_{\mathrm{coh}}} \tag{15.11.13}$$

where τ_{coh} is given by (15.11.10). In this case, therefore, $c\tau_{\mathrm{coh}}$ is just the value of the path separation $|d_2 - d_1|$ at which the visibility drops to e^{-1}, which incidentally illustrates again that there is no sharp boundary between temporal coherence and incoherence, just as there is no sharp boundary between spatial coherence and incoherence.

In the Young two-slit experiment the modulus $|\gamma(S_1, S_2, l/c)|$ appearing in (15.6.6) may be replaced by $|\gamma(S_1, S_2, 0)|$ if

$$l/c \ll \tau \tag{15.11.14}$$

i.e., if the path difference l for the two slits is small compared with the coherence length of the radiation. This condition is frequently well satisfied in practice, even for a "monochromatic" thermal source or a laser operating on many longitudinal modes. In this case the Young experiment gives us a direct measure of spatial coherence, as we assumed in our discussion. This allows us to speak separately of "spatial" and "temporal" coherence. •

15.12   THE PHOTON DEGENERACY FACTOR

Consider a thermal source of quasimonochromatic radiation of bandwidth $\delta\nu$. The radiation from this source is spatially coherent over an area

$$A_{\mathrm{coh}} \sim d_{\mathrm{coh}}^2 \sim \frac{\lambda^2 R^2}{\rho^2} \sim \frac{\lambda^2 R^2}{S} \tag{15.12.1}$$

at a distance R from the source, with S the source area. A is called the *coherence area*. The product of the coherence area and the coherence length $c\tau_{\mathrm{coh}} \sim c/\delta\nu$ defines the *coherence volume*:

$$V_{\mathrm{coh}} = A_{\mathrm{coh}}(c\tau_{\mathrm{coh}}) = \frac{c\lambda^2 R^2}{S\,\delta\nu} \tag{15.12.2}$$

In this section we will ignore geometrical details involving factors of π, 2, 2π, etc. We will only concern ourselves with general orders of magnitude that are independent of source shape—square or circular or whatever.

Let us now think in terms of photons, and consider the number of photons crossing the coherence area in a coherence time. We denote this dimensionless number by δ:

$$\delta = \Phi\, A_{\mathrm{coh}}\tau_{\mathrm{coh}} \tag{15.12.3}$$

where Φ is the photon flux, i.e., the number of photons crossing a unit area per unit time. For blackbody radiation $\rho(\nu)\,\delta\nu$ is the energy per unit volume in the frequency interval from ν to $\nu + \delta\nu$, and so $\rho(\nu)\,\delta\nu/h\nu$ is the number of photons per unit volume in this interval; $\rho(\nu)$ is the spectral energy density (7.7.5). The photon flux from a blackbody source of surface area S is therefore

$$\Phi_\nu \sim c\,\frac{\rho(\nu)\,\delta\nu}{h\nu}\,\frac{S}{4\pi R^2} \tag{15.12.4}$$

at a distance R from the source. The factor $S/4\pi R^2$, the ratio between the emitting surface area and the surface area of a sphere of radius R, takes account of the inverse square law for photon flux. That is, $R^2\Phi_\nu$ must be a constant, independent of R. From (15.12.3) and (15.12.4), therefore,

$$\delta_\nu \sim \left(\frac{c\rho(\nu)\,\delta\nu\,S}{4\pi h\nu R^2}\right)\left(\frac{\lambda^2 R^2}{S}\right)\left(\frac{1}{\delta\nu}\right) = \frac{c\rho(\nu)\,\lambda^2}{4\pi h\nu}$$

$$= \frac{c^3}{4\pi h\nu^3}\,\rho(\nu) \tag{15.12.5}$$

Equation (15.12.5) is the number of photons crossing the coherence area A_{coh} during a coherence time τ_{coh}. Blackbody radiation is unpolarized. The number of photons *of a particular polarization* crossing A_{coh} in a time τ_{coh} is therefore half the value (15.12.5):

$$\delta_\nu = \frac{c^3}{8\pi h\nu^3}\,\rho(\nu) = \frac{1}{e^{h\nu/kT} - 1} \tag{15.12.6}$$

the last step following from the Planck law (7.7.5).

We recognize δ_ν in (15.12.6) from Eq. (7.7.20) as the average number of blackbody photons in a mode of frequency ν. In other words, *the average number of photons crossing an area equal to A_{coh} in a time equal to τ_{coh} is equal to the average number of photons per mode.* This number δ is called the *photon degeneracy factor*, or simply the *degeneracy parameter*. It represents a "degeneracy" in the sense that the δ photons are not distinguished from each other by spatial or temporal labels. For blackbody radiation δ is much smaller than one, being typically 10^{-2} or 10^{-3} (recall the black-dot passage in Section 15.2).

It is clear that the degeneracy parameter is equal to the average number of photons in a volume V_{coh}: we simply write (15.12.3) in the form

$$\delta = \frac{\Phi}{c}\,A_{\text{coh}}(c\tau_{\text{coh}}) = \frac{\Phi}{c}\,V_{\text{coh}} = NV_{\text{coh}} \tag{15.12.7}$$

where N is the density of photons, related to the photon flux Φ by the equation $\Phi = cN$.

On the other hand how many photons per mode are in a laser beam? Our discussion of cavity modes in Chapter 14 assumed a perfectly monochromatic field. Such a field has perfect spatial and temporal coherence; the coherence volume is infinite in this hypothetical limit. Now a single-mode laser is spatially coherent over the entire beam area A, but its temporal coherence is limited by the frequency bandwidth $\delta\nu$. To estimate δ in this case let us assume, as we have proven for blackbody radiation, that δ is equal to the number of photons crossing an area equal to A_{coh} in a time τ_{coh}:

$$\delta = \Phi \, A_{coh} \tau_{coh} \sim \frac{I}{h\nu} A \frac{1}{\delta\nu} = \frac{IA}{hc} \frac{\lambda}{\delta\nu}$$

$$= \frac{P}{hc} \frac{\lambda}{\delta\nu} \qquad (15.12.8)$$

where I is the intensity and $P = IA$ the beam power.

As an example, consider a single-mode, 6328-Å He–Ne laser emitting a power $P = 1$ mW with a bandwidth $\delta\nu = 500$ Hz:

$$\delta = \frac{(10^{-3} \text{ J-sec}^{-1})(6328 \times 10^{-8} \text{ cm})}{(6.625 \times 10^{-34} \text{ J-sec})(3 \times 10^{10} \text{ cm-sec}^{-1})(500 \text{ sec}^{-1})}$$

$$= 6.4 \times 10^{12} \qquad (15.12.9)$$

This is fantastically larger than the photon degeneracy factor of blackbody radiation. Furthermore it is not difficult to exceed (15.12.9) by orders of magnitude in well-stabilized single-mode lasers. And so, although ordinary thermal sources can emit radiation as spatially and temporally coherent as laser radiation, they could produce such huge numbers of degenerate photons only with source temperatures above about 10^{16} K, temperatures unknown in the universe.

• The coherence volume has an interesting interpretation in terms of the Heisenberg uncertainty principle. According to this principle of quantum theory [recall (5.2.25)], there is a fundamental limitation on the accuracy to which the position and momentum of a particle can be simultaneously measured. The uncertainties Δx, Δy, and Δz in the position coordinates of the particle are related to the uncertainties Δp_x, Δp_y, and Δp_z in the momentum components by (still ignoring factors of 2π)

$$\Delta x \, \Delta p_x \geq h, \qquad \Delta y \, \Delta p_y \geq h, \qquad \Delta z \, \Delta p_z \geq h \qquad (15.12.10)$$

The uncertainty $\Delta V = \Delta x \, \Delta y \, \Delta z$ of the volume within which the particle can be localized in a measurement is therefore

$$\Delta V \geq \frac{h^3}{\Delta p_x \, \Delta p_y \, \Delta p_z} \tag{15.12.11}$$

Let us apply (15.12.11) to the case of a photon of light of frequency ν and bandwidth $\delta\nu$ propagating in the z direction. Since the momentum of a photon is $p = h/\lambda = h\nu/c$, we have

$$\Delta p_z = \frac{h}{c} \, \delta\nu \tag{15.12.12}$$

At a distance R from the source of surface area S, the uncertainties in Δp_x and Δp_y are (Figure 15.13)

$$\Delta p_x = \sim \frac{h}{\lambda} \left(\frac{S}{R^2} \right)^{1/2} \tag{15.12.13}$$

Combining (15.12.11)–(15.12.13), we obtain

$$\Delta V = \frac{c\lambda^2 R^2}{S \, \delta\nu} \tag{15.12.14}$$

We recognize the right-hand side as the coherence volume V_{coh} given by (15.12.2).

We can summarize this result as follows. Photons emitted from a given source have linear momenta that can only be determined within a certain tolerance owing to the finite area of the source and the finite bandwidth of the radiation. Associated with this uncertainty in momenta is an uncertainty in the volume within which any measurement can locate the photon. This volume is equal to the coherence volume of the radiation. This connection between the Heisenberg uncertainty principle and coherence, or the ability to produce interference effects, is a general feature of quantum theory. •

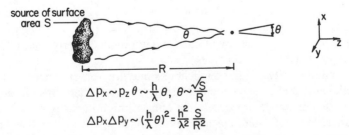

Figure 15.13 Photons from a source of area S. The uncertainties in the momentum components p_x and p_y are directly proportional to the angle θ subtended by the source.

15.13   ORDERS OF COHERENCE

The spatial and temporal coherence properties of radiation determine the degree to which the fields at two points in space, and at two points in time, are found to interfere. We have seen that ordinary sources emit radiation that can be made to have the same degree of spatial and temporal coherence as laser light, even if the laser is well stabilized and oscillating on a single cavity mode. Of course the use of pinholes, lenses, and wavelength filters to increase the coherence of radiation from an ordinary source can severely reduce the light intensity far below that available from lasers.

Aside from this difference in photon number, we might ask whether there is any difference *in principle* between coherent thermal radiation and coherent laser radiation: if they have the same central frequency, the same bandwidth, the same degree of spatial coherence, and the same intensity, can we distinguish thermal radiation from laser radiation?

It seems clear that we cannot distinguish them by measuring the location and visibility of interference fringes in a Young or Michelson experiment. In fact no experiment that divides a wavefront into two parts and then measures in some way the mutual coherence function can show any difference.

However, it turns out that we can distinguish between the two radiation fields if we undertake more sophisticated experiments. The mutual coherence function only characterizes *first-order coherence*. In general we can define *nth-order coherence functions*. A second-order coherence function, for instance, is

$$\Gamma^{(2)}(\mathbf{r}_1, t_1; \mathbf{r}_2, t_2 \,|\, \mathbf{r}_3, t_3; \mathbf{r}_4, t_4)$$

$$= \langle E^*(\mathbf{r}_1, t_1)\, E^*(\mathbf{r}_2, t_2)\, E(\mathbf{r}_3, t_3)\, E(\mathbf{r}_4, t_4) \rangle \qquad (15.13.1)$$

It depends on *four* space-time points. A general nth-order coherence function will likewise be a function of $2n$ space–time variables. Experiments that measure such higher-order coherence functions, and the corresponding higher-order complex degrees of coherence, can distinguish laser radiation from thermal radiation, even if the two fields are identical in their spatial and temporal coherence functions.

The spatial and temporal coherence already described in this chapter depend on the mutual coherence function Γ, which is a measure of first-order coherence. In other words, ordinary *spatial and temporal coherence are only manifestations of first-order coherence*, the ability of radiation to produce interference effects at two spact-time points $(\mathbf{r}_1, t_1)$ and $(\mathbf{r}_2, t_2)$.

In the great majority of laser applications only first-order coherence properties, like directionality and quasimonochromaticity, are important. In some applications not even these coherence properties, but only the high intensity of laser radiation, really matter. We will therefore not devote much space in this book to higher-order coherence, or the types of experiments that measure higher-order coherence properties of radiation. In Section 15.15, however, we will discuss an important example of a second-order coherence function in connection with photon bunching.

It should be mentioned that in laser applications the word "coherence" is used mainly in situations involving only first-order coherence. For instance, a beam may be said to be "coherent" if it produces interference fringes in a Michelson interferometer with a certain path separation, or if it gives interference fringes in a two-slit experiment with a certain slit separation.

15.14   PHOTON STATISTICS OF LASERS AND THERMAL SOURCES

In this chapter we have introduced some of the main ideas of optical coherence theory in terms of the classical wave theory of light. The particle aspect of light was considered only in Section 15.12, where we found it convenient to think in terms of photons in order to elucidate the meaning of the volume of coherence. There we found that lasers and normal thermal sources are drastically different in terms of the number of photons they can put into a single field mode.

As discussed in Chapter 9, it is possible experimentally to *count* photons (or rather, to count the photoelectrons ejected from a photoemissive surface by the absorption of photons). Such experiments very clearly reveal that the difference between lasers and thermal sources is not merely a matter of how many photons can be generated. They show that lasers are different from conventional sources in a *fundamental* way, because their higher-order coherence functions are different.

In Chapter 9 we considered a typical photon-counting experiment, which records the number of photoelectrons ejected by a light beam during a time interval T. By repeating the experiment a large number of times, the probability distribution $P_n(T)$ for the number n of photoelectrons counted during a time T can be determined, and under ideal circumstances we may say that this is the probability distribution for photons counted in a time T. Using our present ensemble-average notation, we found that, if the intensity

$$\langle \bar{I}(t; T) \rangle = \frac{1}{T} \int_{t}^{t+T} \langle I(t') \rangle \, dt' \qquad (15.14.1)$$

averaged over the counting time interval from t to $t + T$ is a constant, independent of t, then $P_n(T)$ is the Poisson distribution:

$$P_n(T) = \frac{\bar{n}^n}{n!} e^{-\bar{n}} \qquad (15.14.2)$$

where $\bar{n}$ is the average number of photons counted in a time interval T.

Equation (15.4.2) may be expected to apply to a cw laser beam. It also applies to a beam of thermal radiation, provided the intensity (15.14.1) is independent of t. This will be true if the counting interval T is large enough, because fluctuations in the thermal-field intensity will be averaged out. In fact T is "large enough" if it is large compared with the coherence time $\tau_{coh} = 1/2\pi\Delta\nu$ of the (quasimonochromatic) thermal radiation.

Suppose, however, that T is small compared with the coherence time of a thermal source. Then the fluctuations in the intensity during a time T are not averaged out, and the Poisson distribution (15.14.2) for the photon counts does not apply. To obtain $P_n(T)$ in this case, let us consider a single-mode thermal field. Such a field is in thermal equilibrium at some temperature T_e, and the probability that there are exactly n photons in the field is given by the Boltzmann law:

$$
\begin{aligned}
p(n) &= \frac{e^{-E_n/kT_e}}{\displaystyle\sum_{m=0}^{\infty} e^{-E_m/kT_e}} \\[2em]
&= \frac{e^{-nh\nu/kT_e}}{\displaystyle\sum_{m=0}^{\infty} e^{-mh\nu/kT_e}}
\end{aligned}
\tag{15.14.3}
$$

since $E_n = nh\nu$ is the energy of a state with n photons of frequency ν. The series can be summed easily:

$$
\sum_{m=0}^{\infty} e^{-mh\nu/kT_e} = \sum_{m=0}^{\infty} x^m = (1 - x)^{-1}
\tag{15.14.4}
$$

where

$$
x = e^{-h\nu/kT_e}
\tag{15.14.5}
$$

Therefore

$$
\begin{aligned}
p(n) &= (1 - x)x^n \\
&= (1 - e^{-h\nu/kT_e}) e^{-nh\nu/kT_e}
\end{aligned}
\tag{15.14.6}
$$

Furthermore the average photon number is

$$
\begin{aligned}
\bar{n} &= \sum_{n=0}^{\infty} np(n) = (1 - x) \sum_{n=0}^{\infty} nx^n \\
&= x(1 - x)^{-1} = (x^{-1} - 1)^{-1} \\
&= (e^{h\nu/kT_e} - 1)^{-1}
\end{aligned}
\tag{15.14.7}
$$

which is just (7.7.20). Using this result in (15.14.6), we may write the probability distribution $p(n)$ in the form

$$p(n) = \frac{\bar{n}^n}{(\bar{n} + 1)^{n+1}} \qquad (15.14.8)$$

Returning now to our photon counting with $T \ll \tau_{coh}$, we observe that for a single mode $\Delta\nu \to 0$ and therefore $\tau_{coh} \to \infty$. We can therefore assume $T \ll \tau_{coh}$, which leads us to suspect that $p(n)$ in (15.14.8) gives the photon-counting probability distribution for a thermal field when the counting interval T is small compared with the coherence time τ_{coh}. This suspicion can be justified by a more rigorous approach, but we will just assume it is true: the probability of counting n photons of a thermal field during a time interval $T \ll \tau_{coh}$ is

$$P_n(T) = \frac{\bar{n}^n}{(\bar{n} + 1)^{n+1}} \qquad \text{(thermal field)} \qquad (15.14.9)$$

where $\bar{n}$ is the average number of photons counted in a time interval T, as opposed to

$$P_n(T) = \bar{n}^n e^{-\bar{n}}/n! \qquad \text{(laser)} \qquad (15.14.10)$$

for a laser beam. Figure 15.14 compares the probability distributions (15.14.9) and (15.14.10) for the case $\bar{n} = 20$. It is clear that a laser and a thermal source, even though they may give the same average number $\bar{n}$ of photons during a counting interval, will nevertheless have completely different photon statistics. The two fields may have the same average intensity, the same frequency and bandwidth, and the same first-order spatial and temporal coherence properties, but they are nevertheless different.

Photon-counting experiments have accurately confirmed the predictions

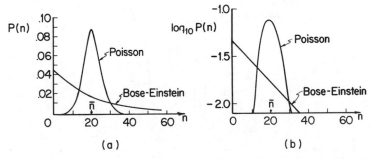

Figure 15.14 Poisson and Bose–Einstein distributions on linear (a) and semilog (b) plots for $\bar{n} = 20$.

(15.14.9) and (15.14.10). They have also shown, again in agreement with theory, that the light from a laser below the threshold for oscillation has the photon statistics of a thermal field. In other words, a laser below threshold is characterized by the photon-counting probability distribution (15.14.9), whereas above threshold it is characterized by the Poisson distribution (15.14.10).

The distribution (15.14.9) is called the *Bose–Einstein distribution*. Since

$$\ln P_n(T) = \ln \frac{\bar{n}^n}{(\bar{n} + 1)^{n+1}}$$

$$= n \ln \frac{\bar{n}}{\bar{n} + 1} - \ln (\bar{n} + 1) \qquad (15.14.11)$$

a plot of $\ln P_n(T)$ vs. n is a straight line. Figure 15.15 shows experimental results for a 6328-Å He–Ne laser, both below and above the threshold for oscillation. Below threshold the observed distribution is given accurately by the straight line (15.14.11), whereas above threshold good agreement is found with the Poisson distribution.

In Section 11.11 we found that a remarkable transition occurs as a laser goes from below threshold to above threshold: the average photon number changes rather abruptly by orders of magnitude. Now we see that the threshold point is also the boundary between two completely different types of photon statistics.

In (9.6.11) we noted that the mean square deviation of the photon number is given by

$$\langle \Delta n^2 \rangle = \langle (n - \bar{n})^2 \rangle = \bar{n} \quad \text{(Poisson distribution)} \quad (15.14.12)$$

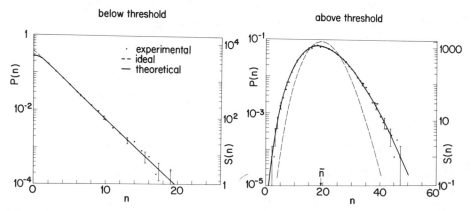

Figure 15.15 Photon statistics of a He–Ne laser above and below threshold. From C. Freed and H. A. Haus, *IEEE Journal of Quantum Electronics* **QE-2**, 190 (1966).

for the Poisson distribution. For the Bose–Einstein distribution it is not difficult to find that a different result holds:

$$\langle \Delta n^2 \rangle = \bar{n}^2 + \bar{n} \quad \text{(Bose–Einstein distribution)} \quad (15.14.13)$$

Thus, whereas the relative rms deviation for the Poisson distribution is

$$\frac{(\Delta n)_{\text{rms}}}{\bar{n}} = \frac{1}{\sqrt{\bar{n}}} \quad \text{(Poisson distribution)} \quad (15.14.14)$$

we have instead

$$\frac{(\Delta n)_{\text{rms}}}{\bar{n}} = \sqrt{1 + \frac{1}{\bar{n}}} \quad \text{(Bose–Einstein distribution)} \quad (15.14.15)$$

for the Bose–Einstein distribution. Whereas the relative rms deviation from the mean for the Poisson distribution decreases toward zero with increasing $\bar{n}$, the corresponding deviation for the Bose–Einstein distribution approaches unity. Thus the fluctuations in the photon number for a thermal field can be much more pronounced than in the case of a coherent laser field.

• Equation (15.14.13) is sometimes referred to as the *Einstein fluctuation formula*. Using Eq. (15.14.7) for $\bar{n}$, and the Planck formula

$$\rho(\nu) = \frac{8\pi h\nu^3}{c^3} \bar{n} \quad (15.14.16)$$

for the spectral energy density of thermal radiation, we obtain

$$\langle \Delta n^2 \rangle = \bar{n}^2 + \bar{n}$$

$$= \frac{c^3}{8\pi h\nu^3} \left(\rho(\nu) + \frac{c^3}{8\pi h\nu^3} \rho(\nu)^2 \right) \quad (15.14.17)$$

The mean square deviation from the average energy for thermal radiation of frequency ν is

$$\langle \Delta E_\nu^2 \rangle = (h\nu)^2 \langle \Delta n^2 \rangle [8\pi\nu^2 V \, d\nu / c^3] \quad (15.14.18)$$

where the factor in brackets is the number of field modes in the volume V and in the frequency interval $[\nu, \nu + d\nu]$, as derived in (2.1.40). Thus, by using (15.14.17) in (15.14.18), we find

$$\langle \Delta E_\nu^2 \rangle = \left[h\nu\rho(\nu) + \frac{c^3}{8\pi\nu^2} \rho(\nu)^2 \right] V \, d\nu \quad (15.14.19)$$

which is the form of the fluctuation formula for thermal radiation found by Einstein in 1909.

The Einstein fluctuation formula is important historically as the first indication of the wave–particle duality of light. The first term inside the brackets in (15.14.19) can be derived from the classical Poisson statistics of distinguishable *particles*, whereas the second term follows from a purely *wave* approach to thermal radiation, as was shown by Lorentz. Thus the fluctuation formula (15.14.19) has both particle and wave contributions. Einstein showed that this formula is a direct consequence of the Planck law for $\rho(\nu)$. •

15.15 BROWN–TWISS CORRELATIONS

We have emphasized that first-order coherence does not in general distinguish between lasers and ordinary, thermal sources of radiation, even though the two have measurably different photon statistics. The results (15.14.12) and (15.14.13) for the variance $\langle \Delta n^2 \rangle$ suggest that lasers and ordinary sources may be distinguished by their *second*-order coherence properties. This is indeed the case, and for this reason we will conclude this chapter with a discussion of an important experiment concerned with second-order coherence properties of radiation.

The experiment is sketched in Figure 15.16. Quasimonochromatic radiation is incident upon a 50:50 beam splitter, so that half the original intensity is directed to a photomultiplier tube PM1, the other half to a second photomultiplier PM2. The responses of the photomultipliers are used to determine the average of the product of $I_1(t)$ and $I_2(t + \tau)$, where I_1 and I_2 are the intensities incident on PM1 and PM2, respectively. We denote this average by $C(\tau)$:

$$C(\tau) = \langle I_1(t) \, I_2(t + \tau) \rangle \tag{15.15.1}$$

The instrument performing this correlation is simply called the "correlator" in Figure 15.16. We assume the field has the property of stationarity, so that $\langle I_2(t + \tau) \rangle = \langle I_1(t) \rangle \equiv I$ for a 50:50 beam splitter. As usual the intensities $I_1(t)$ and $I_2(t)$ are themselves averages over a few optical periods. Note that, from the definition (15.13.1), we have

$$C(\tau) = (c\epsilon_0/2)^2 \langle E^*(\mathbf{r}_1, t) \, E^*(\mathbf{r}_2, t + \tau) \, E(\mathbf{r}_2, t + \tau) \, E(\mathbf{r}_1, t) \rangle \tag{15.15.2}$$

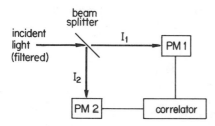

Figure 15.16 Experimental setup to determine second-order coherence properties of light by measuring $C(\tau)$ [Eq. (15.15.1)].

That is, $C(\tau)$ is a second-order coherence function. Here $\mathbf{r}_1$ and $\mathbf{r}_2$ refer to points on PM1 and PM2, respectively. Since spatial variations of the field will not play an important role in our discussion, we will ignore them.

The second-order coherence function $C(\tau)$ was first measured by R. H. Brown and R. Q. Twiss in the mid 1950s. Brown and Twiss used radiation from a mercury arc lamp, and obtained a result like that shown in Figure 15.17. The most significant feature of Figure 15.17 is the positive correlation for small values of the delay time τ: for small τ, $\langle I_1(t)I_2(t + \tau)\rangle$ is *larger* than $\langle I_1(t)\rangle\langle I_2(t + \tau)\rangle$ $= I^2$. This *Brown–Twiss effect* is explainable in either classical or quantum-mechanical terms. We begin with a quantum-mechanical description, focusing our attention on the correlation with zero time delay, the quantity $C(0) = \langle I_1(t)I_2(t)\rangle$.

We will assume that quantum-mechanical expectation values are directly applicable to the interpretation of a photon-counting experiment when the counting time is short compared with the coherence time. The first question we must then address is that of associating a quantum-mechanical expectation value with $C(0)$. Now a measurement of the intensity of a quasimonochromatic field records a quantity proportional to $\langle n\rangle = \bar{n}$, the expectation value of the photon number, and so we might suspect that an experiment like that of Figure 15.16 measures a quantity proportional to $\langle n^2\rangle$. In fact, such an experiment measures not $\langle n^2\rangle$ but $\langle n^2\rangle - \langle n\rangle$. Suppose the field incident on the beam splitter contains one and only one photon, i.e., it is in a stationary state of energy $h\nu$. In this case $\langle n^2\rangle = 1$. But the experiment of Figure 15.16 must give $C(0) = 0$ in this case, because the beam splitter cannot split the incident photon. Thus $C(0)$ cannot be proportional to $\langle n^2\rangle$. On the other hand $\langle n^2\rangle - \langle n\rangle = 0$ in this example, so that $\langle n^2\rangle - \langle n\rangle$ is at least a plausible possibility for the quantum expectation value given by $C(0)$. We will simply outline a derivation of this result. A proof is given in the black-dot section below.

In an experiment like that indicated in Figure 15.16, in which PM1 and PM2 can simultaneously record photon counts, we are interested, according to (15.15.2), in the average value of a product of four fields. If the incident field is in a stationary state of n photons, this expectation value turns out to be proportional to $n(n - 1)$. This result may be thought of as follows. The probability of counting a photon

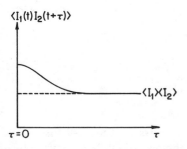

Figure 15.17 Photon bunching or [see R.H. Brown and R.Q. Twiss, Nature **177**, 27 (1956)] Brown–Twiss intensity correlation near $\tau = 0$.

is proportional to n, the number of photons in the field. Thus the probability of then counting a second photon is proportional to $n - 1$, the number of photons remaining after the first is removed by the detection process. The probability of counting the two photons is then proportional to the average product of n and $n - 1$, i.e., to the expectation value $\langle n(n - 1) \rangle = \langle n^2 \rangle - \langle n \rangle$. Since $\langle \Delta n^2 \rangle = \langle n^2 \rangle - \langle n \rangle^2$, it follows that

$$C(0) \propto [\langle n^2 \rangle - \langle n \rangle]$$
$$= \langle \Delta n^2 \rangle + \langle n \rangle^2 - \langle n \rangle$$
$$= \langle \Delta n^2 \rangle + \bar{n}^2 - \bar{n} \tag{15.15.3}$$

For a thermal source $\langle \Delta n^2 \rangle$ is given by (15.14.13), and so

$$C(0) = \bar{n}^2 + \bar{n} + \bar{n}^2 - \bar{n} = 2\bar{n}^2 \tag{15.15.4}$$

Since $C(0) = \langle I_1(t) I_2(t) \rangle$ and $\langle I_1(t) \rangle = \langle I_2(t) \rangle = I \propto \bar{n}$, (15.15.4) implies that

$$\frac{\langle I_1(t) I_2(t) \rangle}{\langle I_1(t) \rangle \langle I_2(t) \rangle} = 2 \quad \text{(thermal radiation)} \tag{15.15.5}$$

According to this result, a thermal field exhibits excess intensity fluctuations (i.e., the ratio is greater than 1) for $\tau = 0$, as shown in Figure 15.17. This is the Brown–Twiss effect for thermal radiation. It is seen from our derivation to be a consequence of the Bose–Einstein statistics of thermal radiation.

This Brown–Twiss correlation is often called *photon bunching*, for it indicates a tendency for photons to arrive simultaneously at PM1 and PM2, i.e., to bunch together. This bunching also occurs if PM1 and PM2 are close together or coincident, in which case we may say that *for thermal radiation there is a statistical tendency for photons to arrive in pairs*.

Suppose, however, that the incident light is from a laser. In this case the photon statistics is Poissonian, and $\langle \Delta n^2 \rangle$ is given by (15.14.12). From (15.15.3) we then have

$$C(0) \propto [\bar{n} + \bar{n}^2 - \bar{n}] = \bar{n}^2 \tag{15.15.6}$$

or

$$\frac{\langle I_1(t) I_2(t) \rangle}{\langle I_1(t) \rangle \langle I_2(t) \rangle} = 1 \quad \text{(laser)} \tag{15.15.7}$$

Thus a field described by Poisson photon statistics shows no Brown–Twiss correlations. There are no excess fluctuations. The photon arrivals are statistically independent. This distinction between laser radiation and thermal radiation has been accurately confirmed experimentally.

- In more advanced analyses it is convenient to define the non-Hermitian operators a and $a^\dagger$ from the coordinate q and momentum p:

$$a = (2\hbar\omega)^{-1/2} (p - i\omega q) \qquad (15.15.8a)$$

$$a^\dagger = (2\hbar\omega)^{-1/2} (p + i\omega q) \qquad (15.15.8b)$$

for a harmonic oscillator of angular frequency ω and unit mass. These operators satisfy the commutation relation

$$aa^\dagger - a^\dagger a = 1 \qquad (15.15.9)$$

as is easily verified from the definitions (15.15.8) and the well-known commutator $[q, p] = i\hbar$. When the operator a acts on a stationary state $|n\rangle$ of oscillator energy $n\hbar\omega$, it converts it to a state $|n - 1\rangle$; a is therefore called a lowering (or annihilation) operator. Similarly $a^\dagger$ acting on the state $|n\rangle$ produces the state $|n + 1\rangle$, and so $a^\dagger$ is called a raising (or creation) operator. In the quantum theory of radiation these operators are called photon annihilation and creation operators, and it may be shown that they represent the quantized complex field amplitudes E and E^*, respectively. Thus we have

$$C(0) \propto \langle a^\dagger a^\dagger a a \rangle \qquad (15.15.10)$$

for a single-mode field. Using (15.15.9), we have

$$a^\dagger a^\dagger a a = a^\dagger (a a^\dagger - 1) a = a^\dagger a a^\dagger a - a^\dagger a \qquad (15.15.11)$$

Now from (15.15.8) it follows that

$$\hbar\omega\, a^\dagger a = \tfrac{1}{2}(p^2 + \omega^2 q^2) - \tfrac{1}{2}$$

$$= H - \tfrac{1}{2}\hbar\omega \qquad (15.15.12)$$

where H is the Hamiltonian operator for the mode "oscillator", with eigenvalues $(n + \tfrac{1}{2})$ $\hbar\omega$. It follows that $a^\dagger a$ is the operator associated with the photon number n, and so from (15.15.10) and (15.15.11) we have

$$C(0) \propto \langle n^2 \rangle - \langle n \rangle \qquad (15.15.13)$$

which is the result used above.

The key step in the derivation of (15.15.13) is the identification (15.15.10) of $C(0)$ with $\langle a^\dagger a^\dagger a a \rangle$. In particular, the ordering of the a, $a^\dagger$ operators is absolutely crucial. This ordering, in which $a^\dagger$'s appear to the left of a's, is called *normal ordering*. For a discussion of the physical motivation for normal ordering we refer the reader to more advanced treatises.[5] •

5. R. Loudon, *The Quantum Theory of Light* (Oxford University Press, second edition, 1983).

The term "photon bunching" for the Brown–Twiss effect might convey the impression that the effect cannot be understood in classical terms. However, it is possible, and instructive, to explain the effect without invoking photons and the quantum theory of radiation. For this purpose we consider a model of a thermal source in which there are N atoms, atom j radiating a field whose time dependence is given by

$$E_j(t) = \mathcal{E}(t) e^{-i\omega t} = \mathcal{E}_0 e^{-i(\omega t + \phi_j)} \tag{15.15.14}$$

where the phase ϕ_j varies randomly in time due to collisions among the atoms. The total complex field is

$$E(t) = \mathcal{E}_0 e^{-i\omega t} \sum_{j=1}^{N} e^{-i\phi_j} \tag{15.15.15}$$

and the total intensity is [recall (15.4.11)]

$$I(t) = \tfrac{1}{2} c \epsilon_0 |\mathcal{E}_0|^2 \sum_{i=1}^{N} \sum_{j=1}^{N} e^{i(\phi_i - \phi_j)} \tag{15.15.16}$$

when we average over a few periods of time $2\pi/\omega$. Now if the randomly varying phases ϕ_i are statistically independent then the statistical average of $e^{i(\phi_i - \phi_j)}$ is zero unless $i = j$, in which case it is unity [recall Problem 12.5b]. Thus the average value of $I(t)$ is

$$\langle I(t) \rangle = \tfrac{1}{2} c \epsilon_0 |\mathcal{E}_0|^2 \sum_{j=1}^{N} 1 = \tfrac{1}{2} c \epsilon_0 N |\mathcal{E}_0|^2 \tag{15.15.17}$$

Note that this is a classical average over the random phases, not a quantum-mechanical expectation value.

Similarly the average square of the intensity corresponding to (15.15.16) is proportional to

$$\langle I(t)^2 \rangle = \left(\tfrac{1}{2} c \epsilon_0 \right)^2 |\mathcal{E}_0|^4 \sum_i \sum_j \sum_l \sum_m \langle e^{i(\phi_i - \phi_j + \phi_l - \phi_m)} \rangle \tag{15.15.18}$$

Because the ϕ_j's are independent, the summand above is nonzero only when $i = j, l = m$ or when $i = m, j = l$. Thus

$$\begin{aligned}
\langle I(t)^2 \rangle &= \left(\tfrac{1}{2} c \epsilon_0 \right)^2 |\mathcal{E}_0|^4 \left(\sum_{i=1}^{N} \sum_{l=1}^{N} 1 + \sum_{i=1}^{N} \sum_{j=1}^{N} 1 \right) \\
&= \left(\tfrac{1}{2} c \epsilon_0 \right)^2 |\mathcal{E}_0|^4 (N^2 + N^2) \\
&= 2 \langle I(t) \rangle^2
\end{aligned} \tag{15.15.19}$$

The factor of 2 leads to the result (15.15.5), i.e., to the photon bunching or Brown–Twiss intensity correlation. Obviously we have here only a very crude model of a thermal source. The point is that the uncorrelated emissions of rapidly fluctuating fields lead naturally to the positive intensity correlations found in the Brown–Twiss experiment.

If the atoms of a source all act coherently to produce a total complex field with a single phase $\phi(t)$,

$$E(t) = N\mathcal{E}_0 \, e^{-i(\omega t + \phi)} \qquad (15.15.20)$$

and therefore a total (cycle-averaged) intensity

$$I = \frac{c\epsilon_0}{2} \left| N\mathcal{E}_0 \right|^2 \qquad (15.15.21)$$

then obviously (15.15.7) is satisfied (i.e., there are no Brown–Twiss correlations) simply because I is constant. Note that this absence of Brown–Twiss correlations applies even if the phase ϕ in (15.15.20) is randomly varying, because (15.15.21) is valid regardless of the variations of a single uniform ϕ. Thus we have a simple, classical explanation for the absence of Brown–Twiss correlations in an ideal laser beam.

• Although classical explanations are possible for many aspects of the coherence properties of optical fields, quantum theory allows for a far richer variety of photon statistics. We have considered only two examples, namely the Poisson statistics of laser radiation and the Bose–Einstein statistics of thermal radiation. The latter exhibits photon bunching; the former does not. To underscore the fact that other kinds of photon statistics can arise, we consider briefly the possibility of *photon antibunching*, i.e., reduced intensity fluctuations (Figure 15.18), or the opposite of the Brown–Twiss correlation. In this case there is a tendency for photons not to arrive in pairs, or even to arrive randomly, but to arrive "well-spaced". In particular, there is zero probability of detecting a second photon immediately after the detection of a first.

Photon antibunching has been observed in the resonance fluorescence of an atom prepared as a two-state system by pumping with circularly polarized light. The two-state atom is driven up and down between the two states by a strong resonant field. In addition to the stimulated transitions, the atom can spontaneously emit a photon when it is in the upper

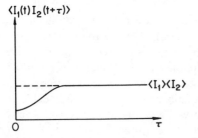

Figure 15.18 Photon antibunching near $\tau = 0$ [see H.J. Kimble, M. Dagenais and L. Mandel, Physical Review Letters **39**, 691 (1977)].

state. After such a spontaneous emission of a photon, however, the atom cannot emit a second photon until it has been pumped by the resonant field back into the upper state. Thus the fluorescence (spontaneous emission) radiation displays a reduced intensity correlation, i.e., photon antibunching. •

A field has first-order coherence if its complex degree of coherence $\gamma(\mathbf{r}_1, \mathbf{r}_2, \tau)$ has modulus unity. This occurs if and only if the mutual coherence function (15.4.18) factors, i.e., if

$$
\begin{aligned}
\Gamma(\mathbf{r}_1, \mathbf{r}_2, \tau) &= \langle \mathcal{E}^*(\mathbf{r}_1, t)\, \mathcal{E}(\mathbf{r}_2, t + \tau) \rangle \\
&= \langle \mathcal{E}^*(\mathbf{r}_1, t) \rangle \langle \mathcal{E}(\mathbf{r}_2, t + \tau) \rangle \quad (15.15.22)
\end{aligned}
$$

This factorization condition for first-order coherence [i.e., for $|\gamma(\mathbf{r}_1, \mathbf{r}_2, \tau)| = 1$] follows easily from the definition (15.6.3) of the complex degree of coherence. Similarly, a field is said to have *second-order coherence* if (15.15.7) is satisfied; this condition is equivalent to a factorization of the second-order coherence function of the field. In general a field is said to have *n*th-order coherence if its *n*th-order coherence function factors. This definition of coherence leads to an elegant formulation of the quantum theory of coherence.

PROBLEMS

15.1   Assume that the sun is a 5800-K blackbody radiator. Calculate the intensity of radiation from the sun.

15.2   In Section 14.6 we obtained the result (14.6.20) for the new waist of a focused Gaussian beam, assuming that the lens intercepts the beam at its waist. Show that (15.2.9) gives the spot size in the focal plane of the lens.

15.3   Assume that the lens of the human eye has a focal length $f = 2.3$ cm, and that the pupil diameter is 2 mm. Suppose a 1-mW laser beam is viewed directly with the eye, and that the eye is approximately a diffraction-limited system. Estimate the size and intensity of the image on the retina. You may assume for simplicity that the beam is of uniform intensity over the pupil aperture. [Note: The retinal damage threshold in the case of continuous illumination can be as low as 2–3 W/cm^2. See A. M. Clarke *et al.*, *Archives of Environmental Health* **18**, 424 (1969).]

15.4   Explain why a nearsighted person sees a speckle pattern moving in the direction opposite to the head motion. (Recall that the brain inverts the image on the retina.) What would a farsighted person see? A person with no visual disorder? [Reference: N. Mohon and A. Rodemann, *Applied Optics* **12**, 783 (1973).] If you wear corrective lenses, remember to remove them when you try this experiment. Explain why this effect of speckle is similar to the following one you can do while sitting at your desk. Focus your eyes on a distant object outside a window, and note the apparent motion of an object on the window sill as you move your head up and down or side to side. Then focus on the object on the sill, and note the apparent motion of the object outside when you move your head.

15.5 In an actual Michelson interferometer using an extended source of light (see Figure 15.11), we observe circular interference fringes when viewing M_1 through BS if M_1 and M_2 are perpendicular (one of the mirrors of the interferometer has tilting screws so that the mirror orientation is adjustable.) Explain the appearance of circular fringes. Do you expect to see circular fringes if the mirrors are not perpendicular?

15.6 Explain how a Michelson interferometer can be used to determine the wavelength of quasimonochromatic radiation.

15.7 Why do we normally add intensities of different frequency components of radiation to obtain the total measured intensity?

15.8 (a) Estimate the bandwidth and coherence time of white light, assuming it comprises wavelengths from 4000 to 7000 Å.

(b) Show that the coherence length of white light is on the order of the wavelength.

(c) Is is possible to observe "white-light fringes"? [See A. Michelson, *Light Waves and Their Uses* (University of Chicago press, Chicago, 1906).]

15.9 Suppose a mercury arc lamp, emitting 5461-Å radiation with a bandwidth of 10^9 Hz, is placed behind a 0.1-mm-diameter circular aperture in an opaque screen. Beyond the screen is a second one with two narrow slits in it, and interference fringes are observed on a third screen a distance 3 m from the second. Calculate the slit separation for which the interference fringes first disappear.

15.10 A piece of transparent ground glass is placed before the two slits of a Young interference experiment and rotated rapidly. It is found that spatially coherent radiation does not produce any interference fringes in this modified two-slit experiment. Explain this observation [Reference: W. Martienssen and E. Spiller, *Naturwissenschaften* **52**, 53 (1965).]

16 SOME LASER APPLICATIONS

16.1 INTRODUCTION

In the early years of its development, the laser was regarded by skeptics as "a solution looking for a problem." Eventually more and more "problems" were found, and lasers have unquestionably become an important part of the science and technology of our time, with applications ranging from medical to military. In this chapter we consider how the special properties of laser radiation may be put to use.

New applications of lasers continue to be found, and some of those described in this chapter may be just the beginnings of major developments. In a single chapter we cannot describe any of these applications more than superficially. Our intention is to emphasize in each instance what specific properties of laser radiation are being exploited.

16.2 DISTANCE AND VELOCITY MEASUREMENTS

One way to determine the distance to an object is to measure the time T taken for a short pulse of laser radiation to reach the object and reflect back to the observer; the distance to the object is then given by $d = cT/2$. Because of its similarity to the radar technique employing radio-frequency waves, this method is known as *optical radar*, or *lidar* (*li*ght *d*etection *a*nd *r*anging). Laser optical radar has been used to determine the distance between points on the earth and the moon to an accuracy of a few inches.

The principal components of such a system are a pulsed laser, a photodetector and timing system, and a telescope such as that indicated in Figure 16.1 to collect the reflected light. Photodetection and timing circuitry typically have resolving times ΔT on the order of a few nanoseconds. This usually limits optical radar to an accuracy of $c \Delta T \approx 15$ cm, which means that the relative accuracy of a distance measurement is greater when large distances (say, > 10 km) are to be measured.

Laser optical radar schemes have the advantage of high directionality of the laser beams and the availability of very short pulses. Furthermore the intensity of the echo pulse can far exceed that obtained from conventional radar sources. For very large distances a suitable retroreflector, such as a *corner cube* (Fig. 16.2)

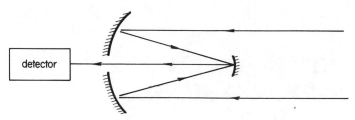

Figure 16.1 A collecting and focusing telescope.

may be necessary to produce a reflected pulse of sufficient intensity for accurate detection with a photomultiplier. A corner cube reflects an incident ray in a direction exactly parallel to its direction of incidence. Arrays of small corner cubes are used in traffic signs and in bicycle reflectors.

An array of 100 corner cubes arranged in an 18-inch square was placed on the moon by Apollo 11 astronauts on 20 July 1969. Pulses from a ruby laser at the Lick Observatory in California were aimed at this retroreflector, and return pulses were first detected on 1 August.

It takes about $2\frac{1}{2}$ seconds for light to travel the 480,000 miles from the earth to the moon and back. After a delay of about $2\frac{1}{2}$ seconds following the firing of each laser pulse, a set of 12 counters in the first lunar ranging experiments were sequentially activated to count pulses registered by a photodetector; the exact delay time was set for each shot according to theoretical predictions for the arrival time. Each counter was active over a time interval that could be adjusted from 0.25 μsec to 2 μsec, and the precise counting sequence was chosen so as to be centered on the predicted arrival time. The distance to the moon could then be accurately computed according to which of the counters "filled up" more rapidly than the rest.

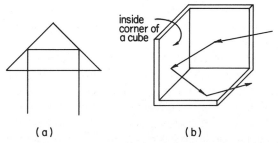

(a)   (b)

Figure 16.2 (a) A *roof prism*, in which any incident ray perpendicular to the face of the prism is reflected parallel to itself. (b) A cube corner prism (or *corner cube*) consisting of three edges intersecting at right angles, a kind of generalization of the roof prism. Any ray incident on the corner cube is reflected parallel to itself.

• The design of the lunar laser reflector was chosen carefully. The smallest diffractive spreading of the echo pulse to be detected on earth is obtained by making the corner cubes as large as weight restrictions would allow. However, the echo pulse is in fact displaced sideways because of velocity aberration, i.e., because the moon and the transmitter on earth are in relative motion. This displaces the center of the return beam about a mile from the point of transmission on earth. In order not to need separate stations for transmission and reception, the corner cubes were chosen small enough to make diffractive spreading of the return beam large enough to illuminate the transmission point. In addition, larger corner cubes would suffer from thermally induced material and optical distortions due to the large temperature variations ($-170°C$ to $+130°C$) on the lunar surface. The $1\frac{1}{2}$-inch corner cubes produce about a 10-mile beam diameter of the return pulse at the earth's surface (Problem 16.1). •

Accurate measurements of the *variations* in time of the earth–moon separation can help to answer important scientific questions, such as how the earth's continents drift, how the lunar mass is distributed, or whether the universal gravitational constant G is changing with time.

In 1976 the U.S. National Aeronautics and Space Administration (NASA) launched the Laser Geodynamics Satellite (LAGEOS). This satellite, a solid sphere of mass 407 kg and diameter 60 cm, is covered with 426 retroreflectors for the purpose of *satellite* laser ranging (SLR), similar in principle to lunar laser ranging. The position of LAGEOS, which orbits the earth at an altitude of 5900 km, can be accurately computed over long periods of time, and so by SLR the earth's rate of rotation can be accurately determined. Laser ranging of LAGEOS over a period of three years revealed small variations in the length of the earth day. These variations are attributed to the exchange of angular momentum between the earth's atmosphere and mantle.

Rangefinding (i.e., distance measurement) has many military applications, for which the lidar technique is used rather extensively. Laser rangefinding weaponry was first used during the war in Vietnam. On a tank, for instance, a lidar system consists of a sighting scope, the laser transmitter and receiver, and a computer to determine the range of the target and perhaps to regulate an automated fire-control system. Portable lidars for ground troops are about the size of a pair of binoculars, weigh only a few pounds, and can measure distances from a few hundred yards to several miles to within an accuracy of a few yards. In laser-guided bombs and missiles a laser beam reflected off a target is sought by a "smart" bomb, which then follows the reflected light to the target.

Laser lidars have numerous other applications, all of which derive from the high directionality and intensity of laser radiation. These include aircraft altimeters, the detection and characterization of fog layers, atmospheric pollution monitoring, and surveying, to name just a few. In many of these applications an important role is also played by the narrow frequency bandwidth of the laser radiation, which permits a high signal-to-noise ratio in discrimination against stray background light.

Lasers may be used in distance measurement in other ways besides the pulse echo technique. *Interferometric methods* are essentially just variations of the Michelson interferometer discussed in Section 15.10. Equations (15.10.8) show that if the arm separation of the interferometer is changed by $\lambda/2$, the interference maxima and minima are interchanged. Thus the magnitude of this change in the arm separation can be determined in terms of the wavelength λ by counting the number of fringe shifts in the interference pattern as the change occurs. The advantage of using laser radiation lies in the fact that much greater coherence lengths are attainable than in the case of ordinary sources. Thus much greater arm separations can be used before the interference fringes get smeared out, and so much larger distances can be measured. Relative accuracies of 10^{-6} are readily achieved, and interferometric methods of distance measurement are used routinely in length calibrations and machining applications. In practice such methods are limited to lengths less than about 100 m because of density fluctuations (and consequent refractive-index fluctuations) in the atmosphere.

Lasers may be used to measure velocities by taking advantage of the Doppler effect. The basic idea is the same as that used in radar-equipped police cars. If a laser beam of frequency ν is directed at a moving target as in Figure 16.3, the reflected radiation has a Doppler-shifted frequency (Section 3.11)

$$\nu' \approx \left(1 - \frac{v}{c}\right)\nu \tag{16.2.1}$$

The Doppler shift $v\nu/c$ can be detected by *optical heterodyning:* two signals, $\mathcal{E}_1 \cos \omega_1 t$ and $\mathcal{E}_2 \cos \omega_2 t$, for instance, are combined at a square-law detector (one sensitive to total light intensity) to produce an intensity

$$
\begin{aligned}
I &= \epsilon_0 c \left(\mathcal{E}_1 \cos \omega_1 t + \mathcal{E}_2 \cos \omega_2 t\right)^2 \\
&= \epsilon_0 c \left(\mathcal{E}_1^2 \cos^2 \omega_1 t + \mathcal{E}_2^2 \cos^2 \omega_2 t + 2\mathcal{E}_1 \mathcal{E}_2 \cos \omega_1 t \cos \omega_2 t\right) \\
&= \epsilon_0 c \left[\tfrac{1}{2} \mathcal{E}_1^2 (1 + \cos 2\omega_1 t) + \tfrac{1}{2} \mathcal{E}_2^2 (1 + \cos 2\omega_2 t)\right. \\
&\quad \left. + \mathcal{E}_1 \mathcal{E}_2 \cos (\omega_1 + \omega_2)t + \mathcal{E}_1 \mathcal{E}_2 \cos (\omega_1 - \omega_2)t\right]
\end{aligned}
\tag{16.2.2}
$$

The oscillations at frequencies $2\omega_1$, $2\omega_2$, and $\omega_1 + \omega_2$ are too rapid to be followed

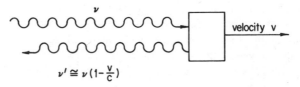

Figure 16.3 The wave reflected from a moving object undergoes a Doppler shift in frequency.

by available detectors. The frequency $|\omega_1 - \omega_2|$, however, is typically in the range of tens of megahertz in velocity measurements, and such oscillations are slow enough to be registered. The detector will thus record the intensity

$$I_D = \epsilon_0 c \left[\tfrac{1}{2} \mathcal{E}_1^2 + \tfrac{1}{2} \mathcal{E}_2^2 + \mathcal{E}_1 \mathcal{E}_2 \cos (\omega_1 - \omega_2)t \right] \qquad (16.2.3)$$

In a velocity measurement the frequency difference is the Doppler shift:

$$|\omega_1 - \omega_2| = 2\pi |\nu_1 - \nu_2| = \frac{2\pi\nu}{c} |v| = \frac{2\pi}{\lambda} |v| \qquad (16.2.4)$$

By measuring the beat frequency between the emitted and reflected laser beams, therefore, we can determine the velocity of the moving target.

An important advantage of lasers over conventional light sources for such velocity measurements is the nearly monochromatic character of laser radiation—bandwidths in perhaps the megahertz range, as opposed to the gigahertz bandwidths typical of nonlaser sources. For accurate heterodyne detection of a Doppler shift $|\nu' - \nu|$, the radiation should have a bandwidth smaller than $|\nu' - \nu|$. This limits conventional light sources to the measurement of velocities greater than about 600 m/sec ($\approx$ 1400 mph) (Problem 16.2). With frequency-stabilized lasers, on the other hand, velocities as small as 10^{-3} cm/sec can be measured. Laser velocimeters have many applications, such as the measurement of fluid flow rates, the speed of aluminum extrusions, and the velocity of aircraft with respect to ground.

16.3 THE LASER GYROSCOPE

The toy gyroscope sketched in Figure 16.4 demonstrates an important property of a mechanical gyroscope: the axis about which the wheel spins maintains its direc-

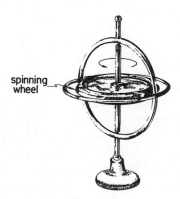

spinning wheel

Figure 16.4 A mechanical gyroscope. The axis about which the wheel spins stays fixed as the frame of the gyroscope is rotated.

tion when the frame of the gyroscope is tilted. A gyroscope on a ship likewise maintains its orientation as the ship rolls in rough water, and thereby provides information on the degree of roll. Gyroscopes are also used on aircraft as compasses and in automatic pilot systems, in torpedoes for steering to the target, and for the inertial guidance of missiles. In a sense the earth itself is a kind of gyroscope: its rotation axis maintains its orientation as the earth orbits about the sun. It is this fixed orientation, of course, that is responsible for changes of season.

In general we can define a gyroscope as a device capable of sensing angular velocity; indeed the word gyroscope itself means "viewing rotation." A gyroscope therefore need not have a spinning wheel, nor need it be a purely mechanical instrument. In this section we will describe how a ring laser (Section 11.6) may be used to sense angular velocity, or in other words how it may be used as a gyroscope.

First let us consider a thought experiment in which we have somehow managed to get two monochromatic waves propagating in a circular path of radius R, one propagating clockwise and the other counterclockwise, as illustrated in Figure 16.5a. The time taken to complete the circular path of circumference $2\pi R$ is

$$t = \frac{2\pi R}{c} \tag{16.3.1}$$

This transit time is the same for the two counterpropagating waves.

Suppose, however, that the plane of the circular path is rotating with angular velocity Ω, as in Figure 16.5b. In this case a point A on the circle is moving with tangential velocity $v = \Omega R$. The light propagating in the same sense as the rotation takes a time t_+ to go from point A to point B in Figure 16.5b, where

$$2\pi R + (\Omega R)t_+ = ct_+ \tag{16.3.2a}$$

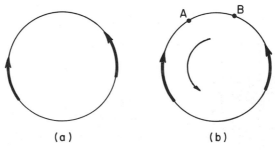

(a)   (b)

Figure 16.5 (a) A thought experiment in which two waves propagate on a circular loop, one clockwise and the other counterclockwise. (b) The case in which the plane of the circular loop is rotating. In this case the two waves take different amounts of time to tranverse the arc between two points A and B.

or

$$t_+ = \frac{2\pi R}{c - \Omega R} \tag{16.3.2b}$$

The left side of (16.3.2a) is the circumference of the circle plus the distance along the circle between points A and B. Thus t_+ is greater than the transit time (16.3.1) when there is no rotation, because during the transit time of the light the reference point A moves from A to B, and so the light must travel the extra distance $\Omega R t_+$ to complete a round trip as reckoned by an observer at rest on the circle. The light progagating in the sense opposite to the rotation, however, takes a shorter time, t_-, to complete a round trip:

$$2\pi R - (\Omega R)t_- = ct_- \tag{16.3.3a}$$

or

$$t_- = \frac{2\pi R}{c + \Omega R} \tag{16.3.3b}$$

According to an observer sitting at a point on the circle, therefore, the two waves have gone different distances in making a round trip; their path difference is

$$\begin{aligned}
\Delta L = c(t_+ - t_-) &= 2\pi Rc\left(\frac{1}{c - \Omega R} - \frac{1}{c + \Omega R}\right) \\
&= 2\pi Rc\frac{2\Omega R}{c^2 - \Omega^2 R^2} \\
&\approx \frac{4\pi R^2\Omega}{c}
\end{aligned} \tag{16.3.4}$$

where the approximation in the last step is based on the assumption that the tangential velocity of a point on the circle is much less than the velocity of light (i.e., $\Omega R \ll c$).

The path difference (16.3.4) corresponds to a phase difference

$$\begin{aligned}
\Delta\phi = 2\pi\frac{\Delta L}{\lambda} &= \frac{8\pi^2 R^2\Omega}{\lambda c} \\
&= \frac{8\pi A\Omega}{\lambda c}
\end{aligned} \tag{16.3.5}$$

where $A = \pi R^2$ is the area enclosed by the circular loop. Actually (16.3.5) is the correct result (to first order in v/c) for a closed loop of *any* shape. For instance,

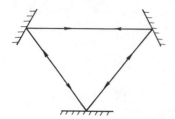 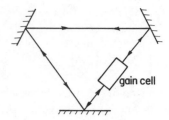

Figure 16.6 A three-mirror ring forming a closed loop for the propagation of light. Equation (16.3.5) applies to *any* such closed loop, whether it is a circle, triangle, square, etc.

Figure 16.7 A three-mirror ring laser that can be used as an optical gyroscope.

it applies to the three-mirror ring arrangement shown in Figure 16.6. And so we can forget the artificial constraints of our thought experiment with a circular loop and consider such a triangular ring, for example. By measuring interferometrically the phase difference $\Delta\phi$ between two counterpropagating light beams at a fixed point on the ring, we can measure the angular velocity Ω. In other words, we can make an *optical gyroscope.*[1]

The ability to sense rotation in this way is called the *Sagnac effect*, after George Sagnac, who in 1913 first demonstrated it using a rapidly rotating table. The main difficulty in using a "Sagnac interferometer" as a gyroscope is that the phase shift $\Delta\phi$ is typically very small, because the path difference ΔL of the counterpropagating waves is small compared with the wavelength λ. After Sagnac's demonstration Michelson used about 5 miles of (evacuated) sewer pipes to construct a Sagnac interferometer for detecting the earth's rotation. Without large loop areas (A) or large angular velocities (Ω), the phase shift (16.3.5) is too small to be measured accurately.

Suppose we put a gain cell inside the three-mirror resonator of Figure 16.6 to make a ring laser (Figure 16.7). In this case the two counterpropagating beams required for the Sagnac interferometer are generated by laser action inside the gain cell. Like the ordinary two-mirror, linear laser resonator, the ring resonator must satisfy the condition that the total round-trip distance L must be equal to an integral number m of wavelengths, or in other words

$$\nu = mc/L \tag{16.3.6}$$

In particular, a small change ΔL in L results in a change $\Delta\nu$ in ν given by (Problem 16.3)

$$\Delta\nu = \frac{\nu\,\Delta L}{L} = \frac{c\,\Delta L}{\lambda L} \tag{16.3.7}$$

---

1. See Dana Z. Anderson, "Optical Gyroscopes," Scientific American **254**, April 1986, p. 94.

Unlike a linear laser resonator, however, the two counterpropagating waves in a ring laser need not have the same frequency. In fact, if the ring is rotating about an axis perpendicular to its plane, the two waves "see" different round-trip distances $L_+ = ct_+$ and $L_- = ct_-$, and they will have a frequency difference (16.3.7) with ΔL given by (16.3.4). That is, the two counterpropagating waves in a rotating ring laser have the frequency difference

$$\Delta \nu = \frac{c}{\lambda L} \left(\frac{\lambda}{2\pi} \Delta \phi \right)$$

$$= \frac{c}{2\pi L} \left(\frac{8\pi A\Omega}{\lambda c} \right)$$

$$= \frac{4A\Omega}{\lambda L} \qquad (16.3.8)$$

where Ω is the angular velocity of rotation and A is the area enclosed by the ring of perimeter L.

Let us consider a numerical example. For a ring resonator of the type shown in Figure 16.6 with side $d = 10$ cm, we have $A = \sqrt{3}d^2/4 \approx 43.3$ cm^2 by simple geometry. If the angular velocity $\Omega = 15$ deg/hr $= 7.3 \times 10^{-5}$ rad/sec, the rotation rate of the Earth, then (16.3.5) gives

$$\Delta \phi = 4.2 \times 10^{-8} \text{ rad} \qquad (16.3.9)$$

for the He–Ne laser wavelength $\lambda = 6328$ Å. This corresponds to a ratio $\Delta L/\lambda = \Delta\phi/2\pi = 6.7 \times 10^{-9}$, and so for this loop area and rotation rate the measurement of $\Delta \phi$ is a formidable task indeed. From (16.3.8) for the ring laser gyro, on the other hand, we compute for the same parameters

$$\Delta \nu = 20 \text{ Hz} \qquad (16.3.10)$$

which is readily measured by optical heterodyning, even though it represents a tiny fraction ($\approx 4 \times 10^{-14}$) of the optical frequency ν.

A great advantage of laser gyroscopes, as compared with conventional spinning-mass gyros, is that they have no moving parts to produce mechanical stress and wear. They can furthermore be operated in the "strapped-down mode" in which they are rigidly fixed to a vehicle without any gimbals. Laser-gyro construction is relatively simple and inexpensive. Typically the ring resonator and a He–Ne gain tube are enclosed in a single drilled-out quartz block small enough to be held in one's hand. Attached to the block is the "readout" apparatus, which uses a partially transmitting mirror to combine parts of the two counterpropagating beams and measure their frequency difference interferometrically. For navigational purposes the measured rotation rate is integrated to give the total angle of rotation over a given time.

A problem with laser gyroscopes is called *lock-in*. It occurs at low rotation rates, and causes low beat frequencies to tend to lock together at a *single* intermediate frequency. (This locking phenomenon often arises in coupled oscillatory systems.) Lock-in obviously destroys the ability to measure small rotation rates. One technique used to alleviate the lock-in problem is to "dither" the gyro, i.e., to rotate it back and forth rapidly enough that lock-in cannot occur. On average the back and forth motions cancel, and what is left after a number of dithering periods is just the rotation rate one wants to measure.

Laser gyroscopes are now used on commercial airliners, and several nations are using them in military vehicles. Another type of optical rotation sensor that uses the same principles of operation is based on an optical fiber loop instead of a ring laser to form a Sagnac interferometer.

16.4 HOLOGRAPHY: THE ESSENTIAL PRINCIPLE

Holography is the art of lensless, three-dimensional photography. The essential principle of holography was discovered by Dennis Gabor in 1947. Before the development of lasers it was difficult (though not impossible) to make holograms. With the advent of the laser it was quickly recognized to be practical, and holography has since become a whole field of research with many applications. In this section we will take up the basic idea underlying the holographic process.

Consider the light scattered by some object illuminated by a monochromatic wave. In some plane ($z = 0$, say) we may represent the electric field of the scattered light as (Figure 16.8)

$$E_O(x, y, t) = \mathcal{E}_O(x, y) e^{-i\omega t} \tag{16.4.1}$$

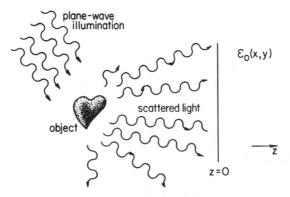

Figure 16.8 The light scattered by an object is represented in a plane $z =$ constant by the electric field $\mathcal{E}_o(x, y)$.

The subscript O labels this field as due to the object. The complex amplitude $\mathcal{E}_O(x, y)$ defined over the plane $z = 0$ of Figure 16.8 may be written as

$$\mathcal{E}_O(x, y) = A_O(x, y) \, e^{i\phi_O(x,y)} \qquad (16.4.2)$$

where $A_O(x, y)$ and $\phi_O(x, y)$ are the (real) amplitude and phase, respectively, in this plane. The real electric field defined by (16.4.1) and (16.4.2) is of course

$$E_O(x, y, t) = A_O(x, y) \cos\left[\omega t - \phi_O(x, y)\right] \qquad (16.4.3)$$

For simplicity we are considering only one scalar component of the electric field vector.

The intensity of the field (16.4.3), averaged over a number of optical periods, is proportional to A_O^2:

$$I_O(x, y) = \tfrac{1}{2}c\epsilon_0 A_O^2(x, y) \qquad (16.4.4)$$

Except for multiplicative constants, this is the intensity distribution measured by developing photographic film placed in the observation plane of Figure 16.8. Such a photograph, of course, is only a flat, two-dimensional record. This record of the object does not look exactly like the real-life object because it registers only the intensity, and not the phase, of the object field. This is where a hologram of the object differs from a photograph: *a hologram contains information about both the intensity and the phase of the object field.*

Suppose that, in addition to the object field (16.4.1), we shine some *reference beam*

$$\begin{aligned} E_R(x, y, t) &= \mathcal{E}_R(x, y) \, e^{-i\omega t} \\ &= A_R(x, y) \, e^{i\phi_R(x,y)} \, e^{-i\omega t} \end{aligned} \qquad (16.4.5)$$

on the observation plane of Figure 16.8, where $A_R(x, y)$ and $\phi_R(x, y)$ are the amplitude and phase of this reference field. Then the total amplitude at the observation plane is

$$\mathcal{E}_T(x, y) = A_O(x, y) \, e^{i\phi_O(x,y)} + A_R(x, y) \, e^{i\phi_R(x,y)} \qquad (16.4.6)$$

and the total intensity is $I_T = \tfrac{1}{2}c\epsilon_0 |\mathcal{E}_T|^2$, where

$$\begin{aligned} \left|\mathcal{E}_T(x, y)\right|^2 &= A_O^2(x, y) + A_R^2(x, y) \\ &\quad + 2A_O(x, y) \, A_R(x, y) \cos\left[\phi_O(x, y) - \phi_R(x, y)\right] \end{aligned} \qquad (16.4.7)$$

Because of the interference of the object and reference waves, this intensity, unlike (16.4.4), does contain information about the phase $\phi_O(x, y)$ of the object field.

The *recording* of this intensity distribution on photographic film is the first step of the holographic process. This intensity distribution does not look anything like the object; to obtain a fully three-dimensional image of the object we must reconstruct the wave front (16.4.2). This *reconstruction* is the second step of the holographic process. The challenge is to use the intensity distribution (16.4.7) on our developed film to reconstruct the object wave front (16.4.2).

Suppose an intensity distribution like (16.4.7) is recorded on photographic film. Exposure of the film to this intensity results in the conversion of silver halide grains in the film emulsion to silver atoms; the density of the silver atoms is greatest where the intensity of the incident light was greatest. Development as a photographic positive reverses the situation: the positive is lightest where the incident intensity was greatest. The transmission function $t(x, y)$ of the positive is therefore largest where the intensity of the incident light was high. In fact, we may assume that $t(x, y)$ is approximately proportional to the intensity $I(x, y)$ of the light to which the film was exposed. Then if the positive is illuminated with a wave front $\mathcal{E}(x, y)$, the transmitted wave front is

$$\mathcal{E}_{\text{trans}}(x, y) = t(x, y)\, \mathcal{E}(x, y)$$

$$= CI(x, y)\, \mathcal{E}(x, y) \tag{16.4.8}$$

where C is the constant of proportionality between t and I.

Now suppose that

$$\mathcal{E}(x, y) = \mathcal{E}_R(x, y) \tag{16.4.9a}$$

and

$$I(x, y) = I_T(x, y) \tag{16.4.9b}$$

where $\mathcal{E}_R$ and I_T are given by (16.4.5) and (16.4.7), respectively. That is, the original intensity distribution to which the film is exposed is the total intensity from the object and reference fields, and after development the incident field $\mathcal{E}(x, y)$ is that of the reference field alone. Then from (16.4.8) we have

$$\mathcal{E}_{\text{trans}} = C\left[A_O^2 + A_R^2 + 2A_O A_R \cos(\phi_O - \phi_R)\right] A_R e^{i\phi_R}$$

$$= C\left[A_O^2 + A_R^2 + A_O A_R\, e^{-i(\phi_O - \phi_R)} + A_O A_R\, e^{i(\phi_O - \phi_R)}\right] A_R e^{i\phi_R}$$

$$= C\left[(A_O^2 + A_R^2)A_R e^{i\phi_R} + A_O A_R^2\, e^{i(2\phi_R - \phi_O)} + A_O A_R^2 e^{i\phi_O}\right]$$

$$\tag{16.4.10}$$

The most interesting part of the transmitted field is the last term of this expression:

$$CA_R^2(x, y)\left[A_O(x, y)^{\,i\phi_O(x,y)}\right] = CA_R^2(x, y)\, \mathcal{E}_O(x, y) \tag{16.4.11}$$

where $\mathcal{E}_O(x, y)$ is the object wave front (16.4.2). If CA_R^2 is a constant independent of (x, y), and we denote this constant by K, then in (16.4.11)

$$CA_R^2 \mathcal{E}_O = K\mathcal{E}_O(x, y) \qquad (16.4.12)$$

is proportional to the object wave front $\mathcal{E}_O(x, y)$. That is, we have managed to reconstruct the complete object wave front, *including both amplitude and phase*.

Let us summarize now in physical terms this process of wave-front reconstruction. First we illuminate the object with a monochromatic field, and superpose the wave front of light scattered by the object on a reference field of the same frequency to produce a total field (16.4.6). This gives a total intensity (16.4.7), which is recorded on photographic film. By developing this film we have recorded a *hologram* ("whole message") of the object. This record appears as a maze of interference fringes that does not look at all like the object, but it contains information about both the amplitude and phase distribution of the object field. When we expose the hologram to the reference beam, the transmitted field has a part that is simply proportional to the object wave front. This reconstructed wave front provides a true, three-dimensional image of the object. Anyone who has seen a hologram can attest to this remarkable property.

It is perhaps worth emphasizing again that the recording of the hologram stores not only the amplitude but also the phase of the object wave front, even though the film (like all detectors of light) responds only to intensity. The essential trick discovered by Gabor is to superpose the object field on a reference beam to produce an interference pattern [Eq. (16.4.7)]. This intensity pattern stores information about the phase of the object field, and this information is recovered when we "read" the hologram [Eq. (16.4.10)].

Actually our treatment of the holographic process has been a bit too restrictive. For one thing, we have assumed that the hologram is a photographic positive, whereas a negative will do just as well. With a negative the transmission function $t(x, y)$ will not be proportional to I but to $B - I$, where B is some constant. In this case the reconstructed object field is simply flipped in sign with respect to the case in which a positive-transparency hologram is used (Problem 16.4). Furthermore the reconstruction of the object wave can be performed without having to illuminate the hologram with the same reference beam used in recording the hologram. In the next section we consider an alternative method, and we will interpret the various terms in the transmitted field that appear in addition to the reconstructed object wave.

16.5 HOLOGRAPHY: PRACTICAL ASPECTS

Figure 16.9 illustrates the "wedge method" of recording a hologram. The object is considered to be a two-dimensional transparency, and the reference beam is described by an electric field of the form

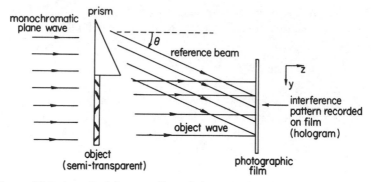

Figure 16.9 A method for recording a hologram of a semitransparent object.

$$E_R = A_R\, e^{-i(\omega t - \mathbf{k} \cdot \mathbf{r})} \tag{16.5.1}$$

where $\mathbf{k}$ lies in the yz plane:

$$\mathbf{k} = k(\hat{\mathbf{y}} \sin \theta + \hat{\mathbf{z}} \cos \theta) \tag{16.5.2}$$

and θ is the angle defined in Figure 16.9. At the plane $z = 0$ of the film we have

$$\mathbf{k} \cdot \mathbf{r} = (k \sin \theta)y \equiv \alpha y \tag{16.5.3}$$

We are assuming the film is so thin that we may effectively take $z = 0$ for all points on the film; this is the case of a *thin hologram*. Using (16.4.5), (16.5.1), and (16.5.3), we may write

$$\phi_R(x, y) = \alpha y \tag{16.5.4}$$

For the sake of simplicity we are ignoring the spatial variable x running perpendicular to the plane of Figure 16.9. Thus the phase ϕ_O of the object field, like ϕ_R, is assumed to be independent of x.

Following our discussion in the preceding section, we now assume that the film is developed to produce a hologram (transparency) with transmission function $t(x, y) = CI_T(x, y)$, where I_T is the total intensity due to the illumination with the object and reference fields. For the recording method of Figure 16.9, I_T is given by (16.4.7) with $\phi_R(x, y)$ defined by (16.5.4). The field transmitted by the hologram when it is illuminated with a uniform plane wave (Figure 16.10)

$$E = A\, e^{-i(\omega t - kz)}\bigg|_{z=0} = A\, e^{-i\omega t} \tag{16.5.5}$$

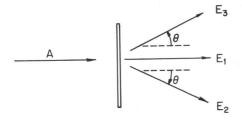

Figure 16.10 Illumination of the hologram as recorded in Figure 16.9 with a normally incident plane wave. E_2 and E_3 are real and virtual images, respectively, of the object, whereas E_1 is an attenuated version of the incident wave.

is then

$$
\begin{aligned}
E_{\text{trans}} &= CI_T(x, y)E \\
&= C\big[A_O^2(y) + A_R^2 + A_O(y)\,A_R\,e^{-i[\phi_O(y)-\alpha y]} \\
&\quad + A_O(y)\,A_R\,e^{i[\phi_O(y)-\alpha y]}\big]Ae^{-i\omega t} \\
&= C\big[A_O^2(y) + A_R^2\big]Ae^{-i\omega t} \\
&\quad + CA_O(y)\,A_R A\,e^{-i[\omega t - \alpha y + \phi_O(y)]} \\
&\quad + CA_O(y)\,A_R A\,e^{-i[\omega t + \alpha y - \phi_O(y)]}
\end{aligned}
\tag{16.5.6}
$$

Both (16.5.5) and (16.5.6) refer to the plane $z = 0$ of the (thin) hologram.

Let us now consider the meaning of the three terms in E_{trans}. Except for the factor $C[A_O(y)^2 + A_R^2]$, the first term on the right side of (16.5.6) is just the same as the uniform plane wave (16.5.5) normally incident on the hologram. This transmitted field is denoted E_1 in Figure 16.10, and is simply a part of the incident field that is transmitted with some change in amplitude because the hologram has a transmission function $t(x, y) \neq 1$.

The second term in (16.5.6) has the phase term

$$
\omega t - \alpha y + \phi_O(y) = \omega t - ky \sin \theta + \phi_O(y) \tag{16.5.7}
$$

If $\phi_O(y)$ were ignored, therefore, the second term in (16.5.6) would have the plane-wave factor $\exp[-i(\omega t - ky\sin\theta)]$ corresponding to propagation at the angle θ with respect to the z axis; this angle defines the direction of the reference beam used in recording the hologram (Figure 16.9). If the phase $\phi_O(y)$ of the object field is only a small correction to αy, we can expect the second term in (16.5.6) to represent a wave propagating at the angle θ as indicated in Figure 16.10, where this wave is labeled E_2.

Similarly the third term in (16.5.6) represents a wave E_3 propagating upwards in Figure 16.10 at the angle θ to the z axis. [The difference between the directions of propagation of E_2 and E_3 stems from the sign difference between the ways αy appears in the second and third terms of (16.5.6).] The wave E_3 has the form

$$E_3 = (CA_RA)\, A_O(y)\, e^{-i[\omega t + ky\sin\theta - \phi_O(y)]} \qquad (16.5.8)$$

whereas the object wave E_O is given by

$$E_O = A_O(y)\, e^{-i[\omega t - \phi_O(y)]} \qquad (16.5.9)$$

in the plane $z = 0$ of the hologram. Thus, except for the factor CA_RA associated with a change in amplitude, and the phase term $ky\sin\theta$ representing a change in direction, *the wave E_3 is the reconstructed object wave*. This image will appear to an observer to be behind the hologram, as illustrated in Figure 16.11. Ideally this image appears to the observer as an exact replica of the object, except for the fact that it is a monochrome.

The wave E_2, on the other hand, has the form

$$E_2 = (CA_RA)\, A_O(y)\, e^{-i[\omega t - ky\sin\theta + \phi_O(y)]} \qquad (16.5.10)$$

at $z = 0$. In addition to the terms CA_RA and $-ky\sin\theta$ representing changes in amplitude and phase, respectively, E_2 differs from the object field (16.5.9) by the sign of the phase $\phi_O(y)$; where the object field has a phase lag, the field E_2 has a phase advance, and vice versa. In other words, if the object wave is diverging, the wave E_2 will be converging to a focus on the opposite side of the hologram from the object (Figure 16.11). This image of the object can be focused onto a screen without the aid of a lens, and is therefore a *real image*.

Lasers are not absolutely crucial for the recording of holograms, but the recording of a hologram does require light with a very high degree of temporal and spatial coherence. Without temporal coherence the path separation of the reference

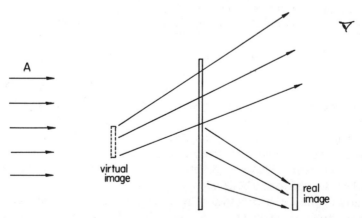

Figure 16.11 The virtual image corresponding to E_3 in Figure 16.10 will appear to an observer to come from behind the hologram. This image consists of diverging rays which can be focused by a lens, such as the lens of an observer's eye. The real image is formed by converging rays.

and object rays, as well as the path separation of light from different parts of the object, would wash out the interference pattern at the film plane. (Recall from Section 15.11 what happens when the arm separation of a Michelson interferometer is extended beyond the coherence length.) Without spatial coherence of the reference and object fields the interference pattern would similarly be smeared out, with information about the object wave front destroyed. (Recall what happens in the Young two-slit experiment, Section 15.3, when the slit separation is made larger than the lateral distance over which the radiation is coherent.) We saw in Chapter 15 that by using spectral filters and pinholes we can obtain a high degree of temporal and spatial coherence with a conventional light source. However, we cannot do this without greatly diminishing the intensity. Lasers, especially when they operate on a single cavity mode, combine coherence with high intensity.

We have assumed the reference and reconstructing waves are perfect plane waves, so that A and A_R are constants in (16.5.8) and (16.5.10). If they depend on y, then the amplitudes of E_2 and E_3 will not vary with y in the same way as the amplitude $A_O(y)$ of the object wave, and consequently we will not obtain a perfect image of the object. In practice it is important that the reference and reconstructing waves be fairly uniform, and this is often achieved by using a beam expander (Section 14.9). The loss in intensity associated with the beam expansion still leaves plenty of light—when a laser is used—to record and read the hologram.

Our discussion has assumed that the object is semitransparent. Figure 16.12 shows a method for recording a hologram of an opaque object. It should be clear that our discussion of recording and reading a hologram carries over essentially *in toto* to the case of an opaque object.

• Holography was conceived in 1947 and first demonstrated in 1948 by Dennis Gabor, whose research at the time was concerned with improving the resolution of electron microscopes. In Gabor's holography a mercury-vapor arc lamp was used to record and view holograms of transparencies with opaque lettering. Most of the light incident on the transparency passed through as a reference beam, but part of it was diffracted near the dark

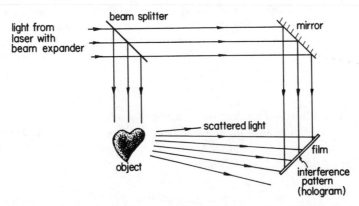

Figure 16.12 A method for recording a hologram of an opaque object.

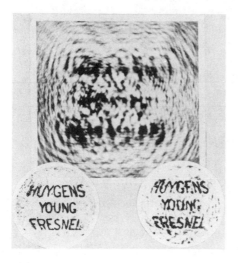

Figure 16.13 One of Gabor's original illustrations of the holographic principle. The object is shown in the lower left and its reconstruction in the lower right. The hologram itself occupies the top of the picture. From D. Gabor, "Microscopy by Reconstructed Wave-Fronts," *Proceedings of the Royal Society, London* **197A,** 454 (1949).

letters and acted as the object wave. That is, the light transmitted by a semitransparent sheet interfered with the light scattered in the same (forward) direction, and the intensity pattern was recorded on film. Figure 16.13 shows one of Gabor's original holograms. In 1971 Gabor was awarded the Nobel prize in physics for his discovery of the holographic principle.

Gabor's holography had a serious limitation: in the reconstruction of the object wave front the real and virtual images were not well separated. An observer focusing his eyes on the virtual image would also see a defocused real image, resulting in a blurring effect. This "twin-image" problem can be solved by introducing an angular offset between the object and reference waves as in Figure 16.9. In this *off-axis holography* the twin images are still present simultaneously, but they are spatially separated as in Figure 16.11.

Non-laser light sources, e.g., a mercury arc lamp, can be used to record and view holograms. Such a lamp typically radiates an intense green line of spectral width ~ 1–2 Å, and the arc itself is only a few millimeters in diameter. The resulting temporal and spatial coherence properties of the light are sufficient for recording holograms of semitransparent sheets, but for three-dimensional objects a laser is essential. An object with a depth of ~ 1 m, for instance, requires light with a coherence length greater than 1m, easily achieved with a single-mode laser, whereas an arc lamp has a coherence length typically ≲ 0.1 mm. The reading (viewing) of a three-dimensional hologram, however, does not require laser light. •

Thus far we have considered only thin holograms. In a *thick hologram* (also called a volume hologram), the interference pattern extends into the photographic emulsion a distance large compared with the spacing of interference fringes (Figure 16.14). If the object wave front is

$$\mathcal{E}_O(\mathbf{r}) = A_O(\mathbf{r}) \, e^{i[\mathbf{k}_O \cdot \mathbf{r} + \phi_O(\mathbf{r})]} \qquad (16.5.11)$$

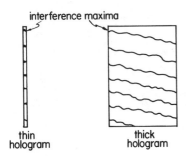

interference maxima

thin
hologram

thick
hologram

Figure 16.14 Comparison of thin and thick holograms. In a thick (or volume) hologram the thickness of the film emulsion is large compared with the fringe spacing.

and the reference field is the plane wave

$$\mathcal{E}_R(\mathbf{r}) = A_R \, e^{i\mathbf{k}_R \cdot \mathbf{r}} \qquad (16.5.12)$$

then the total field at the point $\mathbf{r}$ inside the photosensitive material is

$$\mathcal{E}_T(\mathbf{r}) = A_0(\mathbf{r}) \, e^{i[\mathbf{k}_0 \cdot \mathbf{r} + \phi_0(\mathbf{r})]} + A_R \, e^{i\mathbf{k}_R \cdot \mathbf{r}} \qquad (16.5.13)$$

and the (time-averaged) intensity at $\mathbf{r}$ is proportional to

$$\left| \mathcal{E}_T(\mathbf{r}) \right|^2 = A_R^2 + A_0(\mathbf{r})^2 + 2A_R A_0(\mathbf{r}) \cos\left[(\mathbf{k}_0 - \mathbf{k}_R) \cdot \mathbf{r} + \phi_0(\mathbf{r}) \right] \qquad (16.5.14)$$

Assume as before that the photosensitive material is a film emulsion in which silver halide grains are converted to silver under exposure to light, and that the density of silver atoms is proportional to the light intensity. Then we see from (16.5.14) that the density of silver atoms varies throughout the volume of the emulsion. We can appreciate the significance of the cosine term in this equation by noting that, along the vector $\mathbf{k}_0 - \mathbf{k}_R$, $\cos(\mathbf{k}_0 - \mathbf{k}_R) \cdot \mathbf{r}$ has maxima and minima separated by the distance

$$d = \frac{2\pi}{\left| \mathbf{k}_0 - \mathbf{k}_R \right|} \qquad (16.5.15)$$

Now

$$\begin{aligned}
\left| \mathbf{k}_0 - \mathbf{k}_R \right| &= \left[k_0^2 + k_R^2 - 2\mathbf{k}_0 \cdot \mathbf{k}_R \right]^{1/2} \\
&= \left[k^2 + k^2 - 2k^2 \cos\alpha \right]^{1/2} \\
&= k\left[2(1 - \cos\alpha) \right]^{1/2} \\
&= 2k \sin\frac{\alpha}{2} \qquad (16.5.16)
\end{aligned}$$

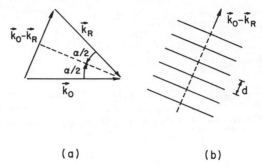

(a)   (b)

Figure 16.15 (a) Geometrical construction showing that $|\mathbf{k}_0 - \mathbf{k}_R| = 2k \sin(\alpha/2)$, where $k = |\mathbf{k}_0| = |\mathbf{k}_R|$ and α is the angle between $\mathbf{k}_0$ and $\mathbf{k}_R$. (b) If $\mathbf{r}$ points in a direction perpendicular to $\mathbf{K} \equiv \mathbf{k}_0 - \mathbf{k}_R$, then $\mathbf{K} \cdot \mathbf{r} = 0$ and $\cos \mathbf{K} \cdot \mathbf{r} = 1$. But if $\mathbf{r}$ is parallel to $\mathbf{K}$, then $\cos \mathbf{K} \cdot \mathbf{r} = \cos Kr$, with maxima (minima) separated by a distance $d = 2\pi/K$. The maxima (minima) of $\cos \mathbf{K} \cdot \mathbf{r}$ thus form a pattern resembling a venetian blind.

where α is the angle between $\mathbf{k}_0$ and $\mathbf{k}_R$ (Figure 16.15). Therefore

$$d = \frac{2\pi}{2k \sin(\alpha/2)} = \frac{\lambda/2}{\sin(\alpha/2)} \qquad (16.5.17)$$

If A_0 and ϕ_0 were constants, independent or $\mathbf{r}$, then the last term in (16.5.14) would vary with $\mathbf{r}$ as $\cos(\mathbf{k}_0 - \mathbf{k}_R) \cdot \mathbf{r}$. That is, there would be a sinusoidal variation of the density of silver atoms, with maxima separated by the distance d given by (16.5.17); d represents the spatial period or "wavelength" of this spatial variation. By analogy with the effect of a sound wave of wavelength λ_s in a medium (Appendix 12.B), we might expect the sinusoidal variation of silver density to produce a sort of "diffraction grating" in which light of wavelength λ' can only be scattered in the direction satisfying the Bragg condition

$$2d \sin \theta_B = \lambda' \qquad (16.5.18)$$

Here θ_B is the Bragg angle indicated in Figure 16.16, and we are assuming for simplicity that the refractive index $n = 1$. Thus the incident light encounters a sort of venetian-blind structure, and diffraction of light of a given wavelength occurs only for a particular angle θ_B. From (16.5.17) and (16.5.18) we have

$$2 \frac{\lambda}{2} \frac{\sin \theta_B}{\sin(\alpha/2)} = \lambda' \qquad (16.5.19)$$

or

$$\theta_B = \alpha/2 \qquad (16.5.20)$$

if we assume $\lambda' = \lambda$.

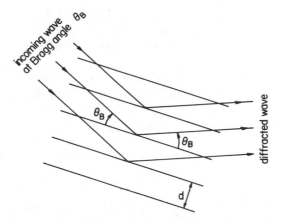

Figure 16.16 An incoming wave will be diffracted by the venetian-blind grating of Figure 16.15b only if the Bragg condition (16.5.18) is satisfied. Otherwise there is no diffracted wave.

Thus, to the extent that $A_0(\mathbf{r})$ and $\phi_0(\mathbf{r})$ in (16.5.14) vary slowly compared with $\cos(\mathbf{k}_0 - \mathbf{k}_R) \cdot \mathbf{r}$, our thick hologram will be essentially a sinusoidal diffraction grating. The hologram will have interference fringes as indicated in Figure 16.14, approximating the idealized venetian-blind case of Figures 16.15b and 16.16. If we illuminate this hologram with a monochromatic wave of wavelength λ', there will be a diffracted wave only in the direction θ_B given by (16.5.19). In fact the diffracted wave represents a virtual image of the object, as indicated in Figure 16.17, where the recording and reconstruction steps for a thick hologram are shown for the case $\lambda' = \lambda$ under consideration. In this case $\theta_B = \alpha/2$, where α is the angular offset between the reference and object waves in the recording step.

A consequence of the Bragg condition for a thick hologram is that only a single image is obtained for an arrangement like that shown in Figure 16.17. Note also

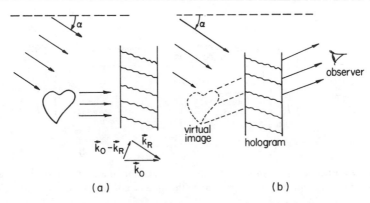

Figure 16.17 Recording (a) and reading (b) a thick hologram.

that whether a hologram is thick or thin is determined not only by its actual thickness (which for photographic film is typically $\sim 10 \ \mu m$) but also by the wavelength and angular offset of the reference and object waves, since these determine the fringe spacing d [Eq. (16.15.17)].

Until now we have considered only (thick or thin) *transmission* holograms, in which one or more images appear on the side of the hologram opposite to the object. That is, the image is contained in light transmitted by the hologram. A *reflection* hologram is a type of thick hologram in which an image appears on the same side of the hologram as the original object. Figure 16.18 illustrates the recording and reconstruction steps for a reflection hologram. Unlike the case of a (thick) transmission hologram, the venetian-blind pattern of interference fringes in a reflection hologram runs approximately parallel to the face of the hologram, as indicated in Figure 16.18 (Problem 16.5). There is a reflected wave (i.e., the reconstructed object wave) only if the Bragg condition is satisfied. In the arrangement shown in Figure 16.18 the reflected wave provides a virtual image of the object. If the reconstructing beam is incident from the object side of the hologram, a real image of the object is obtained (Problem 16.6).

Holograms of the reflection type can be viewed in white light, and can be used to produce three-dimensional *color* images. This was first demonstrated in 1962. The basic principle employed is illustrated in Figure 16.19. The spacing of the interference fringes in this case is

$$d = \lambda/2 \tag{16.5.21}$$

This follows from the Bragg condition (16.5.18) with $\theta_B = \alpha/2 = \pi/2$, since the angle α between the reference and object rays is π. If the hologram is illuminated with white light (Figure 16.19), only the wavelength $\lambda = \lambda_L$ with which the hologram was made will satisfy the Bragg condition and produce a reflected image of the original object. And if the hologram is made with three different wavelengths, using a laser for each of the three primary colors, a full-color image of the object is obtained under white-light illumination by, say, an ordinary lamp.

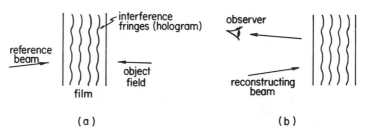

(a)   (b)

Figure 16.18 Recording (*a*) and reading (*b*) of a reflection hologram. In (*b*) the image seen by an observer is a virtual image in which the object appears to be located behind the hologram.

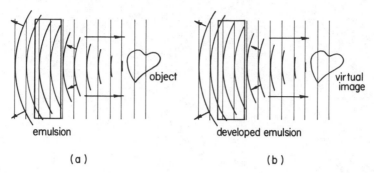

(a)   (b)

Figure 16.19 A color reflection hologram. (*a*) The hologram is recorded by reflecting light from the object, so that the incident and reflected waves produce an interference pattern in a photographic emulsion. (*b*) The hologram is viewed by illuminating it with white light. The reflected image appears at the wavelength satisfying the Bragg condition (16.5.21).

• Such a hologram combines the holographic principle with an old method of color photography invented by Gabriel Lippmann. In the Lippmann process a photographic emulsion plate was coated at its back face by a reflecting layer of mercury, and this resulted after development in a true color picture. Because the path lengths for incident and reflected rays in the plate are about the same for all wavelengths, no special coherence properties of the light were required during the photographing. In 1908 Lippmann was awarded the Nobel Prize in physics for having made the first color photograph. •

Especially bright three-dimensional images are obtained with so-called *rainbow holograms*. The first step in making a rainbow hologram is to make a transmission hologram. If this hologram is viewed in white light, the different wavelengths λ' are diffracted at different angles determined by the Bragg condition, as indicated in Figure 16.20*a* (Problem 16.7). Since a hologram contains many different (three-dimensional) views of the object, and each color is diffracted at a different angle, what we see is a colored blur. In making a rainbow hologram, the transmission hologram is masked except for a horizontal slit, as shown in Figure 16.20*b*, and then a second hologram is made with the real image from this slit as the object. When this second hologram is viewed in white light, the observer sees an image of the object in a single color, depending on the viewing angle. If he moves his head up and down the color changes, because each wavelength gives an image of the slit at a different angle (Figure 16.20*c*). Thus the object is seen in a single color at a time, and in each color of the rainbow as the head of the observer (or the plane of the hologram) is tilted vertically. The image is very bright because it is actually an image of the narrow slit rather than the whole area of the first hologram, and so the light is concentrated on a smaller area.

Because they provide both intensity and phase information, holographic images have parallax, i.e., they appear three-dimensional when viewed from side to side or up and down. The rainbow hologram has horizontal parallax but, because it is

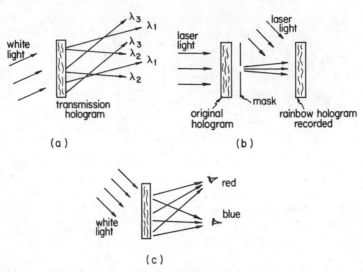

Figure 16.20 (*a*) A transmission hologram produces a colored blur when illuminated with white light. (*b*) A rainbow hologram is recorded by interfering the real image from a slit of the original hologram with a reference beam. (*c*) When viewed in white light, the rainbow hologram produces a colored image of the object, the color depending on the angle of observation.

the image of an object viewed through a (horizontal) slit, it does not offer vertical parallax. That is, as an observer tilts the hologram and sees the rainbow colors, he does not perceive any parallax.

Rainbow holograms are coated at the back with a thin, reflecting aluminum layer, so that they may be viewed by the reflection of light from the back surface. Such holograms are now used on credit cards, and they may eventually become ubiquitous on identification cards, passports, etc. The reader may easily check the rainbow and parallax properties noted above.

Detailed descriptions of practical uses of the holographic principle are readily available in the optics literature. One widely used application is in testing for stresses and structural defects in materials ranging from automobile and aircraft parts and tires to tennis rackets. This may be done by the ''double-exposure'' method in which two holograms of the object, with and without a stress or load, are recorded on the same plate. Deformations due to a stress then appear when the two images are (simultaneously) reconstructed. Compact units for performing ''holographic interferometry'' of this type are commercially available. One of the most exciting possible applications involves X-ray holography, which would provide cytologists with a three-dimensional view of living cells.

• The basic holographic principles have been known since Gabor's work in the late 1940s, but it was the laser that triggered the explosive growth of the field during the 1960s. Nevertheless, it has been possible for some time to produce excellent holographic images without the use of any laser at all. This is done with so-called *composite holograms*. Such holograms are made by the imaging of a series of ordinary photographs providing different perspectives of an object. A collection of (usually contiguous) small holograms then form the composite hologram, which produces the desired image when illuminated.

It is also possible, by using computers, to make holograms of objects that do not actually exist. The computer uses a set of coordinates for the hypothetical object, together with information about the object illumination and reference wave, to construct a transmission function for a hologram, which it then prints out as a magnified version of the hologram; magnification is necessary because of the limited resolution of plotters and other display devices. When the result is then appropriately demagnified and photographed as a transparency, it forms a hologram that can be viewed under illumination with the reference wave. •

16.6 OPTICAL COMMUNICATIONS

Any means of communication, such as smoke signals or speech or telephone, involves a transmitter and a receiver of information. In optical communication systems an electrical signal is converted to a modulated light wave, which then transmits the signal to a receiver. The light signal is converted back to an electrical signal at the receiver, where it might, for example, be "read" as a television picture.

Modulation of the light wave is necessary in order to convey information. A perfectly monochromatic wave, if it could be produced, could not by itself be used to convey information. To transmit information with light we have to turn it on and off, or modulate it in some other way, to have it represent some sort of coded message that is recognizable at the receiver. This sort of thing is familiar from radio wave communication, where amplitude (AM) or frequency (FM) modulation is used to transmit information. The rate of information depends on the rate at which modulations are impressed on the wave, and these modulations (and therefore the rate of information transmission) must always be slow compared with the basic carrier frequency of the wave. The same modulation techniques can be applied to light waves, whose much higher carrier frequencies suggest that they can be used to transmit information at very high rates. This great potential of light waves for communication purposes has been recognized for a long time, but became practical only after the development of lasers.

Even then a major hurdle stood in the way of long-distance optical communication systems: the transmission medium—the earth's atmosphere—introduces uncontrollable signal distortion due to turbulence. And even under the most favorable weather and seeing conditions only direct line-of-sight communication would be possible.

These problems with atmospheric transmission were solved around 1970 with the development of low-loss, single-mode optically transparent fibers that can serve as "cable" for light transmission at near infrared wavelengths, $\lambda \approx 8500$ Å. At about the same time, developments in semiconductor technology made cw, room-

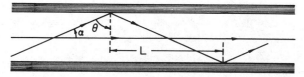

Figure 16.21 An experiment of John Tyndall illustrating the guiding of light by total internal reflection.

temperature laser sources available at this wavelength. By 1977 optical communication systems based on these developments were commercially available.

Optical fibers guide the path of light by total internal reflection. (See Section 4.10 for a brief review of total internal reflection.) The familiar lecture demonstration experiment sketched in Figure 16.21, which was first performed by John Tyndall before the Royal Academy of London in 1870, illustrates the guiding of light by total internal reflection.

Figure 16.22 is an enlarged view of a fiber segment. (The core diameter may be as small as a few micrometers.) We recall that the critical angle for total internal reflection is given by [Eq. (4.10.11)]

$$\theta_c = \sin^{-1}(n_2/n_1) \qquad (16.6.1)$$

Total internal reflection occurs for angle of incidence $\theta \geqq \theta_c$ (Figure 4.27). This leads to a maximum "acceptance angle" for which light injected into the fiber will undergo total internal reflection. Applying Snell's law to the dielectric interface in Figure 16.23, we have

$$n \sin \phi = n_1 \sin \alpha = n_1 \sin \left(\frac{\pi}{2} - \theta \right)$$
$$= n_1 \cos \theta = n_1 \sqrt{1 - \sin^2 \theta} \qquad (16.6.2)$$

Figure 16.22 Cross-sectional view of an optical fiber, showing the path of a mode propagating along the core axis and that of a mode having an angle of incidence θ at the core-cladding interface.

For θ equal to the critical angle θ_c given by (16.6.1), we obtain

$$n \sin \phi = n_1 \sqrt{1 - \frac{n_2^2}{n_1^2}}$$

$$= \sqrt{n_1^2 - n_2^2} \equiv NA \tag{16.6.3}$$

where the number NA is called the *numerical aperture* of the fiber. According to (16.6.2) and (16.6.3) the angle

$$\phi_{max} \equiv \sin^{-1}(NA/n) \tag{16.6.4}$$

is the maximum acceptance angle for total internal reflection. For a fiber in air ($n \approx 1$) with a core-glass refractive index $n_1 = 1.53$ and clad-glass index $n_2 = 1.50$, we calculate $\phi_{max} \approx 18°$.

The numerical aperture is important not only in determining the maximum acceptance angle of an optical fiber, but also in connection with the dispersion of the fiber. If a pulse of light is injected into a fiber, dispersion results in an increase in the width of the pulse, as discussed below. For a string of pulses constituting a signal, this spreading of each pulse may result in overlapping of the pulses at the output end, i.e., successive pulses may become impossible to distinguish. Dispersion thus sets a lower limit on the separation between pulses, or in other words it limits the rate at which information can be transmitted.

One source of dispersion is the fact that an incident beam will be distributed among various modes of the fiber, with different modes making different angles with the core–cladding interface. Figure 16.22 shows the different paths for two modes, one propagating along the core axis and the other having an angle of incidence θ at the core–cladding interface. It is easy to see that the off-axis mode has a total propagation length $L/\cos \alpha$, compared with the length L of the on-axis mode. This results in a difference $\Delta\tau$ in the times taken for fields propagating in these two modes to reach the output end:

$$\Delta\tau = L \frac{(1/\cos \alpha) - 1}{c/n_1}$$

$$\approx n_1 \alpha^2 L/2c \tag{16.6.5}$$

where the approximation is justified if the angle α is very small. For the maximum acceptance angle ϕ_{max} given by (16.6.4), it follows from (16.6.2) that $n_1 \sin \alpha = NA$, or $\alpha \approx NA/n_1$ in the small-angle approximation. Then (16.6.5) becomes

$$\Delta\tau = \frac{(NA)^2}{2n_1 c} \tag{16.6.6}$$

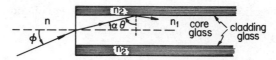

Figure 16.23 Cross-sectional view of an optical fiber.

for the time delay *per unit length of fiber* for the two modes. From this result we can conclude that the amount of spreading in time of an injected pulse will be proportional to $(NA)^2$. In other words, the rate at which the *step-index fiber* of Figure 16.23 can transmit information is inversely proportional to the square of its numerical aperture.

For the numerical example considered earlier ($n_1 = 1.53$, $n_2 = 1.50$), NA $\approx$ 0.3. It is difficult to make NA much smaller (in order to increase the capacity of the fiber for information transmission) because the acceptance angle (16.6.4) decreases with decreasing NA, and so a very small value of NA places stringent requirements on the injection of pulses into the fiber. A better way to decrease $\Delta\tau$ is to use a *single-mode fiber* in which the core diameter is so small (typically ~ 2 μm) that only the on-axis mode can propagate.

The pulse-spreading effect of a multimode fiber is reduced when the fiber is of the *graded-index type* rather than the step-index type considered thus far. In a graded-index fiber the refractive index does not have a sharp, steplike decrease from n_1 to n_2. Instead the index decreases more smoothly from the center of the fiber. An index distribution that is frequently used in practice is described by the formula

$$n^2 = n_c^2(1 - a^2 x^2) \tag{16.6.7}$$

where n_c is the refractive index at the center, x is the distance from the center, and a^2 is a constant. The advantage of a graded-index fiber is a consequence of the following result, which we will not take the time to derive: the pulse spreading ($\Delta\tau$) of a graded-index fiber is proportional to $(NA)^4$, rather than to $(NA)^2$ as in the multimode step-index case. The capacity of a graded-index fiber to transfer information is thus proportional to $1/(NA)^4$, and this gives it an information-transmission capacity about 10–100 times that of a multimode step-index fiber.

The pulse-spreading effect due to multimode propagation in an optical fiber is called *mode dispersion*. Pulse spreading also occurs because of ordinary material dispersion, i.e., because of the dependence of the refractive index on the wavelength. It may be shown (Problem 16.8) that material dispersion leads to a pulse spread given by

$$\Delta\tau \approx \frac{\lambda}{c} \left| \frac{d^2 n}{d\lambda^2} \right| \Delta\lambda \tag{16.6.8}$$

per unit length of fiber, where $\Delta\lambda$ is the spread in wavelengths associated with a pulse. Pulse spreading due to material dispersion occurs even in single-mode fibers, where mode dispersion is absent.

Attenuation in optical fibers occurs as a result of Rayleigh scattering from small-scale refractive index variations, inherent absorption, and absorption due to impurities introduced during fabrication. Losses also occur due to boundary irregularities and imperfect splicing of fiber segments, and because bending of the fiber allows some light rays to have angles of incidence at the core–cladding interface that are too small for total internal reflection. Nevertheless, the attenuation coefficient for single-mode fibers has been made as small as 0.2 dB/km at certain wavelengths.

Fiber attenuation is usually expressed in units of dB/km. If an initial beam power P_0 is reduced to P over a fiber length l, then $[10 \log_{10} (P_0/P)]/l$ is the attenuation coefficient in dB per unit length. Thus 0.2 dB/km corresponds to an attenuation given by $P(z) = P_0 e^{-\alpha z}$, where $\alpha = 0.046$ km^{-1} = 4.6×10^{-7} cm^{-1}. This means that more than 20 km of fiber would be required to reduce the power by a factor of e.

Laser diodes and LEDs are ideal light sources for optical-fiber communications. For one thing, their tiny size facilitates the coupling of light into a hair-thin fiber. And their wavelengths fall in the most propitious regions of low fiber disperson and attenuation. For silica (SiO$_2$) fibers the quantity $d^2n/d\lambda^2$ appearing in Eq. (16.6.8) is near zero at about 1.3 μm, and the attenuation coefficient near this wavelength has a local minimum. GaAsP diode lasers operate in this region, as well as at 1.5–1.6 μm, where single-mode fibers with losses below 0.2 dB/km have been fabricated.

Another important feature of laser diodes is the fact that they can be directly modulated at frequencies reaching onto the gigahertz region. When an injection current is switched on, the laser output, after a short delay, undergoes relaxation oscillations, as sketched in Figure 16.24. The frequency of these relaxation oscillations sets an upper limit to the frequency with which the laser can be modulated by varying the injection current. This maximum modulation frequency is given by Eq. (12.3.16):

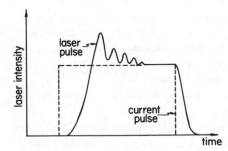

Figure 16.24 Typical laser-diode output when a square current pulse is applied.

$$f_r = T_r^{-1} = \frac{1}{2\pi} \sqrt{\frac{c}{\tau_{21}}(g_0 - g_t)}$$

$$= \frac{1}{2\pi} \sqrt{\frac{cg_t}{\tau_{21}}\left(\frac{g_0}{g_t} - 1\right)}$$

$$= \frac{1}{2\pi} \sqrt{\frac{g_0/g_t - 1}{\tau_p \tau_{21}}} \qquad (16.6.9)$$

where g_0 is the small-signal gain, g_t is the threshold gain, and τ_p and τ_{21} are the photon and upper-level lifetimes, respectively. In the case of a semiconductor laser this expression is applicable if we interpret τ_{21} as the charge-carrier lifetime τ_s at threshold. In addition we can write g_0/g_t as the ratio I/I_{th} of the injection current to its threshold value:

$$f_r = \frac{1}{2\pi} \sqrt{\frac{I/I_{th} - 1}{\tau_p \tau_s}} \qquad (16.6.10)$$

For a direct-gap semiconductor, typical values of τ_s are in the neighborhood of 1 nsec. Then for $\tau_p \approx 3 \times 10^{-12}$ sec and $I/I_{th} \approx 2$, we calculate $f_r \approx 3$ GHz. It might be noted that the carrier frequency ($\sim 10^{14}$ Hz) would in principle allow much higher modulation rates, but in practice only modulation rates less than about 10 GHz have been possible with diode lasers.

Optical communication systems do not make much use of the coherence properties of laser diodes, and in fact LEDs are often preferable to laser diodes because they are less expensive and more reliable in a hostile environment. LEDs do not couple to the tiny cores of single-mode fibers as well as laser diodes, because their light is spatially incoherent and may involve up to 10^3 spatial modes. A one-milliwatt LED emitting uniformly into 10^3 modes, for instance, would only couple about a microwatt into the fiber. (This power is adequate for short-distance communications.) Furthermore, the spectral width of the LED radiation may be as much as 10^2 Å, compared with just a few angstroms for a laser diode, and so pulses from a laser diode will suffer less spreading [recall Eq. (16.6.8)], with the possibility of greater information transmission rates. However, laser diodes have the disadvantage that the turn-on of a current pulse produces laser radiation only after a short delay, and moreover the output radiation itself tends to have rapid oscillations superimposed on each pulse (Figure 16.24).

In addition to light sources (laser diodes or LEDs) and fibers, optical communication systems also require efficient receivers. For the conversion of an optical signal into an electrical one, and for subsequent amplification and processing of the electric current, semiconductor devices are used instead of photomultipliers as detectors in communication systems. This is largely because of their low dark-current noise. *Dark-current noise* refers to random detector current in the absence of any incident radiation. In photocathode surfaces used in photomultipliers, the dark current is due to the thermionic emission of electrons. This causes a random current in any conductor at a temperature $T > 0$ K. For weak signals a preamplification of the signal may be necessary, and the amplifier itself is often the main source of noise in semiconductor junction receivers. Semiconductor devices also have the advantages of small size, lower power consumption, and durability under hostile conditions. We briefly discuss the physical principles behind the *PIN* and *avalanche photodiode* detectors commonly used in optical communication systems.

A PIN diode consists of a *p*-type region, an undoped or *intrinsic* region, and an *n*-type region (Figure 16.25). The diode is reverse-biased, i.e., positive and negative potentials are applied to the *n*-type and *p*-type regions respectively. Thus the externally applied field acts to keep the mobile electrons and holes in the *n*- and *p*-type regions, respectively, opposing their tendency to drift into the intrinsic region. Now photons incident on the intrinsic region can be absorbed in the electron–hole production process in which electrons in the valence band are promoted to the conduction band. The externally applied electric field sweeps the newly freed electrons to the *n* region and the new holes to the *p* region, and this results in a signal current in the external circuit. Silicon is the commonly used material for the intrinsic region because its band-gap energy, corresponding to photons with wavelengths around 9000 Å, allows fast and efficient detection, with low dark-current noise, of 8500-Å radiation from GaAs laser diodes and LEDs. Germanium is used for wavelengths in the 12,000–16,000-Å region. In any case it should be clear that the PIN photodiode is a sort of LED in reverse, converting photon energy to electron energy.

The avalanche photodiode is similar to the PIN photodiode except for an additional *p*-type region. The electrons freed by photon absorption in the intrinsic

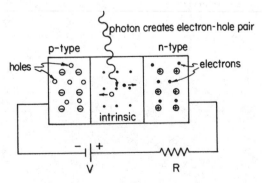

Figure 16.25 Schematic illustration of a reverse-biased PIN photodiode.

region pass through the additional p region before reaching the n region, and in so doing they produce additional electron–hole pairs by a collisional ionization process. These additional charge carriers themselves produce more electron–hole pairs, resulting in a cascade or *avalanche* generation of mobile charge carriers. The avalanche photodiode thus produces a larger electrical signal than a PIN detector. On the other hand it is less stable to variations in applied voltage and temperature, and requires more voltage, typically about 150 V compared with 25 V for a PIN photodiode.

Fiber-optical communication systems have several advantages over the standard electrical systems employing copper cables. Because fiber cables are nonconducting, they are free from stray electromagnetic signals and noise, and also from the possibility of nondestructive or remote electromagnetic tapping. They are also naturally less susceptible to control system failures due to sudden jumps in ground potential or surge currents, and are much safer in environments in which electric sparks might be catastrophic, as in chemical plants. In addition they are generally less susceptible to corrosion. On the other hand, they are less flexible than copper wire, and this makes splicing more complicated.

As noted earlier, commercial optical communication systems have been in operation since 1977. The first systems used AlGaAs laser diodes and multimode fibers to transit signals at around 45 megabits/sec. (This terminology is discussed below.) In the early 1980s LED systems were deployed in "subscriber loops" for local telephone systems, and many telephone companies have deployed 12- and 45-megabit/sec systems at the 13,000-Å wavelength range of InGaAsP LEDs and low-loss multimode fibers. For intercity communications single-mode systems with 1-gigabit/sec capacity are under development. A transatlantic undersea optical cable employing several single-mode fibers at 296 megabits/sec is scheduled for operation in 1988. The transmitted signals will be compensated for loss by repeaters spaced about 50 km and powered from shore. This system is expected to be able to transmit around 40,000 telephone conversations simultaneously.

• In communication theory the concept of information is quantified. Consider an event with two possible outcomes, like heads or tails of a coin flip. We can assign to the outcome of the event a label such as "yes" or "no," or "heads" or "tails," or simply one of the binary digits 0 and 1. Transmission of knowledge of the outcome of the event then requires one binary digit, or *bit* of information. Similarly, if the event has four possible outcomes we require two bits of information to transmit knowledge of the outcome. To see this, imagine the four possible outcomes are represented by four slots in a line. Then one bit of information is required to specify whether the event belongs to, say, one of the left pair of slots or one of the right pair, and one more bit of information then fully specifies the outcome. In general, if there are N possible outcomes, or messages, then the information associated with each message is

$$I = \log_2 N \text{ bits} \tag{16.6.11}$$

If a process has a particular outcome with probability one, then obviously no useful information is conveyed by the outcome. It is useful to define the average information

$$H = \sum_{i=1}^{N} P_i \log_2 \frac{1}{P_i} = -\sum_{i=1}^{N} P_i \log_2 P_i \qquad (16.6.12)$$

where P_i is the probability of the ith outcome (message). If all the outcomes are equally probable, so that $P_i = 1/N$ for all i, then (16.6.12) reduces to (16.6.11). But if any of the $P_i = 1$, then $H = 0$. The expectation value H of information associated with each message is called the *entropy*.

Consider as an example a teletype system in which five binary digits are represented by holes punched in the transmission tape. We might label a punched hole by 1 and the absence of a hole by 0. There are $2^5 = 32$ possible sequences of 0's and 1's, enough to allow for the 26 letters of the English alphabet plus some control signals like start and stop. Each sequence carries $I = \log_2 2^5 = 5$ bits of information.

In communications we are especially concerned with the *rate* of information transfer. If the transmission speed of our teletype system is 60 5-letter words per minute, than the information rate is (60×5 letters/minute) $\times$ (5 bits/letter) or 25 bits/sec.

Consider as another example a black-and-white television signal. The information is contained in small picture elements (pixels) of different light intensity, and if we assume the eye can distinguish 10 intensity gradations, then each pixel contains $H = \log_2 10 = 3.32$ bits of information. Assuming 30 frames/sec, 525 lines/frame, and 500 pixels/line, we have the information rate $(30)(525)(500)(3.32) = 2.61 \times 10^7$ bits/sec = 26.1 megabits/sec.

There are various modulation schemes used to transmit information in communication systems. An especially important one is *digital modulation*, in which information is conveyed by a sequence of pulses corresponding to 0's and 1's depending on whether their amplitudes are less than or greater than some particular value. The major advantage of digital systems is their accuracy, as illustrated in Figure 16.26. A signal of 0's and 1's (Figure 16.26a) is shown degraded by noise in Figure 16.26b. In Figure 16.26c, however,

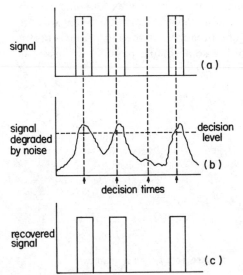

signal

(a)

signal degraded by noise

decision level

(b)

decision times

recovered signal

(c)

Figure 16.26 A digital signal (*a*), degraded by noise (*b*), can be exactly reconstituted (*c*) even in the presence of substantial noise.

the original signal is accurately recovered by letting anything above a certain "decision level" be a 1, and anything below that level a 0.

Laser diodes are well suited for digital transmission because of the relative ease with which their outputs can be switched on and off by varying the injection current. •

16.7 LASERS IN MEDICINE: OPHTHALMOLOGY

The first medical use of lasers was made in ophthalmology, just a few years after the first demonstration of laser oscillation in the early 1960s. This application of lasers for the coagulation of eye tissue has long since become a routine medical procedure throughout the world.

It has been known for centuries that visual loss can result from prolonged, direct viewing of the sun with the naked eye. This occurs due to a burning and consequent scarring of the macula, the central part of the retina responsible for acute vision (Figure 16.27). It was suspected for some time that a localized burning and scarring might actually be useful for the treatment of certain visual disorders, and in the late 1940s G. Meyer-Schwickerath demonstrated that burns produced by white-light sources like the sun and xenon arc lamps could be used to connect the retina to its substratum tissue.

The brightness and monochromaticity of lasers made it immediately apparent that they could be used in ophthalmology. In particular, it was apparent that lasers could be used for photocoagulation of tiny areas of ocular tissue, and that this could be done with very short exposure times. Furthermore the monochromaticity of laser radiation makes it possible to selectively coagulate a particular tissue, since the photocoagulation process is initiated by wavelength-dependent absorption of light.

Laser therapy has replaced conventional surgical procedures for some eye disorders, and lasers have also been used in cases where effective treatment was previously unavailable. Laser therapies may be broadly classified as photoabsorptive or photodisruptive. We will briefly describe some of the physical processes in-

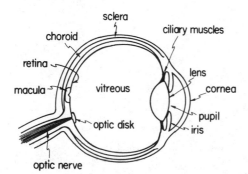

Figure 16.27 The human eye.

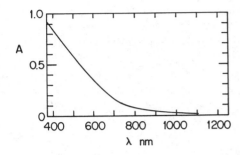

Figure 16.28 Absorption spectrum of melanin. From V.-P. Gabel, "Lasers in Ophthalmology," in *Lasers in Biology and Medicine*, ed. by F. Hillenkamp, R. Pratesi, and C. A. Sacchi (Plenum Press, New York, 1980).

volved in these therapies, and refer the reader to the medical literature for more detailed and up-to-date information.[2]

Large biological molecules, such as proteins, undergo conformational alterations when the temperature is increased sufficiently. These changes are associated with the breaking of hydrogen bonds and the weakening of van der Waals forces. The result is a *thermal denaturation* in which certain biological functions are lost or impaired due, for instance, to changes in cell membranes. Such thermal denaturation is responsible for effects like inflammation and coagulation of tissue.

Coagulation of eye tissue can be beneficial in several ways. Through a normal reparative response it leads to a scar, which can serve to connect tissues like the sensory and pigmented portions of the retina. In other words, coagulation can "glue" disconnected tissue back together. Since coagulation can result in tissue degeneration, it is also useful for making a hole in the iris (iridectomy), which is done in the treatment of glaucoma to relieve the elevated intraocular pressure. In addition, coagulation can occlude and destroy blood vessels. This is used in the treatment of diabetic retinopathy, in which fragile blood vessels may appear on the retina. These abnormal vessels have a tendency to break, and the hemorrhaging can produce severe loss of vision. (This is the leading cause of blindness for people between the ages of 20 and 64 in the United States.)

Photocoagulation begins with absorption of light. This results in electron excitation, vibrational excitation, breaking of molecular bonds, ionization, etc., and ultimately a rise in temperature and coagulation of the irradiated area. Obviously photocoagulation can only occur if there are biomolecules available to absorb the incident light and dissipate it as heat. The primary absorbers of light in the human eye are melanin, hemoglobin, and xanthophyll. The ocular medium itself is transparent between about 380 and 1400 nm. At lower wavelengths the lens and cornea are absorbing, and at the longer wavelengths the primary absorber is water.

An absorption spectrum of melanin is shown in Figure 16.28. Melanin is present in the retinal and iris pigment epithelia and in the choroid. The argon ion laser lines at 488 and 514.5 nm obviously lie in the region of strongest absorption by

---

2. Research reports on laser treatments of ocular disorders have appeared regularly in various journals, such as *Ophthalmology* and *American Journal of Ophthalmology*.

melanin, as does the 568-nm radiation of the krypton ion laser. The blood pigment hemoglobin has absorption maxima near 400 and 600 nm in its oxygenated and deoxygenated forms. Xanthophyll is a yellow pigment present in the macular area of the retina, and its peak absorption occurs between around 420 and 500 nm, again making the argon ion laser attractive for photocoagulation. In fact the argon and krypton ion lasers have been the primary lasers for the treatment of macular degeneration, in which abnormal blood vessels appear below the retinal surface at the macula, leading to local detachment of the retina. Together with the Q-switched or mode-locked Nd:YAG laser, the cw argon and krypton ion lasers have been the principal lasers used in ophthalmology, although the field is by no means confined to these lasers alone.

Lasers allow for a selective energy treatment, in the sense that the total energy of irradiation during the exposure can be accurately controlled. This allows the ophthalmologist to preselect a certain energy that, based on his experience, will produce a minimal degree of photocoagulation. The energy can then be increased gradually until a desirable amount of coagulation has occurred, with minimal damage or side effects.

Exposure time is also a very important consideration. For a given total energy, the temperature rise of the irradiated tissue increases with decreasing laser pulse width, since there is less time available for thermal diffusion to the surrounding tissue. For very short pulses, therefore, the temperature rise can be quite large, and can even lead to vaporization at the irradiated spot; the clinical manifestation of this vaporization is the appearance of gas bubbles near the target. The vaporization can also generate pressure waves strong enough to cause mechanical damage to eye tissue. For this reason cw lasers are generally preferred for photocoagulation.

In contrast to photocoagulation, photodisruptive laser therapies are nonthermal. They are also newer, having been used extensively only since the early 1980s. In addition, laser photodisruption of tissues is generally performed with Q-switched or mode-locked pulses, usually from a 1.064-μm Nd:YAG laser.

Photodisruption is initiated by ionization. For Q-switched pulses the ionization results from the intense heat produced at the target spot, a process called *thermionic emission*. For the much shorter and more intense mode-locked pulses, ionization occurs via multiphoton absorption (Section 18.5). In either case the ionization leads to the formation of a plasma of free electrons and ionized atoms and molecules, and this plasma then absorbs energy from the laser field and becomes very hot. It therefore expands very rapidly, producing a shock wave that can make a hole in an ocular membrane. Such a hole can be several times larger than the waist of the laser beam at the focal point.

One common application of laser photodisruption of tissue is in treating the opacification (i.e., clouding over) of the so-called posterior capsule membrane of the lens, a frequent complication of cataract surgery. In this treatment, called posterior capsulotomy, a Nd:YAG photodisruptor tears a hole in the clouded membrane, opening a path for clear vision. This can be done without damage to the retina by tightly focusing the laser beam to a spot just behind the lens. The diver-

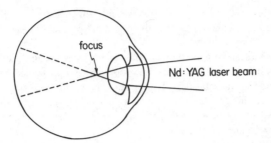

Figure 16.29 In posterior capsulotomy a Nd:YAG laser pulse is focused to a point behind the lens.

gence of the beam beyond the focus then results in a reduced intensity at the retina (Figure 16.29). The attenuation of the beam by the plasma may also help to reduce the intensity at the retina.

Another recent development is the use of lasers for the ablation (removal) of ocular tissue. In this application the ultraviolet radiation of rare-gas excimer lasers, in particular the ArF excimer laser at 193 nm, has been used to make precise incisions in the cornea. The depth of these incisions is determined by the number of laser pulses and the energy of each pulse. Excimer lasers may eventually replace scalpels in radial keratotomy, a procedure for correction of myopia. In this relatively new technique, a pattern of incisions is made, and in the healing process the shape of the cornea changes in a way that alleviates myopia. The basic physics of the photoablation processes is not well understood at present, but it appears that the breaking of molecular bonds by ultraviolet photons plays an important role.

Lasers are now commonplace in the treatment of eye disorders. The vision of hundreds of thousands of patients is aided each year by surgical procedures involving lasers. There are strong indications that lasers will become increasingly important not only in ophthalmology but in other medical fields as well.

• It has been estimated that lasers could eventually be used in two-thirds of all current operating-room procedures. In certain procedures the laser "light knife" has some important advantages, one of the most important being that blood loss may be greatly reduced compared with conventional methods involving steel or electric knives. Together with magnification, this greatly facilitates the removal of tiny structures, and furthermore may reduce morbidity associated with blood loss and the use of tourniquets. These advantages have been demonstrated, for instance, in laryngeal surgery, where, together with an endoscope, the laser makes possible a degree of precision difficult to attain using other methods.

The CO_2 laser was introduced in surgery in the early 1970s, and has remained the leading nonophthalmic surgical laser. This is not very surprising, given that living tissue is about 80% water, and that water strongly absorbs at 10.6 μm. (The absorption coefficient of living tissue is about 200 cm^{-1} at 10.6 μm). Thus CO_2 laser radiation interacts quite strongly with living tissues. A focused CO_2 laser beam can produce an explosive boiling of tissue fluids, resulting in an incision as the focal point is moved. Hand-held CO_2 surgical lasers are now commercially available, and they have been used to remove arterial plaque during open-heart surgery. At present there are no fibers available for guiding CO_2 laser radiation; it is anticipated that the development of such fibers would represent an important advance for laser medicine.

In conventional cancer surgery there is the danger of dislodging cancer cells and so worsening the malignancy. With a laser, the blood and lymph vessels are sealed, thus diminishing the possibility of metastasis. Lasers also appear to facilitate the precise removal of larger numbers of cancer cells than is possible with conventional methods. Although considerable success has been reported with laser surgery for the treatment of cancer, the field is still in a developmental stage.

Aside from ophthalmology, the use of lasers at present seems most well established in dermatology and gynecology. In 1984 the U.S. Food and Drug Administration issued marketing approvals for the use of cw Nd:YAG lasers in certain urological and gastrointestinal procedures, and by 1985 it was estimated that around a fifth of the neurosurgeons in the U.S. used lasers. •

16.8 REMARKS

Like the music of Bach, the great progress in laser applications has a quality of both inevitability and surprise. For instance, it is easy to understand how laser pulses may be used to measure distances, but it is quite remarkable that they can be used to determine distances between the earth and the moon to an accuracy of a few centimeters. And although anyone familiar with Gabor's early efforts could have predicted in the early 1960s that lasers would be used in holography, the degree of realism possible in modern holography is stunning. In medicine the possibilities for lasers were also recognized very early, but probably few people could foresee just how ubiquitous the laser would become in the treatment of visual disorders and perhaps, in the near future, in many other surgical procedures.

In this chapter we have described just a few laser applications chosen, more or less, randomly. Of course lasers have become so much a part of modern technology that in a book of this sort we can hardly even introduce the field of laser applications. Lasers are used to cut the fabric for our clothing, determine our grocery bills, print what we read, and record our music. It seems likely that the growth in laser applications will continue, so much so that some people feel that optics will play a role in the twenty-first century analogous to that of solid-state electronics in the twentieth. Such is the importance of the stimulated emission of radiation discovered by Albert Einstein!

PROBLEMS

16.1 Show that a $1\frac{1}{2}$-inch illuminated corner cube on the moon produces a beam of diameter ≈ 10 miles on the earth.

16.2 Show that the bandwidths of conventional light sources limit velocity measurements by heterodyne detection to velocities greater than about 600 m/sec.

16.3 (a) What properties of the laser are used in the laser gyroscope?

(b) Derive Eq. (16.3.7).

16.4 In our discussion of the holographic principle we assumed the hologram was a photographic positive. Discuss the case in which the hologram is a negative transparency.

16.5 Show that, for the reflection hologram illustrated in Figure 16.18, the interference fringes run approximately parallel to the face of the hologram, as indicated in the figure.

16.6 Suppose the reconstructing beam is incident from the object side of a reflection hologram formed as in Figure 16.18a. Show that a real image of the object is formed.

16.7 Show that when a transmission hologram is illuminated by white light, the different wavelengths are diffracted at different angles determined by the Bragg conditions, as indicated in Figure 16.20a.

16.8 Show that material dispersion in an optical fiber leads to a pulse spread given by (16.6.8). [Hint: Show that the group velocity $v_g = d\omega/dk$ of a pulse may be written as $v_g \approx (c/n) [1 + (\lambda/n) (dn/d\lambda)]$, and then use the expression $\Delta v_g = (dv_g/d\lambda) \Delta\lambda$ for the spread in group velocity. Then use the formula $\tau = L/v_g$ to obtain (16.6.8) for $\Delta\tau$, assuming $(dn/d\lambda)^2 \ll |d^2n/d\lambda^2|$.]

17 INTRODUCTION TO NONLINEAR OPTICS

17.1 INTRODUCTION

When monochromatic light is incident upon a material with sufficient intensity, the transmitted light can contain additional frequencies greatly different from the one present in the incident beam. This is an example of a nonlinear optical effect. With lasers it is easy to reach the intensity levels necessary to observe such effects, and the invention of the laser gave rise to a whole field called *nonlinear optics*.

17.2 NONLINEAR ELECTRON OSCILLATOR MODEL

In the Lorentz oscillator model the equation of motion for an atomic electron in an electric field $E(t)$ is (Table 2.1)

$$\frac{d^2x}{dt^2} + \omega_0^2 x = \frac{e}{m} E(t) \tag{17.2.1}$$

where $x(t)$ is the displacement of the electron from its hypothetical equilibrium position. For simplicity we consider the applied field to be linearly polarized along the x direction, so that only the x component of the electron displacement concerns us. We recall that the elastic restoring force in (17.2.1) is associated with a potential energy

$$V(x) = \tfrac{1}{2} m\omega_0^2 x^2 \tag{17.2.2}$$

That is, the elastic restoring force is

$$F = -\frac{dV}{dx} = -m\omega_0^2 x \tag{17.2.3}$$

A potential energy varying quadratically with x as in (17.2.2) is characteristic of harmonic oscillators, such as a mass at the end of an ideal spring obeying Hooke's law.

Of course a real spring does not follow (17.2.2) exactly. Similarly we should not expect an atomic or molecular oscillator to follow (17.2.2) or (17.2.1) perfectly. The effective potential energy function for the system can be written more realistically in many cases as a power series in x, in which (17.2.2) is the first and most important term:

$$V(x) = \tfrac{1}{2} m\omega_0^2 x^2 + Ax^3 + Bx^4 + \cdots \tag{17.2.4}$$

This may be understood as follows. Whatever the potential energy is, we can expand it in a Taylor series,

$$V(x) = V(0) + x \left(\frac{dV}{dx}\right)_{x=0} + \frac{1}{2!} x^2 \left(\frac{d^2V}{dx^2}\right)_{x=0} + \frac{1}{3!} x^3 \left(\frac{d^3V}{dx^3}\right)_{x=0} + \cdots$$

$$\tag{17.2.5}$$

about the equilibrium point, which we label $x = 0$. Here $V(0)$ is just an additive constant in the total energy; since it does not give rise to any force ($F = -dV/dx$), we may neglect it. Also, at a point of stable equilibrium, there is no force; that is,

$$\left(\frac{dV}{dx}\right)_{x=0} = 0 \tag{17.2.6}$$

as is clear from Figure 17.1. Furthermore

$$\left(\frac{d^2V}{dx^2}\right)_{x=0} > 0 \tag{17.2.7}$$

if $x = 0$ is a point of stable equilibrium [i.e., if $V(x)$ curves upward, as in Figure 17.1]. Thus we may write

$$V(x) = \frac{1}{2} x^2 \left(\frac{d^2V}{dx^2}\right)_{x=0} + \frac{1}{6} x^3 \left(\frac{d^3V}{dx^3}\right)_{x=0} + \cdots \tag{17.2.8}$$

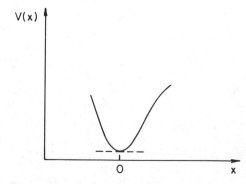

Figure 17.1 A potential $V(x)$ with a point of stable equilibrium at $x = 0$, where $dV/dx = 0$ and $d^2V/dx^2 > 0$.

and since $d^2V/dx^2 > 0$, we can define

$$m\omega_0^2 \equiv \left(\frac{d^2V}{dx^2}\right)_{x=0}$$ (17.2.9)

which shows that it is the curvature of the effective potential function that determines the oscillator frequency ω_0. Similarly we define

$$A \equiv \frac{1}{6}\left(\frac{d^3V}{dx^3}\right)_{x=0}$$ (17.2.10)

Thus the first two terms in (17.2.8) correspond to those in (17.2.4), and indeed we can identify all the terms in (17.2.4) by considering more and more terms in the Taylor series for $V(x)$.

From (17.2.4) we have the force

$$F = -\frac{dV}{dx} = -m\omega_0^2 x - 3Ax^2 - 4Bx^3 - \cdots$$ (17.2.11)

implying the Newton equation of motion

$$\ddot{x} + \omega_0^2 x + \frac{3A}{m}x^2 + \frac{4B}{m}x^3 + \cdots = \frac{e}{m}E(t)$$ (17.2.12)

In the limit of a perfectly elastic restoring force, $A = B = \cdots = 0$ and we recover the familiar oscillator equation (17.2.1).

Equation (17.2.12) is a *nonlinear* differential equation for x, because x enters in powers other than the first. Such nonlinear equations are generally difficult or impossible to solve exactly, and must be treated either approximately or by numerical computation on a computer.

As we shall see, it is precisely the nonlinearities, represented by the terms $3(A/m)x^2, 4(B/m)x^3, \ldots$ in (17.2.12), that are responsible for nonlinear optical effects. Electrons in atoms are really governed by the principles of quantum mechanics, not Newton's equations, of course. However, to a large extent quantum mechanics supplies a formalism in which the nonlinear coefficients $A, B, \ldots$ can be calculated from first principles, without altering the nature of the nonlinear equation (17.2.12). Thus, if we are willing to take the coefficients $A, B, \ldots$ as empirical parameters, at least temporarily, we can study nonlinear optics quite satisfactorily without quantum mechanics.

17.3 PERTURBATIVE SOLUTION OF THE NONLINEAR OSCILLATOR EQUATION

Judging by the success of the linear electron oscillator model considered in Chapters 2 and 3, we should anticipate that the nonlinear terms in (17.2.12) are in some sense "small." More precisely, it is reasonable to suppose that the nonlinear term $3(A/m)x^2$, for instance, is much smaller than the linear term $\omega_0^2 x$ except when x is very large. The ratio of these two terms is $3Ax/m\omega_0^2$, so that we must have

$$x \approx \frac{m\omega_0^2}{3A} \tag{17.3.1}$$

before the nonlinear term is appreciable by comparison with the linear term. If $3A/m \ll \omega_0^2$, then (17.3.1) implies that large displacements of the electron are necessary to make the nonlinearity appreciable. This is consistent with the observation that fairly large field intensities are necessary to realize nonlinear matter-field interactions.

Consider the nonlinear electron oscillator equation

$$\ddot{x} + \omega_0^2 x + ax^2 = \frac{e}{m} E_0 \cos \omega t \tag{17.3.2}$$

obtained from (17.2.12) by defining $a = 3A/m$, neglecting the higher-order non-linear terms, and letting $E(t) = E_0 \cos \omega t$. We will assume that the nonlinear term ax^2 is small in the sense discussed above, and treat it as a small perturbation of the linear equation that is obtained in the limit $a = 0$.

The first approximation to $x(t)$ is obtained by ignoring the nonlinear term in (17.3.2) altogether. Denoting this first approximation by $x^{(1)}(t)$, we have

$$\ddot{x}^{(1)} + \omega_0^2 x^{(1)} = \frac{e}{m} E_0 \cos \omega t \tag{17.3.3}$$

which has the familiar solution (2.3.8):

$$x^{(1)}(t) = \frac{e/m}{\omega_0^2 - \omega^2} E_0 \cos \omega t \tag{17.3.4}$$

We are ignoring the homogeneous solution to equation (17.3.3) in order to concentrate on the electronic response to the field $E_0\cos\omega t$. In any event, given realistic relaxation, the contribution from the homogeneous equation damps to zero. Thus we are in effect considering the steadily driven response.

A better approximation to $x(t)$, labeled $x^{(2)}(t)$, is obtained by solving

$$\ddot{x}^{(2)}(t) + \omega_0^2 x^{(2)}(t) = \frac{e}{m} E_0 \cos \omega t - a\left[x^{(1)}(t)\right]^2 \qquad (17.3.5)$$

That is, to this level of approximation we include the effect of the nonlinear term in (17.3.2), but use the *first* approximation to $x(t)$ in this nonlinear term. Note that, from (17.3.4),

$$\left[x^{(1)}(t)\right]^2 = \left(\frac{e/m}{\omega_0^2 - \omega^2}\right)^2 E_0^2 \cos^2 \omega t$$

$$= \frac{1}{2}\left(\frac{e/m}{\omega_0^2 - \omega^2}\right)^2 E_0^2(1 + \cos 2\omega t) \qquad (17.3.6)$$

since $\cos^2 x = \frac{1}{2}(1 + \cos 2x)$. Thus (17.3.5) becomes

$$\ddot{x}^{(2)}(t) + \omega_0^2 x^{(2)}(t) = \frac{e}{m} E_0 \cos \omega t - \frac{a}{2}\left(\frac{e/m}{\omega_0^2 - \omega^2}\right)^2 E_0^2$$

$$- \frac{a}{2}\left(\frac{e/m}{\omega_0^2 - \omega^2}\right)^2 E_0^2 \cos 2\omega t \qquad (17.3.7)$$

which has the steadily driven solution (Problem 17.1)

$$x^{(2)}(t) = \frac{e/m}{\omega_0^2 - \omega^2} E_0 \cos \omega t - \frac{a}{2\omega_0^2}\left(\frac{e/m}{\omega_0^2 - \omega^2}\right)^2 E_0^2$$

$$- \frac{a}{2}\frac{1}{\omega_0^2 - 4\omega^2}\left(\frac{e/m}{\omega_0^2 - \omega^2}\right)^2 E_0^2 \cos 2\omega t \qquad (17.3.8)$$

An important new effect is evident in (17.3.8). The improved (second) approximation to $x(t)$ shows that the electron displacement has a term oscillating both at the *fundamental* driving field frequency ω, and also a term oscillating at the *second-harmonic* frequency 2ω. There is also a static (or dc) term, the second term on the right side of (17.3.8). Continuing this approximation scheme in like manner, we could compute $x^{(3)}(t)$, $x^{(4)}(t)$, etc. We would find additional harmonics (or overtones) of the fundamental frequency. All of these harmonics arise from the nonlinearity in (17.3.2), i.e., the term ax^2. We must now explore their physical consequences.

Before doing this, we will switch to a different notation that is convenient for nonlinear optics. We will write

$$E(t) = \text{Re} \left(\mathcal{E}_\omega e^{-i\omega t} \right)$$

$$= \tfrac{1}{2} \left(\mathcal{E}_\omega e^{-i\omega t} + \mathcal{E}_\omega^* e^{i\omega t} \right) \tag{17.3.9}$$

instead of $E(t) = \mathcal{E}_0 \cos \omega t$, so that (17.3.2) is replaced by

$$\ddot{x} + \omega_0^2 x + a x^2 = \frac{e}{2m} \left[\mathcal{E}_\omega e^{-i\omega t} + \mathcal{E}_\omega^* e^{i\omega t} \right] \tag{17.3.10}$$

Proceeding as above, we obtain instead of (17.3.8) the expression (Problem 17.2)

$$x^{(2)}(t) = x_0 + \tfrac{1}{2} \left[x_\omega e^{-i\omega t} + x_\omega^* e^{i\omega t} \right]$$

$$+ \tfrac{1}{2} \left[x_{2\omega} e^{-2i\omega t} + x_{2\omega}^* e^{2i\omega t} \right] \tag{17.3.11}$$

where

$$x_0 = \frac{-a}{2\omega_0^2} \left(\frac{e/m}{\omega_0^2 - \omega^2} \right)^2 |\mathcal{E}_\omega|^2 \tag{17.3.12a}$$

$$x_\omega = \frac{e/m}{\omega_0^2 - \omega^2} \mathcal{E}_\omega \tag{17.3.12b}$$

$$x_{2\omega} = \frac{-a}{2} \frac{1}{\omega_0^2 - 4\omega^2} \left(\frac{e/m}{\omega_0^2 - \omega^2} \right)^2 \mathcal{E}_\omega^2 \tag{17.3.12c}$$

Note that, if $\mathcal{E}_\omega$ is real and equal to E_0, then (17.3.11) reduces to (17.3.8), as it must.

• Sum- and difference-frequency generation, of which harmonic generation is a special case, are important phenomena in many areas of science and engineering. In an AM transmitter, for instance, a nonlinear circuit combines a signal with a high-frequency carrier wave to produce a modulated output containing sum and difference frequencies. At the receiver these are fed into a nonlinear circuit to combine sum and difference frequencies and reproduce the original signal. The human ear also has nonlinear response properties, such that sufficiently loud sounds produce the sensation that harmonics and sum and difference frequencies are present.

You may have noticed when driving near a radio-station antenna that your car radio suddenly picks up not only that station but also another one at a different frequency. This *cross modulation* can occur because of a nonlinear response of the receiver to a strong signal. •

17.4 NONLINEAR POLARIZATION AND THE WAVE EQUATION

We recall from Chapter 2 that the polarization density P, the dipole moment per unit volume, is given by

$$P = Nex \qquad (17.4.1)$$

where N is the number of molecules per unit volume. If the electron displacement x is given by (17.3.11), the corresponding polarization is $P^{(2)} = Nex^{(2)}$. Let us assume that the field at the fundamental frequency ω in the medium is a monochromatic plane wave:

$$E = \tfrac{1}{2}\left[\mathcal{E}_\omega(z)\, e^{-i(\omega t - k_\omega z)} + \mathcal{E}_\omega^*(z)\, e^{i(\omega t - k_\omega z)} \right] \qquad (17.4.2)$$

where $k_\omega = n(\omega)\,\omega/c$. By comparison with (17.3.9) we see that this means

$$\mathcal{E}_\omega = \mathcal{E}_\omega(z)\, e^{ik_\omega z} \qquad (17.4.3)$$

in (17.3.12), and therefore in (17.3.11)

$$x_0 = -\frac{a}{2\omega_0^2}\left(\frac{e/m}{\omega_0^2 - \omega^2}\right)^2 \left| \mathcal{E}_\omega(z) \right|^2 \qquad (17.4.4a)$$

$$x_\omega = \frac{e/m}{\omega_0^2 - \omega^2}\, \mathcal{E}_\omega(z)\, e^{ik_\omega z} \qquad (17.4.4b)$$

$$x_{2\omega} = -\frac{a}{2}\,\frac{1}{\omega_0^2 - 4\omega^2}\left(\frac{e/m}{\omega_0^2 - \omega^2}\right)^2 \mathcal{E}_\omega^2(z)\, e^{2ik_\omega z}. \qquad (17.4.4c)$$

Thus the polarization density becomes

$$P^{(2)}(z,\, t) = P_0^{(NL)} + \tfrac{1}{2}\left[P_\omega^{(L)}\, e^{-i(\omega t - k_\omega z)} + P_\omega^{(L)*}\, e^{i(\omega t - k_\omega z)} \right]$$

$$+ \tfrac{1}{2}\left[P_{2\omega}^{(NL)}\, e^{-2i(\omega t - k_\omega z)} + P_{2\omega}^{(NL)*}\, e^{2i(\omega t - k_\omega z)} \right] \qquad (17.4.5)$$

where

$$P_0^{(NL)} = \frac{-Nae^3}{2m^2\omega_0^2(\omega_0^2 - \omega^2)^2}\left| \mathcal{E}_\omega(z) \right|^2 \qquad (17.4.6a)$$

$$P_\omega^{(L)} = \frac{Ne^2/m}{\omega_0^2 - \omega^2}\, \mathcal{E}_\omega(z) \qquad (17.4.6b)$$

$$P_{2\omega}^{(NL)} = \frac{-Nae^3}{2m^2(\omega_0^2 - 4\omega^2)(\omega_0^2 - \omega^2)^2}\, \mathcal{E}_\omega^2(z) \qquad (17.4.6c)$$

The superscripts (L) and (NL) stand for "linear" and "nonlinear." $P_0^{(NL)}$ and $P_{2\omega}^{(NL)}$ vanish if $a = 0$, in which case we are back to the *linear* oscillator model. In other words, the nonlinear terms in the polarization arise from the nonlinear properties of the individual atoms or molecules of the medium.

The second term on the right side of (17.4.5) oscillates at precisely the frequency ω of the applied field. As we saw in Chapter 2, it gives rise to the refractive index [compare (17.4.6b) and (2.3.11)], and, if we include frictional damping in the oscillator model as in Chapter 3, it gives the absorption coefficient of the medium at frequency ω. The term $P_0^{(NL)}$ is a static (i.e., time-independent) polarization, about which we will say more later. For the present we focus our attention on the last term in (17.4.5), which oscillates at the second-harmonic frequency 2ω.

From the Maxwell equation (2.1.13), namely

$$\nabla^2 E - \epsilon_0 \mu_0 \frac{\partial^2 E}{\partial t^2} = \mu_0 \frac{\partial^2 P}{\partial t^2} \qquad (17.4.7)$$

we infer that the second-harmonic term in the polarization P should act as a source for a second-harmonic contribution to the electric field E. In other words, an electric field of frequency ω in a medium in which $a \neq 0$ will generate, via the nonlinear polarization $P_{2\omega}^{(NL)}$, an electric field at the second-harmonic frequency 2ω. This is called *second-harmonic generation*.

We write the second-harmonic field as

$$E = \tfrac{1}{2}[\mathcal{E}_{2\omega}(z)\,e^{-i(2\omega t - k_{2\omega}z)} + \mathcal{E}_{2\omega}^*(z)\,e^{i(2\omega t - k_{2\omega}z)}] \qquad (17.4.8)$$

where $k_{2\omega} = n(2\omega)\,2\omega/c$ and $n(2\omega) = (\epsilon_{2\omega}/\epsilon_0)^{1/2}$ is the refractive index of the medium for radiation of frequency 2ω. Assuming $\mathcal{E}_{2\omega}(z)$ to be slowly varying in z compared with $\exp(ik_{2\omega}z)$, as in Section 8.3, we make the approximation of neglecting second derivatives of $\mathcal{E}_{2\omega}(z)$, to get

$$\nabla^2 E = \frac{\partial^2 E}{\partial z^2} \approx \frac{1}{2}\left(2ik_{2\omega}\frac{d\mathcal{E}_{2\omega}}{dz} - k_{2\omega}^2 \mathcal{E}_{2\omega}\right)e^{-i(2\omega t - k_{2\omega}z)}$$
$$+ \frac{1}{2}\left(-2ik_{2\omega}\frac{d\mathcal{E}_{2\omega}^*}{dz} - k_{2\omega}^2 \mathcal{E}_{2\omega}^*\right)e^{i(2\omega t - k_{2\omega}z)} \qquad (17.4.9)$$

which, together with

$$\epsilon_0 \mu_0 \frac{\partial^2 E}{\partial t^2} = -2\epsilon_0\mu_0\omega^2\,[\mathcal{E}_{2\omega}\,e^{-i(2\omega t - k_{2\omega}z)} + \mathcal{E}_{2\omega}^*\,e^{i(2\omega t - k_{2\omega}z)}]$$

$$(17.4.10)$$

gives the left side of (17.4.7):

$$\nabla^2 E - \epsilon_0 \mu_0 \frac{\partial^2 E}{\partial t^2} \approx \left(ik_{2\omega} \frac{d\mathcal{E}_{2\omega}}{dz} - \tfrac{1}{2} (k_{2\omega}^2 - 4\epsilon_0 \mu_0 \omega^2) \mathcal{E}_{2\omega} \right)$$

$$\times e^{-i(2\omega t - k_{2\omega} z)} + \text{complex conjugate} \qquad (17.4.11)$$

for the second-harmonic field (17.4.8).

The right side of (17.4.7) has both linear and nonlinear contributions at frequency 2ω:

$$P = \tfrac{1}{2} \left[P_{2\omega}^{(L)} e^{-i(2\omega t - k_{2\omega} z)} + P_{2\omega}^{(L)*} e^{i(2\omega t - k_{2\omega} z)} \right.$$

$$\left. + P_{2\omega}^{(NL)} e^{-2i(\omega t - k_{\omega} z)} + P_{2\omega}^{(NL)*} e^{2i(\omega t - k_{\omega} z)} \right] \qquad (17.4.12a)$$

so that

$$\mu_0 \frac{\partial^2 P}{\partial t^2} = -2\mu_0 \omega^2 \left[P_{2\omega}^{(L)} e^{-i(2\omega t - k_{2\omega} z)} + P_{2\omega}^{(L)*} e^{i(2\omega t - k_{2\omega} z)} \right]$$

$$- 2\mu_0 \omega^2 \left[P_{2\omega}^{(NL)} e^{-2i(\omega t - k_{\omega} z)} + P_{2\omega}^{(NL)*} e^{2i(\omega t - k_{\omega} z)} \right] \qquad (17.4.12b)$$

Combining (17.4.11) and (17.4.12), therefore, we get

$$\left(ik_{2\omega} \frac{d\mathcal{E}_{2\omega}}{dz} - \tfrac{1}{2} (k_{2\omega}^2 - 4\epsilon_0 \mu_0 \omega^2) \mathcal{E}_{2\omega} \right) e^{-i(2\omega t - k_{2\omega} z)}$$

$$= -2\mu_0 \omega^2 P_{2\omega}^{(L)} e^{-i(2\omega t - k_{2\omega} z)} - 2\mu_0 \omega^2 P_{2\omega}^{(NL)} e^{-2i(\omega t - k_{\omega} z)} \qquad (17.4.13)$$

From Section 2.3 we know that $P = \epsilon_0 \chi E$, which becomes, at frequency 2ω,

$$P_{2\omega}^{(L)} = \epsilon_0 \chi(2\omega) \mathcal{E}_{2\omega} \qquad (17.4.14)$$

and

$$k_{2\omega}^2 = (2\omega)^2 \epsilon_{2\omega} \mu_0 = 4\omega^2 \epsilon_0 \mu_0 n^2(2\omega) = 4\omega^2 \epsilon_0 \mu_0 [1 + \chi(2\omega)] \qquad (17.4.15)$$

where $\chi(2\omega)$ is the susceptibility for frequency 2ω, taken to be real because we are assuming absorption and other loss processes to be negligible. The use of these relations in (17.4.13) results in the equation

$$ik_{2\omega} \frac{d\mathcal{E}_{2\omega}}{dz} e^{-i(2\omega t - k_{2\omega} z)} = -2\mu_0 \omega^2 P_{2\omega}^{(NL)} e^{-2i(\omega t - k_{\omega} z)}$$

or

$$\frac{d\mathcal{E}_{2\omega}}{dz} = \frac{2i\mu_0\omega^2}{k_{2\omega}} P_{2\omega}^{(\text{NL})} e^{i(2\omega t - k_{2\omega}z - 2\omega t + 2k_\omega z)}$$

$$= i\omega \sqrt{\mu_0/\epsilon_{2\omega}} \, P_{2\omega}^{(\text{NL})} e^{i(2k_\omega - k_{2\omega})z}, \qquad (17.4.16)$$

relating the second-harmonic field to the nonlinear polarization. Finally it is convenient to define a quantity $\bar{d}$ by writing (17.4.6c) as

$$P_{2\omega}^{(\text{NL})} = \bar{d}\mathcal{E}_\omega^2(z) \qquad (17.4.17)$$

Thus we can write (17.4.16) as

$$\frac{d\mathcal{E}_{2\omega}}{dz} = i\omega \sqrt{\frac{\mu_0}{\epsilon_{2\omega}}} \, \bar{d}\mathcal{E}_\omega^2(z) e^{i\Delta kz} \qquad (17.4.18)$$

where

$$\Delta k = 2k_\omega - k_{2\omega} = 2\omega \sqrt{\epsilon_0\mu_0} \left[n(\omega) - n(2\omega) \right] \qquad (17.4.19)$$

The solution of (17.4.18) gives the second-harmonic field amplitude $\mathcal{E}_{2\omega}(z)$. To solve this equation, however, we need to know $\mathcal{E}_\omega(z)$. In the simplest situation there is little attenuation of the fundamental wave, and we make the approximation that $\mathcal{E}_\omega$ is a constant:

$$\mathcal{E}_\omega(z) \approx \mathcal{E}_\omega(0) \qquad (17.4.20)$$

In this case (17.4.18) is easily integrated:

$$\mathcal{E}_{2\omega}(z) \approx i\omega \sqrt{\frac{\mu_0}{\epsilon_{2\omega}}} \, \bar{d}\mathcal{E}_\omega^2(0) \int_0^z e^{i\Delta kz'} \, dz'$$

$$= i\omega \sqrt{\frac{\mu_0}{\epsilon_{2\omega}}} \, \bar{d}\mathcal{E}_\omega^2(0) \left(\frac{1}{i\,\Delta k} (e^{i\Delta kz} - 1) \right) \qquad (17.4.21)$$

Now

$$\frac{1}{i\,\Delta k}\left(e^{i\Delta kz} - 1\right) = \frac{e^{i\Delta kz/2}}{i\,\Delta k}\left(e^{i\Delta kz/2} - e^{-i\Delta kz/2}\right)$$

$$= 2e^{i\Delta kz/2}\frac{1}{\Delta k}\sin\tfrac{1}{2}\Delta k\,z$$

$$= z\,e^{i\Delta kz/2}\left(\frac{\sin\tfrac{1}{2}\Delta k\,z}{\tfrac{1}{2}\Delta k\,z}\right) \qquad (17.4.22)$$

Therefore, in the approximation (17.4.20) of an unattenuated fundamental wave, we have

$$\mathcal{E}_{2\omega}(z) = i\omega\sqrt{\frac{\mu_0}{\epsilon_{2\omega}}}\,\bar{d}\mathcal{E}_\omega^2(0)z\,e^{i\Delta kz/2}\left(\frac{\sin\tfrac{1}{2}\Delta k\,z}{\tfrac{1}{2}\Delta k\,z}\right) \qquad (17.4.23)$$

Equations (17.4.18) and (17.4.23) are the principal results of this section.

Let us briefly summarize the steps leading to (17.4.23). From the nonlinear electron oscillator model we inferred the existence of a contribution to the polarization oscillating at twice the frequency of the applied field. According to the Maxwell wave equation (17.4.7) for the electric field, this polarization acts as the source of a field oscillating at twice the frequency of the applied field. Within the approximation (17.4.9), and the approximation that the applied field is not attenuated inside the medium, this second-harmonic field is given by (17.4.23).

17.5 SECOND-HARMONIC GENERATION

The first observation in the 1960's of various nonlinear optical effects followed quickly after the first successful laser operation. Second-harmonic generation was reported in 1961. Light from a ruby laser ($\lambda = 6943$ Å) incident upon a quartz crystal caused ultraviolet light at half the wavelength of the laser radiation to be generated (Figure 17.2). *Frequency doublers*, acting like this quartz crystal, are now used in many applications.

Suppose the nonlinear crystal indicated in Figure 17.2 is of length L. The second-harmonic electric field at the exit face of the crystal is then given by (17.4.23) with $z = L$. Thus (17.4.23) implies:

$$\left|\mathcal{E}_{2\omega}(L)\right|^2 = \frac{\mu_0\omega^2\bar{d}^2}{\epsilon_{2\omega}}\left|\mathcal{E}_\omega(0)\right|^4 L^2\left(\frac{\sin\tfrac{1}{2}\Delta k\,L}{\tfrac{1}{2}\Delta k\,L}\right)^2 \qquad (17.5.1)$$

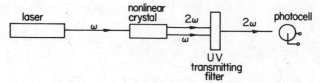

Figure 17.2 Schematic experimental arrangement for detection of second harmonic generation.

The (time-averaged) intensities of the fields at ω and 2ω are given by equation (2.6.8):

$$I_\omega = \tfrac{1}{2} \sqrt{\epsilon_\omega/\mu_0} \, |\mathcal{E}_\omega|^2 \tag{17.5.2a}$$

$$I_{2\omega} = \tfrac{1}{2} \sqrt{\epsilon_{2\omega}/\mu_0} \, |\mathcal{E}_{2\omega}|^2 \tag{17.5.2b}$$

where we have now included the effect of the refractive index in the definition of intensity. It follows from (17.5.1) that

$$I_{2\omega}(L) = \frac{2\mu_0^{3/2}}{\epsilon_\omega \sqrt{\epsilon_{2\omega}}} \, \omega^2 d^2 I_\omega^2(0) L^2 \left(\frac{\sin \tfrac{1}{2} \Delta k \, L}{\tfrac{1}{2} \Delta k \, L} \right)^2$$

$$= 2 \left(\frac{\mu_0}{\epsilon_0} \right)^{3/2} \frac{\omega^2 \overline{d}^2}{n^2(\omega) \, n(2\omega)} \, I_\omega^2(0) L^2 \left(\frac{\sin \tfrac{1}{2} \Delta k \, L}{\tfrac{1}{2} \Delta k \, L} \right)^2 \tag{17.5.3}$$

From this we have the *power conversion efficiency* e_{SHG} for second harmonic generation:

$$e_{\mathrm{SHG}} = \frac{I_{2\omega}(L)}{I_\omega(0)} = 2 \left(\frac{\mu_0}{\epsilon_0} \right)^{3/2} \frac{\omega^2 \overline{d}^2}{n^2(\omega) \, n(2\omega)} \, I_\omega(0) L^2 \left(\frac{\sin \tfrac{1}{2} \Delta k \, L}{\tfrac{1}{2} \Delta k \, L} \right)^2 \tag{17.5.4}$$

If it happens that $n(\omega) = n(2\omega)$, then $\Delta k = 0$ from (17.4.19), and therefore

$$\left(\frac{\sin \tfrac{1}{2} \Delta k \, L}{\tfrac{1}{2} \Delta k \, L} \right)^2 \rightarrow \lim_{x=0} \frac{\sin^2 x}{x^2} = 1 \tag{17.5.5}$$

Thus

$$e_{\mathrm{SHG}} = 2 \left(\frac{\mu_0}{\epsilon_0} \right)^{3/2} \frac{\omega^2 \overline{d}^2}{n^3} \, I_\omega(0) L^2 \tag{17.5.6}$$

where $n = n(\omega) = n(2\omega)$.

As an example, consider second-harmonic generation of 3471-Å radiation in a quartz crystal irradiated by a Q-switched ruby laser. For quartz $\bar{d} \approx 4 \times 10^{-24}$ in mks units, and $n \approx 1.5$. Assuming $I_{\omega}(0) \approx 10^8$ W/cm^2 and $L = 1$ cm, we compute from (17.5.6) the power conversion efficiency (Problem 17.3)

$$e_{SHG} \approx 37\% \tag{17.5.7}$$

This rough computation is misleading in two respects. First, the second-harmonic generation of 3471-Å radiation occurs at the expense of the ruby-laser radiation. That is, the laser radiation is converted to second-harmonic radiation, and for a conversion efficiency as large as (17.5.7) the approximation of no conversion of the fundamental wave is a poor one. Indeed, by taking L large enough in (17.5.6), we would predict $e_{SHG} > 100\%$, in violation of the law of conservation of energy. In other words, the formula (17.5.6) applies only if e_{SHG} is small. Otherwise we would have to go beyond the approximation (17.4.20) and include the depletion of the pump radiation as it is converted to the second harmonic. In such a more accurate analysis, we would have *two* equations of the type (17.4.18) coupling $\mathcal{E}_{\omega}(z)$ and $\mathcal{E}_{2\omega}(z)$. (See Section 17.8.)

But there is a more serious shortcoming of the computation leading to (17.5.7), namely, the assumption that $\Delta k = 0$. For quartz we have $n(6940 \text{ Å}) \approx 1.54$ and $n(3470 \text{ Å}) \approx 1.57$. Then, from (17.4.19) (Problem 17.3),

$$\Delta k \approx 5.4 \times 10^5 \text{ m}^{-1} \tag{17.5.8}$$

and

$$\Delta k L \approx 5.4 \times 10^3 \tag{17.5.9}$$

for $L = 1$ cm. Therefore the last factor in (17.5.4) is

$$\frac{\sin^2 \frac{1}{2} \Delta k L}{(\frac{1}{2} \Delta k L)^2} \approx \frac{\sin^2 2700}{(2700)^2} \sim 10^{-7} \tag{17.5.10}$$

The conversion efficiency (17.5.7) obtained using (17.5.5) is replaced by

$$e_{SHG} \sim (37\%)(10^{-7}) \sim 4 \times 10^{-8} \tag{17.5.11}$$

a far cry from (17.5.7). The principle at work here is phase matching, or mismatching in this case. Phase matching is exceptionally important in nonlinear optics, and we devote the next section to a discussion of it.

• There are many nonlinear optical effects that can occur when laser radiation passes through matter, second-harmonic generation being only one example. Consider, for instance, the dc polarization $P_0^{(NL)}$ appearing in (17.4.5). This gives rise to a static electric field inside the medium. Thus an optical field incident on the material should give rise to a (dc) potential difference. This dc voltage produced by incident light is called *optical rectification*. •

17.6 PHASE MATCHING

The dimensionless number $\Delta k \, L$ in (17.5.9), for example, determines the phase mismatch $\exp(2ik_\omega z - ik_{2\omega}z)$ between the fundamental and second-harmonic waves over the distance $z = L$. It comes from a difference in the indices of refraction for these two frequencies. Our rough estimate of the power conversion efficiency e_{SHG} for ruby laser radiation in quartz indicates that a phase mismatch can be a strongly negative factor in harmonic generation. Indeed, our estimate of e_{SHG} is consistent with the very weak second-harmonic signals observed in early experimental studies of second-harmonic generation. Efficient second-harmonic generation requires the pump and second-harmonic fields to somehow be *phase-matched*.

The reason for the phase mismatch $\Delta k \, L$ between the pump and second-harmonic fields is simple: the second-harmonic field propagates with the phase velocity $c/n(2\omega)$, whereas Eq. (17.4.12a) shows that its nonlinear polarization source has phase velocity $c/n(\omega)$. Because of this difference, the two fields get out of step with each other. Since the medium will only allow the second-harmonic field to have phase velocity $c/n(2\omega)$, there is a reduction in second-harmonic generation determined by $n(2\omega) - n(\omega)$, and this reduction is expressed quantitatively by the factor $(\sin^2 \frac{1}{2} \Delta k \, L)/(\frac{1}{2} \Delta k \, L)^2$.

The function $(\sin^2 x)/x^2$ is plotted in Figure 17.3. At $x = \pi/2$ this function drops to about 40% of its peak value at $x = 0$; therefore we define a distance L_c, called the *coherence length*,[1] by $\frac{1}{2} |\Delta k| L_c = \pi/2$, or

$$L_c = |\pi/\Delta k| \tag{17.6.1}$$

L_c is the distance over which there is significant generation of the second harmonic. If there is perfect phase matching, L_c is effectively infinite, and e_{SHG} increases as the square of the length L of the crystal, as indicated in Eq. (17.5.6), until the depletion of the pump field becomes important. When phase matching is not realized, however, only the coherence length L_c determines the conversion efficiency—increasing L beyond L_c does no good. Without some method of phase matching the pump and second-harmonic fields the coherence length is usually very small, typically on the order of 10 μm.

Figure 17.4 shows experimental data verifying the variation of e_{SHG} with $\Delta k \, L$ according to the function $(\sin^2 \frac{1}{2} \Delta k \, L)/(\frac{1}{2} \Delta k \, L)^2$. Our treatment of second-har-

---

1. There should be no confusion between this coherence length and the coherence length introduced in Chapter 15.

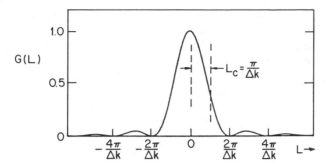

Figure 17.3 The function $G(L) = \sin^2\alpha L/(\alpha L)^2$, where $\alpha \equiv \Delta k/2$. The phase-matching coherence length L_c is indicated.

monic generation using the nonlinear oscillator model gives correct results, as indicated for instance by Figure 17.4. However, we cannot very accurately predict the value of the nonlinear coefficients a or $\bar{d}$; this can be done only with quantum mechanics, and the calculations are rather complicated. Otherwise, our classical approach to nonlinear optics is quite satisfactory.

There are various techniques for phase-matching the pump and second-harmonic waves, i.e., for making $|\Delta k|$ in equation (17.4.19) as small as possible. One common method employs the birefringence of uniaxial crystals used for second-harmonic generation. We will for the present focus our attention on this method, which is sometimes called "angle phase matching," for reasons which will become clear.

As discussed in Section 2.9, a birefringent, or doubly refracting, material is one in which the refractive index depends upon the direction of polarization of a light wave in the medium. However, there exists a direction, called the optic axis, in

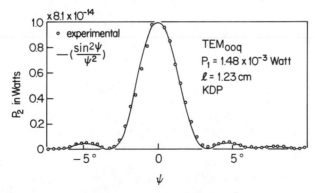

Figure 17.4 Variation of second-harmonic power with $\psi = \frac{1}{2}\Delta k\, L$, as measured for 1.15-$\mu$m pump radiation in the crystal KDP by A. Ashkin, G. D. Boyd, and J. M. Dziedzic, *Physical Review Letters* **11**, 14 (1963).

which the refractive index is independent of the direction of polarization. A uniaxial crystal is one for which there is only one such optic axis. Only two types of wave can propagate in such a crystal, namely, waves plane-polarized perpendicular to the plane formed by the optic axis and the direction of propagation (ordinary waves), and waves plane-polarized parallel to this plane (extraordinary waves). The crystal has different refractive indices for ordinary and extraordinary waves.

The refractive index for ordinary waves is denoted by n_0. The refractive index n_e of an extraordinary wave depends upon its angle of propagation θ relative to the optic axis. According to equation (2.9.9),

$$\frac{1}{n_e(\omega, \theta)^2} = \frac{\cos^2 \theta}{n_0^2(\omega)} + \frac{\sin^2 \theta}{n_e^2(\omega)} \qquad (17.6.2)$$

where $n_e(\omega) = n_e(\omega, \theta = \pi/2)$. This angular dependence of the extraordinary index is the key to angle phase matching.

Consider a uniaxial crystal for which $n_e(\omega) > n_0(\omega)$. Such a crystal is called *positive uniaxial*. [A negative uniaxial crystal is one for which $n_e(\omega) < n_0(\omega)$.] For such a crystal we can phase-match the pump and second-harmonic waves by letting the pump wave of frequency ω be an extraordinary wave propagating at an angle θ_p to the optic axis, while the second-harmonic wave of frequency 2ω is an ordinary wave propagating in this direction. *Phase matching is achieved if θ_p is chosen such that*

$$n_0(2\omega) = n_e(\omega, \theta_p) \qquad (17.6.3a)$$

or

$$\frac{1}{n_0^2(2\omega)} = \frac{1}{n_e^2(\omega, \theta_p)} = \frac{\cos^2 \theta_p}{n_0^2(\omega)} + \frac{\sin^2 \theta_p}{n_e^2(\omega)} \qquad (17.6.3b)$$

Solving for $\sin^2 \theta_p$, we have

$$\sin^2 \theta_p = \frac{n_0(\omega)^{-2} - n_0(2\omega)^{-2}}{n_0(\omega)^{-2} - n_e(\omega)^{-2}} \qquad (17.6.4)$$

Assuming the crystal to be normally dispersive [i.e., $n_0(2\omega) > n_0(\omega)$ and therefore $n_0(\omega)^{-2} - n_0(2\omega)^{-2} > 0$], we see that $n_e(\omega) > n_0(\omega)$ is necessary in order to satisfy the requirement that $\sin^2 \theta_p > 0$, and furthermore

$$n_0(\omega)^{-2} - n_0(2\omega)^{-2} < n_0(\omega)^{-2} - n_e(\omega)^{-2}$$

or

$$n_e(\omega) > n_0(2\omega) \qquad \text{(positive uniaxial)} \qquad (17.6.5)$$

is necessary for $\sin^2 \theta_p < 1$. In other words, this method of phase matching can be used in a positive uniaxial crystal when (17.6.5) is satisfied, since $n_0(2\omega) > n_0(\omega)$ and Eq. (17.6.5) together imply $n_e(\omega) > n_0(\omega)$, as required above. Problem 17.4 concerns the possibility of angle phase-matched second-harmonic generation when 6943-Å ruby-laser radiation is incident on a quartz crystal.

In practice, we send the pump wave into the crystal as an extraordinary wave at the angle θ_p to the optic axis. Then the second-harmonic wave is automatically generated as an ordinary wave propagating in the direction θ_p, because this is the only direction and polarization for which there is phase matching and substantial second-harmonic generation.

For a negative uniaxial crystal the situation is reversed: the angle between the pump and second-harmonic waves is (Problem 17.5)

$$\sin^2 \theta_p = \frac{n_0(\omega)^{-2} - n_0(2\omega)^{-2}}{n_e(2\omega)^{-2} - n_0(2\omega)^{-2}} \tag{17.6.6}$$

and phase matching may be achieved if

$$n_e(2\omega) < n_0(\omega) \qquad \text{(negative uniaxial)} \tag{17.6.7}$$

Consider the example of second-harmonic generation when 6943-Å radiation is incident on the negative uniaxial crystal known as KDP (potassium dihydrogen phosphate, KH_2PO_4). For this crystal $n_0(\omega) = 1.505$, $n_0(2\omega) = 1.534$, $n_e(\omega) = 1.465$, and $n_e(2\omega) = 1.487$. Thus (17.6.7) is satisfied and angle phase matching is achieved with the incident ruby-laser beam linearly polarized as an ordinary wave, incident at the angle θ_p given by (17.6.6):

$$\sin^2 \theta_p = 0.606 \tag{17.6.8}$$

or $\theta_p = 51°$.

There are other methods of phase matching, but the angle method described above is the most common. As we have seen, very high pump-to-second-harmonic conversion efficiencies can be predicted under phase-matched conditions. In such cases, as noted in the preceding section, pump depletion as a result of harmonic generation must be included in a detailed analysis.

17.7 INTRACAVITY SECOND-HARMONIC GENERATION

Equation (17.5.6) indicates that the power conversion efficiency for second-harmonic generation is proportional to the pump intensity. High conversion efficiencies in most instances require the high peak intensities available only from pulsed lasers. However, it is desirable for many applications to generate a cw second-harmonic field. One way to do this is by intracavity second-harmonic generation, i.e., by inserting a nonlinear crystal inside the cavity of a cw laser (Figure 17.5).

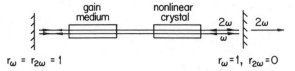

Figure 17.5 Intracavity second-harmonic generation.

The basic advantage of this technique is that it exposes the crystal to a cw pump field of much greater intensity than can be obtained outside the laser in ordinary (extracavity) second-harmonic generation. We recall from Chapter 11 that the output intensity of a laser is typically only a small fraction of the intracavity intensity (recall Eq. 11.2.7).

In Figure 17.5 the ordinary output mirror of the laser is replaced by one that is perfectly reflecting at the laser frequency ω, but perfectly transmitting at the second-harmonic frequency 2ω. The principal loss mechanism for the intracavity laser field is then its conversion to the second harmonic inside the nonlinear crystal. The crystal is oriented at the appropriate phase-matching angle θ_p.

Suppose that the power conversion efficiency for second-harmonic generation inside the crystal can be made equal to the optimal output coupling for the laser without the intracavity crystal. With the crystal in place the loss—due now to second-harmonic generation—is still optimal for the laser, just as it would be with the crystal removed. In this case we still extract the maximum possible power from the laser, but now all this power is at the second-harmonic frequency.

In practice, however, other effects conspire to reduce the second-harmonic output power. The insertion of the nonlinear crystal in the laser cavity represents an additional loss mechanism because of scattering, and so crystals of high optical quality are needed to keep this loss small. Furthermore the crystal may be slightly absorbing at ω and 2ω, thus raising its temperature. The refractive indices n_0 and n_e of the crystal often vary appreciably with temperature, so that even a small temperature rise might seriously reduce the phase matching. Commercially available cw intracavity doubling systems, such as a Nd:YAG laser with an insertable lithium iodate ($LiIO_3$) crystal, may nevertheless have conversion efficiencies of 10% or more.

17.8 THREE-WAVE MIXING

Consider our nonlinear oscillator with a bichromatic applied field

$$E = \tfrac{1}{2}\left[\mathcal{E}_1(z)\, e^{-i(\omega_1 t - k_1 z)} + \mathcal{E}_1^*(z)\, e^{i(\omega_1 t - k_1 z)}\right]$$
$$+ \tfrac{1}{2}\left[\mathcal{E}_2(z)\, e^{-i(\omega_2 t - k_2 z)} + \mathcal{E}_2^*(z)\, e^{i(\omega_2 t - k_2 z)}\right] \tag{17.8.1}$$

where

$$k_i = \frac{n(\omega_i)\omega_i}{c} = \frac{n_i\omega_i}{c} \tag{17.8.2}$$

If we proceed as in Section 17.3 to solve for $x(t)$, we find that $x(t)$ has terms oscillating at various frequencies, including $\omega_1 + \omega_2$ and $|\omega_1 - \omega_2|$. This leads to a nonlinear polarization in the medium with frequency components $\omega_1 + \omega_2$ and $|\omega_1 - \omega_2|$. In other words, the field (17.8.1) gives rise to the generation of fields at frequencies $\omega_1 + \omega_2$ and $|\omega_1 - \omega_2|$. Second-harmonic generation is a special case in which $\omega_1 = \omega_2$.

The field at $\omega_3 = \omega_1 + \omega_2$ may be written

$$E_3 = \tfrac{1}{2}\left[\mathcal{E}_3(z)\,e^{-i(\omega_3 t - k_3 z)} + \mathcal{E}_3^*(z)\,e^{i(\omega_3 t - k_3 z)}\right] \tag{17.8.3}$$

Proceeding as in Section 17.4, we can derive wave equations for the slowly varying amplitudes $\mathcal{E}_1$, $\mathcal{E}_2$, and $\mathcal{E}_3$. Since the derivation, though straightforward, is somewhat lengthy, we simply write the results:

$$\frac{d\mathcal{E}_1}{dz} = i\omega_1 \sqrt{\frac{\mu_0}{\epsilon_1}}\,\overline{d}\mathcal{E}_2^*\mathcal{E}_3\,e^{-i\Delta kz} \tag{17.8.4a}$$

$$\frac{d\mathcal{E}_2}{dz} = i\omega_2 \sqrt{\frac{\mu_0}{\epsilon_2}}\,\overline{d}\mathcal{E}_1^*\mathcal{E}_3\,e^{-i\Delta kz} \tag{17.8.4b}$$

$$\frac{d\mathcal{E}_3}{dz} = i\omega_3 \sqrt{\frac{\mu_0}{\epsilon_3}}\,\overline{d}\mathcal{E}_1\mathcal{E}_2\,e^{i\Delta kz} \tag{17.8.4c}$$

with

$$\Delta k = k_1 + k_2 - k_3 \tag{17.8.5a}$$

$$\epsilon_i = \epsilon_0 n_i^2 \tag{17.8.5b}$$

Equations (17.8.4) couple the three wave amplitudes $\mathcal{E}_1$, $\mathcal{E}_2$, and $\mathcal{E}_3$. That is, they describe the coupling, or *mixing*, of three waves. Nonlinear optical processes described by equations like (17.8.4) are thus examples of *three-wave mixing*, second-harmonic generation being a special example.

Sum-frequency generation, in which waves at ω_1 and ω_2 mix to generate a third wave at $\omega_3 = \omega_1 + \omega_2$, is of particular interest in connection with the detection and measurement of infrared radiation: up-conversion of infrared radiation to the visible by sum-frequency generation allows the use of fast and efficient detectors (e.g., photomultipliers) that are not generally available in the infrared.

Frequency-difference generation, in which two waves mix to produce radiation at their difference frequency, is also described by the coupled wave equations (17.8.4) with $\omega_3 = \omega_1 + \omega_2$. For instance, according to (17.8.4b) we can use pump radiation at ω_3 and signal radiation at ω_1 to generate an *idler* field at ω_2. Or we can mix a field at ω_2 with the pump at ω_3 to generate a field at ω_1; in this case the field at ω_2 is called the signal and that at ω_1 the idler. In other words, what is conventionally called the idler or the signal wave depends on the initial conditions.

From (17.8.4) it follows that (Problem 17.6)

$$\frac{1}{\omega_1}\frac{d}{dz}\left(\sqrt{\frac{\epsilon_1}{\mu_0}}\,|\mathcal{E}_1|^2\right) = \frac{1}{\omega_2}\frac{d}{dz}\left(\sqrt{\frac{\epsilon_2}{\mu_0}}\,|\mathcal{E}_2|^2\right)$$

$$= \frac{-1}{\omega_3}\frac{d}{dz}\left(\sqrt{\frac{\epsilon_3}{\mu_0}}\,|\mathcal{E}_3|^2\right) \qquad (17.8.6a)$$

In other words, for any three-wave mixing process with $\omega_3 = \omega_1 + \omega_2$ we have

$$\frac{1}{\omega_1}\left(\text{rate of change of energy at } \omega_1\right)$$

$$= \frac{1}{\omega_2}\left(\text{rate of change of energy at } \omega_2\right)$$

$$= -\frac{1}{\omega_3}\left(\text{rate of change of energy at } \omega_3\right) \qquad (17.8.6b)$$

Such equations as (17.8.6) are called *Manley–Rowe relations*, and they have a remarkably simple interpretation in terms of photons: each of the three equal terms in (17.8.6) represents the rate of change of the *number of photons* at the corresponding frequency, so that the creation of a photon at ω_3 in sum-frequency generation, for instance, is accompanied by the annihilation of a photon at ω_1 and a photon at ω_2.

The coupled wave equations for second-harmonic generation are found similarly to be

$$\frac{d\mathcal{E}_\omega}{dz} = i\omega\sqrt{\frac{\mu_0}{\epsilon_\omega}}\,\bar{d}\mathcal{E}_\omega^*\mathcal{E}_{2\omega}\,e^{i\Delta kz} \qquad (17.8.7a)$$

$$\frac{d\mathcal{E}_{2\omega}}{dz} = i\omega\sqrt{\frac{\mu_0}{\epsilon_{2\omega}}}\,\bar{d}\mathcal{E}_\omega^2\,e^{i\Delta kz} \qquad (17.8.7b)$$

where Δk is defined by (17.4.19). When we neglect pump depletion in second-harmonic generation, we are simply ignoring (17.8.7a) and replacing $\mathcal{E}_\omega$ by a constant value in (17.8.7b). Thus the analysis reduces in this case to (17.4.18).

Note that second-harmonic generation is just the degenerate case of sum-frequency generation with $\omega_1 = \omega_2 = \omega$ and $\omega_3 = 2\omega$. However, note that (17.8.7) does *not* follow from (17.8.4) by simply letting $\omega_1 = \omega_2 = \omega$, $\mathcal{E}_1 = \mathcal{E}_2 = \mathcal{E}_\omega$, and $\mathcal{E}_3 = \mathcal{E}_{2\omega}$ in the latter. In particular, (17.8.7b) differs from (17.8.4c), when these (improper) substitutions are made, by a factor of $\frac{1}{2}$. To understand this difference, consider the polarization proportional to $E^2(t)$ in a nonlinear medium when $E(t) = E_0 \cos \omega t$. In this case

$$E^2(t) = E_0^2 \cos^2 \omega t = \tfrac{1}{2}[E_0^2 + E_0^2 \cos 2\omega t] \tag{17.8.8}$$

When $E(t) = E_1 \cos \omega_1 t + E_2 \cos \omega_2 t$, however,

$$E^2(t) = (E_1 \cos \omega_1 t + E_2 \cos \omega_2 t)^2$$

$$= \tfrac{1}{2}[E_1^2 + E_2^2 + E_1^2 \cos 2\omega_1 t + E_2^2 \cos 2\omega_2 t]$$

$$+ E_1 E_2 \cos (\omega_1 + \omega_2) t + E_1 E_2 \cos (\omega_1 - \omega_2) t \tag{17.8.9}$$

which follows from the trigonometric identity

$$\cos (x \pm y) = \cos x \cos y \mp \sin x \sin y \tag{17.8.10}$$

We see that the coefficient $(E_1 E_2)$ multiplying $\cos (\omega_1 + \omega_2) t$ in (17.8.9) does not reduce to the coefficient $(E_0^2/2)$ multiplying $\cos 2\omega t$ in (17.8.8) when we let $\omega_1 = \omega_2 = \omega$ and $E_1 = E_2 = E_0$. This is basically the reason why (17.8.4c) does not reduce to (17.8.7b) when we let $\omega_1 = \omega_2 = \omega$, $\mathcal{E}_1 = \mathcal{E}_2 = \mathcal{E}_\omega$, and $\mathcal{E}_3 = \mathcal{E}_{2\omega}$.

This difference can also be appreciated from a different point of view. From (17.8.7) we obtain

$$\frac{1}{\omega} \frac{d}{dz} \left(\sqrt{\frac{\epsilon_\omega}{\mu_0}} \, |\mathcal{E}_\omega|^2 \right) = -2 \frac{1}{2\omega} \frac{d}{dz} \left(\sqrt{\frac{\epsilon_{2\omega}}{\mu_0}} \, |\mathcal{E}_{2\omega}|^2 \right) \tag{17.8.11}$$

We can interpret this Manley–Rowe relation as saying that *two* photons from the field at ω are annihilated to produce *one* photon at 2ω in second-harmonic generation. This correct interpretation, however, does not follow from the Manley–Rowe relation (17.8.6) with $\omega_1 = \omega_2 = \omega$, $\omega_3 = 2\omega$. This is because (17.8.6) is a consequence of the three-wave mixing equations (17.8.4) which, as we have noted, do not reduce trivially to the coupled wave equations (17.8.7) for second-harmonic generation.

17.9 PARAMETRIC AMPLIFICATION

Consider the implication of the Manley–Rowe relation (17.8.6) for frequency-difference generation of light at $\omega_3 - \omega_1 = \omega_2$. In this process waves of frequency ω_3 (the pump) and ω_1 (the signal) mix to produce a wave at the idler frequency ω_2. According to (17.8.6), the decrease in power at ω_3 is accompanied by an increase in power at *both* ω_1 and ω_2. This amplification of light at ω_1 and ω_2 at the expense of the light at ω_3 is called *parametric amplification*.

Parametric amplification is described by the coupled wave equations (17.8.4). Let us assume that the pump wave at ω_3 is approximately undepleted, so that $\mathcal{E}_3(z)$ = constant = $\mathcal{E}_3(0)$; this is a good approximation under the common circumstance that the relative power converted to ω_1 and ω_2 is small. Then (17.8.4a) and (17.8.4b) give

$$\frac{d\mathcal{E}_1}{dz} = i\left[\omega_1 \sqrt{\frac{\mu_0}{\epsilon_1}}\, \bar{d}\mathcal{E}_3(0) \right] \mathcal{E}_2^*(z) = i\sqrt{\frac{\omega_1}{\omega_2}}\, b_1 \mathcal{E}_2^*(z) \qquad (17.9.1a)$$

$$\frac{d\mathcal{E}_2^*}{dz} = -i\left[\omega_2 \sqrt{\frac{\mu_0}{\epsilon_2}}\, \bar{d}\mathcal{E}_3^*(0) \right] \mathcal{E}_1(z) = -i\sqrt{\frac{\omega_2}{\omega_1}}\, b_2^* \mathcal{E}_1(z) \qquad (17.9.1b)$$

where

$$b_i = \left[\omega_1\omega_2(\mu_0/\epsilon_i) \right]^{1/2}\bar{d}\mathcal{E}_3(0), \qquad i = 1, 2 \qquad (17.9.2)$$

In writing (17.9.1) we have assumed perfect phase matching, $\Delta k = 0$, for simplicity; (17.9.1b) is obtained from the complex conjugate of (17.8.4b). We now differentiate (17.9.1a) and use (17.9.1b):

$$\frac{d^2\mathcal{E}_1}{dz^2} = i\sqrt{\frac{\omega_1}{\omega_2}}\, b_1 \frac{d\mathcal{E}_2^*}{dz} = i\sqrt{\frac{\omega_1}{\omega_2}}\, b_1\left(-i\sqrt{\frac{\omega_2}{\omega_1}}\, b_2^* \mathcal{E}_1 \right) = K^2\mathcal{E}_1 \qquad (17.9.3a)$$

where

$$K = \left(\frac{\omega_1\omega_2}{n_1 n_2}\frac{\mu_0}{\epsilon_0} \right)^{1/2}\bar{d}\left| \mathcal{E}_3(0) \right| \qquad (17.9.3b)$$

Similarly we obtain

$$\frac{d^2\mathcal{E}_2}{dz^2} = K^2\mathcal{E}_2 \qquad (17.9.4)$$

The uncoupled equations (17.9.3) and (17.9.4) may be solved in terms of the fields $\mathcal{E}_1(0)$ and $\mathcal{E}_2(0)$ at the input face $z = 0$ of the nonlinear medium (Problem 17.7):

$$\mathcal{E}_1(z) = \mathcal{E}_1(0) \cosh Kz + i \sqrt{\frac{\omega_1}{\omega_2}}\, \mathcal{E}_2^*(0) \sinh Kz \qquad (17.9.5a)$$

$$\mathcal{E}_2(z) = \mathcal{E}_2(0) \cosh Kz + i \sqrt{\frac{\omega_2}{\omega_1}}\, \mathcal{E}_1^*(0) \sinh Kz \qquad (17.9.5b)$$

where we recall that

$$\cosh x = \tfrac{1}{2}(e^x + e^{-x})$$

$$\sinh x = \tfrac{1}{2}(e^x - e^{-x}) \qquad (17.9.5c)$$

If we imagine injecting into the nonlinear medium a pump wave at ω_3 and a signal field at ω_1, then $\mathcal{E}_2(0) = 0$, i.e., there is no idler field at the input face of the medium. In this case the solutions (17.9.5) become

$$\mathcal{E}_1(z) = \mathcal{E}_1(0) \cosh Kz \qquad (17.9.6a)$$

$$\mathcal{E}_2(z) = i \sqrt{\omega_2/\omega_1}\, \mathcal{E}_1^*(0) \sinh Kz \qquad (17.9.6b)$$

or

$$\left|\mathcal{E}_1(z)\right|^2 = \left|\mathcal{E}_1(0)\right|^2 \cosh^2 Kz \qquad (17.9.7a)$$

$$\left|\mathcal{E}_2(z)\right|^2 = \frac{\omega_2}{\omega_1} \left|\mathcal{E}_1(0)\right|^2 \sinh^2 Kz \qquad (17.9.7b)$$

In the limit $Kz \ll 1$ these solutions reduce to (Problem 17.7)

$$\left|\mathcal{E}_1(z)\right|^2 \approx \left|\mathcal{E}_1(0)\right|^2 (1 + K^2z^2) \qquad (17.9.8a)$$

$$\left|\mathcal{E}_2(z)\right|^2 \approx \frac{\omega_2}{\omega_1} \left|\mathcal{E}_1(0)\right|^2 K^2z^2 \qquad (17.9.8b)$$

Let us consider a numerical example to see what to expect for Kz. For the crystal $LiNbO_3$ (lithium niobate), we assume $\bar{d} \approx 4 \times 10^{-23}$ (mks units) and $n_1 \approx n_2 \approx n_3 \approx 1.5$ when the pump is a Nd:YAG laser ($\lambda = 1.06\ \mu m$) and the signal and idler are each at half the pump frequency. We compute (Problem 17.7)

$$K \approx 2 \times 10^{-4} \sqrt{I_3(0)} \text{ cm}^{-1} \qquad (17.9.9)$$

where $I_3(0)$ is the pump intensity in units of W/cm^2. Thus if $I_3(0) = 1 \text{ MW}/\text{cm}^2$ and $z = 1$ cm,

$$Kz \approx 0.2 \qquad (17.9.10)$$

This example suggests that the limiting forms (17.9.8) will be applicable in many circumstances.

Thus far we have assumed perfect phase matching. Without phase matching, however, parametric amplification is drastically reduced. When $\Delta k \neq 0$, and assuming $b_1 \approx b_2$, Eqs. (17.9.1) are replaced by

$$\frac{d\mathcal{E}_1}{dz} = i \sqrt{\frac{\omega_1}{\omega_2}} K \mathcal{E}_2^*(z) e^{-i\Delta kz} \qquad (17.9.11a)$$

$$\frac{d\mathcal{E}_2^*}{dz} = -i \sqrt{\frac{\omega_2}{\omega_1}} K \mathcal{E}_1(z) e^{i\Delta kz} \qquad (17.9.11b)$$

or, by integrating both sides of these equations,

$$\mathcal{E}_1(z) = \mathcal{E}_1(0) + i \sqrt{\frac{\omega_1}{\omega_2}} K \int_0^z \mathcal{E}_2^*(z') e^{-i\Delta kz'} dz' \qquad (17.9.12a)$$

$$\mathcal{E}_2^*(z) = -i \sqrt{\frac{\omega_2}{\omega_1}} K \int_0^z \mathcal{E}_1(z') e^{i\Delta kz'} dz' \qquad (17.9.12b)$$

assuming once again that $\mathcal{E}_2(0) = 0$. A single equation for $\mathcal{E}_2^*(z)$ may be obtained by using the first of these equations in the second:

$$\mathcal{E}_2^*(z) = -i \sqrt{\frac{\omega_2}{\omega_1}} K \int_0^z e^{i\Delta kz'} \left(\mathcal{E}_1(0) + i \sqrt{\frac{\omega_1}{\omega_2}} K \right.$$

$$\times \left. \int_0^{z'} \mathcal{E}_2^*(z'') e^{-i\Delta kz''} dz'' \right) dz'$$

$$= -i \sqrt{\frac{\omega_2}{\omega_1}} K \mathcal{E}_1(0) \int_0^z e^{i\Delta kz'} dz'$$

$$+ K^2 \int_0^z \int_0^{z'} \mathcal{E}_2^*(z'') e^{i\Delta k(z'-z'')} dz' dz''$$

$$= -i \sqrt{\frac{\omega_2}{\omega_1}} \, (Kz) \, \mathcal{E}_1(0) \, e^{i\Delta kz/2} \, \frac{\sin \frac{1}{2}\Delta k \, z}{\frac{1}{2}\Delta k \, z}$$

$$+ K^2 \int_0^z \int_0^{z'} \mathcal{E}_2^*(z'') \, e^{i\Delta k(z'-z'')} \, dz' \, dz'' \qquad (17.9.13)$$

We can now substitute this equation for $\mathcal{E}_2^*(z)$ into the integrand of the second term of the same equation. We can continue by iteration, always substituting an exact expression for $\mathcal{E}_2^*(z)$ into the integrand appearing in the preceding iteration. If $Kz \ll 1$, however, the first term in (17.9.13) is the leading term; all other terms obtained by iterating this equation involve higher powers of Kz than the first. Thus we have the approximation

$$\mathcal{E}_2^*(z) \approx -i \sqrt{\frac{\omega_2}{\omega_1}} \, (Kz) \, \mathcal{E}_1(0) \, e^{i\Delta kz/2} \, \frac{\sin \frac{1}{2}\Delta k \, z}{\frac{1}{2}\Delta k \, z} \qquad (17.9.14)$$

for $Kz \ll 1$, or

$$|\mathcal{E}_2(z)|^2 \approx \frac{\omega_2}{\omega_1} \, (Kz)^2 \, |\mathcal{E}_1(0)|^2 \, \frac{\sin^2 \frac{1}{2}\Delta k \, z}{(\frac{1}{2}\Delta k \, z)^2} \qquad (17.9.15a)$$

Similarly, the first approximation to $|\mathcal{E}_1(z)|^2$ for $Kz \ll 1$ is found to be

$$|\mathcal{E}_1(z)|^2 \approx |\mathcal{E}_1(0)|^2 \left(1 + (Kz)^2 \, \frac{\sin^2 \frac{1}{2}\Delta k \, z}{(\frac{1}{2}\Delta k \, z)^2} \right) \qquad (17.9.15b)$$

For $\Delta k \to 0$ Eqs. (17.9.15) reduce, as they should, to the equations (17.9.8) obtained under the assumption of perfect phase matching. The effect of phase mismatch is thus to introduce a factor $(\sin^2 \frac{1}{2}\Delta k \, z)/(\frac{1}{2}\Delta k \, z)^2$, just as in second-harmonic generation. Now, however, Δk is given by (17.8.5) rather than (17.4.19). When $\omega_1 = \omega_2 = \omega_3/2$, as in second-harmonic generation or degenerate parametric amplification, the two expressions for Δk are the same.

The problem of phase-matching in parametric amplification is basically the same as in second-harmonic generation. For instance, we can have angle phase-matching in parametric amplification in a negative uniaxial crystal by propagating the pump as an ordinary wave and the signal and idler as extraordinary waves.

Although optical parametric amplification is useful in some applications, its most interesting aspect from a practical standpoint is that it suggests the possibility of *parametric oscillation*, just as light amplification by stimulated emission is used to achieve laser oscillation. Parametric oscillation is the subject of the following section.

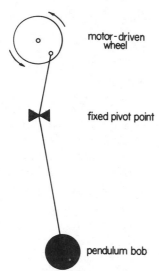

motor-driven wheel

fixed pivot point

pendulum bob

Figure 17.6 Example of a parametric amplification process. A slowly rotating wheel turning at twice the natural pendulum frequency makes the bob oscillate. From A. E. Siegman, *American Journal of Physics* **37**, 843 (1969).

• We have explained that optical parametric amplification occurs in our nonlinear electron oscillator model. However, parametric amplification is a general phenomenon that occurs in many different contexts. A mechanical example of degenerate parametric amplification is provided by the playground swing. A swinger pumps the swing by raising and lowering her center of gravity as she tucks her legs in or extends them. The swing is optimally pumped when the pumping frequency is twice the swing oscillation frequency. By parametric amplification, energy is being fed from the pump at frequency 2ω to the swing oscillation at ω. A similar demonstration of parametric amplification is indicated in Figure 17.6. Parametric amplification is also widely used in electronics.[2]

The parametric resonance principle was well known to nineteenth-century physicists. Lord Rayleigh, for instance, noted several examples, including a pendulum whose point of support vibrates vertically at twice the natural pendulum frequency. (Obviously this example is very similar to that sketched in Figure 17.6.) Such parametric-resonance phenomena are often described by an equation of the type

$$\ddot{x} + \omega^2(t)x = 0 \qquad (17.9.16)$$

where the frequency $\omega(t)$ varies in time according to the formula

$$\omega^2(t) = \omega_0^2(1 + \epsilon \cos \omega't) \qquad (17.9.17)$$

with ϵ small compared with unity. Parametric amplification occurs for $\omega' \approx 2\omega_0$, as may be shown either by perturbation theory or by a numerical solution of the differential equation (17.9.16). Similar equations are encountered in electronic parametric processes when a circuit parameter (e.g., capacitance) is made to vary sinusoidally.

2. See, for instance, W. H. Louisell, *Coupled Mode and Parametric Electronics* (Wiley, New York, 1960).

17.10 PARAMETRIC OSCILLATION

Suppose the signal and idler waves in parametric amplification are propagating back and forth within a resonator containing the nonlinear medium. Because of this feedback, parametric *oscillation* is possible by balancing the gain against transmission loss and whatever other attenuation processes are at work.

Figure 17.7 shows how a parametric oscillator can be designed. The laser oscillator puts out a beam at frequency ω_3, which is focused onto a nonlinear crystal. The crystal is contained in a cavity with mirrors that are transparent to radiation of frequency ω_3, but one of the mirrors allows a small fraction of light at ω_1 and ω_2 ($\omega_1 + \omega_2 = \omega_3$) incident upon it to be transmitted as the output of the parametric oscillator. The waves at ω_1 and ω_2 bounce back and forth inside this cavity, undergoing parametric amplification in the crystal.

The basic idea behind parametric oscillation is quite analogous to that of laser oscillation. In a medium with population inversion and gain over some range of frequencies, there is amplification of an injected signal. If the signal is continually fed back into the gain medium by using a resonator supporting modes within the gain bandwidth, sustained laser oscillation is possible when the small-signal gain exceeds the loss. In the parametric oscillator, however, no population inversion is needed: the signal and idler are amplified at the expense of the energy in the pump wave rather than energy stored in the form of molecular excitation in the nonlinear medium; the medium serves only to mix the pump, signal, and idler waves. Nevertheless we find, as in the laser, a threshold condition for oscillation. As we will see below, the threshold condition for parametric oscillation is for the pump intensity to exceed a certain level.

In parametric amplification we require, in addition to the pump wave, either the signal or idler (or both) to have some initial energy. This is evident, for instance, in (17.9.5). Where does the initial signal or idler energy come from in a parametric oscillator such as that sketched in Figure 17.7? This is quite analogous to asking where the "initial photon" comes from in a laser oscillator, and the answer is the same: the initial radiation triggering the parametric oscillation comes from spontaneous emission. This is not "ordinary" spontaneous emission, however, in which a molecule drops from an excited energy level to a lower one with the emission

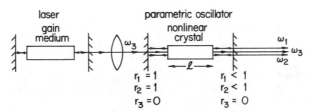

Figure 17.7 Schematic layout of a parametric oscillator with pump frequency $\omega_3 = \omega_1 + \omega_2$.

of a photon. Rather, it is *parametric fluorescence*, in which a molecule exposed to radiation of frequency ω_3 can emit *two* photons, at ω_1 and ω_2, such that $\omega_1 + \omega_2 = \omega_3$. In this process the internal state of the molecule is the same before and after the emission of the two photons; the molecule has, very loosely speaking, acted to split the incident photon into two outgoing photons. Parametric amplification is basically just the *stimulated* emission of the two photons in the presence of a signal photon.

We outline below a derivation of a threshold pump intensity for parameric oscillation, assuming perfect phase matching:

$$I_{3t} = \frac{1}{2} \left(\frac{\epsilon_0}{\mu_0}\right)^{3/2} \frac{n_1 n_2 n_3 (1 - r_1)(1 - r_2)}{\omega_1 \omega_2 \bar{d}^2 l^2} \qquad (17.10.1)$$

where l is the length of the nonlinear crystal and r_1, r_2 are the reflectivities at ω_1, ω_2 of the output mirrors of the parametric oscillator (Figure 17.7). The pump intensity must exceed I_{3t} if parametric oscillation at ω_1 and ω_2 is to occur.

As an example, suppose that ω_1 and ω_2 in Figure 17.7 correspond to wavelengths of approximately 1 μm. Assuming the nonlinear crystal is $LiNbO_3$, and using the values $n_1 = n_2 \approx n_3 \approx 1.5$ and $\bar{d} \approx 4 \times 10^{-23}$ (mks units) as in the preceding section, we compute

$$I_{3t} \approx \frac{6 \times 10^6}{l^2} (1 - r_1)(1 - r_2) \; \text{W/cm}^2 \qquad (17.10.2)$$

with the crystal length l to be expressed in centimeters. For $r_1 = r_2 = 98\%$ and $l = 4$ cm, for instance,

$$I_{3t} \approx 150 \; \text{W/cm}^2 \qquad (17.10.3)$$

This example indicates that parametric oscillation is feasible even with relatively modest pump intensities.

As discussed below, the output intensities of a parametric oscillator oscillating at ω_1 and ω_2 are given by

$$\frac{\omega_3}{\omega_1} I_1^{(\text{out})} = \frac{\omega_3}{\omega_2} I_2^{(\text{out})} = 2 I_{3t} \left(\sqrt{\frac{I_3}{I_{3t}}} - 1\right) \qquad (17.10.4)$$

where I_3 is the input pump intensity. This suggests that high output powers are attainable if the ratio I_3/I_{3t} of input pump intensity to the threshold pump intensity is made large. However, this conclusion is somewhat optimistic.

First, a high pump intensity can seriously damage a nonlinear crystal. Intensities on the order of MW/cm^2, or lower in the case of a cw pump, can induce refractive-index variations that destroy phase matching. Equation (17.10.4) is still applicable when $\Delta k \neq 0$, but the threshold intensity (17.10.1) is increased according to the formula

$$I_{3t} = \frac{1}{2} \left(\frac{\epsilon_0}{\mu_0} \right)^{3/2} \frac{n_1 n_2 n_3 (1 - r_1)(1 - r_2)}{\omega_1 \omega_2 \overline{d}^2 l^2} \left(\frac{\sin \frac{1}{2} \Delta k \, l}{\frac{1}{2} \Delta k \, l} \right)^{-2} \quad (17.10.5)$$

Furthermore, the index variations induced by a high-power field are difficult to predict, and may have complicated spatial variations.

Another difficulty can arise in a parametric oscillator because of the presence of a backward-propagating wave at the pump frequency. In writing the equations (17.8.4) for three-wave mixing we assumed that all three waves propagate in the same direction. In the oscillator case, however, we have signal and idler waves propagating in *both* directions; the backward-propagating waves mix to generate a backward wave at the pump frequency. There is therefore a "reflection" of pump radiation back into the laser oscillator, which means that the reflectivity of the laser output mirror depends, in effect, on the pump intensity it generates. This can lead to a highly unstable and erratic pump intensity because the laser cannot "find" a steady point of oscillation. Such an instability is indeed observed in many lasers when some of the output radiation is reflected back into the laser resonator. The output from such a laser looks chaotic, and the laser may oscillate intermittently or not at all. This can pose a problem for a parametric oscillator of the type shown in Figure 17.7.

Backward waves can be eliminated by using a ring resonator configuration like that shown in Figure 17.8. Furthermore it may be shown that in the absence of backward waves equation (17.10.4) is replaced by

$$\frac{\omega_3}{\omega_1} I_1^{(\text{out})} = \frac{\omega_3}{\omega_2} I_2^{(\text{out})} = 4 I_{3t} \left(\sqrt{\frac{I_3}{I_{3t}}} - 1 \right) \quad (17.10.6)$$

In other words, the parametric oscillator is twice as efficient when backward waves are eliminated.

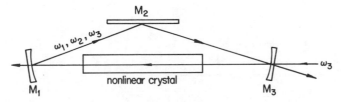

Figure 17.8 Ring-cavity parametric oscillator used by R. L. Byer, A. Kovrigin, and J. F. Young, *Applied Physics Letters* **15**, 136 (1969).

An interesting aspect of a parametric oscillator with bidirectional waves is the *power limiting* of the transmitted pump field. To understand this effect, it is useful to recall the gain clamping that occurs in steady-state laser oscillation: the saturated gain coefficient is exactly equal to the loss coefficient, which is the same as the threshold gain coefficient. That is, the saturated gain is clamped at the value of the threshold gain. Now in a parametric oscillator the effective gain coefficient is a function of the pump intensity; this explains, for instance, why the threshold condition (17.10.1) is a condition on the pump intensity. In steady-state parametric oscillation, therefore, *the pump intensity must be clamped at its threshold value.* It then follows that, regardless of the input pump intensity I_3, the pump intensity emerging from the parametric oscillator cannot exceed I_{3t}. In this sense the parametric oscillator acts as a power limiter for radiation at the pump frequency.

In a unidirectional parametric oscillator like that of Figure 17.8, there is no power limiting. This is evident from the analysis given below.

• Exact solutions are known for the three-wave mixing equations (17.8.4). However, these solutions are complicated, and so to derive an expression for the threshold pump intensity for parametric oscillation we follow a simpler, approximate approach.

As indicated in Figure 17.9, we consider a pump wave of frequency ω_3 and amplitude $\mathcal{E}_3(0)$ incident on a nonlinear crystal of length l. The amplitude of the pump wave at the exit face $z = l$ of the crystal is denoted $\mathcal{E}_3(l)$. We consider for simplicity the "degenerate" case $\omega_1 = \omega_2$ and $\mathcal{E}_1(z) = \mathcal{E}_2(z)$, and consider only waves propagating in the forward direction. Then from (17.8.7) we write the coupled wave equations

$$\frac{d\mathcal{E}_1}{dz} = i\omega_1 \sqrt{\frac{\mu_0}{\epsilon_1}}\, \overline{d}\mathcal{E}_1^* \mathcal{E}_3 \tag{17.10.7a}$$

$$\frac{d\mathcal{E}_3}{dz} = i\omega_1 \sqrt{\frac{\mu_0}{\epsilon_3}}\, \overline{d}\mathcal{E}_1^2 \qquad (\omega_3 = 2\omega_1) \tag{17.10.7b}$$

under the assumption of perfect phase matching ($\Delta k = 0$). We furthermore write

$$\mathcal{E}_j = A_j e^{i\theta_j}, \qquad j = 1, 3 \tag{17.10.8}$$

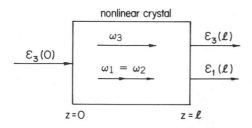

Figure 17.9 Degenerate parametric amplification with pump frequency $\omega_3 = \omega_1 + \omega_2$, $\omega_1 = \omega_2$.

with A_j and θ_j real, and assume that θ_1 and θ_3 are constants, independent of z. Then Eqs. (17.10.7) become

$$\frac{dA_1}{dz} = i\omega_1 \sqrt{\frac{\mu_0}{\epsilon_1}}\ \bar{d}A_1 A_3 e^{i\phi} \qquad (17.10.9a)$$

$$\frac{dA_3}{dz} = i\omega_1 \sqrt{\frac{\mu_0}{\epsilon_3}}\ \bar{d}A_1^2 e^{-i\phi} \qquad (17.10.9b)$$

$$\phi = \theta_3 - 2\theta_1 \qquad (17.10.9c)$$

The approximation we now make is based on the fact that in a laser resonator with small output coupling the intracavity intensity varies little with z. Assuming the same result for the field $\mathscr{E}_1$ in a parametric oscillator, we take A_1 to be approximately constant in (17.10.9b). In this approximation we obtain upon integration

$$A_3(l) \approx A_3(0) + i\omega_1 \sqrt{\mu_0/\epsilon_3}\ \bar{d}lA_1^2 e^{-i\phi} \qquad (17.10.10)$$

where $A_1 = A_1(0) \approx A_1(l)$ by assumption.

The second term in (17.10.10) represents the change in the pump field due to parametric amplification of the signal (idler) wave. This change corresponds to a maximum loss of pump intensity, and therefore a maximum gain of signal intensity, when $ie^{-i\phi} = -1$:

$$A_3(l) \approx A_3(0) - \omega_1 \sqrt{\mu_0/\epsilon_3}\ \bar{d}lA_1^2 \qquad (17.10.11)$$

Therefore we might expect that the parametric oscillation occurs in such a way that $ie^{-i\phi} = -1$, or $\phi = -\pi/2$. This provides some justification of our assumption that θ_1 and θ_3 are constant. It also makes Eqs. (17.10.9) and (17.10.10) real, consistent with our assumption that A_1 and A_3 are real.

From (17.10.11) we have

$$A_3^2(l) \approx A_3^2(0) - 2\omega_1 \sqrt{\frac{\mu_0}{\epsilon_3}}\ \bar{d}lA_1^2 A_3(0) + \omega_1^2 \frac{\mu_0}{\epsilon_3}\left(\bar{d}l\right)^2 A_1^4 \qquad (17.10.12)$$

We can use the Manley–Rowe relation (17.8.6) to relate the input pump intensity to the output pump intensity and the output signal intensity (Figure 17.9):

$$\frac{1}{\omega_3}\sqrt{\frac{\epsilon_3}{\mu_0}}\,A_3^2(0) = \frac{1}{\omega_3}\sqrt{\frac{\epsilon_3}{\mu_0}}\,A_3^2(l) + (1 - r_1)\frac{1}{\omega_1}\sqrt{\frac{\epsilon_1}{\mu_0}}A_1^2$$

or

$$A_3^2(l) = A_3^2(0) - 2(1 - r_1)\sqrt{\epsilon_1/\epsilon_3}\,A_1^2 \qquad (17.10.13)$$

since $\omega_3 = \omega_1 + \omega_2 = 2\omega_1$ in the degenerate case under consideration. Combining (17.10.13) and (17.10.12), we obtain an expression for A_1^2:

$$A_1^2 \approx 2\sqrt{\frac{\epsilon_3}{\mu_0}}\frac{A_3(0)}{\omega_1 \overline{d} l} - 2\sqrt{\frac{\epsilon_1}{\mu_0}\frac{\epsilon_3}{\mu_0}}\frac{1 - r_1}{(\omega_1 \overline{d} l)^2} \tag{17.10.14}$$

Since we must have $A_1^2 > 0$ for parametric oscillation, we obtain the required threshold value for $A_3(0)$ by setting the right side of (17.10.14) equal to zero. Casting this result into an expression for the pump intensity, we find that

$$I_{3t} = \frac{1}{2}\left(\frac{\epsilon_0}{\mu_0}\right)^{3/2}\frac{n_1^2 n_3 (1 - r_1)^2}{\omega_1^2 \overline{d}^2 l^2} \tag{17.10.15}$$

Using (17.10.15) in (17.10.14), we find after some algebra that the output signal intensity is

$$I_1^{(\text{out})} = (1 - r_1)\frac{1}{2}\sqrt{\frac{\epsilon_1}{\mu_0}}A_1^2 = 2I_{3t}\left(\sqrt{\frac{I_3}{I_{3t}}} - 1\right) \tag{17.10.16}$$

where $I_3 = \frac{1}{2}\sqrt{\epsilon_3/\mu_0}\,A_3(0)^2$ is the input pump intensity. These results were derived for the degenerate case $\omega_2 = \omega_1$, $n_2 = n_1$, and $r_2 = r_1$. A similar analysis for the more general case gives (17.10.1) and (17.10.6) in place of (17.10.15) and (17.10.16). When backward-propagating waves are included in the analysis, we obtain (17.10.4) instead of (17.10.6).

Using (17.10.14) in (17.10.13), we also obtain an expression for the intensity of the transmitted pump beam:

$$I_3^{(\text{out})} \approx I_3 - 4I_{3t}\left(\sqrt{I_3/I_{3t}} - 1\right) \tag{17.10.17}$$

This relation also follows from (17.10.13) and (17.10.16). Note that $I_3^{(\text{out})}$ is not limited to a maximum value I_{3t}. Such power limiting is found, however, when backward waves are included, as must be done for a parametric oscillator like that shown in Figure 17.7. •

17.11 TUNING OF PARAMETRIC OSCILLATORS

The parametric oscillator generates output beams at ω_1 and ω_2 such that $\omega_1 + \omega_2 = \omega_3$, where ω_3 is the frequency of the pump beam. For a reasonably strong output at ω_1 and ω_2 the phase-matching condition, $\Delta k = k_1 + k_2 - k_3 = 0$, must also be satisfied. Thus we have the two conditions

$$\omega_1 + \omega_2 = \omega_3 \tag{17.11.1a}$$

and

$$n_1\omega_1 + n_2\omega_2 = n_3\omega_3 \tag{17.11.1b}$$

satisfied by the signal and idler frequencies ω_1 and ω_2.

Now the refractive indices n_i can be changed in various ways. For instance, they generally depend on the temperature of the crystal and, for extraordinary waves, the orientation of the crystal. Any change in the refractive indices will, according to (17.11.1b), change ω_1 and ω_2 when the pump frequency ω_3 is fixed. In others words, *by varying the n_i we can tune the parametric oscillator*. This tunability is an important property of parametric oscillators.

Further consideration of (17.11.1), however, reveals some undesirable features of simultaneous parametric oscillation at two frequencies ω_1 and ω_2. For simultaneous oscillation, both frequencies must be resonant frequencies of the parametric oscillator cavity, i.e., we must have

$$\omega_1 = N_1 \frac{\pi c}{n_1 L} \tag{17.11.2a}$$

and

$$\omega_2 = N_2 \frac{\pi c}{n_2 L} \tag{17.11.2b}$$

where N_1 and N_2 are integers and L is the mirror separation of the cavity containing the nonlinear crystal. In general, however, it is impossible to satisfy both (17.11.2) and (17.11.1). If (17.11.2) is satisfied but not (17.11.1), for instance, oscillation can only occur, if at all, away from the phase-matching condition.

Furthermore other effects that are difficult to control will cause ω_1 and ω_2 to vary randomly. Temperature variations, for instance, will change the refractive indices, while mechanical vibrations can cause L to vary. Slight frequency variations in the pump beam will also be reflected in the variations of the oscillation frequencies.

For these reasons it is often preferable to operate the parametric oscillator so that only one of the two frequencies ω_1 and ω_2 can oscillate. This can be done by introducing a material that absorbs at one but not both frequencies, or by using mirrors that are highly reflecting at one frequency but not the other. In such cases the parametric oscillator is said to be *singly resonant* rather than *doubly resonant*.

Figure 17.10 shows the wavelengths λ_1 and λ_2 satisfying (17.11.1) ($\omega_i = 2\pi c/\lambda_i$) for LiNbO$_3$ with various pump wavelengths λ_3. Such a *tuning curve* gives the wavelengths λ_1 and λ_2 satisfying the phase-matching conditions (17.11.1), and provides valuable information to the designer of a parametric oscillator. With various crystals and pump frequencies it is possible to scan the entire visible range of the electromagnetic spectrum.

One commercially available device employs pump radiation obtained from the frequency doubling of 1.06-μm Nd:YAG laser radiation, $\lambda_p = 0.532$ μm. The nonlinear crystal, LiNbO$_3$, is contained in a small oven with a temperature control. Changing the oven temperature provides oscillation at either λ_1 or λ_2 in Figure 17.10, depending on which of several oscillator mirrors is mounted. The pump wavelength may also be changed to 0.562 or 0.659 μm, giving different tuning

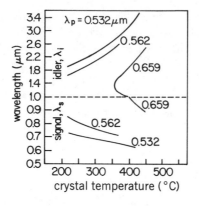

Figure 17.10 Signal and idler wavelengths that satisfy (17.11.1) for LiNbO$_3$ over a range of temperatures. Note the discontinuity in the left scale, marked by the dashed line.

curves of LiNbO$_3$. The different combinations of oscillator mirrors and pump wavelengths, and the temperature control, provide tunability from 0.65 to 3.0 μm. By Q-switching the laser pump, peak output powers (in a TEM$_{00}$ Gaussian mode) on the order of several hundred watts are obtained from this singly resonant parametric oscillator.

• In the preceding section we mentioned that the parametric process of converting a pump photon into two photons can occur spontaneously, i.e., in the absence of signal and idler photons. In the literature this phenomenon goes by several names, including parametric fluorescence, spontaneous parametric emission, parametric luminescence, parametric scattering, and parametric noise. The effect may be usefully employed to determine the nonlinear coefficient of a material, or a tuning curve like that shown in Figure 17.10. At pump powers too low for parametric oscillation, parametric fluorescence still (always) occurs, the fluorescent power being directly proportional to the pump power. •

17.12 NONLINEAR SUSCEPTIBILITIES

From Chapter 2 [Eq. (2.3.27)] we recall that the susceptibility $\chi(\omega)$ is given in the electron oscillator model by the formula

$$\chi(\omega) = \frac{Ne^2/m\epsilon_0}{\omega_0^2 - \omega^2} \qquad (17.12.1)$$

Therefore we have also

$$\chi(2\omega) = \frac{Ne^2/m\epsilon_0}{\omega^2 - 4\omega^2} \qquad (17.12.2)$$

This allows us to write the nonlinear coefficient $\bar{d}$ defined by (17.4.6c) and (17.4.17) as

$$\overline{d} = \frac{ma\epsilon_0^3}{2N^2|e|^3} \chi^2(\omega) \chi(2\omega) \tag{17.12.3}$$

or, using the relation $n^2 = 1 + \chi$ between the susceptibility and the refractive index,

$$\overline{d} = \frac{ma\epsilon_0^3}{2N^2|e|^3} \left[n^2(\omega) - 1 \right]^2 \left[n^2(2\omega) - 1 \right] \tag{17.12.4}$$

This relation suggests that media with large refractive indices will have the largest nonlinear coefficients $\overline{d}$. Such a trend, predicted by our nonlinear oscillator model, is in fact true. However, there is much more than this to the analysis of the nonlinear optical properties of materials. In this section and the next we will introduce some useful concepts and terminology of this field. In particular, we will generalize from the nonlinear coefficient $\overline{d}$ to the notion of a nonlinear susceptibility.

From our study of three-wave mixing with $\omega_3 = \omega_1 + \omega_2$ we know that a nonlinear polarization

$$P_{\omega_3}^{(\mathrm{NL})}(z, t) = \tfrac{1}{2} \left[P^{(\mathrm{NL})}(\omega_3) e^{-i\omega_3 t} e^{i(k_1 + k_2)z} + \text{c.c.} \right] \tag{17.12.5}$$

will be generated in a nonlinear medium by the mixing of the two fields

$$E_1(z, t) = \tfrac{1}{2} \left[\mathcal{E}(\omega_1) e^{-i(\omega_1 t - k_1 z)} + \mathcal{E}^*(\omega_1) e^{i(\omega_1 t - k_1 z)} \right] \tag{17.12.6a}$$

and

$$E_2(z, t) = \tfrac{1}{2} \left[\mathcal{E}(\omega_2) e^{-i(\omega_2 t - k_2 z)} + \mathcal{E}^*(\omega_2) e^{i(\omega_2 t - k_2 z)} \right] \tag{17.12.6b}$$

Here we have changed our notation slightly. We now write the field frequencies explicitly, and suppress the z dependence for convenience, by writing $\mathcal{E}(\omega_i)$ in place of $\mathcal{E}_i(z)$. Furthermore $P_{\omega_3}^{(\mathrm{NL})}(z, t)$ will be proportional to $\mathcal{E}(\omega_1) \mathcal{E}(\omega_2)$, and we denote the constant of proportionality by $\epsilon_0 \chi(-\omega_3, \omega_1, \omega_2)$:

$$P^{(\mathrm{NL})}(\omega_3) = \epsilon_0 \chi(-\omega_3, \omega_1, \omega_2) \mathcal{E}(\omega_1) \mathcal{E}(\omega_2) \tag{17.12.7}$$

We have introduced, for no apparent reason, a minus sign in front of ω_3 in the definition of the *nonlinear susceptibility* $\chi(-\omega_3, \omega_1, \omega_2)$. This conventional way of writing the nonlinear susceptibility turns out to be quite useful.

It is convenient to *define*

$$\mathcal{E}(-\omega_i) = \mathcal{E}^*(\omega_i) \tag{17.12.8}$$

In this notation we can write, in addition to (17.12.7),

$$P^{(\text{NL})}(\omega_2) = \epsilon_0 \chi(-\omega_2, -\omega_1, \omega_3) \, \mathcal{E}(-\omega_1) \, \mathcal{E}(\omega_3) \qquad (17.12.9)$$

for the polarization generated at $\omega_2 = \omega_3 - \omega_1$ by the mixing of fields at ω_1 and ω_2, and

$$P^{(\text{NL})}(\omega_1) = \epsilon_0 \chi(-\omega_1, -\omega_2, \omega_3) \, \mathcal{E}(-\omega_2) \, \mathcal{E}(\omega_3) \qquad (17.12.10)$$

for the polarization generated at $\omega_1 = \omega_3 - \omega_2$ by the mixing of fields at ω_2 and ω_3. Thus we can write the coupled wave equations (17.8.4) for three-wave mixing as (Problem 17.8)

$$\frac{d\mathcal{E}(\omega_1)}{dz} = \frac{i\omega_1}{2n_1 c} \chi(-\omega_1, -\omega_2, \omega_3) \, \mathcal{E}(-\omega_2) \, \mathcal{E}(\omega_3) \, e^{i(k_3 - k_1 - k_2)z} \qquad (17.12.11a)$$

$$\frac{d\mathcal{E}(\omega_2)}{dz} = \frac{i\omega_2}{2n_2 c} \chi(-\omega_2, -\omega_1, \omega_3) \, \mathcal{E}(-\omega_1) \, \mathcal{E}(\omega_3) \, e^{i(k_3 - k_1 - k_2)z} \qquad (17.12.11b)$$

$$\frac{d\mathcal{E}(\omega_3)}{dz} = \frac{i\omega_3}{2n_3 c} \chi(-\omega_3, \omega_1, \omega_2) \, \mathcal{E}(\omega_1) \, \mathcal{E}(\omega_2) \, e^{i(k_1 + k_2 - k_3)z} \qquad (17.12.11c)$$

where we have used the relation $\epsilon_0 \sqrt{\mu_0/\epsilon_i} = 1/cn(\omega_i) \equiv 1/cn_i$.

Thus far all we have done is introduce some notation for the wave amplitudes and the nonlinear susceptibilities. However, we note in comparing (17.12.11) and (17.8.4) that evidently $\bar{d} = \frac{1}{2}\epsilon_0 \chi(-\omega_1, -\omega_2, \omega_3)$ in (17.8.4a), $\bar{d} = \frac{1}{2}\epsilon_0 \chi(-\omega_2, -\omega_1, \omega_3)$ in (17.8.4b), and $\bar{d} = \frac{1}{2}\epsilon_0 \chi(-\omega_3, \omega_1, \omega_2)$ in (17.8.4c). This observation reveals an *approximation* implicit in (17.8.4): we have assumed hitherto that the nonlinear susceptibilities are independent of the values of the frequencies ω_1, ω_2, and ω_3. This turns out for many purposes to be quite an accurate approximation, provided the material is nonabsorbing at ω_1, ω_2, and ω_3.

A possible source of confusion is the factor $\frac{1}{2}$ in the relation

$$\bar{d} = \tfrac{1}{2}\epsilon_0 \chi \qquad (17.12.12)$$

between the nonlinear coefficient and the nonlinear susceptibility. One must be careful in reading the literature to check whether the nonlinear coefficient or the nonlinear susceptibility is used. Experimental researchers and engineers often use $\bar{d}$, whereas their theoretical colleagues generally use χ. The reader should also be cautioned that the factor ϵ_0 is sometimes included in the definition of χ, in which case $\bar{d} = \frac{1}{2}\chi$.

Another source of confusion in the definition of nonlinear susceptibilities is that, whereas (17.12.7) holds for $\omega_3 = \omega_1 + \omega_2$ with $\omega_1 \neq \omega_2$, we have

$$P^{(\text{NL})}(2\omega) = \tfrac{1}{2}\epsilon_0\chi(-2\omega, \omega, \omega)\, \mathcal{E}(\omega)\, \mathcal{E}(\omega) \qquad (17.12.13)$$

when $\omega_1 = \omega_2$. The reason for the factor $\tfrac{1}{2}$ is already implicit in our discussion at the end of Section 17.8, and will be discussed more generally in Section 17.15.

The nonlinear coefficients and susceptibilities of materials are usually expressed in either electrostatic or mks units. For convenience we give here the following formula for converting the nonlinear susceptibility from esu to mks units:

$$\chi(\omega_3, \omega_1, \omega_2)\,(\text{mks}) = \frac{4\pi}{3 \times 10^4}\,\chi(\omega_3, \omega_1, \omega_2)\,(\text{esu}) \qquad (17.12.14)$$

17.13 THE NONLINEAR SUSCEPTIBILITY TENSOR

Two electric fields polarized in different directions in a nonlinear medium will generally mix to generate a field polarized in yet another direction. We must therefore generalize (17.12.7), for instance, to take account of the fact that the polarization at ω_3 might point in a direction different from the field at ω_1 or the field at ω_2. It is convenient to introduce a new subscript $i = 1, 2, 3$ to denote the components x, y, z, respectively, of a vector $\mathbf{P}$ or $\mathbf{E}$, and to write, for $i = 1, 2$, and 3,

$$P_i^{(\text{NL})}(\omega_3) = \epsilon_0 \sum_{j=1}^{3}\sum_{k=1}^{3} \chi_{ijk}(-\omega_3, \omega_1, \omega_2)\, \mathcal{E}_j(\omega_1)\, \mathcal{E}_k(\omega_2) \qquad (17.13.1)$$

as the generalization of (17.12.7). Here χ_{ijk} is called the *nonlinear susceptibility tensor*. Since i, j, k each have three possible values, the nonlinear susceptibility tensor consists of a total of $3 \times 3 \times 3 = 27$ numbers.

It is perhaps worth noting that even the *linear* susceptibility is in general a tensor quantity, with the induced linear polarization taking the form

$$P_i^{(\text{L})}(\omega) = \epsilon_0 \sum_{j=1}^{3} \chi_{ij}(\omega)\, \mathcal{E}_j(\omega) \qquad (17.13.2)$$

Here the matrix $\chi_{ij}(\omega)$ is called the linear susceptibility tensor. In anisotropic media the relation (17.13.2) gives rise to refractive indices that are generally different for different field polarizations, as discussed in Section 2.9.

It is convenient to write (17.13.1) in the abbreviated notation

$$P_i(\omega_3) = P_i^{(\text{NL})}(\omega_3) = \epsilon_0\chi_{ijk}(-\omega_3, \omega_1, \omega_2)\, \mathcal{E}_j(\omega_1)\, \mathcal{E}_k(\omega_2) \qquad (17.13.3)$$

where it is understood that we must sum over the *repeated indices* (they appear twice) j and k on the right. This convention of summing over repeated indices is called the *Einstein summation convention*. It allows us to dispense with the explicit summation symbols Σ_j and Σ_k. In like manner we generalize (17.12.9) and (17.12.10) to

$$P_i(\omega_2) = \epsilon_0 \chi_{ijk}(-\omega_2, -\omega_1, \omega_3) \, \mathcal{E}_j(-\omega_1) \, \mathcal{E}_k(\omega_3) \qquad (17.3.4a)$$

and

$$P_i(\omega_1) = \epsilon_0 \chi_{ijk}(-\omega_1, -\omega_2, \omega_3) \, \mathcal{E}_j(-\omega_2) \, \mathcal{E}_k(\omega_3) \qquad (17.3.4b)$$

The χ_{ijk} have certain symmetry properties. For instance, they satisfy *overall permutation symmetry*, which simply means that subscripts and frequencies together may be freely permuted. Thus

$$\chi_{ijk}(-\omega_A, \omega_B, \omega_C) = \chi_{jik}(\omega_B, -\omega_A, \omega_C)$$

$$= \chi_{kji}(\omega_C, \omega_B, -\omega_A) \qquad (17.13.5)$$

where $\chi_{jik}(\omega_B, -\omega_A, \omega_C)$, for instance, is defined by

$$P_j(-\omega_B) = P_j^*(\omega_B) = \epsilon_0 \chi_{jik}(\omega_B, -\omega_A, \omega_C) \, \mathcal{E}_i(-\omega_A) \, \mathcal{E}_k(\omega_C) \quad (17.13.6)$$

Comparing this with

$$P_i(\omega_A) = \epsilon_0 \chi_{ijk}(-\omega_A, \omega_B, \omega_C) \, \mathcal{E}_j(\omega_B) \, \mathcal{E}_k(\omega_C) \qquad (17.13.7)$$

and (17.13.5) we can see that, loosely speaking, the permutation symmetry has the following meaning: in a three-wave mixing process involving fields of frequency ω_A, ω_B, and ω_C, *the nonlinear susceptibility for the process is the same regardless of which field is being generated and which are doing the generating.* Equation (17.13.6), for example, describes the generation of ω_B by the mixing of ω_A and ω_C, whereas (17.13.7) describes the generation of ω_A by the mixing of ω_B and ω_C. From (17.13.5) it follows that the nonlinear susceptibilities for these two processes are the same.

Another symmetry property is known to hold, at least to an excellent approximation, in many cases. Namely, the subscripts i, j, and k may be freely permuted:

$$\chi_{ijk} = \chi_{jik} = \chi_{kji} = \chi_{ikj} \qquad (17.13.8)$$

This is called *Kleinman's symmetry conjecture*. Combined with the overall permutation symmetry (17.13.5), it states basically that $\chi_{ijk}(-\omega_A, \omega_B, \omega_C)$ is insensitive to the values of ω_A, ω_B, and ω_C. As noted earlier, this is usually an excellent approximation at frequencies for which the material is transparent.

The relation (17.12.3), which may be written

$$\bar{d} = \tfrac{1}{2}\epsilon_0\chi(-2\omega,\omega,\omega) = \frac{ma\epsilon_0^3}{2N^2|e|^3}\chi(2\omega)\chi(\omega)\chi(\omega)$$

$$= \tfrac{1}{2}\epsilon_0 D\,\chi(2\omega)\chi(\omega)\chi(\omega) \qquad (17.13.9)$$

expresses a relation between the nonlinear susceptibility $\chi(-2\omega,\omega,\omega)$ and the linear susceptibility $\chi(\omega)$. As written, it ignores the fact that the susceptibilities are really tensor quantities with various components. A generalization of (17.13.9) is

$$\chi_{ijk}(-\omega_A,\omega_B,\omega_C) = D_{ijk}\,\chi_{ii}(\omega_A)\,\chi_{jj}(\omega_B)\,\chi_{kk}(\omega_C) \qquad (17.13.10)$$

where χ_{ii}, χ_{jj}, and χ_{kk} are the diagonal elements of the linear susceptibility tensor [there is no summation implied on the right of (17.13.10)]. The coefficient D_{ijk} is known to be approximately constant over a considerable range of frequencies for many materials. This empirical result, known as *Miller's rule*, is quite useful in searching for materials with large nonlinear susceptibilities.

From the overall permutation symmetry of χ_{ijk} it follows that $d_{ijk}(-2\omega,\omega,\omega) = d_{ikj}(-2\omega,\omega,\omega)$. This leads to the following *condensed notation* in nonlinear optics:

$d_{ijk} = d_{ikj}$	d_{ijk} in Condensed Notation
d_{i11}	d_{i1}
d_{i22}	d_{i2}
d_{i33}	d_{i3}
d_{i23}	d_{i4}
d_{i13}	d_{i5}
d_{i12}	d_{i6}

In other words, we can label the elements of the nonlinear susceptibility tensor by two subscripts instead of three. From general crystallographic symmetry considerations in specific cases it may be shown that various components d_{ij} are equal, thus reducing the number of independent components still further.

• The nonlinear susceptibility tensor allows us to describe various effects within the same framework. $\chi_{ijk}(0,-\omega,\omega)$, for instance, describes the generation of a static (dc) polarization due to a pump field at ω:

$$P_i(0) = \epsilon_0\chi_{ijk}(0,-\omega,\omega)\,\mathcal{E}_j^*(\omega)\mathcal{E}_k(\omega) \qquad (17.13.11)$$

This will be recognized as the *optical rectification* effect mentioned in Section 17.5. Similarly, $\chi_{ijk}(-\omega,\omega,0)$ describes the generation of a polarization at ω due to the mixing of a field at ω with a static field ($\omega = 0$):

$$P_i(\omega) = \epsilon_0 \chi_{ijk}(-\omega, \omega, 0) \, \mathcal{E}_j(\omega) \, \mathcal{E}_k(0)$$

$$= \epsilon_0 A_{ij}(\omega) \, \mathcal{E}_j(\omega) \tag{17.13.12}$$

where

$$A_{ij}(\omega) = \chi_{ijk}(-\omega, \omega, 0) \, \mathcal{E}_k(0) \tag{17.13.13}$$

Comparing (17.13.12) with (17.13.2), we see that $A_{ij}(\omega)$ acts in effect as a contribution to the *linear* susceptibility tensor at frequency ω. This contribution is linearly proportional to the strength of the applied static field, and is just the *linear electro-optic effect*, or the *Pockel effect*, described in Section 12.5. •

17.14 NONLINEAR MATERIALS

Consider Eq. (17.13.3) with the direction of the electric field reversed, i.e., $\mathcal{E}_j \rightarrow -\mathcal{E}_j$ and $\mathcal{E}_k \rightarrow -\mathcal{E}_k$. We might expect that, upon this reversal of the electric fields, the polarization $P_i(\omega_3)$ induced by them is also reversed. If this is true, then

$$-P_i(\omega_3) = \epsilon_0 \chi_{ijk}(-\omega_3, \omega_1, \omega_2) \left[-\mathcal{E}_j(\omega_1) \right] \left[-\mathcal{E}_k(\omega_2) \right]$$

$$= \epsilon_0 \chi_{ijk}(-\omega_3, \omega_1, \omega_2) \, \mathcal{E}_j(\omega_1) \, \mathcal{E}_k(\omega_2)$$

$$= P_i(\omega_3) \tag{17.14.1}$$

But $P_i(\omega_3) = -P_i(\omega_3)$ means that $P_i(\omega_3) = 0$. A material for which the nonlinear polarization due to three-wave mixing is always zero is called *centrosymmetric*. Evidently three-wave mixing processes do not occur in centrosymmetric media.

Gases and liquids are centrosymmetric. Thus they do not ordinarily exhibit second-harmonic generation, parametric amplification or oscillation, optical rectification, or the linear electro-optic (Pockels) effect. Under ordinary circumstances *no* three-wave mixing process can occur in a gas or a liquid.

However, most crystals, including all crystals with noncubic symmetry, will give rise to three-wave mixing to some degree. A piece of rock candy placed in the path of an invisible 1.06-μm Nd:YAG laser beam, for example, takes on a greenish glow as a consequence of second-harmonic generation. For three-wave mixing applications such noncentrosymmetric crystals are selected according to several criteria.

First, it is desirable for the crystal to have a large nonlinear coefficient $\bar{d}$, which is really a tensor: $\bar{d}_{ijk} = \frac{1}{2}\epsilon_0 \chi_{ijk}$. One thing to bear in mind in a detailed analysis is the effect of the refractive index. The conversion efficiency (17.5.6) for second-harmonic generation, for instance, goes as $\bar{d}^2/n^3$. More generally, three-wave mixing processes depend on $\bar{d}^2/n_1 n_2 n_3$, which is approximately $\bar{d}^2/n^3$ in most instances because $n_1 \approx n_2 \approx n_3 = n$, the "refractive index" of the material. Since Miller's rule tells us that $\bar{d}$ will be largest in crystals with large n, it is $\bar{d}^2/n^3$ rather than just $\bar{d}$ that must be considered the figure of merit in assessing nonlinear crystals. Of course the (ordinary and extraordinary) refractive indices are of interest also for phase-matching considerations. Phase matchability is always a very important criterion in choosing a nonlinear crystal.

Another consideration is the transparency range of a crystal. One reason why $LiNbO_3$ is so useful, for instance, is that it is highly transparent to both visible and infrared light. ADP and KDP are useful in shorter-wavelength applications because they have high transparency in the ultraviolet.

For certain applications a nonlinear crystal must also be able to withstand high radiation powers without suffering any optical damage such as induced refractive-index variations. Another practical consideration is whether the crystal can be grown to reasonably large dimensions (say $\gtrsim 1$ cm) with good optical quality. For tuning of a parametric oscillator, a crystal should have a large refractive-index variation with temperature (or with angle or pressure or electric field). These are all practical considerations that are discussed in the literature on nonlinear optics, especially the literature dealing with the development and improvement of nonlinear crystals.

• Quantum mechanics provides an expression for the nonlinear susceptibility of a material in terms of the transition electric dipole moments μ_{nm} and the transition frequencies ω_{nm} of its atoms or molecules. As an example, we consider the nonlinear susceptibility $\chi_{ijk}(-2\omega, \omega, \omega)$ for second-harmonic generation. Suppose for simplicity that the pump field at ω is linearly polarized along the x direction, and we are interested in the polarization in the x direction:

$$P_1(2\omega) = \epsilon_0 \chi_{111}(-2\omega, \omega, \omega)\, \mathcal{E}_1^2(\omega) \qquad (17.14.2)$$

Quantum mechanics gives the following formula for $\chi(-2\omega, \omega, \omega)$ as computed in lowest (second) order perturbation theory, with $\mu_{nm} = e x_{nm}$:

$$\chi_{111}(-2\omega, \omega, \omega) = \frac{N}{\epsilon_0 \hbar^2} \sum \mu_{gn}\mu_{nm}\mu_{mg}$$

$$\left(\frac{1}{(\omega - \omega_{ng})(2\omega - \omega_{mg})} - \frac{1}{(\omega + \omega_{ng})(2\omega + \omega_{nm})} \right.$$

$$\left. + \frac{1}{(\omega + \omega_{ng})(2\omega + \omega_{mg})} - \frac{1}{(\omega - \omega_{mg})(2\omega + \omega_{nm})} \right) \qquad (17.14.3)$$

where N is the number of molecules per unit volume, and the subscript g refers to the ground state. In writing this formula it is assumed that all the molecules remain in their ground states.

There are certain similarities between (17.14.2) and the expression (17.4.6c) derived with our *classical* nonlinear electron oscillator model. We note, for instance, that our classical theory gives the correct result that $P(2\omega)$ is proportional to Ne^3. Furthermore the classical theory predicts a "resonance enhancement" of the nonlinear susceptibility when $\omega \approx \omega_0$ or $2\omega \approx \omega_0$, while, analogously, (17.14.3) predicts a resonance enhancement when $\omega \approx \omega_{ng}$ or $2\omega \approx \omega_{mg}$ for some transition $n \to g$ of the molecule. Neither (17.4.6c) nor (17.14.3) are valid close to such a resonance, however. Near a resonance we must include the effect of the linewidth of a transition, which has not been done in writing these equations. In most applications the resonances occur only for frequencies in the ultraviolet, and the linewidths may be safely omitted.

All the $\chi_{ijk}(-\omega_3, \omega_1, \omega_2)$ have quantum-mechanical expressions similar to (17.14.3). This provides some insight into why three-wave mixing does not occur in centrosymmetric materials. First let us recall a *selection rule* on the transition $m \to n$ in order that μ_{mn} be nonzero: the states m and n must have different *parity*, which for hydrogenic atoms means that $\Delta l = \pm 1$. With this selection rule it is easy to see that (17.14.3) is zero for hydrogenic atoms: if $\mu_{gn} \neq 0$, then g and n have different parity; if $\mu_{nm} \neq 0$, then n and m have different parity, which means therefore that m and g have the same parity. But then μ_{mg} must be zero, and so the right side of (17.14.3) must vanish. In fact it follows that *the nonlinear susceptibility χ_{ijk} vanishes whenever the medium has states of definite parity.* In other words, three-wave mixing can occur only in materials in which the quantum states do not have a definite parity.

A centrosymmetric material is one in which the quantum states have definite parity. It might be thought that all materials are centrosymmetric. If the molecules are at the lattice sites of a noncubic crystal, however, the local fields arising from neighboring molecules modify the molecular wave functions and destroy their parity. In such a crystal there is no center of inversion symmetry, and the lack of definite parity is a consequence. •

17.15   NOTATIONAL CONVENTION FOR NONLINEAR SUSCEPTIBILITIES

In this chapter we have considered three-wave mixing processes in which two waves mix to generate a third. The nonlinear susceptibility $\chi(\omega_3, \omega_1, \omega_2)$ is called the *second-order* nonlinear susceptibility because it determines the nonlinear polarization up to second order in the electric field. More generally there are n-wave mixing processes in which a wave is generated by the mixing of $n - 1$ other waves. Such higher-order processes are discussed in the following chapter. We conclude this chapter with a summary of the convention we have chosen to follow for the nonlinear susceptibilities.

In Section 17.12 we noted some sources of confusion that often arise in connection with the definition of nonlinear susceptibilities. The convention we have followed for the nonlinear coefficient $\overline{d}$ is a standard one, but by no means universal.[3] The same is true of our definition of the nonlinear susceptibilities. For the purpose of introducing the main concepts of nonlinear optics in this chapter and the next, the precise definition of nonlinear susceptibilities is not terribly important. A precise definition is very important, however, when numerical estimates of intensities, conversion efficiencies, etc., are required.

Let us first recall our definition for the nonlinear susceptibility $\chi(-\omega_3, \omega_1, \omega_2)$. The electric fields at ω_1, ω_2, and ω_3 are written in the form

---

3.   Our definition of $\overline{d}$ follows that of the $\overline{d}$ of A. Yariv, *Quantum Electronics*, second edition (Wiley, New York, 1975), and A. Yariv, *Introduction to Optical Electronics*, second edition (Holt, Rinehart and Winston, 1976).

$$E(\omega_i) = \tfrac{1}{2}\big[\mathcal{E}(\omega_i)\,e^{-i\omega_i t} + \mathcal{E}(-\omega_i)\,e^{i\omega_i t}\big] \tag{17.15.1}$$

with $\mathcal{E}(-\omega_i) \equiv \mathcal{E}^*(\omega_i)$. The polarization at frequency ω_i is similarly written as

$$P(\omega_i) = \tfrac{1}{2}\big[P^{(L)}(\omega_i)\,e^{-i\omega_i t} + P^{(L)}(\omega_i)^* e^{i\omega_i t}\big]$$

$$+ \tfrac{1}{2}\big[P^{(NL)}(\omega_i)\,e^{-i\omega_i t} + P^{(NL)}(\omega_i)^* e^{i\omega_i t}\big] \tag{17.15.2}$$

where (L) and (NL) denote the linear and nonlinear components, respectively, of the induced polarization.

Consider the special case of second-harmonic generation, and assume for simplicity that the electric fields and polarization have only one Cartesian component. For this three-wave process it is the square of the electric field at the fundamental frequency ω that determines the nonlinear polarization:

$$\big[\tfrac{1}{2}\mathcal{E}(\omega)\,e^{-i\omega t} + \tfrac{1}{2}\mathcal{E}(-\omega)\,e^{i\omega t}\big]^2 = \tfrac{1}{4}\mathcal{E}^2(\omega)\,e^{-2i\omega t} + \tfrac{1}{2}\mathcal{E}(\omega)\,\mathcal{E}(-\omega)$$

$$+ \tfrac{1}{4}\mathcal{E}^2(-\omega)\,e^{2i\omega t} \tag{17.15.3}$$

and the polarization at the second-harmonic frequency is

$$P(2\omega) = \tfrac{1}{2}P^{(L)}(2\omega)\,e^{-2i\omega t} + \tfrac{1}{2}P^{(NL)}(2\omega)\,e^{-2i\omega t} + \text{c.c.} \tag{17.15.4}$$

The nonlinear susceptibility $\chi(-2\omega, \omega, \omega)$ is *defined* by writing

$$\tfrac{1}{2}P^{(NL)}(2\omega) = \epsilon_0 \chi(-2\omega, \omega, \omega)\,\tfrac{1}{4}\mathcal{E}^2(\omega)$$

or

$$P^{(NL)}(2\omega) = \tfrac{1}{2}\epsilon_0 \chi(-2\omega, \omega, \omega)\,\mathcal{E}^2(\omega) \tag{17.15.5}$$

which is Eq. (17.12.13).

In the case of sum-frequency generation with $\omega_3 = \omega_1 + \omega_2$, but $\omega_1 \neq \omega_2$, the square of the total electric field acting to generate the field at frequency ω_3 is

$$\big[\tfrac{1}{2}\mathcal{E}(\omega_1)\,e^{-i\omega_1 t} + \tfrac{1}{2}\mathcal{E}(\omega_2)\,e^{-i\omega_2 t} + \text{c.c.}\big]^2 = \tfrac{1}{2}\mathcal{E}(\omega_1)\,\mathcal{E}(\omega_2)\,e^{-i(\omega_1 + \omega_2)t} + \cdots$$

$$\tag{17.15.6}$$

and the polarization at $\omega_3 = \omega_1 + \omega_2$ is given by (17.15.4) with 2ω replaced by ω_3. We define $\chi(-\omega_3, \omega_1, \omega_2)$ analogously to $\chi(-2\omega, \omega, \omega)$ above, as the ratio between $\frac{1}{2}P^{(NL)}(\omega_3)$ and the factor multiplying $e^{-i\omega_3 t}$ in (17.15.6):

$$\tfrac{1}{2}P^{(NL)}(\omega_3) = \epsilon_0 \chi(-\omega_3, \omega_1, \omega_2)\, \tfrac{1}{2}\mathcal{E}(\omega_1)\, \mathcal{E}(\omega_2)$$

or

$$P^{(NL)}(\omega_3) = \epsilon_0 \chi(-\omega_3, \omega_1, \omega_2)\, \mathcal{E}(\omega_1)\, \mathcal{E}(\omega_2) \qquad (17.15.7)$$

which is Eq. (17.12.7). Comparison of (17.15.5) and (17.15.7) shows that the "degenerate" case $\omega_1 = \omega_2$ requires special consideration in order to avoid a double counting of contributions to the nonlinear polarization. Obviously we have simply rephrased the discussion near the end of Section 17.8. However, the same considerations, and often confusion, also arise in n-wave mixing processes.

Consider the example of four-wave mixing, which is taken up in the following chapter. In this case three waves at frequencies ω_1, ω_2, and ω_3 mix to generate a fourth wave at $\omega_4 = \omega_1 + \omega_2 + \omega_3$. (As in the case of three-wave mixing, one or more of the frequencies ω_1, ω_2, and ω_3 may be negative.) The nonlinearity responsible for four-wave mixing is a cubic one, i.e., the induced nonlinear polarization depends on the *third* power of the electric field. The cube of the total electric field of the waves at ω_1, ω_2, and ω_3 is

$$\left[\tfrac{1}{2}\mathcal{E}(\omega_1)\, e^{-i\omega_1 t} + \tfrac{1}{2}\mathcal{E}(\omega_2)\, e^{-i\omega_2 t} + \tfrac{1}{2}\mathcal{E}(\omega_3)\, e^{-i\omega_3 t} + \text{c.c.}\right]^3$$

$$= \tfrac{6}{8}\mathcal{E}(\omega_1)\, \mathcal{E}(\omega_2)\, \mathcal{E}(\omega_3)\, e^{-i(\omega_1 + \omega_2 + \omega_3)t} \qquad (17.15.8)$$

$$+ \tfrac{6}{8}\mathcal{E}(\omega_1)\, \mathcal{E}(\omega_2)\, \mathcal{E}(-\omega_3)\, e^{-i(\omega_1 + \omega_2 - \omega_3)t} + \cdots$$

while the polarization at $\omega_4 = \omega_1 + \omega_2 + \omega_3$ is

$$P(\omega_4) = \tfrac{1}{2}P^{(L)}(\omega_4)\, e^{-i\omega_4 t} + \tfrac{1}{2}P^{(NL)}(\omega_4)\, e^{-i\omega_4 t} + \text{c.c.} \qquad (17.15.9)$$

Proceeding exactly as in the three-wave case, we define the *third-order* nonlinear susceptibility $\chi(-\omega_4, \omega_1, \omega_2, \omega_3)$ by writing

$$\tfrac{1}{2}P^{(NL)}(\omega_4) = \epsilon_0 \chi(-\omega_4, \omega_1, \omega_2, \omega_3)\, \tfrac{6}{8}\, \mathcal{E}(\omega_1)\, \mathcal{E}(\omega_2)\, \mathcal{E}(\omega_3)$$

or

$$P^{(NL)}(\omega_4) = \tfrac{3}{2}\, \epsilon_0 \chi(-\omega_4, \omega_1, \omega_2, \omega_3)\, \mathcal{E}(\omega_1)\, \mathcal{E}(\omega_2)\, \mathcal{E}(\omega_3) \qquad (17.15.10)$$

In the fully degenerate case $\omega_1 = \omega_2 = \omega_3 \equiv \omega$ and $\omega_4 = 3\omega$, and the cube of the field inducing the nonlinear polarization at ω_4 is

$$\left[\tfrac{1}{2}\mathcal{E}(\omega)e^{-i\omega t} + \tfrac{1}{2}\mathcal{E}(-\omega)e^{i\omega t}\right]^3 = \tfrac{1}{8}\mathcal{E}^3(\omega)e^{-3i\omega t}$$

$$+ \tfrac{3}{8}\mathcal{E}^2(\omega)\mathcal{E}(-\omega)e^{-i(2\omega-\omega)t} + \cdots$$

$$(17.15.11)$$

and so $\chi(-3\omega, \omega, \omega, \omega)$ is defined by writing

$$\tfrac{1}{2}P^{(\mathrm{NL})}(3\omega) = \epsilon_0\chi(-3\omega, \omega, \omega, \omega)\tfrac{1}{8}\mathcal{E}^3(\omega)$$

or

$$P^{(\mathrm{NL})}(3\omega) = \tfrac{1}{4}\epsilon_0\chi(-3\omega, \omega, \omega, \omega)\,\mathcal{E}^3(\omega) \qquad (17.15.12)$$

Similarly, if $\omega_1 = \omega_2 \neq \omega_3$, then $\chi(-\omega_4, \omega_1, \omega_1, \omega_3)$ is defined by

$$P^{(\mathrm{NL})}(\omega_4) = \tfrac{3}{4}\epsilon_0\chi(-\omega_4, \omega_1, \omega_1, \omega_3)\,\mathcal{E}^2(\omega_1)\mathcal{E}(\omega_3) \qquad (17.15.13)$$

as the reader may easily verify.

These same considerations apply more generally when we allow for the tensor character of the nonlinear susceptibilities, and may easily be extended to wave-mixing process higher than fourth order in the field strength. We will not take the time here to generalize to these cases, for it should be clear that the confusion that has often arisen about factors like $\tfrac{1}{4}$, $\tfrac{3}{4}$, $\tfrac{3}{2}$, etc. in nonlinear optics may be traced to the definition of the nonlinear susceptibility and the electric fields.[4]

PROBLEMS

17.1 Show that the expression (17.3.8) satisfies Eq. (17.3.7).

17.2 Show that the expression (17.3.11) is a solution of Eq. (17.3.10) valid up to second order in the electric field strength.

17.3 (a) Derive (17.5.7).
(b) Derive (17.5.8).

4. For a more extensive discussion see D. C. Hanna, M. A. Yuratich, and D. Cotter, *Nonlinear Optics of Free Atoms and Molecules* (Springer-Verlag, Berlin, 1979), Chapter 2. A different convention for the nonlinear susceptibilities is employed by Y. R. Shen, *The Principles of Nonlinear Optics* (Wiley, New York, 1984).

17.4 For quartz the refractive indices for the frequency ω corresponding to $\lambda = 6943$ Å are $n_0(\omega) \approx 1.5408$ and $n_e(\omega) \approx 1.5498$, while at the second-harmonic frequency $n_0(2\omega) \approx 1.5664$ and $n_e(2\omega) \approx 1.5774$. Discuss the possibility of angle phase matching for second-harmonic generation in quartz with ruby-laser radiation as the pump. What is the angle θ_p between the pump and second-harmonic waves for angle phase matching?

17.5 Derive the expression (17.6.6) for angle phase matching in a negative uniaxial crystal. Does the second-harmonic field propagate as an ordinary or an extraordinary wave?

17.6 Derive the Manley–Rowe relations (17.8.6).

17.7 (a) Verify (17.9.5).

(b) Show that (17.9.7) reduces to (17.9.8) in the limit $Kz \ll 1$.

(c) Verify (17.9.9).

17.8 Derive the coupled wave equations (17.12.11) for three-wave mixing.

18 NONLINEAR OPTICS: HIGHER-ORDER PROCESSES

18.1 INTRODUCTION

The nonlinear optical processes considered in the preceding chapter are all three-wave mixing processes in which two waves mix to produce a third. In this chapter we take up nonlinear processes of higher order, especially four-wave mixing processes in which three waves combine to produce a fourth wave at a frequency which may be different from any of the three pump frequencies. The existence of such higher-order nonlinear processes may be inferred from our classical nonlinear electron oscillator model of Section 17.2. Four-wave mixing, for instance, is predicted when the term proportional to x^3 is retained in the Newton equation (17.2.12).

In this chapter we also take up the possibility of resonance enhancement of nonlinear susceptibilities, which leads us to consider multiphoton absorption and emission processes. Then we turn our attention to Raman and Brillouin scattering, two of the most important nonlinear optical processes. Finally we describe how these and other nonlinear wave mixing processes can produce so-called phase-conjugate waves, and we discuss how the phase conjugation process can be put to use in various laser applications.

18.2 FOUR-WAVE MIXING

In four-wave mixing, three waves of frequency ω_1, ω_2, and ω_3 are coupled to produce a fourth wave at ω_4. For instance, we can have sum-frequency generation, in which $\omega_4 = \omega_1 + \omega_2 + \omega_3$. The degenerate case $\omega_1 = \omega_2 = \omega_3 = \omega$ and $\omega_4 = 3\omega$ is called *third-harmonic generation*, or frequency tripling.

Unlike three-wave mixing, *four-wave mixing can occur in any medium, even a liquid or gas.*

Four-wave mixing is described by four coupled wave equations, just as three-wave mixing is described by the three coupled wave equations (17.8.4). Because three waves mix to generate a fourth, the nonlinear polarization associated with four-wave mixing is proportional to a product of *three* fields. In the case of third-harmonic generation, for example, the nonlinear polarization that is the source of the third-harmonic electric field may be written

$$P_{3\omega}^{(\mathrm{NL})}(z, t) = \tfrac{1}{2}\big[P(3\omega; z)\, e^{-3i\omega t}\, e^{i(k_1 + k_2 + k_3)z} + \text{c.c.}\big] \qquad (18.2.1)$$

671

where

$$P(3\omega; z) = \tfrac{1}{4}\epsilon_0\chi(-3\omega, \omega, \omega, \omega)\,\mathcal{E}(\omega; z)\,\mathcal{E}(\omega; z)\,\mathcal{E}(\omega; z) \quad (18.2.2)$$

and the three electric fields E_1, E_2, and E_3 are defined as in (17.12.6) by

$$E_i(z, t) = \tfrac{1}{2}\big[\mathcal{E}(\omega_i; z)\,e^{-i(\omega_i t - k_i z)} + \mathcal{E}(-\omega_i; z)\,e^{i(\omega_i t - k_i z)}\big] \quad (18.2.3)$$

$\chi(-3\omega, \omega, \omega, \omega)$ is called a *third-order nonlinear susceptibility*, because the corresponding polarization (18.2.2) is proportional to a product of three electric fields.[1]

The origin of a polarization like (18.2.1) may be explained in terms of our nonlinear electron oscillator model with a *cubic* nonlinearity, x^3, in the equation of motion for the electron displacement (Problem 18.1). The argument closely parallels that in Section 17.3 for second-harmonic generation, except that in place of $\cos^2 \omega t = \tfrac{1}{2}(1 + \cos 2\omega t)$ we use

$$\cos^3 \omega t = \tfrac{1}{2}\cos \omega t + \tfrac{1}{2}\cos \omega t \cos 2\omega t$$

$$= \tfrac{3}{4}\cos \omega t + \tfrac{1}{4}\cos 3\omega t \quad (18.2.4)$$

in order to see explicitly the third harmonic of the fundamental driving frequency ω.

As in the case of three-wave mixing, there are various processes that fall within the framework of four-wave mixing. In the next few sections we will describe just a few of these four-wave processes.

• Expressions analogous to (17.14.3) may be derived for third-order nonlinear susceptibilities. For instance, quantum mechanics in third order perturbation theory provides the following expression for $\chi(-3\omega, \omega, \omega, \omega)$:

$$\chi(-3\omega, \omega, \omega, \omega) = \frac{N}{\epsilon_0 \hbar^3} \sum_{l,m,n} \mu_{gl}\mu_{lm}\mu_{mn}\mu_{ng} A_{lmn} \quad (18.2.5a)$$

where

$$A_{lmn} = \frac{1}{(\omega_{lg} - 3\omega)(\omega_{mg} - 2\omega)(\omega_{ng} - \omega)} + \frac{1}{(\omega_{lg} + \omega)(\omega_{mg} + 2\omega)(\omega_{ng} + 3\omega)}$$

$$+ \frac{1}{(\omega_{lg} + \omega)(\omega_{mg} + 2\omega)(\omega_{ng} - \omega)} + \frac{1}{(\omega_{lg} + \omega)(\omega_{mg} - 2\omega)(\omega_{ng} - \omega)}$$

$$(18.2.5b)$$

1. Sometimes the third-order nonlinear susceptibility is denoted $\chi^{(3)}$ or χ_3. Similarly the *second-order nonlinear* polarization associated with three-wave mixing is sometimes denoted $\chi^{(2)}$ or χ_2.

As with (17.14.3), it is assumed that the electric fields are all linearly polarized in the same direction, and that the molecules remain in their ground states with probability ≈ 1. In the general case the third-order susceptibilities are tensors with components χ_{ijkl}.

Comparing (18.2.5) with (17.14.3), we note that the former has products of four μ's instead of three, an extra factor of $\hbar^{-1}$, and products of three frequency differences in denominators instead of two. These differences make third-order nonlinear susceptibilities generally smaller than second-order ones, unless a resonance enhancement occurs in which one of the frequency differences in a denominator is very small. Such a resonance enhancement is responsible for the *Raman effect*, as we will see in Section 18.6. •

18.3   THIRD-HARMONIC GENERATION

Coupled wave equations for four-wave mixing are generally more complicated than those for three-wave mixing. This is due mainly to the fact that there are more processes of the four-wave type. For instance, we could have $\omega_4 = \omega_1 + \omega_2 + \omega_3$, $\omega_4 = \omega_1 + \omega_2 - \omega_3$, $\omega_4 = -\omega_1 + \omega_2 + \omega_3$, etc. In this section we will consider one of the simplest examples of four-wave mixing, namely third-harmonic generation, in which $\omega_1 = \omega_2 = \omega_3 = \omega$ and $\omega_4 = 3\omega$. Using (18.2.2) and (18.2.3), we obtain the following equation for the amplitude $\mathcal{E}(3\omega, z) \equiv \mathcal{E}(3\omega)$ of the third-harmonic field:

$$\frac{d}{dz}\mathcal{E}(3\omega) = \frac{3i\omega}{8cn_3}\chi^{(3)}(3\omega)\,\mathcal{E}^3(\omega)e^{i\Delta kz} \qquad (18.3.1)$$

where $\chi^{(3)}(3\omega) \equiv \chi(-3\omega, \omega, \omega, \omega)$ and

$$\Delta k = 3k_\omega - k_{3\omega} = \frac{3\omega}{c}\left[n(\omega) - n(3\omega)\right] \qquad (18.3.2)$$

Equation (18.3.1) is quite analogous to the equation (17.4.18) for second-harmonic generation. In (18.3.1), however, the amplitude $\mathcal{E}(\omega)$ on the right side of the equation is raised to the third power, because third-harmonic generation is a four-wave process. Moreover, additional terms on the right-hand side such as $\mathcal{E}(3\omega)|\mathcal{E}(3\omega)|^2$ and $\mathcal{E}(3\omega)|\mathcal{E}(\omega)|^2$ will arise as soon as the third harmonic begins to act as a pump field itself. These can be ignored only if the pump wave is much stronger than the third-harmonic wave:

$$\left|\mathcal{E}(\omega)\right| \gg \left|\mathcal{E}(3\omega)\right| \qquad (18.3.3)$$

We should also write alongside (18.3.1) an equation for the change with z of $\mathcal{E}(\omega)$.

In the approximation in which pump depletion is neglected and (18.3.3) holds, however, (18.3.1) accurately describes the generation of the third-harmonic field, just as (17.4.18) describes second-harmonic generation approximately. Assuming $\mathcal{E}(\omega) = $ constant, then, we have

$$\mathcal{E}(3\omega) \approx \frac{3i\omega}{8cn_3} \chi^{(3)}(3\omega)\, \mathcal{E}^3(\omega) z\, e^{i\Delta k z/2} \frac{\sin\,(\Delta k\, z/2)}{\Delta k\, z/2} \qquad (18.3.4)$$

which is obviously analogous to (17.4.23). In particular, the effect of phase mismatch (i.e., $\Delta k \neq 0$) appears in the same way in the two equations.

From (18.3.4) we obtain the power conversion efficiency for third-harmonic generation:

$$\begin{aligned}
e_{\text{THG}} &= \frac{I_{3\omega}(L)}{I_\omega(0)} \\[1em]
&= \frac{9\omega^2}{16\epsilon_0^2 c^4 n_1^3 n_3} \left| \chi^{(3)}(3\omega) \right|^2 I_\omega^2(0) L^2 \left(\frac{\sin\,(\Delta k\, L/2)}{\Delta k\, L/2} \right)^2 \qquad (18.3.5)
\end{aligned}$$

The dependence of e_{THG} on Δk and L is the same as in (17.5.4) for e_{SHG}. Note, however, that e_{THG} is proportional to the *square* of the pump intensity.

Third-harmonic generation was reported for the first time in the 1960's. It continues to be a subject of much interest for the generation of ultraviolet radiation. For instance, a 6943-Å ruby-laser beam focused to a high intensity in He and other inert gases, or H_2, O_2, CO_2, N_2, or air, gives rise to the third harmonic at 2314 Å. Such short-wavelength coherent radiation is sometimes difficult to produce by laser oscillation for a number of reasons. First, spontaneous emission rates increase rapidly with decreasing wavelength, making it difficult to achieve and maintain a population inversion on a transition of short wavelength. And there are practical difficulties, such as the fabrication of high-quality mirrors for short wavelengths.

One of the major considerations in third-harmonic generation, as in second-harmonic generation, is phase matching. We have seen that a phase mismatch can drastically reduce the power conversion efficiency of second-harmonic generation, and the same is true for third-harmonic generation. We will now describe two phase-matching techniques that have been employed for third-harmonic generation.

In our simplified treatment of nonlinear optical effects we have restricted ourselves to the propagation of all fields in one direction (z). If the various waves have propagation wave vectors $\mathbf{k}_i$ in different directions, however, we should represent each wave by an expression like

$$E_i(\mathbf{r},\, t) = \tfrac{1}{2}\mathcal{E}(\omega_i)\, e^{-i(\omega t - \mathbf{k}_i \cdot \mathbf{r})} + \tfrac{1}{2}\mathcal{E}*(\omega_i)\, e^{i(\omega_i t - \mathbf{k}_i \cdot \mathbf{r})} \qquad (18.3.6)$$

instead of (18.2.3). Consider, for instance, third-harmonic generation in which two input beams of frequency ω have the same propagation wave vector $\mathbf{k}_1$, whereas the third input beam has wave vector $\mathbf{k}_1'$, with $|\mathbf{k}_1'| = |\mathbf{k}_1| = n(\omega)\,\omega/c$. In this case the phase-matching condition $\Delta k = 0$ involves the directions as well as the magnitudes of the different wave vectors. It is not difficult to show that the phase-matching condition is (Problem 18.2)

$$2\mathbf{k}_1 + \mathbf{k}_1' = \mathbf{k}_3 \qquad (18.3.7)$$

If $|2\mathbf{k}_1| + |\mathbf{k}_1'| > |\mathbf{k}_3|$, i.e., if the medium is negatively dispersive (the refractive index decreases with increasing frequency), then we can satisfy (18.3.7) by an appropriate choice of the angle between $\mathbf{k}_1$ and $\mathbf{k}_1'$ (Figure 18.1). This angle is determined by the refractive indices $n(\omega)$ and $n(3\omega)$ (Problem 18.3). A disadvantage of such *phase matching by noncollinear beams* is that larger and more intense pump beams are required to offset the decrease in conversion efficiency resulting from the decreased spatial overlap of the beams.

Another phase-matching technique, involving *collinear* beams, can be used in some instances. In the case of collinear beams (18.3.7) reduces to $n(\omega) = n(3\omega)$. In general, of course, this phase-matching condition will not be satisfied, because the nonlinear medium will be dispersive. But in a gaseous medium we can *force* the equality of $n(\omega)$ and $n(3\omega)$ by adding another gas to the vapor cell. Suppose that gas A is to be used for third-harmonic generation and it is normally dispersive, so that $n_A(3\omega) > n_A(\omega)$. By adding a suitable amount of negatively dispersive gas B, for which $n_B(3\omega) < n_B(\omega)$, we can make $n_{A+B}(\omega) = n_{A+B}(3\omega)$. Now let n_p and n_n be the refractive indices of the two gases, which are positively and negatively dispersive, respectively, and let f_p and f_n be the fractional concentrations of the two species. Then

$$n(\omega) = f_n n_n(\omega) + f_p n_p(\omega) \qquad (18.3.8a)$$

and

$$n(3\omega) = f_n n_n(3\omega) + f_p n_p(3\omega) \qquad (18.3.8b)$$

are the refractive indices of the mixture. Obviously $n(\omega) = n(3\omega)$ requires that

$$f_p n_p(\omega) + f_n n_n(\omega) = f_p n_p(3\omega) + f_n n_n(3\omega)$$

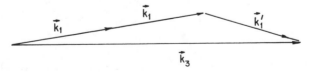

Figure 18.1 Phase matching ($\mathbf{k}_3 = 2\mathbf{k}_1 + \mathbf{k}_1'$) by noncollinear beams.

or

$$\frac{f_n}{f_p} = \frac{n_p(3\omega) - n_p(\omega)}{n_n(\omega) - n_n(3\omega)} \tag{18.3.9}$$

for the ratio of the species concentrations giving $n(\omega) = n(3\omega)$. A similar expression may be derived for the case in which the gas generating the third harmonic is negatively dispersive and the buffer gas is positively dispersive. This technique requires the relative concentrations to be accurately fixed, as well as a high degree of homogeneity of the mixture.

• At present the best way to obtain *tunable* coherent ultraviolet radiation is by frequency doubling and summing of tunable dye-laser radiation in nonlinear crystals. However, it is very difficult to generate wavelengths down to 2000 Å in nonlinear crystals, because at such low wavelengths their birefringence tends to be too weak for phase matching, and furthermore they usually absorb in the ultraviolet.

It has been possible to obtain coherent 887-Å radiation by third-harmonic generation with 2660-Å radiation in a cell containing Ar gas at a pressure of a few Torr. The 2660-Å beam was obtained by twice frequency-doubling the 1.064-μm radiation of a Nd:YAG laser. Although the third-harmonic conversion efficiency was very small ($\sim 10^{-7}$), it is possible to generate wavelengths below 200 Å by fifth-, seventh-, or even higher-order harmonic generation. •

18.4 RESONANCE ENHANCEMENT OF NONLINEAR SUSCEPTIBILITIES

We mentioned in Section 17.14 that nonlinear susceptibilities may be resonantly enhanced near certain frequencies. The second-order nonlinear susceptibility (17.14.3), for instance, shows the possibility of one-photon ($\omega \approx \omega_{ng}$) and two-photon ($2\omega \approx \omega_{mg}$) resonance enhancement. The third-order susceptibility (18.2.5) has a three-photon resonance ($3\omega \approx \omega_{lg}$) as well as one- and two-photon resonances.

Such "multiphoton resonances" in nonlinear susceptibilities have been used to enhance various nonlinear optical effects. Consider the four-wave mixing process in which frequencies ω_1 and ω_2 are mixed to generate a frequency $\omega_3 = 2\omega_1 + \omega_2$. This four-wave mixing process is characterized by a third-order nonlinear susceptibility $\chi(-\omega_3, \omega_1, \omega_1, \omega_2)$, which is found to have certain multiphoton resonances. In particular, there is a resonance enhancement of $\chi(-\omega_3, \omega_1, \omega_1, \omega_2)$ if $2\hbar\omega_1 \approx E_a - E_g$ and $\hbar\omega_2 \approx E_b - E_a$, as indicated in Figure 18.2a. Figure 18.2b shows how such a double resonance enhancement has been realized using sodium vapor as the nonlinear medium. The frequencies ω_1 and ω_2 corresponded to wavelengths of 6856 Å (from a parametric oscillator pumped by a pulsed Nd:YAG laser) and 10.6 μm (from a cw CO_2 laser), respectively, and the generated field (at $\omega_3 = 2\omega_1 + \omega_2$) was at 3321 Å.

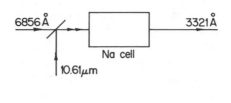

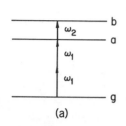

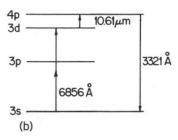

(a)   (b)

Figure 18.2 (a) $\chi(-\omega_3, \omega_1, \omega_1, \omega_2)$ is resonantly enhanced if $2\hbar\omega_1 \approx E_a - E_g$ and $\hbar\omega_2$ $\approx E_b - E_a$. (b) Realization of the two-photon resonance enhancement scheme of (a) in sodium vapor. Form D. M. Bloom, J. T. Yardley, J. F. Young and S. E. Harris, Applied Physics Letters **24**, 427 (1974).

Frequency up-conversion processes of this type are of interest in connection with the detection of infrared radiation, as noted in Section 17.8. Given the pump radiation at 6856 Å in the experiment indicated in Figure 18.2, the detection of 3321-Å radiation signals the presence of 10.6-μm radiation, for in the absence of the latter no 3321-Å radiation is generated. (The power generated at 3321 Å is proportional to the power at 10.6 μm and the square of the power at 6856 Å.) Such a process can therefore convert an infrared signal to one of much shorter wavelength, for which efficient photodetectors may be employed.

18.5 MULTIPHOTON ABSORPTION AND EMISSION

Of course resonance enhancement is not peculiar to nonlinear optics: the *linear* susceptibility $\chi(\omega)$ is resonantly enhanced when ω is near a transition frequency ω_0 of the medium. Near such a "one-photon resonance" at $\omega = \omega_0$ we must include the effect of damping processes, which, as we saw in Chapter 3, give rise to an imaginary part of the polarizability (and susceptibility), corresponding to the *absorption* of radiation near the resonance frequency $\omega = \omega_0$. That is, absorption is associated with the possibility of a one-photon resonance enhancement of the linear susceptibility $\chi(\omega)$.

With this in mind, let us consider the nonlinear susceptibility $\chi(-\omega, \omega, -\omega, \omega)$, corresponding to an induced third-order polarization

$$P^{(NL)}(\omega) = \tfrac{3}{4}\epsilon_0\chi(-\omega, \omega, -\omega, \omega)\, \mathcal{E}(\omega)\, \mathcal{E}^*(\omega)\, \mathcal{E}(\omega)$$

$$= \tfrac{3}{4}\epsilon_0\chi(-\omega, \omega, -\omega, \omega)|\mathcal{E}(\omega)|^2\mathcal{E}(\omega) \qquad (18.5.1)$$

at frequency ω in an applied field of the same frequency. $\chi(-\omega, \omega, -\omega, \omega)$ is given by an expression analogous to (18.2.5). In particular, like (18.2.5), $\chi(-\omega, \omega, -\omega, \omega)$ may be resonantly enhanced if $\omega_{ag} \approx 2\omega$ for some level a of the atom or molecule (which is assumed to be in the ground state, denoted by g). Under such a two-photon resonance enhancement $\chi(-\omega, \omega, -\omega, \omega)$ is given by

$$
\chi(-\omega, \omega, -\omega, \omega) \approx \frac{2N}{3\epsilon_0\hbar^3} \sum_{m,n} \frac{\mu_{gm}\mu_{ma}\mu_{an}\mu_{ng}}{(\omega_{mg} - \omega)(\omega_{ag} - 2\omega)(\omega_{ng} - \omega)}
$$

$$
= \frac{2N}{3\epsilon_0\hbar^3} \frac{1}{\omega_{ag} - 2\omega} \sum_m \frac{\mu_{gm}\mu_{ma}}{\omega_{mg} - \omega} \sum_n \frac{\mu_{an}\mu_{ng}}{\omega_{ng} - \omega}
$$

$$
= \frac{2N}{3\epsilon_0\hbar^3} \frac{1}{\omega_{ag} - 2\omega} \left| \sum_n \frac{\mu_{gn}\mu_{na}}{\omega_{ng} - \omega} \right|^2 \tag{18.5.2}
$$

In writing this expression the notation and assumptions are the same as in (18.2.5) for $\chi(-3\omega, \omega, \omega, \omega)$. Here again $\mu_{mn} = ex_{mn}$.

When $\omega_{ag} = 2\omega$, of course, the expression (18.5.2) is meaningless, for we cannot divide by zero. This blow-up at resonance does not arise when the theory on which (18.5.2) is based is modified to include damping terms. This modification leads to the replacement

$$
\omega_{ag} - 2\omega \rightarrow \omega_{ag} - 2\omega - i\beta_{ag}, \tag{18.5.3}
$$

where β_{ag} is a damping rate associated with the molecular oscillation at frequency ω_{ag}. We will not bother to justify this modification of (18.5.2). It should be clear, however, that this modification of a nonlinear susceptibility near a resonance is quite analogous to the modification of the linear susceptibility [or polarizability $\alpha(\omega) = \epsilon_0\chi(\omega)/N$] near a resonance [cf. equation (3.4.8)]. Indeed, the damping rate β_{ag} in (18.5.3) is associated with the same physical effects—collisions and spontaneous emission, for instance—as in the theory of the linear susceptibility.

Using (18.5.3) in (18.5.2), we may write $\chi(-\omega, \omega, -\omega, \omega)$ in terms of its real and imaginary parts, χ_r and χ_i, respectively:

$$
\chi(-\omega, \omega, -\omega, \omega) = \chi_r + i\chi_i \tag{18.5.4a}
$$

$$
\chi_r = \frac{2N}{3\epsilon_0\hbar^3} \frac{\omega_{ag} - 2\omega}{(\omega_{ag} - 2\omega)^2 + \beta_{ag}^2} \left| \sum_n \frac{\mu_{gn}\mu_{na}}{\omega - \omega_{ng}} \right|^2 \tag{18.5.4b}
$$

$$
\chi_i = \frac{2N}{3\epsilon_0\hbar^3} \frac{\beta_{ag}}{(\omega_{ag} - 2\omega)^2 + \beta_{ag}^2} \left| \sum_n \frac{\mu_{gn}\mu_{na}}{\omega - \omega_{ng}} \right|^2 \tag{18.5.4c}
$$

From (18.5.1) we may derive in the now familiar way the equation for the propagation of the amplitude of the field of frequency ω:

$$\frac{d}{dz}\mathcal{E}(\omega) = \frac{3i\omega}{8cn(\omega)}(\chi_r + i\chi_i)|\mathcal{E}(\omega)|^2\mathcal{E}(\omega) \qquad (18.5.5)$$

The effect of the linear susceptibility is included in the usual way in the definition of $n(\omega)$. It follows from (18.5.5) that

$$\frac{d}{dz}|\mathcal{E}(\omega)|^2 = -\frac{3\omega}{4cn(\omega)}\chi_i|\mathcal{E}(\omega)|^4$$

or

$$\frac{dI}{dz} = -a_2 I^2 \qquad (18.5.6)$$

where I is the intensity ($I = \frac{1}{2}\sqrt{\epsilon_\omega/\mu_0}\,|\mathcal{E}(\omega)|^2$) and

$$a_2 = \frac{3\omega\chi_i}{2\epsilon_0 c^2 n(\omega)^2} \qquad (18.5.7)$$

The solution of the differential equation (18.5.6) is

$$I(z) = \frac{I(0)}{1 + a_2 z I(0)} \qquad (18.5.8)$$

which shows that the intensity decreases with increasing z at a rate determined by the coefficient a_2.

In other words, near a two-photon resonance ($2\omega \approx \omega_{ag}$) the field diminishes in intensity [eq. (18.5.8)] as it propagates. The coefficient a_2 is called the *two-photon absorption coefficient*, because it is associated with the process in which two photons are absorbed from the field and a molecule goes from state g—in our example the ground state—to a state of higher energy (Figure 18.3). It is not difficult to express (18.5.6) in a form giving the *rate* at which molecules are lifted from state g to state a by two-photon absorption (Problem 18.4).

In general the two photons can have different frequencies ω_1 and ω_2, with the resonance condition taking the form $\omega_1 + \omega_2 = \omega_{ag}$ for these two-photon processes. One way in which two-photon absorption can be put to use is in *Doppler-free spectroscopy*. In this technique the two fields propagate in opposite directions, so that an atom in a gas sees Doppler-shifted frequencies $\omega_1 + \Delta$ and $\omega_2 - \Delta$, say, where Δ represents the Doppler shift. Then since $(\omega_1 + \Delta) + (\omega_2 - \Delta) =$

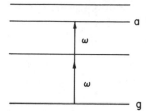

Figure 18.3 Two-photon absorption occurs if $2\hbar\omega \approx E_a - E_g$ and the transition $g \leftrightarrow a$ is forbidden to occur by one-photon absorption or emission.

$\omega_1 + \omega_2$, the Doppler shift is effectively removed from the two-photon resonance condition (Problem 18.5).

Except for the fact that two photons are involved in changing the state of the molecule, the two-photon absorption process is quite analogous to ordinary, one-photon absorption. For instance, it is characterized by a Lorentzian lineshape function under the assumption made here of homogeneous line broadening. Furthermore, there is associated with two-photon absorption the possibility of *two-photon stimulated emission* if we have a population inversion on the two-photon transition $a \rightarrow g$. And not surprisingly, there is also the two-photon spontaneous emission process, in which the atom (molecule) jumps from state a to state g with the simultaneous emission of two photons.

It is worth noting that the selection rules for two-photon transitions are different from those for one-photon transitions. In particular, two-photon transitions $a \rightarrow g$ are allowed only if $a \rightarrow g$ is forbidden in the usual sense of a one-photon transition. This is easily inferred from the sum over states appearing in (18.5.2), using an argument similar to that used in Section 17.14 to show that $\chi(-2\omega, \omega, \omega)$ vanishes when the molecular states have definite parity.

It should also be noted that summations over intermediate states n as in (18.5.2) are generally difficult to perform and various approximations must be used.[2] For one thing, we require not only the $|\mu_{gn}|$ and $|\mu_{na}|$, which can be computed from transition oscillator strengths, if they are known, but also the relative phases of the matrix elements. These are exactly calculable only for atomic hydrogen.

In general we must add to (18.5.6) a contribution from one-photon (ordinary) absorption:

$$\frac{dI}{dz} = -a_1 I - a_2 I^2 \tag{18.5.9}$$

where a_1 is the ordinary absorption coefficient. The solution of (18.5.9) is

$$I(z) = \frac{a_1 I(0) e^{-a_1 z}}{a_1 + a_2(1 - e^{-a_1 z}) I(0)} \tag{18.5.10}$$

2. R. B. Miles and S. E. Harris, "Optical Third-Harmonic Generation in Alkali Metal Vapors," *IEEE Journal of Quantum Electronics* **QE-9**, 470 (1973).

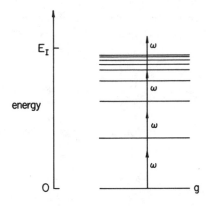

Figure 18.4 Multiphoton ionization. In the example shown $4\hbar\omega > E_I$, the ionization energy.

which, by use of l'Hospital's rule, may be shown to reduce to (18.5.8) in the limit $a_1 \to 0$. In the limit $a_2 \to 0$, (18.5.10) reduces to the familiar exponential decay law (Beer's law).

Two-photon emission and absorption are the simplest of *multiphoton* emission and absorption processes, in which n photons can be simultaneously emitted or absorbed in a single atomic (molecular) transition. These are typically *forbidden* transitions in that they violate the selection rules governing ordinary, one-photon transitions. The rates for spontaneous multiphoton emission are much smaller than the rates for one-photon emission. The two-photon transition $2^2 S_{1/2} \to 1^2 S_{1/2}$ in hydrogen, for instance, occurs at a rate of around 8 sec^{-1}. The n-photon spontaneous emission processes with $n > 2$ are entirely negligible.

Stimulated multiphoton absorption or emission processes can be significant at the high intensities available with lasers, since an n-photon absorption or stimulated emission process has a rate proportional to I^n, where I is the intensity of the stimulating field (assumed to be of a single frequency). Nevertheless the degree of excitation is usually small, so that sensitive detection techniques are required for multiphoton spectroscopy.

In addition to multiphoton absorption processes involving bound electron states, there are multiphoton *ionization* transitions in which one or more electrons are freed from the atom or molecule (Figure 18.4). Multiphoton ionization will occur in any atom or molecule when the laser radiation is sufficiently intense. Thus any transparent medium will suffer *optical breakdown* under sufficiently intense irradiation.[3] Optical breakdown refers to the formation of a high-temperature plasma due to ionization of the medium. Usually there is a visible spark, which can be

---

3. Optical breakdown is distinct from *thermal breakdown*, which arises from absorption of laser radiation and a consequent heating of the medium. In a gas the temperature rise can lead to collisional ionization and plasma formation, whereas in a solid it can lead to melting, vaporization, and surface plasma formation.

quite large. The plasma can be highly opaque, resulting in a blockage of the laser beam. In pure STP air, for instance, the threshold intensity for optical breakdown is

$$I_{BD} \sim \frac{3 \times 10^{11}}{\lambda^2} \, W/cm^2 \qquad (18.5.11)$$

where λ is the wavelength in micrometers. The λ^{-2} dependence can be understood on the basis of a simple classical model (Problem 18.6).

• Optical breakdown is an avalanche process in which electrons already present in the medium take up energy from the field and so become able to produce more electrons by impact ionization of atoms and molecules. Avalanche ionization and breakdown occurs if, among other things, the rate of free-electron production exceeds the rate of loss due to recombination and diffusion of electrons out of the interaction region (e.g., the focal volume of the laser radiation). The avalanche process requires some primary electrons to be present initially. Multiphoton ionization is believed to be important in creating the primary electrons, and so setting the stage for avalanche ionization and optical breakdown.

The presence of particulate matter (aerosols) such as dust results in a dramatic lowering of the threshold intensity for optical breakdown of gases. Dirty air, for instance, will have breakdown threshold intensities several orders of magnitude lower than that given by Eq. (18.5.11). •

18.6 RAMAN SCATTERING

In the early 1920s the Indian physicist C. V. Raman discovered that the light scattered by a gas, liquid, or solid ensemble of molecules can have a wavelength slightly different from that of the incident light. From a modern perspective we can say that this *Raman effect* is the spontaneous version of a four-wave mixing process that can be either stimulated or spontaneous. To understand this effect we will consider the process described by the third-order nonlinear susceptibility $\chi(-\omega_s, \omega_p, -\omega_p, \omega_s)$. In this process light at the pump frequency ω_p mixes with light at the *Raman* frequency ω_s to generate additional light at ω_s.

Quantum mechanics gives the following expression for the Raman susceptibility $\chi(-\omega_s, \omega_p, -\omega_p, \omega_s) \equiv \chi^{(R)}$:

$$\chi^{(R)} = \frac{N}{6\epsilon_0 \hbar^3} \frac{P_2 - P_1}{\omega_p - \omega_s - \omega_{21}} \left| \sum_n \mu_{2n}\mu_{n1} \left(\frac{1}{\omega_p + \omega_{n2}} - \frac{1}{\omega_p - \omega_{n1}} \right) \right|^2 \qquad (18.6.1)$$

where P_i is the probability that a molecule is in the state labeled i. As in previous examples, we assume the fields at ω_p and ω_s are linearly polarized in the same direction. In addition we assume that the molecular states 1 and 2 are such that

$$\omega_p - \omega_s \approx \omega_{21} = \frac{1}{\hbar}(E_2 - E_1) \tag{18.6.2}$$

and we have used this resonance assumption in writing the sum over n in (18.6.1) as a function only of ω_p rather than both ω_p and ω_s. Actually we should allow the possibility that 1 and 2 are degenerate energy levels, but in the interest of simplicity we will ignore this possibility.

When $\omega_p - \omega_s = \omega_{21}$, (18.6.1) is of course meaningless, and again we must include the effect of damping near a resonance:

$$\omega_p - \omega_s - \omega_{21} \rightarrow \omega_p - \omega_s - \omega_{21} - i\beta_{21} \tag{18.6.3}$$

With this modification of (18.6.1) we have

$$\chi^{(R)} = \chi_r^{(R)} + i\chi_i^{(R)} \tag{18.6.4a}$$

where

$$\chi_r^{(R)} = \frac{N}{6\epsilon_0 \hbar^3} \frac{(P_2 - P_1)(\omega_p - \omega_s - \omega_{21})}{(\omega_p - \omega_s - \omega_{21})^2 + \beta_{21}^2} |S_{12}|^2 \tag{18.6.4b}$$

$$\chi_i^{(R)} = \frac{N}{6\epsilon_0 \hbar^3} \frac{(P_2 - P_1)\beta_{21}}{(\omega_p - \omega_s - \omega_{21})^2 + \beta_{21}^2} |S_{12}|^2 \tag{18.6.4c}$$

and S_{12} is the sum over n appearing in (18.6.1). In the case of exact resonance ($\omega_p - \omega_s = \omega_{21}$) we have

$$\chi_r^{(R)} = 0 \tag{18.6.5a}$$

$$\chi_i^{(R)} = \frac{N}{6\epsilon_0 \hbar^3 \beta_{21}}(P_2 - P_1)|S_{12}|^2 \tag{18.6.5b}$$

We will see that the Raman effect changes the state of a molecule. For this reason we are carrying along the probabilities P_1 and P_2, instead of just assuming that the molecules remain in their ground states with high probability, as we have done in our previous discussions of nonlinear susceptibilities.

Coupled wave equations for the field amplitudes $\mathcal{E}_p$ and $\mathcal{E}_s$ are easily derived. In particular, for the case of exact resonance ($\chi_r^{(R)} = 0$) we obtain from

$$P(\omega_s) = \tfrac{3}{2}\epsilon_0 \chi(-\omega_s, \omega_p, -\omega_p, \omega_s)|\mathcal{E}_p|^2 \mathcal{E}_s \tag{18.6.6}$$

the equation

$$\frac{d\mathcal{E}_s}{dz} = \tfrac{1}{2} g_s \mathcal{E}_s \tag{18.6.7}$$

where

$$g_s = -\frac{3\omega_s}{2cn_s} \chi_i^{(R)} \left| \mathcal{E}_p \right|^2$$

$$= \frac{N\omega_s I_p}{2\epsilon_0^2 c^2 \hbar^3 n_p n_s \beta_{21}} (P_1 - P_2) \left| S_{12} \right|^2 \tag{18.6.8}$$

In the approximation in which pump depletion is neglected, therefore, g_s is constant and

$$\left| \mathcal{E}_s(z) \right|^2 = \left| \mathcal{E}_s(0) \right|^2 e^{g_s z} \tag{18.6.9}$$

This implies there is gain (i.e., amplification) at the Raman frequency ω_s if $g_s > 0$, or loss (i.e., attenuation) if $g_s < 0$.

Figure 18.5 shows how the condition (18.6.2) for the resonance enhancement of $\chi(-\omega_s, \omega_p, -\omega_p, \omega_s)$ can be realized. Although Eq. (18.6.9) indicates that the field at the frequency difference $\omega_s = \omega_p - \omega_{21}$ between the pump radiation and the transition frequency ω_{21} is amplified only if $\mathcal{E}_s(0) \neq 0$ (i.e., only if some radiation at frequency ω_s is already present), there is also the possibility of *spontaneous* scattering of radiation at frequency ω_s when $\mathcal{E}_s(0) = 0$. That is, if radiation of frequency ω_p is incident on the medium, then some of the scattered radiation will have the down-shifted frequency $\omega_s = \omega_p - \omega_{21}$.

Equation (18.6.9) describes a stimulated process in which radiation at frequency ω_s stimulates the generation of further radiation at ω_s in the presence of a pump field of frequency ω_p. As in ordinary stimulated emission, the stimulated radiation is in the same direction (z) as the stimulating radiation. In the spontaneous version of the process, however, the radiation at frequency ω_s goes off in all directions as in ordinary spontaneous emission. And as in ordinary spontaneous

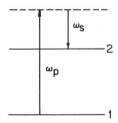

Figure 18.5 Resonance enhancement of $\chi(-\omega_2, \omega_p, -\omega_p, \omega_s)$ occurs if $\omega_p = \omega_{21} + \omega_s$. The process indicated corresponds to the Raman effect in which pump radiation at ω_p leads to Stokes radiation at the frequency $\omega_s = \omega_p - \omega_{21}$. The molecule is initially in level 1 and is left in the excited level 2 by the scattering process.

Figure 18.6 Raman scattering of anti-Stokes radiation at frequency $\omega_a = \omega_p + \omega_{21}$. In this case the molecule is initially in level 2 and is left in level 1 by the scattering process.

emission, the "seed" radiation for the process [e.g., the field $\mathcal{E}_s$ in (18.6.6)] can be thought of as arising from the zero-point field predicted by the quantum theory of radiation (recall Sec. 12.13).

We have assumed that the molecules are initially in the ground state, labeled 1. The spontaneous scattering of a photon of frequency ω_s (Figure 18.5) implies, by energy conservation, that a molecule makes an upward transition to state 2. On the other hand if a molecule starts out in level 2, then it can drop to the ground state 1 when there is incident radiation of frequency ω_p, while spontaneously scattering radiation at the *up*-shifted frequency $\omega_a = \omega_p + \omega_{21}$ (Figure 18.6). The Raman-shifted frequencies

$$\omega_s = \omega_p - \omega_{21} \qquad (18.6.10a)$$

$$\omega_a = \omega_p + \omega_{21} \qquad (18.6.10b)$$

corresponding to the processes indicated in Figures 18.5 and 18.6 are referred to as *Stokes* and *anti-Stokes lines*, respectively.

Under ordinary circumstances (e.g., thermal equilibrium) most molecules will be in the lower of the two states indicated in Figures 18.5 and 18.6. This means that the radiation scattered at the anti-Stokes frequency will normally be much weaker in intensity than the radiation scattered at the Stokes frequency (Figure 18.7).

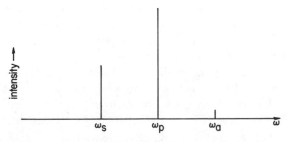

Figure 18.7 For molecules in thermal equilibrium the anti-Stokes lines are usually much weaker than the Stokes lines in Raman scattering.

• Raman's experiments began around 1921. He and his students gradually realized that the small shifts in wavelength observed in the light scattered from different materials signified a fundamental process analogous to the Compton effect. The earliest observations were made using filtered sunlight as the pump radiation. Raman announced the discovery in 1928 at a meeting of the South Indian Science Association at Bangalore. In 1930 he was awarded the Nobel Prize in Physics for this work.

Shortly after Raman's discovery, the Russian scientists G. Landsberg and L. Mandelstam independently discovered the effect in crystals. (In the Russian literature the Raman effect is called combination scattering or Mandelstam scattering.) Actually the effect had been anticipated theoretically, by analogy with the recently discovered Compton effect, by Smekal in 1923. •

Although (spontaneous) Raman scattering is weak (typically several orders of magnitude weaker in intensity than Rayleigh scattering), it is relatively easy to observe and is an extremely valuable tool in mapping out the vibrational and rotational structure of chemical species in the gas, liquid, or solid phase.[4] For one thing, ω_p and ω_s are typically in the visible or ultraviolet, whereas the material frequency ω_{21} is in the infrared or far infrared. This allows *infrared* spectroscopy (i.e., the determination of frequencies ω_{21}) to be done by looking at frequency differences between *optical* signals.

Another major advantage of spontaneous Raman scattering as a spectroscopic technique is revealed by looking more carefully at $\chi^{(R)}$. The factors $\mu_{2n}\mu_{n1}$ vanish unless $2 \leftrightarrow n$ and $1 \leftrightarrow n$ are allowed transitions; but if these are allowed, then $1 \rightarrow 2$ is forbidden. In other words, Raman scattering occurs only when the transition $1 \leftrightarrow 2$ of Figures 18.5 and 18.6 is forbidden to occur by ordinary (one-photon) absorption or emission. The frequency ω_{21} might therefore be difficult to probe by an ordinary spectroscopic technique based on absorption or fluorescence. Thus Raman scattering has been invaluable for spectroscopic studies of molecular vibrations that are not *infrared-active*, i.e., that do not have allowed one-photon transitions between vibrational levels. This is the case, for instance, in all homonuclear diatomic molecules (e.g., H_2, O_2, and N_2). Such molecules have no permanent electric dipole moments and therefore no vibrational–rotational or purely rotational spectra.

Anti-Stokes lines in rotational Raman spectra are often nearly as intense as the Stokes lines, in contrast to the vibrational case. This is because the rotational energy level spacings are much smaller than the vibrational spacings, and so the Boltzmann factors $\exp(-\Delta E/kT)$ in the rotational case are much larger. In other words, in the rotational case the Raman scattering is nearly as likely to occur off the upper level (Figure 18.6) as off the lower level (Figure 18.5) of the transition.

The Stokes and anti-Stokes lines are very weak compared with the elastic (Rayleigh) line, and this has always been the main experimental difficulty in Raman spectroscopy. For this reason Raman spectroscopy of polyatomic molecules was for some time confined mainly to liquids, which of course scatter more Raman

4. See, for instance, G. Herzberg, *Infrared and Raman Spectra of Polyatomic Molecules* (Van Nostrand, New York, 1945).

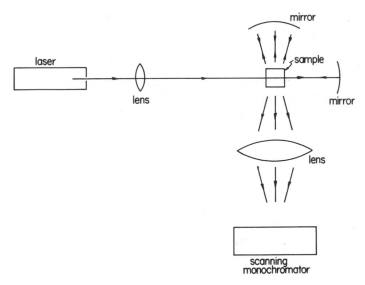

Figure 18.8 An experimental arrangement for laser Raman spectroscopy.

radiation than gases. In liquids, however, the molecules do not freely rotate, and so Raman scattering could not be used to study rotational structure. During the 1950s, special techniques were devised to facilitate the Raman spectroscopy of low-pressure gases, and so to measure rotational Raman spectra.

Lasers are now widely used as light sources for Raman spectroscopy. Figure 18.8 shows schematically a typical experimental arrangement for laser Raman spectroscopy. The intensity and tunability possible with laser sources has also introduced new aspects of Raman scattering, such as *resonance Raman scattering* and *hyper-Raman scattering*. Resonance Raman scattering refers to the situation in which the pump radiation is resonant with some transition involving level 1 or 2, say $\omega_p \approx \omega_{n1}$ for some level n in (18.6.1). This results in a resonance enhancement of the Stokes and anti-Stokes lines. Hyper-Raman scattering results from an interplay between a second-harmonic type of nonlinearity and Raman scattering, so that a pump at ω_p produces lines at $2\omega_p \pm \omega_{21}$.[5]

18.7 STIMULATED RAMAN SCATTERING

When the pump field is sufficiently intense, the generation of Raman-shifted radiation can proceed by stimulated as well as spontaneous scattering. Equation (18.6.9) describes the growth of Stokes radiation of frequency $\omega_s = \omega_p - \omega_{21}$ by stimulated Raman scattering due to the presence of both a Stokes field $\mathcal{E}_s(0)$ and a pump field $\mathcal{E}_p$. The exponential growth of the Stokes field with distance z predicted by this equation suggests that the Stokes line will be much more intense

5. See, for instance, D. A. Long, *Raman Spectroscopy* (McGraw-Hill, New York, 1977).

than in the case of spontaneous Raman scattering. The form of (18.6.9) also suggests the possibility of making a Raman laser, with a gain coefficient g_s proportional to the intensity of the pump field. Both of these suggestions can be realized.

The initial (seed) Stokes field $\mathcal{E}_s(0)$ can be externally applied from a laser, or it can build up from spontaneous Stokes radiation when the pump field is strong enough. In other words, if the pump is sufficiently strong there can be enough Stokes radiation generated spontaneously to make the stimulated Raman process dominate the spontaneous scattering along the direction of the pump beam. This is the situation usually implied by the term "stimulated Raman scattering."

Unlike spontaneous Raman scattering, the stimulated Stokes wave propagates in the direction of the pump beam as a laser-like beam in its own right. The stimulated Raman effect can therefore be used to extend the wavelength range of laser sources by varying the Raman shift ω_{21}, i.e., by choosing different materials.

The growth in intensity of the Stokes wave results in the generation of additional Stokes lines, for the Stokes wave at $\omega_p - \omega_{21}$ can itself act as a pump for a second Stokes line at

$$(\omega_p - \omega_{21}) - \omega_{21} = \omega_p - 2\omega_{21} \qquad (18.7.1)$$

Furthermore, this second Stokes component can act as a pump for a third Stokes component at $\omega_p - 3\omega_{21}$, and so forth. This suggests that the Stokes lines at $\omega_p - n\omega_{21}$ are born in succession: $\omega_p - \omega_{21}$ followed by $\omega_p - 2\omega_{21}$, which is followed by $\omega_p - 3\omega_{21}$, etc., and this sequence can in fact be observed in careful experiments.

The stimulated Raman process also produces anti-Stokes radiation. But unlike spontaneous Raman scattering, the anti-Stokes radiation is not associated with a downward transition from an excited level. What happens instead is that the pump and Stokes radiation are mixed via the third-order nonlinear susceptibility $\chi(-\omega_a, \omega_p, \omega_p, -\omega_s)$ to generate a field at the anti-Stokes frequency

$$\omega_a = 2\omega_p - \omega_s = 2\omega_p - (\omega_p - \omega_{21}) = \omega_p + \omega_{21} \qquad (18.7.2)$$

This first anti-Stokes line can then mix with the pump and first Stokes lines via the nonlinear susceptibility $\chi(\omega_a', \omega_p, \omega_a, -\omega_s)$ to generate a second anti-Stokes line at

$$\omega_a' = \omega_p + \omega_a - \omega_s = \omega_p + 2\omega_{21} \qquad (18.7.3)$$

By further mixing of Stokes and anti-Stokes lines, higher-order anti-Stokes lines at $\omega_p + n\omega_{21}$ can be generated.

In spontaneous Raman scattering the higher-order Stokes and anti-Stokes lines at $\omega_p \pm n\omega_{21}$, $n \geq 2$, do not appear. Furthermore, the anti-Stokes line is usually (but not always) much weaker in intensity than the Stokes. In stimulated Raman scattering, however, the Stokes and anti-Stokes lines are often of comparable intensity, even at low temperatures. This is because, as noted earlier, excited states are not necessary for the generation of anti-Stokes radiation.

A complete theoretical description of stimulated Raman scattering involves wave equations coupling all the various frequency components in the chain of Stokes and anti-Stokes lines. Formulas like (18.6.9) are valid only in the small-signal regime in which only a small portion of the pump is converted to Stokes radiation. If the pump is not too intense, the gain for the first Stokes line may not be sufficient to seed a substantial chain of Stokes and anti-Stokes lines.

A detailed understanding of stimulated Raman scattering also requires consideration of phase matching. In spontaneous Raman scattering there is no phase-matching requirement. This should by now be evident from the way the frequencies ω_p and ω_s appear in $\chi(-\omega_s, \omega_p, -\omega_p, \omega_s)$: all $\exp(i\mathbf{k}_p \cdot \mathbf{r})$ and $\exp(i\mathbf{k}_s \cdot \mathbf{r})$ factors cancel out, because the pump field enters as $|\mathcal{E}_p|^2$, not $\mathcal{E}_p^2$ or $\mathcal{E}_p^{*2}$. In stimulated Raman scattering, however, where many frequency components can be generated, this is not the case. Consider, for instance, the first anti-Stokes wave at $\omega_a = \omega_p + \omega_{21}$, with wave vector

$$\mathbf{k}_a = 2\mathbf{k}_p - \mathbf{k}_s \qquad (18.7.4)$$

Here $\mathbf{k}_p$ and $\mathbf{k}_s$ are the wave vectors of the pump and first Stokes fields, respectively. Equation (18.7.4) may be regarded as a consequence of the conservation of linear momentum, since photons associated with a plane wave $\exp[i(\mathbf{k} \cdot \mathbf{r} - \omega t)]$ carry energy and linear momentum $\hbar\omega$ and $\hbar\mathbf{k}$, respectively. In terms of photons, therefore, we can say that an anti-Stokes photon is created by scattering two pump photons and annihilating one Stokes photon; conservation of energy and linear momentum then imply that $\omega_a = 2\omega_p - \omega_s = \omega_p + \omega_{21}$ and $\mathbf{k}_a = 2\mathbf{k}_p - \mathbf{k}_s$, respectively. Equation (18.7.4) may also be regarded, in classical wave language, as just another (four-wave) phase-matching condition, fully analogous to (18.3.7) (Problem 18.7). This equation is shown graphically in Figure 18.9.

The smaller the phase mismatch

$$\Delta k = \left| \mathbf{k}_a - (2\mathbf{k}_p - \mathbf{k}_s) \right| \qquad (18.7.5)$$

the stronger is the coupling between the Stokes and anti-Stokes waves. For $\Delta k = 0$ the Stokes and anti-Stokes waves couple so strongly that they become, in effect, parts of a single mode of propagation. The solution of the coupled wave equations shows that for small Δk there is a large reduction in the gain coefficient, due in effect to strong competition between the Stokes and anti-Stokes waves. In gases, where $n(\omega_p) \approx n(\omega_s) \approx n(\omega_a) \approx 1$, the phase-matching direction is essentially

Figure 18.9 The condition $\mathbf{k}_a = 2\mathbf{k}_p - \mathbf{k}_s$ for the wave vectors of the anti-Stokes, pump, and Stokes waves in stimulated Raman scattering.

the forward direction defined by the direction of propagation of the pump beam. In this case, therefore, there is a "dark spot" in the scattered field in the forward direction. Conditions analogous to (18.7.4) hold also for higher-order Stokes and anti-Stokes waves, and result in different anti-Stokes components being emitted at different angles about the forward direction; the anti-Stokes output then appears as a series of concentric bright rings about the forward direction.

However, such predictions of the coupled plane-wave equations are not always well borne out experimentally. The differences stem mainly from self-focusing, which, as discussed in Section 18.12 below, is of considerable importance in the propagation of intense laser beams.

• Stimulated Raman scattering was first observed, accidentally, in 1962. Q-switching of a ruby laser with a nitrobenzene Kerr cell produced in the output beam a component downshifted by 1345 cm^{-1} from the ruby laser frequency. This downshift corresponded to the strong Raman vibrational frequency of nitrobenzene. The effect was soon reported in a large number of other liquids, and also in gases and solids. •

18.8   COHERENT ANTI-STOKES RAMAN SCATTERING (CARS)

The four-wave mixing process described by the nonlinear susceptibility $\chi(-\omega, \omega_1, \omega_1, -\omega_2)$ generates radiation at frequency

$$\omega = 2\omega_1 - \omega_2 \qquad (18.8.1)$$

through the mixing of radiation at ω_1 and ω_2. If the difference $\omega_1 - \omega_2$ is equal to a Raman transition frequency ω_{21}, then the generated radiation has frequency

$$\omega = \omega_1 + \omega_1 - \omega_2 = \omega_1 + \omega_{21} = \omega_a \qquad (18.8.2)$$

In other words, if the frequency difference $\omega_1 - \omega_2$ is equal to a Raman frequency ω_{21} of the medium, then the generated radiation has the anti-Stokes frequency $\omega_a = \omega_1 + \omega_{21}$. Since $\chi(-\omega, \omega_1, \omega_2, -\omega_2)$ is resonantly enhanced when $\omega_1 - \omega_2 = \omega_{21}$, exactly as in Section 18.6 with $\omega_1 = \omega_p$ and $\omega_2 = \omega_s$, the intensity of generated radiation is itself resonantly enhanced when $\omega_1 - \omega_2 = \omega_{21}$.

This resonance enhancement of the generated "signal" suggests the following spectroscopic technique. A laser beam of fixed frequency ω_1 is incident on a medium, together with a second laser beam of *variable* frequency ω_2, and the radiation at frequency $\omega = 2\omega_1 - \omega_2$ produced by four-wave mixing is the signal that is monitored. When this signal exhibits a resonant peak, then the frequency difference $\omega_1 - \omega_2$ is equal to a Raman frequency ω_{21} of the material.

Because this is just an example of stimulated Raman scattering, the generated radiation at the anti-Stokes frequency ω_a is not only quasimonochromatic, but also highly directional. That is, the anti-Stokes radiation at the resonance emerges as a coherent beam. For this reason this spectroscopic technique is called *coherent anti-Stokes Raman scattering* (CARS). This process was first studied in 1965, but it became widely practicable only with the advent of tunable dye lasers some time later.

Some important features of the CARS process can be understood using a simple theoretical model that ignores any depletion of the input fields at ω_1 and ω_2. The nonlinear polarization at frequency ω_a is given by

$$P_{\omega_a}^{(NL)} = \tfrac{1}{2}\left[P(\omega_a)\, e^{-i\omega_a t}\, e^{i(2k_1 - k_2)z} + \text{c.c.}\right] \tag{18.8.3}$$

where

$$P(\omega_a) = \tfrac{3}{4}\epsilon_0 \chi(-\omega_a, \omega_1, \omega_1, -\omega_2)\, \mathcal{E}^2(\omega_1)\, \mathcal{E}*(\omega_2) \tag{18.8.4}$$

The anti-Stokes amplitude satisfies the equation (Problem 18.9)

$$\frac{d\mathcal{E}(\omega_a)}{dz} = \frac{3i\omega_a}{8cn_a}\, \chi_a \mathcal{E}^2(\omega_1)\, \mathcal{E}*(\omega_2)\, e^{i\Delta k z} \tag{18.8.5}$$

where $\chi_a \equiv \chi(-\omega_a, \omega_1, \omega_1, -\omega_2)$ and $\Delta k \equiv 2k_1 - k_2$. The solution of (18.8.5) gives

$$|\mathcal{E}(\omega_a)|^2 = \frac{9\omega_a^2}{64c^2 n_a^2}\, |\chi_a|^2\, |\mathcal{E}_1|^4\, |\mathcal{E}_2|^2\, z^2 \left(\frac{\sin \Delta k\, z/2}{\Delta k\, z/2}\right)^2 \tag{18.8.6}$$

which shows the familiar dependence on the phase mismatch Δk. Note also that the intensity of the CARS signal will be proportional to the intensity of the tunable laser, and to the *square* of the intensity of the fixed-frequency laser. The nonlinear susceptibility χ_a is of the form

$$\chi_a = \frac{A}{\omega_1 - \omega_2 - \omega_{21} - i\beta_{21}} \tag{18.8.7}$$

where A is a constant proportional to a sum over states like S_{12} in Section 18.6. The CARS signal should therefore have a Lorentzian dependence on the detuning $\omega_1 - \omega_2 - \omega_{21}$, i.e.,

$$|\mathcal{E}_a|^2 \sim |\chi_a|^2 \sim \frac{1}{(\omega_1 - \omega_2 - \omega_{21})^2 + \beta_{21}^2} \tag{18.8.8}$$

Actually this Lorentzian applies only if nonresonant contributions to the third-order susceptibility are negligible. More generally (18.8.7) is replaced by

$$\chi_a = \chi_{NR} + \frac{A}{\omega_1 - \omega_2 - \omega_{21} + i\beta_{21}} \tag{18.8.9}$$

where χ_{NR} is the (real) nonresonant contribution to χ_a. In general, therefore,

$$|\mathcal{E}_a|^2 \propto |\chi_a|^2 \propto \left(\chi_{NR} + \frac{A(\omega_1 - \omega_2 - \omega_{21})}{(\omega_1 - \omega_2 - \omega_{21})^2 + \beta_{21}^2} \right)^2$$
$$+ \frac{A^2 \beta_{21}^2}{(\omega_1 - \omega_2 - \omega_{21})^2 + \beta_{21}^2} \tag{18.8.10}$$

This expression provides a good fit to measured CARS signals.[6]

CARS offers high-resolution Raman spectra superior in several ways to those obtained by traditional (spontaneous) Raman spectroscopy. For one thing, the CARS signal is much more intense than that obtained in ordinary Raman spectra. This makes possible the detection of trace molecules in gases, for instance. (Note in this regard that the CARS signal is proportional to the square of the number N of scatterers, since χ_a itself is proportional to N.) The high degree of directionality and coherence of the CARS beam is also put to use in many applications. In studies of combustion processes, for example, the CARS signal can be effectively distinguished from background fluorescence by employing a spatial filter, i.e., by using a screen with a small hole through which passes the CARS beam, but very little of the diffusely emitted thermal radiation from the hot sample.

18.9   STIMULATED BRILLOUIN SCATTERING

In Raman scattering a light wave has its frequency changed by an amount equal to some frequency ω_{21} characteristic of the molecules of the medium. It is also possible for a light wave to have its frequency shifted by an amount determined by bulk characteristics of the medium, such as the frequency ω_s of a sound wave. In this case the scattering process associated with the frequency shift is called *Brillouin scattering*.

A sound wave of frequency ω_s and wave vector $\mathbf{k}_s$ may be described in terms of units of energy and momentum called *phonons*, much as a light wave may be described in terms of photons. Then Brillouin scattering may be described as a

---

6. See, for instance, Y. R. Shen, *The Principles of Nonlinear Optics* (Wiley, New York, 1984, Chapter 15).

scattering of a photon by a phonon, and conservation of energy and linear momentum in this process demands that

$$\omega_1 = \omega_2 + \omega_s \tag{18.9.1a}$$

$$\mathbf{k}_1 = \mathbf{k}_2 + \mathbf{k}_s \tag{18.9.1b}$$

where the subscript s (for sound) must not be confused with the Stokes subscript of Raman scattering, and where the subscripts 1 and 2 refer to the incident and scattered photons, respectively. The second equation implies that

$$k_s^2 = k_1^2 + k_2^2 - 2\mathbf{k}_1 \cdot \mathbf{k}_2$$

$$\approx k_1^2(1 + 1 - 2\cos\theta)$$

$$= 4k_1^2 \sin^2 \frac{\theta}{2}$$

or

$$k_s = |\mathbf{k}_s| = 2k_1 \sin \frac{\theta}{2} \tag{18.9.2}$$

where θ is the angle between $\mathbf{k}_1$ and $\mathbf{k}_2$, i.e., the scattering angle. We have used the approximation $k_1 \approx k_2$, which is an excellent one because the Brillouin shift, though hypersonic (Table 18.1), is small compared with an optical frequency. Since $k_s = \omega_s/v_s$, where v_s is the speed of sound in the medium, the Brillouin shift is

$$\omega_s = 2k_1 v_s \sin \frac{\theta}{2} = 2n\omega_1 \frac{v_s}{c} \sin \frac{\theta}{2} \tag{18.9.3}$$

TABLE 18.1 Frequency Shift $v_s = \omega_s/2\pi$, Linewidth Γ_B, and Measured Gain $g_B^{(0)}$ for Stimulated Brillouin Scattering in Some Typical Liquids.

Liquid	Frequency Shift v_s (MHz)	Gain $g_B^{(0)}$ (cm/MW)	Linewidth Γ_b (MHz)
CS_2	5850	0.13	52.3
Acetone	4600	0.020	224
Benzene	6470	0.018	289
Methanol	4250	0.013	250
Toluene	5910	0.013	579
CCl_4	4390	0.006	520
H_2O	5690	0.0048	317

After I. L. Fabellinskii, *Molecular Scattering of Light* (Plenum, New York, 1968).

We can see that the relative shift ω_s/ω_1 must be very small, as it is proportional to the ratio of the speed of sound to the speed of light. It is also clear from (18.9.3) that the frequency shift ω_s is zero when the fields at ω_1 and ω_2 propagate in the same direction, and has its maximum, $2n\omega_1(v_s/c)$, when they propagate in opposite directions ($\theta = \pi$).

Equation (18.9.2) will be recognized as a Bragg condition for the diffraction of light by sound (Appendix 12.B). In the case of *stimulated* Brillouin scattering the sound wave that scatters light is produced by the light itself. To see how this can happen, let us recall that an acoustic (sound) wave can be described by a density variation ρ' satisfying the wave equation

$$\nabla^2\rho' - \frac{1}{v_s^2}\frac{\partial^2\rho'}{\partial t^2} = 0 \qquad (18.9.4)$$

The prime is used to emphasize that this density variation represents a small *perturbation* that is added to a constant density ρ_0. That is, the total mass per unit volume is $\rho = \rho_0 + \rho'$, with ρ' small compared with ρ_0. Since we wish to include the possibility of damping of the sound wave, we replace (18.9.4) by

$$v_s^2\nabla^2\rho' - \frac{\partial^2\rho'}{\partial t^2} + \eta\nabla^2\frac{\partial\rho'}{\partial t} = 0 \qquad (18.9.5)$$

where η is a viscosity coefficient. The way this viscosity term enters the wave equation may not be familiar, but the reader may easily check that it does indeed lead to a damping of a sound wave. One more modification is necessary: we must include a source term for the forces that produce the density variation ρ'. Denoting the local force per unit volume by $\mathbf{f}$, we write

$$v_s^2\nabla^2\rho' + \eta\nabla^2\frac{\partial\rho'}{\partial t} - \frac{\partial^2\rho'}{\partial t^2} = \nabla\cdot\mathbf{f} \qquad (18.9.6)$$

We will not take the time to derive this equation more rigorously. The reader should at least convince himself that the source term has been added in a dimensionally consistent way.

We are interested in the density variation induced by light, and therefore need to know the force $\mathbf{f}$ per unit volume due to light in the medium. It is reasonable to suppose that, in a charge-neutral medium, this force involves spatial variations of the square of the electric field. In fact we show below that, under certain simplifying assumptions,

$$\mathbf{f} = \frac{\gamma_e}{2}\nabla E^2 \qquad (18.9.7)$$

where γ_e is a constant that is characteristic of the medium. Thus we write

$$v_s^2 \nabla^2 \rho' + \eta \nabla^2 \frac{\partial \rho'}{\partial t} - \frac{\partial^2 \rho'}{\partial t^2} = \frac{\gamma_e}{2} \nabla \cdot \nabla E^2 = \frac{\gamma_e}{2} \nabla^2 E^2 \qquad (18.9.8)$$

for the density variation ρ' due to an electric field $\mathbf{E}$.

Now the polarization $\mathbf{P}$ of the medium is given by $\mathbf{P} = (\epsilon - \epsilon_0) \mathbf{E}$, and the permittivity ϵ will be a function of the density of matter in the medium. Keeping in mind that $\rho' \equiv \rho - \rho_0$ represents a small variation on the unperturbed density ρ_0, we write

$$\mathbf{P} \approx \left(\epsilon(\rho_0) + \rho' \frac{\partial \epsilon}{\partial \rho} - \epsilon_0 \right) \mathbf{E}$$

$$= [\epsilon(\rho_0) - \epsilon_0] \mathbf{E} + \rho' \frac{\partial \epsilon}{\partial \rho} \mathbf{E}$$

$$= \mathbf{P}_L + \mathbf{P}_{NL} \qquad (18.9.9)$$

where $\partial \epsilon / \partial \rho$ is understood to be evaluated at ρ_0 and

$$\mathbf{P}_{NL} = \rho' \frac{\partial \epsilon}{\partial \rho} \mathbf{E} \qquad (18.9.10)$$

is the nonlinear contribution to $\mathbf{P}$, in the sense that it is proportional to ρ', which according to (18.9.8) varies nonlinearly with $\mathbf{E}$. The remaining contribution to $\mathbf{P}$ is assumed to vary linearly with $\mathbf{E}$. Thus $\mathbf{P}_L$ leads to the ordinary (linear) refractive index, and for our purposes may just as well be ignored. For $\mathbf{E}$, therefore, we write the wave equation

$$\nabla^2 \mathbf{E} - \frac{1}{c^2} \frac{\partial^2 \mathbf{E}}{\partial t^2} = \mu_0 \frac{\partial^2 \mathbf{P}_{NL}}{\partial t^2}$$

$$= \mu_0 \frac{\partial^2}{\partial t^2} [\rho'(\partial \epsilon / \partial \rho) \mathbf{E}] \qquad (18.9.11)$$

The constant γ_e introduced in (18.9.7) is shown below to be given by the formula

$$\gamma_e = \rho_0 \frac{\partial \epsilon}{\partial \rho} \qquad (18.9.12)$$

This allows us to write (18.9.11) in the form

$$\nabla^2 \mathbf{E} - \frac{1}{c^2} \frac{\partial^2 \mathbf{E}}{\partial t^2} = \frac{\mu_0 \gamma_e}{\rho_0} \frac{\partial^2 (\rho' \mathbf{E})}{\partial t^2} \qquad (18.9.13)$$

• To understand (18.9.7) and (18.9.12) we consider for simplicity a charge-neutral fluid described at every point by a density ρ and a velocity $\mathbf{u}$. The energy in the electric field in this medium is

$$W = \frac{1}{2} \int \mathbf{E} \cdot \mathbf{D} \, dV = \frac{1}{2} \int \frac{1}{\epsilon} D^2 \, dV \tag{18.9.14}$$

We now compute dW/dt. The contribution from the time variation of D vanishes since the medium is charge-neutral, and we find

$$\frac{dW}{dt} = -\frac{1}{2} \int \frac{1}{\epsilon^2} \frac{d\epsilon}{dt} D^2 \, dV = -\frac{1}{2} \int E^2 \frac{d\epsilon}{dt} \, dV$$

$$= -\frac{1}{2} \int E^2 \frac{d\epsilon}{d\rho} \frac{d\rho}{dt} \, dV \tag{18.9.15}$$

for $\epsilon = \epsilon(\rho)$. Now the *total* rate of change of ρ at a given point in the medium is given by the "convective" derivative

$$\frac{d\rho}{dt} = \frac{\partial \rho}{\partial t} + \mathbf{u} \cdot \nabla \rho \tag{18.9.16}$$

while the equation of continuity expressing the conservation of mass is

$$\frac{\partial \rho}{\partial t} + \nabla \cdot (\rho \mathbf{u}) = 0 \tag{18.9.17}$$

Thus

$$\frac{d\rho}{dt} = -\nabla \cdot (\rho \mathbf{u}) + \mathbf{u} \cdot \nabla \rho = -\rho \nabla \cdot \mathbf{u} \tag{18.9.18}$$

and

$$\frac{dW}{dt} = \frac{1}{2} \int E^2 \frac{d\epsilon}{d\rho} \rho \nabla \cdot \mathbf{u} \, dV$$

$$= -\frac{1}{2} \int \mathbf{u} \cdot \nabla \left(\rho \frac{d\epsilon}{d\rho} E^2 \right) dV \tag{18.9.19}$$

where the second line follows from an integration by parts. Since we can also write

$$\frac{dW}{dt} = -\int \mathbf{f} \cdot \mathbf{u} \, dV \tag{18.9.20}$$

where $\mathbf{f}$ is the force per unit volume exerted by the field on the fluid, we see from (18.9.19) that

$$\mathbf{f} = \tfrac{1}{2} \nabla \left(\rho \, \frac{d\epsilon}{d\rho} \, E^2 \right) \tag{18.9.21}$$

In the derivation of this formula ρ is assumed to be the total mass per unit volume. If the largest portion of this density is just a constant value ρ_0, then

$$\mathbf{f} \approx \tfrac{1}{2} \gamma_e \nabla E^2 \tag{18.9.22}$$

where $\gamma = \rho_0 \, d\epsilon/d\rho$ is called the electrostrictive coefficient of the medium. •

Equations (18.9.8) and (18.9.13) are coupled nonlinear equations for the density perturbation ρ' and the electric field $\mathbf{E}$. We suppose that the electric field is of the form

$$\begin{aligned}
E = \tfrac{1}{2} \big[&\mathcal{E}_1(z) \, e^{-i(\omega_1 t - k_1 z)} + \mathcal{E}_1^*(z) \, e^{i(\omega_1 t - k_1 z)} \\
&+ \mathcal{E}_2(z) \, e^{-i(\omega_2 t + k_2 z)} + \mathcal{E}_2^*(z) \, e^{i(\omega_2 t + k_2 z)} \big]
\end{aligned} \tag{18.9.23}$$

where $\mathcal{E}_1$ and $\mathcal{E}_2$ are slowly varying compared with $\exp{(ik_1 z)}$ and $\exp{(ik_2 z)}$. By taking $\mathcal{E}_1$ and $\mathcal{E}_2$ to be functions of z, but not t, we are restricting ourselves to a steady-state regime. We have also assumed that the fields at ω_1 and ω_2 are linearly polarized, in the same direction, and that they propagate in opposite directions. The latter assumption is consistent with the fact that there is no frequency shift in the forward direction defined by $\mathbf{k}_1$ [Eq. (18.9.3)], and will be justified later.

From (18.9.23) we have

$$\begin{aligned}
E^2 = \tfrac{1}{2} \big(&|\mathcal{E}_1|^2 + |\mathcal{E}_2|^2 + \mathcal{E}_1 \mathcal{E}_2^* \, e^{-i(\omega_1 - \omega_2)t} \, e^{i(k_1 + k_2)z} \\
&+ \mathcal{E}_1^* \mathcal{E}_2 \, e^{i(\omega_1 - \omega_2)t} \, e^{-i(k_1 + k_2)z} \big)
\end{aligned} \tag{18.9.24}$$

if we average over the rapidly oscillating contributions at frequencies $2\omega_1$, $2\omega_2$, and $\omega_1 + \omega_2$. Again defining

$$k_s = |\mathbf{k}_1 - \mathbf{k}_2| = |k_1 \hat{\mathbf{z}} + k_2 \hat{\mathbf{z}}| = k_1 + k_2 \tag{18.9.25}$$

we write (18.9.24) as

$$E^2 = \tfrac{1}{2} \big(|\mathcal{E}_1|^2 + |\mathcal{E}_2|^2 + \mathcal{E}_1 \mathcal{E}_2^* \, e^{-i[(\omega_1 - \omega_2)t - k_s z]} + \mathcal{E}_1^* \mathcal{E}_2 \, e^{i[(\omega_1 - \omega_2)t - k_s z]} \big) \tag{18.9.26}$$

Thus

$$\nabla^2 E^2 \rightarrow -\tfrac{1}{2} k_s^2 \big[\mathcal{E}_1 \mathcal{E}_2^* \, e^{-i[(\omega_1 - \omega_2)t - k_s z]} + \mathcal{E}_1^* \mathcal{E}_2 \, e^{i[(\omega_1 - \omega_2)t - k_s z]} \big] \tag{18.9.27}$$

when we ignore the contributions from $|\mathcal{E}_1|^2$ and $|\mathcal{E}_2|^2$. It is clear from (18.9.8) that these terms cannot produce a propagating sound wave, because they do not vary in time.

To solve (18.9.8) with the source term determined by (18.9.27), we write

$$\rho' = \tfrac{1}{2}\left(Ae^{-i[(\omega_1-\omega_2)t-k_sz]} + A^*e^{i[(\omega_1-\omega_2)t-k_sz]}\right) \qquad (18.9.28)$$

and solve for A, obtaining

$$A = \frac{(\gamma_e k_s^2/2)\,\mathcal{E}_1\mathcal{E}_2^*}{k_s^2 v_s^2 - (\omega_1-\omega_2)^2 - ik_s^2\eta(\omega_1-\omega_2)} \qquad (18.9.29)$$

We now use this solution for ρ' in the equation (18.9.13) for the electric field (18.9.23). Using the assumption that $\mathcal{E}_1$ and $\mathcal{E}_2$ are slowly varying, we obtain the coupled wave equations (Problem 18.10)

$$\frac{d\mathcal{E}_1}{dz} = -K|\mathcal{E}_2|^2\mathcal{E}_1 \qquad (18.9.30a)$$

$$\frac{d\mathcal{E}_2}{dz} = -K|\mathcal{E}_1|^2\mathcal{E}_2 \qquad (18.9.30b)$$

where

$$K \equiv \frac{-i\omega_1 c\mu_0\gamma_e^2 k_s^2/8n\rho_0}{k_s^2 v_s^2 - (\omega_1-\omega_2)^2 - 2i\Gamma_B(\omega_1-\omega_2)} \qquad (18.9.31)$$

and we follow a convention by defining

$$\Gamma_B \equiv k_s^2\eta/2 \qquad (18.9.32)$$

We see from (18.9.31) that the coupling of the fields at ω_1 and ω_2 is strongest when $\omega_1 - \omega_2 = k_s v_s$, which is just the result (18.9.1a) deduced from the photon–phonon point of view. Assuming $\omega_1 - \omega_2 \approx \omega_s = k_s v_s = 2\omega_1 n v_s/c$, we have (Problem 18.11)

$$K \approx \frac{i\omega_1^2\mu_0\gamma_e^2/8\rho_0 v_s}{\omega_1 - \omega_2 - \omega_s + i\Gamma_B} \qquad (18.9.33)$$

and, exactly at resonance,

$$K = \frac{\omega_1^2\mu_0\gamma_e^2}{8\rho_0 v_s\Gamma_B} \qquad (18.9.34)$$

The coupled wave equations (18.9.30) imply the following equations for the intensities I_1 and I_2:

$$\frac{dI_1}{dz} = -g_B I_1 I_2 \qquad (18.9.35a)$$

$$\frac{dI_2}{dz} = -g_B I_1 I_2 \qquad (18.9.35b)$$

where the gain factor for stimulated Brillouin scattering is given by (Problem 18.11)

$$g_B = \frac{\omega_1^2 (\gamma_e/\epsilon_0)^2}{2 n \rho_0 v_s c^3} \frac{\Gamma_B}{(\omega_1 - \omega_2 - \omega_s)^2 + \Gamma_B^2} \qquad (18.9.36)$$

In the approximation that the pump intensity I_1 is constant we have, from (18.9.35b),

$$\frac{dI_2}{dz} = -G I_2 \qquad (18.9.37)$$

with $G \equiv g_B I_1$. Since the Brillouin-scattered "Stokes" field at ω_2 has been assumed to propagate in the $-z$ direction, Eq. (18.9.37) describes an *increase* in I_2 as the field propagates. The gain coefficient for this stimulated amplification of the backward wave is

$$G = g_B I_1 \equiv g_B^{(0)} I_1 \frac{\Gamma_B^2}{(\Delta\omega)^2 + \Gamma_B^2} \qquad (18.9.38)$$

with $\Delta\omega \equiv \omega_1 - \omega_2 - \omega_s$. Table 18.1 lists the quantities ω_s, $g_B^{(0)}$, and Γ_B for a few liquid media.

It is clear from the form of (18.9.37) that we are dealing here with a stimulated scattering process. The backward, amplified wave may be injected into the Brillouin cell or, more commonly, it can be seeded from spontaneous Brillouin scattering in which I_2 is initially zero. The assumption of negligible pump depletion, of course, breaks down when I_2 becomes comparable in magnitude to the pump intensity I_1, and this is not uncommon in stimulated Brillouin scattering experiments. Analytic solutions of the coupled intensity equations (18.9.35) are not difficult to derive.

In deriving (18.9.36) we used the relation $k_s = 2k_1$ for the case of backward scattering ($\theta = \pi$). In general we have $k_s = 2k_1 \sin(\theta/2)$, and furthermore the decay coefficient Γ_B given by (18.9.32) also depends on the angle between $\mathbf{k}_1$ and $\mathbf{k}_2$:

$$\Gamma_B = \frac{k_s^2 \eta}{2} = 2 k_1^2 \eta \sin^2 \frac{\theta}{2} \qquad (18.9.39)$$

This leads to the result that the maximum gain factor $g_B^{(0)}$ at resonance ($\Delta\omega = 0$) is independent of the light frequency and is inversely proportional to $\sin(\theta/2)$ (Problem 18.11). It should be noted that the values of ω_s, $g_B^{(0)}$, and Γ_B listed in Table 18.1 are for the backward direction, $\theta = \pi$. Furthermore the tabulated parameters ω_s and Γ_B are themselves dependent on the frequency of the pump radiation:

$$\omega_s(\theta = \pi) = \frac{2n\upsilon_s}{c}\omega_1 \tag{18.9.40a}$$

$$\Gamma_B(\theta = \pi) = \frac{2\eta n^2}{c^2}\omega_1^2 \tag{18.9.40b}$$

The values given in Table 18.1 are for the ruby-laser wavelength $\lambda_1 = 6943$ Å, and must be scaled according to (18.9.40) for other wavelengths.

We should also include in (18.9.35) the effect of the absorption coefficient α of the medium:

$$\frac{dI_1}{dz} = -g_B I_1 I_2 - \alpha I_1 \tag{18.9.41a}$$

$$\frac{dI_2}{dz} = -g_B I_1 I_2 + \alpha I_2 \tag{18.9.41b}$$

We are assuming that the absorption coefficient is approximately the same at the two frequencies ω_1 and ω_2. Note that, because the fields at ω_1 and ω_2 propagate in the $+z$ and $-z$ directions, respectively, the absorption terms enter with different signs in (18.9.41).

Thus far we have not really explained why the stimulated Brillouin scattering is strongest in the backward direction, opposite to the direction of the pump beam. There are actually two aspects to the answer to this question. The first is that, since the gain for the process is proportional to the pump intensity, the buildup of the "Stokes" field at ω_2 is only significant along the axis defined by the pump beam. Therefore we can expect this field to propagate either in the same direction as the pump or opposite to it, or perhaps in both directions. The second consideration involves the damping coefficient Γ_B given by (18.9.39). If $\mathbf{k}_1$ and $\mathbf{k}_2$ are parallel, then $\Gamma_B = 0$, and so the density perturbation ρ' induced by the fields cannot relax to the form (18.9.28). When $\mathbf{k}_2 = -\mathbf{k}_1$, however, Γ_B has its largest possible value, given by (18.9.40b), and the medium can quickly respond to the fields and form a scattering "grating" of the form (18.9.28). In practice stimulated Brillouin scattering is therefore almost always found only in the backward direction.

As already noted, stimulated Brillouin scattering is normally seeded by spontaneous Brillouin scattering. That is, a beam at ω_1 is sent into a medium, and a beam at ω_2 is found to emerge in the backward direction. The density fluctuations needed for spontaneous Brillouin scattering are always available, even at thermal equilibrium. If the gain coefficient $G = g_B I_1$ is large enough, the spontaneously scattered photons that happen to be going in the backward direction can create replicas of themselves by the stimulated scattering process, much as laser oscillation is initiated by spontaneously emitted photons. Since the gain factors $g_B^{(0)}$ listed in Table 18.1 are given in units of cm/MW, it is clear that fairly large pump intensities are required to produce appreciable gain for stimulated scattering. CS_2, for instance, has one of the largest known values of $g_B^{(0)}$, and it requires a pump intensity of 1 MW/cm^2 to give a gain coefficient G equal to $(0.13 \text{ cm/MW}) \times (1 \text{ MW/cm}^2) = 0.13 \text{ cm}^{-1}$.

Since spontaneous Brillouin scattering is normally very weak, high pump intensities are required to produce a signal comparable in intensity to the pump. Based on typical pump geometries and spontaneous scattering cross sections, one estimates that a total gain factor

$$e^{Gz} = e^{g_B I_1 z} \sim e^{30} \tag{18.9.42}$$

is typically necessary to give $I_2 \sim I_1$. Thus a reduction in the pump intensity by just 25% will reduce I_2 by a factor $\exp{(0.25 \times 30)} \approx 1800$. In practice, therefore, it is found that the pump intensity must exceed a certain threshold determined by (18.9.42) before stimulated Brillouin scattering is observed. (This is true too for stimulated Raman scattering, which is also found to have a gain requirement of $\sim e^{30}$.)

We have discussed stimulated scattering from induced density variations. Stimulated scattering from induced temperature and entropy fluctuations is also possible, although it is generally not as useful as the stimulated scattering from sound waves. Stimulated Brillouin scattering may be observed not only in liquids but also in gases and solids. This process was first reported in 1964. A backward-scattered signal was observed when quartz and sapphire were irradiated with a Q-switched ruby laser. One of the most useful applications of stimulated Brillouin scattering is in optical phase conjugation, to which we now turn our attention.

18.10 OPTICAL PHASE CONJUGATION

An especially interesting application of nonlinear optical processes is in optical phase conjugation. We discuss this application in the following section. In this section we explain what optical phase conjugation is and why it is useful.

Suppose a uniform plane wave enters a medium in which the refractive index has irregular spatial variations. Different parts of the wave will then be retarded or advanced by different amounts, and the emergent wave will have an irregular profile, as indicated in Figure 18.10. Such distortion of a wave front is detrimental

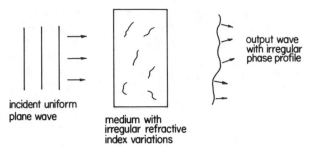

Figure 18.10 A uniform plane wave incident on a medium with irregular refractive-index variations emerges as a wave with an irregular phase profile.

to many applications, because it results in a spreading of the wave by an amount much larger than the diffraction-limited value associated with a uniform wave front. Phase distortions are therefore of great concern in applications involving, for instance, the propagation of laser beams in the atmosphere. In such applications phase distortions associated with atmospheric turbulence can make it impossible to obtain high peak intensities at large distances from the laser.

It is possible, however, for a uniform wave front to emerge from a phase-distorting medium, provided the incident wave front is appropriately "tailored." Consider for simplicity a medium that imparts a little blip to a wave front, as indicated in Figure 18.11a. If a wave whose phase is exactly the reverse of the distorted wave front of Figure 18.11a is incident on the medium, it will be distorted in just the right way as to emerge as a uniform wave front, as shown in Figure 18.11b. Of course the blip of Figure 18.11 is used only to illustrate the idea of wave-front reversal as simply as possible. More generally, a wave with a uniform phase profile might emerge from a distorting medium as a wave of the form

$$E(\mathbf{r}, t) = \mathcal{E}(\mathbf{r}) \cos \left[\omega t - kz + \phi(\mathbf{r})\right] \qquad (18.10.1)$$

In this case a *time-reversed* wave

$$E_{\text{rev}}(\mathbf{r}, t) = C\mathcal{E}(\mathbf{r}) \cos \left[\omega t + kz - \phi(\mathbf{r})\right] \qquad (18.10.2)$$

propagating in the opposite direction will emerge with a uniform phase profile, i.e., as a wave without any phase variations $\phi(\mathbf{r})$. In (18.10.2) C is a constant. Note that $E_{\text{rev}}(\mathbf{r}, t) = CE(\mathbf{r}, -t)$, so that, loosely speaking, the wave (18.10.2) is essentially a time-reversed replica of the wave transmitted by the distorting medium. When $E_{\text{rev}}(\mathbf{r}, t)$ propagates through the medium in the opposite direction, therefore, it will emerge as a time-reversed replica of the wave that was originally incident on the medium. Figure 18.11 is just a simple illustration of this undoing of phase distortions.

Figure 18.11 (a) An idealized model in which a phase-distorting medium imparts a blip to an incident uniform plane wave. The blip represents a phase advance. (b) If a wave whose phase is exactly the reverse of the transmitted wave in (a) is incident on the same medium, a uniform plane wave emerges.

If we write (18.10.1) and (18.10.2) as the real parts of

$$E(\mathbf{r}, t) = \mathcal{E}(\mathbf{r}) e^{-i\omega t} e^{i[kz - \phi(\mathbf{r})]} \qquad (18.10.3)$$

and

$$E_{\text{rev}}(\mathbf{r}, t) = C\mathcal{E}(\mathbf{r}) e^{-i\omega t} e^{-i[kz - \phi(\mathbf{r})]} \qquad (18.10.4)$$

respectively, then it becomes clear that E_{rev} may be called the phase conjugate of E. That is, if we write

$$E(\mathbf{r}, t) = \mathcal{E}(\mathbf{r}) e^{-i\omega t} e^{i\Phi(\mathbf{r})} \qquad (18.10.5)$$

with $\mathcal{E}^*(\mathbf{r}) = \mathcal{E}(\mathbf{r})$, then E_{rev} is obtained by replacing $e^{i\Phi(\mathbf{r})}$ by its complex conjugate:

$$E_{\text{rev}}(\mathbf{r}, t) = C\mathcal{E}(\mathbf{r}) e^{-i\omega t} e^{-i\Phi(\mathbf{r})} \qquad (18.10.6)$$

For this reason the process of wave-front reversal is called *phase conjugation*.

It is easy to phase-conjugate a uniform plane wave. All we need to do is reflect it with a plane mirror. Similarly a diverging spherical wave may be phase-conjugated by reflecting it with a concave mirror whose center of curvature coincides with the point source of the spherical wave (Problem 18.12). In general, however, a mirror does not perform phase conjugation. This is obvious from Figure 18.12.

Figure 18.13 illustrates a typical application of optical phase conjugation. A laser beam is amplified by passing it through a gain medium which, unfortunately, also introduces distortions in the wave front of the amplified beam. By employing a phase conjugator as shown, however, a distortion-free amplified beam may be produced.

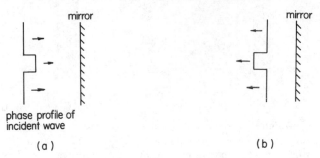

Figure 18.12 The phase profile of a wave (*a*) before and (*b*) after reflection by an ordinary mirror, showing that a mirror does not phase-conjugate an incident wave.

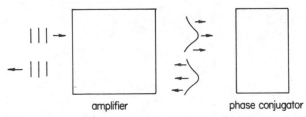

Figure 18.13 A laser beam passing through an amplifier undergoes phase distortion. After reflection from a phase conjugator and a second pass through the amplifier, the distortions due to the amplifier are removed.

Complicated and expensive phase-correction techniques involving deformable mirror arrays have been developed for various laser applications requiring good optical-beam quality. In 1972, however, a group of scientists at the Lebedev Physical Institute in Moscow reported the observation of optical phase conjugation by stimulated Brillouin scattering, and by the late 1970s phase conjugation had been observed by researchers in a variety of nonlinear optical processes.[7] We now turn our attention to such methods of phase conjugation.

18.11 PHASE CONJUGATION BY NONLINEAR OPTICAL PROCESSES

Consider the experiment indicated in Figure 18.14. Two counterpropagating pump waves of the form

$$E_1(\mathbf{r}, t) = \mathcal{E}_1 e^{-i(\omega t - \mathbf{k}_1 \cdot \mathbf{r})} \tag{18.11.1a}$$

$$E_2(\mathbf{r}, t) = \mathcal{E}_2 e^{-i(\omega t + \mathbf{k}_1 \cdot \mathbf{r})} \tag{18.11.1b}$$

---

7. See, for example, V. V. Shkunov and B. Ya. Zel'dovich, Optical Phase Conjugation, Scientific American **253**, December 1985, p. 54, and D. M. Pepper, Applications of Optical Phase Conjugation, ibid., **254**, January 1986, p. 74.

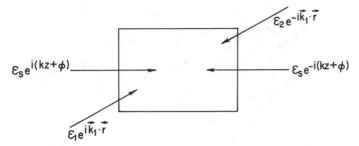

Figure 18.14 Phase conjugation by degenerate four-wave mixing. The two counterpropagating pump waves $\mathcal{E}_1 \exp(i\mathbf{k}_1 \cdot \mathbf{r})$ and $\mathcal{E}_2 \exp(-i\mathbf{k}_1 \cdot \mathbf{r})$ mix with the signal $\mathcal{E}_s \exp[i(kz + \phi)]$ to generate the phase conjugate of the signal.

are incident on a material, together with a "signal" wave

$$E_s(\mathbf{r}, t) = \mathcal{E}_s(\mathbf{r}) e^{-i(\omega t - kz)}$$

$$= \mathcal{E}_s e^{-i[\omega t - kz + \phi(\mathbf{r})]} \qquad (18.11.2)$$

The signal wave has been passed through an "aberrator" of some sort that imparts phase distortions to the wave front. The phase distortions are contained in the function $\phi(\mathbf{r})$. Note that the pump and signal fields have the same frequency.

As a result of the third-order nonlinear susceptibility $\chi(-\omega, \omega, \omega, -\omega)$ of the material, a polarization

$$P_\omega(\mathbf{r}, t) = \tfrac{3}{2} \epsilon_0 \chi(-\omega, \omega, \omega, -\omega) E_1 E_2 E_s^*$$

$$= \tfrac{3}{2} \epsilon_0 \chi(-\omega, \omega, \omega, -\omega) \left[\mathcal{E}_1 e^{-i(\omega t - \mathbf{k}_1 \cdot \mathbf{r})}\right] \left[\mathcal{E}_2 e^{-i(\omega t + \mathbf{k}_1 \cdot \mathbf{r})}\right]$$

$$\times \left[\mathcal{E}_s(\mathbf{r}) e^{-i(\omega t - kz)}\right]*$$

$$= \tfrac{3}{2} \epsilon_0 \chi(-\omega, \omega, \omega, -\omega) \mathcal{E}_1 \mathcal{E}_2 \mathcal{E}_s^*(\mathbf{r}) e^{-i(\omega t + kz)}$$

$$= \tfrac{3}{2} \epsilon_0 \chi(-\omega, \omega, \omega, -\omega) \mathcal{E}_1 \mathcal{E}_2 \mathcal{E}_s^* e^{-i[\omega t + kz - \phi(\mathbf{r})]}$$

$$= K \mathcal{E}_s e^{-i[\omega t + kz - \phi(\mathbf{r})]} \qquad (18.11.3)$$

is produced in the material. Here $K \equiv \tfrac{3}{2} \epsilon_0 \chi(-\omega, \omega, \omega, -\omega) \mathcal{E}_1 \mathcal{E}_2$ and is assumed for simplicity to be independent of $\mathbf{r}$ and t. Equation (18.11.3) says that the polarization induced by the nonlinear mixing of the pump and signal waves is proportional to the phase conjugate of the signal wave. Thus this polarization acts as a source for the spontaneous buildup of a backward propagating wave that is a phase conjugate of the signal wave. There is a striking simplicity to this method of phase conjugation by degenerate four-wave mixing. [The four-wave mixing via the susceptibility $\chi(-\omega, \omega, \omega, -\omega)$ is said to be *degenerate* because all four fields have the same frequency.] The phase conjugation is performed automatically—independently of the form of the incident signal wave.

By now it should be clear to the reader how the polarization (18.11.3) leads, via the wave equation for the electric field, to a field that is a phase conjugate of the original signal wave (Problem 18.13). There are other contributions to $P_\omega(\mathbf{r}, t)$ in addition to (18.11.3), such as terms proportional to

$$\chi(-\omega, \omega, \omega, -\omega) E_1 E_s E_s^*$$
$$= \chi(-\omega, \omega, \omega, -\omega) \, \mathcal{E}_1 \left| \mathcal{E}_s \right|^2 e^{-i(\omega t - \mathbf{k}_1 \cdot \mathbf{r})} \qquad (18.11.4)$$

and

$$\chi(-\omega, \omega, \omega, -\omega) E_1 E_s E_2^* = \chi(-\omega, \omega, \omega, -\omega) \, \mathcal{E}_1 \mathcal{E}_2^* \mathcal{E}_s \, e^{-i\omega t} \, e^{i(2\mathbf{k}_1 \cdot \mathbf{r} + kz)}$$
$$(18.11.5)$$

The contribution (18.11.4) produces a wave in the same direction as the pump field E_1, whereas the field generated by (18.11.5) propagates in the direction $2\mathbf{k}_1 + k\hat{\mathbf{z}}$. Such contributions would have to be included in a detailed analysis of degenerate four-wave mixing, but the main point for our purposes here is simply that there is a term (18.11.3) in the polarization that leads to a phase conjugate of what we have called the "signal" wave.

It might be suspected that our idealized description of phase conjugation by degenerate four-wave mixing does not apply to the real world, or that phase conjugation by nonlinear wave mixing is only possible in the special case of degenerate four-wave mixing. In fact, however, there have been many observations of phase conjugation not only in four-wave mixing but in a variety of other nonlinear optical processes. Indeed, it seems that "Nature surely loves the phase-conjugate beam."[8]

Like other nonlinear optical processes, four-wave mixing can be resonantly enhanced. This can occur, for instance, when ω is close to a one-photon resonance (i.e., near an absorption line of the medium).

One of the most frequently used nonlinear processes for phase conjugation is stimulated Brillouin scattering (SBS). Unlike the four-wave mixing process described above, it is certainly not obvious how SBS can lead to a phase-conjugate wave. In fact the origin of phase conjugation by SBS is rather subtle. To understand it we return to the coupled wave equations (18.9.30), this time including the effect of the transverse Laplacian in the wave equation. Thus we improve on the approximate (plane-wave) equation (18.9.30b) for the backward scattered wave by writing instead the paraxial wave equation (Section 14.4),

$$\frac{i}{2k} \nabla_T^2 \mathcal{E}_2 + \frac{\partial \mathcal{E}_2}{\partial z} = -K \left| \mathcal{E}_1 \right|^2 \mathcal{E}_2 \qquad (18.11.6)$$

8. R. W. Hellwarth, in *Optical Phase Conjugation*, edited by R. A. Fisher (Academic Press, New York, 1983).

We define the quantity

$$P_2(z) = \int \left| \mathcal{E}_2(\mathbf{r}, z) \right|^2 d^2\mathbf{r} \qquad (18.11.7)$$

where $\mathbf{r} = x\hat{\mathbf{x}} + y\hat{\mathbf{y}}$ and $d^2\mathbf{r} \equiv dx\, dy$. Thus $P_2(z)$ is proportional to the power in the backward-scattered wave as a function of the distance z. From (18.11.6) it follows that

$$\frac{dP_2}{dz} = -2K \int \left| \mathcal{E}_1(\mathbf{r}, z) \right|^2 \left| \mathcal{E}_2(\mathbf{r}, z) \right|^2 d^2\mathbf{r}$$

$$\equiv -g_2(z)\, P_2(z) \qquad (18.11.8)$$

where

$$g_2(z) \equiv 2K \int \left| \mathcal{E}_1(\mathbf{r}, z) \right|^2 \left| \mathcal{E}_2(\mathbf{r}, z) \right|^2 d^2\mathbf{r} \left(\int \left| \mathcal{E}_2(\mathbf{r}, z) \right|^2 d^2\mathbf{r} \right)^{-1} \qquad (18.11.9)$$

This result is derived below. $g_2(z)$ acts as a gain coefficient for the power in the backward-scattered SBS wave. [Again we remind the reader that, since the backward SBS wave goes in the direction of negative z in our convention, Eq. (18.11.8) implies a growth of the scattered power P_2 when $g_2 > 0$.]

Equation (18.11.8) is important for an understanding of how SBS produces a backward-scattered wave that is a phase-conjugate of the forward-propagating pump wave. In general the pump intensity, proportional to $\left| \mathcal{E}_1(\mathbf{r}, z) \right|^2$, will vary with $\mathbf{r}$ in a complicated way, and the gain coefficient $g_2(z)$ will obviously be largest when the local maxima of $\left| \mathcal{E}_2(\mathbf{r}, z) \right|^2$ coincide with the local maxima of $\left| \mathcal{E}_1(\mathbf{r}, z) \right|^2$. One way for this to occur is for $\mathcal{E}_2(\mathbf{r}, z)$ to be proportional to $\mathcal{E}_1^*(\mathbf{r}, z)$, for then the corresponding intensities are proportional and their maxima are coordinated. In fact, since both fields will vary spatially as they propagate, this phase conjugacy is the *only* way of keeping the intensities correlated in such a way as to maximize the gain coefficient for the SBS power. If the pump wave is converging towards a waist, the overlap integral (18.11.9) is largest when the backward SBS wave is *diverging* from the waist, i.e., when the backward-scattered wave is phase-conjugate to the pump wave.

Finally we recall that the backward-scattered wave is seeded by the very weak *spontaneous* Brillouin scattering. We can expect, then, that the phase-conjugate wave will develop in preference to other possible wave structures by "survival of the fittest," i.e., by virtue of the fact that it has the largest gain. This expectation is well borne out experimentally.

• To derive (18.11.8) we simply differentiate (18.11.7) with respect to z and then use (18.11.6):

$$
\begin{aligned}
\frac{dP_2}{dz} &= \int \frac{\partial}{\partial z} \left[\mathcal{E}_2^*(\mathbf{r}, z)\, \mathcal{E}_2(\mathbf{r}, t) \right] d^2\mathbf{r} \\
&= \int \left(\mathcal{E}_2^* \frac{\partial \mathcal{E}_2}{\partial z} + \mathcal{E}_2 \frac{\partial \mathcal{E}_2^*}{\partial z} \right) d^2\mathbf{r} \\
&= \int \left(-K |\mathcal{E}_1|^2 |\mathcal{E}_2|^2 - \frac{i}{2k} \mathcal{E}_2^* \nabla_T^2 \mathcal{E}_2 \right. \\
&\qquad \left. - K |\mathcal{E}_1|^2 |\mathcal{E}_2|^2 + \frac{i}{2k} \mathcal{E}_2 \nabla_T^2 \mathcal{E}_2^* \right) d^2\mathbf{r} \\
&= -2K \int |\mathcal{E}_1|^2 |\mathcal{E}_2|^2 \, d^2\mathbf{r} \\
&\qquad + \frac{i}{2k} \int \left[\mathcal{E}_2 \nabla_T^2 \mathcal{E}_2^* - \mathcal{E}_2^* \nabla_T^2 \mathcal{E}_2 \right] d^2\mathbf{r}
\end{aligned}
\tag{18.11.10}
$$

An integration by parts shows that the second and third terms on the right cancel for any wave of finite cross-sectional area. Then we obtain Eq. (18.11.8). •

As already noted, nonlinear phase conjugation has been observed using a variety of processes since the first report of its observation (in stimulated Brillouin scattering) in 1972. Figure 18.15 shows some quite dramatic results obtained using

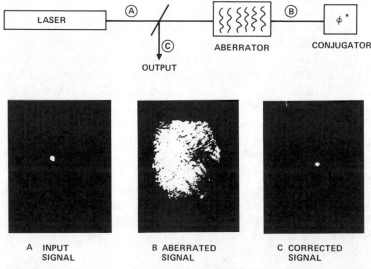

Figure 18.15 Correction of aberrations by degenerate four-wave mixing in semiconductor-doped glass. From R. K. Jain and R.C. Lind, *Journal of the Optical Society of America* **73**, 647 (1983).

a glass sheet doped with a semiconductor as the nonlinear medium. Figure 18.15*a* is a photograph of the far-field output from a *Q*-switched ruby laser, and Figure 18.15*b* shows what happens to the spot size when the laser beam is passed through an aberrator consisting of an etched glass plate. Obviously the beam quality has been seriously degraded. When the aberrated beam is phase-conjugated and then passed through the aberrator again, however, the result shown in Figure 18.15*c* is obtained. Needless to say, the phase conjugation has undone the wave-front distortions produced by the etched plate.

18.12 SELF-FOCUSING

The polarization induced by a single applied field may be expressed as a power series in the field. In particular, the induced polarization at the field frequency ω has the form

$$P(\omega) = \tfrac{1}{2}\big[P^{(L)}(\omega) + P^{(NL)}(\omega)\big] e^{-i\omega t} + \text{c.c.}$$

$$= \tfrac{1}{2}\Big[\epsilon_0 \chi(\omega)\, \mathcal{E}(\omega) + \tfrac{3}{4}\epsilon_0 \chi(-\omega, \omega, \omega, -\omega)\big|\mathcal{E}(\omega)\big|^2 \mathcal{E}(\omega)$$

$$+ \cdots\Big] e^{-i\omega t} + \text{c.c.} \tag{18.12.1a}$$

which can be rewritten in condensed notation with $\mathcal{E}(\omega)$ factored out:

$$P(\omega) \equiv \tfrac{1}{2}\epsilon_0\big[\chi_1 + \chi_3\big|\mathcal{E}(\omega)\big|^2 + \cdots\big]\mathcal{E}(\omega)\, e^{-i\omega t} + \cdots \tag{18.12.1b}$$

It is often useful to identify the entire bracket in (18.12.1b) as a field-dependent susceptibility:

$$P(\omega) = \tfrac{1}{2}\epsilon_0 \chi\big(\big|\mathcal{E}(\omega)\big|^2\big)\mathcal{E}e^{-i\omega t} + \text{c.c.} \tag{18.12.1c}$$

Here $\chi_1 \equiv \chi(\omega)$ is the linear susceptibility related to the refractive index via the familiar formula

$$n_0^2 = 1 + \chi_1 \tag{18.12.2}$$

where the subscript 0 is used in this section to denote the usual *linear* refractive index. In general the coefficient of proportionality between $P(\omega)$ and $\mathcal{E}(\omega)$ determines the total refractive index, and in general we have, from (18.12.1c),

$$n^2 = 1 + \chi\big(\big|\mathcal{E}\big|^2\big) \tag{18.12.3}$$

In general, therefore, the refractive index will be a function of the light intensity.

It is often a good approximation to retain only the first nonlinear contribution to $\chi(|\mathcal{E}|^2)$, i.e.,

$$\chi(|\mathcal{E}|^2) \cong \chi_1 + \chi_3|\mathcal{E}|^2 \qquad (18.12.4)$$

For simplicity we will use this approximation, in which case the refractive index is given by

$$n = \left(1 + \chi_1 + \chi_3|\mathcal{E}|^2\right)^{1/2} = \left(n_0^2 + \chi_3|\mathcal{E}|^2\right)^{1/2}$$

$$= n_0\left(1 + \frac{\chi_3|\mathcal{E}|^2}{n_0^2}\right)^{1/2}$$

$$\approx n_0 + \frac{\chi_3|\mathcal{E}|^2}{2n_0} \equiv n_0 + \frac{n_2}{2}|\mathcal{E}|^2 \qquad (18.12.5)$$

We have introduced a factor of $\frac{1}{2}$ here in order to conform to a notational convention in which n is written as

$$n = n_0 + n_2E^2, \qquad E = \tfrac{1}{2}\left[\mathcal{E}(\omega)e^{-i\omega t} + \text{c.c.}\right] \qquad (18.12.6)$$

Since the average of E^2 over an optical period is just $|\mathcal{E}|^2/2$, (18.12.6) is equivalent to (18.12.5) for practical purposes. More generally n can be written as a power series:

$$n = n_0 + n_2E^2 + n_4E^4 + \cdots \qquad (18.12.7)$$

but (18.12.6) will suffice for our discussion, and indeed it is often an excellent approximation.

One consequence of (18.12.6) is immediately obvious. Suppose $n_2 > 0$. Then (18.12.6) says that the refractive index is largest where the intensity is largest. A Gaussian laser beam, for instance, experiences the largest refractive index at the center of the beam, and the index monotonically decreases away from the beam axis (Figure 18.16). In effect, then, the medium acts as a (positive) lens tending

Figure 18.16 Self-focusing. A Gaussian beam in a medium with refractive index $n = n_0 + (n_2/2)|\mathcal{E}|^2$, $n_2 > 0$, leads to a refractive-index variation as shown. This index acts to focus the beam.

to focus the beam to a small spot size. Since the beam itself induces the nonlinear polarization and therefore the focusing, this effect is called *self-focusing*.

Self-focusing is an important consideration in many nonlinear optical processes. This is no surprise, for if a field is intense enough to produce nonlinear effects like harmonic generation or stimulated Raman scattering, then the nonlinear contribution to the refractive index is also likely to be significant.

In many instances self-focusing does not profoundly alter a particular nonlinear process of interest, but it does significantly modify the predictions of theoretical analyses that leave out self-focusing. For example, self-focusing in liquids can reduce the incident intensity required for stimulated Raman scattering by a factor ~ 100. The large power densities resulting from self-focusing can also cause optical breakdown and material damage at lower incident intensities than might otherwise be expected. Self-focusing therefore sets limitations on the design of various high-power laser systems.

To attain a more quantitative understanding of self-focusing, we consider the wave equation for the electric field when the index of refraction is given by (18.12.6):

$$\nabla^2 E - \frac{n^2}{c^2}\frac{\partial^2 E}{\partial t^2} \approx \nabla^2 E - \frac{1}{c^2}(n_0^2 + 2n_0 n_2 E^2)\frac{\partial^2 E}{\partial t^2} = 0 \quad (18.12.8)$$

for $n_2 \ll n_0$. Using the paraxial approximation (Section 14.4), and averaging over an optical period, which results in the replacement of E^2 by $|\mathcal{E}|^2/2$, we obtain (Problem 18.15)

$$\nabla_T^2 \mathcal{E} + 2ik\frac{\partial \mathcal{E}}{\partial z} + \frac{k^2 n_2}{n_0}|\mathcal{E}|^2 \mathcal{E} = 0 \quad (18.12.9)$$

where $k \equiv n_0 \omega / c$. The term $\nabla_T^2 \mathcal{E}$ accounts for diffraction; without it (18.12.9) describes only plane-wave propagation, and cannot account for the decrease in beam diameter due to self-focusing. If the cross section is characterized by some radius a_0, then

$$\nabla_T^2 \mathcal{E} \sim a_0^{-2} \mathcal{E} \quad (18.12.10)$$

We expect that self-focusing can compete with diffraction if the last term on the left side of (18.12.9) is comparable in magnitude to the first, i.e., if

$$\frac{k^2 n_2}{n_0}|\mathcal{E}|^2 \mathcal{E} \sim a_0^{-2}\mathcal{E}$$

or

$$a_0^2 |\mathcal{E}|^2 \sim \frac{n_0}{k^2 n_2} \quad (18.12.11)$$

Since the beam intensity $I = (n_0 c \epsilon_0 / 2) |\mathcal{E}|^2$, we expect, based on (18.12.11), that a critical beam power on the order of

$$
\begin{aligned}
P_{\text{cr}} \sim (\pi a_0^2) I &= \frac{\pi n_0 c \epsilon_0}{2} a_0^2 |\mathcal{E}|^2 \\
&= \frac{\pi n_0 c \epsilon_0}{2} \frac{n_0}{k^2 n_2} \\
&= \frac{\pi n_0^2 c \epsilon_0}{2 k^2 n_2} = \frac{c \epsilon_0 \lambda^2}{8 \pi n_2}
\end{aligned}
\tag{18.12.12}
$$

is required for self-focusing to overcome the diffractive spreading of the beam. This result for P_{cr} is in reasonably good agreement with results obtained by numerical integration of the nonlinear partial differential equation (18.12.9).

Consider as an example the liquid CS_2, which has a rather large n_2 value of around 10^{-20} (mks units). For the ruby-laser wavelength $\lambda = 6943$ Å, we estimate from (18.12.12) a critical power $P_{\text{cr}} \approx 5$ kW. This shows that self-focusing can be significant even at modest beam powers.

Note that it is the beam *power* that must exceed a certain threshold for self-focusing, not the beam intensity. Thus if a beam of a certain power $P < P_{\text{cr}}$ is focused with a lens to create a very large intensity, self-focusing will not occur, simply because the reduction in beam diameter increases the diffractive spreading that must be overcome to realize a net focusing.

We outline below an approximate approach to the solution of (18.12.9) that supports the estimate (18.12.12) for the critical power necessary for self-focusing, and also provides an estimate of the focal length.

Resonance enhancement of n_2 can occur near an absorption line. In this case the intensity dependence of the refractive index may be regarded as a consequence of the saturation of the resonant refractive index (or "anomalous dispersion") with intensity. This intensity dependence can give rise to self-focusing or defocusing ($n_2 < 0$), depending on whether the laser frequency is above or below the central frequency of the absorption line. Resonance enhancement of n_2 near a two-photon absorption line is believed to be responsible for the coefficient n_2 near the ruby-laser frequency of many liquids.

• Let us write

$$
\mathcal{E}(\mathbf{r}) = A(\mathbf{r}) e^{ikS(\mathbf{r})}
\tag{18.12.13}
$$

where A and S are both real functions of $\mathbf{r}$. This approach is used frequently in classical optics; the function $S(\mathbf{r})$ is called the *eikonal*, after the Greek word for "image." Using (18.12.13) in (18.12.9), and equating the real and imaginary parts of both sides of the resulting equation, we obtain the two (real) equations

$$\frac{\partial A^2}{\partial z} + \nabla_T \cdot (A^2 \nabla_T S) = 0 \tag{18.12.14a}$$

$$2\frac{\partial S}{\partial z} + (\nabla_T S)^2 = \frac{\nabla_T^2 A}{k^2 A} + \frac{n_2 A^2}{n_0} \tag{18.12.14b}$$

Assume for A a Gaussian beam form

$$A(\mathbf{r}) = \frac{A_0 w_0}{w(z)} e^{-r^2/w^2(z)} \tag{18.12.15}$$

Using this assumption in (18.12.14a), and the replacements $\nabla_T \to \hat{\mathbf{r}} \, \partial/\partial r$, $\nabla_T^2 \to \partial^2/\partial r^2 + (1/r)\,\partial/\partial r$ for the case of cylindrical symmetry, we obtain for S the equation

$$\frac{\partial^2 S}{\partial r^2} + \frac{1}{r}\left(1 - \frac{4r^2}{w^2}\right)\frac{\partial S}{\partial r} + \frac{2}{w}\left(\frac{2r^2}{w^2} - 1\right)\frac{dw}{dz} = 0 \tag{18.12.16}$$

It is easily checked that this equation has a solution of the form

$$S(r, z) = \frac{r^2}{2w}\frac{dw}{dz} \tag{18.12.17}$$

[Actually we can add to (18.12.17) any function of z and still satisfy (18.12.16), but (18.12.17) is adequate for our purposes here.] Next we use (18.12.17), together with (18.12.15), in Eq. (18.12.14b), and the result is the following equation for $w(z)$:

$$r^2 \frac{d^2 w}{dz^2} = \frac{2}{k^2 w}\left(\frac{2r^2}{w^2} - 1\right)$$
$$+ \frac{n_2 A_0^2 w_0^2}{n_0 w} e^{-2r^2/w^2} \tag{18.12.18}$$

Equation (18.12.18) reveals a flaw in the assumption (18.12.15) that a Gaussian wave form is maintained during propagation in the nonlinear medium: w is supposed to depend only on z, but (18.12.18) implies it varies also with r. It turns out that the "aberrationless approximation" (18.12.15) is valid only near the z axis. Based on this consideration, we write

$$e^{-2r^2/w^2} \approx 1 - \frac{2r^2}{w^2} \tag{18.12.19}$$

in (18.12.18), and then equate coefficients of r^2 on both sides of the resulting equation. This leads to the following equation for w:

$$\frac{d^2 w}{dz^2} = \frac{4/k^2 - 2n_2 A_0^2 w_0^2/n_0}{w^3} \tag{18.12.20}$$

which is consistent with the assumption that w is independent of r. Assuming $w(0) = w_0$ and $w'(0) = 0$, we have the following solution to (18.12.20):

$$w(z) = w_0 \left[1 - \left(\frac{P_0}{P_{cr}} - 1 \right) \frac{z^2}{z_0^2} \right]^{1/2} \tag{18.12.21}$$

where

$$P_0 = \pi n_0 c \epsilon_0 w_0^2 A_0^2 / 4 \tag{18.12.22a}$$

is the beam power,

$$P_{cr} = c \epsilon_0 \lambda^2 / 8 \pi n_2 \tag{18.12.22b}$$

and

$$z_0 = \frac{k w_0^2}{2} = \frac{n_0 \pi w_0^2}{\lambda} \tag{18.12.22c}$$

From (18.12.21) we see that, for $P_0 \ll P_{cr}$,

$$w(z) \approx w_0 \left(1 + \frac{z^2}{z_0^2} \right)^{1/2} \tag{18.12.23}$$

which corresponds to the growth in spot size of a Gaussian beam, as in Chapter 14. The critical power (18.12.22b) agrees with our rough estimate (18.12.12). For a beam power $P_0 = P_{cr}$ we see that $w(z) = w_0$, so that diffraction and self-focusing cancel each other and the beam neither spreads nor focuses as it propagates. This is called self-trapping.

At a focusing distance

$$z_f \equiv z_0 \left(\frac{P_0}{P_{cr}} - 1 \right)^{1/2} \tag{18.12.24}$$

the beam spot size vanishes, $w(z) = 0$. Of course this cannot occur, for it corresponds to infinite intensity. In practice various other nonlinear processes, which are left out of our analysis here, come into play to prevent the avalanching of self-focusing. These include, for instance, optical breakdown and stimulated Raman and Brillouin scattering. The distance z_f is nevertheless a useful gauge of the distance required for substantial self-focusing. •

PROBLEMS

18.1 Show that the polarization (18.2.1) follows from a cubic nonlinearity (x^3) in the classical electron oscillator model.

18.2 Derive the phase-matching condition (18.3.7).

18.3 **(a)** Consider phase matching by noncollinear beams for third-harmonic generation in a negatively dispersive gas. Derive an expression for the angle between $\mathbf{k}_1$ and $\mathbf{k}'_1$ that allows the phase-matching condition (18.3.7) to be satisfied.

(b) Is it possible to phase-match using noncollinear beams if the medium is positively dispersive, i.e., if $n(3\omega) > n(\omega)$?

18.4 Using (18.5.6), derive an expression for the two-photon absorption rate in terms of a_2.

18.5 Consider two counterpropagating fields in a gas, one at frequency ω_1 and the other at ω_2. Show that the Doppler shift does not affect the two-photon resonance condition $\omega_1 + \omega_2 = \omega_{ag}$.

18.6 Consider an electron with equation of motion

$$\ddot{x} + \beta\dot{x} = \frac{e}{m}\mathcal{E}\cos\omega t$$

in an electric field $\mathcal{E}\cos\omega t$. Here β is a damping rate associated with collisions.

a) Show that the rate of absorption of energy by the electron is given by the expression

$$\frac{dW}{dt} = \frac{\beta e^2 \mathcal{E}^2}{2 m\omega^2}$$

for $\omega \gg \beta$.

b) Suppose the "free" electrons of part (a) are in a gas of atoms or molecules that can be ionized by collisions with the electrons. This collisional ionization increases the number of free electrons while other processes, such as electron–ion recombination, decrease the number. If optical breakdown is to occur, the rate of gain of energy by the electrons from their interaction with the field should exceed the product of the effective ionization rate times the energy E_I required for ionization of the atoms or molecules. Show that this implies a threshold intensity for optical breakdown that varies as $1/\lambda^2$, where λ is the wavelength.

18.7 Based on coupled wave equations for stimulated Raman scattering, derive the expression (18.7.4) for the wave vector of the first anti-Stokes wave.

18.8 Show that the condition (18.7.4) cannot in general be satisfied when all three waves propagate in the same direction in a normally dispersive medium.

18.9 Derive Eq. (18.8.5) for the anti-Stokes amplitude in coherent anti-Stokes Raman scattering.

18.10 Assuming the amplitudes $\mathcal{E}_1$ and $\mathcal{E}_2$ are slowly varying, derive the coupled wave equations (18.9.30) for stimulated Brillouin scattering.

18.11 (a) Derive (18.9.33) from (18.9.31), assuming $\omega_1 - \omega_2 \approx \omega_s$.

(b) Derive the expression (18.9.36) for the gain factor in stimulated Brillouin scattering.

(c) Show that the maximum gain factor in stimulated Brillouin scattering is independent of frequency and inversely proportional to $\sin(\theta/2)$.

18.12 Consider a diverging spherical wave reflected by a concave mirror whose center of curvature coincides with the point source of the spherical wave. Show that the reflected wave is the phase conjugate of the incident wave.

18.13 Show that the polarization (18.11.3) generates a wave that is the phase conjugate of the signal wave.

18.14 Why is there no term proportional to $E(\mathbf{r})^2$ or $|\mathcal{E}(\mathbf{r})|^2$ in the polarization (18.12.1) giving rise to self-focusing?

18.15 Using the paraxial-wave approximation, derive (18.12.9) from (18.12.8).

INDEX